Updates in Hypertension and Cardiovascular Protection

Series Editors
Giuseppe Mancia, Milan, Italy
Enrico Agabiti-Rosei, Brescia, Italy

The aim of this series is to provide informative updates on both the knowledge and the clinical management of a disease that, if uncontrolled, can very seriously damage the human body and is still among the leading causes of death worldwide. Although hypertension is associated mainly with cardiovascular, endocrine, and renal disorders, it is highly relevant to a wide range of medical specialties and fields – from family medicine to physiology, genetics, and pharmacology. The topics addressed by volumes in the series *Updates in Hypertension and Cardiovascular Protection* have been selected for their broad significance and will be of interest to all who are involved with this disease, whether residents, fellows, practitioners, or researchers.

Maria Dorobantu • Victor Voicu
Guido Grassi
Enrico Agabiti-Rosei • Giuseppe Mancia
Editors

Hypertension and Heart Failure

Epidemiology, Mechanisms and Treatment

Second Edition

European Society of Hypertension

Springer

Editors
Maria Dorobantu
Carol Davila University of Medicine and Pharmacy
Bucharest, Bucuresti, Romania

Guido Grassi
University of Milano-Bicocca
Milano, Italy

Giuseppe Mancia
University of Milano-Bicocca
Milano, Italy

Victor Voicu
Department of Pharmacology, Toxicology and Clinical Psychopharmacology Faculty of Medicine
Carol Davila University of Medicine and Pharmacy
Bucharest, Romania

Enrico Agabiti-Rosei
Dept of Clinical and Experimental Sci
University of Brescia
Brescia, Italy

ISSN 2366-4606 ISSN 2366-4614 (electronic)
Updates in Hypertension and Cardiovascular Protection

ISBN 978-3-031-39314-3 ISBN 978-3-031-39315-0 (eBook)
https://doi.org/10.1007/978-3-031-39315-0

This Springer imprint is published by the registered company Springer Nature Switzerland AG
The registered company address is: Gewerbestrasse 11, 6330 Cham, Switzerland

Paper in this product is recyclable

Introduction

We are thrilled to introduce the second edition of the book *Hypertension and Heart Failure* published by Springer Nature which is devoted to two conditions of major importance for public health which are closely related. Arterial hypertension is a major public health burden, affecting a significant portion of the general population worldwide. The World Health Organization (WHO) estimates that hypertension is responsible for over nine million deaths annually, making it the leading cause of death globally. Furthermore, the burden of hypertension is increasing, and it is projected to continue to rise in the coming decades due to aging populations and lifestyle changes. Understanding the facts and numbers, as well as individualized treatment options for hypertension, is crucial for healthcare professionals who interact with patients with hypertension. Heart failure (HF) is a condition responsible for the largest number of hospitalizations in the elderly fraction of the population. It has an elevated mortality and a very high negative impact on healthcare services and costs. Its prevalence is also on the rise. Blood pressure elevation is the number one risk factor for incident HF, and blood pressure control through treatment is a most important measure to prevent this event.

The first edition of this book, published in 2019, was met with great success among healthcare professionals. It was accessed 28,000 times on Springer's website, emphasizing the need for comprehensive and up-to-date information on hypertension and HF. The feedback we received from our readers was overwhelmingly positive, and we are grateful for the support and encouragement we received. However, since the publication of the first edition, there have been important novelties and concepts in hypertension and HF. These include advances in hypertension management, new insights into the mechanisms of hypertension-induced HF, and the emergence of novel therapeutic strategies. The second edition of our book reflects these new developments and provides a comprehensive overview of the current state of knowledge in this so medically important area.

The 2021 ESC guidelines on HF have introduced a new classification system based on left ventricular ejection fraction. The new classification system includes three different categories, i.e., HF with preserved, mildly reduced, and reduced ejection fraction. The new classification system takes into account the different pathophysiology and prognosis of each category. This will help healthcare professionals to better identify the proper treatment for each category of HF, hopefully also with an earlier diagnostic identification and a more timely therapeutic intervention. The new guidelines also provide a more detailed description of the management of each category of HF and the use of different drugs and treatment options. They offer a more detailed description of the management of HF in special populations such as the elderly, women, and patients with diabetes. Most importantly, they also focus on HF with preserved ejection fraction for which hypertension is by far the most common precursor, and new knowledge on the protection associated with treatment has been recently made available.

Recent drug developments for the treatment of HF have focused on targeting specific pathways in the heart and vasculature to improve cardiac function and reduce symptoms. One promising new development is the use of angiotensin receptor-neprilysin inhibitors (ARNIs), such as sacubitril/valsartan, which have been shown to improve survival and reduce hospitalization in patients with HF. Another new drug class is the SGLT-2 inhibitors, which have been shown not only to improve glucose control in patients with diabetes but also to reduce the risk of HF and cardiovascular events in this population, a protective effect that appears to extend to renal disease and include nondiabetic patients.

Hyperuricemia is a prevalent condition in the general population and can contribute to hypertension and its complications, as well as having an adverse prognostic effect in patients with HF with reduced and preserved ejection fraction. Recent research has shown that the reduction of uric acid levels by Xanthine Oxidase inhibitors and certain medications for HF (SGLT-2 inhibitors, ARNI, and Vericiguat) can contribute to the clinical benefit observed in patients with HF.

In the future, there may come new opportunities for fixed drug combinations to improve cost-effectiveness and compliance of diabetes treatment when antihypertensive, lipid-lowering and glucose-lowering drugs are combined.

One of the most notable features of this second edition is the inclusion of new chapters that cover a wide range of topics. New topics are HF in hypertensive patients across the age spectrum, patients with cancer, and HF in pregnancy. New pathophysiological aspects have also been added such as the role of heart rate variability, the renal-cardiac interrelationship in hypertension, uric acid in hypertension, myocardial ischemia—as an intermediate step between hypertension and HF, the role of endothelial dysfunction in the progression from hypertension to HF as well as the panendothelitis associated with the SARS-CoV infection, its relationship with hypertension and its possible role in HF. These chapters offer a wealth of information that was not available in the first edition.

There is increasing prevalence of high blood pressure in children and adolescents. The prevalence of pediatric hypertension is difficult to establish due to regional differences, but a recent meta-analysis reported a pooled prevalence ranging between 4% and 15.3% among obese children. High blood pressure in childhood increases the risk of premature cardiovascular disease, and effective preventive interventions need to be applied early in life to modify the disease progression. Furthermore, the assessment of organ damage needs to be optimized by looking for early markers.

Hypertensive disorders of pregnancy are one of the leading causes of maternal, fetal, and neonatal morbidity and mortality. They can induce HF in the ante-, intra- or postpartum period. Strategies for prevention and treatment of HDP-associated HF must cover both pregnancy and the postpartum period, including first-trimester screening, lifestyle changes, aspirin administration, early referral to specialist care, structured cardiovascular follow-up, limitation of sodium intake, and regular assessment and management of cardiovascular risk.

This edition of the book has also updated several chapters included in the first edition, thus providing the latest information on various topics related to hypertension and HF. One of the most notable updates in this edition is the chapter on the genetics and epigenetics of arterial hypertension and HF. Other updated chapters deal with blood pressure variability, sympathetic activity, new markers of HF, cardiac imaging, and treatment of cardiac arrhythmias associated with HF and hypertension.

In summary, the second edition of *Hypertension and Heart Failure* is an excellent resource for healthcare professionals and researchers working in the field of hypertension and HF. It covers the latest research findings, guidelines, and practical recommendations for the management of hypertension and heart failure.

We are grateful to the contributors for writing their chapters Their expertise and dedication have made this second edition possible.

We would like to express our gratitude to Springer Nature for their support and guidance throughout the production of this book.

We hope that readers will find this book useful and consider it one of the reference materials to look at when discussing clinical and research problems in this area.

Giuseppe Mancia
Maria Dorobanțu

Contents

Part I

Epidemiology of Hypertension and Hypertensive Heart Disease

1 Epidemiological Aspects: Prevalence and Risk of Heart Failure Related to Blood Pressure

Peter Wohlfahrt and Renata Cífková

1.1 The Prevalence and Incidence of Heart Failure (HF)

1.1.1 Heart Failure Definition and Classification

According to the European Society of Cardiology (ESC) guidelines, HF is a clinical syndrome consisting of cardinal symptoms (e.g., breathlessness, ankle swelling, and fatigue) that may be accompanied by signs (e.g., elevated jugular venous pressure, pulmonary crackles, and peripheral edemas) caused by a structural and/or functional cardiac abnormality, resulting in elevated intracardiac pressures and/or inadequate cardiac output at rest and/or during exercise [1]. Based on left ventricular ejection fraction (LVEF), heart failure patients can be categorized as having HF with preserved ejection fraction (HFpEF, ejection fraction (EF) ≥50%), HF with mildly reduced ejection fraction (HFmrEF, EF 41–49%), or HF with reduced ejection fraction (HFrEF, EF ≤40%). Because patients with HFmrEF benefit from similar therapies as do those with LVEF ≤40%, [2–4], the previously used term

P. Wohlfahrt (✉)
Department of Preventive Cardiology, Institute for Clinical and Experimental Medicine, Prague, Czech Republic

Centre for Cardiovascular Prevention, Charles University Medical School I and Thomayer Hospital, Prague, Czech Republic

Department of Medicine II, Charles University in Prague, First Faculty of Medicine, Prague, Czech Republic

R. Cífková
Centre for Cardiovascular Prevention, Charles University Medical School I and Thomayer Hospital, Prague, Czech Republic

Department of Medicine II, Charles University in Prague, First Faculty of Medicine, Prague, Czech Republic
e-mail: renata.cifkova@ftn.cz

M. Dorobantu et al. (eds.), *Hypertension and Heart Failure*, Updates in Hypertension and Cardiovascular Protection,
https://doi.org/10.1007/978-3-031-39315-0_1

"mid-range EF" was replaced by "mildly reduced EF" in the latest guidelines. These three entities differ in terms of their epidemiological profiles, presentation, mechanisms, and response to HF therapy.

Compared to patients with HFrEF, those with HFpEF are older and more commonly have hypertension and atrial fibrillation, whereas a history of myocardial infarction is less common. In several studies, HFpEF was more common in females than in males. This may be partly explained by the sex distribution in the highest age groups. However, in an analysis of elderly individuals of similar age and prevalence of other HF risk factors, women were not at a higher risk of HFpEF than were men but were at a lower risk of HFrEF [5]. Heart failure with a mildly reduced ejection fraction is an intermediate phenotype, with the prevalence of coronary heart disease (CHD) similar to that of HFrEF, whereas other demographic characteristics, symptom profiles, comorbidities, laboratory values, and short-term outcomes are closer to those with HFpEF [6].

Although these categories, which are based on left ventricular ejection fraction, are widely accepted, some have criticized this approach due to oversimplification of a complex syndrome and the use of arbitrary cutoffs [7]. Novel classification recognizing the spectral nature of the syndrome and including new technologies may surpass the current classification.

1.1.2 The Prevalence of Heart Failure

Because HF is a heterogeneous syndrome and several criteria (Framingham, 2016 ESC, Gothenburg, Boston) have been used for its diagnosis, summarizing its epidemiology is challenging. According to the 2017 Global Burden of Disease Study, HF remains a rising global epidemic with an estimated prevalence of 64.3 million worldwide [8]. The reported prevalence of HF in developed countries varies between studies and affects 1.2–4.2% of the adult population [9–12], with a steep rise with aging to ≥10% among people over 70 years of age (Fig. 1.1). The chance that a 30-year-old person will develop HF during the rest of his or her life is 5%, rising to 7.8% in a person with arterial hypertension [13]. In individuals aged 55 years, one in three will develop HF during their remaining life span [14]. In the community,

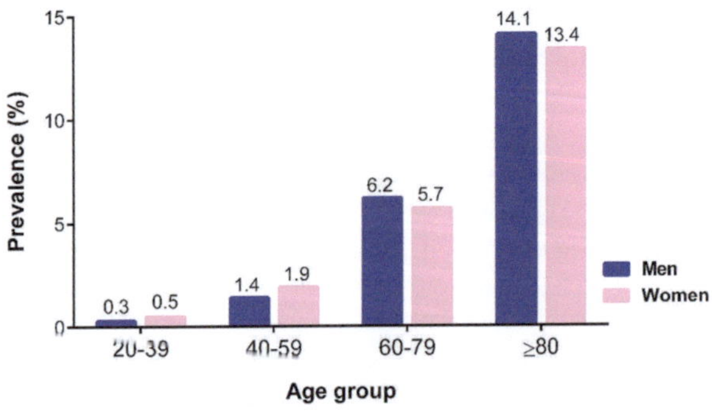

Fig. 1.1 Prevalence of heart failure by gender and age in the United States according to the National Health and Nutrition Examination Survey 2011–2014

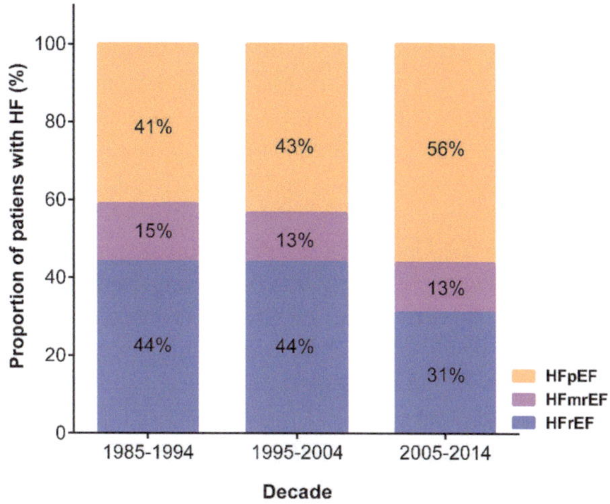

Fig. 1.2 Changes in the proportion of heart failure subtypes over three decades in the Framingham Heart Study. *HF* heart failure, *HFpEF* heart failure with preserved ejection fraction, *HFmrEF* heart failure with mid-range ejection fraction, *HFrEF* heart failure with reduced ejection fraction. (Adapted from Vasan et al. [17])

more than half of patients with HF have preserved EF [15]. The incidence and prevalence of HFpEF increase more sharply with age as compared to those of HFrEF. In the Olmsted County (Minnesota, USA) population, a decrease in HFrEF prevalence and an increase in HFpEF rate were noted during a 15-year period [16]. Similar trends were observed over a three-decade period in the Framingham Heart Study [17], with a lower prevalence of asymptomatic left ventricular systolic dysfunction accompanied by a shift in the HF phenotype toward a preponderance of HFpEF (56% vs. 31% for HFpEF vs. HFrEF in the 2005–2014 decade, respectively) (Fig. 1.2). The prevalence of HFmrEF did not change over a 30-year period. Temporal trends in the risk factors for HF, with a lower prevalence of CHD and rising hypertension rates among those with HF, explain 75% of the shift toward the greater prevalence of HFpEF.

1.1.3 The Incidence of Heart Failure

The incidence of HF varies between 3 and 29 per 1000 person-years, reflecting differences in ascertainment and adjustment between studies. In the Atherosclerosis Risk in Communities (ARIC) study, the HF incidence rates varied by ethnicity, with the highest risk in African Americans, intermediate risk in Whites and Hispanics, and the lowest risk in Chinese Americans [18]. This ethnic disparity was related to differences in hypertension and diabetes prevalence.

1.1.4 Secular Trends in Heart Failure Epidemiology

Because most of the literature on HF epidemiology comes from North America and western European countries, estimation of HF trends in the global population is scarce and unreliable. A recent literature review on the secular trends in HF epidemiology has indicated that the standardized incidence of HF in White populations has decreased since the middle of the twentieth century and that the HF epidemic in these countries is largely caused by improved HF survival rates and aging of the population [19]. Improvements in public health in high-income countries have shifted HF demographics toward the aging population with a high prevalence of chronic diseases. Elderly individuals live longer with HF, which leads to a great increase in hospitalizations with >1 million HF hospitalizations each year in both Europe and the United States. Assuming a stable prevalence of HF, the number of individuals with HF in the United States will increase by 46% over the next 20 years [20]. The total direct medical costs of HF will increase from $21 billion in 2012 to $53 billion by 2030, with hospitalizations accounting for up to three-quarters of these costs [20]. In addition, for demographic changes, an expected 54% increase in diabetes prevalence by 2030 will further increase HF prevalence in the United States in the near future [21]. Similar trends are anticipated in other high-income countries.

Although the majority of HF burden falls on the elderly, recent studies have suggested an increase in HF incidence in the young. In a Danish national sample of hospitalized patients, the incidence of HF declined among older individuals but increased among younger individuals [22]. Similarly, a study from Sweden observed a 50% and 43% increase in HF incidence in individuals aged 18–34 and 35–44 years, respectively [23]. On the other hand, since the mid-1990s, there has been a decrease in HF incidence in individuals over 45 years of age in Sweden. One of the explanations for this disturbing trend is the increase in obesity and obesity-related comorbidities, such as arterial hypertension, type 2 diabetes, and atrial fibrillation, which are occurring in the young.

1.1.5 Heart Failure Epidemiology in Middle- and Low-Income Countries

Although there is an estimated 80% cardiovascular disease burden in middle- and low-income countries, we lack data on HF epidemiology and trends. Available data show several differences from high-income regions. In the PURE (Prospective Urban Rural Epidemiology) study, the burden of cardiovascular risk factors was the lowest in low-income countries, whereas the rates of major cardiovascular disease and death were substantially higher than those in high-income countries [24]. Case fatality rates for HF increase with decreasing national income, being 2.6-times higher in middle-income and 3.7-times higher in low-income countries as compared to high-income countries. This likely reflects the lower availability of heart failure therapies in low-income countries.

The International Congestive Heart Failure (INTER-CHF) study evaluating prospectively enrolled HF patients from low- and middle-income regions showed several demographic differences compared to HF patients coming from high-income regions [25]. The mean age of HF patients in Africa and India was 53 and 56 years, respectively, whereas the mean age of HF patients from high-income regions was at least 10 years higher. Furthermore, the 20% prevalence of ischemic heart disease (IHD) in HF patients in Africa and the 25% prevalence in South America are substantially lower than in other regions of the world. Similar findings were noted in a systematic review of HF in low- and middle-income countries [26], in which the human development index integrating a country's life expectancy, education, and gross national income per capita positively correlated with age at admission for HF. Furthermore, IHD was the main reported cause of HF in all regions, except Africa and the Americas, where hypertension was the predominant cause. According to the Global Burden of Disease Study, causes of HF differ by region, with a high incidence of preventable causes of HF such as hypertensive heart disease and rheumatic heart disease in low-income countries [27]. Thus, tailoring of policies to population-specific risks and the underlying etiologies is required.

Although reliable estimates for middle- and low-income countries are lacking, evidence from the literature suggests that HF is the fastest growing cardiovascular condition globally. The expected HF prevalence increase in middle- and low-income regions will be driven by population aging and the increasing burden of hypertension and other cardiovascular risk factors. Although the age-standardized prevalence of hypertension decreased by 2.6% in high-income countries from 2000 to 2010, there was a 7.7% increase in low- and middle-income countries [28]. Thus, global hypertension control is an important target to decrease the global epidemic of HF.

1.1.6 Mortality in Heart Failure

In developed countries, HF survival improved substantially during the early 1990s and early to mid-2000s, likely due to evidence-based medications [29, 30]. Lately, no significant change in mortality has been observed in the community-based cohort from Olmsted County between the years 2000 and 2010, with reported 20% age-adjusted mortality rates for incident HF at 1 year and 56% at 5 years [31]. Similar trends were observed in other high-income countries [32]. This may be explained by the shift in the HF phenotype toward a preponderance of HFpEF and the increasing comorbidity burden of HF. Although the proportion of IHD decreased over the 10-year period in patients with HFrEF, there was a significant increase in the rates of hypertension, diabetes, and hyperlipidemia among subjects with HFpEF [31]. Moreover, more than half of the deaths were due to non-cardiovascular causes, with respiratory issues, neoplasms, and mental or behavioral health being the most common. The proportion of non-cardiovascular deaths is higher in HFpEF than in HFrEF. In the Framingham Heart Study, during the 1985–2014 period, the prognosis of HFrEF improved, whereas those of HFmrEF and HFpEF remained unchanged [17]. This may be due to the absence of specific therapy influencing the prognosis of patients with HFpEF.

In the INTER-CHF study, marked regional differences in HF mortality were noted in low- and middle-income countries, with the highest rates documented in Africa and India, intermediate in southern Asia, and the lowest in China, South America, and the Middle East. A higher threshold for case definition, greater disease severity, and limited availability of evidence-based therapies might explain these differences.

The recent shift in cardiovascular mortality in the United States has been highly disturbing. Although heart disease mortality declined between 1980 and 2000 by more than 40%, and slowed down between 2011 and 2014, there has been a mortality increase since 2015 [33]. This increase in cardiovascular mortality is mainly attributable to the increase in HF mortality, with deaths due to ischemic heart disease continuing to decline. The increase in HF mortality at the population level contrasts with the stable HF mortality observed in HF cohorts. This can be explained by increasing HF burden on the population. Although a similar increase in cardiovascular mortality has not been observed in European countries, unfavorable trends in some HF risk factors such as diabetes and obesity may also increase cardiovascular mortality in Europe.

1.2 The Epidemiology of Hypertension and Heart Failure

Hypertension is the most prevalent modifiable risk factor for HF [34]. In the Framingham Heart Study cohort, hypertension antedated the development of HF in 91% of subjects, whereas in the Cardiovascular Health Study, the proportion was 82% [35]. After adjusting for age and other heart failure risk factors, hypertension increases the risk of HF twofold in men and threefold in women. Hypertension accounts for 39% of cases of HF in men and 59% in women. Together with arrhythmias and CHD, arterial hypertension belongs to comorbidities with the largest attributable risk of HF development. The risk of HF attributable to hypertension is twofold higher in patients with HFpEF than in those with HFrEF [36]. The lifetime risk of HF for individuals with blood pressure >160/90 mmHg is double that of those with blood pressure <140/90 mmHg. An analysis of the Cardiovascular Health Study and Health ABC study in the elderly (mean age 73 years), who were not receiving antihypertensive therapy, has shown that the risk of incident HF over a 10-year follow-up increases with increasing systolic blood pressure, with individuals with blood pressure <120 mmHg having the lowest HF risk [37]. However, 38% of all incident HF events occurred in individuals with systolic blood pressure between 120 and 130 mmHg due to the highest proportion of individuals being in this group (Fig. 1.3).

In the Multi-Ethnic Study of Atherosclerosis (MESA), late systolic hypertension, defined as the ratio of late (last one-third of systole) to early (first two-thirds of systole) pressure–time integrals (PTIs) of the aortic pressure waveform, was an independent predictor of incident HF [38]. Late systolic hypertension was more predictive than the presence of hypertension. In another study [39], increased aortic pressure wave pulsatility and a greater decrease in pulsatility on treatment were associated with functional improvement in patients with HFrEF receiving aggressive vasodilator titration. These differences were not identifiable using brachial cuff pressures. This suggests that central waveform analysis may provide additional prognostic information to traditional brachial blood pressure.

Several interventional studies have highlighted the importance of hypertension control to decrease HF risk. In subjects aged ≥60 years, antihypertensive therapy reduced the risk of HF by 36–68% and had a greater impact on HF prevention than on any other major cardiovascular outcome (Table 1.1). In the Systolic Blood

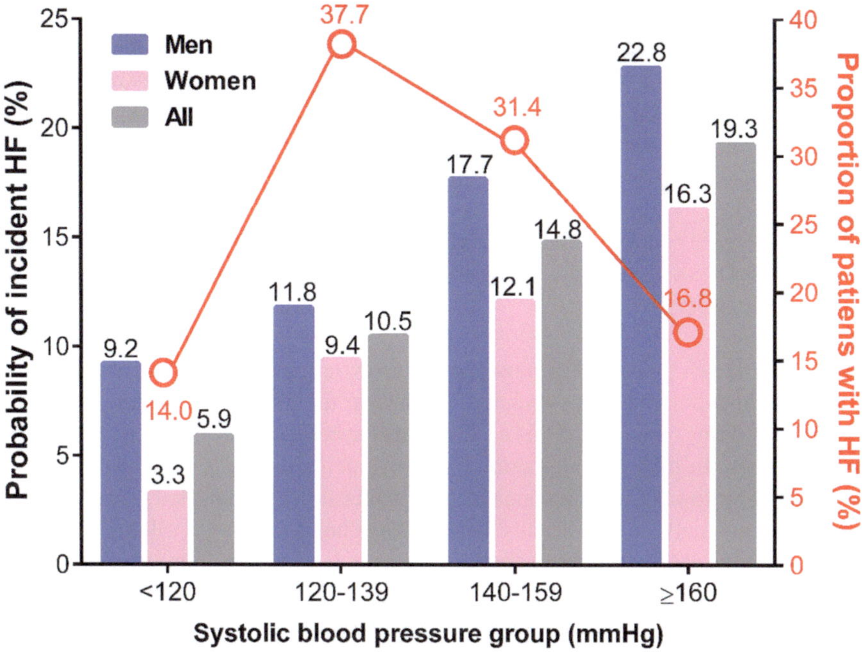

Fig. 1.3 Incident heart failure and proportion of patients developing heart failure over 10 years, by systolic blood pressure. Among elderly subjects not receiving antihypertensive therapy, the risk of incident HF over a 10-year follow-up increases with increasing systolic blood pressure, with subjects with blood pressure <120 mmHg having the lowest HF risk. However, 38% of all incident HF events occurred in subjects with systolic blood pressure between 120 and 130 mmHg (red marks) due to the highest proportion of subjects in this group. (Adapted from Butler et al. [37])

Table 1.1 Cardiovascular risk reduction by antihypertensive trials in subjects aged ≥60 years

Trial	*N*	Age (years)	Stroke (%)	CHD (%)	HF (%)	All CVDs (%)
STOP-HTN	1627	70–84	47	13	51	40
SHEP	4736	≥60	33	27	55	32
Syst-Eur	4695	≥60	42	26	36	31
STONE	1632	60–79	57	6	68	60
Syst-China	2394	≥60	38	33	38	37
HYVET	3845	≥80	30	28	64	34
SPRINT	9361	68	11	17	38	25

CHD coronary heart disease, *HF* heart failure, *CVD* cardiovascular disease, *HYVET* Hypertension in the Very Elderly Trial, *SHEP* Systolic Hypertension in the Elderly Program, *STONE* Shanghai Trial of Nifedipine in the Elderly, *STOP-HTN* Swedish Trial in Old Patients with Hypertension, *Syst-China* Systolic Hypertension in China, *Syst-Eur* Systolic Hypertension in Europe

Pressure Intervention Trial (SPRINT) among subjects at an increased cardiovascular risk without diabetes or prior stroke and baseline systolic blood pressure >130/80 mmHg, target systolic blood pressure <120 mmHg was associated with a 46% HF risk reduction and an overall decrease in cardiovascular death. In a meta-analysis, the risk of heart failure decreased by 24% for each 5-mmHg reduction in systolic blood pressure [40].

The absence of hypertension, obesity, and diabetes substantially prolongs HF-free survival. Subjects free of these three risk factors at age 45 have up to an 85% lower risk of incident HF, greater than 10 years longer HF-free survival, and live up to 13 years longer than those with all three risk factors [41]. This suggests that primordial prevention of hypertension, diabetes, and obesity leads not only to overall longer survival but also to a shortened period of chronic illness at the end of life.

1.3 Conclusions

Worldwide, heart failure is the most rapidly growing cardiovascular disease, which will consequently lead to a substantial burden on the global health-care system in the coming years. Sources of this epidemic differ between developed and developing countries. In high-income nations, improvements in public health are shifting demographics toward an aging population. Furthermore, an expected increase in the prevalence of diabetes and obesity will increase heart failure prevalence in developed countries. Although the proportion of heart failure with a preserved ejection fraction is increasing in these countries, no therapy has been shown to affect the prognosis of this heart failure subtype. In developing countries, preventable causes of heart failure such as hypertensive heart disease are responsible for most of the heart failure cases. Affected patients are younger than those in the developed countries. The switch toward a Western lifestyle, with an increase in the age-standardized prevalence of hypertension and other cardiovascular risk factors, is expected to drive the heart failure epidemic in developing countries. To combat this heart failure pandemic, improvement in the global control of heart failure risk factors will be required.

Main Points

- Heart failure is the most rapidly growing cardiovascular disease with a substantial burden on the global health-care system in the near future.
- Hypertension is the most prevalent risk factor for heart failure.
- Hypertension is among the three risk factors with the highest population-attributable risk of heart failure, particularly for heart failure with a preserved ejection fraction.
- In high-income countries, the increase in heart failure prevalence is mostly related to aging of the population, whereas in low- and middle-income countries, there is also a contribution of the increasing burden of hypertension.
- Treatment of hypertension substantially reduces the risk of developing heart failure.

Funding This work was supported by the Ministry of Health of the Czech Republic, Grant Number NV 19-09 00125.

References

1. McDonagh TA, et al. 2021 ESC guidelines for the diagnosis and treatment of acute and chronic heart failure: developed by the Task Force for the diagnosis and treatment of acute and chronic heart failure of the European Society of Cardiology (ESC) With the special contribution of the Heart Failure Association (HFA) of the ESC. Eur Heart J. 2021;42(36):3599–726.
2. Cleland JG, et al. The perindopril in elderly people with chronic heart failure (PEP-CHF) study. Eur Heart J. 2006;27(19):2338–45.
3. Abdul-Rahim AH, et al. Effect of digoxin in patients with heart failure and mid-range (borderline) left ventricular ejection fraction. Eur J Heart Fail. 2018;20(7):1139–45.
4. Cleland JGF, et al. Beta-blockers for heart failure with reduced, mid-range, and preserved ejection fraction: an individual patient-level analysis of double-blind randomized trials. Eur Heart J. 2018;39(1):26–35.
5. Silverman MG, et al. Impact of race, ethnicity, and multimodality biomarkers on the incidence of new-onset heart failure with preserved ejection fraction (from the Multi-Ethnic Study of Atherosclerosis). Am J Cardiol. 2016;117(9):1474–81.
6. Hsu JJ, Ziaeian B, Fonarow GC. Heart failure with mid-range (borderline) ejection fraction: clinical implications and future directions. JACC Heart Fail. 2017;5(11):763–71.
7. Triposkiadis F, et al. The continuous heart failure spectrum: moving beyond an ejection fraction classification. Eur Heart J. 2019;40(26):2155–63.
8. Global, regional, and national incidence, prevalence, and years lived with disability for 354 diseases and injuries for 195 countries and territories, 1990-2017: a systematic analysis for the Global Burden of Disease Study 2017. Lancet. 2018;392(10159):1789–858.
9. Benjamin EJ, et al. Heart disease and stroke statistics-2018 update: a report from the American Heart Association. Circulation. 2018;137(12):e67–e492.
10. Störk S, et al. Epidemiology of heart failure in Germany: a retrospective database study. Clin Res Cardiol. 2017;106(11):913–22.
11. Conrad N, et al. Temporal trends and patterns in heart failure incidence: a population-based study of 4 million individuals. Lancet. 2018;391(10120):572–80.
12. van Riet EES, et al. Epidemiology of heart failure: the prevalence of heart failure and ventricular dysfunction in older adults over time. A systematic review. Eur J Heart Fail. 2016;18(3):242–52.
13. Rapsomaniki E, et al. Blood pressure and incidence of twelve cardiovascular diseases: lifetime risks, healthy life-years lost, and age-specific associations in 1·25 million people. Lancet. 2014;383(9932):1899–911.
14. Bleumink GS, et al. Quantifying the heart failure epidemic: prevalence, incidence rate, lifetime risk and prognosis of heart failure: the Rotterdam study. Eur Heart J. 2004;25(18):1614–9.
15. Bursi F, et al. Systolic and diastolic heart failure in the community. JAMA. 2006;296(18):2209–16.
16. Owan TE, et al. Trends in prevalence and outcome of heart failure with preserved ejection fraction. N Engl J Med. 2006;355(3):251–9.
17. Vasan RS, et al. Epidemiology of left ventricular systolic dysfunction and heart failure in the Framingham study: an echocardiographic study over 3 decades. JACC Cardiovasc Imaging. 2017;11(1):1–11.
18. Bahrami H, et al. Differences in the incidence of congestive heart failure by ethnicity: the Multi-Ethnic Study of Atherosclerosis. Arch Intern Med. 2008;168(19):2138–45.
19. Roger VL. Epidemiology of heart failure: a contemporary perspective. Circ Res. 2021;128(10):1421–34.
20. Heidenreich PA, et al. Forecasting the impact of heart failure in the United States: a policy statement from the American Heart Association. Circ Heart Fail. 2013;6(3):606–19.
21. Rowley WR, et al. Diabetes 2030: insights from yesterday, today, and future trends. Popul Health Manag. 2017;20(1):6–12.
22. Christiansen MN, et al. Age-specific trends in incidence, mortality, and comorbidities of heart failure in Denmark, 1995 to 2012. Circulation. 2017;135(13):1214–23.

23. Barasa A, et al. Heart failure in young adults: 20-year trends in hospitalization, aetiology, and case fatality in Sweden. Eur Heart J. 2014;35(1):25–32.
24. Yusuf S, et al. Cardiovascular risk and events in 17 low-, middle-, and high-income countries. N Engl J Med. 2014;371(9):818–27.
25. Dokainish H, et al. Global mortality variations in patients with heart failure: results from the International Congestive Heart Failure (INTER-CHF) prospective cohort study. Lancet Global Health. 2017;5(7):e665–72.
26. Callender T, et al. Heart failure care in low- and middle-income countries: a systematic review and meta-analysis. PLoS Med. 2014;11(8):e1001699.
27. Vos T, et al. Years lived with disability (YLDs) for 1160 sequelae of 289 diseases and injuries 1990-2010: a systematic analysis for the Global Burden of Disease Study 2010. Lancet. 2012;380(9859):2163–96.
28. Mills KT, et al. Global disparities of hypertension prevalence and control: a systematic analysis of population-based studies from 90 countries. Circulation. 2016;134(6):441–50.
29. Jhund PS, et al. Long-term trends in first hospitalization for heart failure and subsequent survival between 1986 and 2003: a population study of 5.1 million people. Circulation. 2009;119(4):515–23.
30. Yeung DF, et al. Trends in the incidence and outcomes of heart failure in Ontario, Canada: 1997 to 2007. CMAJ. 2012;184(14):E765–73.
31. Gerber Y, et al. A contemporary appraisal of the heart failure epidemic in Olmsted County, Minnesota, 2000-2010. JAMA Intern Med. 2015;175(6):996–1004.
32. Taylor CJ, et al. Trends in survival after a diagnosis of heart failure in the United Kingdom 2000–2017: population based cohort study. BMJ. 2019;364:l223.
33. Sidney S, et al. Heterogeneity in national U.S. mortality trends within heart disease subgroups, 2000-2015. BMC Cardiovasc Disord. 2017;17(1):192.
34. Dunlay SM, et al. Risk factors for heart failure: a population-based case-control study. Am J Med. 2009;122(11):1023–8.
35. Levy D, et al. The progression from hypertension to congestive heart failure. JAMA. 1996;275(20):1557–62.
36. Chamberlain AM, et al. Risk factors for heart failure in the community: differences by age and ejection fraction. Am J Med. 2020;133(6):e237–48.
37. Butler J, et al. Systolic blood pressure and incident heart failure in the elderly. The Cardiovascular Health Study and the Health, Ageing and Body Composition Study. Heart. 2011;97(16):1304–11.
38. Chirinos JA, et al. Late systolic central hypertension as a predictor of incident heart failure: the Multi-ethnic Study of Atherosclerosis. J Am Heart Assoc. 2015;4(3):e001335.
39. Wohlfahrt P, et al. Aortic waveform analysis to individualize treatment in heart failure. Circ Heart Fail. 2017;10(2):e003516.
40. Verdecchia P, et al. Blood pressure reduction and renin-angiotensin system inhibition for prevention of congestive heart failure: a meta-analysis. Eur Heart J. 2009;30(6):679–88.
41. Ahmad FS, et al. Hypertension, obesity, diabetes, and heart failure-free survival: the cardiovascular disease lifetime risk pooling project. JACC Heart Fail. 2016;4(12):911–9.

2 An Update on the Prevalence of Heart Failure

Tomasz Zdrojewski

2.1 Introduction

Heart failure (HF) is a global public health problem. An estimated 64.3 million people are affected worldwide, making it a disease of epidemic outreach with severe medical and economic consequences [1, 2]. Older Americans and Europeans are hospitalized more for HF than for any other medical condition. Heart failure is a multifaceted and life-threatening syndrome characterized by significant morbidity and mortality, poor functional capacity and quality of life, and immense health-care costs. The lifetime risk of developing HF is high: approximately one in five people [2]. Mortality remains high, and the estimated rates of 30-day mortality are 2–3%, 1-year mortality 15–30%, 3-year mortality 30–50%, and 5-year mortality 50–75% [3]. In a population-based cohort study conducted in Olmsted County, Minnesota (USA), in the years 2007–2017, the median time from advanced HF diagnosis to death was 12.2 months and the mean rate of hospitalization was 2.91 per person-year in the first year after advanced HF was diagnosed [4].

Regardless of the criteria used, HF has shown a progressive rise in prevalence over the past few decades, attributable to both increased efficacy of treatment methods and increased age of the population. Therefore, attempts to decrease its medical, social, and economic burden have become a major global public health priority. The socioeconomic burden of HF is especially worrisome in the low- and middle-income regions of Africa, South America, and the Middle East, where the prevalence of HF is rising rapidly and the clinical characteristics, treatment patterns, and outcomes vary substantially. Unfortunately, good quality statistical data are mainly available

T. Zdrojewski (✉)
Department of Preventive Medicine and Education, Medical University of Gdansk, Gdansk, Poland
e-mail: tomasz.zdrojewski@gumed.edu.pl

M. Dorobantu et al. (eds.), *Hypertension and Heart Failure*, Updates in Hypertension and Cardiovascular Protection,
https://doi.org/10.1007/978-3-031-39315-0_2

only from developed countries. For this reason, in this chapter, we mainly summarize the data from industrialized countries and regions. In those countries, the prevalence of known heart failure is generally estimated at 1–3% of the general adult population [3]. Although in the Western world, the incidence of HF has stabilized, the prevalence is increasing. This increase in HF prevalence is due to aging of the population, improved treatment of and survival with ischemic heart disease, and availability of effective evidence-based therapies that aid in prolonging the life of patients with HF. The heart failure syndrome was first described as an emerging epidemic about 25 years ago and remains a progressive epidemic due to an increase in prevalence and hospitalizations, particularly among the elderly [3]. Clearly, effective action cannot be achieved without adequate monitoring of the prevalence of HF and its changes.

2.2 Definition of HF and Its Changes

Heart failure is difficult to define as a separate clinical entity, as its features comprise a constellation of signs and symptoms. Due to the nature of the disease process and development of diagnostic laboratory tests and echocardiography, multiple diagnostic criteria have arisen. For this reason, the published data on the prevalence of HF were highly dependent on the definition used and should be interpreted cautiously, in accordance with the respective criteria.

The first widely accepted definition of HF was introduced more than 50 years ago, based on the Framingham Heart Study [5]. However, that definition did not comprise echocardiographic examination as a diagnostic test, resulting in poor sensitivity, particularly when formulating a diagnosis for early HF. The contemporary *2022 ACC/AHA/HFSA Guideline for the Management of Heart Failure: A Report of the American College of Cardiology/American Heart Association Joint Committee on Clinical Practice Guidelines* by the American College of Cardiology/American Heart Association/Heart Failure Society of America (ACC/AHA/HFSA) defines HF as a complex clinical syndrome with symptoms and signs that result from any structural or functional impairment of ventricular filling or ejection of blood [6]. The writing committee recognized that asymptomatic stages with structural heart disease or cardiomyopathies are not included under the above definition as having HF. Such asymptomatic stages are considered at-risk of HF (stage A) or pre-HF (stage B). Patients with current or previous symptoms (due to effective therapy) of HF are considered to be at stage C, whereas those with marked HF symptoms that interfere with daily activities and with recurrent hospitalizations despite attempts to optimize guideline-related medical therapy are at stage D. It is worth stressing that from the health policy point of view, the new ACC/AHA/HFSA 2022 guidelines and the classification of HF stages (A–D) introduced therein are aimed at prevention or delay of the development of overt HF and prevention of death before the onset of symptoms.

In 2021, major scientific organizations proposed a consensus on the universal definition and classification of HF [7]. HF was defined as a clinical syndrome with symptoms and/or signs caused by a structural and/or functional cardiac abnormality with elevated natriuretic peptide levels and/or objective evidence of pulmonary or systemic congestion. Based on left ventricular ejection fraction (LVEF), HF was classified into three ejection fraction (EF) categories: HF with reduced ejection fraction (HFrEF), HF with mildly reduced ejection fraction (HFmrEF), and HF with preserved ejection fraction (HFpEF), according to the EF ranges ≤40%, 41–49%, and ≥50%, respectively. Additionally, based on the trajectories of EF over time, a new entity of HF with improved EF was introduced and defined as HF with a baseline EF >a 10-point EF increase from baseline.

This revised diagnostic approach was published in the European Society of Cardiology (ESC) 2021 guidelines for the diagnosis and treatment of heart failure [8]. In comparison with the 2016 edition of the same guidelines, HF with mid-range EF (HFmrEF 41–49%) was redefined as a phenotype closer to HFrEF than to HFpEF. This was underpinned by renaming the abbreviation HFmrEF to "HF with mildly reduced EF" instead of "HR with mid-range EF." This redefinition of HFmrEF resulted in N-terminal pro-B-type natriuretic peptide (NT-proBNP) and echocardiographic abnormalities beyond a mildly reduced EF no longer being considered as a diagnostic marker for HFmrEF. The removal of elevated NT-proBNP from the 2021 ESC HF guidelines as a requirement for diagnosing HF with mildly reduced EF does offer new treatment options for this group of patients.

Thus, on one hand, the new diagnostic algorithm reflected a growing eligibility for evidence-based therapies in patients with HFrEF and HFmrEF, but, on the other, the impact of this redefinition on the prevalence of HF and its subtypes (HFrEF, HFmrEF (mildly reduced), and HFpEF) was not known. To resolve this issue, the authors of the Hamburg City Health Study [9] compared the prevalence of HF and its subtypes based on the diagnostic criteria published in the 2016 and 2021 ESC HF guidelines in a representative, completely phenotyped sample of residents of Hamburg: 7074 participants aged 45–78 years (51.5% women). Compared with using the 2016 edition, HF prevalence increased when applying the 2021 ESC HF guidelines from 4.31% to 4.83%. This increase was driven by a higher number of subjects with HFrEF (0.47–0.52%) and HFmrEF (1.37–2.12%), whereas the number of subjects with HFpEF decreased from 2.46% to 2.19%. The authors concluded that the application of the new guidelines at the population level resulted in a 12% net increase in HF prevalence. This was mainly driven by a large increase of 54% of HFmrEF. Importantly, this did not impact the known risk factor profiles of the phenotypes [9].

Some of the differences between the new 2022 ACC/AHA/HFSA and 2021 ESC HF guidelines are a subject of current debate [10]. Certainly, both sets of guidelines provide valuable guidance for clinicians in clinical decision-making based on the results of large trials. Although they have a similar structure, each of the guidelines have elaborated more on specific topics. The point of view in the United States and Europe can contribute to these differences, as the American guidelines focus more on the patient's quality of life, cost-effectiveness, and quality of care, whereas the ESC guidelines provide more specific suggestions on the management of certain comorbidities and special populations.

2.2.1 Lifetime Risk

Although chronic HF has been described as an emerging cardiovascular disease epidemic, limited studies are available on the lifetime risk of developing heart failure. An important analysis regarding the lifetime risk was performed in the Framingham Heart Study cohort [11]. The follow-up of 3757 men and 4472 women free from HF at baseline lasted from 1971 to 1996, a period of development of invasive cardiology strategies and the introduction of angiotensin-converting enzyme inhibitors to the medical management of HF. At the age of 40 years, the lifetime risk of chronic HF was 21.0% in men and 20.3% in women, i.e., approximately one in five. As evidenced by these data, HF is overwhelmingly a disease of the elderly and its increasing incidence with advancing age outpaces the increase in mortality from competing causes.

The first European study to assess lifetime HF risk was the Rotterdam survey [12], a prospective population-based cohort study performed in participants aged ≥55 years. The baseline examination of 7734 subjects who were free from HF at the baseline began in 1989 with follow-up until a de novo diagnosis of HF. Data collection lasted until the year 2000, and the definition of HF from the 2001 ESC HF guidelines was applied. At the age of 55 years, the lifetime HF risk in the Rotterdam Study was 33% for men and 28.5% for women.

In the Cardiovascular Lifetime Risk Pooling project [13], data from several American cohorts (Cardiovascular Health Study, the Atherosclerosis Risk in Communities (ARIC) study, and the Chicago Heart Association Cohort) were pooled in 2013, including nearly 40,000 individuals. At age 45 years, the lifetime risk of heart failure through age 95 years ranged from 20% to 46%. The lifetime risk of HF increased with higher blood pressure and body mass index at all ages in both Blacks and Whites. In a recent study from the UK, linking electronic health records of 1.25 million individuals from primary care, national registries, and hospitals, the overall chance that a 30-year-old person develops HF during the rest of his or her life was 5% and in people with hypertension 7.8% [14]. The lower estimates in this British study were due to exclusion of all subjects who had experienced any type of cardiovascular disease before the outcome. This observation confirms that HF is strongly associated with age and preexisting cardiovascular disease.

2.2.2 Prevalence

The clinical epidemiology of HF has not been clearly appreciated in the context of its burden on health-care systems around the world. First, HF is likely to have changed due to changes in demography, risk factors, diagnostic procedures, and medical care. This is why several research and clinical criteria have been proposed to diagnose heart failure, each with its own advantages and disadvantages. Second, the prevalence and incidence are ideally estimated in repeated examinations of a random sample of the general population, using validated criteria that remain unchanged over time, and include objective methods such as natriuretic peptide measurement and echocardiography to evaluate cardiac dysfunction or overload. A generally accepted "gold standard" for diagnosing HF is lacking and substantial observer variability occurs even at the expert level of test interpretation like echocardiography. As a result, the prevailing information on the clinical epidemiology of HF is in part outdated, incomplete, and, to some extent, contradictory. The major epidemiological studies on the prevalence of HF were conducted in western Europe and the United States, using samples of mostly Caucasians. Their results were well-summarized in the respective chapters of the 2021 ESC [7] and the 2022 ACC/AHA/HFSA guidelines [5] as well as in the latest reviews [2, 3] dedicated to the epidemiology of HF.

The results of the studies on HF prevalence described below and undertaken already in the 1990s should be interpreted cautiously due to changes in the diagnosis and treatment that have occurred since then and differences in population characteristics. Thus, comparisons between the various studies are quite difficult. The most important surveys were the Rotterdam [12] and EPICA (EPidemiologia da Insuficiência Cardiaca e Aprendizagem) [15] studies in Europe and a survey performed in Olmsted County (Minnesota) in the United States [16]. In the Framingham Heart Study [11], clinical criteria did not include evidence of cardiac dysfunction on echocardiography, and, therefore, the true prevalence of HF might have been underestimated.

In the Rotterdam Study [12], the prevalence of HF according to the 2001 ESC HF criteria (with 40% of participants assessed by echocardiography) was 8.0% in men and 6.0% in women. The point prevalence increased from 0.9% in subjects aged 55–64 years to 4.0% in subjects aged 65–74 years, 9.7% in those aged 75–84 years, and 17.4% in those aged 85 years or older. Only 35% of patients survived 5 years after the initial diagnosis of HF.

The Olmsted County study conducted in the years 1997–2000 in randomly selected residents aged ≥45 years was the first population-based study on the prevalence of both systolic and diastolic left ventricular (LV) dysfunction based on Doppler echocardiography in relation to the symptoms and signs of HF [16]. Heart failure was diagnosed using the modified Framingham criteria. Of the 4203 invited eligible residents, 2042 (47%) participated in this cross-sectional survey. The prevalence of validated HF was 2.2% (with LVEF higher than 50% in 44% of participants), increasing from 0.7% in persons aged 45–54 years to 8.4% in those aged 75 years or older. The prevalence of LVEF ≤40% was 2.0%, and moderate or severe diastolic dysfunction was present in 7.3% of participants.

The EPICA study [15], carried out in Portugal in 1998, was the first community-based, well-designed epidemiological survey aimed to estimate the nationwide prevalence of HF. This study included a large sample size, direct evaluation by a physician, and the use of two-dimensional (2D) echocardiography. This survey included subjects attending primary care centers selected by a two-stage sampling and stratification procedure. Heart failure cases were identified based on the 1995 ESC HF guidelines. The health-care centers were randomly selected from every district by sampling in proportion to the population of the district. Subjects >25 years of age who attended primary care centers were recruited consecutively, stratified by age and examined. Overall, 5434 eligible subjects were evaluated by 365 general practitioners. The overall prevalence of chronic HF in Portugal was 4.4%. The prevalence was similar in males (4.3%) and in females (4.4%). Age-specific HF prevalence was 1.4% in the age group 25–49 years, 2.9% in the age group 50–59 years, 7.6% in the age group 60–69 years, 12.7% in the age group 70–79 years, and 16.2% among those aged 80 years or older.

It is noteworthy that the authors of the EPICA and Rotterdam studies defined LV systolic dysfunction (LVSD) as fractional shortening below 28%, which was assumed to be equivalent to LVEF below 45%. The Rotterdam, Olmsted, and EPICA studies evaluated the prevalence of HF in large Caucasian populations in Europe and the United States. However, due to the methodological differences with regard to the definition of cases, sample selection, population coverage, geographic location, and/or ethnic background, the results of these three studies cannot be directly compared. Nevertheless, their agreement regarding the prevalence of symptomatic LV systolic dysfunction was remarkable.

Ten years later, in 2008, Sanches et al. [17] reported on the prevalence of HF in the Spanish general population aged >45 years. The Prevalencia de Insuficiencia Cardiaca en España (PRICE) study was a population-based survey with participation of 15 health-care centers throughout Spain. In each area, a random sample of residents was invited and examined ($n = 1772$, 44% of males, mean age 64 ± 12 years) by their primary care physicians who used the modified Framingham criteria. Subjects ($n = 242$) who met the criteria of HF were referred for Doppler echocardiography by a cardiologist. The weighted prevalence of HF was 6.8%, similar in men (6.5%) and in women (7%). Age-specific prevalence was 1.3% in those aged 45–54 years, 5.5% in those aged 55–64 years, 8% in those aged 65–74 years, and 16.1% in those aged >74 years.

In more recent analyses, the National Health and Nutrition Health Survey (NHANES) investigators have estimated heart failure prevalence in the United States to be 2.5% based on self-reported data [18]. In Germany, an analysis of healthcare claims that data between 2009 and 2013 highlighted a crude HF prevalence of 4.0%, which was 3.9% after adjustments for age, sex, and region, with estimates increasing with age [19]. Based on primary care data from 48 health-care centers in Belgium in 2015, the prevalence of heart failure was 1.2% in men and 1.3% in women [20].

However, one should be cautious while interpreting results based on registries from official health-care claims. Liew et al. [21] undertook a retrospective cohort study based on the analysis of anonymized medical records of adult patients under the care of 43 Australian general practices between 2013 and 2018. Data were extracted from coded and uncoded fields in electronic medical records. Only 19% of patients with definite HF had this formally captured as a diagnosis in their medical record. The authors concluded that the majority of HF patients did not have this diagnosis optimally noted in their primary care medical records. Another important finding in this study was that HF was more frequent among patients of lower socioeconomic status.

In a population-based study by Conrad et al., using routine primary care data from the UK Clinical Practice Research Datalink of four million individuals, the prevalence was 1.6% [22]. The prevalence of HF standardized by age and sex was stable between 2002 and 2014, ranging between 1.5% and 1.6%, but the absolute number of patients with HF increased by 23%. In Sweden, between 2010 and 2014, the age-adjusted prevalence of HF increased from 1.61% to 1.72%, according to data from national health registries, and from 2.15% to 2.18%, according to electronic health records [23]. In a study considering data from 2.1 million inhabitants from the Stockholm region in Sweden in 2010, the crude and demographic composition-adjusted prevalence of HF was 1.8% and 2.2%, respectively, similar in women and in men [24]. A nationwide registry study in Norway showed that crude HF prevalence increased from 1.98% in 2013 to 2.42% in 2016, being higher in males than in females in 2016 (2.88% vs. 1.97%) [25]. Age-standardized prevalence was 2.3% in 2013 and 2.8% in 2016 [25]. In Italy, in a primary care setting between 2002 and 2013, the prevalence of HF was 1.25% [26].

Additionally, in a large registry-based study in a Belgian general practice between 2000 and 2015, the age-adjusted prevalence of HF stage A (at risk: patients at a high risk of developing HF but no functional/structural heart disorder) increased from 27% to 35%, similarly in males and in females, whereas for HF stage B (pre-HF: structural heart disorder but no symptoms), it increased from 7.6% to 11%. Females had a steeper increase compared to males. For HF stages C/D (structural abnormalities and symptoms managed with medical treatment or advanced HF requiring hospital-based support), a general downward trend was observed [20].

In the 2019 Heart Failure Association (HFA) Atlas project, the median prevalence of HF was estimated as 17 per 1000 persons across 13 European countries, which reported data, and ranging from ≤12 in Greece and Spain to >30 per 1000 persons in Lithuania and Germany [27]. The data from the HFA Atlas were thoroughly discussed by Ziaeian and Fonarow [28]. They stressed drastic variation in the reported burden of HF between countries. In their opinion, the data sources were of variable, if not questionable, quality and not entirely nationally representative. Thus, some of the striking variability observed may reflect differences in data quality, sources, and definitions, rather than true intercountry variability. With respect to prevalence, the HFA Atlas reports that Germany leads the included nations with a prevalence of 39 per 1000 person-years compared to 10 per 1000 person-years for Greece. Why there might be a 3.9-fold difference in HF burden is not clear for two nations with similar age distributions. Understanding both reporting differences and the underlying factors in these estimates should hopefully improve future epidemiological reporting practices.

Since none of these large studies included neither previously unrecognized cases nor removed misclassified cases, they cannot account for all heart failure patients in the population at large. A meta-analysis based on echocardiographic screening studies in the general population taking into analysis previously unrecognized cases showed that the prevalence of "all-type" heart failure in developed countries is around 11.8% in those aged ≥65 years [29]. This would account for a calculated prevalence of 4.2% in the general population. This figure is twice as high as some of the reported prevalence values based on registries containing only established cases and seems to be a more realistic estimate. This means that even half of HF cases may be undetected. Especially, HFpEF is easily missed: up to 76% of the unrecognized cases of HF are patients with a preserved ejection fraction [30]. The reasons for not recognizing heart failure are multiple and include misclassification as chronic obstructive pulmonary disease, deconditioning, aging, or obesity due to a similarity in symptoms and unavailability of echocardiography in primary care [30].

The current consensus is that the absolute numbers of patients living with heart failure have been increasing as a result of aging of the population, global population growth, and improved survival after diagnosis [31]. However, part of this increase might be caused by introduction of new diagnostic tools and criteria, e.g., the recognition of HFpEF as an important HF subtype.

2.3 Heart Failure with a Reduced Versus a Preserved Ejection Fraction

Data on the prevalence of the different HF phenotypes (HFrEF, HF mid-range EF, and HFpEF) were for a long time limited by the lack of echocardiographic systolic and diastolic parameters in many data sources. In addition, there might be high variability of measurements of echocardiographic parameters, particularly of the diastolic function. Prior to the introduction of the 2021 ESC HF guidelines, another difficulty was that in quite many patients with echocardiographic criteria of HFpEF, normal natriuretic peptide levels were observed.

The proportion of patients with HFpEF widely ranges from 16% to 73% depending on the definition used, clinical settings (primary care, hospital clinic, hospitalized patients), age and sex distribution of the studied population, previous myocardial infarction, and the year of publication. In all consecutive patients hospitalized with decompensated heart failure in the years 1987–2001 at the Mayo Clinic Hospitals in Olmsted County, Minnesota, 53% had reduced EF and 47% had preserved EF [32].

It is also worth stressing that the proportion of patients with the diagnosis of HFpEF increased over time due to a significant increase in the prevalence rates of hypertension, atrial fibrillation, and diabetes among patients with HF. In the cross-sectional, selective, screening study performed in the years 2010–2012 in patients ≥65 years presenting to primary care with shortness of breath on exertion without earlier diagnosis of HF according to the 2012 and 2016 ESC HF guidelines (54.5% of the 585 participants were female), overall 15.7% participants had HF, 2.9% had HFrEF (LVEF ≤45%), 12.0% had HFpEF, and 0.9% had isolated right-sided HF [33]. In a cross-sectional study in 2010 in the south west of the Netherlands comprising 581 patients aged ≥60 years with type 2 diabetes with no prior diagnosis of HF, 27.7% were found to have previously unrecognized HF, 4.8% with reduced EF, and 22.9% with preserved EF [34].

In the large ESC Long-Term Registry, 60% of patients were classified as HFrEF, 24% as HFmrEF, and 16% as HFpEF [35], whereas in the Swedish HF registry, the estimates were 56%, 21%, and 23%, respectively [36]. In patients hospitalized for acute HF in Spain, the figures were 32% for HFrEF, 16% for HFmrEF, and 52% for

HFpEF [37]. The corresponding estimates in the Get With The Guidelines-Heart Failure study in the Medicare population were 39%, 14%, and 47% [38]. The investigators form the Global Congestive Heart Failure (G-CHF) registry, based on analyses of 23,047 patients with newly detected or a prior diagnosis of HF from 40 countries in eight major geographic regions of the world, reported 54% of patients with HFrEF, 21% with HFmrEF, and 24% with HFpEF [39].

The data published so far support the thesis that the epidemic of HF seems to be changing. Data from a review published by Riet et al. [29] in 2016, which included 28 papers from 25 different study populations ≥60 years of age, showed that in older patients, the prevalence of diastolic dysfunction was on the rise and higher than that of systolic dysfunction. The prevalence of the latter seems to have decreased in the twenty-first century. This thesis is confirmed by the echocardiographic study presenting changes of epidemiology of left ventricular systolic dysfunction (LVSD) and HF during three decades in the Framingham Study. The purpose of this study was to describe the trends in the prevalence of LVSD in individuals without and with heart failure (HF). Vasan et al. [40] compared the frequency, correlates, and prognosis of LVSD <50% among Framingham Study participants in three periods: 1985–1994, 1995–2004, and 2005–2014. They concluded that the profile of HF in the community has changed with a lower prevalence of LVSD and an increased frequency of HFpEF, presumably due to concomitant risk factor trends. Even though in the Olmsted County Study [32], patients with advanced HFpEF were at a lower risk of cardiovascular mortality compared with advanced HFrEF (hazard ratio (HR): 0.79; 95% confidence interval (95% CI): 0.65–0.97), improving care and treatment of HFpEF should be given priority since this is expected to be the major player in the next years.

2.4 Asymptomatic Left Ventricular Dysfunction (ALVD)

By definition, HF is characterized by its symptoms and signs, and, thus, ALVD is not equivalent to HF. Population-based echocardiographic studies have demonstrated that about 50% of participants with LV systolic dysfunction (LVEF <40%) have no symptoms or signs of HF [16]. It has appeared to be important to establish the role of ALVD in the natural history of HF. Such an analysis was undertaken for the first time in a cohort of 4257 Framingham Study participants (1860 men) who underwent routine 2D echocardiography [41]. The authors showed that subjects with only mild LVEF impairment (≤50%) were at a high risk of congestive heart failure (CHF) and death. After adjustment for cardiovascular disease risk factors, ALVD was associated with a hazard ratio for HF of 4.7 compared to individuals without ALVD. An increased risk of HF was observed even in individuals with ALVD and no history of myocardial infarction or valvular disease, with an adjusted hazard ratio of 6.5.

At present, it is widely accepted that individuals may progress through the asymptomatic phase of LV systolic dysfunction before developing overt HF. The potential for any preventive measures is stressed in the new ACC/AHA/HFSA

2022 guidelines, which introduced two stages for asymptomatic patients, i.e., those being at risk of HF (stage A) and those with structural heart disorder but no symptoms (stage B) (pre-HF). Two contemporary trends (aging of industrialized populations and improvements in survival after a myocardial infarction) are expected to cause a substantial increase in the prevalence of HF. Thus, preventing HF by targeting its preclinical stages and treating known risk factors may be the best strategy. A more comprehensive review and discussion of the current research findings on ALVD and incidence rates of HF are beyond the scope of this chapter.

To summarize, in developed countries, the prevalence of HF ranges between 1% and 3% of the general adult population. Estimates from low-income countries are scarce. HF is particularly prevalent among older individuals. According to registry data from Western countries, up to roughly 50% of HF patients have HFrEF. This subtype of HF is most likely to remain stable or even decline due to improved treatments of ischemic heart disease, whereas the prevalence of HFpEF is steadily increasing and may become the most common form of HF in the future. The total number of HF patients is still continuing to rise due to a growing and aging population, improved availability of diagnostic tools, better recognition of heart failure, and progress in medical treatments that aid in prolonging life. These findings, together with the currently insufficient decline in HF mortality rates, indicate that the worldwide epidemic of HF may progress even more rapidly than assumed so far, particularly when the fight against communicable diseases in low-income countries becomes more effective.

2.5 Future Directions

The epidemiological data highlight the need for better HF treatment strategies to improve outcomes. It is also important to establish the role of asymptomatic left ventricular dysfunction in the natural history of HF and the potential for its prevention.

References

1. GBD 2017 Disease and Injury Incidence and Prevalence Collaborators. Global, regional, and national incidence, prevalence, and years lived with disability for 354 diseases and injuries for 195 countries and territories, 1990–2017: a systematic analysis for the Global Burden of Disease Study 2017. Lancet. 2018;392(10159):1789–858. https://doi.org/10.1016/S0140-6736(18)32279-7. Epub 2018 Nov 8. Erratum in: Lancet. 2019 Jun 22;393(10190):e44. PMID: 30496104; PMCID: PMC6227754.
2. Groenewegen A, Rutten FH, Mosterd A, Hoes AW. Epidemiology of heart failure. Eur J Heart Fail. 2020;22(8):1342–56. https://doi.org/10.1002/ejhf.1858. Epub 2020 Jun 1. PMID: 32483830; PMCID: PMC7540043.
3. Savarese G, Becher PM, Lund LH, Seferovic P, Rosano GMC, Coats AJS. Global burden of heart failure: a comprehensive and updated review of epidemiology. Cardiovasc Res. 2023;118(17):3272–87. https://doi.org/10.1093/cvr/cvac013. PMID: 35150240.

4. Dunlay SM, Roger VL, Killian JM, Weston SA, Schulte PJ, Subramaniam AV, Blecker SB, Redfield MM. Advanced heart failure epidemiology and outcomes: a population-based study. JACC Heart Fail. 2021;9(10):722–32. https://doi.org/10.1016/j.jchf.2021.05.009. Epub 2021 Aug 11. PMID: 34391736; PMCID: PMC8487916.
5. McKee P, Castelli W, McNamara P, et al. The natural history of congestive heart failure: the Framingham study. N Engl J Med. 1971;285(26):1441–6. https://doi.org/10.1056/nejm197112232852601.
6. Heidenreich PA, Bozkurt B, Aguilar D, Allen LA, Byun JJ, Colvin MM, Deswal A, Drazner MH, Dunlay SM, Evers LR, Fang JC, Fedson SE, Fonarow GC, Hayek SS, Hernandez AF, Khazanie P, Kittleson MM, Lee CS, Link MS, Milano CA, Nnacheta LC, Sandhu AT, Stevenson LW, Vardeny O, Vest AR, Yancy CW. 2022 AHA/ACC/HFSA guideline for the management of heart failure: a report of the American College of Cardiology/American Heart Association joint committee on clinical practice guidelines. Circulation. 2022;145(18):e895–1032. https://doi.org/10.1161/CIR.0000000000001063. Epub 2022 Apr 1. Erratum in: Circulation. 2022 May 3;145(18):e1033. Erratum in: Circulation. 2022 Sep 27;146(13):e185. PMID: 35363499.
7. Bozkurt B, Coats AJS, Tsutsui H, Abdelhamid CM, Adamopoulos S, Albert N, Anker SD, Atherton J, Böhm M, Butler J, Drazner MH, Michael Felker G, Filippatos G, Fiuzat M, Fonarow GC, Gomez-Mesa JE, Heidenreich P, Imamura T, Jankowska EA, Januzzi J, Khazanie P, Kinugawa K, Lam CSP, Matsue Y, Metra M, Ohtani T, Francesco Piepoli M, Ponikowski P, Rosano GMC, Sakata Y, Seferović P, Starling RC, Teerlink JR, Vardeny O, Yamamoto K, Yancy C, Zhang J, Zieroth S. Universal definition and classification of heart failure: a report of the Heart Failure Society of America, Heart Failure Association of the European Society of Cardiology, Japanese Heart Failure Society and Writing Committee of the Universal Definition of Heart Failure: Endorsed by the Canadian Heart Failure Society, Heart Failure Association of India, Cardiac Society of Australia and New Zealand, and Chinese Heart Failure Association. Eur J Heart Fail. 2021;23(3):352–80. https://doi.org/10.1002/ejhf.2115. Epub 2021 Mar 3. PMID: 33605000.
8. McDonagh TA, Metra M, Adamo M, Gardner RS, Baumbach A, Böhm M, Burri H, Butler J, Čelutkienė J, Chioncel O, Cleland JGF, Coats AJS, Crespo-Leiro MG, Farmakis D, Gilard M, Heymans S, Hoes AW, Jaarsma T, Jankowska EA, Lainscak M, Lam CSP, Lyon AR, McMurray JJV, Mebazaa A, Mindham R, Muneretto C, Francesco Piepoli M, Price S, Rosano GMC, Ruschitzka F, Kathrine Skibelund A, ESC Scien-tific Document Group. 2021 ESC guidelines for the diagnosis and treatment of acute and chronic heart failure. Eur Heart J. 2021;42(36):3599–726. https://doi.org/10.1093/eurheartj/ehab368. Erratum in: Eur Heart J. 2021 Oct 14; PMID: 34447992.
9. Wenzel JP, Nikorowitsch J, Bei der Kellen R, Magnussen C, Bonin-Schnabel R, Westermann D, Twerenbold R, Kirchhof P, Blankenberg S, Schrage B. Heart failure in the general population and impact of the 2021 European Society of Cardiology Heart Failure Guidelines. ESC Heart Fail. 2022;9(4):2157–69. https://doi.org/10.1002/ehf2.13948. Epub 2022 Apr 20. PMID: 35445582; PMCID: PMC9288760.
10. Behnoush AH, Khalaji A, Naderi N, Ashraf H, von Haehling S. ACC/AHA/HFSA 2022 and ESC 2021 guidelines on heart failure comparison. ESC Heart Fail. 2022;10(3):1531–44. https://doi.org/10.1002/ehf2.14255. PMID: 36460629.
11. Lloyd-Jones DM, Larson MG, Leip EP, Beiser A, D'Agostino RB, Kannel WB, Murabito JM, Vasan RS, Benjamin EJ, Levy D, Framingham Heart Study. Lifetime risk for developing congestive heart failure: the Framingham Heart Study. Circulation. 2002;106(24):3068–72. https://doi.org/10.1161/01.cir.0000039105.49749.6f. PMID: 12473553.
12. Bleumink GS, Knetsch AM, Sturkenboom MC, Straus SM, Hofman A, Deckers JW, Witteman JC, Stricker BH. Quantifying the heart failure epidemic: prevalence, incidence rate, lifetime risk and prognosis of heart failure The Rotterdam Study. Eur Heart J. 2004;25(18):1614–9. https://doi.org/10.1016/j.ehj.2004.06.038. PMID: 15351160.
13. Huffman MD, Berry JD, Ning H, Dyer AR, Garside DB, Cai X, Daviglus ML, Lloyd-Jones DM. Lifetime risk for heart failure among white and black Americans: cardiovascular lifetime

risk pooling project. J Am Coll Cardiol. 2013;61(14):1510–7. https://doi.org/10.1016/j.jacc.2013.01.022. PMID: 23500287; PMCID: PMC3618527.
14. Rapsomaniki E, Timmis A, George J, Pujades-Rodriguez M, Shah AD, Denaxas S, White IR, Caulfield MJ, Deanfield JE, Smeeth L, Williams B, Hingorani A, Hemingway H. Blood pressure and incidence of twelve cardiovascular diseases: lifetime risks, healthy life-years lost, and age-specific associations in 1.25 million people. Lancet. 2014;383(9932):1899–911. https://doi.org/10.1016/S0140-6736(14)60685-1. PMID: 24881994; PMCID: PMC4042017.
15. Ceia F, Fonseca C, Mota T, Morais H, Matias F, de Sousa A, Oliveira A, EPICA Investigators. Prevalence of chronic heart failure in Southwestern Europe: the EPICA study. Eur J Heart Fail. 2002;4(4):531–9. https://doi.org/10.1016/s1388-9842(02)00034-x. PMID: 12167394.
16. Redfield MM, Jacobsen SJ, Burnett JC Jr, Mahoney DW, Bailey KR, Rodeheffer RJ. Burden of systolic and diastolic ventricular dysfunction in the community: appreciating the scope of the heart failure epidemic. JAMA. 2003;289(2):194–202. https://doi.org/10.1001/jama.289.2.194. PMID: 12517230.
17. Anguita Sánchez M, Crespo Leiro MG, de Teresa Galván E, Jiménez Navarro M, Alonso-Pulpón L, Muñiz García J, PRICE Study Investigators. Prevalence of heart failure in the Spanish general population aged over 45 years. The PRICE Study. Rev Esp Cardiol. 2008;61(10):1041–9. https://doi.org/10.1016/s1885-5857(09)60007-4. PMID: 18817680.
18. Benjamin EJ, Virani SS, Callaway CW, Chamberlain AM, Chang AR, Cheng S, Chiuve SE, Cushman M, Delling FN, Deo R, de Ferranti SD, Ferguson JF, Fornage M, Gillespie C, Isasi CR, Jiménez MC, Jordan LC, Judd SE, Lackland D, Lichtman JH, Lisabeth L, Liu S, Longenecker CT, Lutsey PL, Mackey JS, Matchar DB, Matsushita K, Mussolino ME, Nasir K, O'Flaherty M, Palaniappan LP, Pandey A, Pandey DK, Reeves MJ, Ritchey MD, Rodriguez CJ, Roth GA, Rosamond WD, Sampson UKA, Satou GM, Shah SH, Spartano NL, Tirschwell DL, Tsao CW, Voeks JH, Willey JZ, Wilkins JT, Wu JH, Alger HM, Wong SS, Muntner P, American Heart Association Council on Epidemiology and Prevention Statistics Committee and Stroke Statistics Subcommittee. Heart disease and stroke statistics-2018 update: a report from the American Heart Association. Circulation. 2018;137(12):e67–492. https://doi.org/10.1161/CIR.0000000000000558. Epub 2018 Jan 31. Erratum in: Circulation. 2018 Mar 20;137(12):e493. PMID: 29386200.
19. Störk S, Handrock R, Jacob J, Walker J, Calado F, Lahoz R, Hupfer S, Klebs S. Epidemiology of heart failure in Germany: a retrospective database study. Clin Res Cardiol. 2017;106(11):913–22. https://doi.org/10.1007/s00392-017-1137-7. Epub 2017 Jul 26. PMID: 28748265; PMCID: PMC5655572.
20. Smeets M, Vaes B, Mamouris P, Van Den Akker M, Van Pottelbergh G, Goderis G, Janssens S, Aertgeerts B, Henrard S. Burden of heart failure in Flemish general practices: a registry-based study in the Intego database. BMJ Open. 2019;9(1):e022972. https://doi.org/10.1136/bmjopen-2018-022972. PMID: 30617099; PMCID: PMC6326340.
21. Liew D, Audehm RG, Haikerwal D, Piazza P, Neville AM, Lim K, Parsons RW, Sindone AP. Epidemiology of heart failure: Study of Heart failure in the Australian Primary carE setting (SHAPE). ESC Heart Fail. 2020;7(6):3871–80. https://doi.org/10.1002/ehf2.12979. Epub 2020 Sep 9. PMID: 32902206; PMCID: PMC7754764.
22. Conrad N, Judge A, Tran J, Mohseni H, Hedgecott D, Crespillo AP, Allison M, Hemingway H, Cleland JG, McMurray JJV, Rahimi K. Temporal trends and patterns in heart failure incidence: a population-based study of 4 million individuals. Lancet. 2018;391(10120):572–80. https://doi.org/10.1016/S0140-6736(17)32520-5. Epub 2017 Nov 21. PMID: 29174292; PMCID: PMC5814791.
23. Lindmark K, Boman K, Olofsson M, Törnblom M, Levine A, Castelo-Branco A, Schlienger R, Bruce Wirta S, Stålhammar J, Wikström G. Epidemiology of heart failure and trends in diagnostic work-up: a retrospective, population-based cohort study in Sweden. Clin Epidemiol. 2019;11:231–44. https://doi.org/10.2147/CLEP.S170873. PMID: 30962724; PMCID: PMC6435223.
24. Zarrinkoub R, Wettermark B, Wändell P, Mejhert M, Szulkin R, Ljunggren G, Kahan T. The epidemiology of heart failure, based on data for 2.1 million inhabitants in Sweden. Eur J Heart

Fail. 2013;15(9):995–1002. https://doi.org/10.1093/eurjhf/hft064. Epub 2013 May 3. PMID: 23645498.
25. Ødegaard KM, Hallén J, Lirhus SS, Melberg HO, Halvorsen S. Incidence, prevalence, and mortality of heart failure: a nationwide registry study from 2013 to 2016. ESC Heart Fail. 2020;7(4):1917–26. https://doi.org/10.1002/ehf2.12773. Epub 2020 Jun 12. PMID: 32530563; PMCID: PMC7373922.
26. Piccinni C, Antonazzo IC, Simonetti M, Mennuni MG, Parretti D, Cricelli C, Colombo D, Nica M, Cricelli I, Lapi F. The burden of chronic heart failure in primary care in Italy. High Blood Press Cardiovasc Prev. 2017;24(2):171–8. https://doi.org/10.1007/s40292-017-0193-4. Epub 2017 Mar 21. PMID: 28324597.
27. Seferović PM, Vardas P, Jankowska EA, Maggioni AP, Timmis A, Milinković I, Polovina M, Gale CP, Lund LH, Lopatin Y, Lainscak M, Savarese G, Huculeci R, Kazakiewicz D, Coats AJS. National Heart Failure Societies of the ESC member countries (see Appendix). The Heart Failure Association Atlas: heart failure epidemiology and management statistics 2019. Eur J Heart Fail. 2021;23(6):906–14. https://doi.org/10.1002/ejhf.2143. Epub 2021 Mar 13. PMID: 33634931.
28. Ziaeian B, Fonarow GC. Making heart failure count. Eur J Heart Fail. 2021;23(6):917–8. https://doi.org/10.1002/ejhf.2148. Epub 2021 Mar 18. PMID: 33713517; PMCID: PMC8286334.
29. van Riet EE, Hoes AW, Wagenaar KP, Limburg A, Landman MA, Rutten FH. Epidemiology of heart failure: the prevalence of heart failure and ventricular dysfunction in older adults over time. A systematic review. Eur J Heart Fail. 2016;18(3):242–52. https://doi.org/10.1002/ejhf.483. Epub 2016 Jan 4. PMID: 26727047.
30. Caruana L, Petrie MC, Davie AP, McMurray JJ. Do patients with suspected heart failure and preserved left ventricular systolic function suffer from "diastolic heart failure" or from misdiagnosis? A prospective descriptive study. BMJ. 2000;321(7255):215–8. https://doi.org/10.1136/bmj.321.7255.215. PMID: 10903655; PMCID: PMC27439.
31. Roth GA, Forouzanfar MH, Moran AE, Barber R, Nguyen G, Feigin VL, Naghavi M, Mensah GA, Murray CJ. Demographic and epidemiologic drivers of global cardiovascular mortality. N Engl J Med. 2015;372(14):1333–41. https://doi.org/10.1056/NEJMoa1406656. PMID: 25830423; PMCID: PMC4482354.
32. Owan TE, Hodge DO, Herges RM, Jacobsen SJ, Roger VL, Redfield MM. Trends in prevalence and outcome of heart failure with preserved ejection fraction. N Engl J Med. 2006;355(3):251–9. https://doi.org/10.1056/NEJMoa052256. PMID: 16855265.
33. van Riet EE, Hoes AW, Limburg A, Landman MA, van der Hoeven H, Rutten FH. Prevalence of unrecognized heart failure in older persons with shortness of breath on exertion. Eur J Heart Fail. 2014;16(7):772–7. https://doi.org/10.1002/ejhf.110. Epub 2014 May 26. PMID: 24863953.
34. Boonman-de Winter LJ, Rutten FH, Cramer MJ, Landman MJ, Liem AH, Rutten GE, Hoes AW. High prevalence of previously unknown heart failure and left ventricular dysfunction in patients with type 2 diabetes. Diabetologia. 2012;55(8):2154–62. https://doi.org/10.1007/s00125-012-2579-0. Epub 2012 May 18. PMID: 22618812; PMCID: PMC3390708.
35. Chioncel O, Lainscak M, Seferovic PM, Anker SD, Crespo-Leiro MG, Harjola VP, Parissis J, Laroche C, Piepoli MF, Fonseca C, Mebazaa A, Lund L, Ambrosio GA, Coats AJ, Ferrari R, Ruschitzka F, Maggioni AP, Filippatos G. Epidemiology and one-year outcomes in patients with chronic heart failure and preserved, mid-range and reduced ejection fraction: an analysis of the ESC Heart Failure Long-Term Registry. Eur J Heart Fail. 2017;19(12):1574–85. https://doi.org/10.1002/ejhf.813. Epub 2017 Apr 6. PMID: 28386917.
36. Koh AS, Tay WT, Teng THK, Vedin O, Benson L, Dahlstrom U, Savarese G, Lam CSP, Lund LH. A comprehensive population-based characterization of heart failure with mid-range ejection fraction. Eur J Heart Fail. 2017;19(12):1624–34. https://doi.org/10.1002/ejhf.945. Epub 2017 Sep 25. PMID: 28948683.
37. Santas E, de la Espriella R, Palau P, Miñana G, Amiguet M, Sanchis J, Lupón J, Bayes-Genís A, Chorro FJ, Núñez Villota J. Rehospitalization burden and morbidity risk in patients with heart failure with mid-range ejection fraction. ESC Heart Fail. 2020;7(3):1007–14. https://doi.org/10.1002/ehf2.12683. Epub 2020 Mar 25. PMID: 32212327; PMCID: PMC7261530.

38. Cheng RK, Cox M, Neely ML, Heidenreich PA, Bhatt DL, Eapen ZJ, Hernandez AF, Butler J, Yancy CW, Fonarow GC. Outcomes in patients with heart failure with preserved, borderline, and reduced ejection fraction in the Medicare population. Am Heart J. 2014;168(5):721–30. https://doi.org/10.1016/j.ahj.2014.07.008. Epub 2014 Jul 22. PMID: 25440801.
39. Joseph P, Dokainish H, McCready T, Budaj A, Roy A, Ertl G, Gomez-Mesa JE, Leong D, Ezekowitz J, Hage C, Lanas F, Maggioni AP, Sliwa K, Zhu J, Rouleau J, Balasubramanian K, Yusuf S, G-CHF Investigators. A multinational registry to study the characteristics and outcomes of heart failure patients: the global congestive heart failure (G-CHF) registry. Am Heart J. 2020;227:56–63. https://doi.org/10.1016/j.ahj.2020.06.002. Epub 2020 Jun 6. PMID: 32679282.
40. Vasan RS, Xanthakis V, Lyass A, Andersson C, Tsao C, Cheng S, Aragam J, Benjamin J, Larson MG. Epidemiology of left ventricular systolic dysfunction and heart failure in the Framingham Study: an echocardiographic study over 3 decades. JACC Cardiovasc Imaging. 2018;11(1):1–11. https://doi.org/10.1016/j.jcmg.2017.08.007. Epub 2017 Oct 5. PMID: 28917679; PMCID: PMC5756128.
41. Wang TJ, Evans JC, Benjamin EJ, Levy D, LeRoy EC, Vasan RS. Natural history of asymptomatic left ventricular systolic dysfunction in the community. Circulation. 2003;108(8):977–82. https://doi.org/10.1161/01.CIR.0000085166.44904.79. Epub 2003 Aug 11. PMID: 12912813.

Heart Failure in Severely Hypertensive Children

3

Empar Lurbe and Francisco Aguilar

3.1 Introduction

High blood pressure (BP) is a clearly established, but modifiable, risk factor for early disability and death. Although most of the adverse outcomes occur in adulthood, it has become clear that high BP is a life course problem that can become evident in early life. While few would dispute the importance of taking effective steps to identify and manage this condition in middle-aged and older people, relatively little attention has been paid to the problem of high BP in children and adolescents.

Although long considered a disease of aging, its prevalence is increasing in children and adolescents [1, 2]. Importantly, the negative and potentially severe consequences of high BP are not limited to adulthood. Evidence of target organ damage, such as left ventricular hypertrophy and pathological vascular changes, has been found even in young children and already in children with newly diagnosed high BP. Without intervention, high BP in childhood will increase the risk of premature cardiovascular disease; consequently, effective preventive interventions that are applied early in life will modify disease progression [3].

The definition of hypertension (HTN) in children and adolescents is still based on the normal distribution of BP in healthy children and not on the cardiovascular morbidity and mortality or on the risk of developing early organ damage associated with a certain level of BP. Clinicians can use the available pediatric normative BP data to determine whether BP is within the normal range or is at a level that warrants attention or preventive intervention.

E. Lurbe (✉) · F. Aguilar
Pediatric Department, Consorcio Hospital General, University of Valencia, Valencia, Spain

CIBER Fisiopatología Obesidad y Nutrición (CB06/03), Instituto de Salud Carlos III, Madrid, Spain
e-mail: empar.lurbe@uv.es; francisco.aguilar@uv.es

M. Dorobantu et al. (eds.), *Hypertension and Heart Failure*, Updates in Hypertension and Cardiovascular Protection,
https://doi.org/10.1007/978-3-031-39315-0_3

The prevalence of pediatric HTN worldwide is difficult to establish because of the regional differences in the definition, the distribution of reference BP data, and the methods of BP measurement. A recent meta-analysis, including 47 studies, has reported the pooled prevalence of HTN in 4%, among obese 15.3%, and overweight 4.9% versus 1.9% in normal-weight children [4]. The data are worrying because HTN in childhood is usually already linked to cardiovascular risk factors and intermediate markers of target organ damage (e.g., left ventricular hypertrophy or thickening of the carotid artery vessel wall) [1, 2].

Besides BP values, better knowledge about the natural history of early organ damage is necessary. The assessment of organ damage needs to be optimized, looking for early markers. High BP-induced organ damage can evolve concurrently in the heart, carotid artery walls, the kidneys, and in the eye vessels; however, the most common approach is to study damage in one or two areas only.

3.2 Severe Hypertension and Heart Failure

Unlike in adults, hypertension is not a common cause of heart failure in children and adolescents [1], even though the presence of severe hypertension is a condition in which congestive heart failure can be present. Severe hypertension has not been rigor ously defined, which has led to some confusion with respect to the terminology. For the purpose of this chapter, severe hypertension is defined as a BP elevation that fulfills and usually exceeds the definition of stage 2 hypertension (Table 3.1) and that is accompanied by severe symptoms [1, 2].

Severe hypertension has traditionally been divided into hypertensive emergencies and hypertensive urgencies, where the former is associated with life-threatening symptoms and/or target organ injury and the latter is associated with less significant symptoms and no target injury. The distinction is arbitrary and depends somewhat on clinical judgment; therefore, what is most important is a careful assessment of patients presenting with severe hypertension, followed by the application of meticulous treatment and monitoring [5–8].

Table 3.1 2016 ESH guidelines: classification of HTN

	0–15 years	16 years and older
Category	SBP and/or DBP percentile	SBP and/or DBP values
Normal	<90th	<130/85 mmHg
High–normal	≥90th to <95th percentile	130–139/85–89 mmHg
Hypertension	≥95th percentile	≥140/90 mmHg
Stage 1 hypertension	95th percentile to the 99th percentile plus 5 mmHg	140–159/90–99 mmHg
Stage 2 hypertension	>99th percentile plus 5 mmHg	160–179/100–109 mmHg
Isolated systolic hypertension	SBP ≥ 95th percentile and DBP <90th percentile	≥140/<90 mmHg

Lurbe E et al. *J Hypertens*. 2016; 34:1887–1920 [1]. *ESH* European Society of Hypertension, *SBP* systolic blood pressure, *DBP* diastolic blood pressure

3.2.1 The Etiology of Severe Hypertension in Children and Adolescents

Secondary hypertension is still the major cause of severe hypertension. Conditions like coarctation of the aorta, renal vein or artery thrombosis, and polycystic kidney disease predominate in neonates [6]. Renal parenchymal diseases, endocrine conditions, autoimmune diseases, and medications, on the other hand, are important etiologies in older children and adolescents [6]. In adolescents, substance abuse, specifically with cocaine, amphetamines, hallucinogens, and nonsteroidal anti-inflammatory drugs, should be strongly considered as a cause of severe hypertension [6]. More causes of secondary hypertension are included in Table 3.2.

3.2.2 Clinical Presentation

Special attention should be paid to the presence of acute severe hypertension that may result in life-threatening complications and congestive heart failure, among others. Acute severe hypertension is a relatively rare condition in children, defined as a significant BP elevation with or without evidence of acute hypertensive target organ damage. Although there has not been extensive research in this area in children and adolescents, it requires prompt therapy to prevent and/or ameliorate further target organ damage.

In children and adolescents, acute severe HTN may present with life-threatening symptoms or signs and symptoms of end organ damage. Seizure-induced altered mental status, lethargy, and visual disturbances are often present in hypertensive encephalopathy. Patients may also present with tachypnea, pulmonary edema, or a galop rhythm or a new heart murmur may suggest congestive heart failure. An echocardiogram might show left ventricular hypertrophy if hypertension is chronic. A fundoscopic examination might reveal papilledema or hypertensive retinopathy. Younger children may have nonspecific symptoms such as lethargy and poor feeding. Severe HTN may also be present with milder symptoms [1, 2, 5–7].

Table 3.2 Causes of acute severe hypertension in children and adolescents

Kidney disease	Vascular	Malignancies	Others
Acute glomerulonephritis Reflux nephropathy Acute kidney injury Polycystic kidney disease Hemolytic uremic syndrome End-stage kidney disease at presentation	Aortic coarctation (thoracic, abdominal) Renal artery stenosis Vasculitis (Takayasu's, polyarteritis nodosa, others)	Pheochromocytoma Wilms tumor Neuroblastoma	Medication noncompliance in a patient with known hypertension Drug-induced, especially abrupt withdrawal of clonidine or beta-blockers Monogenic hypertension Illicit substances (cocaine, DDMA) Primary hypertension (unusual)

Although children and adolescents can become symptomatic from hypertension at lesser degrees of BP elevation than adults, in general, patients who present with acute severe HTN will have a BP elevation well above the stage 2 hypertension threshold (Table 3.1). Clinicians should be concerned about the development of acute target organ effects when a child's BP increases 30 mmHg or more above the 95th percentile [1, 2]. The presence of encephalopathy, seizures, renal failure, or congestive heart failure signifies the presence of hypertensive emergency regardless of the degree of elevation of BP and requires immediate evaluation and treatment. Although some patients may have no identifiable cause, the majority of affected patients have secondary causes of HTN [5–8] (Table 3.2). Thus, for patients who present with acute severe HTN, an evaluation for secondary causes is appropriate and should be conducted expediently [1, 2].

3.2.3 Initial Investigations

A laboratory evaluation should include serum electrolytes, blood urea nitrogen, creatinine, liver enzymes, a complete blood count, drug screen, and urinalysis. Females of reproductive age should always have a pregnancy test done. These initial laboratory tests might reveal or increase the suspicion for a secondary cause of HTN. Hypokalemia and metabolic alkalosis may be seen in aldosteronism and some cases of monogenic HTN. High serum creatinine indicates either acute or chronic renal failure. Blood and protein in the urine could suggest glomerulonephritis. Low hemoglobin and platelets may suggest hemolytic uremic syndrome [1, 2, 7].

3.2.4 Imaging Studies

Imaging studies should include brain imaging if there are any concerns about the mental status of the patient or any focal neurological deficits. Chest X-rays should be done to look for cardiomegaly and pulmonary edema. An echocardiogram could reveal left ventricular hypertrophy, decreased function in patients with congestive heart failure, and coarctation of the aorta as a cause of HTN. A renal ultrasound should be done in any hypertensive child to look for renal asymmetry, scarring, cystic kidney disease, and hydronephrosis. Additionally, target organ effects should be assessed with renal function, echocardiography, and central nervous system imaging [1, 2, 5–8].

3.2.5 Mechanisms

As the majority of hypertensive crises in the pediatric population are attributed to secondary hypertension, the mechanism of severe hypertension is directly related to the nature of the underlying condition. However, regardless of the cause, these are some common pathways that may relate to the severity of hypertension itself and not to the processes leading to symptoms and end organ damage.

The mechanisms of hypertension leading to the development of hypertensive emergencies often involve the renin–angiotensin–aldosterone system [5, 6]. High renin and aldosterone are often found in renovascular and other renal causes of hypertension. Activation of this system leads to vasoconstriction via angiotensin II production and sodium retention through the effects of aldosterone on the kidneys. Angiotensin II may also promote endothelial dysfunction and increased expression of proinflammatory cytokines such as nuclear factor kappa B (NF-kB). Other mechanisms leading to severe blood pressure elevation may include fluid overload, as may occur in acute kidney injury or chronic kidney disease, activation of the sympathetic nervous system by secretion of vasoactive substances, as in a pheochromocytoma, and medications [9].

3.2.6 Treatment

Although there has not been extensive research in this area in children, and recommendations are mostly derived from adult data, in the last few years, there have been more pediatric studies on the safety and effectiveness of antihypertensives. Treatment of a child with severe hypertension should be done concomitantly with the evaluation. Treatment of acute severe HTN requires admission in the pediatric intensive care unit with meticulous patient monitoring to prevent excessively rapid lowering of BP [1, 2, 5–8]. In situations in which the underlying condition causing HTN has been identified, treatment should be targeted at the underlying pathophysiology of HTN [1, 2, 5–8].

In patients with the most significant target organ involvement, BP should be lowered no more than 25% of the planned reduction over the first 8 h to prevent compromise of organ perfusion (e.g., coronary, cerebral, or renal ischemia). The ultimate BP target in the acute phase is usually the 95th percentile and can be achieved over the next 24–48 h [1, 2, 5–8]. A faster normalization must be strictly avoided as it can cause more harm than severe HTN itself. Therefore, careful neurological and CV assessment should be undertaken throughout the initial treatment. In most patients, this will require administration of intravenous antihypertensive medications, and, in less severely affected patients, oral medications may be considered. Drug dosing for patients with acute severe HTN has been described in detail in a recent paper published by Baracco [7]; suggested drug doses are presented in Tables 3.3 and 3.4.

Table 3.3 Useful for severely hypertensive patients with life-threatening symptoms

Drug and class	Dose	Route	Onset	Side effects and precautions
Esmolol [10] β-Adrenergic blocker	100–500 μg/kg/ loading dose 50–150 μg/kg/min Maximum: 1000 μg/kg/min	IV infusion	1 min	Contraindication in asthma May cause profound bradycardia Do not use in catecholamine excess or CHF
Hydralazine [11] Direct vasodilator	0.1–0.2 mg/kg/dose Up to 0.6 mg/kg/ dose	IV, IM	10 min	Tachycardia, sodium retention
Labetalol [12] α- and β-adrenergic blockers	Bolus 0.20–1.0 mg/kg/ dose, up to 40 mg/ dose Infusion: 0.25–3.0 mg/kg/h	IV bolus or infusion	2–5 min	Do not use in asthma and congestive heart failure
Nicardipine [13, 14] Calcium channel blocker	1–3 μg/kg/min	IV infusion	1–2 min	May cause reflex tachycardia, hypotension, flushing, phlebitis
Sodium nitroprusside [15] Direct vasodilator	0.3–1 μg/kg/min Maximum: 10 mcg/ kg/min	IV infusion	1–2 min	Risk of cyanide toxicity after 24–48 h of use and in renal failure Avoid in infants

IM intramuscular, *IV* intravenous, *kg* kilogram, *μg* microgram, *mg* milligram, *CHF* congestive heart failure

Table 3.4 Useful for severely hypertensive patients with less significant symptoms

Drug and class	Dose	Route	Onset	Side effects and precautions
Captopril [7] ACE inhibitor	0.3–0.5 mg/kg/ dose	PO	15 min	Angioedema Avoid in patients with CHF, volume depletion, and bilateral renal artery stenosis
Clonidine [16] Central α-agonist	0.05–0.1 mg/dose in older children	Po	15–30 min	Side effects include dry mouth and drowsiness
Isradipine [17] Calcium channel blocker	0.05–0.1 mg/kg/ dose Maximum: 5 mg/ dose 0.05 mg/kg/dose in <2-year-olds	Po	30–60 min	Hypotension, headache, flushing Avoid in patients receiving azole antifungals
Minoxidil [18] Direct vasodilator	0.1–0.2 mg/kg/ dose Maximum: 10 mg/ dose	Po	30 min	Fluid retention and hirsutism
Nifedipine [19, 20] Calcium channel blocker	0.1–0.25 mg/kg/ dose	Po	5–15 min	Hypotension, tachycardia, and flushing

CHF congestive heart failure, *ACE* angiotensin-converting enzyme, *BP* blood pressure, *IM* intramuscular, *IV* intravenous, *kg* kilogram, *μg* microgram, *mg* milligram, *PO* oral

3.3 Conclusions

- Unlike in adults, HTN is not a common cause of heart failure in children and adolescents.
- Tachypnea, pulmonary edema, a galop rhythm, or a new heart murmur may suggest congestive heart failure.
- The presence of acute severe HTN may result in life-threatening complications such as congestive heart failure and/or hypertensive encephalopathy.
- Acute severe HTN should be evaluated emergently, and treatment should be instituted without delay to reduce risks such as severe elevation of BP in a child, which may lead to devastating complications.
- Immediate treatment with short-acting antihypertensive medication should be initiated in children and adolescents with acute severe HTN and life-threatening symptoms.
- Several available drugs have been shown to be safe and effective in children with severe HTN, although more research is needed to guide the management of these patients.DisclaimersNone.

References

1. Lurbe E, Agabiti-Rosei E, Cruickshank JK, Dominiczak A, Erdine S, Hirth A, et al. 2016 European Society of Hypertension guidelines for the management of high blood pressure in children and adolescents. J Hypertens. 2016;34:1887–920. https://doi.org/10.1097/HJH.0000000000001039.
2. Flynn JT, Kaelber DC, Baker-Smith CM, Blowey D, Carroll AE, Daniels SR, SUBCOMMITTEE ON SCREENING AND MANAGEMENT OF HIGH BLOOD PRESSURE IN CHILDREN, et al. Clinical practice guideline for screening and management of high blood pressure in children and adolescents. Pediatrics. 2017;140:e20171904. https://doi.org/10.1542/peds.2017-1904.
3. Falkner B. Hypertension in children and adolescents: epidemiology and natural history. Pediatr Nephrol. 2010;25:1219–24. https://doi.org/10.1007/s00467-009-1200-3.
4. Song P, Zhang Y, Yu J, Zha M, Zhu Y, Rahimi K, Rudan I. Global prevalence of hypertension in children: a systematic review and meta-analysis. JAMA Pediatr. 2019;173:1154–63. https://doi.org/10.1001/jamapediatrics.2019.3310.
5. Flynn JT, Tullus K. Severe hypertension in children and adolescents: pathophysiology and treatment. Pediatr Nephrol. 2009;24:1101–12.
6. Seeman T, Hamdani G, Mitsnefes M. Hypertensive crisis in children and adolescents. Pediatr Nephrol. 2019;34:2523–37. https://doi.org/10.1007/s00467-018-4092-2.
7. Baracco R. A practical guide to the management of severe hypertension in children. Pediatr Drugs. 2020;22:13–20. https://doi.org/10.1007/s40272-019-00377-z.
8. Larkins N, Lurbe E, Flynn JT. Evaluation and management of hypertension in children. In: Emma F, Goldstein S, Bagga A, Bates CM, Shroff R, editors. Pediatric nephrology. Berlin, Heidelberg: Springer; 2021. https://doi.org/10.1007/978-3-642-27843-3_123-1.
9. Belsha C. Management of hypertensive emergencies. In: Pediatric hypertension. Portman Humana Press; 2013. p. 557–71. https://doi.org/10.1007/978-1-62703-490-6_37.
10. Tabbutt S, Nicolson SC, Adamson PC, Zhang X, Hofman ML, Wells W, et al. The safety, efficacy, and pharmacokinetics of esmolol for blood pressure control immediately after repair of coarctation of the aorta in infants and children: a multicenter, double-blind, randomized trial. J Thorac Cardiovasc Surg. 2008;136:321–8. https://doi.org/10.1016/j.jtcvs.2007.09.086.

11. Flynn JT, Bradford MC, Harvey EM. Intravenous hydralazine in hospitalized children and adolescents with hypertension. J Pediatr. 2016;168:88–92. https://doi.org/10.1016/j.jpeds.2015.07.070.
12. Thomas CA, Mofett BS, Wagner JL, Mott AR, Feig DI. Safety and efcacy of intravenous labetalol for hypertensive crisis in infants and small children. Pediatr Crit Care Med. 2011;12:28–32. https://doi.org/10.1097/PCC.0b013e3181e328d8.
13. Tenney F, Sakarcan A. Nicardipine is a safe and efective agent in pediatric hypertensive emergencies. Am J Kidney Dis. 2000;35:E20. https://doi.org/10.1016/s0272-6386(00)70285-x.
14. Flynn JT, Mottes TA, Brophy PD, Kershaw DB, Smoyer WE, Bunchman TE. Intravenous nicardipine for treatment of severe hypertension in children. J Pediatr. 2001;139:38–43. https://doi.org/10.1067/mpd.2001.114030.
15. Drover DR, Hammer GB, Barrett JS, Cohane CA, Reece T, Zajicek A, et al. Evaluation of sodium nitroprusside for controlled hypotension in children during surgery. Front Pharmacol. 2015;6:136. https://doi.org/10.3389/fphar.2015.00136.
16. Stein DR, Ferguson MA. Evaluation and treatment of hypertensive crises in children. Integr Blood Press Control. 2016;16:49–58. https://doi.org/10.2147/IBPC.S50640.
17. Miyashita Y, Peterson D, Rees JM, Flynn JT. Isradipine for treatment of acute hypertension in hospitalized children and adolescents. J Clin Hypertens. 2010;12:850–5. https://doi.org/10.1111/j.1751-7176.2010.00347.x.
18. Halling SE, Asling-Monemi K, Herthelius M, Celsi G, Vavilis G, Kuru NK, et al. Minoxidil therapy in children and young adult patients with renal disease and refractory hypertension: value when multidrug regimens have failed to achieve blood pressure control. J Hum Hypertens. 2010;24:552–4. https://doi.org/10.1038/jhh.2010.40.
19. Egger DW, Deming DD, Hamada N, Perkin RM, Sahney S. Evaluation of the safety of short-acting nifedipine in children with hypertension. Pediatr Nephrol. 2002;17:35–40. https://doi.org/10.1007/s004670200006.
20. Yiu V, Orrbine E, Rosychuk RJ, MacLaine P, Goodyer P, Girardin C, et al. The safety and use of short-acting nifedipine in hospitalized children. Pediatr Nephrol. 2004;19:644–50. https://doi.org/10.1007/s00467-004-1444-x.

Part II

Mechanisms of Heart Failure in Arterial Hypertension

Genetics of Hypertension and Heart Failure

4

Sandosh Padmanabhan, Clea du Toit,
and Anna F. Dominiczak

The high prevalence of hypertension and its significant health and economic burden on the individual and population highlights the importance of understanding its causal pathways and developing effective therapeutic, diagnostic, and primary prevention measures [1]. Heart failure (HF) is the leading cause of death worldwide [2], resulting typically from myocardial disease with impairment of left ventricular (LV) function and manifesting with either a reduced (HFrEF) or a preserved ejection fraction (HFpEF). In this review, HF refers to HFrEF. The etiology of HF comprises several cardiovascular and systemic disorders, which frequently co-occur, most notably coronary artery disease (CAD), obesity, and hypertension [2]. In this chapter, we will describe the genetic underpinnings of hypertension and HF and the common pathways that affect both as part of the cardiovascular continuum [3]. The genetics of other causes of HF, such as ischemic heart disease, is beyond the scope of this chapter.

4.1 Heritability of Hypertension and Heart Failure

Family studies have consistently demonstrated a genetic component influencing blood pressure (BP) and HF. The heritabilities of clinic systolic BP (SBP) and diastolic BP (DBP) are around 15–40% and 15–30%, respectively, whereas they are

S. Padmanabhan · C. du Toit
Institute of Cardiovascular and Medical Sciences, University of Glasgow, Glasgow, UK
e-mail: Sandosh.Padmanabhan@glasgow.ac.uk; Clea.DuToit@glasgow.ac.uk

A. F. Dominiczak (✉)
Institute of Cardiovascular and Medical Sciences, University of Glasgow, Glasgow, UK

Medicine, University of Glasgow, BHF Glasgow Cardiovascular Research Centre, Glasgow, UK
e-mail: Anna.Dominiczak@glasgow.ac.uk

M. Dorobantu et al. (eds.), *Hypertension and Heart Failure*, Updates in Hypertension and Cardiovascular Protection,
https://doi.org/10.1007/978-3-031-39315-0_4

higher for ambulatory nighttime systolic and diastolic BP at 32–70% and 32–50%, respectively [4–10]. The sibling recurrent risk of hypertension is around 1.2–1.5 [11], and a large Swedish adoption study estimated a heritability of ~26% for HF [12]. The Framingham Heart Study found that the occurrence of HF (before 75 years of age) in at least one parent was a significant predictor of the HF phenotype in the offspring even after accounting for echocardiographically determined LV mass (LVM), thus suggesting a heritable trait [13]. A sizeable proportion of the HF burden is attributable to coronary heart disease (CHD) and myocardial infarction (MI). This is evidenced by the reduction from a one in five lifetime risk of HF in all men to one in nine in the absence of an antecedent MI [14]. However, the familial aggregation of HF is not explained solely by the presence or absence of CHD but also by qualitative differences in disease burden, for instance, the type of coronary lesion. Heritability is high for hazardous coronary lesion locations, such as disease of the left main artery or proximal vessels and coronary calcification, but low for distal disease [15, 16]. HFpEF is a highly heterogeneous disease, and genetic effects can be expected to be extremely limited as the disease is of late onset and associated with multiple environmental triggers [17, 18].

4.2 Genetics of Hypertension and Heart Failure

Linkage analysis of pedigrees exhibiting a Mendelian pattern of inheritance has been highly successful in the identification of rare mutations in genes causing monogenic syndromes. In contrast, the standard method for the dissection of the genetics of complex traits is a genome-wide association study (GWAS), which is based on the common disease/common variant hypothesis. A GWAS involves testing hundreds of thousands of single nucleotide polymorphisms (SNPs) across the DNA of large numbers of individuals to find variants statistically associated with a specific trait or disease. Currently, the genetic architecture of hypertension encompasses approximately 30 monogenic rare variants and > 1500 single nucleotide variants comprising both low-frequency and common variants reflecting the continuous spectrum of risk alleles influencing BP and, consequently, hypertension [19–22].

Similar to hypertension, the genetic architecture of HF comprises both rare and common genetic variants. Monogenic hypertrophic and dilated cardiomyopathy (DCM) syndromes account for a small proportion of the disease burden, and it is likely that polygenic contribution will play a major role in the predisposition to HF. However, the complex etiology of HF has made HF GWASs more challenging and has so far identified 14 SNPs [23–28]. The top GWAS signals for BP and HF are summarized in Fig. 4.1 and Table 4.1.

4.2.1 Sodium and Intravascular Volume

4.2.1.1 Monogenic Hypertension Syndromes

Sodium homeostatic pathways are the main pathways accounting for the majority of the monogenic hypertension and hypotension syndromes (Fig. 4.1) [29]. Sodium

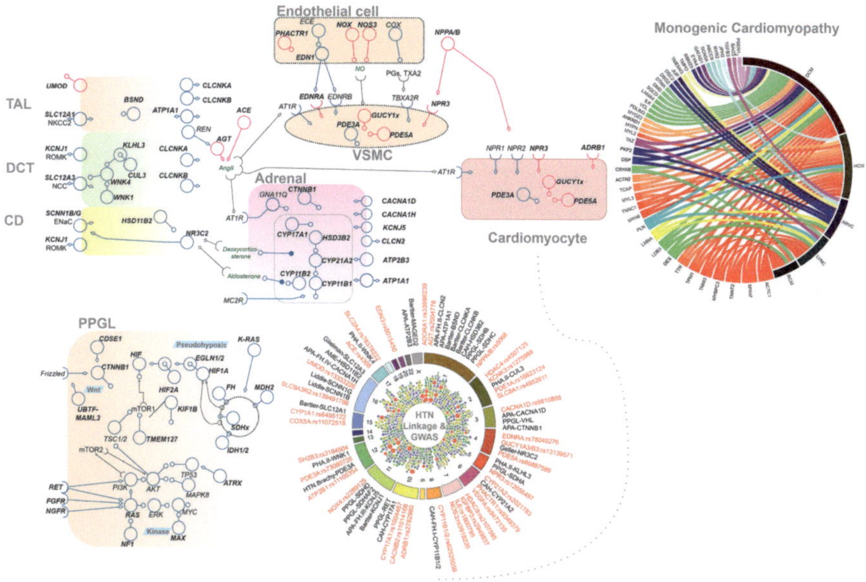

Fig. 4.1 Genetic pathways involved in blood pressure and cardiomyopathy. Genes containing known mutations that cause high or low blood pressure or cardiomyopathy in the context of the circulatory system. Plausible genes linked to the GWAS loci are presented near the site of their likely phenotype effect. *CD* collecting duct, *DCT* distal convoluted tubule, *TAL* thick ascending limb of the loop of Henle, *PPGL* phaeochromocytoma paraganglioma, *VSMC* vascular smooth muscle cell, *ARVC* arrhythmogenic right ventricular cardiomyopathy, *DCM* dilated cardiomyopathy, *HCM* hypertrophic cardiomyopathy, *LVNC* left ventricular non-compaction cardiomyopathy, *RCM* restrictive cardiomyopathy

and blood volume are also crucial in the etiology and manifestation of HF. Glucocorticoid-remediable aldosteronism or familial hyperaldosteronism (FH) type 1 (Online Mendelian Inheritance in Man, OMIM #103900) is an autosomal dominant syndrome in which hypertension is caused by increased aldosterone secretion that is driven by the adrenocorticotropic hormone (ACTH). The fusion of the 5′-regulatory sequences of 11β-hydroxylase (*CYP11B1*) with the distal coding sequences of aldosterone synthase (*CYP11B2*) leads to a chimeric gene, which results in ACTH becoming the main controller of aldosterone secretion, instead of angiotensin II or potassium [30]. FH type 2 (OMIM #605635) is an autosomal dominant syndrome caused by hyperplasia or adenoma of the aldosterone-producing adrenal cortex (APA), with the genetic cause localized to chromosome 7p22 [31]. The syndrome of apparent mineralocorticoid excess (OMIM # 218030) accompanied by hypokalemia and metabolic alkalosis is due to the absence or reduced activity of 11β-hydroxysteroid dehydrogenase (*HSD11B2*), resulting in hypertension in which cortisol acts as if it were a potent mineralocorticoid [32]. It is estimated that ≤40% of APAs harbor somatic mutations in *KCNJ5*, *ATP1A1*, *ATP2B3*, *ATP2A2*, and *CACNA1D* [33]. Congenital adrenal hyperplasia is a group of autosomal recessive disorders caused by defects in the enzymes of cortisol biosynthesis [34]. In some of these syndromes, plasma ACTH will increase to produce cortisol, thus

Table 4.1 GWAS signals for blood pressure, heart failure, and heart failure-related traits. *BP* blood pressure

Locus	Phenotype	SNP	Nearest gene(s)
1p36.2	BP/hypertension	rs880315	*CASZ1*
1p36.13	Dilated cardiomyopathy	rs10927875	*ZBTB17*
1p36.22	BP/hypertension	rs17367504	*MTHFR, CLCN6, NPPA, NPPB*
		rs5068	
1p34.2	HF founder population	rs16830359	*SLC2A1*
1p21.2	Aortic root size	rs7543130	*PALMD*
1p13.3	HF GWAS meta-analysis	rs660240	*CELSR2*
1p13.2	BP/hypertension	rs2932538	*SLC16A1, CAPZA1, ST7L, MOV10*
		rs17030613	
		rs10745332	
1q32.1		rs2169137	*MDM4*
1q41	HF founder population	rs12757165	*ESRRG*
1q42.2	BP/hypertension	rs2004776	*AGT*
2p23.2		rs1275988	*KCNK3*
2q11.2		rs7599598	*FER1L5*
2q24.3		rs1446468	*FIGN*
		rs13002573	*FIGN*
		rs16849225	*FIGN*
		rs6749447	*STK39*
2q32.1		rs16823124	*PDE1A*
3p25.3		rs347591	*HRH1-ATG7*
3p24.1		rs13082711	*SLC4A*
		rs820430	
3p22.3	Heart failure	rs12638540	CMTM7
3p22.1	BP/hypertension	rs9815354	*ULK4*
		rs3774372	
		rs1717027	
3p21.31		rs319690	*MAP4*
		rs7651237	
3p21.1		rs9810888	*CACNA1D*
3q26.1		rs16833934	*MIR1263*
3q26.2		rs419076	*MECOM*
4q12		rs871606	*CHIC2*
4q21.21		rs16998073	*FGF5*
		rs1458038	
4q24		rs13107325	*SLC39A8*
4q25		rs6825911	*ENPEP, PITX2*
4q25	HF GWAS meta-analysis	rs17042102	*PITX2, FAM241A*
4q32.1	BP/hypertension	rs13139571	*GUCY1A3-GUCY1B3*
5p13.3	BP/hypertension	rs1173771	*NPR3-C5orf23*
		rs7733331	
		rs1173766	
5q23.2	Aortic root size	rs17470137	*CCDC100*
5q31.2	HF GWAS meta-analysis	rs11745324	*KLHL3*
5q33.3	BP/hypertension	rs11953630	*EBF1*
6p25.3	HF GWAS meta-analysis	rs55730499	*LPA*
6p25.3		rs140570886	*LPA*
6p22.2	BP/hypertension	rs1799945	*HFE*
		rs198823	
6p22.1	HF founder population	rs10947055	*TRIM38*
6p21.33	BP/hypertension	rs805303	*BAG1*
		rs2021783	*CYP21A2*
	Dilated cardiomyopathy	rs9262636	*HGC22*

Table 4.1 (continued)

Locus	Phenotype	SNP	Nearest gene(s)
6p21.32	BP/hypertension	rs2854275	*HLA-DQB1*
6p21.2	HF GWAS meta-analysis	rs4135240	*CDKN1A*
6p21.1		rs10948071	*CRIP3*
6p24.1		rs9349379	*PHACTR1*
6q22.31	LV internal dimension	rs89107	*SLC35F1*
		rs11153768	*PLN*
6q22.33	BP/hypertension	rs13209747	*RSPO3*
6q25.1		rs17080102	*PLEKHG1*
7p15.2		rs17428471	*EVX1-HOXA*
7p12.3		rs2949837	*IGFBP3*
7q21.2		rs2282978	*CDK6*
7q22.3		rs17477177	*PIK3CG*
		rs12705390	
7q36.1		rs3918226	*NOS3*
8p23.1		rs4841569	*BLK-GATA4*
		rs2898290	
8q11.21	LV mass	rs4552930	*UBE2V2*
8q24.12	BP/hypertension	rs2071518	*NOV*
9p21.3	HF GWAS meta-analysis	rs1556516	*CDKN2B-AS1*
9q34.2	HF GWAS meta-analysis	rs600038	*ABO, SURF1*
10p12.31	BP/hypertension	rs11014166	*CACNB2*
		rs1813353	
		rs4373814	
		rs12258967	
10q21.2		rs1530440	*c10orf107*
		rs4590817	
		rs12244842	
		rs7070797	
10q22.2		rs4746172	*VCL*
10q22.2	HF GWAS meta-analysis	rs4746140	*SYNPO2L, AGAP5*
10q23.33	BP/hypertension	rs932764	*PLCE1*
10q24.32		rs1004467	*CYP17A1-NT5C2*
		rs11191548	
		rs12413409	
		rs4409766	
		rs3824755	
10q25.1	HF founder population	rs1320448	*COL17A1*
10q25.3	BP/hypertension	rs2782980	*ADRB1*
		rs7076938	
		rs1801253	
10q26.11	HF GWAS meta-analysis	rs17617337	*BAG3*
10q26.11	Dilated cardiomyopathy	rs2234962	*BAG3*
10q26.12	Interventricular Septum (IVS) wall thickness	rs1571099	*PPAPDC1A*
11p15.5	BP/hypertension	rs661348	*LSP1-TNNT3*
11p15.4		rs7129220	*ADM*
11p15.1		rs381815	*PLEKHA7*
		rs757081	*PIK3C2A, NUCB2, NCR3LG1*
11p15.2		rs2014408	*SOX6*
		rs4757391	
11q13.1		rs4601790	*EHBP1L1*
		rs3741378	*RELA*
11q22.1		rs633185	*FLJ32810-TMEM133*
11q24.3		rs11222084	*ADAMTS8*

(continued)

Table 4.1 (continued)

Locus	Phenotype	SNP	Nearest gene(s)
12p12.2	Aortic root size	rs10770612	*PDE3A*
12q14.1	Heart failure	rs11172782	*LRIG3*
12q14.3	Aortic root size	rs4026608	*HMGA2*
12q13.13	BP/hypertension	rs7297416	*HOXC4*
12q21.33		rs11105354	*ATP2B1*
		rs2681492	
		rs2681472	
		rs17249754	
12q22	Dilated cardiomyopathy	rs10859313	*CLLU1*
12q24.12	HF GWAS meta-analysis	rs4766578	*ATXN2*
12q24.12	BP/hypertension	rs3184504	*SH2B3*
		rs653178	
12q24.13		rs11066280	*RPL6-ALDH2*
12q24.21		rs35444	*TBX5-TBX3*
		rs2384550	
		rs10850411	
		rs1991391	
		rs11067763	*MED13L*
13q22.1	Ejection fraction	rs9530176	*KLF5*
15q12	HF founder population	rs17636733	*UBE3A*
15q21.1	BP/hypertension	rs1036477	*FBN1*
15q22.2	Heart failure	rs10519210	*USP3*
15q24.1	BP/hypertension	rs6495122	*CYP1A1-ULK3*
		rs1378942	
15q24.2		rs11072518	*COX5A*
		rs1133323	
15q26.1		rs2521501	*FURIN-FES*
16p12.3		rs13333226	*UMOD*
16p12.2	HF GWAS meta-analysis	rs56094641	*FTO*
16q22.1	BP/hypertension	rs33063	*NFAT5*
17p13.3	Aortic root size	rs10852932	*SMG6*
		rs4523957	*SRR*
		rs413016	*TSR1*
17q21.31	BP/hypertension	rs12946454	*PLCD3*
17q21.32		rs17608766	*GOSR2*
17q21.33		rs12940887	*ZNF652*
		rs16948048	
17q24.2	LV internal dimension	rs7213314	*WIPI1*
20p12.2	BP/hypertension	rs1327235	*JAG1*
		rs1887320	
20q13.32		rs6015450	*GNAS-EDN3*
		rs6092743	*C20orf174*

causing the accumulation of aberrant products, some of which lead to hypertension. Enzyme mutations that are associated with hypertension include (in order of frequency) 11β-hydroxylase (OMIM #202010; *CYP11B1*), 3β-hydroxysteroid dehydrogenase (OMIM #613890; *HSD3B2*), 17α-hydroxylase (OMIM #609300; *CYP17A1*), and cholesterol desmolase (OMIM #118485; *CYP11A1*). Furthermore, a common SNP near *CYP17A1* has emerged in multiple large BP GWAS meta-analyses [35–38]. Pseudohypoaldosteronism type II (Gordon's syndrome; familial hyperkalemia; OMIM #145260) is an autosomal dominant low-renin hypertension syndrome, resulting from either gain-of-function mutations in *WNK1* or loss-of-function mutations in *WNK4* and mutations in Kelch-like 3 (*KLHL3*) and Cullin 3 (*CUL3*) genes [39]. Liddle's syndrome (OMIM # 177200) is an autosomal dominant condition with a clinical picture of hypertension and aldosterone excess but with extremely low aldosterone and renin levels. This is caused by gain-of-function mutations in the genes coding the β- or γ-subunits of epithelial sodium channels (*SCNN1B*, *SCNN1G*) [40, 41].

4.2.1.2 Monogenic Hypotension Syndromes

Two examples of low BP syndromes include Bartter (*SLC12A1*, *KCNJ1*, *CLCNKB*, *BSND*, *CaSR*, *ClCK-A*) and Gitelman (*SLC12A3*) syndromes, which are associated with mutations that reduce salt retention, tend to lower BP, and protect against the development of hypertension [29, 42]. A mutation in the *CLCNKB* gene (1p36), encoding a basolateral chloride channel ClCKb, has been identified as the most frequent cause of the classic Bartter syndrome. Mutations in *SLC12A1* and *KCNJ1* cause the classic, less severe phenotype.

4.2.1.3 Polygenic Pathways

In addition to the kidneys, the heart secretes a family of vasodilatory and natriuretic hormones in response to increased wall stress: atrial and B-type natriuretic peptides (*NPPA* and *NPPB*, respectively). Knockout of one copy of *NPPA* in mice increases BP, and overexpression of *NPPA* lowers BP [43, 44]. There is now convincing evidence in humans for common variations at the *NPPA–NPPB* locus, influencing both levels of natriuretic peptides and BP in opposite directions [36–38, 45]. Furthermore, an SNP near the natriuretic peptide clearance receptor (*NPR3*) also showed genome-wide significant association in European, African, and Japanese BP GWASs [36, 37, 46]. Interestingly, there is evidence of convergence between studies in animal models of hypertension and their emerging signals from human BP GWAS analyses [47–49]. Levels of N-terminal pro-brain natriuretic peptide (NT-proBNP) increase with myocardial wall stress and are associated with HF. Although natriuretic peptide gene variants have not been directly associated with HF, they have been associated with increased susceptibility to LV hypertrophy [50].

In line with the monogenic syndromes described above, there is a functional variant within a promoter region of the uromodulin gene (*UMOD*), which is almost exclusively expressed in the thick ascending limb of the loop of Henle in the kidneys, identifying a potentially novel pathway of BP regulation through an effect on sodium homeostasis [51]. Moreover, independent studies have identified SNPs highly correlated with the hypertension SNP near *UMOD* to be associated with chronic kidney disease [52]. Trudu et al. showed that furosemide treatment significantly enhanced natriuresis and reduced BP levels both in transgenic mice and in hypertensive individuals homozygous for the *UMOD* increasing allele, thus making this a potentially interesting locus for both hypertension and HF [53].

4.2.2 Vascular Mechanisms

Vascular diameter and compliance of resistance arteries are important determinants of arterial pressure. Vascular changes in hypertension may result from either abnormal extracellular stimuli or an altered intracellular signaling cascade, leading to enhanced vasoconstriction, blunted vasodilation, and vascular wall hypertrophy or remodeling, all of which contribute to elevated peripheral vascular resistance [54].

Vascular endothelial function also modulates vascular tone. The vascular endothelium synthesizes and releases a spectrum of vasoactive substances, including nitric oxide (NO), a potent vasodilator. NO exerts vasodilating and antiproliferative effects on vascular smooth muscle cells (VSMCs) and inhibits thrombocyte aggregation and leukocyte adhesion. Other factors that regulate vascular reactivity include endothelins and prostacyclin (PGI2). Endothelin-1 (*EDN1*) activates specific endothelin ETA receptors (*EDNRA*) on VSMCs to cause vasoconstriction and cell proliferation. In contrast, endothelial ETB receptors (*EDNRB*) mediate vasodilatation via release of NO and PGI2 [55]. GWASs of CAD showed the SNP rs9349379 G allele (frequency 36%) was associated with an increased risk of CAD and coronary calcification but a decreased risk of four conditions (migraine headache, cervical artery dissection, fibromuscular dysplasia, and hypertension) [56, 57]. This SNP is located within the third intron of the gene encoding phosphatase and actin regulatory protein 1 (*PHACTR1*). However, functional analysis of this variant shows that it regulates expression of endothelin 1 (*EDN1*), a gene located 600 kb upstream of *PHACTR1*. Interestingly, variants in the *PHACTR1* gene have been associated with fibromuscular dysplasia, a nonatherosclerotic vascular disease leading to stenosis, dissection, and aneurysm mainly affecting the renal and cerebrovascular arteries [58]. Variants in the gene for vascular endothelial growth factor A (*VEGFA*), which induces proliferation and migration of vascular endothelial cells and stimulates angiogenesis, is one of the replicated signals from a GWAS [22].

Ion transport by VSMCs may contribute to hypertension-associated abnormalities of vascular tone and vascular growth, both of which are modulated by intracellular pH. Increased expression of the sodium/hydrogen exchanger (*SLC9A1*) can stimulate vascular tone and cell growth by increasing sodium reabsorption in renal proximal tubule cells [59]. Other ion transporters considered are the sodium/bicarbonate cotransporter (*SLC4A10)* and the sodium/calcium exchanger NCX (*SLC8A1*). One hypothesis for the mechanism by which excess salt intake elevates BP is through the observed rise in cardiotonic steroids such as ouabain in response to salt intake [60]. Ouabain inhibits the plasma membrane sodium/potassium ATPase, leading to an increase in cytosolic sodium concentration and cytosolic calcium concentration through the involvement of the sodium/calcium exchanger NCX (*SLC8A1*), thereby increasing contraction in the vascular or heart muscle [61].

4.2.3 The Autonomic Nervous System

Increased sympathetic nervous system activity has a role in causing hypertension [62], which may be the result of background genetic susceptibility interacting with chronic psychogenic stress, obesity, or high sodium intake. Hypertension could also arise or be sustained by defects in baroreceptor function [63]. Furthermore, catecholamines may exert a direct toxic effect on the myocardium through enhanced lipid mobility, calcium overload, free radical production, or increased sarcolemmal permeability, thus leading to HF. Rare syndromes caused by mutations affecting the function of the autonomic system and causing hypertension are exemplified by pheochromocytomas (PCCs) and paragangliomas (PGLs). Up to 10% of genetically determined PCC/PGLs are due to mutations in the *SDHD*, *SDHC*, *SDHB*, *SDHA*, and *SDHAF2* (or *SDH5*) genes [64]. Autosomal dominantly inherited PCCs are due to a variety of *RET* proto-oncogene mutations. Other PCC-susceptibility genes include *RET* (multiple endocrine neoplasia syndrome type 2), the tumor suppressor gene *VHL* observed in families with Von Hippel–Lindau syndrome, and the genes that encode succinate dehydrogenase subunits B and D (*SDHB* and *SDHD*, respectively) that cause familial PGL [64].

Polymorphisms in the α2c-adrenergic receptor and the β1-adrenergic receptor are associated with an increased risk of HF in African American individuals via increased synaptic norepinephrine release [65]. The simultaneous occurrence of both gene mutations had a synergistic effect with a ten-fold increase in the risk of HF in double homozygotes [65].

4.2.4 Cardiomyopathies

Cardiomyopathies (intrinsic diseases of the heart muscle), including dilated, hypertrophic, and restrictive forms, can all lead to HF, although DCM is the leading global cause of heart transplantation [66]. Idiopathic DCM is a form of HF defined by the presence of LV dilatation and LV systolic dysfunction in the absence of an obvious etiology, such as CAD, hypertension, valvular disease, or congenital defects. It affects ~1/2500 adults, and it is more common in men than in women [67]. There are more than 50 currently recognized genes associated with this condition, most of which encode proteins in the cardiomyocyte sarcomere (Fig. 4.1 and Table 4.2).

GWASs have identified two loci for DCM. The first, rs10927875, maps to a region on chromosome 1p36.13, which encompasses several genes among which *HSPB7* has been formerly suggested to be implicated in DCM. The second identified locus involves rs2234962, a non-synonymous SNP (c.T757C, p. C151R) located within the sequence of *BAG3* on chromosome 10q26 [68]. Other GWASs of systolic HF have identified rs10519210 (nearest gene *USP3*) in Europeans and rs11172782 (nearest gene *LRIG3*) in Africans [27]. Two intronic SNPS, rs1739843 (*HSPB7*) and rs6787362 (*FRMD4B*), were identified in a case–control study of advanced HF [23].

Among patients with thickening of the LV wall that was not explained by hypertension or valvular disease, nearly one in five individuals had a mutation in known sarcomeric (e.g., β-myosin heavy chain, myosin-binding protein C, troponin T, troponin I, and myosin light chain) or storage (e.g., α-galactosidase A) genes [69].

Table 4.2 Genes associated with dilated cardiomyopathy

Gene	Protein	OMIM	Percentage of familial DCM cases (%)
LMNA	Lamin A/C	150,330	6%
MYH7	β-Myosin heavy chain	160,760	4.2%
MYPN	Myopalladin	608,517	3.5%
TNNT2	Cardiac troponin T	191,045	2.9%
SCN5A	Sodium channel	600,163	2.6%
MYBPC3	Myosin-binding protein C	600,958	2%
RBM20	RNA-binding protein 20	613,171	1.9%
TMPO	Thymopoietin	188,380	1.1%
LAMA4	Laminin a-4	600,133	1.1%
VCL	Metavinculin	193,065	1%
LDB3	Cypher/ZASP	605,906	1%
TCAP	Titin-cap or telethonin	604,488	1%
PSEN1/2	Presenilin 1/2	104,311/600759	1%
ACTN2	α-Actinin-2	102,573	0.9%
CRYAB	Alpha B-crystalin	123,590	0.7%
TPM1	α-Tropomyosin	191,010	0.6%
ABCC9	SUR2A	601,439	0.6%
ACTC1	Cardiac actin	102,540	0.5%
PDLIM3	PDZ LIM domain protein 3	605,889	0.5%
ILK	Integrin-linked kinase	602,366	0.5%
TNNC1	Cardiac troponin C	191,040	0.4%
TNNI3	Cardiac troponin I	191,044	0.4%
PLN	Phospholamban	172,405	0.4%
DES	Desmin	125,660	0.3%
SGCD	δ-Sarcoglycan	601,411	0.3%
CSRP3	Muscle LIM protein	600,824	0.3%
MYH6	α-Myosin heavy chain	160,710	Unknown
TTN	Titin	188,840	Unknown
EYA4	Eyes absent 4	603,550	Unknown
ANKRD1	Ankyrin repeat domain-containing protein 1	609,599	Unknown
DMD	Dystrophin	300,377	Unknown
GATAD1	GATA zinc finger domain-containing 1	614,518	Unknown
BAG3	BCL2-associated athanogene 3	603,883	Unknown
TAZ/G4.5	Tafazzin	300,394	Unknown

4.2.5 Mendelian Randomization (MR) Studies

Mendelian randomization (MR) uses genetic variation as a natural experiment to investigate the causal relationships between potentially modifiable risk factors and diseases or phenotypes in observational data. The idea behind MR is that as the genotype is randomly allocated at conception and is invariant over the lifetime, it can be used as genetic proxies unaffected by confounding or reverse causation to infer causality [70].

The causal relationship between high BP and cardiovascular disease (CVD) and non-CVD outcomes has been the subject of MR studies confirming a causal relationship between BP and CAD, stroke, HF, and atrial fibrillation (AF) but not for estimated glomerular filtration rate (eGFR) [71–73]. For HF, the causal effects of several risk factors were confirmed through MR studies: a 10-mmHg higher DBP or SBP shows a 30% and 18% higher risk of HF, respectively, whereas a 4.4 kg/m^2 (men) and 5.4 kg/m^2 (women) body mass index (BMI) was causally associated with a 74% higher HF risk [28]. An MR study using 3 sets of BP GWAS SNPs specific for BP traits (242 independent SNPs associated with both SBP and DBP, 120 SBP-exclusive SNPs, and 80 DBP-exclusive SNPs) was carried out to infer SBP- and DBP-specific effects of cardiovascular outcomes. It showed that while SBP was the causal driver of CAD, stroke, and ischemic stroke, DBP was exclusively associated with small vessel stroke and SBP was exclusively causally associated with HF, AF, and type 2 diabetes mellitus (T2DM) [74].

4.2.6 Future Directions

Although the technological and analytic aspects of genomics have been highly successful in discovering DNA sequence variants associated with BP and HF, the functional significance of the vast number of these variants in the human genome remains unknown. The *UMOD* loci from GWASs is now the basis of a clinical trial (clinicaltrials.gov NCT03354897) to reposition a loop diuretic in the hypertension care pathway. Advances in the genetics of DCM and HF have highlighted numerous rare variants linked to DCM and fewer common variants linked to DCM and HF. Unraveling functional genomics of HF is challenging and requires great care in phenotypic definition and characterization for our collective efforts to be fruitful. Ultimately, understanding the genetic underpinnings of both hypertension and heart failure has a common goal: early detection and precision treatment.

References

1. Krousel-Wood MA, Muntner P, He J, Whelton PK. Primary prevention of essential hypertension. Med Clin North Am. 2004;88(1):223–38.
2. Ziaeian B, Fonarow GC. Epidemiology and aetiology of heart failure. Nat Rev Cardiol. 2016;13(6):368–78.
3. Dzau VJ, Antman EM, Black HR, Hayes DL, Manson JE, Plutzky J, et al. The cardiovascular disease continuum validated: clinical evidence of improved patient outcomes: part II: clinical trial evidence (acute coronary syndromes through renal disease) and future directions. Circulation. 2006;114(25):2871–91.
4. Bochud M, Bovet P, Elston RC, Paccaud F, Falconnet C, Maillard M, et al. High heritability of ambulatory blood pressure in families of East African descent. Hypertension. 2005;45(3):445–50.
5. Havlik RJ, Garrison RJ, Feinleib M, Kannel WB, Castelli WP, McNamara PM. Blood pressure aggregation in families. Am J Epidemiol. 1979;110(3):304–12.

6. Hottenga JJ, Boomsma DI, Kupper N, Posthuma D, Snieder H, Willemsen G, et al. Heritability and stability of resting blood pressure. Twin Res Hum Genet. 2005;8(5):499–508.
7. Kupper N, Willemsen G, Riese H, Posthuma D, Boomsma DI, de Geus EJ. Heritability of daytime ambulatory blood pressure in an extended twin design. Hypertension. 2005;45(1):80–5.
8. Snieder H, Harshfield GA, Treiber FA. Heritability of blood pressure and hemodynamics in African-and European-American youth. Hypertension. 2003;41(6):1196–201.
9. Fagard R, Brguljan J, Staessen J, Thijs L, Derom C, Thomis M, et al. Heritability of conventional and ambulatory blood pressures. A study in twins. Hypertension. 1995;26(6 Pt 1):919–24.
10. Fava C, Burri P, Almgren P, Groop L, Hulthen UL, Melander O. Heritability of ambulatory and office blood pressure phenotypes in Swedish families. J Hypertens. 2004;22(9):1717–21.
11. Caulfield M, Munroe P, Pembroke J, Samani N, Dominiczak A, Brown M, et al. Genome-wide mapping of human loci for essential hypertension. Lancet (London, England). 2003;361(9375):2118–23.
12. Lindgren MP, PirouziFard M, Smith JG, Sundquist J, Sundquist K, Zoller B. A Swedish Nationwide adoption study of the heritability of heart failure. JAMA Cardiol. 2018;3(8):703–10.
13. Post WS, Larson MG, Myers RH, Galderisi M, Levy D. Heritability of left ventricular mass: the Framingham heart study. Hypertension. 1997;30(5):1025–8.
14. Lloyd-Jones DM, Larson MG, Leip EP, Beiser A, D'Agostino RB, Kannel WB, et al. Lifetime risk for developing congestive heart failure: the Framingham heart study. Circulation. 2002;106(24):3068–72.
15. Nasir K, Michos ED, Rumberger JA, Braunstein JB, Post WS, Budoff MJ, et al. Coronary artery calcification and family history of premature coronary heart disease: sibling history is more strongly associated than parental history. Circulation. 2004;110(15):2150–6.
16. Fischer M, Broeckel U, Holmer S, Baessler A, Hengstenberg C, Mayer B, et al. Distinct heritable patterns of angiographic coronary artery disease in families with myocardial infarction. Circulation. 2005;111(7):855–62.
17. Lewis GA, Schelbert EB, Williams SG, Cunnington C, Ahmed F, McDonagh TA, et al. Biological phenotypes of heart failure with preserved ejection fraction. J Am Coll Cardiol. 2017;70(17):2186–200.
18. Zakeri R, Cowie MR. Heart failure with preserved ejection fraction: controversies, challenges and future directions. Heart. 2018;104(5):377–84.
19. Surendran P, Feofanova EV, Lahrouchi N, Ntalla I, Karthikeyan S, Cook J, et al. Discovery of rare variants associated with blood pressure regulation through meta-analysis of 1.3 million individuals. Nat Genet. 2020;52(12):1314–32.
20. Padmanabhan S, Dominiczak AF. Genomics of hypertension: the road to precision medicine. Nat Rev Cardiol. 2021;18(4):235–50.
21. Giri A, Hellwege JN, Keaton JM, Park J, Qiu C, Warren HR, et al. Trans-ethnic association study of blood pressure determinants in over 750,000 individuals. Nat Genet. 2019;51(1):51–62.
22. Evangelou E, Warren HR, Mosen-Ansorena D, Mifsud B, Pazoki R, Gao H, et al. Genetic analysis of over 1 million people identifies 535 new loci associated with blood pressure traits. Nat Genet. 2018;50(10):1412–25.
23. Cappola TP, Li M, He J, Ky B, Gilmore J, Qu L, et al. Common variants in HSPB7 and FRMD4B associated with advanced heart failure. Circ Cardiovasc Genet. 2010;3(2):147–54.
24. Parsa A, Chang YP, Kelly RJ, Corretti MC, Ryan KA, Robinson SW, et al. Hypertrophy-associated polymorphisms ascertained in a founder cohort applied to heart failure risk and mortality. Clin Transl Sci. 2011;4(1):17–23.
25. Vasan RS, Glazer NL, Felix JF, Lieb W, Wild PS, Felix SB, et al. Genetic variants associated with cardiac structure and function: a meta-analysis and replication of genome-wide association data. JAMA. 2009;302(2):168–78.
26. Meder B, Ruhle F, Weis T, Homuth G, Keller A, Franke J, et al. A genome-wide association study identifies 6p21 as novel risk locus for dilated cardiomyopathy. Eur Heart J. 2014;35(16):1069–77.

27. Smith NL, Felix JF, Morrison AC, Demissie S, Glazer NL, Loehr LR, et al. Association of genome-wide variation with the risk of incident heart failure in adults of European and African ancestry: a prospective meta-analysis from the cohorts for heart and aging research in genomic epidemiology (CHARGE) consortium. Circ Cardiovasc Genet. 2010;3(3):256–66.
28. Shah S, Henry A, Roselli C, Lin H, Sveinbjornsson G, Fatemifar G, et al. Genome-wide association and mendelian randomisation analysis provide insights into the pathogenesis of heart failure. Nat Commun. 2020;11(1):163.
29. Lifton RP, Gharavi AG, Geller DS. Molecular mechanisms of human hypertension. Cell. 2001;104(4):545–56.
30. Lifton RP, Dluhy RG, Powers M, Rich GM, Cook S, Ulick S, et al. A chimaeric 11 beta-hydroxylase/aldosterone synthase gene causes glucocorticoid-remediable aldosteronism and human hypertension. Nature. 1992;355(6357):262–5.
31. Lafferty AR, Torpy DJ, Stowasser M, Taymans SE, Lin JP, Huggard P, et al. A novel genetic locus for low renin hypertension: familial hyperaldosteronism type II maps to chromosome 7 (7p22). J Med Genet. 2000;37(11):831–5.
32. Cerame BI, New MI. Hormonal hypertension in children: 11beta-hydroxylase deficiency and apparent mineralocorticoid excess. J Pediatr Endocrinol Metab. 2000;13(9):1537–47.
33. Fernandes-Rosa FL, Boulkroun S, Zennaro MC. Somatic and inherited mutations in primary aldosteronism. J Mol Endocrinol. 2017;59(1):R47–63.
34. New MI, Levine LS. Hypertension of childhood with suppressed renin. Endocr Rev. 1980;1(4):421–30.
35. Levy D, Ehret GB, Rice K, Verwoert GC, Launer LJ, Dehghan A, et al. Genome-wide association study of blood pressure and hypertension. Nat Genet. 2009;41(6):677–87.
36. Kato N, Takeuchi F, Tabara Y, Kelly TN, Go MJ, Sim X, et al. Meta-analysis of genome-wide association studies identifies common variants associated with blood pressure variation in East Asians. Nat Genet. 2011;43(6):531–8.
37. International Consortium for Blood Pressure Genome-Wide Association S, Ehret GB, Munroe PB, Rice KM, Bochud M, Johnson AD, et al. Genetic variants in novel pathways influence blood pressure and cardiovascular disease risk. Nature. 2011;478(7367):103–9.
38. Newton-Cheh C, Johnson T, Gateva V, Tobin MD, Bochud M, Coin L, et al. Genome-wide association study identifies eight loci associated with blood pressure. Nat Genet. 2009;41(6):666–76.
39. Louis-Dit-Picard H, Barc J, Trujillano D, Miserey-Lenkei S, Bouatia-Naji N, Pylypenko O, et al. KLHL3 mutations cause familial hyperkalemic hypertension by impairing ion transport in the distal nephron. Nat Genet. 2012;44(4):456–60. S1-3
40. Hansson JH, Nelson-Williams C, Suzuki H, Schild L, Shimkets R, Lu Y, et al. Hypertension caused by a truncated epithelial sodium channel gamma subunit: genetic heterogeneity of Liddle syndrome. Nat Genet. 1995;11(1):76–82.
41. Shimkets RA, Warnock DG, Bositis CM, Nelson-Williams C, Hansson JH, Schambelan M, et al. Liddle's syndrome: heritable human hypertension caused by mutations in the beta subunit of the epithelial sodium channel. Cell. 1994;79(3):407–14.
42. Ji W, Foo JN, O'Roak BJ, Zhao H, Larson MG, Simon DB, et al. Rare independent mutations in renal salt handling genes contribute to blood pressure variation. Nat Genet. 2008;40(5):592–9.
43. John SW, Krege JH, Oliver PM, Hagaman JR, Hodgin JB, Pang SC, et al. Genetic decreases in atrial natriuretic peptide and salt-sensitive hypertension. Science. 1995;267(5198):679–81.
44. Schillinger KJ, Tsai SY, Taffet GE, Reddy AK, Marian AJ, Entman ML, et al. Regulatable atrial natriuretic peptide gene therapy for hypertension. Proc Natl Acad Sci U S A. 2005;102(39):13789–94.
45. Newton-Cheh C, Larson MG, Vasan RS, Levy D, Bloch KD, Surti A, et al. Association of common variants in NPPA and NPPB with circulating natriuretic peptides and blood pressure. Nat Genet. 2009;41(3):348–53.
46. Zhu X, Young JH, Fox E, Keating BJ, Franceschini N, Kang S, et al. Combined admixture mapping and association analysis identifies a novel blood pressure genetic locus on 5p13: contributions from the CARe consortium. Hum Mol Genet. 2011;20(11):2285–95.

47. Brosnan MJ, Clark JS, Jeffs B, Negrin CD, Van Vooren P, Arribas SM, et al. Genes encoding atrial and brain natriuretic peptides as candidates for sensitivity to brain ischemia in stroke-prone hypertensive rats. Hypertension. 1999;33(1):290–7.
48. Jeffs B, Clark JS, Anderson NH, Gratton J, Brosnan MJ, Gauguier D, et al. Sensitivity to cerebral ischaemic insult in a rat model of stroke is determined by a single genetic locus. Nat Genet. 1997;16(4):364–7.
49. Ye P, West MJ. Cosegregation analysis of natriuretic peptide genes and blood pressure in the spontaneously hypertensive rat. Clin Exp Pharmacol Physiol. 2003;30(12):930–6.
50. Rubattu S, Bigatti G, Evangelista A, Lanzani C, Stanzione R, Zagato L, et al. Association of atrial natriuretic peptide and type a natriuretic peptide receptor gene polymorphisms with left ventricular mass in human essential hypertension. J Am Coll Cardiol. 2006;48(3):499–505.
51. Padmanabhan S, Melander O, Johnson T, Di Blasio AM, Lee WK, Gentilini D, et al. Genome-wide association study of blood pressure extremes identifies variant near UMOD associated with hypertension. PLoS Genet. 2010;6(10):e1001177.
52. Kottgen A, Glazer NL, Dehghan A, Hwang SJ, Katz R, Li M, et al. Multiple loci associated with indices of renal function and chronic kidney disease. Nat Genet. 2009;41(6):712–7.
53. Trudu M, Janas S, Lanzani C, Debaix H, Schaeffer C, Ikehata M, et al. Common noncoding UMOD gene variants induce salt-sensitive hypertension and kidney damage by increasing uromodulin expression. Nat Med. 2013;19(12):1655–60.
54. Heagerty AM, Heerkens EH, Izzard AS. Small artery structure and function in hypertension. J Cell Mol Med. 2010;14(5):1037–43.
55. Dhaun N, Webb DJ. Endothelins in cardiovascular biology and therapeutics. Nat Rev Cardiol. 2019;16(8):491–502.
56. Nikpay M, Goel A, Won HH, Hall LM, Willenborg C, Kanoni S, et al. A comprehensive 1,000 genomes-based genome-wide association meta-analysis of coronary artery disease. Nat Genet. 2015;47(10):1121–30.
57. Gupta RM, Hadaya J, Trehan A, Zekavat SM, Roselli C, Klarin D, et al. A genetic variant associated with five vascular diseases is a distal regulator of Endothelin-1 gene expression. Cell. 2017;170(3):522–33. e15
58. Kiando SR, Tucker NR, Castro-Vega LJ, Katz A, D'Escamard V, Treard C, et al. PHACTR1 is a genetic susceptibility locus for fibromuscular dysplasia supporting its complex genetic pattern of inheritance. PLoS Genet. 2016;12(10):e1006367.
59. Xiang M, Feng M, Muend S, Rao R. A human Na+/H+ antiporter sharing evolutionary origins with bacterial NhaA may be a candidate gene for essential hypertension. Proc Natl Acad Sci. 2007;104(47):18677–81.
60. Blaustein MP. Physiological effects of endogenous ouabain: control of intracellular Ca2+ stores and cell responsiveness. Am J Phys. 1993;264(6 Pt 1):C1367–87.
61. Iwamoto T, Kita S, Zhang J, Blaustein MP, Arai Y, Yoshida S, et al. Salt-sensitive hypertension is triggered by Ca2+ entry via Na+/Ca2+ exchanger type-1 in vascular smooth muscle. Nat Med. 2004;10(11):1193–9.
62. Esler M, Rumantir M, Kaye D, Lambert G. The sympathetic neurobiology of essential hypertension: disparate influences of obesity, stress, and noradrenaline transporter dysfunction? Am J Hypertens. 2001;14(6 Pt 2):139S–46S.
63. Izzo JL Jr, Taylor AA. The sympathetic nervous system and baroreflexes in hypertension and hypotension. Curr Hypertens Rep. 1999;1(3):254–63.
64. Welander J, Soderkvist P, Gimm O. Genetics and clinical characteristics of hereditary pheochromocytomas and paragangliomas. Endocr Relat Cancer. 2011;18(6):R253–76.
65. Small KM, Wagoner LE, Levin AM, Kardia SL, Liggett SB. Synergistic polymorphisms of beta1- and alpha2C-adrenergic receptors and the risk of congestive heart failure. N Engl J Med. 2002;347(15):1135–42.
66. McKenna WJ, Judge DP. Epidemiology of the inherited cardiomyopathies. Nat Rev Cardiol. 2021;18(1):22–36.

67. Codd MB, Sugrue DD, Gersh BJ, Melton LJ 3rd. Epidemiology of idiopathic dilated and hypertrophic cardiomyopathy. A population-based study in Olmsted County, Minnesota, 1975-1984. Circulation. 1989;80(3):564–72.
68. Villard E, Perret C, Gary F, Proust C, Dilanian G, Hengstenberg C, et al. A genome-wide association study identifies two loci associated with heart failure due to dilated cardiomyopathy. Eur Heart J. 2011;32(9):1065–76.
69. Morita H, Larson MG, Barr SC, Vasan RS, O'Donnell CJ, Hirschhorn JN, et al. Single-gene mutations and increased left ventricular wall thickness in the community: the Framingham heart study. Circulation. 2006;113(23):2697–705.
70. Holmes MV, Ala-Korpela M, Smith GD. Mendelian randomization in cardiometabolic disease: challenges in evaluating causality. Nat Rev Cardiol. 2017;14(10):577–90.
71. Nazarzadeh M, Pinho-Gomes AC, Bidel Z, Canoy D, Dehghan A, Smith Byrne K, et al. Genetic susceptibility, elevated blood pressure, and risk of atrial fibrillation: a mendelian randomization study. Genome Med. 2021;13(1):38.
72. Malik R, Georgakis MK, Vujkovic M, Damrauer SM, Elliott P, Karhunen V, et al. Relationship between blood pressure and incident cardiovascular disease: linear and nonlinear mendelian randomization analyses. Hypertension. 2021;77(6):2004–13.
73. Higgins H, Mason AM, Larsson SC, Gill D, Langenberg C, Burgess S. Estimating the population benefits of blood pressure lowering: a wide-angled mendelian randomization study in UK biobank. J Am Heart Assoc. 2021;10(17):e021098.
74. Le NN, Tran TQB, Lip S, McCallum L, McClure J, Dominiczak AF, et al. Unravelling the distinct effects of systolic and diastolic blood pressure using mendelian randomisation. Genes. 2022;13(7):1226.

5 High Heart Rate: A Predictor of Heart Failure in Healthy Subjects and a Risk Factor for Adverse Outcome in Patients with Heart Failure

Paolo Palatini

5.1 Introduction

Congestive heart failure (HF) is a common and progressive cardiovascular clinical disease affecting a large number of people. Coronary artery disease, myocardial infarction, systemic and pulmonary hypertension, valvular heart diseases, and cardiomyopathies are the most frequent causes of HF in adults [1, 2]. HF is characterized by neurohumoral activation of the sympathetic nervous system [3, 4], which is accompanied by increased heart rate (HR), myocardial contractility, vascular tone, and volume expansion, which, in the long run, lead to a progressive deterioration of left ventricular performance. In fact, chronic sympatho-adrenergic activation causes deleterious effects on the myocardium characterized by cardiomyocyte hypertrophy, ventricular dilatation, and pump failure [5]. These conditions are accompanied by impairment of left ventricular filling and cause inadequate perfusion and oxygenation to the tissues [6].

According to the degree of impairment of the left ventricular pump function, HF is now classified into two main conditions: HF with preserved ejection fraction (HFpEF) and HF with reduced ejection fraction (HFrEF) [7, 8]. HFrEF is characterized by left ventricular systolic dysfunction, reduced ejection of blood, and a left ventricular ejection fraction (LVEF) $\leq 40\%$. HFpEF is known as diastolic HF with a $\geq 50\%$ LVEF. However, recent studies have suggested a more heterogeneous pathophysiology, including the stiffening of the ventricular tissue, atrial dysfunction, ventricular systolic and diastolic reserve abnormalities, or endothelial dysfunction [9, 10]. Between these two syndromes, there is a "gray area," including patients with an LVEF in the range of 40–49%, characterized by signs of HF with modest systolic dysfunction and with features of diastolic dysfunction.

P. Palatini (✉)
Department of Medicine, Studium Patavinum, University of Padova, Padua, Italy
e-mail: palatini@unipd.it

M. Dorobantu et al. (eds.), *Hypertension and Heart Failure*, Updates in Hypertension and Cardiovascular Protection,
https://doi.org/10.1007/978-3-031-39315-0_5

The compensatory increase in heart rate associated with HF, which causes increased myocardial oxygen consumption, decreased coronary blood flow, and reduced myocardial efficiency, is probably one of the main mechanisms that promotes the progression of left ventricular dysfunction, leading to increased mortality in HF [9, 11]. However, not only is a high heart rate deleterious in patients with overt HF but it is also a risk factor for HF in healthy individuals. Thus, in this chapter, I will first describe the epidemiological association between heart rate and HF in apparently healthy individuals and will then focus on the role of high HR in patients with HF.

5.2 High Heart Rate As A Risk Factor for the Development of Heart Failure

The first studies to show an association between heart rate and adverse cardiovascular outcomes were performed more than 50 years ago and have been described thoroughly in preview review papers [12–16]. In almost all of the studies, a graded association was found between heart rate and mortality from cardiovascular disease, ischemic heart disease, stroke, and HF.

In a recent meta-analysis by Aune et al. [17] who took into account 87 studies, for each 10-bpm increase in the resting heart rate, the risk was 1.07 (95% confidence interval (95% CI): 1.05–1.10) for coronary heart disease, 1.09 (95% CI: 1.00–1.18) for sudden cardiac death, 1.06 (95% CI: 1.02–1.10) for total stroke, and 1.18 (95% CI: 1.10–1.27) for HF. Thus, among the cardiovascular events, the strongest association was found for HF (Fig. 5.1). A more recent meta-analysis by Shi

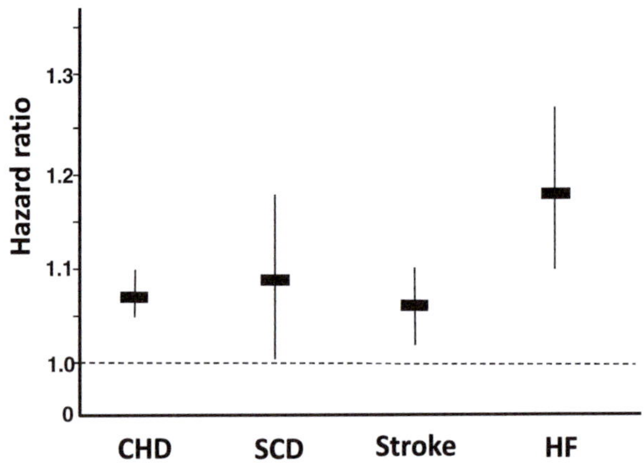

Fig. 5.1 The results of the meta-analysis by Aune et al. (17) who took into account 87 studies on the relationship between heart rate and future cardiovascular events in noncardiac subjects. Hazard ratios and 95% confidence intervals for each 10-bpm increase in the resting heart rate are shown. This study showed that the baseline heart rate is a predictor of future cardiovascular events and that the strongest association is present for incident heart failure. *CHD* coronary heart disease, *SCD* sudden cardiac death, *HF* heart failure

et al. [18] has focused more specifically on HF. In this chapter, which identified 17 cohort studies, restricted cubic spline analysis was used to model the resting heart rate–heart failure risk association. In fact, it is important to establish whether the relationship between HR and HF is linear or J- or U-shaped because this will help identify the optimal level to which the heart rate should be lowered with cardiac-slowing treatment. In agreement with Aune's meta-analysis, the risk of HF was 1.19 (95% confidence interval: 1.13–1.25) for each 10-bpm increment in the resting heart rate. The results provided no evidence of a nonlinear association of an elevated resting heart rate with HF development. The association was consistent in most subgroups, except for gender, because no significant association was observed among women.

An analysis of the Framingham Heart Study [19] evaluated possible nonlinear relationships by fitting restricted cubic splines with 3 knots at the 25th, 50th, and 75th percentiles. The baseline heart rate was associated with incident cardiovascular disease, and the association was particularly strong for HF (hazard ratio 1.32; 95% CI: 1.18–1.48). A lower heart rate appeared to be associated with a lower risk of HF even at extremely low heart rates (<60 bpm). Also in the Framingham study, spline analyses could not identify a threshold beyond which the benefit of a lower heart rate abated or increased.

An association between elevated resting heart rate and increased risk of incident HF was also observed in the MESA (Multi-Ethnic Study of Atherosclerosis) study [20], in 6766 asymptomatic participants followed up for 7 years. For each 1-bpm increase in the resting heart rate, there was a 4% greater adjusted relative risk of incident HF. An interesting finding of the MESA study was that a high heart rate was related to the development of regional and global left ventricular dysfunction.

In conclusion, the results of the literature indicate that (1) a high heart rate is consistently associated with the risk of future HF, (2) the strength of the association of heart rate with HF is greater than those for other cardiovascular outcomes, and (3) the association is linear without any flattening or upturn of the heart rate–heart failure risk relationship.

5.2.1 High Heart Rate: A Risk Factor or a Marker of Disease?

Some detractors of the clinical value of heart rate claim that the association with adverse cardiovascular outcomes was due to it being a marker of high sympathetic activity or of an underlying chronic disease. However, it should be pointed out that in many epidemiological studies, the relationship between heart rate and outcomes was also shown after long follow-ups, which, in some studies, lasted for 20 years or more [12–16]. In addition, the heart rate–event associations persisted when a large variety of comorbid conditions were taken into account. Besides traditional risk factors for atherosclerosis, several other clinical variables were taken into account, such as physical activity, indexes of pulmonary function, inflammatory markers, level of hemoglobin, parameters derived from analysis of heart rate variability, and use of beta-blockers or other antihypertensive drugs. Reverse causality is another important issue when dealing with heart rate and cardiovascular risk because a preexisting disease such as incipient HF might be associated with increased heart rate. To rule out this possibility, events occurring during the first years after baseline assessment were excluded in many large studies such as the Framingham, the Finland, Italy, Netherland, Elderly (FINE), the Cooper Clinic, or the Valsartan Antihypertensive Long-Term Use Evaluation trial (VALUE) studies [21–25]. In all of the abovementioned studies, the association with outcome remained significant even when events occurring during the first years after baseline assessment were excluded. Another argument against reverse causality is the association of heart rate with outcomes found in young individuals after extremely long follow-ups. This was shown in a large Swedish study in 1,008,485 young men enrolled in 1968–2005 and followed up until 2014 [26]. In this cohort, the heart rate measured in adolescence was associated with an increased risk of incident HF and death from all causes. In 502,534 individuals from the UK Biobank study, the associations of heart rate with adverse outcomes was even of greater magnitude at younger than at older ages [27].

5.3 Association of High Heart Rate with Adverse Outcomes in Patients with Congestive Heart Failure

If high heart rate is a precursor of HF in apparently healthy subjects or in uncomplicated hypertensive patients, then it may also have deleterious effects in patients with overt HF. This has already been observed in the "pre-beta-blocker era" by Aaronson et al. who evaluated the clinical characteristics of a group of patients with end-stage HF waiting for heart transplant [28]. These authors found that the resting heart rate was one of the most powerful predictors of urgent transplant or death. The prognostic value of an elevated resting heart rate in patients with HF was confirmed in several other studies such as the Assessment of Treatment with Lisinopril and Survival (ATLAS) [29] or the Cardiac Insufficiency Bisoprolol Study II (CIBIS II) trial [30]. Also in the Carvedilol Or Metoprolol European Trial (COMET), patients with a high heart rate had increased mortality risk at 4 months with relative risk for heart rate > 68 bpm of 1.33 (95% CI: 1.15–1.54) [31]. In the Systolic Heart Failure Treatment with the I(f) Inhibitor Ivabradine Trial (SHIFT) [32], cardiovascular outcomes were assessed in the placebo ($n = 3264$) and ivabradine arms ($n = 3241$). In the placebo group, patients with heart rates ≥87 bpm were at more than a twofold higher risk of cardiovascular mortality or hospital admission for worsening HF than were those in the lowest heart rate group (hazard ratio 2.34; 95% CI: 1.84–2.98). The highest risk differential was observed for events related to HF. In the ivabradine group, there was a direct association between heart rate achieved after 4 weeks and future cardiovascular outcomes. In the EVEREST (Efficacy of Vasopressin Antagonism in Heart Failure: Outcome Study With Tolvaptan), [33] the association of baseline heart rate with all-cause mortality did not achieve the level of statistical significance, but the 1-week post-discharge heart rate was independently associated with increased all-cause mortality (hazard ratio 1.13; 95% CI: 1.05–1.22), as was the 4-week post-discharge heart rate (hazard ratio 1.12; 95% CI: 1.05–1.19). Similar results were obtained by Habal et al. in the Enhanced Feedback For Effective Cardiac Treatment (EFFECT-HF) study [34]. A significant increase in all-cause mortality was found with an adjusted odds ratio of 1.56 (95% CI: 1.13–2.16) for heart rates >90 bpm when compared with the reference heart rate group (61–70 bpm).

In conclusion, a large body of evidence has shown that an elevated resting heart rate is associated with worse outcomes in HFrEF patients. Whether this association is also present in HFpEF is less known. This issue has been explored in the CHARM (Candesartan in Heart failure: Assessment of Reduction in Mortality and morbidity) study in HF patients stratified according to whether their left ventricular ejection fraction was ≤40% or > 40% [35]. In the group as a whole, patients in the highest heart rate tertile had worse composite cardiovascular end point than did those in the lowest heart rate tertile (hazard ratio 1.23; 95% CI: 1.11–1.36). A concordant increment in death rates was seen with increasing heart rate both in HFrEF and HFpEF patients. For each additional 10-bpm heart rate increase, there was a 6% and 5% adjusted risk accrual in patients with reduced and preserved ejection fraction, respectively (Fig. 5.2). Thus, these findings indicate that the strong association between elevated heart rate and mortality in HF is also present in patients with a preserved ejection fraction.

Whether the presence of atrial fibrillation alters the relationship between heart rate and outcomes in HF patients is less known. In the CHARM study [35], when heart rhythm was taken into account, the association between a higher heart rate and

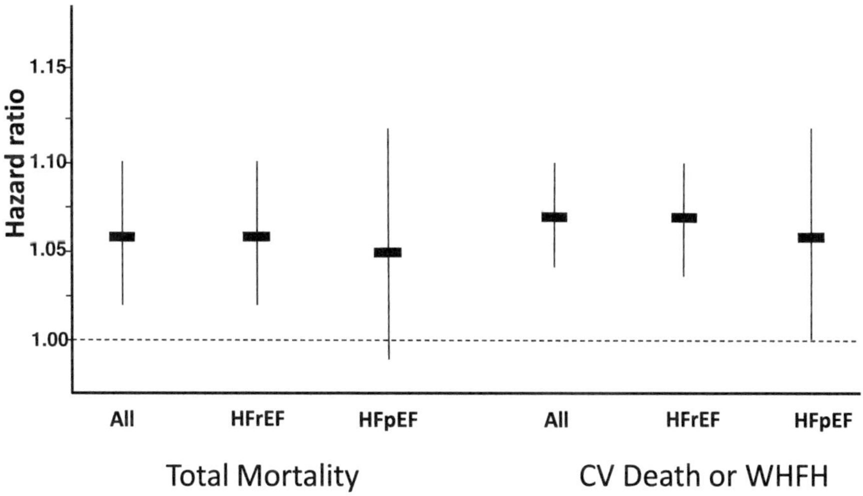

Fig. 5.2 The results of the CHARM study (35) showing the relationship between baseline heart rate and outcomes in 7597 patients with heart failure. Adjusted hazard ratios and 95% confidence intervals for each 10-bpm increase in resting heart rate are shown. This study showed that the baseline heart rate was a predictor of total mortality, cardiovascular mortality, and hospital stay for worsening heart failure in patients with heart failure. The association was present for both patients with reduced (HFrEF) and with a preserved (HFpEF) left ventricular ejection fraction. However, due to the smaller number of patients, the association was of borderline statistical significance in the HFpEF patients. *CV* cardiovascular, *WHFH* hospital stay for worsening heart failure

the risk of death was still present in patients in sinus rhythm but not in those with atrial fibrillation (p for interaction <0.0001). In the American Heart Association's Get With The Guidelines-HF (GWTG-HF) study, 145,221 HF patients were enrolled [36]. Moreover, in this study, the association of admission heart rate with in-hospital mortality was evaluated by heart rhythm and the heart rate–mortality relationship differed between patients presenting in sinus rhythm and atrial fibrillation. For heart rates >100 bpm, the mortality curve tended to plateau in patients with atrial fibrillation, whereas it continued to rise in those with sinus rhythm.

A recent meta-analysis has investigated whether heart rate is a primary predictor of mortality and/or hospitalization in adult ambulatory patients with HF [37]. A total of 62 studies encompassing 163,445 patients were considered. A 10-bpm increase in heart rate was significantly associated with an increased risk of all-cause mortality with a hazard ratio of 1.10, (95% confidence interval: 1.08–1.13), and a strongly positive interaction with age was present. Thus, this meta-analysis showed that an increasing resting heart rate is a significant predictor of all-cause mortality in ambulatory patients with HF on optimal medical therapy.

5.4 The Effect of Heart Rate Reduction in Patients with Heart Failure

The mechanisms for the detrimental effects of a high heart rate on the failing heart have been described extensively in previous reviews [1–9]. Briefly, an elevated heart rate increases myocardial workload and oxygen consumption and impairs diastolic filling time and oxygen delivery. These mechanisms can cause myocardial ischemia, necrosis, and apoptosis of the myocytes with consequent left ventricular remodeling and hypertrophy. Thus, tachycardia can both indirectly and directly facilitate ventricular fibrillation as well as other malignant arrhythmias and sudden cardiac death. This suggests that a cardiac-slowing treatment can be beneficial in failing hearts. The first evidence was provided by the DIG (Digitalis Investigation Group) trial in which digoxin reduced hospital admission for worsening heart failure [38]. However, a stronger evidence for the beneficial effect of heart rate reduction on HF was provided by trials performed with beta-blockers.

In the US Carvedilol Program, the safety committee terminated the study early because of a significant risk reduction for mortality in patients treated with carvedilol (3.2% and 7.8% in the carvedilol and placebo groups, respectively) [39]. In the Metoprolol CR/XL Randomised Intervention Trial in Congestive Heart Failure (MERIT-HF), extended-release metoprolol succinate was used to a maximum target dose of 200 mg/day [40]. Moreover, this trial was terminated prematurely because metoprolol significantly reduced all-cause mortality with a 34% decrease in risk of all-cause mortality, a 19% decrease in the risk of total mortality or all-cause hospitalization, and a 41% risk reduction for sudden death. Three large mortality trials investigated the effects of bisoprolol in patients with stable HF [41–43]. In the Cardiac Insufficiency Bisoprolol (CIBIS) trial, bisoprolol caused a risk reduction of mortality and of the rate of hospitalization for worsening HF [41]. A more dramatic decrease in mortality was observed in the CIBIS-II trial, which was terminated prematurely after

1.3 years of follow-up because of the significant mortality benefit associated with bisoprolol (hazard ratio 0.66; 95% CI: 0.54–0.81) [42]. In the third CIBIS study (CIBIS III), the main aim was to compare the effect of initial monotherapy with either bisoprolol or enalapril, followed by combination therapy [43]. No significant difference between the two strategies was found, although there was a tendency for a lower risk of death in the bisoprolol-first group than in the enalapril-first group (hazard ratio 0.88; 95% CI: 0.63–1.22). In the Carvedilol Prospective Randomized Cumulative Survival (COPERNICUS) trial, the effect of carvedilol was evaluated in patients with severe heart failure and an LVEF ≤25% [44]. Moreover, this trial was stopped early because of the mortality benefits observed with carvedilol. The aim of the Carvedilol Post-Infarction Survival Control in LV Dysfunction (CAPRICORN) study [45] was to investigate the effect of carvedilol in patients with previous myocardial infarction. In this study, a significant 23% reduction in all-cause mortality was observed. At variance with the above studies, in the Beta-Blocker Evaluation Survival Trial (BEST) study, a clinical trial in which HF patients were randomly assigned to bucindolol or placebo, no significant differences in mortality were observed between the two treatment groups [46]. This was probably due to the fact that bucindolol has some intrinsic sympathomimetic activity and thus reduces heart rate to a lower extent than other beta-blockers. The Study of the Effects of Nebivolol Intervention on Outcomes and Rehospitalisation in Seniors with Heart Failure (SENIORS) study was performed to assess the effects of the beta-blocker nebivolol, a third-generation beta-1-selective blocker with vasodilating properties, in patients ≥70 years [47]. The primary outcome (a composite of all-cause mortality or cardiovascular hospital admission) occurred in 332 patients (31.1%) on nebivolol compared with 375 (35.3%) on placebo (hazard ratio 0.86; 95% CI: 0.74–0.99). However, the difference in all-cause mortality was not statistically significant (Hazard ratio (HR) 0.88; 95% CI: 0.71–1.08; $P = 0.21$). On the basis of these results, in the year 2012, the European Society of Cardiology (ESC) guidelines recommended only bisoprolol, carvedilol, metoprolol succinate CR/XL, and nebivolol for the treatment of patients with HF [48].

The risk of sudden cardiac death is extremely high among patients with HF, in whom the 1-year absolute risk is between 4% and 13% [4, 5]. In a meta-analysis by Al-Gobari et al. [49], 30 trials, which randomized 24,779 patients to beta-blockers or placebo/control, were identified. The rate of sudden death was 5.27% in patients treated with beta-blockers and was 7.69% in those untreated (odds ratio 0.69; 95% CI: 0.62–0.77; $P < 0.00001$). Based on their data, the authors calculated that 43 patients must be treated with a beta-blocker to prevent 1 sudden death.

On the basis of the evidence from the abovementioned clinical trials, today, beta-blockers are recommended in addition to an renin–angiotensin system (RAS) blocker, for all patients with HFrEF [6, 8]. Although HF guidelines recommend that beta-blockers be titrated to achieve the target doses shown to be effective in major clinical trials, little information is available regarding the doses applied in routine clinical practice. Beta-blocker underdosing has been shown to occur frequently in clinical practice as shown by The Organized Program to Initiate Lifesaving Treatment in Hospitalized Patients with Heart Failure (OPTIMIZE-HF) [50]. This study revealed that the mean total daily dose of beta-blockers before admission was much lower than that currently recommended. After 3 months of follow-up, only

17.5% and 7.9% of patients were being treated with the recommended target doses of carvedilol and metoprolol, respectively.

Although treatment with beta-blockers has become a cornerstone of HF therapy, it should be borne in mind that in some patients heart rate remains elevated in spite of treatment with full doses. In this event, a different therapeutic strategy should be sought using ivabradine, a specific inhibitor of the I_f current in the sinoatrial node [51–53]. The aim of the SHIFT study was to establish whether ivabradine, on top of standard therapy including beta-blockers, may improve prognosis in HF patients with a resting heart rate ≥ 70 [32]. During a median follow-up of 22.9 months, ivabradine was able to reduce the occurrence of the primary composite end point of cardiovascular death or HF hospitalizations by 18% (hazard ratio 0.82; 95% CI: 0.75–0.90). The SHIFT study confirmed both the important role of heart rate in the pathophysiology of HF and the beneficial effect of heart rate reduction.

5.4.1 The Effect of Beta-Blockers in Patients with a Preserved Left Ventricular Function

Many patients with clinical signs of HF have a normal or a near-normal LVEF, but mortality in this clinical condition may be as high as in HF with left ventricular dysfunction [54]. However, whether beta-blockers may also be beneficial in patients with HFpEF is not well-known because the data from the literature are inconclusive [10]. In the SENIORS study, reduced combined all-cause mortality or cardiovascular hospitalization was seen both in patients with an LVEF ≤35% and in patients with an LVEF >35% [47]. The results from the OPTIMIZE-HF registry did not show any association between beta-blocker treatment and outcome in HFpEF patients [55, 56], whereas in two small studies [57, 58], a beneficial effect of beta-blockers was seen. In the Swedish Heart Failure Registry, which included 41,976 patients, beta-blockers were associated with reduced all-cause mortality (hazard ratio 0.93; 95% CI: 0.86–0.996; $P = 0.04$) in HFpEF patients also, thus suggesting that beta-blockade can be beneficial in this setting as well [54]. A meta-analysis by Liu et al., including 12 clinical studies and 21,206 patients with an LVEF ≥40%, demonstrated that treatment with a beta-blocker was associated with a 9% reduction in relative risk of all-cause death (95% CI: 5%–13%; $P < 0.001$) [59].

Although observational data suggest that use of beta-blockers in HFpEF patients is associated with a lower risk of mortality, these data are challenged by the differential criteria of the LVEF cutoff value (ranging from 35% to 55%) used in these clinical studies. The cutoff values of LVEF ≥50% and LVEF ≥40% were mainly used as the diagnostic criteria, but in the recent American College of Cardiology/American Heart Association (ACC/AHA) guidelines, the range of 40–50% has been defined as an intermediate condition [6]. In the study by Lund et al., HF patients with an LVEF from 40% to 50% were characteristically and prognostically similar to those with an LVEF ≥50% [54]. According to some authors, pharmacological HR lowering almost certainly has no benefit in most patients with HFpEF and may even be deleterious [60]. However, there are selected indications for which pharmacological HR lowering may be beneficial, including atrial fibrillation with a high ventricular rate, prophylaxis of tachyarrythmias, and symptomatic treatment of effort angina, even though the long-term effects of treatment in these clinical settings are not known.

There has been some conflict in the literature about the optimal doses of beta-blockers in patients with HF. The results of most clinical trials suggest that the achieved heart rate rather than the beta-blocker dose is the main determinant of beta-blocker effect on outcomes. This has been clearly shown in a meta-analysis of 35 trials, including a total of 22,926 patients, in which a close relation between reduction in all-cause annualized mortality rate and reduction in resting heart rate was found ($p = 0.004$) [61]. The importance of heart rate reduction has been confirmed by a more recent meta-analysis encompassing 19,209 HF patients in which for every heart rate reduction of 5 bpm with beta-blocker treatment, a 18% reduction (95% CI: 6–29%) in the risk of total mortality was observed [62]. In contrast, no association between all-cause mortality and beta-blocker dosing was observed. These data indicate that the most important contributor to the clinical benefits of beta-blocker therapy in systolic HF is the heart rate-lowering effect of these agents. However, it should be pointed out that beta-blockers may have additional beneficial effects not related to heart rate reduction, as these drugs may block the deleterious effects of sympathetic activity on myocardial oxygen consumption and cardiac myocyte survival [63, 64].

5.4.2 Ivabradine as a Heart Rate-Lowering Agent in Patients with HF

Ivabradine is a heart rate-lowering drug that inhibits the pacemaker ("funny") current (I_f) and was first used for the treatment of chronic stable angina pectoris [51–53]. As mentioned above, in patients with HFrEF, ivabradine proved to be an effective heart rate-lowering agent, associated with a lower risk of cardiovascular death and hospital readmissions [32]. Data from the SHIFT study, in which both beta-blockers and ivabradine were used, confirmed that the heart rate-lowering effect of pharmacological treatment is more important for improving outcomes than the actual dose of the drugs [65]. Thus, increasing the beta-blocker dose to the maximal dosing level may not be necessary in many HF patients. Conversely, patients who have little or no reduction of heart rate with beta-blocking treatment may benefit from a higher dose of the beta-blocker or the addition of ivabradine. However, the results from the Study assessInG the morbidity-mortality beNefits of the If inhibitor ivabradine in patients with coronarY artery disease (SIGNIFY) study suggest that caution should be exercised when treating HF with ivabradine on top of beta-blockers [53]. It should be noted that ivabradine tends to increase the risk of atrial fibrillation as shown by a meta-analysis, including eight randomized controlled trials (RCTs) [66]. Indeed, the European Society of Cardiology guidelines recommend ivabradine treatment only for symptomatic patients in sinus rhythm, with an LVEF ≤35% and a resting heart rate ≥ 70 bpm, in spite of treatment with an evidence-based dose of a beta-blocker and an RAS inhibitor [8].

Some studies suggest that ivabradine fails to improve outcomes in patients with HFpEF, in spite of effective heart rate reduction [67]. However, there is no study directly comparing the ivabradine effects between HFpEF and HFrEF. A meta-analysis by Toth et al. investigated the effect of ivabradine in HFpEF compared to HFrEF patients focusing on mortality, cardiovascular events, and HF hospitalization [68]. Ivabradine caused a decrease in heart rate in both HFrEF (−17.646 beats/min) and HFpEF (−11.434 beats/min) patients, although the bradycardic effect was stronger in HFrEF ($p = 0.094$) and significantly improved left ventricular performance in HFrEF. These differences may account for the beneficial effects of ivabradine treatment in patients with HFrEF and the neutral effects observed in those with HFpEF.

5.4.3 Use of Beta-Blockers in HF Patients with Atrial Fibrillation

HF patients with atrial fibrillation have higher mortality and hospital admission rates compared to patients in sinus rhythm. However, few studies have examined the relationship of heart rate with outcomes among patients with atrial fibrillation. As mentioned above, at heart rates above 100 bpm, the mortality curve for atrial fibrillation plateaued, whereas that for sinus rhythm continued to increase [36]. In a meta-analysis of individual patient data by Kotecha et al., taking into account more than 18,000 patients randomly assigned to beta-blockers or placebo, death rates over a mean follow-up of 1.5 years were lower in patients with sinus rhythm (16%) than in patients with atrial fibrillation (21%) [69]. After beta-blocker therapy, a significant reduction in all-cause mortality was observed in patients with sinus rhythm (hazard ratio 0.73; 95% CI: 0.67–0.80; $p < 0.001$), but not in patients with atrial fibrillation (HR 0.97; 95% CI: 0.83–1.14; $p = 0.73$), with a significant interaction with baseline rhythm ($p = 0.002$). These data indicate that although heart rate is a predictor of adverse outcome in patients with atrial fibrillation also, beta-blockers are not beneficial in patients with concomitant HF and atrial fibrillation. However, beta-blockers can be helpful in atrial fibrillation also because they protect from extremely high heart rates and improve exercise capacity.

5.5 Summary and Conclusions

Several randomized controlled trials performed with beta-blockers have provided undisputed evidence for a beneficial effect of heart rate reduction in patients with HFrEF. In the MERIT-HF trial, metoprolol succinate significantly reduced all-cause mortality by 34%. In the CIBIS-II trial, similar results were obtained with bisoprolol. The COPERNICUS and CAPRICORN trials showed that the beneficial effects of carvedilol can also be achieved in patients with severe HF and with prior acute myocardial infarction (AMI), respectively. In the SENIORS study, nebivolol showed an improvement in the primary outcome in patients ≥70 years, regardless of the ejection fraction. These results led the ESC guidelines to recommend only bisoprolol, carvedilol, metoprolol succinate CR/XL, and nebivolol for the treatment of patients with HF. There is general agreement that beta-blockers should be titrated to achieve the target doses shown to be effective in major clinical trials. However, beta-blockers are frequently underdosed in HF as shown by the OPTIMIZE-HF registry of patients hospitalized for HF [55].

Ivabradine, on top of standard beta-blocking therapy, improved prognosis in the HF patients from the SHIFT study. However, this effect was observed in those patients with a heart rate ≥ 75 bpm but not in those with a lower heart rate [32, 65]. These findings emphasize the importance of identifying HF patients with a high heart rate because they can benefit the most from pharmacological heart rate reduction.

Whether beta-blockers may also be beneficial in patients with HFpEF is not well-known. A recent meta-analysis has shown that treatment with beta-blockers was associated with a significant reduction in the relative risk of adverse outcomes only in patients with HFrEF [60]. The efficacy of beta-blockers has also been shown to be limited in patients with atrial fibrillation, thus indicating that beta-blockers do not confer prognostic advantage on patients with concomitant HF and atrial fibrillation.

Most clinical trials suggest that the achieved heart rate rather than the beta-blocker dose is the main determinant of beta-blocker effect on HF. However, it is not known whether the lower the heart rate, the better the outcome, and the current guidelines do not provide specific suggestions about this issue. Recent results have suggested that the relationship between posttreatment heart rate and mortality in HF is not linear in patients with a heart rate of 58–64 bpm having the best prognosis [64]. Although it is difficult to establish an optimal evidence-based heart rate level that should be pursued when treating patients with HF, some data suggest that there is a J-shaped relationship between posttreatment heart rate and mortality in these patients. Some authors recommend titrating beta-blockers until a decrease in resting heart rate in the 60-bpm range is achieved [70, 71].Conflicts of InterestThere are no conflicts of interest.

References

1. Inamdar AA, Inamdar AC. Heart failure: diagnosis, management and utilization. J Clin Med. 2016;5(7):62.
2. Chow J, Senderovich H. It's time to talk: challenges in providing integrated palliative care in advanced congestive heart failure. A narrative review. Curr Cardiol Rev. 2018;14:128–37.
3. Hasking GJ, Esler MD, Jennings GL, et al. Norepinepherine spillover to plasma in patients with congestive heart failure: evidence of increased overall and cardiorenal sympathetic nervous activity. Circulation. 1986;73:615–21.
4. Kaye DM, Lefkovits J, Jennings GL, et al. Adverse consequences of high sympathetic nervous activity in the failing human heart. J Am Coll Cardiol. 1995;26:1257–63.
5. Milo-Cotter O, Cotter-Davison B, Lombardi C, et al. Neurohormonal activation in acute heart failure: results from VERITAS. Cardiology. 2011;119:96–105.
6. Writing Committee, Maddox TM, Januzzi JL Jr, Allen LA, et al. 2021 Update to the 2017 ACC expert consensus decision pathway for optimization of heart failure treatment: answers to 10 pivotal issues about heart failure with reduced ejection fraction: a report of the American College of Cardiology Solution Set Oversight Committee. J Am Coll Cardiol. 2021;77:772–810.

7. Biegus J, Niewinski P, Josiak K, et al. Pathophysiology of advanced heart failure: what knowledge is needed for clinical management? Heart Fail Clin. 2021;17:519–31.
8. Seferovic PM, Ponikowski P, Anker SD, et al. Clinical practice update on heart failure 2019: pharmacotherapy, procedures, devices and patient management. An expert consensus meeting report of the Heart Failure Association of the European Society of Cardiology. Eur J Heart Fail. 2019;21:1169–86.
9. Hsu S, Fang JC, Borlaug BA. Hemodynamics for the heart failure clinician: a State-of-the-art review. J Card Fail. 2022;28:133–48.
10. Omote K, Verbrugge FH, Borlaug BA. Heart failure with preserved ejection fraction: mechanisms and treatment strategies. Annu Rev Med. 2022;73:321–37.
11. Metra M. Tachycardia after a heart failure hospitalization: another piece of the puzzle? JACC Heart Fail. 2013;1:497–9.
12. Palatini P, Casiglia S, Julius S, et al. Heart rate, a risk factor for cardiovascular mortality in elderly men. Arch Int Med. 1999;159:585–92.
13. Palatini P, Julius S. Review article: heart rate and the cardiovascular risk. J Hypertens. 1997;15:3–17.
14. Palatini P, Julius S. Association of tachycardia with morbidity and mortality: pathophysiological considerations. J Hum Hypertens. 1997;11(S1):19–27.
15. Fox K, Borer JS, Camm AJ, et al. Resting heart rate in cardiovascular disease. J Am Coll Cardiol. 2007;50:823–30.
16. Cook S, Togni M, Schaub MC, et al. High heart rate: a cardiovascular risk factor? Eur Heart J. 2006;27:2387–93.
17. Aune D, Sen A, O'Hartaigh B, et al. Resting heart rate and the risk of CV disease, total cancer, and all-cause mortality - a systematic review and dose-response meta-analysis of prospective studies. Nutr Metab Cardiovasc Dis. 2017;27:504–17.
18. Shi Y, Zhou W, Liu X, et al. Resting heart rate and the risk of hypertension and heart failure: a dose-response meta-analysis of prospective studies. J Hypertens. 2018;36:995–1004.
19. Ho JE, Larson MG, Ghorbani A, et al. Long-term cardiovascular risks associated with an elevated heart rate: the Framingham heart study. J Am Heart Assoc. 2014;3:e000668.
20. Opdahl A, Ambale G, Venkatesh B, et al. Resting heart rate as predictor for left ventricular dysfunction and heart failure: MESA (multi-ethnic study of atherosclerosis). J Am Coll Cardiol. 2014;63:1182–9.
21. Menotti A, Mulder I, Nissinen A, et al. Cardiovascular risk factors and 10-year all-cause mortality in elderly European male populations; the FINE study. Finland, Italy, Netherlands, elderly. Eur Heart J. 2001;22:573–9.
22. Legeai C, Jouven X, Tafflet M, et al. Resting heart rate, mortality and future coronary heart disease in the elderly: the 3C study. Eur J Cardiovasc Prev Rehabil. 2011;18:488–97.
23. Hartaigh BO, Allore HG, Trentalange M, et al. Elevations in time-varying resting heart rate predict subsequent all-cause mortality in older adults. Eur J Prev Cardiol. 2015;22:527–34.
24. Saxena A, Minton D, Lee DC, et al. Protective role of resting heart rate on all-cause and cardiovascular disease mortality. Mayo Clin Proc. 2013;88:1420–6.
25. Julius S, Palatini P, Kjeldsen S, et al. Tachycardia predicts cardiovascular events in the VALUE trial. Am J Cardiol. 2012;109:685–92.
26. Lindgren M, Robertson J, Adiels M, et al. Resting heart rate in late adolescence and long term risk of CV disease in Swedish men. Int J Cardiol. 2018;259:109–15.
27. Raisi-Estabragh Z, Cooper J, Judge R, et al. Age, sex and disease-specific associations between resting heart rate and cardiovascular mortality in the UK BIOBANK. PLoS One. 2020;15:e0233898.
28. Aaronson KD, Schwartz JS, Chen TM, et al. Development and prospective validation of a clinical index to predict survival in ambulatory patients referred for cardiac transplant evaluation. Circulation. 1997;95:2660–7.
29. Poole-Wilson PA, Uretsky BF, Thygesen K, Atlas Study Group, et al. Mode of death in heart failure: findings from the ATLAS trial. Heart. 2003;89:42–8.
30. Lechat P, Hulot JS, Escolano S, et al. Heart rate and cardiac rhythm relationships with bisoprolol benefit in chronic heart failure in CIBIS II trial. Circulation. 2001;103:1428–33.

31. Metra M, Torp-Pedersen C, Swedberg K, et al. Influence of heart rate, blood pressure, and beta-blocker dose on outcome and the differences in outcome between carvedilol and metoprolol tartrate in patients with chronic heart failure: results from the COMET trial. Eur Heart J. 2005;26:2259–68.
32. Swedberg K, Komajda M, Bohm M, et al. Ivabradine and outcomes in chronic heart failure (SHIFT): a randomised placebo-controlled study. Lancet. 2010;376:875–85.
33. Greene SJ, Vaduganathan M, Wilcox JE, EVEREST Trial Investigators, et al. The prognostic significance of heart rate in patients hospitalized for heart failure with reduced ejection fraction in sinus rhythm: insights from the EVEREST (efficacy of vasopressin antagonism in heart failure: outcome study with Tolvaptan) trial. JACC Heart Fail. 2013;1:488–96.
34. Habal MV, Liu PP, Austin PC, et al. Association of heart rate at hospital discharge with mortality and hospitalizations in patients with heart failure. Circ Heart Fail. 2014;7:12–20.
35. Castagno D, Skali H, Takeuchi M, CHARM Investigators, et al. Association of heart rate and outcomes in a broad spectrum of patients with chronic heart failure: results from the CHARM (candesartan in heart failure: assessment of reduction in mortality and morbidity) program. J Am Coll Cardiol. 2012;59:1785–95.
36. Bui AL, Grau-Sepulveda MV, Hernandez AF, et al. Admission heart rate and in-hospital outcomes in patients hospitalized for heart failure in sinus rhythm and in atrial fibrillation. Am Heart J. 2013;165:567–74.
37. Lau K, Malik A, Foroutan F, et al. Resting heart rate as an important predictor of mortality and morbidity in ambulatory patients with heart failure: a systematic review and meta-analysis. J Card Fail. 2021;27:349–63.
38. Ahmed A, Rich MW, Love TE, et al. Digoxin and reduction in mortality and hospitalization in heart failure: a comprehensive post hoc analysis of the DIG trial. Eur Heart J. 2006;27:178–86.
39. Packer M, Bristow MR, Cohn JN, et al. The effect of carvedilol on morbidity and mortality in patients with chronic heart failure: U.S. carvedilol heart failure study group. N Engl J Med. 1996;334:1349–55.
40. MERIT Investigators. Effect of metoprolol CR/XL in chronic heart failure: metoprolol CR/XL randomised intervention trial in congestive heart failure (MERIT-HF). Lancet. 1999;353:2001–7.
41. CIBIS Investigators. A randomised trial of beta-blockade in heart failure: the cardiac insufficiency Bisoprolol study (CIBIS). Circulation. 1994;90:1765–73.
42. CIBIS II Investigators and Committees. The cardiac insufficiency Bisoprolol study II (CIBIS II): a randomised trial. Lancet. 1999;353:9–13.
43. Willenheimer R, van Veldhuisen DJ, Silke B, et al. Effect on survival and hospitalization of initiating treatment for chronic heart failure with bisoprolol followed by enalapril, as compared with the opposite sequence: results of the randomized cardiac Insuffi ciency Bisoprolol study (CIBIS) III. Circulation. 2005;112:2426–35.
44. Packer M, Coats A. Fowler M, and carvedilol prospective randomized cumulative survival study group: effect of carvedilol on survival in severe chronic heart failure (COPERNICUS). N Engl J Med. 2001;344:1651–8.
45. Dargie HJ. Effect of carvedilol on outcome after myocardial infarction in patients with left-ventricular dysfunction: the CAPRICORN randomised trial. Lancet. 2001;357:1385–90.
46. Investigators BEST. A trial of the beta-blocker bucindolol in patients with advanced chronic heart failure. N Engl J Med. 2001;344:1659–67.
47. Flather M, Shibata M, Coats A, et al. Randomized trial to determine the effect of nebivolol on mortality and cardiovascular hospital admission in elderly patients with heart failure (SENIORS). Eur Heart J. 2005;26:215–25.
48. McMurray JJ, Adamopoulos S, Anker SD, et al. ESC guidelines for the diagnosis and treatment of acute and chronic heart failure 2012: the task force for the diagnosis and treatment of acute and chronic heart failure 2012 of the European Society of Cardiology. Developed in collaboration with the heart failure association (HFA) of the ESC. Eur J Heart Fail. 2012;14:803–69.
49. Al-Gobari M, El Khatib C, Pillon F, et al. β-Blockers for the prevention of sudden cardiac death in heart failure patients: a meta-analysis of randomized controlled trials. BMC Cardiovasc Disord. 2013;13:52–61.

50. Fonarow GC, Abraham WT, Albert NM, et al. Dosing of beta-blocker therapy before, during, and after hospitalization for heart failure (from organized program to initiate lifesaving treatment in hospitalized patients with heart failure). Am J Cardiol. 2008;1(102):1524–9.
51. Tardif JC, Ponikowski P, Kahan T, ASSOCIATE study investigators. Efficacy of the if current inhibitor ivabradine in patients with chronic stable angina receiving beta-blocker therapy: a 4 month, randomized, placebo-controlled trial. Eur Heart J. 2009;30:540–8.
52. Fox K, Ford I, Steg PG, et al. Ivabradine for patients with stable coronary artery disease and left-ventricular systolic dysfunction (BEAUTIFUL): a randomised, double-blind, placebo-controlled trial. Lancet. 2008;372:807–16.
53. Fox K, Ford I, Steg PG, et al. Ivabradine in stable coronary artery disease without clinical heart failure. N Engl J Med. 2014;371:1091–9.
54. Lund LH, Benson L, Dahlström U, et al. Association between use of β-blockers and outcomes in patients with heart failure and preserved ejection fraction. JAMA. 2014;312:2008–18.
55. Hernandez AF, Hammill BG, O'Connor CM, et al. Clinical effectiveness of beta-blockers in heart failure: findings from the OPTIMIZE-HF (organized program to initiate lifesaving treatment in hospitalized patients with heart failure) registry. J Am Coll Cardiol. 2009;53:184–92.
56. Patel K, Fonarow GC, Ekundayo OJ, et al. Beta-blockers in older patients with heart failure and preserved ejection fraction: class, dosage, and outcomes. Int J Cardiol. 2014;173:393–401.
57. Nevzorov R, Porath A, Henkin Y, et al. Effect of beta blocker therapy on survival of patients with heart failure and preserved systolic function following hospitalization with acute decompensated heart failure. Eur J Intern Med. 2012;23:374–8.
58. El-Refai M, Peterson EL, Wells K, et al. Comparison of β-blocker effectiveness in heart failure patients with preserved ejection fraction versus those with reduced ejection fraction. J Card Fail. 2013;19:73–9.
59. Liu F, Chen Y, Feng X, et al. Effects of beta-blockers on heart failure with preserved ejection fraction: a meta-analysis. PLoS One. 2014;9:e90555.
60. Meyer M, LeWinter MM. Heart rate and heart failure with preserved ejection fraction: time to slow beta-blocker use? Circ Heart Fail. 2019;12:e006213.
61. Flannery G, Gehrig-Mills R, Billah B, et al. Analysis of randomized controlled trials on the effect of magnitude of heart rate reduction on clinical outcomes in patients with systolic chronic heart failure receiving beta-blockers. Am J Cardiol. 2008;101:865–9.
62. McAlister FA, Wiebe N, Ezekowitz JA, et al. Meta-analysis: beta-blocker dose, heart rate reduction, and death in patients with heart failure. Ann Intern Med. 2009;150:784–94.
63. Sarraf M, Francis GS. It is all about heart rate. Or is it? J Am Coll Cardiol. 2012;59:1946–7.
64. Cullington D, Goode KM, Clark AL, et al. Heart rate achieved or beta-blocker dose in patients with chronic heart failure: which is the better target? Eur J Heart Fail. 2012;14:737–47.
65. Swedberg K, Komajda M, Böhm M, for the SHIFT Investigators, et al. Effects on outcomes of heart rate reduction by ivabradine in patients with congestive heart failure: is there an influence of betablocker dose? Findings from the SHIFT (systolic heart failure treatment with the if inhibitor ivabradine trial) study. J Am Coll Cardiol. 2012;59:1938–45.
66. Tanboğa İH, Topçu S, Aksakal E, et al. The risk of atrial fibrillation with ivabradine treatment: a meta-analysis with trial sequential analysis of more than 40000 patients. Clin Cardiol. 2016;39:615–20.
67. Komajda M. Ivabradine. Handb Exp Pharmacol. 2017;243:167–75.
68. Tóth N, Soós A, Váradi A, et al. Effect of ivabradine in heart failure: a meta-analysis of heart failure patients with reduced versus preserved ejection fraction. Can J Physiol Pharmacol. 2021;99:1159–74.
69. Kotecha D, Holmes J, Krum H, et al. Beta-blockers in heart failure collaborative group. Efficacy of β blockers in patients with heart failure plus atrial fibrillation: an individual-patient data meta-analysis. Lancet. 2014;384:2235–43.
70. Porapakkham P, Krum H. Is target dose of beta-blocker more important than achieved heart rate or heart rate change in patients with systolic chronic heart failure? Cardiovasc Ther. 2010;28:93–100.
71. Huang RL, Listerman J, Goring J, et al. Beta-blocker therapy for heart failure: should the therapeutic target be dose or heart rate reduction? Congest Heart Fail. 2006;12:206–10.

Blood Pressure Variability and Heart Failure: Pathophysiological and Clinical Aspects

6

Gianfranco Parati and Juan Eugenio Ochoa

6.1 Introduction

Blood pressure (BP) values change significantly over time as a result of the interaction between extrinsic environmental and behavioral factors and intrinsic cardiovascular regulatory mechanisms. Although these variations represent a continuous phenomenon, definitions and classification of BP variability (BPV) have been proposed on the basis of the time window over which this phenomenon is assessed. BPV can indeed be measured over different time windows: from beat to beat (**very short-term** BP variability (BPV)), within 24 h (from minute to minute, hour to hour, and day to night; **short-term** BPV), over different days (**mid-term** BPV), or between clinic visits performed over weeks, months, seasons, and years (**long-term** BPV) [1] (Fig. 6.1).

Although the clinical significance and prognostic implications of different measures of BPV may substantially differ, in general, a large body of evidence supports the concept that increasing values of BPV are associated with a higher risk of preclinical organ damage and with the development of cardiac and cerebrovascular events and cardiovascular and all-cause mortality [2] (see Fig. 6.1). Only few studies, however, have provided evidence on the contribution of BPV to development of heart failure (HF) [3] or on its prognostic impact on patients with HF. It should be emphasized that most of the evidence on the role of BPV in relation to HF comes

G. Parati (✉)
Department of Cardiology, Istituto Auxologico Italiano, IRCCS, San Luca Hospital, Milan, Italy

Department of Medicine and Surgery, University of Milano-Bicocca, Milan, Italy
e-mail: gianfranco.parati@unimib.it

J. E. Ochoa
Department of Cardiology, Istituto Auxologico Italiano, IRCCS, San Luca Hospital, Milan, Italy

M. Dorobantu et al. (eds.), *Hypertension and Heart Failure*, Updates in Hypertension and Cardiovascular Protection,
https://doi.org/10.1007/978-3-031-39315-0_6

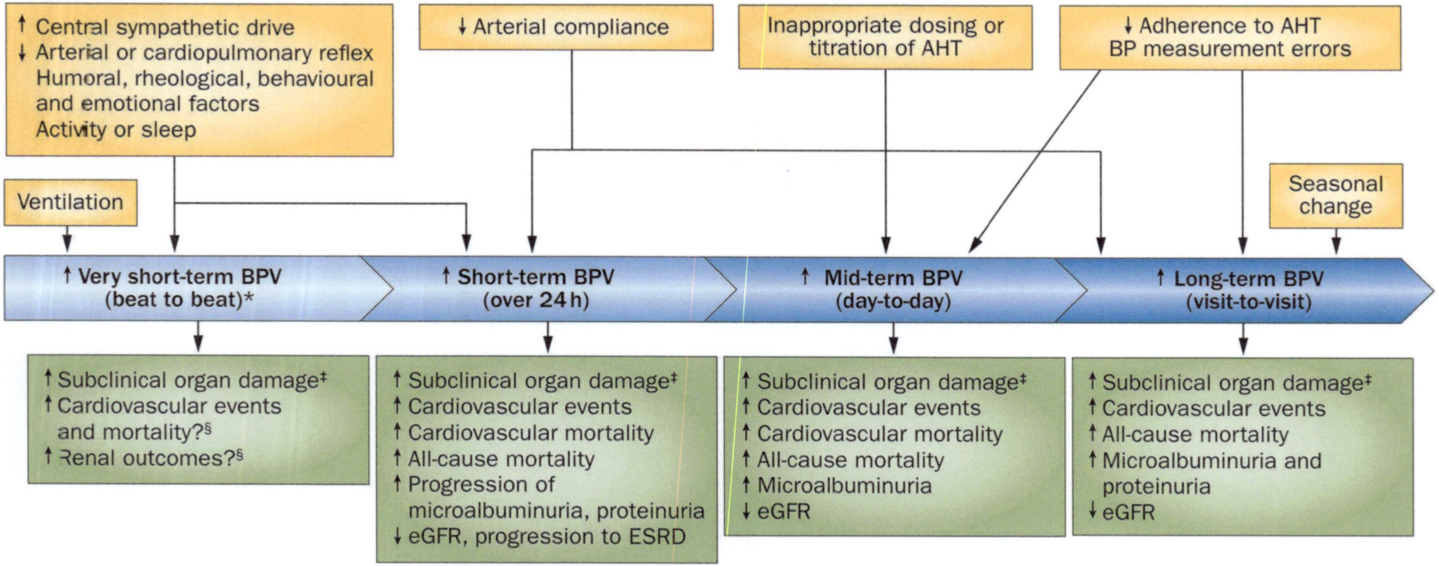

Fig. 5.1 Different types of blood pressure (BP) variability (BPV), their determinants, and prognostic relevance. (Taken from Parati et al. [2] with permission). *Assessed in laboratory conditions; ‡cardiac, vascular, and renal subclinical organ damage; §BPV on a beat-to-beat basis has not been routinely measured in population studies. *AHT* antihypertensive treatment, *BP* blood pressure, *BPV* blood pressure variability, *ESRD* end-stage renal disease, *eGFR* estimated glomerular filtration rate

mainly from *post hoc* analyses of clinical trials in hypertension or from HF cohorts in which HF was part of the composite primary outcome or a secondary outcome. On the other hand, understanding the actual prognostic impact of BPV for the progression of HF or for the development of cardiovascular outcomes in HF patients has been made difficult by the complex pathophysiology of HF. Indeed, HF may present with different phenotypes based on the characteristics of the left ventricle (LV), which are different from a pathophysiological and a clinical viewpoint. For instance, while in patients with HF with preserved ejection fraction (HFpEF), higher values of BPV have been shown to be associated with a poor prognosis, in patients with HF with reduced ejection fraction (HFrEF) lower and not just higher BPV values have been predictive of adverse events.

In its first part, this chapter will address the role of BPV in promoting cardiac organ damage and development of HF, also discussing the potential pathophysiological mechanisms for this association. The second part will address the clinical relevance of BPV in HF patients according to their clinical phenotype. Among other issues, we will also deal with the paradoxical relationship reported between a number of clinical variables and HF with reduced ejection fraction.

6.2 Relationship of BPV with Preclinical Cardiac Organ Damage

A number of studies, either in the general population or in different clinical settings, have provided evidence that preclinical cardiac organ damage is associated with a higher incidence of cardiovascular events and HF independently from traditional risk factors.

Left ventricular hypertrophy (LVH), estimated by electrocardiography or more accurately by echocardiographic examination, is a recognized landmark of hypertensive heart disease. Patients with LVH have been reported to have a significantly higher risk of cardiovascular events compared to those without LVH, with the magnitude of risk depending on the demographic and clinical characteristics of the samples [4]. Furthermore, at variance from other markers of organ damage, regression of LVH has been largely documented to be associated with significant reductions in cardiovascular risk. Elevated average BP levels have been classically considered a major determinant of LV function and structure, and, in recent decades, a number of studies have also provided evidence that increasing values of BPV are associated with an increased risk of cardiac organ damage independently of the average BP values.

6.2.1 Very Short-Term BPV

Studies addressing the relationship of very short-term BPV (estimated by means of intra-arterial beat-to-beat BP recordings) with cardiac organ damage in hypertensive subjects have reported a higher prevalence and severity of hypertension-mediated cardiac organ damage in individuals with a higher 24-h BPV [5, 6]. For nearly any level of the 24-h mean BP (MBP), subjects in whom the 24-h BPV (i.e., assessed as standard deviations of systolic BP (SBP), mean BP, and diastolic BP (DBP)) was low had a lower prevalence and severity of organ damage than did those in whom the 24-h BPV was high. In the long-term follow-up, after an average of 7.4 years, a group of hypertensive patients participating in the original study were reexamined, including echocardiographic assessment of LV mass index (LVMI). Interestingly, the same indices of variability predicted future development and progression of LVH [6].

The results of these studies provided evidence that the level of BP achieved by treatment and the degree of organ damage at the time of initial evaluation are important determinants of future organ damage related to hypertension. They also constitute the first longitudinal evidence that the cardiovascular complications of hypertension may depend on the degree of 24-h BPV [5, 6].

6.2.2 Short-Term BPV

A number of studies have also provided evidence that **short-term BPV**, evaluated on the basis of intermittent 24-h ambulatory blood pressure monitoring (ABPM) recordings, is associated with preclinical cardiac organ damage. In a study by Tatasciore et al. [7], in a population of individuals referred for suspected hypertension, LVMI progressively increased across tertiles of standard deviation (SD) of awake systolic (S) ambulatory BP independent of the average BP levels [7]. Such relationship in patients with recently detected high BP values suggest that BPV and not just elevated BP levels may promote changes in LV structure early in the natural history of hypertension.

A meta-analysis of 12 selected studies evaluating the correlation of LVMI with various indices of short-term BPV (estimated from intermittent ABPM recordings) showed that estimates of BPV such as the SD of 24-h SBP, the SD of daytime SBP, weighted SD (wSD) of 24-h SBP, and the average real variability (ARV) of 24-h SBP were all associated with higher values of LV mass [8].

Evidence has also been provided that specific alterations of the physiological circadian BP profile, characterized by loss (non-dipping) or inversion (rising) of nocturnal BP fall, are associated with concomitant cardiac subclinical organ damage [4].

In studies conducted in older hypertensive patients, recently diagnosed with essential hypertension, and in chronic kidney disease (CKD) and in HF patients, the prevalence rates of LVH (defined by Sokolow–Lyon or Romhilt–Estes criteria and Cornell voltage product) have been reported to be not significantly higher in reverse dippers than in dippers, non-dippers, or extreme dippers [9, 10].

Findings from echocardiographic studies have offered more consistent evidence on the increased risk of LVH associated with the reverse dipping pattern. In a population of 375 untreated and treated middle-aged hypertensive patients, Ivanovic

Table 6.1 Summary of echocardiographic data on LVH and/or LVMI in extreme dipper, dipper, non-dipper, and reverse dipper patients provided by four echocardiographic studies

	Extreme dippers		Dippers		Nondippers		Reverse dippers	
Author (reference)	LVH, %	LVMI, g/h$^{2.7}$	LVH, %	LVMI, g/h$^{2.7}$	LVH, %	LVMI, g/h$^{2.7}$	LVH, %	LVMI, g/h$^{2.7}$
Pogue 14	NA	NA	62	NA	69	NA	72	NA
Ivanovic 11	5	46 ± 8	9	47 ± 9	17	51 ± 10	31*	54 ± 10*
Komori 12	NA	NA	NA	150 ± 51 g/m^2	NA	163 ± 51 g/m^2	NA	164 ± 48 g/m^2**
Wang 13	NA	NA	NA	45 ± 16	NA	52 ± 24	NA	58 ± 19***

Taken from Cuspidi et al. with permission [4]. *LVH* left ventricular hypertrophy, *LVMI* left ventricular mass index, *NA* not available. *$P < 0.05$ reverse dippers vs. non-dippers, dippers, and extreme dippers. **$P = 0.05$ reverse dippers vs. dippers. ***$P < 0.05$ reverse dippers vs. non-dippers and dippers

et al. [11] reported a progressive and significant increase in LVH rates from extreme dippers (5%) to dippers (9%), non-dippers (17%), and reverse dippers (31%) [11]. A multicenter Japanese study performed in 334 hospitalized patients with congestive heart Failure (CHF) documented that the LV mass indexed to body surface area was significantly higher in non-dippers and reverse dippers than in dippers [12]. Similarly, Wang et al. [13] examined a cohort of 540 patients with CKD and reported that the LV mass indexed to height was significantly greater in reverse dipper than in non-dipper and dipper participants [13].

In contrast to the abovementioned echocardiographic studies, in the African American Study of Kidney Disease and Hypertension Study (AASK Trial), only marginal differences in LVH prevalence among various ambulatory day/night BP phenotypes were reported [14] (see Table 6.1).

Among 508 hospitalized patients with congestive heart failure (46% with a preserved ejection fraction (EF)), a paradoxical increase in BP during sleep was significantly more common in those with a preserved LV EF (29%) compared to those with a reduced systolic function (20%) [19]. Reverse dipping pattern was shown to be an independent correlate of HF with a preserved systolic function (hazard ratio (HR) 1.73; 95% confidence interval (95% CI): 1.02–2.91; $P = 0.04$) after adjusting for several covariates [19]. Some evidence has also been provided that nocturnal BP surge within seconds is associated with the LVMI [20].

6.2.3 Mid-Term BPV

Concerning mid-term BPV, a few studies have provided evidence that increased day-to-day BPV identified by home BP monitoring (HBPM) is associated with increased prevalence and severity of cardiac organ damage. In a population of 356 never-treated hypertensive subjects, Matsui et al. [21] investigated the relationship between the indices of day-by-day BPV and LVMI. From the various indices evaluated, the maximum home SBP (defined as the maximum mean triplicate BP reading in a 14-day period for each individual) was significantly correlated with LVMI ($r = 0.51$, $P < 0.001$). It should be noted that the correlation coefficient reported for LVMI was significantly larger for the maximum home SBP than for the mean home SBP. In multivariate regression analyses, the maximum home SBP was independently associated with LVMI regardless of the average home BP level. In the prediction of LVH and carotid atherosclerosis, the goodness of fit of the model was significantly improved when the maximum home SBP was added to the sum of the mean office and home BPs ($P = 0.002$ and $P < 0.001$, respectively). These findings indicate that assessment of the maximum home SBP, in addition to the mean home SBP, might increase the predictive value for cardiac organ damage. However, despite this evidence, no index of cardiac organ damage has shown consistent and independent relationships with the indices of mid-term BPV [15, 21–28].

6.2.4 Long-Term BPV

Finally, long-term visit-to-visit BPV has also been shown to be associated with cardiac organ damage [29, 30]. In a group of treated hypertensive patients, Masugata et al. evaluated the relationship between visit-to-visit BPV (estimated as the SD of BP measures performed every 1 or 2 months over a 1-year period or as the difference between the maximum and minimum SBP) and the echocardiographic indices of LV diastolic function (quantified by the ratio (E/A) of early (E) and late (A) diastolic transmitral flow, early diastolic mitral annular velocity (e'), and the ratio (E/e') of E to e') [29]. LV diastolic function assessed by the E/A ratio only correlated with the SD of SBP, whereas e’ correlated with the SD of SBP and the maximum–minimum SBP difference. E/e' correlated with the SD of SBP, the maximum–minimum SBP difference, and the mean value of SBP. In multiple regression analysis, only the maximum–minimum SBP difference was independently associated with E/e' ($\beta = 0.410$, $P = 0.009$). It should be noted that the visit-to-visit variability of SBP showed better correlation with LV diastolic dysfunction than the mean values of SBP. Based on these findings, the authors concluded that a high visit-to-visit variability of SBP is associated with LV diastolic dysfunction and may represent a risk factor for diastolic HF in hypertensive patients [29].

In a study by Okada et al., the ratio of early ventricular filling velocity to early diastolic mitral annular velocity (E/e'), ejection fraction, and LVMI were compared between the highest (high variability) and lowest (low variability) tertiles of systolic BP SD. E/e' and maximum intima media thickness (max-IMT) were significantly greater in the high variability group than in the low variability group after adjusting for age, sex, baseline systolic BP, and other covariates. Despite the significant differences in the indices of LV function, no significant differences were observed in LVMI or LV EF [30].

6.3 Relationship of BPV with Development of Heart Failure

Increasing values of systolic and diastolic BPV have shown to be associated with important cardiac, vascular, and hemodynamic alterations, all of which may promote development of CHF. On the basis of this evidence, a series of recent studies have been conducted in order to explore whether increasing values of BPV may be associated with a higher risk of developing HF. It should be mentioned, however, that despite the large body of evidence reporting on the associations between BPV and cardiovascular events and mortality, only very few studies have been conducted specifically to address the association between BPV and development of CHF. The evidence on this relationship mainly comes from *post hoc* analyses of clinical trials or cohort studies in which the relationship between BPV and CHF was not predefined. In most of these studies, HF was indeed part of the composite end point or a secondary outcome. Moreover, most previous studies have not reported on the association between BPV and development of different HF subtypes based on the characteristics of the LV, by separately considering HF with preserved ejection fraction (HFpEF) and HF with reduced ejection fraction (HFrEF), which are pathophysiologically and clinically different. For instance, patients with HFpEF typically exhibit a stiffer myocardium characterized by an increased presence of the fibrotic tissue that is particularly sensitive to preload increases. By contrast, acute afterload elevations in the setting of ventricular-arterial stiffening can deteriorate myocardial relaxation, leading to significant increases in cardiac filling pressures during stress. Both increases in preload and afterload can thus predispose patients to cardiac decompensation, and this is the rationale for aiming at tight volume and BP control in these patients [16–18, 31]. This suggests that BPV may be more sensitive than the mean BP levels as an indicator of arterial BP dysregulation associated with preload and afterload disturbances during the evolution of HF. Identifying BP dysregulation early in the preclinical stages of HF might improve the prevention of its clinical manifestations. Finally, while higher systolic BPV has been frequently reported to be associated with a greater risk of developing CHF, the evidence is less regarding the prognostic role of diastolic BPV.

6.3.1 Long-Term BPV

In a number of studies, increasing values of long-term BPV have been associated with a higher risk of developing HF or with hospitalization due to CHF [32–37]. However, not all studies have consistently confirmed this relationship [38, 39]. For instance, BPV was not associated with incident HF in the Antihypertensive and Lipid-Lowering Treatment to Prevent Heart Attack Trial (ALLHAT) [38] nor in the Systolic Blood Pressure Intervention Trial (SPRINT) [39].

In a *post hoc* analysis of the PROspective Study of Pravastatin in the Elderly at Risk (PROSPER) in **older subjects at risk of cardiovascular disease (CVD_ events**, systolic long-term visit-to-visit BPV was predictive for the risk of HF hospitalizations and cardiovascular and total mortality [35]. In this study, higher values of SD and CV (estimated from BP measurements obtained from visits 1–9 over 24 months) were associated with an increased risk of HF hospitalization (HR 1.4; 95% confidence interval: 1.1–1.8) after 7.1 years of follow-up. This association was independent of the average BP levels, age, gender, in-trial treatment group (pravastatin or placebo), and other common cardiovascular risk factors [35].

A *post hoc* analysis of the Valsartan Antihypertensive Long-term Use Evaluation (VALUE) trial analyzed data of 13,803 hypertensive patients at different risk levels in order to explore whether visit-to-visit BPV was associated with an increased risk of cardiovascular events and death [34]. Overall, after a mean of 4.2 years of follow-up, compared to patients with the lowest values of BPV, those in the highest quintile of SBP SD showed a significantly higher risk of cardiovascular events (HR 2.1; 95% CI: 1.7–2.4). More specifically, increased values of systolic visit-to-visit BPV were associated with a higher risk of cardiac events (HR 2.3; 95% CI: 1.9–2.8), stroke (HR 1.5; 95% CI: 1.1–2.1), myocardial infarction (HR 3.2; 95% CI: 2.3–4.3), ischemic stroke (HR 1.9; 95% CI: 1.3–2.7), and CHF (HR 3.1; 95% CI: 2.2–4.3). Associations were stronger among younger patients and in patients with lower systolic BP and were similar between patients with different baseline risks, except for the higher risk of death among patients with an established cardiovascular disease [34].

An analysis of a large dataset of 28,790 patients recruited for the Ongoing Telmisartan Alone and in Combination with Ramipril Global Endpoint Trial (ONTARGET) and the Telmisartan Randomized Assessment Study in ACE Intolerant Subjects with Cardiovascular Disease (TRANSCEND) trial explored whether on-treatment visit-to-visit systolic BPV versus mean SBP and the two measures together were predictors of cardiovascular events [33]. It should be noted that in this analysis, quantification of the indices of visit-to-visit BPV was standardized using the same number of visits ($n = 5$) over the same time interval (2 years) in the vast majority (>90%) of the patients. In both trials, the primary outcome was a composite of mortality from cardiovascular causes, nonfatal myocardial infarction, nonfatal stroke, and hospitalization for HF. Secondary outcomes were cardiovascular mortality, all-cause mortality, and cause-specific outcomes (myocardial infarction, stroke, and HF) separately considered. The main result of this report was the finding of an association between increasing values of the CV of the mean SBP with fatal events. Although, on-treatment mean SBP provided an overall better prediction of cardiovascular risk than did the visit-to-visit SBP-CV, prediction of the primary outcome improved by the combined use of both measures (HR 1.42; 95% CI: 1.20–1.68) compared with SBP-CV (HR 1.13; 95% CI: 1.01–1.27) and mean SBP (HR 1.24; 95% CI: 1.11–1.40) separately considered. Based on these results, authors concluded that although mean on-treatment systolic BP is superior to systolic BPV in predicting the overall risk of cardiovascular events, the combined use of the two measures improves event prediction, thereby improving the assessment of the protective effect of antihypertensive treatment [33]. Regarding the time-related incidence of HF, there was a progressive increase in the risk of HF from the lowest to the highest SBP-CV quintiles. It should be noted that compared with the lowest SBP-CV quintile, the adjusted risk usually exhibited a significant increase of 30% only in the highest SBP-CV quintile (see Fig. 6.2).

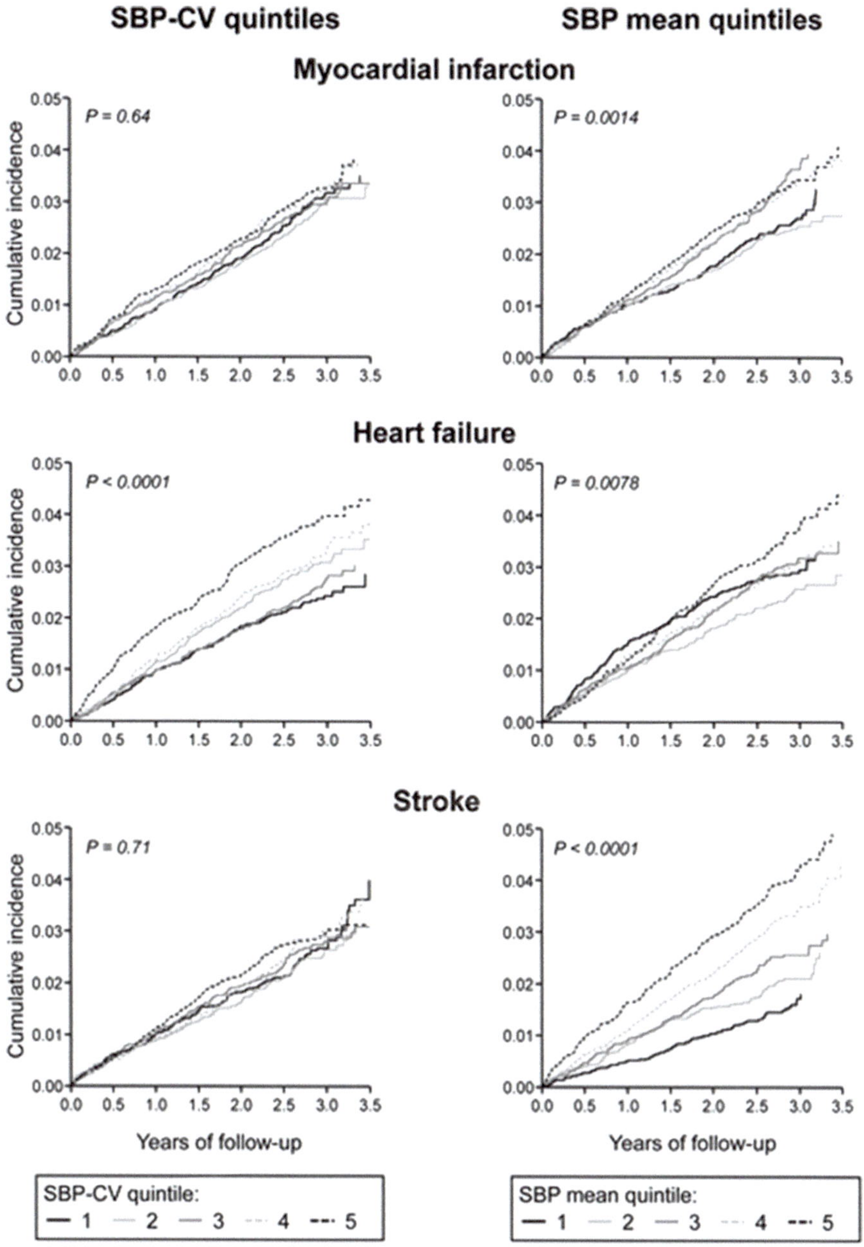

Fig. 6.2 Time-relatedincidence (Kaplan–Meier curves) of myocardial infarction, heart failure, and stroke for the different SBP-CV and mean on-treatment SBP quintiles. *CV* coefficient of variation, *SBP* systolic blood pressure. (Taken from Mancia et al. [33] with permission)

SBP-CV and mean on-treatment SBP quintiles together resulted in an outcome prediction that reflected in an additive manner the contribution of these factors when they were examined individually. However, this was not the case for HF, which remained exclusively related to SBP-CV whose relative importance exceeded that of the mean SBP, thus confirming the independent prognostic role of visit-to-visit BPV in HF [33].

It should be noted that while systolic visit-to-visit BPV did not predict the risk of stroke and myocardial infarction (the prediction of which was entirely accounted for by mean SBP), it did predict, to a greater degree than the average systolic BP levels, the risk of cardiovascular and all-cause mortality [33]. This raises the possibility for visit-to-visit BPV to be a preferential predictor of serious events, which would be in line with the observation that systolic BPV also predicted a clinical condition associated with high mortality, such as HF [40, 41]. This might occur because unstable BP may have a particularly damaging effect on the structure and function of a vital organ such as the heart. It is, nevertheless, also possible that the neurohumoral activation that precedes and favors HF [41] promotes BP instability. Thus, an increase in BPV would reflect an impairment of BP homeostasis associated with a condition in which the mechanisms of autonomic and hemodynamic modulation are impaired. Future studies will have to address to what extent an increased BPV represents a risk factor for developing HF or a marker of an especially severe clinical condition such as HF.

Evidence on the relationship between visit-to-visit BPV and risk of HF has also been provided in patients with **type 2 diabetes** in whom HF is a common comorbidity [42]. Most of this evidence comes from large cohorts of patients with type 2 diabetes who had participated in clinical trials of glucose and other cardiovascular risk factor management. In a *post hoc* analysis of the Action to Control Cardiovascular Risk in Diabetes (ACCORD) trial [43], the CV and ARV of both SBP and DBP were associated with an increased risk of HF. It should be noted that these effects were related to dips, not elevations, in BP. Indeed, ACCORD patients with progressively lower baseline average BP levels demonstrated a stepwise increase in the risk of HF with higher CV-SBP, ARV-SBP, and CV-DBP [43]. Similar findings were also observed in the Veterans Affairs Diabetes Trial (VADT), in which the predictive value for HF was limited, however, to CV and ARV of DBP only [43]. It should be noted that in both ACCORD and VADT studies, BPV was a significant and robust predictor of HF even in the setting of well-controlled management of other type 2 diabetes risk factors and in those undergoing additional BP-lowering treatment [43].

Additional evidence on the relationship between visit-to-visit BPV and incidence of HF in type 2 diabetes was provided in a report of the Look AHEAD (Action for Health in Diabetes) study, conducted in a large community-based sample of 4200 adults [44]. In all individuals, indices of long-term BPV were determined on the basis of BP measures performed at four visits (baseline and at 12, 24, and 36 months). Participants free of HF duringthe first 36 months were followed up for HF events. Over a median follow-up of 6.7 years, higher values of systolic and diastolic BPV were independently associated with a greater risk of incident HF after accounting for other common HF risk factors, including average BP. Specifically, the HR of incident HF for the highest vs. lowest quartile of the SD of SBP was 1.77 (95% CI: 1.01–3.09), whereas the HR for the highest vs. lowest quartile of variability independent of the mean (VIM) of SBP was 1.29 (95% CI: 0.78–2.14). The adjusted HR for participants in the highest compared with the lowest quartile of the SD of DBP was 1.61 (95% CI: 1.01–2.59), and the adjusted HR for VIM of DBP was 1.65 (95% CI: 1.03–2.65) [44]. This study emphasized the importance of stable and consistent BP control over time in the reduction of the incidence of HF among patients with type 2 diabetes.

It should be noted that recent studies have been conducted in order to explore the relationship between long-term BPV and the incidence of CHF after also considering the specific subtypes of HF (i.e., HFpEF and HFrEF).

A recent analysis of a large multiethnic cohort of postmenopausal women ($n = 23{,}918$), enrolled in the Women's Health Initiative Hormone Therapy Trials, has explored the relationship between BPV and HF sub-phenotypes. Specifically, the authors explored the association between annual visit-to-visit BPV (assessed on the basis of BP measures performed at baseline from 1993 to 1998 and then annually up to 2005) and incident HF with either a preserved or a reduced ejection fraction [45]. After approximately 16 years of follow-up, the authors found increasing values of systolic visit-to-visit BPV (defined either as the SD of the mean BP across visits (SDm) or the SD of the participant's regression line for BP across visits (Sdreg)) to be associated with a higher risk of HFpEF and of coronary events during HF hospitalization, independent of the usual mean BP. Specifically, women in the highest versus lowest quartile of the SD of SBP were at a statistically significantly higher risk of HFpEF (HR 1.61; 95% CI: 1.12–2.31) but not of HFrEF (HR 1.18; 85% CI: 0.70–1.96). Conversely, the risk for the highest versus lowest quartile of DBP SD was 1.56 (0.89, 2.74) for HFrEF and 1.19 (0.85, 1.65) for HFpEF [45] (see Fig. 6.3).

It should be noted that the association between diastolic BPV and the risk of HF differed in women with and without preexisting LV hypertrophy. Although diastolic visit-to-visit BPV was not significantly associated with HF risk, an interaction of diastolic BPV with LV hypertrophy on the risk of HFrEF was significant [45]. Women who had LV hypertrophy had a twofold higher risk of HFrEF associated with each 5-mmHg increment in the SDm of DBP and a 93% higher risk of HFrEF with each 5-mmHg in the SDreg of DBP. This potentially implies stronger risk associations among those with long-standing hypertension, especially if poorly treated. Interestingly, the authors reported a stronger association between higher SDm of SBP and the risk of HF in younger women (50–59 years) and a stronger association between higher systolic BPV and the risk of HFrEF among those with a mean SBP <120 mmHg. A possible explanation for this is that younger women, often with lower SBP and lower cardiovascular risk, might be more susceptible to BPV than older women in whom other risk factors are potentially more important [32, 34, 46].

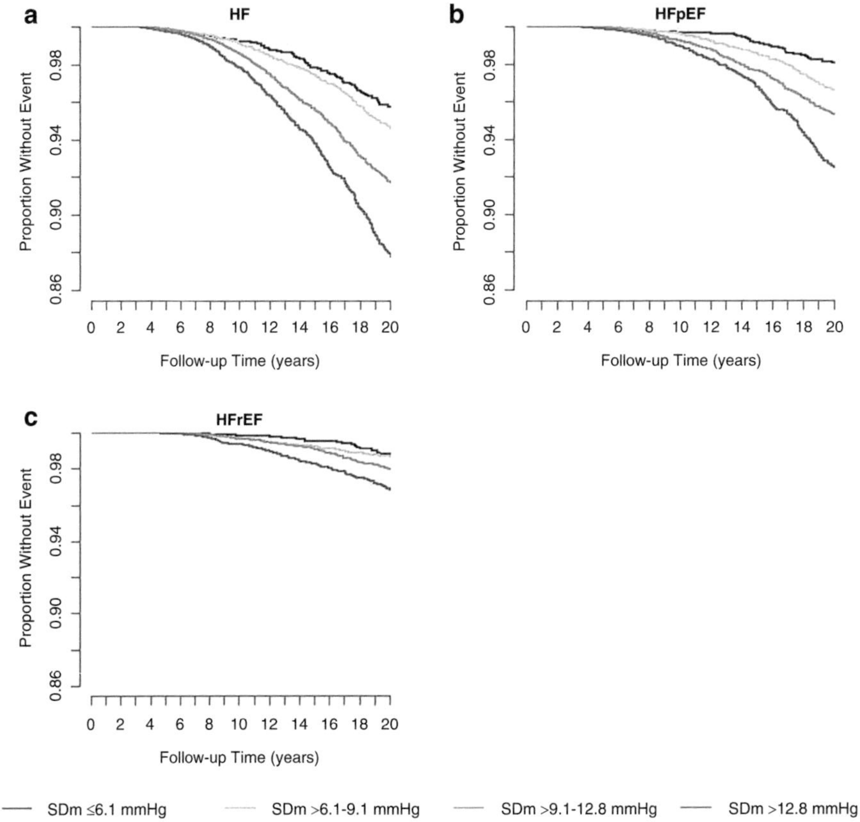

Fig. 6.3 Kaplan–Meier curves of HF outcomes associated with the quartiles of the variability of systolic blood pressure. (**a**) A Kaplan–Meier curve of HF. (**b**) A Kaplan–Meier curve of HFpEF. (**c**) A Kaplan–Meier curve of HFrEF. *HF* heart failure, *HFpEF* heart failure with preserved ejection fraction, *HFrEF* heart failure with reduced ejection fraction, *SDm* SD of the participant's mean blood pressure. (Taken from Haring et al. [45] with permission)

6.4 The Relationship of BPV with Outcomes in CHF Patients

Most of the studies in non-HF populations have indicated that in general increasing values of BPV are associated with an increased risk of events. However, the evidence on the predictive value of BPV in CHF populations either for cardiovascular events or for progression of CHF is somehow limited. Besides, it is not clear whether increasing values of BPV have the same prognostic impact on patients with HFpEF and HFrEF.

6.4.1 Very Short-Term BPV

Gibelin et al. [47] comparatively evaluated the prognostic value of very short-term BPV (i.e., assessed on the basis of continuous noninvasive BP recordings) in patients with CHF vs. healthy controls. Overall, compared to healthy controls, patients with CHF not only showed significantly lower values of BPV but also a worse prognosis. Moreover, among patients with CHF, decreased BPV was related to LV dysfunction. Overall, the total spectral amplitude (TA) for DBP, very low frequencies over total amplitude ratio (VLF/TA) for DBP and for mean BP (MBP), and very low over high frequencies ratio (VLF/HF) for DBP were the significant predictors of death. In a multivariate analysis, BPV predicted survival independently of EF or peak VO2. In Kaplan–Meier survival curves, VLF/TA < 55% for DBP, MBP, and SBP was shown to be a significant predictor of risk. It should be noted that 1-year survival in patients with DBP VLF/TA < 55% was significantly reduced compared to subjects with DBP VLF/TA > 55% (i.e., 53% versus 95%, respectively). Based on these results, the authors concluded that in patients with HF, an analysis of BPV can identify patients at an increased risk of cardiac death [47].

6.4.2 Short-Term BPV

Most of the evidence on the prognostic impact of short-term BPV has been obtained from *post hoc* analyses of clinical trials in hypertension [48] showing that BPV is an independent predictor of cardiovascular events and mortality [49]. The few studies exploring this issue in HF patients have also indicated a predictive value for BPV in this population, also reporting a paradoxical prognostic impact of short-term BPV on cardiovascular event occurrence in the case of patients with systolic HF (see below).

A study by Berry et al. evaluated the prognostic impact of short-term BPV (i.e., assessed on the basis of 24-h ABPM) on a population of patients with systolic HF (i.e., with a mean LV ejection fraction of 28 ± 9%). After a mean follow-up of 4.4 years, patients with a systolic daytime BPV below a cutoff value of 19 mmHg had the poorest prognosis with a significant odds ratio (OR) for cumulative cardiac events (cardiac death and/or cardiac transplantation) of 1.65 (1.1–2.7) [50] (see Fig. 6.4).

In the same line, the mean daytime systolic BP below a cutoff value of 110 mmHg was associated with a reduction in survival (Fig. 6.5).

This study also explored the relative impact of BPV on patients with systolic HF (HFrEF) according to the New York Heart Association (NYHA) stage. Overall, the lower values of daytime BPV were associated with lower survival rates either in patients with NYHA I/II or in patients with NYHA III/IV. It should be noted that the association between NYHA and BPV produced a better prognosis (see Fig. 6.6).

Evidence has also been provided that alterations in day-to-night BP profiles among HF patients may be associated with an adverse cardiovascular prognosis. A study by Shin et al. [51] evaluated whether alterations in diurnal BP profiles were predictors of adverse outcomes such as hospitalization attributable to HF exacerbation or death in patients with symptomatic LV systolic dysfunction. Overall, compared to dippers, subjects with a reverse dipper profile showed a significantly higher risk of presenting the main outcome of the study. Another study in a population of

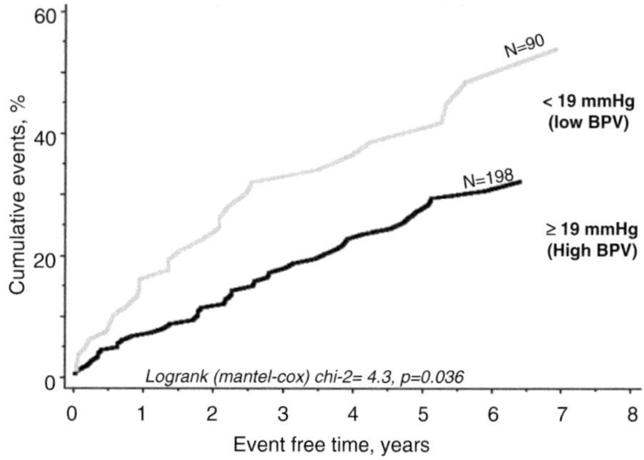

Fig. 6.4 The impact of blood pressure variability with a cutoff value of 19 mmHg on survival in patients with systolic heart failure. (Taken from Berry et al. [50] with permission)

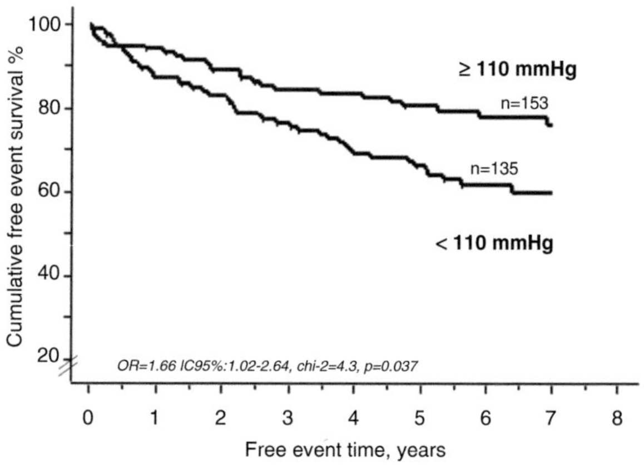

Fig. 6.5 The mean daytime systolic blood pressure level and survival. This figure shows the impact of daytime systolic blood pressure level with a cutoff value of 110 mmHg on survival in patients with systolic heart failure. (Taken from Berry et al. [50] with permission)

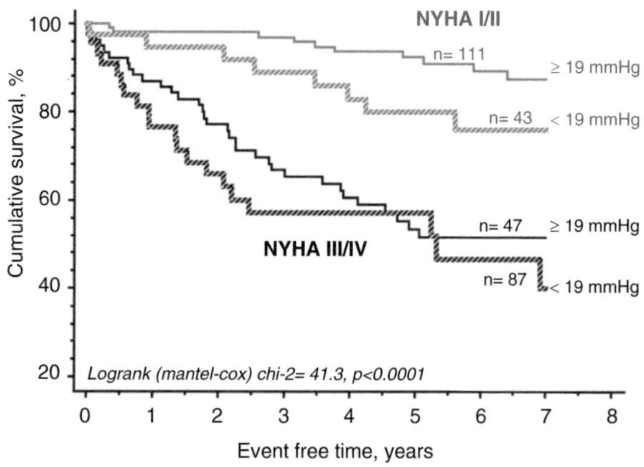

Fig. 6.6 The impact of daytime BPV and functional status on survival

hospitalized HF patients and preserved ejection fraction showed reverse dipping to be associated with a significantly higher incidence of the composite outcome of all-cause mortality and cardiovascular events [52].

6.4.3 Long-Term BPV

A recent analysis conducted in 1988 patients with HFpEF from the TOPCAT (Treatment of Preserved Cardiac Function Heart Failure with an Aldosterone Antagonist) trial has evaluated whether visit-to-visit BPV is associated with the primary end point of cardiovascular death, hospitalization for HF, or aborted cardiac arrest occurring after the first 12 months after randomization [53]. After a median follow-up of 35 months, visit-to-visit systolic BPV (i.e., assessed as the SD of the six measurements performed at baseline and at months 1, 2, 4, 8, and 12 after randomization) was independently associated with the primary outcome of the study and with a higher risk of HF hospitalization. Compared to other quartiles, patients in the upper quartile of SBP-SD showed a significantly higher risk for the primary outcome and HF hospitalization (HR 1.63; 95% CI: 1.07–2.49 and HR 2.44; 95% CI: 1.22–4.85, respectively). When considered as a continuous variable (per 5-mmHg increase), SBP-SD was associated with a significant 23% increase in the risk of the primary outcome and HF hospitalization (HR 1.23; 95% CI: 1.06–1.43 and HR 1.35; 95% CI: 1.10–1.65, respectively). Based on these results, the authors concluded that higher values of systolic visit-to-visit BPV in patients with HFpEF are associated with an increased risk of adverse outcomes and of HF hospitalization over and beyond the average BP levels [53].

More recently, evidence from studies conducted in HFrEF patients has indicated that not only higher but also lower values of BPV are associated with a higher risk of outcomes. These findings differ from most prior investigations in other disease states where the vast majority of studies reported an increased risk of adverse outcomes with higher rather than lower values of BPV.

Evidence in this regard was provided by a report of the Eplerenone in Mild Patients Hospitalization and Survival Study in Heart Failure (EMPHASIS-HF) in patients with HFrEF and mild symptoms, in whom both lower and higher values of SBP-CV were associated with worse outcomes [54]. It also seemed that lower and higher values of SBP-CV had stronger prognostic implications in the context of lower and higher SBP values, with less of an “effect” of SBP-CV when the SBP was within the range of 110–140 mmHg. As with SBP-CV, patients in the lowest third of DBP-CV had the highest risk of outcomes (see Fig. 6.7).

The Systolic Heart Failure Treatment with the If inhibitor ivabradine Trial (SHIFT) study, conducted in a large population of patients with HFrEF ($n = 6505$) [55], also provided evidence that a lower visit-to-visit BPV is a predictor of cardiovascular outcomes with additive effects on risk in HF patients. In this report, the authors evaluated the relationship of visit-to-visit BPV (assessed as CV) with the primary composite end point (cardiovascular mortality or HF hospitalization), and its components, all-cause mortality, and all-cause hospitalization. Patients in the

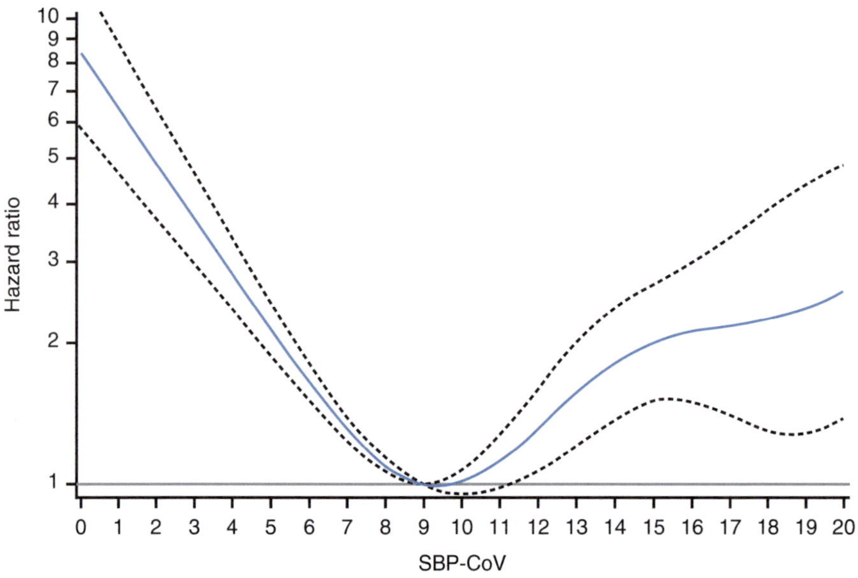

Fig. 6.7 Spline' graphical representation of the prognostic implications of the SBP coefficient of variation. Blue line: hazard ratio; dashed lines: 95% confidence interval. (Taken from Monzo et al. [54] with permission)

lowest thirds of mean SBP and SBP-CV had the highest risk. It should be noted that the combination of a high heart rate and a low heart rate CV had an additive deleterious effect on risk, as did that of low SBP and low SBP-CV [55]. Low SBP-CV was more likely to be associated with adverse outcomes in patients with both low and high SBP, whereas the SBP within the "normal" range showed less relevance. It is noteworthy that patients in the lowest tertile of DBP-CV had the worse outcome. This "interaction" may be potentially attributable to two linked mechanisms: on one hand, hypertension is a well-known risk factor for worse cardiovascular outcomes in most populations, whereas, on the other, pump failure and poor contractile reserve may lead to low SBP in the setting of LV systolic dysfunction. This U-shaped relationship appeared to mainly influence HF hospitalizations that may be more susceptible to variations in BP values. This evidence seems to support the hypothesis by which in HF, inverse epidemiology occurs: high SBP-CV is associated with a higher risk in hypertension but with a lower risk in systolic HF patients [55].

6.5 The Paradoxical Prognostic Value of BPV in CHF

Traditional cardiovascular risk factors (i.e., body mass index, serum cholesterol, and BP), which are associated with poor clinical outcome and mortality in the general population, have also been found to be related to adverse outcomes in CHF patients but in an opposite direction [56]. Indeed, in several studies, obesity, hypercholesterolemia, and elevated BP values have been reported to be associated with greater survival rates among CHF patients. These findings are in contrast to the well-known adverse prognosis associated with these risk factors in the general population. The association between traditional cardiovascular risk factors and an adverse clinical outcome in CHF patients is referred to as "reverse epidemiology." This phenomenon has also been reported in other populations, including patients with end-stage renal disease receiving dialysis, those with advanced malignancies, and individuals with advanced age.

For the last three decades, several studies have provided evidence demonstrating that in patients with CHF, and in particular in those with systolic HF, the usual cardiovascular risk factors become good prognostic markers. For instance, high body mass index [57], hypercholesterolemia, or high BP have been reported to be associated with survival in a CHF population [58]. While in hypertensive patients and in general populations, increasing values of BPV have been associated with a worse prognosis, [59] very low BPV in pathological conditions such as CHF has been associated with a worse prognosis. This relationship, however, should still be better explored in future research studies, taking into account the complexity of a condition such as CHF and the possibility of confounding factors [56].

6.6 Mechanisms of the Relationship Between BPV and CHF

It remains unclear whether BPV itself is a pathogenic risk factor for CHF or whether it is a biomarker of underlying biological alterations. However, on the basis of the evidence provided by some clinical and experimental studies, several mechanisms have been proposed to account for the association between increasing values of BPV and the risk of developing HF (Fig. 6.8).

On the basis of the evidence from outcome studies linking BPV to the development of HF, BPV could be considered as a potential risk factor for development of CHF. Evidence in this regard has been provided by a series of animal and human studies in which BPV was associated with several alterations, which characterize the course of HF. In animal studies, increasing values of BPV were reported to influence myocardial remodeling, by inducing chronic inflammation and fibrosis in the myocardium, leading to adverse cardiac remodeling and impaired systolic function independently of the mean BP values [60]. A series of studies have also suggested that increased BPV could lead to endothelial damage [61] and decreased endothelial function [62], which, in turn, negatively affects coronary blood flow reserve, leading to myocardial hypertrophy and diastolic dysfunction [63, 64]. Increased BPV has also been reported to be associated with arterial stiffness [65] and with decreased aortic distensibility [66], which, in turn, may contribute to the development of HF. An additional possibility is that BPV may directly increase the risk of myocardial ischemia. As coronary blood flow peaks during diastole, repeated transient declines in DBP over time may put the cardiac tissue at an increased risk of relative hypoperfusion. Moreover, extrinsic factors such as daily activities, stress, smoking,

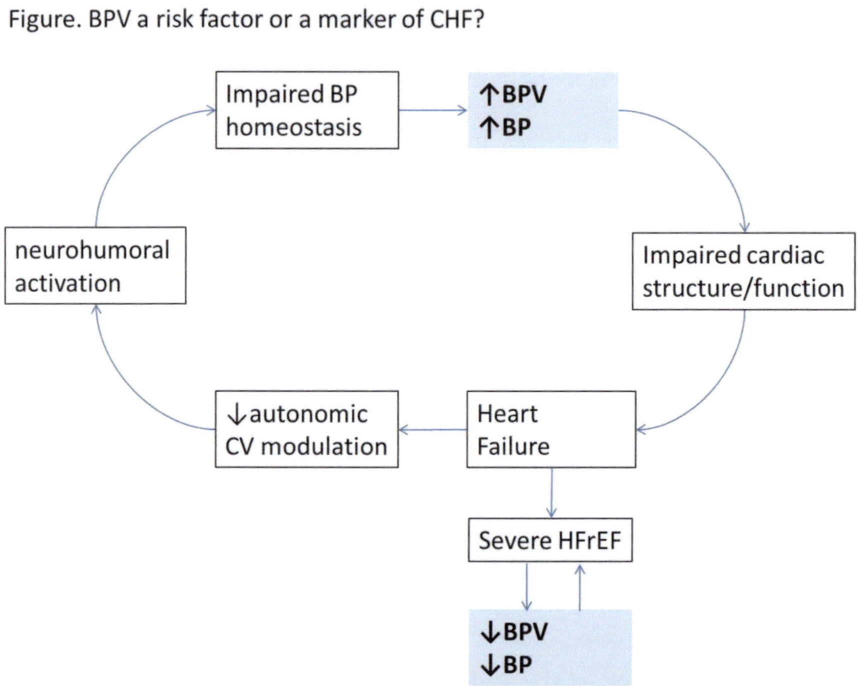

Fig. 6.8 BPV: a risk factor or a marker of CHF? *BP* blood pressure, *BPV* blood pressure variability, *HFrEF* heart failure with reduced ejection fraction, *CV* cardiovascular

and antihypertensive medication nonadherence may participate in increasing BPV and promoting development of HF [67].

It is, nevertheless, also possible that the neurohumoral activation that precedes and favors HF [41] promotes BP instability. Impaired autonomic cardiovascular modulation (i.e., reduced baroreceptor sensitivity) and neuroendocrine activation (i.e., increased sympathetic drive) are known to occur in patients with CHF. Studies comparing patients with LV dysfunction with and without CHF have provided evidence that the neurohumoral activation that precedes and favors HF is further increased as overt HF ensues [41] (see Fig. 6.9).

In turn, impaired autonomic modulation and unopposed sympathetic activation may lead to alterations in BP homeostasis and to BP instability, reflected in an increased BPV. From this perspective, BPV would seem to be the consequence (a marker) of a number of alterations, which develop during the course of HF.

Future studies will have to address to what extent an increased BPV represents a risk factor for developing HF or a marker of an especially severe clinical condition such as HF.

As discussed above, evidence by some studies in patients with HFrEF has indicated that lower and not just higher values of BPV are associated with worse outcomes [54]. It also seemed that lower and higher values of BPV have stronger

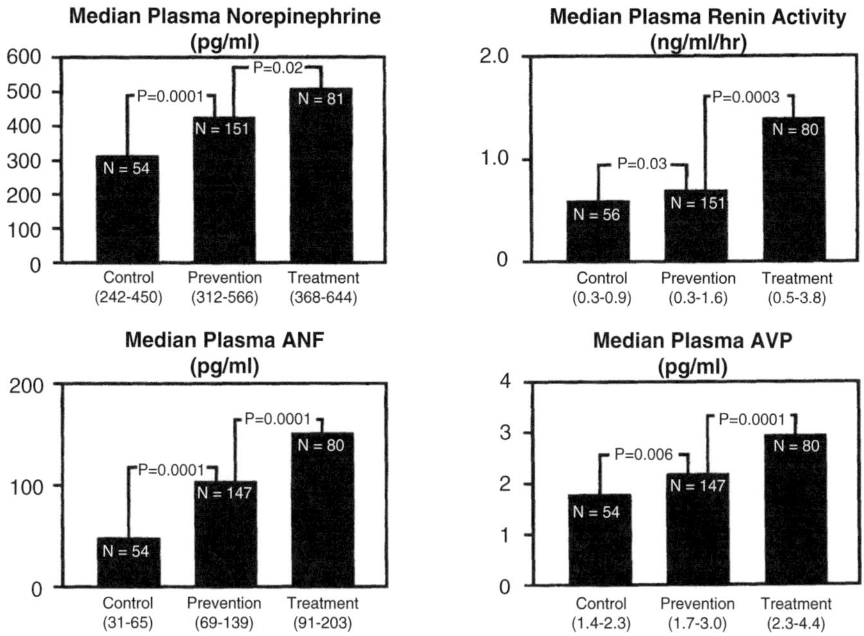

Fig. 6.9 Bar graphs of incremental and significant increase in plasma norepinephrine, plasma renin activity, plasma arginine vasopressin (AlP), and plasma atrial natriuretic factor (ANF) in control subjects, prevention patients, and treatment patients. Treatment values are higher than control values for each neurohormone. Median values and interquartile ranges: 25%–75%. (Taken from Francis et al. [41] with permission)

prognostic implications in the context of lower and higher average BP values, respectively, with less of an "effect" BPV when the SBP is within the range of 110–140 mmHg [54].

Although several explanations have been proposed for the association between high visit-to-visit BPV and worse outcomes, the potential mechanisms underlying the association between low visit-to-visit BPV and poor outcomes reported in some studies are still debated.

Decreased BPV in patients with CHF is well-described. Radaelli et al. [68] demonstrated significant reductions in BPV during exercise in patients with CHF compared to healthy controls (i.e., BPV of 15 vs. 41 mmHg, respectively) [68].

While reduced BP values in HF patients may be explained by a number of factors (i.e., myocardial stiffening, reduced myocardial contractile reserve, reduced cardiac output, reduced preload volume, changes in blood vessel elasticity and in systemic vascular resistance), the mechanisms responsible for reduced values of BPV are not well-understood. It is likely that lower BPV values in patients with HFrEF may be related to alterations in the mechanisms involved in BP control such as impaired autonomic cardiovascular modulation [69], reduced baroreflex sensibility [70], and circadian alterations in neurohormonal systems [71]. Collectively, the dysregulation of these pathways may identify sicker patients at a higher risk of adverse events.

Box 6.1 Potential Mechanisms for Alterations in BPV in CHF Patients

Intrinsic factors

- Myocardial hypertrophy and stiffening.
- Diastolic dysfunction.
- Reduced myocardial contractile reserve.
- Reduced cardiac output and preload volume.
- Alterations in systemic vascular resistance.
- Changes in blood vessel elasticity.
- Increased arterial stiffness.
- Impaired autonomic cardiovascular modulation.
- Reduced baroreflex sensibility.
- Circadian alterations in neurohormonal systems.
- Decreased endothelial function.
- Endothelial damage.

Extrinsic factors

- Daily activities.
- Stress and smoking changes in blood vessel elasticity.
- Antihypertensive medication nonadherence.

6.7 Implications for Treatment

In addition to the lack of understanding of the pathogenic mechanisms, it also remains unclear whether and how individuals with a higher systolic BPV should be specifically targeted. Although previous studies have shown that various antihypertensive drug classes have differential effects on BPV [72, 73], no interventional study has yet explored the key question of whether a reduction in BPV by treatment translates into a better outcome. Future studies aiming at investigating whether certain therapeutic strategies can reduce BPV and improve the outcomes of HF incidence are warranted. Although some reports have indicated that a greater systolic BPV is associated with a higher risk of HFpEF [45], thus suggesting that control of systolic BPV may be a relevant target for the prevention of HFpEF, further research is needed to determine whether the reduction of systolic BPV influences the risk of HF.

On the other hand, the reverse epidemiology of BPV (i.e., that also lower values of BPV are associated with poor outcomes) and other conventional cardiovascular risk factors reported in patients with systolic HF may have important implications for the management of these patients. However, this is an issue that still needs to be further investigated.

A more precise quantification of the benefit of treatment by considering both on-treatment mean BP and visit-to-visit BPV may improve the ability of future studies to determine the benefits of treatment and compare the prognostic impact of different treatment strategies.

Although restoring alterations in circadian BP profiles is considered a therapeutic target for improving cardiovascular prognosis in HF patients, information on treatments aimed at restoring an abnormal circadian BP pattern remains extremely limited, in particular regarding HF populations.

Although there is evidence that drugs with a long-lasting antihypertensive effect and elevated smoothness index are effective in reducing nocturnal BP and in restoring alterations in circadian BP profiles [74–77], evidence on whether normalization of nocturnal BP translates into reductions in cardiovascular risk is still extremely limited. One of the few studies addressing this issue is the Monitorizacion Ambulatoria de la Presion Arterial y Eventos Cardiovasculares (MAPEC) study [78], which assessed whether bedtime administration of antihypertensive drugs is able to control BP and reduce cardiovascular risk more effectively than conventional therapy administration (i.e., all antihypertensive drugs taken in the morning hours) [78]. After 5.6 years of follow-up, participants taking one or more BP-lowering medications at bedtime had a lower nighttime BP as well as a reduced prevalence of non-dipping/reverse dipping status compared to patients taking all drugs upon awakening. The tighter control in nocturnal BP levels was associated with a substantial decrease in the risk of cardiovascular events. Although on the basis of this evidence, it seems appropriate to consider chronotherapy in order to improve cardiovascular prevention, further studies to assess the beneficial effects on the cardiovascular prognosis of therapeutic interventions aimed to normalize circadian BP patterns are needed.

6.8 Conclusions

Accumulating evidence supports the concept that BPV may contribute to cardiovascular risk prediction over and above the impact of the average BP levels. The role of BPV in relation to HF has also been supported by some studies, in which increasing values of BPV have been associated with a higher frequency of cardiac organ damage and an increased incidence of HF. In some reports, the predictive value of BPV for the incidence of HF has exceeded that of the mean SBP, thus confirming the independent prognostic role of BPV in HF [33]. In particular, visit-to-visit BPV has been shown to be an independent prognostic factor for the risk of fatal events, as well as for HF, which suggests that it may predict the risk (or reflect the existence) of particularly severe clinical conditions. In patients already affected by HF, BPV has been associated with an increased risk of cardiovascular events and complications and with a higher incidence of HF hospitalizations. However, the predictive value of BPV in HF patients may depend on the HF phenotype. For instance, while in patients with HFpEF, increasing values of BPV have been associated with a higher risk of adverse outcomes, in patients with systolic HF (HFrEF), poor prognosis has also been associated with lower values of BPV. Given the complexity of these associations, future studies should establish what is the relative prognostic impact of BPV on different phenotypes of HF, i.e., preserved versus reduced ejection fraction.

Although the evidence reported above suggests the possible usefulness of assessing BPV in clinical practice and of considering an elevated BPV as a potential target for treatment in order to prevent development of HF and its complications, currently available studies are characterized by several limitations in order to draw conclusions in this regard. In particular, the heterogeneity among studies addressing the prognostic value of BPV, in terms of study design (i.e., variable follow-up duration) and methodology applied for assessment of BPV (i.e., diversity of protocols used to estimate indices of BPV), has so far not allowed to adequately answer a number of practical questions nor to clarify several important issues related to the clinical implementation of BPV assessment. In addition, although many indices of BPV have been shown to be of prognostic value, no interventional longitudinal outcome study has yet been conducted to specifically address what BPV levels should be regarded as normal and which BPV level should be achieved as target for antihypertensive treatment. Similarly, no intervention study has yet explored the key question of whether a reduction in BPV by treatment translates into a better outcome.

References

1. Parati G, Ochoa JE, Salvi P, Lombardi C, Bilo G. Prognostic value of blood pressure variability and average blood pressure levels in patients with hypertension and diabetes. Diabetes Care. 2013;36(Suppl 2):S312–24.
2. Parati G, Ochoa JE, Lombardi C, Bilo G. Assessment and management of blood-pressure variability. Nat Rev Cardiol. 2013;10(3):143–55.
3. Ernst ME, Chowdhury EK, Beilin LJ, Margolis KL, Nelson MR, Wolfe R, et al. Long-term blood pressure variability and risk of cardiovascular disease events among community-dwelling elderly. Hypertension. 2020;76(6):1945–52.
4. Cuspidi C, Sala C, Tadic M, Gherbesi E, De Giorgi A, Grassi G, et al. Clinical and prognostic significance of a reverse dipping pattern on ambulatory monitoring: an updated review. J Clin Hypertens (Greenwich). 2017;19(7):713–21.
5. Parati G, Pomidossi G, Albini F, Malaspina D, Mancia G. Relationship of 24-hour blood pressure mean and variability to severity of target-organ damage in hypertension. J Hypertens. 1987;5(1):93–8.
6. Frattola A, Parati G, Cuspidi C, Albini F, Mancia G. Prognostic value of 24-hour blood pressure variability. J Hypertens. 1993;11(10):1133–7.
7. Tatasciore A, Renda G, Zimarino M, Soccio M, Bilo G, Parati G, et al. Awake systolic blood pressure variability correlates with target-organ damage in hypertensive subjects. Hypertension. 2007;50(2):325–32.
8. Madden JM, O'Flynn AM, Fitzgerald AP, Kearney PM. Correlation between short-term blood pressure variability and left-ventricular mass index: a meta-analysis. Hypertens Res. 2016;39(3):171–7.
9. Garcia-Ortiz L, Gomez-Marcos MA, Martin-Moreiras J, Gonzalez Elena LJ, Recio-Rodriguez JI, Castano-Sanchez Y, et al. Pulse pressure and nocturnal fall in blood pressure are predictors of vascular, cardiac and renal target organ damage in hypertensive patients (LOD-RISK study). Blood Press Monit. 2009;14(4):145–51.

10. Cha RH, Kim S, Ae Yoon S, Ryu DR, Eun OJ, Han SY, et al. Association between blood pressure and target organ damage in patients with chronic kidney disease and hypertension: results of the APrODiTe study. Hypertens Res. 2014;37(2):172–8.
11. Ivanovic BA, Tadic MV, Celic VP. To dip or not to dip? The unique relationship between different blood pressure patterns and cardiac function and structure. J Hum Hypertens. 2013;27(1):62–70.
12. Komori T, Eguchi K, Saito T, Nishimura Y, Hoshide S, Kario K. Riser blood pressure pattern is associated with mild cognitive impairment in heart failure patients. Am J Hypertens. 2016;29(2):194–201.
13. Wang C, Zhang J, Liu X, Li C, Ye Z, Peng H, et al. Reversed dipper blood-pressure pattern is closely related to severe renal and cardiovascular damage in patients with chronic kidney disease. PLoS One. 2013;8(2):e55419.
14. Pogue V, Rahman M, Lipkowitz M, Toto R, Miller E, Faulkner M, et al. Disparate estimates of hypertension control from ambulatory and clinic blood pressure measurements in hypertensive kidney disease. Hypertension. 2009;53(1):20–7.
15. Liu Z, Zhao Y, Lu F, Zhang H, Diao Y. Day-by-day variability in self-measured blood pressure at home: effects on carotid artery atherosclerosis, brachial flow-mediated dilation, and endothelin-1 in normotensive and mild-moderate hypertensive individuals. Blood Press Monit. 2013;18(6):316–25.
16. Borlaug BA, Kass DA. Ventricular-vascular interaction in heart failure. Heart Fail Clin. 2008;4(1):23–36.
17. Borlaug BA. The pathophysiology of heart failure with preserved ejection fraction. Nat Rev Cardiol. 2014;11(9):507–15.
18. Pfeffer MA, Shah AM, Borlaug BA. Heart failure with preserved ejection fraction in perspective. Circ Res. 2019;124(11):1598–617.
19. Komori T, Eguchi K, Saito T, Hoshide S, Kario K. Riser pattern: another determinant of heart failure with preserved ejection fraction. J Clin Hypertens (Greenwich). 2016;18(10):994–9.
20. Kokubo A, Kuwabara M, Ota Y, Tomitani N, Yamashita S, Shiga T, et al. Nocturnal blood pressure surge in seconds is a new determinant of left ventricular mass index. J Clin Hypertens (Greenwich). 2022;24(3):271–82.
21. Matsui Y, Ishikawa J, Eguchi K, Shibasaki S, Shimada K, Kario K. Maximum value of home blood pressure: a novel indicator of target organ damage in hypertension. Hypertension. 2011;57(6):1087–93.
22. Veloudi P, Blizzard CL, Head GA, Abhayaratna WP, Stowasser M, Sharman JE. Blood pressure variability and prediction of target organ damage in patients with uncomplicated hypertension. Am J Hypertens. 2016;29(9):1046–54.
23. Wei FF, Li Y, Zhang L, Xu TY, Ding FH, Wang JG, et al. Beat-to-beat, reading-to-reading, and day-to-day blood pressure variability in relation to organ damage in untreated Chinese. Hypertension. 2014;63(4):790–6.
24. Stergiou GS, Ntineri A, Kollias A, Ohkubo T, Imai Y, Parati G. Blood pressure variability assessed by home measurements: a systematic review. Hypertens Res. 2014;37(6):565–72.
25. Shibasaki S, Hoshide S, Eguchi K, Ishikawa J, Kario K, Japan Morning Surge-Home Blood Pressure Study G. Increase trend in home blood pressure on a single occasion is associated with B-type natriuretic peptide and the estimated glomerular filtration rate. Am J Hypertens. 2015;28(9):1098–105.
26. Ushigome E, Fukui M, Hamaguchi M, Tanaka T, Atsuta H, Mogami S, et al. Maximum home systolic blood pressure is a useful indicator of arterial stiffness in patients with type 2 diabetes mellitus: post hoc analysis of a cross-sectional multicenter study. Diabetes Res Clin Pract. 2014;105(3):344–51.
27. Matsui Y, O'Rourke MF, Hoshide S, Ishikawa J, Shimada K, Kario K. Combined effect of angiotensin II receptor blocker and either a calcium channel blocker or diuretic on day-by-day variability of home blood pressure: the Japan combined treatment with Olmesartan and a Calcium-Channel blocker versus Olmesartan and diuretics randomized efficacy study. Hypertension. 2012;59(6):1132–8.

28. Ushigome E, Fukui M, Hamaguchi M, Senmaru T, Sakabe K, Tanaka M, et al. The coefficient variation of home blood pressure is a novel factor associated with macroalbuminuria in type 2 diabetes mellitus. Hypertens Res. 2011;34(12):1271–5.
29. Masugata H, Senda S, Murao K, Inukai M, Hosomi N, Iwado Y, et al. Visit-to-visit variability in blood pressure over a 1-year period is a marker of left ventricular diastolic dysfunction in treated hypertensive patients. Hypertens Res. 2011;34(7):846–50.
30. Okada R, Okada A, Okada T, Nanasato M, Wakai K. Visit-to-visit blood pressure variability is a marker of cardiac diastolic function and carotid atherosclerosis. BMC Cardiovasc Disord. 2014;14:188.
31. Redfield MM. Heart failure with preserved ejection fraction. N Engl J Med. 2016;375(19):1868–77.
32. Rothwell PM, Howard SC, Dolan E, O'Brien E, Dobson JE, Dahlof B, et al. Prognostic significance of visit-to-visit variability, maximum systolic blood pressure, and episodic hypertension. Lancet. 2010;375(9718):895–905.
33. Mancia G, Schumacher H, Bohm M, Redon J, Schmieder RE, Verdecchia P, et al. Relative and combined prognostic importance of on-treatment mean and visit-to-visit blood pressure variability in ONTARGET and TRANSCEND patients. Hypertension. 2017;70(5):938–48.
34. Mehlum MH, Liestol K, Kjeldsen SE, Julius S, Hua TA, Rothwell PM, et al. Blood pressure variability and risk of cardiovascular events and death in patients with hypertension and different baseline risks. Eur Heart J. 2018;39:2243.
35. Poortvliet RK, Ford I, Lloyd SM, Sattar N, Mooijaart SP, de Craen AJ, et al. Blood pressure variability and cardiovascular risk in the PROspective study of pravastatin in the elderly at risk (PROSPER). PLoS One. 2012;7(12):e52438.
36. Suchy-Dicey AM, Wallace ER, Mitchell SV, Aguilar M, Gottesman RF, Rice K, et al. Blood pressure variability and the risk of all-cause mortality, incident myocardial infarction, and incident stroke in the cardiovascular health study. Am J Hypertens. 2013;26(10):1210–7.
37. Wu C, Shlipak MG, Stawski RS, Peralta CA, Psaty BM, Harris TB, et al. Visit-to-visit blood pressure variability and mortality and cardiovascular outcomes among older adults: the health, aging, and body composition study. Am J Hypertens. 2017;30(2):151–8.
38. Muntner P, Whittle J, Lynch AI, Colantonio LD, Simpson LM, Einhorn PT, et al. Visit-to-visit variability of blood pressure and coronary heart disease, stroke, heart failure, and mortality: a cohort study. Ann Intern Med. 2015;163(5):329–38.
39. Chang TI, Reboussin DM, Chertow GM, Cheung AK, Cushman WC, Kostis WJ, et al. Visit-to-visit office blood pressure variability and cardiovascular outcomes in SPRINT (systolic blood pressure intervention trial). Hypertension. 2017;70(4):751–8.
40. Velagaleti RS, Pencina MJ, Murabito JM, Wang TJ, Parikh NI, D'Agostino RB, et al. Long-term trends in the incidence of heart failure after myocardial infarction. Circulation. 2008;118(20):2057–62.
41. Francis GS, Benedict C, Johnstone DE, Kirlin PC, Nicklas J, Liang CS, et al. Comparison of neuroendocrine activation in patients with left ventricular dysfunction with and without congestive heart failure. A substudy of the studies of left ventricular dysfunction (SOLVD). Circulation. 1990;82(5):1724–9.
42. Jorsal A, Wiggers H, McMurray JJV. Heart failure: epidemiology, pathophysiology, and management of heart failure in diabetes mellitus. Endocrinol Metab Clin North Am. 2018;47(1):117–35.
43. Nuyujukian DS, Koska J, Bahn G, Reaven PD, Zhou JJ, Investigators V. Blood pressure variability and risk of heart failure in ACCORD and the VADT. Diabetes Care. 2020;43(7):1471–8.
44. Kaze AD, Santhanam P, Erqou S, Bertoni AG, Ahima RS, Echouffo-Tcheugui JB. Long-term variability of blood pressure and incidence of heart failure among individuals with type 2 diabetes. ESC Heart Fail. 2021;8(4):2959–67.
45. Haring B, Hunt RP, Manson JE, LaMonte MJ, Klein L, Allison MA, et al. Blood pressure variability and heart failure hospitalization: results from the Women's Health Initiative. Am J Prev Med. 2022;63(3):410–8.

46. Shimbo D, Newman JD, Aragaki AK, LaMonte MJ, Bavry AA, Allison M, et al. Association between annual visit-to-visit blood pressure variability and stroke in postmenopausal women: data from the Women's Health Initiative. Hypertension. 2012;60(3):625–30.
47. Gibelin P, Spillner E, Bonnan S, Chevallier T. Non-invasive blood pressure variability in chronic heart failure: characteristics and prognostic value. Arch Mal Coeur Vaiss. 2003;96(10):955–62.
48. Clement DL, De Buyzere ML, De Bacquer DA, de Leeuw PW, Duprez DA, Fagard RH, et al. Prognostic value of ambulatory blood-pressure recordings in patients with treated hypertension. N Engl J Med. 2003;348(24):2407–15.
49. Rothwell PM. Limitations of the usual blood-pressure hypothesis and importance of variability, instability, and episodic hypertension. Lancet. 2010;375(9718):938–48.
50. Berry M, Lairez O, Fourcade J, Roncalli J, Carrie D, Pathak A, et al. Prognostic value of systolic short-term blood pressure variability in systolic heart failure. Clin Hypertens. 2016;22:16.
51. Shin J, Kline S, Moore M, Gong Y, Bhanderi V, Schmalfuss CM, et al. Association of diurnal blood pressure pattern with risk of hospitalization or death in men with heart failure. J Card Fail. 2007;13(8):656–62.
52. Komori T, Eguchi K, Saito T, Hoshide S, Kario K. Riser pattern is a novel predictor of adverse events in heart failure patients with preserved ejection fraction. Circ J. 2017;81(2):220–6.
53. Zhang Q, Zhou B, Ma Y, Hu Y, Li X, Cong H. Blood pressure visit-to-visit variability and outcomes in patients with heart failure with preserved ejection fraction. ESC Heart Fail. 2021;8(5):3984–96.
54. Monzo L, Ferreira JP, Abreu P, Szumski A, Bohm M, McMurray JJV, et al. Visit-to-visit blood pressure variation and outcomes in heart failure with reduced ejection fraction: findings from the Eplerenone in patients with systolic heart failure and mild symptoms trial. J Hypertens. 2020;38(3):420–5.
55. Bohm M, Robertson M, Borer J, Ford I, Komajda M, Mahfoud F, et al. Effect of visit-to-visit variation of heart rate and systolic blood pressure on outcomes in chronic systolic heart failure: results from the systolic heart failure treatment with the If Inhibitor Ivabradine Trial (SHIFT) Trial. J Am Heart Assoc. 2016;5(2):e002160.
56. Kalantar-Zadeh K, Block G, Horwich T, Fonarow GC. Reverse epidemiology of conventional cardiovascular risk factors in patients with chronic heart failure. J Am Coll Cardiol. 2004;43(8):1439–44.
57. Davos CH, Doehner W, Rauchhaus M, Cicoira M, Francis DP, Coats AJ, et al. Body mass and survival in patients with chronic heart failure without cachexia: the importance of obesity. J Card Fail. 2003;9(1):29–35.
58. Canesin MF, Giorgi D, Oliveira MT Jr, Wajngarten M, Mansur AJ, Ramires JA, et al. Ambulatory blood pressure monitoring of patients with heart failure. A new prognosis marker. Arq Bras Cardiol. 2002;78(1):83–9.
59. Chen AF. Blood pressure variability reduction and organ protection in hypertension treatment. Hypertens Res. 2008;31(4):587–8.
60. Kudo H, Kai H, Kajimoto H, Koga M, Takayama N, Mori T, et al. Exaggerated blood pressure variability superimposed on hypertension aggravates cardiac remodeling in rats via angiotensin II system-mediated chronic inflammation. Hypertension. 2009;54(4):832–8.
61. Diaz KM, Veerabhadrappa P, Kashem MA, Thakkar SR, Feairheller DL, Sturgeon KM, et al. Visit-to-visit and 24-h blood pressure variability: association with endothelial and smooth muscle function in African Americans. J Hum Hypertens. 2013;27(11):671–7.
62. Diaz KM, Veerabhadrappa P, Kashem MA, Feairheller DL, Sturgeon KM, Williamson ST, et al. Relationship of visit-to-visit and ambulatory blood pressure variability to vascular function in African Americans. Hypertens Res. 2012;35(1):55–61.
63. Low Wang CC, Hess CN, Hiatt WR, Goldfine AB. Clinical update: cardiovascular disease in diabetes mellitus: atherosclerotic cardiovascular disease and heart failure in type 2 diabetes mellitus—mechanisms, management, and clinical considerations. Circulation. 2016;133(24):2459–502.

64. Seddon M, Looi YH, Shah AM. Oxidative stress and redox signalling in cardiac hypertrophy and heart failure. Heart. 2007;93(8):903–7.
65. Lee HT, Lim YH, Kim BK, Lee KW, Lee JU, Kim KS, et al. The relationship between ambulatory arterial stiffness index and blood pressure variability in hypertensive patients. Korean Circ J. 2011;41(5):235–40.
66. Shimbo D, Shea S, McClelland RL, Viera AJ, Mann D, Newman J, et al. Associations of aortic distensibility and arterial elasticity with long-term visit-to-visit blood pressure variability: the multi-ethnic study of atherosclerosis (MESA). Am J Hypertens. 2013;26(7):896–902.
67. Mancia G. Visit-to-visit blood pressure variability: an insight into the mechanisms. Hypertension. 2016;68(1):32–3.
68. Radaelli A, Perlangeli S, Cerutti MC, Mircoli L, Mori I, Boselli L, et al. Altered blood pressure variability in patients with congestive heart failure. J Hypertens. 1999;17(12 Pt 2):1905–10.
69. Grassi G, Bombelli M, Seravalle G, Dell'Oro R, Quarti-Trevano F. Diurnal blood pressure variation and sympathetic activity. Hypertens Res. 2010;33(5):381–5.
70. Lanfranchi PA, Somers VK. Arterial baroreflex function and cardiovascular variability: interactions and implications. Am J Physiol Regul Integr Comp Physiol. 2002;283(4):R815–26.
71. Giles TD, Roffidal L, Quiroz A, Sander G, Tresznewsky O. Circadian variation in blood pressure and heart rate in nonhypertensive congestive heart failure. J Cardiovasc Pharmacol. 1996;28(6):733–40.
72. Rothwell PM, Howard SC, Dolan E, O'Brien E, Dobson JE, Dahlof B, et al. Effects of beta blockers and calcium-channel blockers on within-individual variability in blood pressure and risk of stroke. Lancet Neurol. 2010;9(5):469–80.
73. Muntner P, Levitan EB, Lynch AI, Simpson LM, Whittle J, Davis BR, et al. Effect of chlorthalidone, amlodipine, and lisinopril on visit-to-visit variability of blood pressure: results from the antihypertensive and lipid-lowering treatment to prevent heart attack trial. J Clin Hypertens (Greenwich). 2014;16(5):323–30.
74. Kario K, Schwartz JE, Pickering TG. Changes of nocturnal blood pressure dipping status in hypertensives by nighttime dosing of alpha-adrenergic blocker, doxazosin: results from the HALT study. Hypertension. 2000;35(3):787–94.
75. Hermida RC, Calvo C, Ayala DE, Fernandez JR, Covelo M, Mojon A, et al. Treatment of non-dipper hypertension with bedtime administration of valsartan. J Hypertens. 2005;23(10):1913–22.
76. Kario K, Nariyama J, Kido H, Ando S, Takiuchi S, Eguchi K, et al. Effect of a novel calcium channel blocker on abnormal nocturnal blood pressure in hypertensive patients. J Clin Hypertens (Greenwich). 2013;15(7):465–72.
77. Gosse P, Schumacher H. Effect of telmisartan vs. ramipril on 'dipping' status and blood pressure variability: pooled analysis of the PRISMA studies. Hypertens Res. 2014;37(2):151–7.
78. Hermida RC, Ayala DE, Mojon A, Fernandez JR. Influence of circadian time of hypertension treatment on cardiovascular risk: results of the MAPEC study. Chronobiol Int. 2010;27(8):1629–51.

Hemodynamic Markers in the Progression from Hypertension to Heart Failure

7

Nicholas Cauwenberghs and Tatiana Kuznetsova

7.1 Introduction

The normal heart is an efficient muscle designed to serve both as a pump and an integrator of two independent vascular systems: the pulmonary and systemic circulation. The capacity of the body to augment cardiac output, regulate systemic blood pressure (BP), and respond appropriately to changes in heart rate and pre- and afterload depends on the properties of both the heart and the vasculature into which the left ventricle ejects blood [1]. Two components of systemic BP can be identified: the steady component, represented by the mean arterial pressure (MAP), and the pulsatile component, represented by pulse pressure. MAP is determined by peripheral arterial resistance, which depends on the physical characteristics of the arterial tree and on the volume of blood that the left ventricle ejects. On the other hand, left ventricular (LV) stroke volume and aortic compliance are the major determinants of pulse pressure. In the absence of aortic stenosis, conventionally measured brachial BP provides a clinically useful estimate of LV afterload.

When a high afterload opposes LV ejection, reduction of the LV stroke volume could be observed in the short term. This reduction is further compensated by shifting the LV pump function to a higher energy level (the Frank–Starling mechanism) and by activating an autoregulatory mechanism (the Anrep response). However, the long-term increased afterload and, consequently, the chronically increased cardiac performance lead to adverse LV remodeling and dysfunction and increased LV oxygen requirements and, eventually, cause symptomatic heart failure (HF). Because the process of myocardial remodeling/dysfunction starts long before the onset of symptomatic HF, it is of importance to better understand the pathophysiological

N. Cauwenberghs · T. Kuznetsova (✉)
Research Unit Hypertension and Cardiovascular Epidemiology,
Department of Cardiovascular Sciences, University of Leuven, Leuven, Belgium
e-mail: nicholas.cauwenberghs@kuleuven.be; tatiana.kouznetsova@kuleuven.be

M. Dorobantu et al. (eds.), *Hypertension and Heart Failure*, Updates in Hypertension and Cardiovascular Protection,
https://doi.org/10.1007/978-3-031-39315-0_7

mechanisms leading to subclinical (asymptomatic) LV maladaptation and the timely identification of patients who are at risk of developing overt cardiac events. In this chapter, we discuss the distinct aspects of cardiac maladaptive responses to a chronically increased load and highlight the noninvasive markers reflecting the hemodynamic impact on the heart.

7.2 LV Remodeling and Hypertrophy

For a better understanding of cardiac mechanics, it is important to describe the relationship between pressure, volume, and wall stress of the left ventricle. The law of Laplace is commonly used as a mathematical model to predict LV wall stress from given pressure and geometry [2]. According to this equation, LV wall stress is directly proportional to LV pressure and radius and is inversely proportional to the wall thickness of the left ventricle (Fig. 7.1). Higher pressure can cause thickening of LV walls in order to accommodate an increased load and maintain normal wall tension. Indeed, hypertension induces a compensatory thickening of the ventricular wall, so-called concentric hypertrophy, in order to normalize wall stress. Thus, patients with hypertensive heart disease usually present with concentric remodeling or concentric LV hypertrophy but have a normal-sized LV chamber and a normal ejection fraction [3]. On the other hand, a hypertrophic left ventricle is stiffer, so it requires elevated pressures to fill it, leading to a condition known as diastolic dysfunction.

Cardiac maladaptation such as LV hypertrophy in response to an increased hemodynamic load is not a benign condition and is associated with an increased risk of cardiovascular outcome, including incident HF [4, 5]. Several studies have

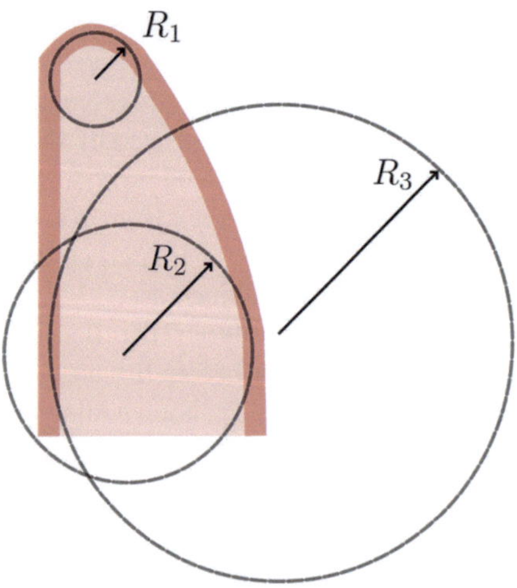

$$\sigma = \frac{PR}{2h}, \qquad h \ll R$$

Wall stress (σ) related to:

Pressure	$P\uparrow$	$\rightarrow$	$\uparrow\sigma$
Shape, cavity size	$R\uparrow$	$\rightarrow$	$\uparrow\sigma$
Wall thickness	$h\uparrow$	$\rightarrow$	$\downarrow\sigma$

Fig. 7.1 Schematic representation of the law of Laplace, which states that wall tension is proportionate to the pressure (P) times radius (R). Therefore, wall stress (σ) is wall tension divided by wall thickness (h). At a given intraventricular pressure, wall stress increases with an increase in the radius of the ventricular cavity

documented the relationship between LV hypertrophy (increased LV mass index) detected by electrocardiography and echocardiography and an adverse prognosis. A meta-analysis combined 48,545 subjects from 20 prospective studies and showed that the adjusted risk of future cardiovascular morbidity associated with baseline LV hypertrophy ranged from 1.5 to 3.5, with a weighted mean risk ratio of 2.3 for all studies combined [6]. Several mechanisms may explain why adverse LV remodeling/hypertrophy is a harbinger of adverse cardiovascular outcomes. First, LV hypertrophy or remodeling may lead to diastolic filling abnormalities that predispose to symptomatic HF. Second, maladaptive LV remodeling may lead to dysfunction of the autonomic nervous system, reduce coronary reserve, and increase LV oxygen requirements. Third, it may predispose to ventricular arrhythmias and a greater risk of sudden death.

7.3 LV End-Diastolic Filling Pressure (Diastolic Function)

LV diastolic function tends to worsen over the adult life course in patients with hypertension in parallel to changes in cardiac geometry and systolic performance [7]. Diastolic dysfunction refers to a condition in which abnormalities in LV function are present during diastole.

The early stage of LV diastolic dysfunction, as impaired myocardial relaxation, is characterized by decreased transmitral early (*E* peak) and enhanced atrial (*A* peak) LV filling as well as less vigorous mitral annulus motion (e' peak) during early diastole. The more advanced stage of diastolic dysfunction is typically presented by increased LV end-diastolic filling pressure in response to increased LV stiffness. We might estimate LV filling pressure noninvasively by combining early transmitral blood flow velocity with early mitral annular velocity (E/e' ratio) [8]. It should be noted that an accurate prediction of LV filling pressures for an individual patient requires further characterization of the intermediate E/e' group ("gray zone"), for instance, by measurement of left atrial volume and deformation (strain), tricuspid regurgitation velocity, and blood flow in the pulmonary vein [9].

Previous studies demonstrated that elderly women appear to be the most susceptible to the detrimental effects of increased pulsatile load on LV diastolic function [10–12]. This observation might be explained by the higher aortic pulsatile load and aortic stiffness, enhanced LV systolic performance, and lower LV compliance in women as compared to men [11, 13–15].

7.4 LV Deformation

LV deformation (strain) is determined by the fiber structure and curvature of the myocardium and its interaction with local wall stress at the beginning of ejection, which decreases from the endocardium to epicardium and from the LV base to the apex [16]. During LV ejection, longitudinal deformation of the heart results from contraction of longitudinally oriented subendocardial and subepicardial fibers, whereas radial LV wall thickening mainly originates from contraction of circumferential fibers located in the mid-wall [17]. Gould et al. assessed the relation between the directional components of LV contraction and ejection fraction in 122 subjects with or without heart disease, using angiocardiography [18]. The contribution of the longitudinal and radial components to total cardiac work was 14% and 40%, respectively [18]. It was suggested that separate analysis of the various components of LV systolic deformation might help us understand the progression of LV systolic dysfunction at different stages of heart disease [19, 20].

7.4.1 Changes in LV Longitudinal Strain in Response to an Increased Afterload

Theoretical computer models of the heart predicted that LV systolic deformation in a longitudinal direction increases with a higher cardiac output (preload) and decreases with increasing mean arterial pressure (MAP) (afterload) (Fig. 7.2) [21]. Because high systemic arterial pressure leads to increased LV wall stress, deformation of the longitudinally oriented and less curved myofibers in the subendocardium is particularly impaired. Thus, longitudinal systolic dysfunction could already be observed at early stages of progressive myocardial maladaptation related to a chronically increased hemodynamic load [22].

Along similar lines, an experimental study involving an aortic banding model showed disparate changes in longitudinal and radial myocardial strain in response to acute alternation in LV afterload [23]. Longitudinal systolic function dramatically fell as the afterload increased, whereas LV fractional shortening and radial strain were still preserved after mild banding [23]. Previous clinical studies reported depressed LV longitudinal function even in asymptomatic patients with hypertension as compared to normotensives controls [24–26]. In the presence of hypertension, subclinical impairment of LV longitudinal deformation appears to be worse if the high BP is long-standing and uncontrolled [27]. In a large-scale general

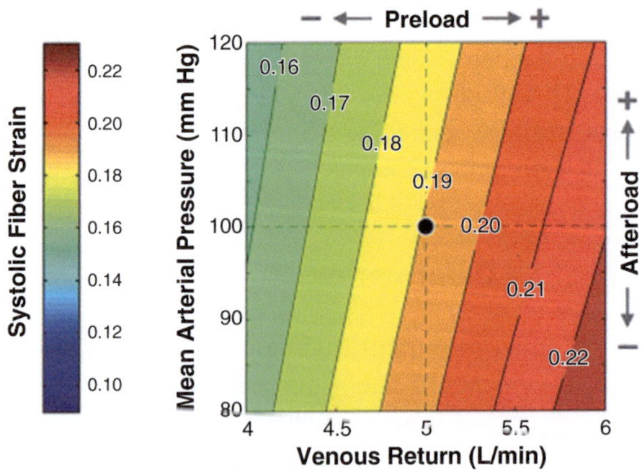

Fig. 7.2 Dependence of LV global LS (color gradient) to preload (venous return) and afterload (mean arterial pressure) in a normal heart. (Reproduced from Lumens et al. with permission [21])

population study, global longitudinal strain decreased significantly with higher MAP (by 0.29% per 10 mmHg, $P < 0.01$) [6]. Moreover, LV longitudinal deformation declined strongly over time in individuals with higher MAP at baseline and stronger increase in MAP during follow-up [28].

7.4.2 Changes in LV Radial Strain in Response to an Increased Afterload

At systole, the heart ejects a volume of blood into the aorta and generates a forward pressure wave that is reflected at various sites in the arterial system. If the aorta is compliant, then the aortic walls elastically expand to accommodate the ejected blood. An elastic aorta, therefore, dampens pulsatility and maintains a continuous blood flow from the heart to the periphery. Hypertension accelerates the age-related stiffening of the large arteries, including the aorta, which plays a key role in the development of HF due to the additional mechanical load on the heart [29]. Indeed, to expand stiffened arteries, the heart needs to produce greater pressure and therefore its energy expenditure increases [30–32]. In addition, increased pulse wave velocity (PWV) in stiffening arteries leads to an early return of the reflected wave, which might, in turn, also augment the late systolic LV load. On the other hand, as suggested recently, entrapment of reflected waves in the periphery might limit the influence of the peripheral reflected waves on central hemodynamic and late systolic load [11].

Population studies have shown that measures reflecting increased aortic stiffness such as a higher aortic PWV is associated with a higher risk of cardiovascular events, including HF beyond traditional cardiovascular risk factors [33–36]. For instance, in an analysis including 7283 Framingham Study participants, greater aortic stiffness as reflected by increased aortic PWV was associated with multiple health outcomes [36]. In multivariable-adjusted analyses, a 1-standard deviation (SD) increase in aortic PWV was associated with 20% and 21% higher risk of incidence of composite cardiovascular events and HF, respectively.

Along these lines, we previously reported an age-dependent relationship between changes in LV radial systolic deformation and early and late systolic load in a general population [14]. Radial strain increased significantly with higher central pulse pressure and PWV in middle-aged participants (50–60 years) only, whereas it decreased with these indexes in older subjects (above 70 years). These findings suggest that a chronic rise in pulse pressure increases LV load and enhances LV radial systolic performance but eventually might lead to adverse LV remodeling and impaired radial deformation. This mechanism probably contributes to the development and progression of LV dysfunction in hypertensive patients. However, future serial imaging studies should clarify the progression of LV radial strain changes in response to chronically increased LV loading due to arterial stiffening.

Figure 7.3 summarizes the changes in LV function components during the transition from subclinical to symptomatic HF in arterial hypertension.

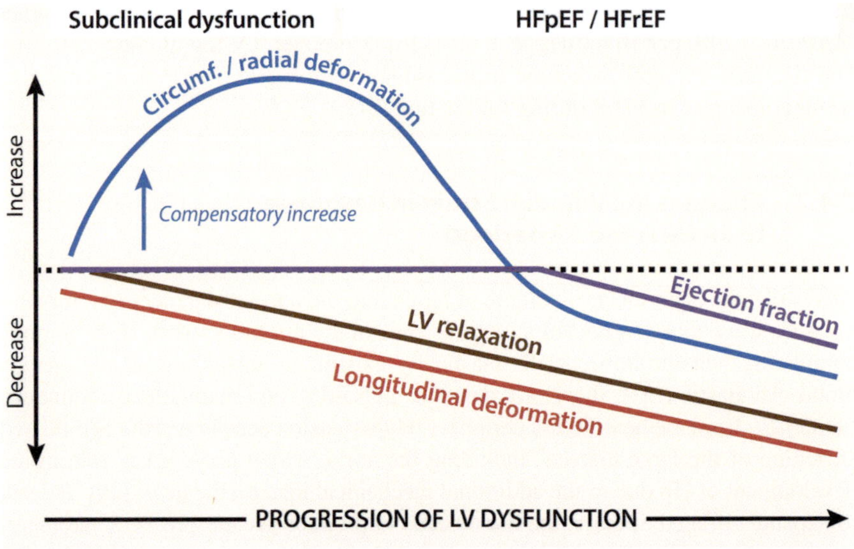

Fig. 7.3 The transition from subclinical to overt heart failure with preserved ejection fraction (HFpEF) or reduced ejection fraction (HFrEF) in arterial hypertension. Initially, hypertension-mediated changes in left ventricular (LV) function encompass reduced longitudinal shortening and compensatory increased circumferential shortening to preserve LV ejection fraction. In the long run, circumferential/radial deformation progressively deteriorates, thereby gradually impairing the overall LV performance. (Modified from Shah and Solomon [53])

7.5 The Correlation Network of BP and LV Traits

Figure 7.4 illustrates a complex network of interactions between the multivariable-adjusted components of BP and the echocardiographic indexes of LV systolic and diastolic function and structure. The network was constructed from population data of 791 participants randomly recruited within the Flemish Study on Environment, Genes and Health Outcomes (FLEMENGHO) (the mean age was 50.9 years, 51.8% were women, and 41.2% had hypertension) [37]. The figure represents a partial regression diagram, including multivariable-adjusted components of BP such as MAP and pulse pressure and echocardiographically measured LV phenotypes. This approach fits covariance selection models, estimating the correlation between two components of the network adjusted for the correlations of these two components with all other variables in the network (i.e., partial correlations).

While accounting for all BP and LV traits interactions, partial regression analysis confirmed the relationships between hemodynamics variables and LV phenotypes as described in previous sections. That is, a lower global longitudinal strain is significantly correlated with a higher MAP and an increased LV mass index. Moreover, we observed a strong relation between higher pulse pressure and increased LV filling pressure (as estimated by *E*/*e*′ ratio) and LV mass index (Fig. 7.4). Lower radial strain was related to higher relative wall thickness (index of concentric remodeling)

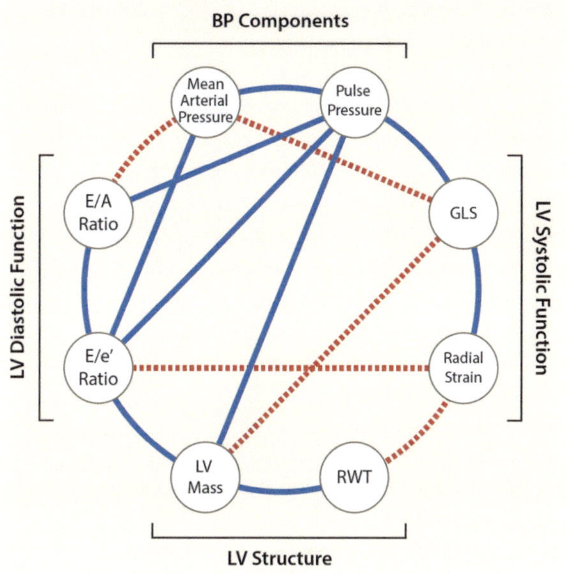

Fig. 7.4 A partial correlation diagram between blood pressure components and the echocardiographic indexes of LV function and structure based on the data of 791 participants from the FLEMENGHO population study [24]. The full blue lines represent direct (positive) correlations, and the dashed red lines represent inverse (negative) correlations ($P < 0.05$ for all). All indexes were adjusted for age, sex, heart rate, body height, and body weight. Models for LV mass index did not include body height and weight. *BP* blood pressure, *GLS* global longitudinal strain, *LV* left ventricular, *RWT* relative wall thickness

and greater LV filling pressure (by *E*/*e*′ ratio). Thus, higher BP evokes a complex network of functional and structural changes in the heart. As such, early detection and effective management of BP may prevent or delay the development of subclinical LV remodeling and dysfunction preceding symptomatic HF. Preventive strategies might tackle the rising contribution of poorly controlled BP to the HF epidemic.

7.6 Assessment of LV and Arterial Elastance

As highlighted in the previous section, the interaction of the heart with the systemic vasculature, or ventricular-arterial coupling, is a key determinant of cardiovascular performance. Therefore, simultaneous measurement of arterial and LV stiffness or elastance is important to better understand the hemodynamic mechanisms leading to HF. The measurements of elastance could be derived from invasive or noninvasive registration of LV pressures and LV volumes (pressure–volumes curves), which might also be recorded over a wide range of LV loading conditions (Fig. 7.5). By definition, elastance reflects volume change per unit of pressure change; it is the reciprocal of compliance. For instance, effective arterial elastance (E_a) could be

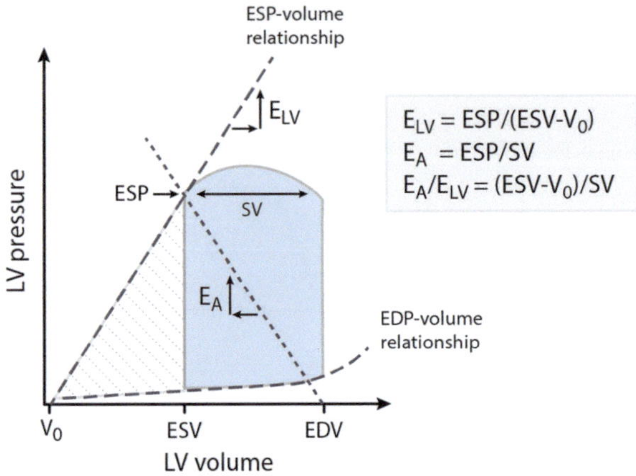

Fig. 7.5 Schematic representations of a pressure–volume loop and derived ventricular-arterial indexes. *EDV* end-diastolic volume, *ESV* end-systolic volume, *ESP* end-systolic pressure, E_A arterial elastance, E_{LV} left ventricular end-systolic elastance, *LV* left ventricular, *SV* stroke volume

calculated as the ratio of LV end-systolic pressure to stroke volume and it reflects the net arterial load imposed on the left ventricle [38].

LV end-systolic elastance (E_{lv}) provides an estimate of the overall LV performance and is calculated by measuring the slope of the end-systolic pressure–volume relations registered over a range of LV loads [38]. Alternatively, E_{lv} could be calculated as a ratio of LV end-systolic pressure to LV end-systolic volume (Fig. 7.5). Therefore, E_{lv} as an index of myocardial performance reflects the ability of the left ventricle to eject blood opposed to a given pressure. An increase in E_{lv} is associated with enhanced myocardial contractility. On the other hand, to some degree, E_{lv} also reflects chamber geometry and passive myocardial stiffening, which could be altered in hypertensive patients. Indeed, numerous clinical studies [39–42] show that effective arterial and LV elastance based on invasive or noninvasive determination of LV volumes and end-systolic pressure are both increased in hypertensive patients. For instance, Cohen-Solal et al. [42] showed that E_a and E_{lv} as measured by angiography were significantly higher in 19 hypertensive patients as compared to 25 normotensive men. Later, Saba et al. [39] confirmed this finding using echocardiography for measurement of LV volumes and carotid pressure waveforms for assessment of ESP in 81 normotensive and 174 hypertensive patients. Furthermore, Chantler et al. [40] reported that the ventricular-arterial coupling ratio (E_a/E_{lv}) was about 25% lower in hypertensive compared with normotensive women. Along similar lines, we also found that in hypertensive patients, the ventricular-arterial coupling ratio at rest was 16.4% lower compared to that in normotensive subjects [41]. The lower ventricular-arterial coupling ratio in hypertensive patients was due to a disproportionate increase in E_{lv} compared with E_a (32% vs. 10%) [41]. Therefore, in asymptomatic hypertensive patients, a higher E_{lv} at rest might not only mean greater myocardial performance but also reflect geometrical and passive structural changes in hypertensive hearts.

7.7 Noninvasive Indexes of LV Myocardial Work from Pressure–Strain Loops

Additional information about changes in LV performance due to increased loading conditions might also be derived from a simultaneous consideration of pulse wave and myocardial deformation (strain) curves. Recent developments in the construction and quantification of pressure–strain loops have illustrated the pressing need for a noninvasive clinical tool that integrates both LV deformation and its loading conditions [43].

In an analogy to pressure–volume curves, one can construct LV pressure–strain loops by plotting myocardial deformation (derived from two-dimensional speckle tracking [44]) against the pulse waveform. The pulse waveform can either be measured by applanation tonometry [45] or estimated from a reference curve standardized to brachial BP and cardiac cycle events [46]. From LV pressure–strain loops, surrogate markers of myocardial work can then be calculated as quantitative measures of regional LV performance [41, 46].

For instance, the area of pressure–strain loops during ejection is called the LV ejection work density as it represents the cumulative work done by the cardiac muscle in order to instantaneously shorten a given amount (i.e., change in strain) at a given instantaneous resistance (i.e., pressure) [41]. As such, ejection work density might better reflect LV systolic performance than peak LV deformation, as it integrates the instantaneous deformation as well as the instantaneous afterload against which the myocardium has to shorten [43]. In a random cohort of 148 subjects, higher arterial load in asymptomatic hypertensive patients matched with enhanced LV myocardial performance as ejection work density was 24% higher in hypertensive subjects as compared to normotensives [41].

Inspired by the work of Russel et al. [46], GE Healthcare added a feature to their echocardiographic post-processing software to construct whole pressure–strain loops from a 2D LV strain tracing and an estimated LV pressure waveform retrieved from a reference pressure curve that is standardized to brachial BP and the timing of aortic and mitral valve events (Fig. 7.6). Surrogate LV myocardial work indexes derived from pressure–strain loops include global constructive work and global wasted work, from which the global work index and mean global work efficiency can be calculated (Fig. 7.6). The reference values for these indices have been published recently [47]. In the "Characteristics and Course of Heart Failure Stages A–B and Determinants of Progression" (STAAB) study, higher global wasted work and lower global work efficiency was independently associated with hypertension [48] as well as with abnormal LV geometry profiles such as LV enlargement and LV hypertrophy [49]. Moreover, peak oxygen uptake (and thus functional capacity) was found to decrease with a higher global myocardial work index ($P_{adjusted}$ = 0.042; [50]). However, in contrast to measured pressure waves, this approach does not account for interindividual differences in the pressure wave shape. In fact, Hubert

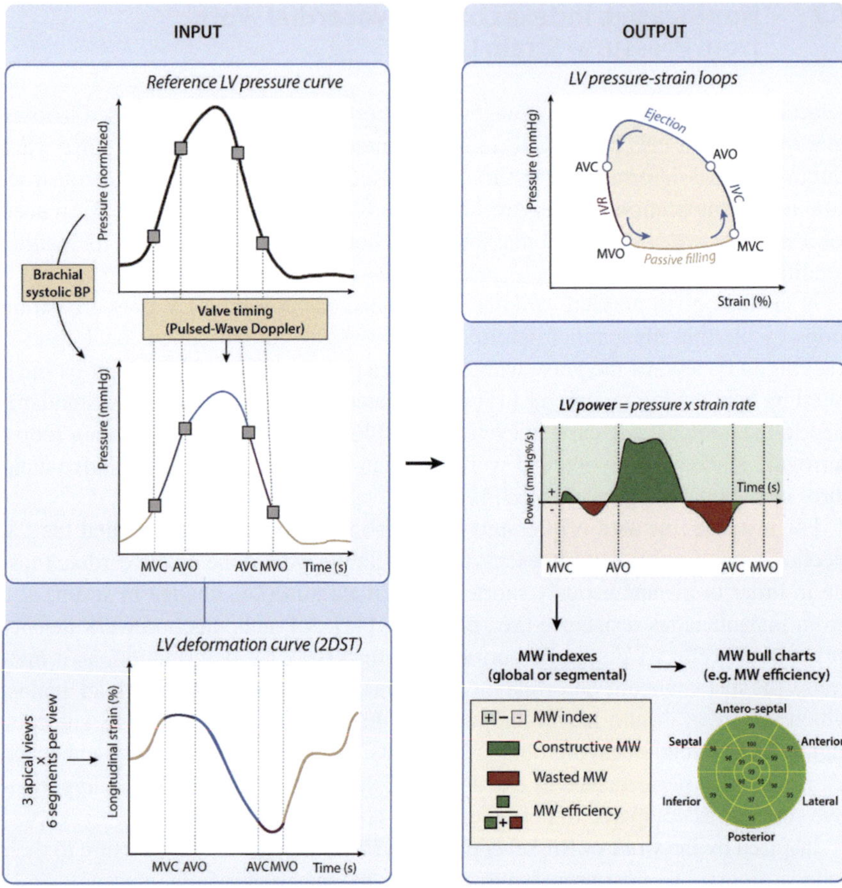

Fig. 7.6 Noninvasive assessment of left ventricular (LV) myocardial work (MW) using echocardiographic deformation imaging and LV pressure estimation. First, an LV pressure waveform is constructed from a reference pressure curve fitted to brachial systolic BP and echo-derived timing of valve events. The LV pressure waveform can be combined with global or segmental LV deformation curves to plot LV pressure–strain loops for visual inspection. In addition, LV power curves can be constructed (=pressure × strain rate) from which global and segmental MW indexes can be derived. Segmental MW values can be presented conjointly in bull charts. *2DST* two-dimensional speckle tracking, *AVC* aortic valve closure, *AVO* aortic valve opening, *BP* blood pressure, *IVC* isovolumic contraction, *IVR* isovolumic relaxation, *MVC* mitral valve closure, *MVO* mitral valve opening, *MW* myocardial work

et al. observed substantial shortcomings in the estimation of the pressure curve, particularly at greater pressures and loop areas [51]. To date, larger multicenter studies are warranted to further validate the diagnostic and prognostic value of the noninvasive indexes of LV performance in patients at risk of cardiac maladaptation [52].

7.8 Conclusions

As shown by epidemiological studies, hypertension is one of the most important modifiable risk factors for the development of symptomatic HF. Identifying patients at asymptomatic stages of HF would allow the institution of more aggressive risk management strategies and will likely decrease the progression to symptomatic disease. In this regard, a better understanding of the pathophysiological mechanisms of cardiac maladaptation in patients with hypertension is crucial for early detection of this condition. The chronically increased cardiac performance due to high afterload leads to LV concentric remodeling and hypertrophy, impairment of LV systolic deformation, and diastolic dysfunction. LV systolic and diastolic dysfunction coexist to varying degrees and appear early in the course of hypertensive heart disease. Community-based studies have revealed a higher-than-hitherto expected prevalence of LV systolic and diastolic dysfunction and their independent prognostic significance.

Hypertension also accelerates the age-related stiffening of the large arteries, which plays a key role in the development of HF due to the additional mechanical load on the heart. The heart typically adapts to confront higher systolic loads by both hypertrophy and LV stiffening. Increased vascular loading on the heart also contributes to LV dysfunction. Ventricular-arterial coupling disease must be further explored in subjects with subclinical LV dysfunction.

References

1. De Tombe PP, Jones S, Burkhoff D, Hunter WC, Kass DA. Ventricular stroke work and efficiency both remain nearly optimal despite altered vascular loading. Am J Physiol. 1993;264(6 Pt 2):H1817–24.
2. Moriarty TF. The law of Laplace. Its limitations as a relation for diastolic pressure, volume, or wall stress of the left ventricle. Circ Res. 1980;46(3):321–31.
3. Devereux RB, Roman MJ. Left ventricular hypertrophy in hypertension: stimuli, patterns, and consequences. Hypertens Res. 1999;22(1):1–9.
4. Zile MR, Gaasch WH, Patel K, Aban IB, Ahmed A. Adverse left ventricular remodeling in community-dwelling older adults predicts incident heart failure and mortality. JACC Heart Fail. 2014;2(5):512–22.
5. Velagaleti RS, Gona P, Pencina MJ, Aragam J, Wang TJ, Levy D, et al. Left ventricular hypertrophy patterns and incidence of heart failure with preserved versus reduced ejection fraction. Am J Cardiol. 2014;113(1):117–22.
6. Vakili BA, Okin PM, Devereux RB. Prognostic implications of left ventricular hypertrophy. Am Heart J. 2001;141(3):334–41.
7. Kuznetsova T, Thijs L, Knez J, Cauwenberghs N, Petit T, Gu YM, et al. Longitudinal changes in left ventricular diastolic function in a general population. Circ Cardiovasc Imaging. 2015;8(4):e002882.
8. Ommen SR, Nishimura RA, Appleton CP, Miller FA, Oh JK, Redfield MM, et al. Clinical utility of Doppler echocardiography and tissue Doppler imaging in the estimation of left ventricular filling pressures: a comparative simultaneous Doppler-catheterization study. Circulation. 2000;102(15):1788–94.

9. Kuznetsova T, Cauwenberghs N, Sabovčik F, Kobayashi Y, Haddad F. Evaluation of diastole by echocardiography for detecting early cardiac dysfunction: an outcome study. ESC Heart Fail. 2022;9:1775.
10. Shim CY, Park S, Choi D, Yang WI, Cho IJ, Choi EY, et al. Sex differences in central hemodynamics and their relationship to left ventricular diastolic function. J Am Coll Cardiol. 2011;57(10):1226–33.
11. Coutinho T, Borlaug BA, Pellikka PA, Turner ST, Kullo IJ. Sex differences in arterial stiffness and ventricular-arterial interactions. J Am Coll Cardiol. 2013;61(1):96–103.
12. Cauwenberghs N, Knez J, D'hooge J, Thijs L, Yang W-Y, Wei F-F, et al. Longitudinal changes in LV structure and diastolic function in relation to arterial properties in general population. JACC Cardiovasc Imaging. 2017;10(11):1307–16.
13. Borlaug BA. Sex, load, and relaxation: are women more susceptible to load-dependent diastolic dysfunction? J Am Coll Cardiol. 2011;57(10):1234–6.
14. Cauwenberghs N, Knez J, Tikhonoff V, D'hooge J, Kloch-Badelek M, Thijs L, et al. Doppler indexes of left ventricular systolic and diastolic function in relation to the arterial stiffness in a general population. J Hypertens. 2016;34(4):762–71.
15. Hayward CS, Kalnins WV, Kelly RP. Gender-related differences in left ventricular chamber function. Cardiovasc Res. 2001;49(2):340–50.
16. Choi HF, D'hooge J, Rademakers FE, Claus P. Distribution of active fiber stress at the beginning of ejection depends on left-ventricular shape. Conf Proc Annu Int Conf IEEE Eng Med Biol Soc IEEE Eng Med Biol Soc Annu Conf. 2010;2010:2638–41.
17. Spotnitz HM. Macro design, structure, and mechanics of the left ventricle. J Thorac Cardiovasc Surg. 2000;119(5):1053–77.
18. Gould KL, Kennedy JW, Frimer M, Pollack GH, Dodge HT. Analysis of wall dynamics and directional components of left ventricular contraction in man. Am J Cardiol. 1976;38(3):322–31.
19. Sengupta PP. Left ventricular transmural mechanics: tracking opportunities in-depth. J Am Soc Echocardiogr. 2009;22(9):1022–4.
20. Cheng S, Larson MG, McCabe EL, Osypiuk E, Lehman BT, Stanchev P, et al. Age- and sex-based reference limits and clinical correlates of myocardial strain and synchrony. Circ Cardiovasc Imaging. 2013;6(5):692–9.
21. Lumens J, Prinzen FW, Delhaas T. Longitudinal strain: "think globally, track locally". JACC Cardiovasc Imaging. 2015;8(12):1360–3.
22. Sengupta PP, Narula J. Reclassifying heart failure: predominantly subendocardial, subepicardial, and transmural. Heart Fail Clin. 2008;4(3):379–82.
23. Donal E, Bergerot C, Thibault H, Ernande L, Loufoua J, Augeul L, et al. Influence of afterload on left ventricular radial and longitudinal systolic functions: a two-dimensional strain imaging study. Eur J Echocardiogr. 2009;10(8):914–21.
24. Tadic M, Sala C, Carugo S, Mancia G, Grassi G, Cuspidi C. Myocardial strain in hypertension: a meta-analysis of two-dimensional speckle tracking echocardiographic studies. J Hypertens. 2021;39(10):2103–12.
25. Ballo P, Quatrini I, Giacomin E, Motto A, Mondillo S. Circumferential versus longitudinal systolic function in patients with hypertension: a nonlinear relation. J Am Soc Echocardiogr. 2007;20(3):298–306.
26. Baltabaeva A, Marciniak M, Bijnens B, Moggridge J, He FJ, Antonios TF, et al. Regional left ventricular deformation and geometry analysis provides insights in myocardial remodelling in mild to moderate hypertension. Eur J Echocardiogr. 2008;9(4):501–8.
27. Soufi Taleb Bendiab N, Meziane-Tani A, Ouabdesselam S, Methia N, Latreche S, Henaoui L, et al. Factors associated with global longitudinal strain decline in hypertensive patients with normal left ventricular ejection fraction. Eur J Prev Cardiol. 2017;24(14):1463–72.
28. Kuznetsova T, Nijs E, Cauwenberghs N, Knez J, Thijs L, Haddad F, et al. Temporal changes in left ventricular longitudinal strain in general population: clinical correlates and impact on cardiac remodeling. Echocardiography. 2019;36(3):458–68.
29. Kawaguchi M, Hay I, Fetics B, Kass DA. Combined ventricular systolic and arterial stiffening in patients with heart failure and preserved ejection fraction: implications for systolic and diastolic reserve limitations. Circulation. 2003;107(5):714–20.

30. Davies JE, Baksi J, Francis DP, Hadjiloizou N, Whinnett ZI, Manisty CH, et al. The arterial reservoir pressure increases with aging and is the major determinant of the aortic augmentation index. Am J Physiol Heart Circ Physiol. 2010;298(2):H580–6.
31. Wang JJ, O'Brien AB, Shrive NG, Parker KH, Tyberg JV. Time-domain representation of ventricular-arterial coupling as a windkessel and wave system. Am J Physiol Heart Circ Physiol. 2003;284(4):H1358–68.
32. Schultz MG, Davies JE, Hardikar A, Pitt S, Moraldo M, Dhutia N, et al. Aortic reservoir pressure corresponds to cyclic changes in aortic volume: physiological validation in humans. Arterioscler Thromb Vasc Biol. 2014;34(7):1597–603.
33. Vlachopoulos C, Aznaouridis K, Stefanadis C. Prediction of cardiovascular events and all-cause mortality with arterial stiffness. A systematic review and meta-analysis. J Am Coll Cardiol. 2010;55(13):1318–27.
34. Ben-Shlomo Y, Spears M, Boustred C, May M, Anderson SG, Benjamin EJ, et al. Aortic pulse wave velocity improves cardiovascular event prediction: an individual participant meta-analysis of prospective observational data from 17,635 subjects. J Am Coll Cardiol. 2014;63(7):636–46.
35. Tsao CW, Lyass A, Larson MG, Levy D, Hamburg NM, Vita JA, et al. Relation of central arterial stiffness to incident heart failure in the community. J Am Heart Assoc. 2015;4(11):e002189.
36. Vasan RS, Pan S, Xanthakis V, Beiser A, Larson MG, Seshadri S, et al. Arterial stiffness and long-term risk of health outcomes: the Framingham heart study. Hypertension. 2022;79(5):1045–56.
37. Kuznetsova T, Cauwenberghs N, Knez J, Yang W-Y, Herbots L, D'hooge J, et al. Additive prognostic value of left ventricular systolic dysfunction in a population-based cohort. Circ Cardiovasc Imaging. 2016;9(7):e004661.
38. Sunagawa K, Sagawa K, Maughan WL. Ventricular interaction with the vascular system in terms of pressure-volume relationships. In: Ventriculo-vascular coupling: clinical, physiologic, and engineering aspects. New York: Springer Verlag; 1987. p. 210–39.
39. Saba PS, Ganau A, Devereux RB, Pini R, Pickering TG, Roman MJ. Impact of arterial elastance as a measure of vascular load on left ventricular geometry in hypertension. J Hypertens. 1999;17(7):1007–15.
40. Chantler PD, Melenovsky V, Schulman SP, Gerstenblith G, Becker LC, Ferrucci L, et al. The sex-specific impact of systolic hypertension and systolic blood pressure on arterial-ventricular coupling at rest and during exercise. Am J Physiol Heart Circ Physiol. 2008;295(1):H145 53.
41. Kuznetsova T, D'hooge J, Kloch-Badelek M, Sakiewicz W, Thijs L, Staessen JA. Impact of hypertension on ventricular-arterial coupling and regional myocardial work at rest and during isometric exercise. J Am Soc Echocardiogr. 2012;25(8):882–90.
42. Cohen-Solal A, Caviezel B, Himbert D, Gourgon R. Left ventricular-arterial coupling in systemic hypertension: analysis by means of arterial effective and left ventricular elastances. J Hypertens. 1994;12(5):591–600.
43. Cauwenberghs N, Tabassian M, Thijs L, Yang W-Y, Wei F-F, Claus P, et al. Area of the pressure-strain loop during ejection as non-invasive index of left ventricular performance: a population study. Cardiovasc Ultrasound. 2019;17(1):15.
44. Geyer H, Caracciolo G, Abe H, Wilansky S, Carerj S, Gentile F, et al. Assessment of myocardial mechanics using speckle tracking echocardiography: fundamentals and clinical applications. J Am Soc Echocardiogr. 2010;23(4):351–5.
45. Urheim S, Rabben SI, Skulstad H, Lyseggen E, Ihlen H, Smiseth OA. Regional myocardial work by strain Doppler echocardiography and LV pressure: a new method for quantifying myocardial function. Am J Physiol Heart Circ Physiol. 2005;288(5):H2375–80.
46. Russell K, Eriksen M, Aaberge L, Wilhelmsen N, Skulstad H, Remme EW, et al. A novel clinical method for quantification of regional left ventricular pressure-strain loop area: a non-invasive index of myocardial work. Eur Heart J. 2012;33(6):724–33.
47. Truong VT, Vo HQ, Ngo TNM, Mazur J, Nguyen TTH, Pham TTM, et al. Normal ranges of global left ventricular myocardial work indices in adults: a meta-analysis. J Am Soc Echocardiogr. 2022;35(4):369–377.e8.

48. Sahiti F, Morbach C, Cejka V, Tiffe T, Wagner M, Eichner FA, et al. Impact of cardiovascular risk factors on myocardial work—insights from the STAAB cohort study. J Hum Hypertens. 2022;36(3):235–45.
49. Sahiti F, Morbach C, Cejka V, Albert J, Eichner FA, Gelbrich G, et al. Left ventricular remodeling and myocardial work: results from the population-based STAAB cohort study. Front Cardiovasc Med. 2021;8:669335.
50. Tadic M, Cuspidi C, Pencic B, Vukomanovic V, Taddei S, Grassi G, et al. Association between myocardial work and functional capacity in patients with arterial hypertension: an echocardiographic study. Blood Press. 2021;30(3):188–95. https://doi.org/10.1080/08037051.2021.1902267.
51. Hubert A, Le Rolle V, Leclercq C, Galli E, Samset E, Casset C, et al. Estimation of myocardial work from pressure-strain loops analysis: an experimental evaluation. Eur Heart J Cardiovasc Imaging. 2018;19(12):1372–9.
52. Papadopoulos K, Özden Tok Ö, Mitrousi K, Ikonomidis I. Myocardial work: methodology and clinical applications. Diagnostics. 2021;11(3):573.
53. Shah AM, Solomon SD. Phenotypic and pathophysiological heterogeneity in heart failure with preserved ejection fraction. Eur Heart J. 2012;33(14):1716–7.

Sympathetic Activity in Hypertension and Heart Failure

8

Dagmara Hering, Gino Seravalle, Guido Grassi, and Krzysztof Narkiewicz

8.1 Introduction

Hypertension is a leading risk factor for the development of heart failure (HF), with nearly two-thirds of patients with HF having a history of elevated blood pressure (BP). The lifetime risk of HF remains high given the aging of the population and the rapidly increasing prevalence of hypertension, diabetes, and obesity. Over the past decades, two different HF phenotypes have been identified: HF with reduced ejection fraction (HFrEF) and HF with preserved ejection fraction (HFpEF). Pharmacological treatment for HFrEF is a cornerstone and is recommended to initiate therapy with four drugs that modulate the renin–angiotensin–aldosterone system (RAAS) and the sympathetic nervous system (SNS), namely, angiotensin-converting enzyme inhibitors or angiotensin receptor/neprilysin inhibitors, beta-blockers, mineralocorticoid receptor antagonists, and sodium-glucose cotransporter-2 inhibitors. With the goal of reducing mortality, preventing recurrent hospitalizations due to worsening HF, and improving the clinical status, functional capacity, and quality of life [1]. The diagnosis of HFpEF remains challenging; the average time to diagnose HFpEF is significantly longer compared to that to diagnose HFrEF and is often delayed due to nonspecific disease symptoms. Currently, only symptoms and

D. Hering (✉) · K. Narkiewicz
Department of Hypertension and Diabetology, Medical University of Gdansk, Gdansk, Poland
e-mail: hering@gumed.edu.pl; knark@gumed.edu.pl

G. Seravalle
IRCCS Istituto Auxologico Italiano, Cardiology Department, San Luca Hospital, Milan, Italy
e-mail: g_seravalle@yahoo.com

G. Grassi
Clinica Medica, Department of Medicine and Surgery, University Milano-Bicocca, Milan
e-mail: guido.grassi@unimib.it

M. Dorobantu et al. (eds.), *Hypertension and Heart Failure*, Updates in Hypertension and Cardiovascular Protection,
https://doi.org/10.1007/978-3-031-39315-0_8

comorbidity management are available for HFrEF due to the absence of proven treatments targeting the underlying HFpEF-related disease mechanisms.

8.2 Sympathetic Nervous System Activation in Human Hypertension

Hemodynamic characteristics of the initial phase of essential hypertension (EH) are induced by increased peripheral resistance or raised cardiac output [2]. Approximately one-third of younger adults with borderline and/or mild hypertension present with hyperkinetic hypertension with increased heart rate (HR), cardiac output, forearm blood flow, and plasma noradrenaline (NA) levels [3]. Other abnormalities present in the hyperkinetic state include increased renal blood flow and plasma renin activity [4]. The hyperkinetic state is likely to indicate increased sympathetic activity. The activation of the SNS in human hypertension has been demonstrated through the use of two state-of-the-art methods, namely, the isotope dilution method for quantifying NA spillover rates and postganglionic efferent sympathetic nerve recording with the microneurography technique. NA is the main neurotransmitter of sympathetic nerves, and the rate of its release from nerve terminals allows for the measurement of sympathetic nerve activity. Quantifying NA kinetics with the spillover technique was a breakthrough in studies on the sympathetic nervous system. Although the overall (total body) NA spillover in hypertension is moderate, a selective increase of sympathetic outflow from the heart and the kidneys characterizes EH, likely contributing to established hypertension [5, 6]. Augmented NA from the renal sympathetic nerves is evident in untreated patients with EH, mostly adults below the age of 40, and is a prime mover for BP rise [5, 7]. Increased cardiac NA spillover and decreased NA neuronal reuptake further potentiate sympathetic activation in maintaining elevated BP [8]. Compared to younger hypertensives, NA release from sympathetic nerves was found to be lower due to an age-dependent fall in NA plasma clearance in EH [9]. With aging and disease progression, cardiac output generally becomes normal in uncomplicated hypertension; however, a shift toward increased vascular resistance potentiates sympathetic activation, which is a hallmark of established hypertension, leading to vascular remodeling, organ damage, and adverse cardiovascular (CV) complications [10].

Notably, at rest, in the healthy human heart, the amount of adrenaline released from the sympathetic nerves is negligible as the majority (80%) of adrenaline is secreted from the adrenal medulla into the bloodstream, with only a slight amount released from the sympathetic nerve endings. In untreated EH, augmented adrenaline release from cardiac sympathetic nerves enhances cardiac NA release through adrenaline co-transmission [11]. Mental stress is another factor to potentiate adrenaline plasma concentrations, and it increases the amount of neuronal uptake of adrenaline in cardiac sympathetic nerve endings, leading to the subsequent release of adrenaline and potentiating NA co-release, thus acting as a co-transmitter mechanism and supporting the "adrenaline hypothesis" [11]. Adrenaline contribution to the pathogenesis of hypertension has been demonstrated by other findings documenting an 18-h elevation in BP following a 6-h adrenaline infusion [12]. These data suggest that adrenaline exerts a delayed pressor effect during sympathetic stimulation through the activation of presynaptic β-2-adrenoreceptors.

Microneurography studies have demonstrated increased muscle sympathetic nerve activity (MSNA) in high-normal BP [13, 14], indicating that sympathetic activation precedes overt arterial hypertension even in very low-risk subjects with high-normal BP [15]. MSNA has been independently linked to arterial mechanical properties assessed by pulse wave velocity (PWV), contributing to arterial stiffness in normotensive subjects [16]. Tonic sympathetic activation evident in patients with prehypertension appears to contribute to a time-related increase in BP and the development of sustained hypertension and asymptomatic arterial stiffness [17]. In this context, interesting are findings linking long-term aircraft noise exposure to arterial stiffness, [18] suggesting that environmental noise appears to be an important non-CV risk factor for the development of hypertension. On the contrary, even short-term noise reduction, as experienced during the coronavirus disease of 2019 (COVID-19) lockdown, reduces BP and PWV, suggesting the potential reversal effects of environmental noise on arterial compliance [19].

Persistent sympathetic activation is commonly found in patients with uncontrolled hypertension and resistant hypertension (RH). High levels of multiunit MSNA (often 50–70 bursts/min) with burst activity synchronized with every heartbeat typically characterize patients with RH when compared to healthy controls (<20 bursts/min) and patients with untreated EH (25–40 bursts/min). Patients with RH display markedly elevated activity of single-unit muscle vasoconstrictor fibers, including firing rate, firing probability, and incidence of multiple spikes within a cardiac cycle [20]. These findings indicate that a high sympathetic drive is a hallmark of RH despite the use of antihypertensive drug classes [20, 21], which are supposed to target the pathophysiology of hypertension. An increased prevalence of conditions associated with sympathetic activation, including diabetes, obesity, obstructive sleep apnea, and chronic kidney disease, are commonly found in RH patients [22, 23], thereby further potentiating a high sympathetic drive and drug resistance.

8.2.1 Neural Reflex Mechanisms in the Pathogenesis of Hypertension

Tonic sympathetic activation is generated by the neurons residing in the rostral ventrolateral medulla (RVLM), which integrates neural reflex mechanisms from afferent arterial baroreceptors, afferent arterial chemoreceptors, and cardiopulmonary mechanoreceptors [24]. The arterial baroreceptor reflex acts as an important inhibitory regulatory mechanism induced by BP changes mediated by the parasympathetic and sympathetic components. Reduced baroreflex sensitivity was found in patients with hypertension and in subjects with a family history of hypertension and normal BP levels [25]. Increased cardiopulmonary baroreflex control of sympathetic activity, unrelated to the attenuation of the arterial baroreflex, has also been demonstrated in hypertension compared to normotensive controls [26]. A further important causative mechanism of sympathetic activation, contributing to the pathogenesis of hypertension, is the potentiated sensitivity of arterial chemoreceptors in response to hypoxia [27–29]. These findings support the concept of disturbed peripheral regulatory mechanisms associated with sympathetic activation in human hypertension.

8.2.2 Interaction Between the Sympathetic Nervous System and the Renin–Angiotensin System

The SNS is directly related to the renin–angiotensin system (RAS), with both playing an important role in BP and blood volume control, acting at different levels on the central nervous system (CNS) and peripheral sites and leading to the synthesis and release of angiotensin II and NA, the fundamental neurotransmitters implicated in the development and progression of the disease [30]. Sympathetic activation increases the activity of the RAS via direct effects on the three renal neuroeffectors: (1) the juxtaglomerular granular cells enhancing renin release (via stimulation of β_1-adrenoreceptors), (2) the renal tubular epithelial cells increasing sodium reabsorption (via stimulation of α_{1B}-adrenoreceptors), and (3) the renal vasculature (via stimulation of α_{1A}-adrenoreceptors), thus reducing renal blood flow. These result in increased efferent renal sympathetic nerve activity (ERSNA) [6]. Notably, in physiological conditions, ERSNA stimulates afferent renal sensory nerves, which, via the renorenal reflex mechanism, leads to a subsequent reduction in ERSNA and resultant diuresis and natriuresis. In the presence of elevated BP, increased signals arising from the injured and/or ischemic kidney via afferent renal sensory nerves are projected to the central integrative structure in the brainstem (RVLM), thus potentiating sympathetic outflow to the periphery and causing end organ damage and adverse CV complications.

8.2.3 Central Neural Activation in Hypertension

There is evidence to indicate that sympathetic activation in EH is of CNS origin and contributes to the efferent sympathetic outflow to the periphery. Studies assessing the combined overflow of brain NA have demonstrated increased NA release within the CNS [31, 32]. Direct blood sampling from the internal jugular veins and concomitant cerebral blood flow scans revealed that suprabulbar noradrenergic projections from the brainstem to the hypothalamus play a key role in neurogenic hypertension [33]. Subcortical NA turnover in brain regions (not the cerebral cortex) was found to be significantly higher in EH when compared to healthy subjects and directly related to neurochemical indices of the SNS activity and renal NA spillover in patients with EH [33].

The association between "hypertension" and the "brain" is reciprocal. Hypertension alone leads to cerebral hypoperfusion and a reduction in cortical thickness independent of the use of antihypertensive drugs [34]. Further proof linking functional neural reorganization early in the course of hypertension development comes from functional magnetic resonance imaging studies, demonstrating substantial variations in brain activation in a reasonably controlled hypertensive cohort compared to age-matched normotensive controls [35].

8.2.4 The Link Between Hypertension and Heart Failure

Chronically elevated BP critically contributes to the progression of hypertensive heart disease, resulting in a constellation of abnormalities in the myocardium, coronary vasculature, heart function, and conduction [36]. Changes in cardiac chamber geometry induced by hypertension include asymptomatic left ventricular (LV) hypertrophy, ischemic heart disease, systolic and diastolic dysfunction, and resultant clinical manifestations, including arrhythmia and HF. The relationship of antecedent hypertension to the development of HF has been documented in a large study cohort (n = 5143) followed over a 20-year duration [37]. Hypertension accounted for 39% of new HF cases in men and 59% in women. Among hypertensive subjects, myocardial infarction, diabetes, LV hypertrophy, and valvular heart disease predicted an increased risk of congestive HF in both females and males. Outcomes following the onset of hypertension-related HF were poor with only 24% of men and 31% of women surviving for 5 years [37].

The pathophysiology of hypertensive heart disease involves a complex interplay of hemodynamic, structural, cellular, neurohormonal, molecular, and genetic factors. Sympathetic activation plays a critical role in cardiac, arterial, and vascular remodeling and further progression. High levels of NA release from cardiac and renal sympathetic nerves accompanying increased MSNA have been linked to LV remodeling in untreated hypertension, including LV hypertrophy and LV dysfunction [38–40]. Activation of the RAS is a further major contributor to the development of myocardial hypertrophy. The associated sympathetically mediated rise in BP and induced organ damage are likely to contribute to adverse CV events.

8.3 Sympathetic Nervous System Activation in Heart Failure

Neurohormonal activation, including increased sympathetic activity and withdrawal of vagal activity, are hallmarks of HF. Altered sympathetic CV regulation is at the core of disease development, progression, and associated increased CV morbidity and mortality. Augmented MSNA is already evident in early-stage HF and is directly related to the clinical severity of congestive HF as classified by the NYHA (the New York Heart Association) [41]. Moreover, the potentiated high sympathetic drive in HF has been associated with markedly impaired baroreflex control when compared to healthy subjects [41]. Higher levels of MSNA have been linked to end organ damage, including reduced LV function, LV stroke work index, stroke volume, and cardiac chamber size [42], indicating that resting sympathetic activity parallels impairment of cardiac performance in HF. Support for sympathetic activation in HF has been shown in studies applying NA spillover. Increased NA release from the renal and cardiac sympathetic nerves are typical features in patients with congestive HF [43].

Previous studies found that in acute HF, high levels of catecholamines augment ventricular contractility to maintain sufficient cardiac output. Heightened NA released from the sympathetic nerves predominantly stimulates β_1-adrenergic receptors (ARs) with its higher binding affinity to β_1-ARs than to β_2-ARs, irrespective of the physiological and pathophysiological conditions [44]. In the healthy heart, the concentration of β_1-AR is four times higher than that of β_2-AR, whereas in the chronically failing heart, β_1-AR is selectively downregulated, thus shifting the proportion of β-AR and resulting in high levels of circulating NA. In the failing heart, NA was found to be exceptionally cardiotoxic, producing cardiac myocyte injury via NA-mediated cell toxicity through β- rather than α-AR [44]. Additionally, high cardiac NA levels contribute to the selective decrease of β_1-AR density in HF. Although initially this is a protective compensatory mechanism preventing a deleterious surge in intracellular cyclic adenosine monophosphate (cAMP) and mediating the catecholaminergic control of HR and contractility, it ultimately leads to a loss of inotropy and damage to the failing heart [45].

8.3.1 The Role of the Kidneys in the Pathophysiology of Heart Failure

Neural control of the kidneys is of particular relevance in HF with a subsequent increase in renin release, sodium reabsorption, and reduction in renal perfusion [7]. The resulting increase in venous return triggers a rise in end-diastolic volume and stroke volume in the heart. Excessive and prolonged exposure to sympathetic activation produces adverse systemic effects on various organs, critically contributing to the progression of HF, including myocardial ischemia and remodeling, with enlargement of the ventricular chamber [46] and a decrease in ventricular fibrillation threshold, thus resulting in sudden cardiac death and worsened prognosis [7]. An increased elevated cardiac NA spillover rate independently predicted mortality in congestive HF patients [47]. Additionally, congestive HF patients with higher levels of renal NA spillover had substantially reduced transplantation-free survival rates [48]. Chronic kidney disease has predicted repeated HF hospitalizations [49]. Stimulation of the RAAS accompanying sympathetic activation is an additional fundamental contributor to the progression of congestive HF [50, 51], producing adverse systemic consequences on the myocardium, blood vessels, and peripheral vasoconstriction and resulting in increased afterload, reduced cardiac output, and renal perfusion.

8.3.2 Neural Reflex Mechanisms in the Pathogenesis of Heart Failure

Altered reflex and neurohumoral and metabolic factors substantially contribute to the initiation and maintenance of increased sympathetic activation in HF. This includes an impairment of arterial baroreceptor function in mediating changes in tonic sympathetic nerve activity in the heart and blood vessels in HF [52–54]. While in healthy subjects, the arterial baroreflex exerts a powerful inhibitory influence on the SNS and arterial chemoreceptors [55], baroreflex impairment with subsequent loss of sympathetic inhibition in HF is likely to increase chemoreflex sensitivity and further potentiate sympathetic activation, which is unopposed by inhibitory influences from reflex mechanisms. HF patients display a marked and selective augmentation of the central chemoreflex (located in the brainstem on the ventral surface of the medulla) in response to hypercapnia [56, 57]. An increase in central chemoreflex sensitivity may contribute to the development of central sleep apnea (CSA) in HF. Direct intraneural recordings of MSNA showed conflicting results in HF patients regarding chemoreflex activation. High resting levels of MSNA evident in HF patients were not influenced by hyperoxia (deactivation of peripheral chemoreceptors), indicating that increased efferent MSNA is unrelated to tonic activation of excitatory chemoreflex afferents even in the presence of mild hypoxemia [56]. This has been further confirmed in another study demonstrating that congestive HF patients, despite higher ventilation (which inhibits sympathetic activity), are characterized by a selective potentiation of ventilatory and sympathetic responses to central chemoreceptor activation induced by hypercapnia but not hypoxia or cold pressor test when compared to healthy subjects [57]. In contrast, another study demonstrated reduced MSNA in response to hyperoxia in patients with HF-associated anemia [58]. Moreover, the deactivation of peripheral chemoreceptors led to a reduction in MSNA and was closely associated with impaired baroreflex function in HF [59]. Nevertheless, CSA and cyclic episodes of hyperpnea–apnea breathing (C-S (Cheyne–Stokes) respiration) are commonly present in HF patients [60], eliciting a further augmentation of sympathetic activation [56] and likely contributing to CV events, bradyarrhythmia, and worsening prognosis. It has been shown that increased peripheral chemoreflex sensitivity is an independent prognostic factor in ambulatory patients with HF [61]. An additional important contributor to sympathetic activation in HF is the reduced ability of the inhibitory influences from arterial and cardiopulmonary mechanoreceptors to exert tonic restraint on sympathetic outflow [62]. Chronic sympathetic activation with reduced parasympathetic tone present in HF is critical in the development and progression of disease, including ventricular remodeling and arrhythmia, resulting in increased CV morbidity and mortality [63, 64].

8.4 Device-Based Therapeutic Interventions

Inhibition of the neurogenic pathways has been a major target for the management of hypertension and HF. Therapeutic modulation of the SNS activity via interruption of afferent signaling arising from (1) the kidneys, (2) carotid arterial baroreceptors, and (3) carotid arterial chemoreceptors projected to the RVLM (the center for BP and sympathetic outflow control) has been demonstrated in the treatment of uncontrolled hypertension and HF.

8.4.1 Renal Denervation (RDN) in Hypertension

Since 2009, the publication of the first-in-man proof-of-concept Symplicity HTN-1 trial, most interventional advancements for the treatment of RH have focused on renal denervation (RDN) [65]. Evidence from the unblinded studies applying catheter-based RDN demonstrated a significant decrease of ~10–15 mmHg in mean ambulatory systolic BP [66–68]. However, the first prospective, randomized, double-blind, sham-controlled Symplicity HTN-3 study has not confirmed the BP reduction for treatment difference between the RDN and control groups in patients with RH [69]. Conflicting results come from other sham-controlled RDN studies that show no significant changes in the primary efficacy end point between the RDN and the sham-controlled groups [70] or comparable reductions in daytime systolic BP between both groups [71]. However, the effectiveness in BP reduction achieved with RDN was higher when a standardized, stepped-care, antihypertensive treatment was added to the procedure when compared to the same medication alone [72].

The uncertainty of the BP reduction following RDN has been addressed in multicenter sham-controlled trials, with adequate blinding of patients and outcome assessors, with BP assessment by 24-h ambulatory BP monitoring as the primary efficacy end point, study completion as planned, and the use of advanced RDN systems and procedural techniques [73]. The second-generation randomized controlled trials with the use of radiofrequency energy or high-focused ultrasound energy have demonstrated a significant reduction in office and ambulatory BP in hypertensive patients without and with antihypertensive medication [74–78]. In the SPYRAL HTN-OFF Med Proof-of-Concept Study, patients who were drug-naive or discontinued their antihypertensive medications with an office systolic BP of 150 mmHg or greater and less than 180 mmHg, office diastolic BP of 90 mmHg or greater, and a mean 24-h ambulatory systolic BP of 140 mmHg or greater and less than 170 mmHg with eligible renal anatomy were randomly assigned to RDN ($n = 38$) or sham control ($n = 42$) groups. At 3 months follow-up, the mean difference in 24-h ambulatory blood pressure monitoring (ABPM) between the groups was −5.0 mmHg for 24-h systolic BP and −4.4 mmHg for diastolic BP, favored RDN, and had no major adverse events in either group [74]. Further analysis of combined evidence from the pilot and Pivotal trials that included a larger number of patients in RDN ($n = 166$) or a sham procedure ($n = 165$), it was found that the treatment difference between the two groups for 24-h systolic BP was −3.9 mmHg, thus suggesting the superiority of catheter-based RDN compared with a sham procedure in the absence of antihypertensive medications [76].

In parallel to SPYRAL HTN-OFF MED, the safety and BP response to RDN was observed in patients with uncontrolled hypertension on antihypertensive medications (SPYRAL HTN-ON MED) with drug adherence testing [75]. In this international, randomized, single-blind, sham-controlled, proof-of-concept trial, patients with uncontrolled hypertension on one to three antihypertensive drugs with stable doses for at least 6 weeks were randomly assigned to RDN ($n = 38$) or sham control ($n = 42$) groups. At 6 months follow-up, the mean baseline-adjusted treatment differences in 24 h were − 7.0 mmHg for systolic and −4.3 mmHg for diastolic

BP. This study confirmed that RDN in the main renal arteries and branches significantly reduced BP compared with sham control, with no major safety events. Incomplete medication adherence was common, accounting for ~60%, and varied for individual patients throughout the study [75].

Long-term efficacy and safety of the SPYRAL HTN-ON MED trial have been recently reported to demonstrate significantly lower BP reduction in RDN compared to sham control groups at 24 and 36 months, despite a similar treatment intensity of antihypertensive drugs. At 36 months, the ambulatory systolic BP reduction was −18.7 mmHg in the RDN ($n = 30$) and −8.6 mmHg in the sham control ($n = 32$) groups. The medication burden at 36 months was 2.13 ± 1.15 in the RDN group and 2.55 ± 2.19 in the sham control group. In this study, 24 (77%) of 31 patients in the RDN and 25 (93%) of 27 patients in the sham control groups adhered to medication at 36 months. There were no short- or long-term safety issues associated with renal denervation [79].

A further RADIANCE-HTN SOLO trial examined the safety and efficacy of alternative RDN technology using endovascular ultrasound in patients with moderate hypertension in the absence of antihypertensive medications [77]. In this multicenter, international, single-blind, randomized, sham-controlled study, hypertensive patients with ambulatory BP ≥135/85 mmHg and <170/105 mmHg after 4-week discontinuation of up to 2 antihypertensive medications and suitable renal artery anatomy were randomized to RDN (n = 74) with the Paradise system or a sham procedure ($n = 72$) consisting of only renal angiography. At 2 months post-procedure, the change in daytime ambulatory systolic BP was greater with RDN (−8.5 ± 9.3 mmHg) than in the sham procedure (−2.2 ± 10.0 mmHg) with the baseline-adjusted difference between groups being −6.3 mmHg, with no major adverse events reported in either group [77].

In this study, patients were followed over a longer time post-procedure. Between 2 and 5 months post procedure, patients received standardized medication escalation protocol at physicians' discretion if BP was elevated [80]. A total of 65 of 74 RDN patients and 67 of 72 sham patients had 12-month ambulatory BP measurements. The proportion of patients on ≥2 medications (27.7% vs. 44.8%), the number of medications (1.0 vs. 1.4), and the defined daily dose (1.4 vs. 2.2) were reduced with RDN versus sham. There was a decrease in daytime ambulatory systolic BP from baseline by −16.5 ± 12.9 mmHg in the RDN group, which remained stable at 12 months post-procedure. The RDN versus sham-adjusted difference at 12 months was −2.3 mmHg for daytime systolic BP [80].

Endovascular ultrasound RDN has demonstrated efficacy and safety in patients with hypertension resistant to three or more antihypertensive medications by reducing daytime ambulatory systolic BP by −8.0 mmHg compared to −3.0 mmHg in the sham procedure at 2 months follow-up, indicating that RDN might be an alternative therapeutic approach to the addition of further antihypertensive medications in patients with RH [78].

Currently, available data indicate a biological proof of principle for the reduction of BP with a vascular approach and a sustained reduction in MSNA out to 12 months post-procedure [21] as shown in Fig. 8.1, with a recent meta-analysis of 11 studies,

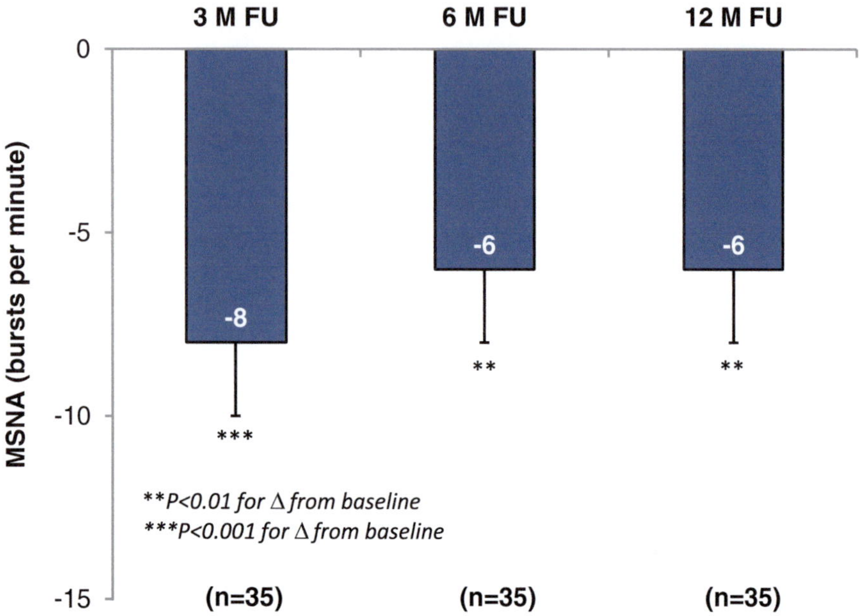

Fig. 8.1 Mean fall in muscle sympathetic nerve activity (MSNA) at 3, 6, and 12 month (M) follow-up (FU) after renal denervation. (Adapted with permission from Hering et al. [21])

including more than 400 patients with RH, thus confirming previous findings of significant, although modest, central inhibition after RDN, with MSNA unrelated to the BP-lowering effects post-procedure [81]. Although these findings are promising, the broad clinical utility of RDN for the management of hypertension is limited by the lack of diagnostic tools and markers to predict BP response after RDN and the lack of reliable markers to confirm the completeness of renal nerve ablation.

8.4.2 Renal Denervation in Heart Failure

With a proven safety profile and substantial continued BP reduction in a large proportion of patients with uncontrolled hypertension, the safety and efficacy of bilateral RDN have been demonstrated in the Renal Artery Denervation in Chronic Heart Failure (REACH) trial, which included 7 HF patients with reduced ejection fraction (HFrEF) [82] and the Symplicity HF Feasibility Study with a total of 39 HFrEF patients [83]. Importantly, the procedure was well-tolerated and no major drops in BP were observed [82]. No changes in left ventricular ejection fraction (LVEF) and cardiac function were observed in the REACH trial at 6 months follow-up and in the Symplicity HF trial 12 months post-procedure. A significant improvement in the 6-min walk test was found in the REACH trial but not in the Symplicity HF trial [83]. The effectiveness and risks of bilateral RDN in patients with HF with an LVEF <50% were assessed in a systematic review of clinical trials [84]. This meta-analysis found a decrease in the mean NYHA class and N-terminal pro-B-type natriuretic peptide (NT-proBNP) levels, improvements in the 6-min walk test and LV structure, and function, including increases in LVEF and a decrease in left atrium diameter after RDN with safety profile on BP levels without affecting renal function.

The clinical and hemodynamic effects of RDN have been retrospectively analyzed in a single-center study in patients with HF with preserved ejection fraction (HFpEF) [85]. Patients undergoing RDN were retrospectively classified as HFpEF (n = 99) and no HF (n = 65). While RDN decreased systolic BP comparable in patients with HFpEF and no-HF patients, there was a reduction in stroke volume index and increases in aortic distensibility and systolic stiffness in HFpEF patients. LV diastolic stiffness, LV filling pressures, and NT-proBNP (N-terminal pro-B-type natriuretic peptide) decreased after RDN in patients with HFpEF. These findings suggest that RDN exerts safety beneficial effects on cardiac, vascular, and functional capacity performance in patients with uncontrolled hypertension and HFpEF.

Although there are potential benefits of RDN in patients with HF, no randomized controlled and long-term data are yet available.

8.4.3 Baroreflex Activation Therapy (BAT) in Hypertension

Electric stimulation of carotid sinus baroreceptors is a further attractive approach for the treatment of uncontrolled hypertension. The initial proof-of-concept study (DEBuT-HT (Device-Based Therapy in Hypertension Trial)) has demonstrated the safety and efficacy of the CVRx Rheos System device in producing a substantial and durable BP-lowering effect [86, 87]. The implantation of the first-generation system was associated with procedure-related serious adverse events and short-term battery life, which limited its utility. The next-generation minimally invasive Barostim neo led to a significant BP reduction in RH patients at 3- and 6-month follow-up, even in a subset of patients (n = 6) previously treated with RDN, and was associated with fewer device-related side effects [88]. Further proof for additive BP lowering and antiproteinuric effects achieved by BAT was demonstrated in a cohort of 28 patients who presented with elevated BP despite previous RDN performed 5 months prior [89]. Two recent case reports that have demonstrated the beneficial clinical utility of BAT in acute clinical scenarios deserve to be mentioned. Generally, the BAT device is activated 2–4 weeks after surgical implantation to allow the site to heal; however, its immediate activation in a young male with hypertensive crisis following aortic dissection due to RH that was unresponsive to sympatholytic agents resulted in a rapid, significant, and sustained reduction in BP out to 12 months post-procedure with no further incidence of hypertensive crisis [90]. The clinical utility of the second-generation Barostim neo has been also successfully demonstrated in the first-in-man treatment of severe BP variability >30 mmHg in a patient diagnosed with a progressive central and peripheral dysautonomia secondary to Sjogren's syndrome and impaired CV reflex regulation [91]. This case report further highlights the clinical applicability of BAT in severe forms of difficult-to-manage hypertension. Although the application of BAT in emergency cases may not be feasible, it emphasizes the potential of this treatment option.

The long-term efficacy and safety of BAT are based on data available from the previous 3 trials of 383 patients with RH [92]. In this study, 143 patients completed 5 years of follow-up and 48 patients completed 6 years of follow-up. In the entire cohort, office systolic BP fell from 179 ± 24 mmHg to 144 ± 28 mmHg, whereas office diastolic BP dropped from 103 ± 16 mmHg to 85 ± 18 mmHg. The number of medications was reduced from a median of 6 to 3 drugs in approximately one-fourth of patients. These data demonstrate that BAT maintains its efficacy for persistent reduction of office BP in patients with RH without major safety issues out to 6 years of follow-up.

8.4.4 Mechanical Endovascular Baroreflex Amplification

The endovascular device-based implant, which increases wall strain and mechanically activates carotid baroreceptors, is an alternative therapeutic approach that modulates the autonomic nervous system. The MobiusHD (Vascular Dynamics, Mountain View, CA, USA) is delivered by a catheter via the femoral artery and is a nitinol self-expanding device implanted in the internal carotid artery to amplify the carotid baroreceptor signal. The CALM-FIM_EUR (Controlling and Lowering Blood Pressure with MobiusHD First-in-Man) prospective study evaluated the safety and performance of the MobiusHD system in 30 patients with RH [93]. Unilateral implantation of the MobiusHD device in the internal carotid artery decreased office BP from 184 ± 18 mmHg for systolic and 109 ± 14 mmHg at baseline by 24/12 mmHg at 6 months. The mean baseline 24-h ambulatory BP was 166/100 mmHg (17/14) at baseline and was reduced by 21/12 mmHg at 6 months for systolic and diastolic BP. Five serious adverse events (two hypotension, one worsening hypertension, one intermittent claudication, and one wound infection) occurred 6 months post-procedure. While endovascular baroreceptor amplification with the MobiusHD system substantially reduced BP with an acceptable safety profile, the currently ongoing CALM-2 randomized, double-blind, sham-controlled multicenter pivotal study will determine the safety and effectiveness of this approach for the treatment of RH.

8.5 Baroreflex Activation Therapy in Heart Failure

The beneficial results of BAT in patients with RH have led to further proof-of-concept studies in HF patients characterized by impaired baroreflex function [94, 95]. In one of these studies, out of the 11 HF patients with NYHA III class and an average LVEF of 31 ± 7% receiving optimal drug therapy, 7 had a previously implanted cardioverter defibrillator and 1 with a pacemaker underwent BAT and were followed up for 6 months [94]. The reduction in MSNA was more pronounced between 1 and 3 months post-procedure and remained at this level in 8 out of 11 patients at 6 months follow-up (Fig. 8.2) [94]. The sympathetic inhibition was accompanied by progressively increasing improvement in baroreflex sensitivity

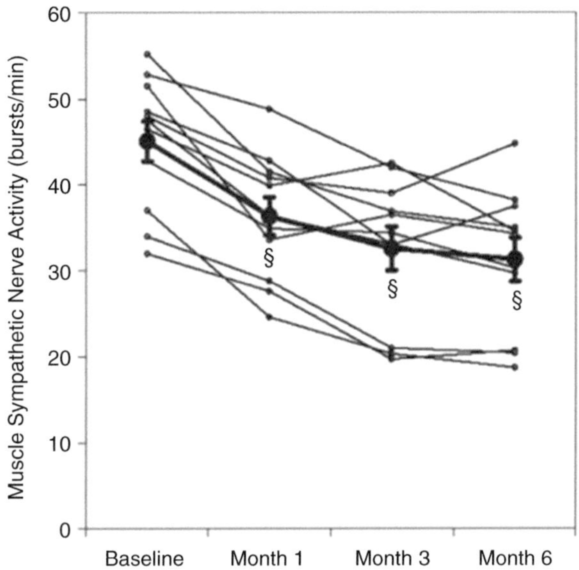

Fig. 8.2 Change in muscle sympathetic nerve activity (MSNA) during treatment with baroreflex activation therapy. (Adapted with permission from Gronda et al. [94])

from 1-month post-procedure through to 6 months follow-up. BAT was associated with improved clinical symptoms as demonstrated by increased exercise tolerance and LVEF and reduced NYHA class and hospitalization rate in addition to improved quality of life, with no significant changes in BP or symptoms of orthostatic hypotension [94]. Two patients died throughout the study due to causes deemed to be unrelated to the procedure. Nevertheless, the remaining nine patients who experienced clinical benefits with BAT at 6 months continued to maintain the improvements over the longer term at 12 and 21.5 ± 4.2 months post-procedure [95]. Given the causal link between baroreflex dysfunction and its prognostic significance in HF, these preliminary findings appear reassuring for the potential applicability of BAT in patients with a reduced LVEF. However, larger clinical trials need to confirm these clinical observations and assess whether BAT may complement current management in the treatment of HF, both in patients with preserved and reduced LVEF.

8.6 Carotid Body Denervation in Hypertension

The association between potentiated tonic chemoreflex sensitivity and increased sympathetic activation in the pathophysiology of hypertension has encouraged the initiation of studies investigating the feasibility and efficacy of a therapeutic intervention directed at the modulation of peripheral arterial chemoreceptors located in the carotid body. Results from the first-in-human proof-of-concept study of a total of 15 patients with RH who underwent unilateral ($n = 4$ on left side, $n = 11$ on right) carotid body removal found that 8 out of 15 patients, who underwent carotid body removal on the right side, experienced a reduction in daytime and nighttime systolic

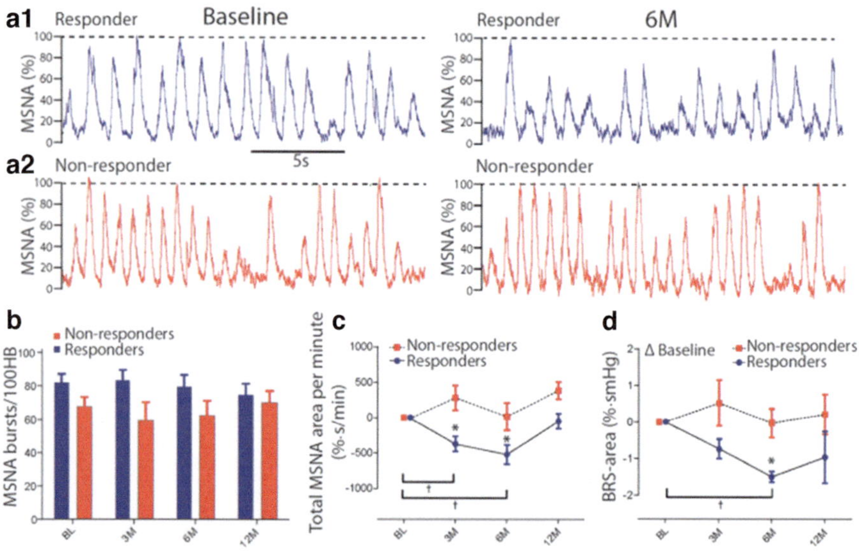

Fig. 8.3 Representative recordings of muscle sympathetic nerve activity (MSNA) for a responder ($\mathbf{a}_1$) and a nonresponder ($\mathbf{a}_2$) at baseline (left trace) and at 6-month follow-up (right trace). There was no difference in MSNA burst incidence between the responders and nonresponders (**b**). However, the MSNA area was reduced after carotid body removal in the responders but not in the nonresponders (**c**), and spontaneous MSNA area baroreflex gain improved in the responders but not in the nonresponders (**d**). Two-way repeated measures analysis of variance was used (within groups $*p < 0.05$; between groups $\dagger p < 0.05$), bursts/100 HB = bursts per 100 heartbeats. (Adapted with permission from Narkiewicz et al. [96])

BP at 3 (−23 ± 3 mmHg, −20 ± 4 mmHg), 6 (−26 ± 4 mmHg, −16 ± 5 mmHg), and 12 (−12 ± 8 mmHg, −15 ± 6 mmHg) months follow-up [96]. Lowering the BP was accompanied by a reduction in MSNA and improved baroreflex sensitivity in responders but not in nonresponders (Fig. 8.3). In a patient previously treated with RDN and removal of the left carotid body, no short- and long-term BP changes were noted post-procedure. Another novel finding is the association between the BP responses and increased chemoreflex sensitivity, indicating the underlying contributing mechanism of RH [96].

8.7 Carotid Body Denervation in Heart Failure

The prognostic significance of augmented peripheral chemoreflex sensitivity in HF patients has prompted the extension of therapeutic removal of carotid body in patients with systolic CHF [97]. Surgical carotid body removal was well-tolerated in a treated patient and reduced peripheral chemoreceptor sensitivity whilst improving exercise tolerance, LVEF, sleep disorder breathing, and quality of life. Although the patient's medication remained unchanged, no symptoms of hypotension following unilateral carotid body removal were observed. Whether surgical or endovascular carotid body denervation may be offered as a therapeutic approach to improve CV outcomes in HF merits further investigation. The effect of carotid body removal on the sympathetic activity in HF is unknown.

Although this treatment approach seems promising, there are no further clinical trials on carotid body modulation for the treatment of hypertension or HF.

8.8 Future Directions

Hypertension is the leading and most preventable risk factor for cardiovascular disease. Despite the wide range of available pharmacological approaches, there is a proportion of patients with uncontrolled BP. Innovative device-based and procedural interventions that directly manipulate the mechanisms underlying hypertension have been successful in demonstrating their ability to lower BP in the vast majority of patients. The high variability in BP response to these therapies indicates that the pathophysiology, including complex neural reflexes that may trigger a disease in an individual, is still not completely understood.

HF and its progression remain a challenging clinical problem. Preliminary findings from HF pilot studies with recently introduced therapeutic interventions appear promising; however, their long-term outcomes, including the rate of hospital admission for HF deterioration and associated death, need to be confirmed in larger clinical trials.

Disclosure The authors declare no conflict of interest.

References

1. McDonagh TA, Metra M, Adamo M, et al. Corrigendum to: 2021 ESC guidelines for the diagnosis and treatment of acute and chronic heart failure: developed by the task force for the diagnosis and treatment of acute and chronic heart failure of the European Society of Cardiology (ESC) with the special contribution of the Heart Failure Association (HFA) of the ESC. Eur Heart J. 2021;42:4901.
2. Folkow B. Physiological aspects of primary hypertension. Physiol Rev. 1982;62:347–504.
3. Julius S, Krause L, Schork NJ, et al. Hyperkinetic borderline hypertension in Tecumseh, Michigan. J Hypertens. 1991;9:77–84.
4. Messerli FH, Ventura HO, Reisin E, et al. Borderline hypertension and obesity: two prehypertensive states with elevated cardiac output. Circulation. 1982;66:55–60.
5. Esler M, Lambert G, Jennings G. Regional norepinephrine turnover in human hypertension. Clin Exp Hypertens A. 1989;11(Suppl 1):75–89.
6. DiBona GF, Esler M. Translational medicine: the antihypertensive effect of renal denervation. Am J Physiol Regul Integr Comp Physiol. 2010;298:R245–53.
7. Parati G, Esler M. The human sympathetic nervous system: its relevance in hypertension and heart failure. Eur Heart J. 2012;33:1058–66.
8. Schlaich MP, Lambert E, Kaye DM, et al. Sympathetic augmentation in hypertension: role of nerve firing, norepinephrine reuptake, and angiotensin neuromodulation. Hypertension. 2004;43:169–75.

9. Esler M, Jennings G, Biviano B, Lambert G, Hasking G. Mechanism of elevated plasma noradrenaline in the course of essential hypertension. J Cardiovasc Pharmacol. 1986;8(Suppl 5):S39–43.
10. Julius S, Nesbitt S. Sympathetic overactivity in hypertension. A moving target. Am J Hypertens. 1996;9:113S–20S.
11. Rumantir MS, Jennings GL, Lambert GW, Kaye DM, Seals DR, Esler MD. The 'adrenaline hypothesis' of hypertension revisited: evidence for adrenaline release from the heart of patients with essential hypertension. J Hypertens. 2000;18:717–23.
12. Blankestijn PJ, Man in't Veld AJ, Tulen J, et al. Support for adrenaline-hypertension hypothesis: 18 hour pressor effect after 6 hours adrenaline infusion. Lancet. 1988;2:1386–9.
13. Greenwood JP, Stoker JB, Mary DA. Single-unit sympathetic discharge: quantitative assessment in human hypertensive disease. Circulation. 1999;100:1305–10.
14. Seravalle G, Lonati L, Buzzi S, et al. Sympathetic nerve traffic and baroreflex function in optimal, normal, and high-normal blood pressure states. J Hypertens. 2015;33:1411–7.
15. Hering D, Kara T, Kucharska W, Somers VK, Narkiewicz K. High-normal blood pressure is associated with increased resting sympathetic activity but normal responses to stress tests. Blood Press. 2013;22:183–7.
16. Swierblewska E, Hering D, Kara T, et al. An independent relationship between muscle sympathetic nerve activity and pulse wave velocity in normal humans. J Hypertens. 2010;28:979–84.
17. Hering D, Kara T, Kucharska W, Somers VK, Narkiewicz K. Longitudinal tracking of muscle sympathetic nerve activity and its relationship with blood pressure in subjects with prehypertension. Blood Press. 2016;25:184–92.
18. Rojek M, Rajzer MW, Wojciechowska W, et al. Relationship among long-term aircraft noise exposure, blood pressure profile, and arterial stiffness. J Hypertens. 2019;37:1350–8.
19. Wojciechowska W, Januszewicz A, Drozdz T, et al. Blood pressure and arterial stiffness in association with aircraft noise exposure:long-term observation and potential effect of COVID-19 lockdown. Hypertension. 2022;79:325–34.
20. Hering D, Lambert EA, Marusic P, et al. Substantial reduction in single sympathetic nerve firing after renal denervation in patients with resistant hypertension. Hypertension. 2013;61:457–64.
21. Hering D, Marusic P, Walton AS, et al. Sustained sympathetic and blood pressure reduction 1 year after renal denervation in patients with resistant hypertension. Hypertension. 2014;64:118–24.
22. Calhoun DA, Jones D, Textor S, et al. Resistant hypertension: diagnosis, evaluation, and treatment. A scientific statement from the American Heart Association Professional Education Committee of the Council for high blood pressure research. Hypertension. 2008;51:1403–19.
23. Bertoia ML, Waring ME, Gupta PS, Roberts MB, Eaton CB. Implications of new hypertension guidelines in the United States. Hypertension. 2012;60:639–44.
24. Narkiewicz K, Grassi G, Mancia G, Hedner T. The sympathetic nervous system and cardiovascular disease. Gdansk: Gdańsk Via Medica; 2008.
25. Parmer RJ, Cervenka JH, Stone RA. Baroreflex sensitivity and heredity in essential hypertension. Circulation. 1992;85:497–503.
26. Rea RF, Hamdan M. Baroreflex control of muscle sympathetic nerve activity in borderline hypertension. Circulation. 1990;82:856–62.
27. Trzebski A, Tafil M, Zoltowski M, Przybylski J. Increased sensitivity of the arterial chemoreceptor drive in young men with mild hypertension. Cardiovasc Res. 1982;16:163–72.
28. Somers VK, Mark AL, Abboud FM. Potentiation of sympathetic nerve responses to hypoxia in borderline hypertensive subjects. Hypertension. 1988;11:608–12.
29. Sinski M, Lewandowski J, Przybylski J, et al. Tonic activity of carotid body chemoreceptors contributes to the increased sympathetic drive in essential hypertension. Hypertens Res. 2012;35:487–91.
30. Mancia G, Dell'Oro R, Quarti-Trevano F, Scopelliti F, Grassi G. Angiotensin-sympathetic system interactions in cardiovascular and metabolic disease. J Hypertens Suppl. 2006;24:S51–6.
31. Ferrier C, Esler MD, Eisenhofer G, et al. Increased norepinephrine spillover into the jugular veins in essential hypertension. Hypertension. 1992;19:62–9.

32. Lambert GW, Kaye DM, Thompson JM, et al. Internal jugular venous spillover of noradrenaline and metabolites and their association with sympathetic nervous activity. Acta Physiol Scand. 1998;163:155–63.
33. Ferrier C, Jennings GL, Eisenhofer G, et al. Evidence for increased noradrenaline release from subcortical brain regions in essential hypertension. J Hypertens. 1993;11:1217–27.
34. Alosco ML, Gunstad J, Xu X, et al. The impact of hypertension on cerebral perfusion and cortical thickness in older adults. J Am Soc Hypertens. 2014;8:561–70.
35. Naumczyk P, Sabisz A, Witkowska M, et al. Compensatory functional reorganization may precede hypertension-relatedbrain damage and cognitive decline: a functional magnetic resonance imaging study. J Hypertens. 2017;35:1252–62.
36. Drazner MH. The progression of hypertensive heart disease. Circulation. 2011;123:327–34.
37. Levy D, Larson MG, Vasan RS, Kannel WB, Ho KK. The progression from hypertension to congestive heart failure. JAMA. 1996;275:1557–62.
38. Mancia G, Grassi G, Giannattasio C, Seravalle G. Sympathetic activation in the pathogenesis of hypertension and progression of organ damage. Hypertension. 1999;34:724–8.
39. Schlaich MP, Kaye DM, Lambert E, Sommerville M, Socratous F, Esler MD. Relation between cardiac sympathetic activity and hypertensive left ventricular hypertrophy. Circulation. 2003;108:560–5.
40. Grassi G, Seravalle G, Quarti-Trevano F, et al. Sympathetic and baroreflex cardiovascular control in hypertension-related left ventricular dysfunction. Hypertension. 2009;53:205–9.
41. Grassi G, Seravalle G, Cattaneo BM, et al. Sympathetic activation and loss of reflex sympathetic control in mild congestive heart failure. Circulation. 1995;92:3206–11.
42. Ferguson DW, Berg WJ, Sanders JS. Clinical and hemodynamic correlates of sympathetic nerve activity in normal humans and patients with heart failure: evidence from direct microneurographic recordings. J Am Coll Cardiol. 1990;16:1125–34.
43. Hasking GJ, Esler MD, Jennings GL, Burton D, Johns JA, Korner PI. Norepinephrine spillover to plasma in patients with congestive heart failure: evidence of increased overall and cardiorenal sympathetic nervous activity. Circulation. 1986;73:615–21.
44. Bristow MR. Beta-adrenergic receptor blockade in chronic heart failure. Circulation. 2000;101:558–69.
45. Lamba S, Abraham WT. Alterations in adrenergic receptor signaling in heart failure. Heart Fail Rev. 2000;5:7–16.
46. Colucci WS. The effects of norepinephrine on myocardial biology: implications for the therapy of heart failure. Clin Cardiol. 1998;21:I20–4.
47. Kaye DM, Lefkovits J, Jennings GL, Bergin P, Broughton A, Esler MD. Adverse consequences of high sympathetic nervous activity in the failing human heart. J Am Coll Cardiol. 1995;26:1257–63.
48. Petersson M, Friberg P, Eisenhofer G, Lambert G, Rundqvist B. Long-term outcome in relation to renal sympathetic activity in patients with chronic heart failure. Eur Heart J. 2005;26:906–13.
49. Setoguchi S, Stevenson LW. Hospitalizations in patients with heart failure: who and why. J Am Coll Cardiol. 2009;54:1703–5.
50. Schrier RW, Abraham WT. Hormones and hemodynamics in heart failure. N Engl J Med. 1999;341:577–85.
51. Weber KT. Mechanisms of disease—aldosterone in congestive heart failure. N Engl J Med. 2001;345:1689–97.
52. Ferguson DW, Abboud FM, Mark AL. Selective impairment of baroreflex-mediated vasoconstrictor responses in patients with ventricular dysfunction. Circulation. 1984;69:451–60.
53. Mark AL. Sympathetic dysregulation in heart failure: mechanisms and therapy. Clin Cardiol. 1995;18:I3–8.
54. Mortara A, La Rovere MT, Pinna GD, et al. Arterial baroreflex modulation of heart rate in chronic heart failure: clinical and hemodynamic correlates and prognostic implications. Circulation. 1997;96:3450–8.
55. Somers VK, Mark AL, Abboud FM. Interaction of baroreceptor and chemoreceptor reflex control of sympathetic nerve activity in normal humans. J Clin Invest. 1991;87:1953–7.

56. van de Borne P, Oren R, Anderson EA, Mark AL, Somers VK. Tonic chemoreflex activation does not contribute to elevated muscle sympathetic nerve activity in heart failure. Circulation. 1996;94:1325–8.
57. Narkiewicz K, Pesek CA, van de Borne PJ, Kato M, Somers VK. Enhanced sympathetic and ventilatory responses to central chemoreflex activation in heart failure. Circulation. 1999;100:262–7.
58. Franchitto N, Despas F, Labrunee M, et al. Tonic chemoreflex activation contributes to increased sympathetic nerve activity in heart failure-related anemia. Hypertension. 2010;55:1012–7.
59. Despas F, Lambert E, Vaccaro A, et al. Peripheral chemoreflex activation contributes to sympathetic baroreflex impairment in chronic heart failure. J Hypertens. 2012;30:753–60.
60. Yumino D, Wang H, Floras JS, et al. Prevalence and physiological predictors of sleep apnea in patients with heart failure and systolic dysfunction. J Card Fail. 2009;15:279–85.
61. Ponikowski P, Chua TP, Anker SD, et al. Peripheral chemoreceptor hypersensitivity: an ominous sign in patients with chronic heart failure. Circulation. 2001;104:544–9.
62. Shepherd JT. Heart failure: role of cardiovascular reflexes. Cardioscience. 1990;1:7–12.
63. Cohn JN, Levine TB, Olivari MT, et al. Plasma norepinephrine as a guide to prognosis in patients with chronic congestive heart failure. N Engl J Med. 1984;311:819–23.
64. Brunner-La Rocca HP, Esler MD, Jennings GL, Kaye DM. Effect of cardiac sympathetic nervous activity on mode of death in congestive heart failure. Eur Heart J. 2001;22:1136–43.
65. Krum H, Schlaich M, Whitbourn R, et al. Catheter-based renal sympathetic denervation for resistant hypertension: a multicentre safety and proof-of-principle cohort study. Lancet. 2009;373:1275–81.
66. Krum H, Schlaich MP, Sobotka PA, et al. Percutaneous renal denervation in patients with treatment-resistant hypertension: final 3-year report of the Symplicity HTN-1 study. Lancet. 2014;383:622–9.
67. Esler MD, Bohm M, Sievert H, et al. Catheter-based renal denervation for treatment of patients with treatment-resistant hypertension: 36 month results from the SYMPLICITY HTN-2 randomized clinical trial. Eur Heart J. 2014;35:1752–9.
68. Bohm M, Mahfoud F, Ukena C, et al. First report of the global SYMPLICITY registry on the effect of renal artery denervation in patients with uncontrolled hypertension. Hypertension. 2015;65:766–74.
69. Bhatt DL, Kandzari DE, O'Neill WW, et al. A controlled trial of renal denervation for resistant hypertension. N Engl J Med. 2014;370:1393–401.
70. Desch S, Okon T, Heinemann D, et al. Randomized sham-controlled trial of renal sympathetic denervation in mild resistant hypertension. Hypertension. 2015;65:1202–8.
71. Mathiassen ON, Vase H, Bech JN, et al. Renal denervation in treatment-resistant essential hypertension. A randomized, SHAM-controlled, double-blinded 24-h blood pressure-based trial. J Hypertens. 2016;34:1639–47.
72. Azizi M, Sapoval M, Gosse P, et al. Optimum and stepped care standardised antihypertensive treatment with or without renal denervation for resistant hypertension (DENERHTN): a multicentre, open-label, randomised controlled trial. Lancet. 2015;385:1957–65.
73. Barbato E, Azizi M, Schmieder RE, et al. Renal denervation in the management of hypertension in adults. A clinical consensus statement of the ESC Council on hypertension and the European Association of Percutaneous Cardiovascular Interventions (EAPCI). Eur Heart J. 2023;44:1313.
74. Townsend RR, Mahfoud F, Kandzari DE, et al. Catheter-based renal denervation in patients with uncontrolled hypertension in the absence of antihypertensive medications (SPYRAL HTN-OFF MED): a randomised, sham-controlled, proof-of-concept trial. Lancet. 2017;390:2160–70.
75. Kandzari DE, Bohm M, Mahfoud F, et al. Effect of renal denervation on blood pressure in the presence of antihypertensive drugs: 6-month efficacy and safety results from the SPYRAL HTN-ON MED proof-of-concept randomised trial. Lancet. 2018;391:2346–55.

76. Bohm M, Kario K, Kandzari DE, et al. Efficacy of catheter-based renal denervation in the absence of antihypertensive medications (SPYRAL HTN-OFF MED pivotal): a multicentre, randomised, sham-controlled trial. Lancet. 2020;395:1444–51.
77. Azizi M, Schmieder RE, Mahfoud F, et al. Endovascular ultrasound renal denervation to treat hypertension (RADIANCE-HTN SOLO): a multicentre, international, single-blind, randomised, sham-controlled trial. Lancet. 2018;391:2335–45.
78. Azizi M, Sanghvi K, Saxena M, et al. Ultrasound renal denervation for hypertension resistant to a triple medication pill (RADIANCE-HTN TRIO): a randomised, multicentre, single-blind, sham-controlled trial. Lancet. 2021;397:2476–86.
79. Mahfoud F, Kandzari DE, Kario K, et al. Long-term efficacy and safety of renal denervation in the presence of antihypertensive drugs (SPYRAL HTN-ON MED): a randomised, sham-controlled trial. Lancet. 2022;399:1401–10.
80. Azizi M, Daemen J, Lobo MD, et al. 12-month results from the unblinded phase of the RADIANCE-HTN SOLO trial of ultrasound renal denervation. JACC Cardiovasc Interv. 2020;13:2922–33.
81. Biffi A, Dell'Oro R, Quarti-Trevano F, et al. Effects of renal denervation on sympathetic nerve traffic and correlates in drug-resistant and uncontrolled hypertension: a systematic review and meta-analysis. Hypertension. 2023;80:659–67.
82. Davies JE, Manisty CH, Petraco R, et al. First-in-man safety evaluation of renal denervation for chronic systolic heart failure: primary outcome from REACH-pilot study. Int J Cardiol. 2013;162:189–92.
83. Hopper I, Gronda E, Hoppe UC, et al. Sympathetic response and outcomes following renal denervation in patients with chronic heart failure: 12-month outcomes from the Symplicity HF feasibility study. J Card Fail. 2017;23:702–7.
84. Xia Z, Han L, Pellegrino PR, et al. Safety and efficacy of renal denervation in patients with heart failure with reduced ejection fraction (HFrEF): a systematic review and meta-analysis. Heliyon. 2022;8:e08847.
85. Kresoja KP, Rommel KP, Fengler K, et al. Renal sympathetic denervation in patients with heart failure with preserved ejection fraction. Circ Heart Fail. 2021;14:e007421.
86. Scheffers IJ, Kroon AA, Schmidli J, et al. Novel baroreflex activation therapy in resistant hypertension: results of a European multi-center feasibility study. J Am Coll Cardiol. 2010;56:1254–8.
87. Bakris GL, Nadim MK, Haller H, Lovett EG, Schafer JE, Bisognano JD. Baroreflex activation therapy provides durable benefit in patients with resistant hypertension: results of long-term follow-up in the Rheos pivotal trial. J Am Soc Hypertens. 2012;6:152–8.
88. Hoppe UC, Brandt MC, Wachter R, et al. Minimally invasive system for baroreflex activation therapy chronically lowers blood pressure with pacemaker-like safety profile: results from the Barostim neo trial. J Am Soc Hypertens. 2012;6:270–6.
89. Wallbach M, Halbach M, Reuter H, et al. Baroreflex activation therapy in patients with prior renal denervation. J Hypertens. 2016;34:1630–8.
90. Weipert KF, Most A, Dorr O, et al. Rescue baroreflex activation therapy after Stanford B aortic dissection due to therapy-refractory hypertension. J Am Soc Hypertens. 2016;10:490–2.
91. Floyd CN, Adeel MY, Wolff CB, et al. First-in-man treatment of severe blood pressure variability with baroreflex activation therapy. Int J Cardiol. 2016;220:577–9.
92. de Leeuw PW, Bisognano JD, Bakris GL, et al. Sustained reduction of blood pressure with baroreceptor activation therapy: results of the 6-year open follow-up. Hypertension. 2017;69:836–43.
93. Spiering W, Williams B, Van der Heyden J, et al. Endovascular baroreflex amplification for resistant hypertension: a safety and proof-of-principle clinical study. Lancet. 2017;390:2655–61.
94. Gronda E, Seravalle G, Brambilla G, et al. Chronic baroreflex activation effects on sympathetic nerve traffic, baroreflex function, and cardiac haemodynamics in heart failure: a proof-of-concept study. Eur J Heart Fail. 2014;16:977–83.

95. Gronda E, Seravalle G, Trevano FQ, et al. Long-term chronic baroreflex activation: persistent efficacy in patients with heart failure and reduced ejection fraction. J Hypertens. 2015;33:1704–8.
96. Narkiewicz K, Ratcliffe LE, Hart EC, et al. Unilateral carotid body resection in resistant hypertension: a safety and feasibility trial. JACC Basic Transl Sci. 2016;1:313–24.
97. Niewinski P, Janczak D, Rucinski A, et al. Carotid body removal for treatment of chronic systolic heart failure. Int J Cardiol. 2013;168:2506–9.

Renal and Heart Failure Interrelationship in Hypertension

9

Josep Redon

9.1 Introduction

There is mounting evidence that heart failure (HF) and chronic kidney disease (CKD) are conditions with a huge impact on healthcare [1]. A direct association exists between the development of HF and CKD, and both can be the cause and consequence of each other [2] (Fig. 9.1). Furthermore, well-known factors contribute to the development and progression of cardiac and renal dysfunction, such as age, hypertension (HTN) [3, 4], diabetes [5], and atherosclerotic vascular disease

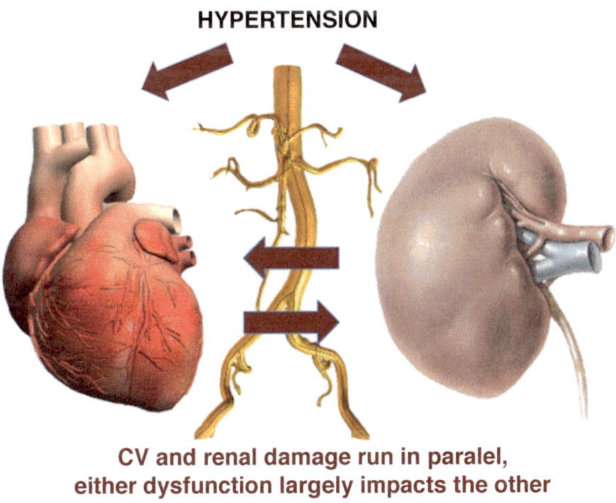

Fig. 9.1 Cardiac and renal cross talk

J. Redon (✉)
Cardiorenal Risk Research Group, INCLIVA Research Institute, University of Valencia, Valencia, Spain
e-mail: josep.redon@uv.es

M. Dorobantu et al. (eds.), *Hypertension and Heart Failure*, Updates in Hypertension and Cardiovascular Protection,
https://doi.org/10.1007/978-3-031-39315-0_9

[6, 7]. Increased urinary albumin excretion (UAE) and/or a reduced glomerular filtration rate (GFR), ventricular dysfunction, and atrial fibrillation all increase the risk of progression to the final stages of both HF and CKD. Heart and kidney functions are tightly connected, and an interplay between these two organs occurs through a vast array of dynamic and bidirectional mechanisms: hemodynamic parameters, neurohumoral signaling, endothelial injury, immune imbalance, inflammatory cascades, and oxidative stress.

Hypertension is a major risk factor for the development and progression of CKD, irrespective of the cause of CKD. In patients with CKD, resistant HTN, masked HTN, and elevated nighttime blood pressure (BP) are common and are associated with a lower GFR, higher levels of albuminuria, tubular dysfunction, and atrophy and interstitial fibrosis [8, 9]. Likewise, HTN is the leading risk factor for the development of HF [10], and most patients with HF will have an antecedent history of HTN. Heart failure may be a consequence of coronary artery disease, which results in HF with reduced ejection fraction (HFrEF). Hypertension also causes left ventricular hypertrophy (LVH), which impairs LV relaxation, so-called diastolic dysfunction, and is a potent predictor of HF even when LV systolic function is normal and there is no preceding myocardial infarction (HF with preserved ejection fraction (HFpEF)). Hypertension-dependent fibrosis and structural alteration of both the large and small arteries (microvascular disease) also contribute.

Likewise, it is relevant to mention diabetes in this scenario. Diabetes is the main cause of CKD and is a risk factor for their progression to end-stage renal disease (ESRD). Patients with diabetes and CKD have an extremely high risk of cardiovascular (CV) disease, as this is the main cause of morbidity and mortality.

The name "cardiorenal syndrome" indicates the bidirectional link between the heart and kidneys and includes five different subtypes based on the organ that initially suffers from claudication and also on whether this is acute or chronic. Although the criteria defining the five subtypes have been outlined, sometimes it is difficult to assign one specific class and overlaps exist. This chapter will help to provide a link between HTN, HF, and CKD.

9.2 Epidemiology

The estimated HF prevalence affects 3% of the general population aged more than 18 years, and, CKD, defined by an estimated glomerular filtration rate (eGFR) <60 mL/min/1.73 m^2, is around 7.5%, but both depend on age with increasing prevalence rates predominantly seen in the elderly. When both conditions occur together, 20–30% of individuals with HF have an eGFR <60 mL/min/1.73 m^2. Importantly, at least 30% of the population in the presence of decompensated HF has an eGFR <60 mL/min/1.73 m^2. On the other side, patients with stage 3 CKD, with an eGFR <60 mL/min/1.73 m^2, are at a threefold higher risk of developing HF. Acute kidney injury (AKI), defined as an abrupt decrease in kidney function, which encompasses both structural injury and loss of function [11], is identified in around 20%, and it has been recognized that AKI is a strong independent predictor of both in-hospital and 1-year mortality. When both HF and CKD are present, some comorbidities are more frequent, thus resulting in weight loss and cachexia.

9.3 Mechanisms of Heart Failure and Kidney Dysfunction

Several mechanisms contribute to damage in the heart and kidneys and accelerate progressive organ dysfunction. These mechanisms are the result of an interplay between hemodynamic, neurohormonal, and cardiovascular disease-associated mechanisms as well as local and systemic inflammation, activation of cellular immunity, oxidative stress, bone and mineral metabolism and acid–base metabolism, anemia, and sarcopenia.

9.3.1 Hemodynamics

Hypertension is one of the leading causes of renal and heart damage [12] (Fig. 9.2). The longer the duration and the higher the values, the more increased the risk. Heart failure impacts kidney damage through the so-called "forward failure" [13] or "congestion" [14]. The forward failure is produced when the ventricular function is not able to supply the necessary blood flow to maintain glomerular filtration, with the consequent activation of mechanisms of autoregulation, and the reduction on oxygen apport to the renal tubules leading to acute tubular necrosis. Hypoperfusion of

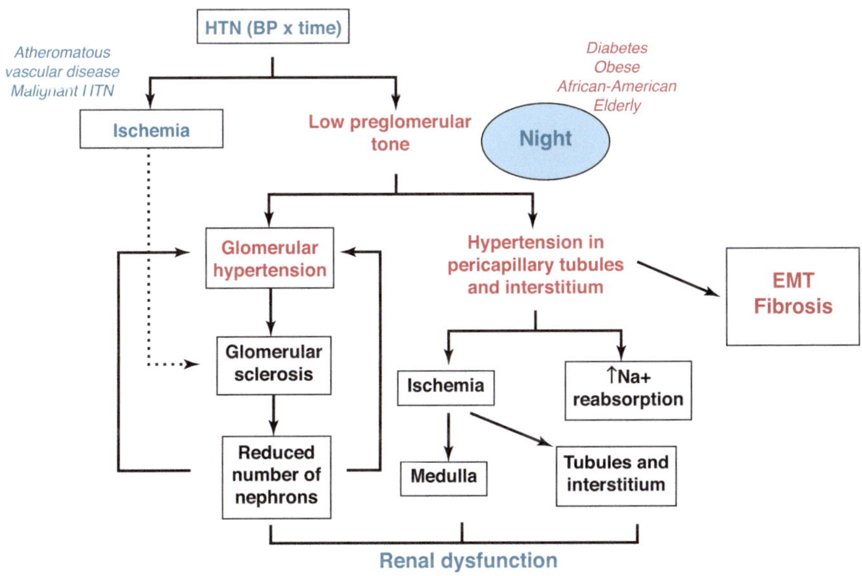

Fig. 9.2 Mechanisms of renal damage induced by hypertension. *EMT* endothelial mesenchymal transition

the kidneys leads to activation of the renin–angiotensin–aldosterone system (RAAS), thus resulting in vasoconstriction that affects the glomerulus and tubular structures. Congestion, defined as an increment in central venous pressure, acts as the outflow pressure of the blood through the kidneys and increases the pressure in them. This leads to a decline in the GFR as a consequence of decreasing transrenal perfusion, thus increasing the intraglomerular hydrostatic pressure and reducing the GFR. If the increased pressure is >15 mmHg, then the peritubular capillaries and intratubular pressures are also impacted.

In addition, a progressive decline in the GFR is one of the main factors increasing BP since both sodium handling and the renal–angiotensin system are disturbed. Hypertension induced under such circumstances causes left ventricular overload, thus inducing hypertrophy, fibrosis, and dysfunction.

9.3.2 Neurohumoral Mechanisms

Both the RAAS and the sympathetic nervous system (SNS) are activated in the presence of HF and/or CKD (Fig. 9.3) [15]. Stimulation of the RAAS increases sodium retention and vasoconstriction in the kidneys as well as leads to an increment in end-diastolic ventricular pressure, augmented oxidative stress, increased production of pro-inflammatory cytokines, phenotypic changes in cell-mediated immunity, and acceleration of renal and cardiac remodeling via mechanisms that include the promotion of fibrosis. At the same time, macrophage-derived galectin-3 is stimulated, which reinforces the impact on cardiac and renal fibrosis. Sodium retention increases volume and the secretion of arginine vasopressin that inhibits free water clearance. Overactivity of the SNS at the systemic level and in the renal nerves together with increment of asymmetric dimethylarginine (ADMA) are other mechanisms that contribute to the progressive decline of both heart and kidney functions.

Fig. 9.3 Hypertension in ventricular dysfunction

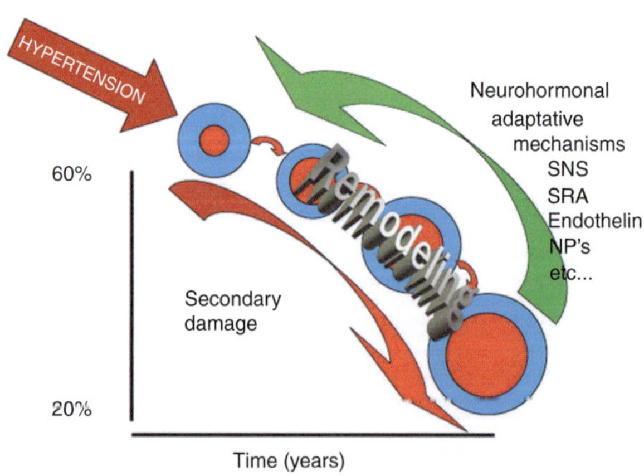

9.3.3 Cardiovascular-Associated Mechanisms

Among the factors outlined above, chronic inflammation, anemia, and bone and mineral disorders play central roles. "Chronic low-grade inflammation" causes damage to both cardiovascular structures and the kidneys and is a disease marker. The factors that trigger inflammation are multiple, including endothelial activation, oxidative stress, obesity, smoking dietary, microbiota, and lifestyle-related factors together with those that arise from renal dysfunction such as uremic toxins and dialysis. Consequently, circulating biomarkers progressively increase when the renal function declines. However, changes in innate and adaptive immunity also play a role in both vascular and renal damage. Besides the impact on the organs, it also has an impact on nutritional status and body composition with progressive development of sarcopenia. "Cardiorenal anemia," a form of normochromic normocytic anemia, is induced by a reduction in erythropoietin production, is not infrequent in patients with renal dysfunction, and contributes to the progression of ventricular dysfunction, decrease in the quality of life, and increase in mortality. Bone and mineral disorders are frequent in patients with stage 3 CKD, often induced by a reduction in phosphate excretion. Hyperphosphatemia causes an increase in parathyroid hormone secretion and fibroblast growth factor 23 and also vitamin D deficiency. All of these reduce intestinal calcium absorption and are associated with cardiovascular and left ventricular mass toxicity.

Finally, genetic factors may predispose to renal and heart dysfunction; however, epigenetics also plays a pivotal role. Although the cross talk between the heart and kidneys is clearly evidenced with the abovementioned mechanism, there is a lack of studies simultaneously analyzing cellular mechanisms predisposed to reduce function overtime.

9.4 Clinically Relevant Issues

9.4.1 Comorbidities

The importance of comorbidities in the evolution and prognosis of HF has been emphasized [16], and several studies have tried to identify the factors related to hospitalization and prognosis [17, 18]. Besides ischemic heart disease, other factors such as noncompliance to the prescribed treatments or diet, infections, arrhythmias, uncontrolled hypertension, anemia, and renal dysfunction are among the most common reasons for AHF, progression, and mortality [19, 20].

9.4.2 Atrial Fibrillation (AF) and Progression in Renal Dysfunction

The association of AF with a reduction of eGFR by >20% was analyzed in a large cohort of HF patients [21]. Among 128,000 patients with HF, 18,513 had significantly reduced eGFR values, and the incidence was 66.2/1000 patients/year with

AF increasing the risk (hazard ratio (HR) 1.13; 95% confidence interval (95% CI): 1.09–1.18; $p < 0.001$) more so than the presence of diabetes, HTN, dyslipidemia, and coronary heart disease. The association with risk increase was also demonstrated in a cohort of patients with recently diagnosed HF. Among 55,000 patients, the number of subjects with decreased eGFR values was 7067, the incidence was 94.0/1000 patients/year with AF increasing the risk (HR 1.22; 95% CI: 1.14–1.31; $p < 0.001$) independent of the other potential factors. For a subgroup of patients classified according to the Kidney Disease Improving Global Outcomes (KDIGO) risk guidelines, 10,425 (27.8%) of them in the entire cohort and 3297 (23.2%) with a recent diagnosis, the risk category increased by at least one level.

The mechanism in reducing renal function in AF may be related due the hemodynamic changes associated to AF which will impact by reducing the renal blood flow or by congestion in the venous territory.

9.4.3 Acute Kidney Injury

Acute kidney injury (AKI), defined as an abrupt decrease in kidney function, is a cardiorenal syndrome, which encompasses both structural injury and loss of function [22]. It is a syndrome of multiple etiologies associated with an increased risk of hospitalization and high mortality [23]. In patients with HF, AKI is a frequent event in which the hemodynamic status, low cardiac output or congestive status, and the impact of drugs, mainly diuretics and renin–angiotensin system blockade, are relevant factors. As expected, the incidence increases in patients with an eGFR <60 mL/min/1.73 m^2 owing to the lower renal functional reserve caused by the HF itself or owing to age and comorbidities that increase the risk of renal functional deterioration. The more reduced eGFR increases not only the incidence but also the severity and reduces the lag period of time free of AKI. It has been recognized that even patients who have complete or near-complete kidney function recovery are at an increased risk of progressive CKD and that superimposition of AKI on CKD is associated with acceleration in the rate of progression to end-stage renal disease (ERSD) [24].

The incidence and impact of AKI have been mainly reported in subjects hospitalized with acute heart failure, with the prevalence of AKI around 20%; thus, it is now recognized that AKI is a strong independent predictor of both in-hospital and 1-year mortality [25, 26].

A meta-analysis of cohorts, registries, and post hoc studies concluded that CKD and "worsening of renal function (WRF)" [12], a term that has been used instead of AKI in HF patients [27], are frequently observed in patients with acute and chronic heart failure. However, the authors acknowledge heterogeneity due to different inclusion criteria, selection biases, different definitions, and criteria used to qualify AKI. Indeed, the incidence, prevalence, and consequences of AKI in patients with HF are not well-established.

9.4.4 Anemia

The relevance of anemia in the natural history and prognosis of HF has been recognized [28, 29]. Well-established as a prognostic factor, the prevalence is more than 25% in all patients with HF and even greater in older adults [30]. Moreover, a progressive reduction of hemoglobin (Hb) implies a poor prognosis [31, 32]. The origin of anemia in HF is multifactorial and includes fluid retention, CKD, undernutrition, relative low levels of erythropoietin, inflammation, and drugs blocking the renin–angiotensin system that reduces erythropoietin, erythroid progenitors, and the degradation of an inhibitor of erythropoiesis [30, 33]. In addition, blood loss can contribute to significant reduction in Hb levels. Antiplatelet and anticoagulant treatment, with or without gastrointestinal lesions [34], as well as AKI, induce episodes of rapid Hb reduction [26]. This Hb reduction not only impacts the comorbidities that are frequently present in HF patients but also increases myocardial workload [35].

It is well-known that anemia is the worst predictor in patients with HF, with reduced [36] or preserved [37] ejection fraction. Likewise, anemia reduces the exercise capacity of HF patients and has been linked to more elevated levels of proN-BNP and high-sensitivity troponin (hs-troponin) [37]. Episodes of Hb reduction increase the risk of AHF hospitalization and all-cause mortality [38].

9.5 Treatment

In individuals with both HF and renal dysfunction, a combined treatment for both the conditions should be administered, although in those in whom acute or chronic heart disease predominates, the main treatment should address the heart disease; in contrast, if the principal condition is acute or chronic renal disease, then this should be the first focus of treatment.

Overall, "prevention" is the first step. One should take care to avoid nephrotoxic drugs (certain antibiotics, contrast media, nonsteroidal anti-inflammatory drugs), control BP and lipid levels, prevent diabetes and anemia, avoid malnutrition and smoking, and prevent obesity. Salt reduction is recommended to less than 5 g/day, even though there is no agreement in different studies. Treatment of HTN is a cornerstone in the prevention of both HF and CKD. The RAAS is recommended since renal and heart protection has been described with their use. A combination of a RAAS blocker with a calcium channel blocker and/or a diuretic is recommended to reduce BP levels to <130/80 mmHg but not 120/70 mmHg. If a reduction in the GFR is observed at the time of starting treatment, the percentage of the reduction should be taken into account. Reductions below 20% indicate a decrease in the filtration fraction, but it is beneficial in the long term. Whatever the case, after initiation of treatment, both the GFR and potassium levels should be monitored [39].

The most frequently used drugs for HF treatment are beta-blockers, angiotensin-converting enzyme inhibitors (ACEis), angiotensin receptor blockers (ARBs), and sacubitril/valsartan. Recently, sodium-glucose cotransporter-2 inhibitors (SGLT2is), a class of glucosuric drugs initially introduced in the treatment of diabetes, have demonstrated renal protection and improve function and survival in HF patients with HFrEF and HFpEF [40, 41]. Diuretics, which reduce the expansion of extracellular fluid volume, are a cornerstone of treatment, mainly in those at stage 4 with a severely reduced renal function; however, diuretic-resistant states are a big issue in the treatment of patients. Loop diuretics, thiazides or thiazide-like, alone or in combination, are the most used [42]; but they are central. Despite their wide use, they have steep dose–response curves, reducing the effect when the threshold is reached. Approaches such a combination with a diuretic of proximal tubule action or a subcutaneous administration of furosemide may improve the diuretic action.

Mineralocorticoid receptor antagonists (MRAs) with or without a steroidal structure are drugs that can be used to provide protection against HF and CKD, mainly in diabetic patients. Steroidal blockers of MRAs have demonstrated benefits in reducing proteinuria and CV events in HFrEF [43], but hyperkaliemia and hormonal effects are limiting factors in clinical use. Furthermore, they play a role in the treatment of resistant HTN [44]. A non-steroidal potent MRAs blocker, finerenone, inhibits recruitment of transcriptional cofactors binding less hypertrophy, proinflammatory and profibrotic gene expression and induces a dose-dependent moderately increased of K^+ retention. Phase II program demonstrated efficacy and safety to reduce HF cardiac biomarkers, ARTS-HF, and albuminuria in T2DM, ARTS-DM [45]. Phase III has been developed with the FIDELIO [46] and FIGARO [47] trials. The first released, FIDELIO, demonstrated a favorable effect on cardiorenal outcomes in T2DM patients with predominantly stages 3 and 4 CKD.

FIGARO was designed to explore the beneficial impact in patients with T2DM to reduce CV outcomes, at stages 1 and 2, with severely elevated albuminuria, or at stages 3 and 4 with moderately elevated albuminuria. This is a double-blind clinical trial, in which 7437 patients were randomized to receive finerenone or placebo on top of renin–angiotensin system blockers, adjusted to the maximal dose without unacceptable side effects. In an average of 3.4-years of follow-up, the primary outcome (death for a CV event, nonfatal myocardial infarction, nonfatal stroke, or hospitalization due to heart failure) was significantly reduced (HR 0.87; $p = 0.03$; absolute risk −1.8%), mainly toward HF hospitalization expenses. The impact on the secondary end point (kidney failure, sustained decrease from the baseline of at least 40% in the eGFR, or death for renal causes) was not significant (HR 0.87; p = NS; absolute risk 1.3%) [48], although a sustained decrease below 56% was significant. FIDELITY, a pooled analysis of the two trials, FIDELIO and FIGARO, offered a wide perspective of the impact of finerenone on the T2DM population with CKD [49]. Finerenone, a high-capacity blocker of MRAs, can be incorporated into the armamentarium of a highly prevalent disease such as diabetic nephropathy, mainly to reduce HF incidence and/or to decompensate, although the complex scenario in the mechanisms and therapeutic interventions requires a careful evaluation and close follow-up of patients when used [50].

Treatment of anemia is also an important part in the control of acidosis. Infusion of carboxymaltose iron improves cardiac dysfunction, although the impact on hemoglobin values is not relevant. At stages 4 and 5 of CKD, treatment with recombinant erythropoietin can be administered when the levels of hemoglobin are <9 g/dL [48, 51].

References

1. Tuegel C, Bansal N. Heart failure in patients with kidney disease. Heart. 2017;103:1848–53.
2. Schefold JC, Filippatos G, Hasenfuss G, Anker SD, von Haehling S. Heart failure and kidney dysfunction: epidemiology, mechanisms and management. Nat Rev Nephrol. 2016;12:610–23.
3. Lurbe E, Redon J, Kesani A, Pascual JM, Tacons J, Alvarez V, et al. Increase in nocturnal blood pressure and progression to microalbuminuria in type 1 diabetes. N Engl J Med. 2002;347:797–805.
4. Redon J, Tellez-Plaza M, Orozco-Beltran D, Gil-Guillen V, Pita Fernandez S, Navarro-Pérez J, et al. Impact of hypertension on mortality and cardiovascular disease burden in patients with cardiovascular risk factors from a general practice setting: the ESCARVAL-risk study. J Hypertens. 2016;34:1075–83.
5. Redon J, Mancia G, Sleight P, Schumacher H, Gao P, Pogue J, et al. Safety and efficacy of low blood pressures among patients with diabetes: subgroup analyses from the ONTARGET (ONgoing Telmisartan alone and in combination with Ramipril global endpoint trial). J Am Coll Cardiol. 2012;59:74–83.
6. Dubin RF, Whooley M, Pico A, Ganz P, Schiller NB, Meyer C. Proteomic analysis of heart failure hospitalization among patients with chronic kidney disease: the heart and soul study. PloS One. 2018;13:e0208042.
7. Kieswich JE, Chen J, Alliouachene S, Caton PW, McCafferty K, Thiemermann C, et al. A novel model of Reno-cardiac syndrome in the C57BL/ 6 mouse strain. BMC Nephrol. 2018;19:346.
8. Drawz PE, Alper AB, Anderson AH, Brecklin CS, Charleston J, Chen J, et al. Chronic renal insufficiency cohort study investigators. Masked hypertension and elevated nighttime blood pressure in CKD: prevalence and association with target organ damage. Clin J Am Soc Nephrol. 2016;11:642–52.
9. Rossignol P, Massy ZA, Azizi M, Bakris G, Ritz E, Covic A, et al. ERA-EDTA EURECA-m working group, red de Investigacion renal (REDINREN) network, cardiovascular and renal clinical Trialists (F-CRIN INI-CRCT) network. The double challenge of resistant hypertension and chronic kidney disease. Lancet. 2015;386:1588–98.
10. Rapsomaniki E, Timmis A, George J, Pujades-Rodriguez M, Shah AD, Denaxas S, et al. Blood pressure and incidence of twelve cardiovascular diseases: lifetime risks, healthy life-years lost, and age-specific associations in 1.25 million people. Lancet. 2014;383:1899–911.
11. Heywood JT, Fonarow GC, Costanzo MR, Mathur VS, Wigneswaran JR, Wynne J. High prevalence of renal dysfunction and its impact on outcome in 118,465 patients hospitalized with acute decompensated heart failure: a report from the ADHERE database. J Card Fail. 2007;13:422–30.
12. Fortrie G, de Geus HRH, Betjes MGH. The aftermath of acute kidney injury: a narrative review of long-term mortality and renal function. Crit Care. 2019;23:24.
13. Nohria A, Hasselblad V, Stebbins A, Pauly DF, Fonarow GC, Shah M, et al. Cardiorenal interactions: insights from the ESCAPE trial. J Am Coll Cardiol. 2008;51:1268–74.
14. Mullens W, Abrahams Z, Francis GS, Sokos G, Taylor DO, Starling RC, et al. Importance of venous congestion for worsening of renal function in advanced decompensated heart failure. J Am Coll Cardiol. 2009;53:589–96.
15. Clementi A, Virzi GM, Battaglia GG, Ronco C. Neurohormonal, endocrine, and immune dysregulation and inflammation in Cardiorenal syndrome. Cardiorenal Med. 2019;9:265–73.
16. Damman K, Perez AC, Anand IS, Komajda M, McKelvie RS, Zile MR, et al. Worsening renal function and outcome in heart failure patients with preserved ejection fraction and the impact of angiotensin receptor blocker treatment. J Am Coll Cardiol. 2014;64:1106–13.

17. Mentz RJ, O'Connor CM. Pathophysiology and clinical evaluation of acute heart failure. Nat Rev Cardiol. 2016;13:28–35.
18. Martínez F, Martínez-Ibañez L, Pichler G, Ruiz A, Redon J. Multimorbidity and acute heart failure in internal medicine. Int J Cardiol. 2017;232:208–15.
19. Murton M, Goff-Leggett D, Bobrowska A, Garcia Sanchez JJ, James G, Wittbrodt E, et al. Burden of chronic kidney disease by KDIGO categories of glomerular filtration rate and albuminuria: a systematic review. Adv Ther. 2021;38:180–200.
20. Löfman I, Szummer K, Dahlström U, Jernberg T, Lund LH. Associations with and prognostic impact of chronic kidney disease in heart failure with preserved, mid-range, and reduced ejection fraction. Eur J Heart Fail. 2017;19:1606–14.
21. Diaz J, Martinez F, Calderon JM, Fernandez A, Sauri I, Uso R, et al. Incidence and impact of atrial fibrillation in heart failure patients: real-world data in a large community. ESC Heart Fail. 2022;9:4230. https://doi.org/10.1002/ehf2.14124.
22. Schytz PA, Blanche P, Nissen AB, Torp-Pedersen C, Gislason GH, Nelveg-Kristensen KE, et al. Acute kidney injury and risk of cardiovascular outcomes: a nationwide cohort study. Nefrologia (Engl Ed). 2022;42:338–46.
23. Gansevoort RT, Correa-Rotter R, Hemmelgarn BR, Jafar TH, Heerspink HJ, Mann JF, et al. Chronic kidney disease and cardiovascular risk: epidemiology, mechanisms, and prevention. Lancet. 2013;382:339–52.
24. Go AS, Hsu CY, Yang J, Tan TC, Zheng S, Ordonez JD, Liu KD. Acute kidney injury and risk of heart failure and atherosclerotic events. Clin J Am Soc Nephrol. 2018;13:833–41.
25. Pickering JW, Blunt IRH, Than MP. Acute kidney injury and mortality prognosis in acute coronary syndrome patients: a meta-analysis. Nephrology. 2018;23:237–46.
26. Makris K, Spanou L. Acute kidney injury: definition, pathophysiology and clinical phenotypes. Clin Biochem Rev. 2016;37:85–98.
27. Roger VL. Epidemiology of heart failure. Circ Res. 2013;113:646–59.
28. Ponikowski P, Voors AA, Anker SD, Bueno H, Cleland JGF, Coats AJS, et al. ESC guidelines for the diagnosis and treatment of acute and chronic heart failure: the task force for the diagnosis and treatment of acute and chronic heart failure of the European Society of Cardiology (ESC). Developed with the special contribution of the heart failure association (HFA) of the ESC. Eur Heart J. 2016;37:2129–200.
29. Anand IS, Gupta P. Anemia and iron deficiency in heart failure. Circulation. 2018;138:90–8.
30. Goodnough LT, Schrier SL. Evaluation and management of anemia in the elderly. Am J Hematol. 2014;89:88–96.
31. Tang YD, Katz SD. The prevalence of anemia in chronic heart failure and the impact on the clinical outcomes. Heart Fail Rev. 2008;13:387–92.
32. van der Meer P, Postmus D, Ponikowski P, Cleland JG, O'Connor CM, Cotter G, et al. The predictive value of short-term changes in hemoglobin concentration in patients presenting with acute decompensated heart failure. J Am Coll Cardiol. 2013;61:1973–81.
33. Magrì D, De Martino F, Moscucci F, Agostoni P, Sciomer S. Anemia and iron deficiency in heart failure: clinical and prognostic role. Heart Fail Clin. 2019;15:359–69.
34. Lopez A, Cacoub F, Macdougall C, Peyrin-Broulet L. Iron deficiency anaemia. Lancet. 2016;387:907–16.
35. Chung YJ, Luo A, Park KC, Loonat AA, Lakhal-Littleton S, Robbins PA, Swietach P. Iron-deficiency anemia reduces cardiac contraction by downregulating RyR2 channels and suppressing SERCA pump activity. JCI Insight. 2019;4:4.
36. Komajda M, Anker SD, Charlesworth A, Okonko D, Metra M, Di Lenarda A, et al. The impact of new onset anaemia on morbidity and mortality in chronic heart failure: results from COMET. Eur Heart J. 2006;27:1440–6.
37. Parcha V, Patel N, Kalra R, Bhargava A, Prabhu SD, Arora G, Arora P. Clinical, demographic, and imaging correlates of anemia in heart failure with preserved ejection fraction (from the RELAX trial). Am J Cardiol. 2020;125:1870–8.
38. Lopez C, Holgado JL, Fernandez A, Sauri I, Uso R, Trillo JL, et al. Impact of acute hemoglobin falls in heart failure patients: a population study. J Clin Med. 2020;9:1869.

39. Williams B, Mancia G, Spiering W, Agabiti Rosei E, Azizi M, Burnier M, et al. Practice guidelines for the management of arterial hypertension of the European Society of Cardiology and the European Society of Hypertension. Blood Press. 2018;27:314–40.
40. McCullough PA, Rahimi G, Tecson KM. Ambulatory worsening of renal function in heart failure with preserved ejection fraction. J Am Coll Cardiol. 2021;77:1222–4.
41. Zhou J, Liew D, Kaye DM, Zoungas S, Stub D. Cost-effectiveness of Empagliflozin in patients with heart failure and preserved ejection fraction. Circ Cardiovasc Qual Outcomes. 2022;15:e008638.
42. Jentzer JC, Bihorac A, Brusca SB, Del Rio-Pertuz G, Kashani K, et al. Contemporary management of severe acute kidney injury and refractory cardiorenal syndrome: JACC council perspectives. J Am Coll Cardiol. 2020;76:1084–101.
43. Pitt B, Zannad F, Remme WJ, Cody R, Castaigne A, Perez A, Palensky J, Wittes J. The effect of spironolactone on morbidity and mortality in patients with severe heart failure. Randomized Aldactone evaluation study investigators. N Engl J Med. 1999;341:709–17.
44. Williams B, MacDonald TM, Morant S, Webb DJ, Sever P, McInnes G, et al. British hypertension Society's PATHWAY studies group. Spironolactone versus placebo, bisoprolol, and doxazosin to determine the optimal treatment for drug-resistant hypertension (PATHWAY-2): a randomised, double-blind, crossover trial. Lancet. 2015;386:2059–68.
45. Agarwal R, Anker SD, Bakris G, Filippatos G, Pitt B, Rossing P, et al. FIDELIO-DKD and FIGARO-DKD investigators. Investigating new treatment opportunities for patients with chronic kidney disease in type 2 diabetes: the role of finerenone. Nephrol Dial Transplant. 2022;37:1014–23.
46. Bakris GL, Agarwal R, Anker SD, Pitt B, Ruilope LM, Rossing P, et al. Effect of Finerenone on chronic kidney disease outcomes in type 2 diabetes. N Engl J Med. 2020;383:2219–29.
47. Pitt B, Filippatos G, Agarwal R, Anker SD, Bakris GL, Rossing P, et al. Cardiovascular events with Finerenone in kidney disease and type 2 diabetes. N Engl J Med. 2021;385:2252–63.
48. Cattadori G, Agostoni P, Corrà U, Sinagra G, Veglia F, Salvioni E, et al. Heart failure and anemia: effects on prognostic variables. Eur J Intern Med. 2017;37:56–63.
49. Agarwal R, Filippatos G, Pitt B, Anker SD, Rossing P, Joseph A, et al. Cardiovascular and kidney outcomes with finerenone in patients with type 2 diabetes and chronic kidney disease: the FIDELITY pooled analysis. Eur Heart J. 2022;43:474–84.
50. Redon J. New insights of cardiovascular and renal protection in diabetic chronic kidney disease with finerenone. Cardiovasc Res. 2022;118:e36–7.
51. Ambrosy AP, Gurwitz JH, Tabada GH, Artz A, Schrier S, Rao SV, et al. Incident anaemia in older adults with heart failure: rate, aetiology, and association with outcomes. Eur Heart J Qual Care Clin Outcomes. 2019;5:361–9.

The Role of Uric Acid in Hypertension and Heart Failure

10

Claudio Borghi, Alexander Adorno, Ilaria Gallelli, and Massimo Dall'Olio

10.1 The Pathophysiology of Hyperuricemia and Cardiovascular Diseases

Uric acid (UA) is the final product of the degradation of purines. It was considered a potent antioxidant agent acting as a free radical scavenger, until the discovery that its increased serum level (SUA) and tissue deposition as a crystal was the cause of gout, the most prevalent form of adult arthritis [1]. The possible causative role of elevated SUA level in the risk of cardiovascular diseases (CVDs) has been known for more than a century [2]. Very recently, the 2018 ESC/ESH Guidelines for the management of hypertension as well as the 2020 ISH Guidelines for Hypertension have included UA among the additional cardiovascular (CV) risk factors in patients with hypertension suggesting the importance of measuring urate levels for the quantification of CV risk [3]. Hyperuricemia affects approximately 20% of the general population and more than 25% of unselected hospitalized patients [4, 5]. A recent paper on the results of the EMPEROR REDUCED study has reported a prevalence exceeding 50% in patients with heart failure [6]. Plasma UA level is the result of a balance between food intake, endogenous purine degradation, and the UA excretion that involves the gut and kidney. In terms of physiology, 25% of UA clearance occurs in the intestine and 75% of UA elimination is via the kidney. Under normal

C. Borghi (✉)
Department of Medical and Surgical Sciences, IRCCS Azienda Ospedaliero-Universitaria di Bologna, University of Bologna, Bologna, Italy

UOC Medicina Cardiovascolare, Ospedale S.Orsola-Malpighi, Bologna, Italy
e-mail: claudio.borghi@unibo.it

A. Adorno · I. Gallelli · M. Dall'Olio
Department of Medical and Surgical Sciences, IRCCS Policlinico S.Orsola, University of Bologna, Bologna, Italy
e-mail: ilaria.gallelli@aosp.bo.it; massimo.dallolio@aosp.bo.it

M. Dorobantu et al. (eds.), *Hypertension and Heart Failure*, Updates in Hypertension and Cardiovascular Protection,
https://doi.org/10.1007/978-3-031-39315-0_10

physiological circumstances, only 5–10% of the filtered UA is excreted [7, 8]. Serum uric acid levels are mainly related to cellular turnover that provides a large quantity of nucleic acids involved in the metabolism of UA. Dietary intake of protein- and fructose-rich foods can contribute to 10–12% of circulating UA. Xanthine oxidoreductase (XOR) is the key enzyme that converts hypoxanthine to xanthine as well as xanthine to UA. Reactive oxygen species (ROS) are produced during this conversion, and this supports the role of XOR activity as one of the main mechanisms involved in the development of CVD in patients with gout or asymptomatic hyperuricemia [1, 9]. A recent paper has described a close relationship between the increase in the XOR activity and the RR of major cardiovascular diseases [10]. This is more evident in patients with heart failure with a significant increase in mortality and HF hospitalization in the presence of higher levels of circulating XO [11]. Hyperuricemia may, therefore, result from an increased production of UA or by an impaired excretion or by the combination of both mechanisms. Reduced urinary excretion of UA is a more prevalent etiology of H-SUA in patients with gout, while the mechanisms of overproduction of UA can be particularly active in patients where who complain CV disease [9]. This mechanism is crucial in patients with heart failure (HF), according to the evidence of an overexpression of XO activity associated with impaired left ventricular function [12, 13]. An additional contribution to cardiovascular disease can be offered by the cellular entry of UA levels that is mediated by two main carrier systems: the urate transporter 1 (URAT1) and the glucose transporter 9 (GLUT9) that are involved in the renal tubular handling as well as in the intracellular transport of urate. This results in a further activation of pro-oxidative pathways leading to endothelium-mediated vasoconstriction and cardiac dysfunction [9, 14–16]. Figure 10.1 illustrates the pathophysiological mechanisms linking hyperuricemia and CV disease.

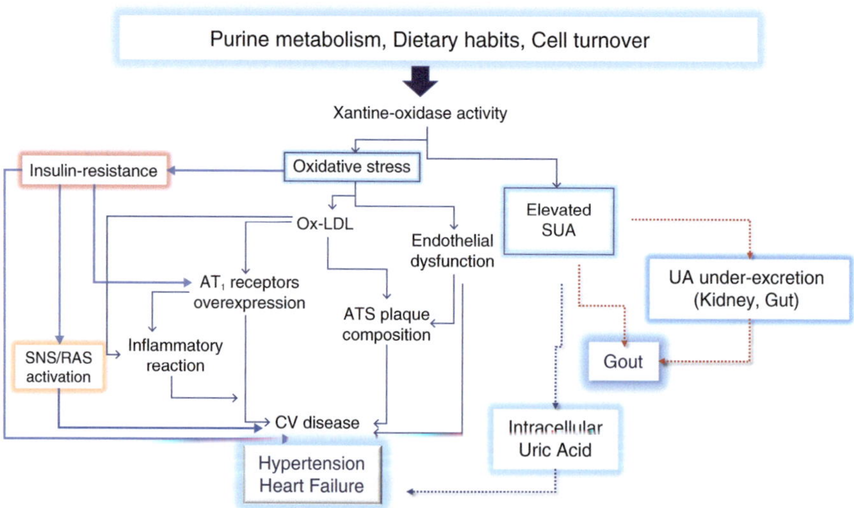

Fig. 10.1 Pathophysiological mechanisms linking hyperuricemia and CV disease

10.2 The Role of Inflammation in Gout and Cardiovascular Diseases

The most remarkable consequence of hyperuricemia is inflammation that is leading primarily to the development of gouty arthritis and can contribute to cardiovascular disease. Uric acid levels >6.5 mg/dL can promote deposit of monosodium urate (MSU) crystals responsible for and inflammatory response involving the macrophages through the activation of the NOD-, LRR-, and pyrin domain-containing protein 3 (NLRP3) inflammasome protein complex (interleukin-1 beta, tumor necrosis factor-alpha, interleukin-6, and interleukin-8) [17]. The proinflammatory response is also relevant in the development of cardiovascular diseases including hypertension and heart failure. Inflammation is associated with every step of the atherothrombotic process; patients with elevated levels of high-sensitivity C-reactive protein are known to be of a greater vascular risk. The increased levels of SUA can directly (inflammasome) or indirectly (oxidative stress) promote inflammatory response and negatively affect left ventricular function and the development of HF. An increase in the level of inflammatory response has been described in patients with heart failure [18], particularly in those with preserved ejection fraction where the activation of the inflammatory pathway has been considered among the causes of left ventricular stiffness leading to increased filling pressure and pulmonary congestion. The same inflammatory mechanism can cause a remodeling of the arterial wall and destabilizes atherosclerotic plaque [19] and may contribute to cardiovascular diseases in patients with hyperuricemia and hypertension. The level of oxidative stress and inflammatory response can be reduced in patients with symptomatic or asymptomatic hyperuricemia with administration of XO inhibitors [20], and this effect can contribute to CV prevention in combination with the recommended treatment strategies.

10.3 Hyperuricemia, Hypertension, Heart Failure, and Cardiovascular Disease

The presence of hyperuricemia can directly or indirectly promote the development of cardiometabolic mechanisms that are involved in the pathogenesis and clinical profile of CV diseases including HF. Elevated SUA levels are common in patients with metabolic syndrome that is associated with a high risk of CV diseases [21, 22]. Increased UA levels often precede the onset of metabolic syndrome [23] as the consequence of the negative impact of hyperuricemia over insulin resistance [24] with a reduction of the renal tubular secretion/reabsorption balance of UA. However, the larger amount of evidence relating elevated SUA and CV diseases including HF is that involving the pathogenesis of hypertension. A large meta-analysis reported an association between hyperuricemia and hypertension, independent of other risk factors for elevated blood pressure. A 13% increase in the risk of an incidence of hypertension was reported per 1 mg/dL increase in SUA level. There was a linear relationship between UA levels and hypertension with no cut-point or threshold. This correlation was more prominent in young individuals and females [25]. Potential mechanisms linking hyperuricemia and the development of cardiometabolic diseases include oxidative stress, reduced NO availability, and alterations of the renin angiotensin axis [26, 27]. UA is also reported as an independent risk factor for a decrease in renal function and end-stage renal disease (ESRD) that may negatively affect the natural history of HF [28, 29].

10.4 The Role of Hyperuricemia in Hypertension and Implications for Heart Failure

Uric acid and its increased serum levels have been directly involved in the pathogenesis and natural history of hypertension. The relationship between elevated uric acid and hypertension has been proven in both animals and humans, and its relevance is already evident in childhood and adolescent population [30]. The mechanism responsible for blood pressure increase in hyperuricemic subjects involves both oxidative stress and intracellular urate activity with a primary involvement of XOR activity [30]. An increase in the relative risk of hypertension has been confirmed by genetic data and by large meta-analyses of epidemiological data [31]. The effects of urate lowering treatment on blood pressure control in patients with elevated SUA has been investigated in a small number of studies with a large heterogeneity of patient populations and study designs. However, two large meta-analyses suggest a significant effect of urate lowering treatment on blood pressure particularly in patients with history of CV disease, thus confirming the significant relationship between high serum urate and blood pressure [31]. These data support the hypothesis that a reduction in urate levels can result in a significant decrease in the residual cardiovascular risk beyond blood pressure control. The future research should be focused on a more appropriate identification of hypertensive patients with "cardiovascular" hyperuricemia by considering the correct cardiovascular threshold of serum urate, the time course of uricemia fluctuations, and the identification of reliable markers of urate overproduction that could significantly clarify the clinical and therapeutic implications of the interaction between serum uric acid and hypertension (Table 10.1).

Table 10.1 New proposed approach for the identification of serum urate overproduction

1. Combined assessment of serum and urinary levels of uric acid
2. SUA/sCreatinine ratio
3. SUA/eGFR ratio
4, SUA/XOR activity

10.5 The Role of Hyperuricemia in Heart Failure

Elevated levels of UA are a common finding in patients with HF. Among 1869 participants of the Japanese Cardiac Registry of Heart Failure in Cardiology (JCARE-CARD), SUA concentration was ≥7.0 mg/dL in 56% of outpatients with HF [32] and this strictly corresponds with the prevalence of elevated SUA reported in the population of the EMPEROR Reduced study (55%) [6]. Many studies have revealed a relationship between increased UA levels and symptoms of HF according to the New York Heart Association (NYHA) functional class, reduced exercise tolerance, systemic congestion, and impaired cardiac function [33, 34]. Moreover, current evidence suggests that UA could be involved both as a potential determinant in the pathogenesis of HF and/or as an indicator of poor prognosis [35] with a threshold value for the risk of incident HF lower that usually considered [36]. In terms of the severity of the disease, we reported a linear inverse correlation between UA levels and EF% in a population of elderly patients with HF [37]. Amin et al. found a significant correlation between UA levels and NYHA class, levels of mean pulmonary arterial pressure, and pulmonary capillary wedge pressure [38]. Likewise, a post-hoc analysis of the EVEREST trial revealed a significant correlation between high SUA levels and NYHA functional class [39]. Another study reported that increased UA levels are associated with an impaired right ventricular systolic function and reduced left atrium function in patients with newly diagnosed HF [35]. Hyperuricemia has also been described in patients with diastolic dysfunction in chronic HF [40]. Elevated levels of UA have been associated with an increased all-cause mortality and rate of hospital admission in patients with HF with reduced [41] and preserved ejection fraction [42] after adjustment for the possible confounders including renal function. Palazzuoli et al. recently found a strong association between increased UA levels and risk of death and re-hospitalization during the 6-month follow-up period in patients hospitalized for acute HF [43].

Despite the relevance of the evidence, some of the aspects of the clinical and prognostic role of increased levels of UA in patients with HF needs to be better defined. Since most of the patients with HF complain some degree of renal dysfunction, it remains to be elucidated whether UA is a primary and independent promoter of the development and progression of HF or if it is just a marker of the underlying degree of renal impairment. In other words, the pathogenesis of hyperuricemia in HF may depend on the increased activity of XOR, with enhanced oxidative stress impairing left ventricular function, or it may depend on an increase in the renal tubular reabsorption or decreased urate excretion. Despite the evidence supporting that the level of hyperuricemia correlates with the degree of renal dysfunction in patients with HF [44], the results published by Filippatos et al. [41] and Vaduganathan et al. [39] reported a higher mortality rate and hospitalization in patients with elevated SUA associated with a better preserved renal function (estimated glomerular filtration rate >30 mL/min) thus excluding the primary role of renal dysfunction as responsible for the negative prognostic impact of elevated SUA in patients with HF. The same trend has been reported by Palazzuoli et al. in patients with acute HF [43]. This is clearly supported by an interesting paper measuring the levels of plasma XO in patients with HF [11] and showing a lower event-free survival in the presence of a higher serum concentration of the enzyme. The same mechanism(s) could be involved in both patients with HFrEF and HFpEF where the pathogenetic role of serum urate levels vs. XOR activation and oxidative stress warrants further investigations.

Among the most important problems in patients with HF is the effect of the extensive use of diuretic treatment on serum urate levels. According to their mechanism of action, diuretics can significantly reduce the renal excretion of uric acid leading to a significant increase in the levels of SUA, often in the range of hyperuricemia [45]. This is more common in patients with reduced GFR regardless of the functional nature of renal dysfunction or the presence of overt CKD. The use of diuretics, mainly loop diuretics, is recommended by guidelines to control symptoms of HF. In patients with HF, the diuretic-induced increase in SUA is more frequently responsible for gout and tissue urate deposition, while the CV impact of hyperuricemia has been less investigated. A recent paper by Maloberti et al. [46] carried out in a cohort of subjects with or at risk of CV disease has summarized the experience of the URRAH study group, suggesting that hyperuricemia carries a similar risk of cardiovascular events and all-cause mortality independently of diuretic treatment confirming the multiple nature of the urate-related CV complications in patients with HF (Fig. 10.2).

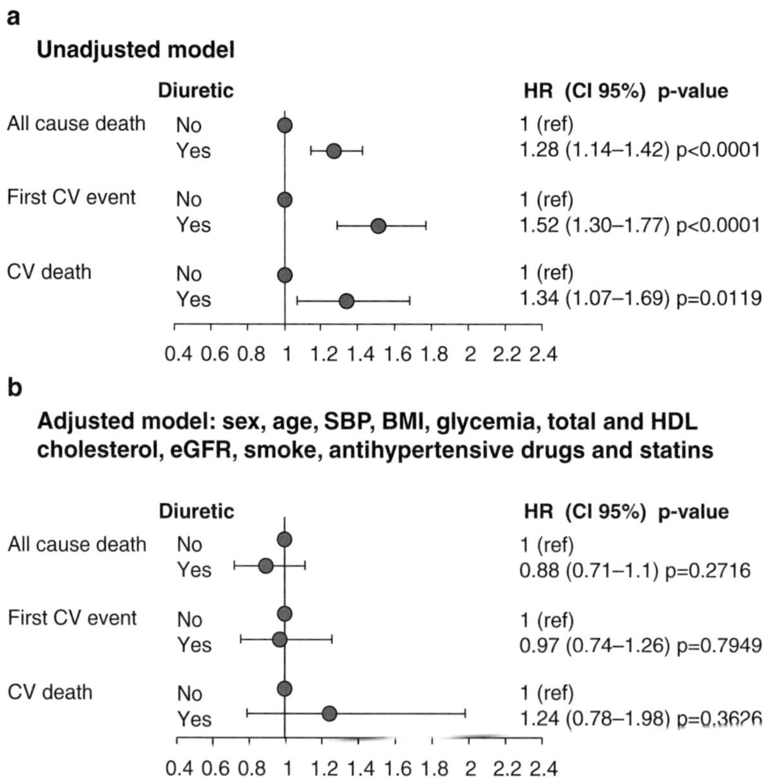

Fig. 10.2 Unadjusted (a) and adjusted (b) models of the effects of urate induced hyperuricemia on the relative risk of cardiovascular disease in the URRAH cohort (From [45])

10.6 The Cardiovascular Benefits of Treating Hyperuricemia

10.6.1 Xanthine Oxidase Inhibitors

The effects of treatment with urate lowering drugs have been investigated in a mixed population of patients with and without cardiovascular disease [47]. The most studied drugs are XO inhibitors such as allopurinol, oxypurinol, and febuxostat. More recently, topiroxostat, a more selective inhibitor of XO, has been studied in patients with hypertension [48] and in those with HFpEF [49] reporting interesting results. Allopurinol treatment is associated with a decreased CV risk in the general population of subjects with elevated serum uric acid [50]. A large meta-analyses has reported a significant decrease in systolic blood pressure values in response to allopurinol [51], and these results have been confirmed by a more recent paper dealing with the effects of urate lowering treatment on blood pressure [31]. In particular, the blood pressure lowering effect of XO inhibitors (febuxostat and topiroxostat) has been reported to be more evident in patients with preserved renal function (GFR > 90 mL/min/1.73 m^2) [52] and in the presence of a stronger blockade of the XOR [48] again supporting the hypothesis that the cardiovascular protective effects of XO inhibitors are significantly influenced by the level of activity of the enzyme.

As far as the trials comparing the effects of allopurinol and febuxostat, they have been carried out in patients with gout complicated by hypertension and previous cardiovascular diseases (including heart failure). The results of the CARES trial [53], based on large-scale, randomized study design, demonstrated a comparable effect of the two drugs on the primary CV endpoint with a modest increase in the rate of cardiovascular events with febuxostat that could be largely explained by the poor quality of the study design, the huge rate of drug discontinuation (>50%), and the high rate of lost to follow-up (8%). Conversely, the recent publication of the results of the Febuxostat versus Allopurinol Streamlined Trial (FAST) [54] does not support the finding of an increased cardiovascular risk of febuxostat, despite using higher dosages compared to the CARES trial. In 6128 patients (with previous cardiovascular disease), the incidence of the primary endpoint (a composite of hospitalization for non-fatal myocardial infarction or biomarker-positive acute coronary syndrome, non-fatal stroke, or cardiovascular death) was slightly reduced (formally noninferior) to allopurinol (adjusted HR 0·85 [95% CI 0·70–1·03], $p < 0{\cdot}000$). No excess of deaths was observed in any subgroup of patients in FAST study with a substantial trend toward a generalized favorable outcome in patients treated with febuxostat.

As far as the population of patients with HF, the results of studies with urate lowering treatment provided mixed results. While recent reports have raised the possibility of preventing HF using UA-lowering therapy, the literature is still conflicted on whether the reduction in UA will result in a significant clinical benefit [55]. In a randomized placebo-controlled trial in HF, Gavin et al. found that allopurinol decreased B-type natriuretic peptide (BNP) levels despite no improvement in exercise tolerance [56]. The La Plata study found that the treatment with oxypurinol improved the left ventricular ejection fraction (LVEF) by approximately 7%

in patients with a baseline LVEF <40% [57]. Other studies showed that endothelium-dependent vasodilation improvement after allopurinol or oxypurinol therapy in patients with CAD and chronic HF. Conversely, the uricosuric agent, benzbromarone, had no significant effect on BNP levels, left ventricular function, or NYHA class, suggesting no clinical benefit in left ventricular dysfunction [58]. One possible explanation is the central role of XO in patients with impaired left ventricular function and elevated SUA. Among prospective interventional studies on uric acid-lowering therapy in patients with hyperuricemia and HF, few trials have reported favorable results for uric acid-lowering therapy [59, 60]. The OPT-CHF study [61] investigated the efficacy of oxipurinol administered to NYHA 3–4 patients with HF. The study reported no improvement in the clinical symptoms, all-cause mortality, CV disease mortality, hospitalization, or hospitalization for HF in the overall population. However, a sub-analysis involving the subgroup with SUA value ≥9.5 mg/dL showed a decrease in the total number of all-cause and CV deaths with reduction of the clinical symptoms. To confirm these results, a prospective placebo-controlled double-blind study (EXACT-HF) [62] was conducted to assess the effects of 600 mg allopurinol/day in a population of about 300 patients with symptomatic severe HF with LVEF ≤40% and SUA acid value ≥9.5 mg/dL. In the group treated with allopurinol, a significant decrease in the SUA level occurred without significant changes in the clinical symptoms, all-cause mortality, 6-min walk test, and LVEF between the two groups. However, the Kaplan–Meier curves for hospital admission showed a remarkable difference that did not reach the level of statistical significance because of the small sample size. This suggests the need of an appropriate methodological approach to the efficacy of XO inhibition treatment and a better characterization of patients with HF. More recently, a US cohort study, involving approximately 100,000 Medicare participants with mild to moderate renal impairment and early-stage chronic HF patient, showed no difference in the cumulative CV events and CV deaths between patients treated with febuxostat or allopurinol. Less worsening of HF was observed in patients who were administered febuxostat [63]. In the CARES trial, no difference was observed in the incidence and mortality for HF heart failure in patients with gout and previous CV disease and treated with allopurinol or febuxostat [53]. In the Japanese FREED [64] and FEATHER [65] studies, the administration of febuxostat did not increase the prevalence of CV events/HF or worsen patient prognosis compared with placebo or small-dose administration of allopurinol. More recently, the results of the FAST study [54] reported no differences in the rate of hospitalization for heart failure (3.0% vs. 3.3%; $p = 0.441$) between patients treated with febuxostat and allopurinol that suggest the primary role of XO inhibition over single drugs. More information is expected from the results of the ongoing LEAF-CHF [66] and EXCITED-UA [49] studies comparing the effect of uric acid-lowering agent, febuxostat and topiroxostat, to conventional treatment in chronic HF patients with hyperuricemia.

10.6.2 Sodium Glucose Cotransporter 2 Inhibitors (SGLT2-I)

Sodium glucose cotransporter 2 (SGLT2) inhibitors are a group of medications used in the treatment of type 2 diabetes mellitus (DM2) and heart failure. Dapagliflozin, empagliflozin, and canagliflozin are the drugs approved by the FDA and EMA for use in the USA and European countries. Besides they primary effect on glucose control, SGLT2 inhibitors decrease the serum levels of uric acid by increasing the urinary excretion [67]. This was the consequence of the effects of SGLT2 inhibitors on the GLUT9 isoform 2 that exchanges glucose for UA in the renal proximal tubule and contributes to the UA reabsorption at the collecting ducts of the renal tubules. The effects of *Canagliflozin* (in comparison to placebo) have been tested in patients with DM2, reporting a decrease in SUA levels of 13% that was confirmed in the subset of patients with hyperuricemia [68]. This was associated with a lower number of patients who experienced a primary outcome event (defined as the composite of nonfatal myocardial infarction, nonfatal stroke, or death from CV causes) in the combined analysis of the CANVAS and CANVAS-R trials [69]. Patients who were treated with canagliflozin were also at a lower risk for HF hospitalization. Zinman et al. reported a reduction in the SUA level of patients with DM2 treated with *Empagliflozin* during the EMPAREG-OUTCOME trial [70]. Administration of empagliflozin in addition to standard treatment resulted in a lower rate of death from CV causes, modest weight loss, and a decrease in systolic and diastolic blood pressure. Additionally, empagliflozin treatment leads to a considerably lowered risk of HF hospitalization. A more recent multivariable analysis published by Inzucchi et al. [71] found that changes in variables related to glycemia and urate metabolism had smaller but highly significant mediating effects that include patients with HF.

The clinical efficacy of SGLT2 inhibitors has been tested in patients with both HFrEF and HFpEF reporting successful results in terms of clinical outcome and serum urate control (Table 10.2). In the DAPA-HF study [72], the effects of *Dapagliflozin* on the primary composite outcome of cardiovascular death or worsening heart failure were examined in 3119 patients according to tertiles of

Table 10.2 Effects of nonurate-lowering drugs on SUA in randomized controlled trials in heart failure

Drug	Study	Patients	Uric acid	Clinical outcome
Empagliflozin	EMPEROR-REDUCED	HFrEF	Reduced (−1.11 mg/dL vs. Placebo)	Improved-MACE
Empagliflozin	EMPEROR-PRESERVED	HFpEF	Reduced (−0.8 mg/dL vs. Placebo)	Improved-MACE
Dapagliflozin	DAPA-HF	HfrEF	Reduced vs. placebo (−0.84 mg/dL)	Improved-MACE
Sacubitril-Valsartan	PARADIGM-HF	HFrEF	Reduced (−0.24 mg/dL vs. ACEi	Improved-MACE
Sacubitril-Valsartan	PARAGON-HF	HfpEF	Reduced (−0.32 mg/dL vs. valsartan)	Partially improved-MACE
Vericiguat	SOCRATES-HF	HFrEF	Reduced (−6.5 to −10.3%) vs. Placebo	Improved-BNP

UA and after adjustment for other prognostic variables. Higher UA was associated with a greater risk of the primary outcome and the risk of heart failure hospitalization and cardiovascular death increased by 7% and 6%, respectively, per 1 mg/dL unit increase of UA. Compared with placebo, dapagliflozin reduced UA and improved outcomes, irrespective of baseline UA concentration. In the EMPEROR-REDUCED study [6], hyperuricemia was associated with advanced severity of HF and with a worst outcome in multivariate adjusted analyses. SUA was reduced throughout follow-up in patients treated with empagliflozin, with a similar reduction in all prespecified subgroups. Empagliflozin reduced the primary CV endpoint with a benefit that was confirmed regardless of baseline SUA and of the change in SUA at 4 weeks. A comparable reduction in the levels of SUA has also been observed in the EMPEROR-PRESERVED Study [73] supporting a significant interaction between SUA changes and treatment benefit of empagliflozin on mortality (cardiovascular and all-cause mortality) in patients in elevated SUA. These data support the therapeutic role of SGLT2 inhibitors in patients with heart failure complicated by elevated serum urate and suggest the possibility that their uricosuric mechanism of action could be synergistically integrated with XOI.

10.6.3 ARNI

Angiotensin-II/neprilysin inhibitors (ARNIs) are a new class of CV drugs whose mechanism of action is based on the blockade of angiotensin-II receptors and inhibition of degradation of natriuretic peptides. At present, sacubitril-valsartan is the only drug available in his class, and the results of the PARADIGM-HF study have demonstrated his capacity of improving mortality and hospitalization for HF in patients with HFrEF in comparison with enalapril [74]. A sub-analysis of the PARADIGM-HF study showed that high SUA levels are an independent prognostic factor for HFrEF and that lowering SUA level by treatment with sacubitril-valsartan contribute to the favorable outcome observed in the PARADIGM-HF study [75]. A significant reduction of serum urate levels has also been reported in the PARAGON-HF study where sacubitril-valsartan was compared with Valsartan alone in patients with HFpEF [76]. This was associated with a nonsignificant trend toward a reduction of the primary outcome (death + hospitalization for HF) that achieved a formal statistical significance in some subgroup of patients (EF >40 < 55%, female gender, CKD). This opens the next important question about the importance of serum urate control in a large proportion of patients with HF and provides some alternative or additional solutions to the treatment with XOI that should be considered only for those patients whose serum uric acid levels remains elevated after the treatment with the recommended modifying disease drugs.

10.6.4 Vericiguat

The effects of vericiguat vs. placebo on serum uric acid (SUA) were assessed in patients with HFrEF end enrolled in the Phase 2 SOCRATES-REDUCED study [77]. Changes from baseline SUA values at 12 weeks with placebo and vericiguat were assessed. Baseline-adjusted mean percentage changes in SUA were 5.0%, −1.3%, −1.1%, −3.5%, and − 5.3% in the placebo, and vericiguat 1.25 mg, 2.5 mg, 5.0 mg, and 10.0 mg groups, respectively. Significance vs. placebo was observed in the 5.0 mg and 10.0 mg groups ($P = 0.02$ and $P = 0.004$, respectively). Estimated probability for an end-of-treatment SUA value of <7.0 mg/dL was higher with vericiguat compared with placebo. In the same study, the higher doses of Vericiguat were associated with a greater reduction in the NT-proBNP level ($P < 0.02$) that suggested a dose–response relationship and a better neurohumoral outcome when vericiguat is used at a dose associated with a reduction of serum levels of UA.

10.7 Conclusions

In summary, hyperuricemia is a common feature that can result from an increase in the synthesis of UA or from a decrease in the amount of UA excreted through the kidney and the intestine. Hyperuricemia can lead to the development of gout through the deposition of MSU crystals in the joints with local inflammation. In addition, an overwhelming amount of evidence demonstrates a direct correlation between increased SUA levels and CV problems including hypertension and heart failure.

The main conclusions of this review are that (1) hyperuricemia was very common in patients with hypertension (prevalence >25%) and HFrEF (prevalence >50%) where it was associated with more advanced disease state, as reflected by NYHA class, the risk of hospitalization for HF, NT-proBNP, and LVEF; (2) elevated SUA was a strong and independent predictor of increased mortality (all-cause and cardiovascular mortality), morbidity, and of hospitalization for HF; (3) the negative effects of elevated uric acid on the CV system (and particularly HF) are more relevant in patients with normal renal function supporting a mechanistic role for urate production through the pro-oxidative enzyme XOR, (4) the administration of XO inhibitors (allopurinol and febuxostat) favorably affects blood pressure control and could improve the clinical outcome of patients with HF depending on the baseline characteristics of the patients (overproducers vs. underexcretors), (5) SGLT-2 inhibitors with uricosuric effects (empagliflozin and dapagliflozin), ARNI (sacubitril-valsartan), and vericiguat produce an early and sustained decrease in SUA in HF that was maintained in several prespecified subgroups, with a beneficial effects in terms of neurohumoral profile (vericiguat) and the incidence of clinically relevant HF events (SGLT-2 inhibitors and ARNI); and (6) the effect of non-XO inhibitor drugs to reduce the risk of HF outcomes was related to serum UA decrease and independent of SUA levels.

References

1. Ndrepepa G. Uric acid and cardiovascular disease. Clin Chim Acta. 2018;484:150–63.
2. Wu XW, Lee CC, Muzny DM, Caskey CT. Urate oxidase: primary structure and evolutionary implications. Proc Natl Acad Sci. 1989;86:9412–6.
3. Unger T, Borghi C, Charchar F, Khan NA, Poulter NR, Prabhakaran D, Ramirez A, Schlaich M, Stergiou GS, Tomaszewski M, Wainford RD, Williams B, Schutte AE. 2020 International Society of Hypertension global hypertension practice guidelines. Hypertension. 2020;75(6):1334–57.
4. Culleton BF, Larson MG, Kannel WB, Levy D. Serum uric acid and risk for cardiovascular disease and death: the Framingham heart study. Ann Intern Med. 1999;131:7.
5. Borghi C, Agabiti-Rosei E, Johnson RJ, Kielstein JT, Lurbe E, Mancia G, Redon J, Stack AG, Tsioufis KP. Hyperuricaemia and gout in cardiovascular, metabolic and kidney disease. Eur J Intern Med. 2020;80:1–11.
6. Doehner W, Anker SD, Butler J, Zannad F, Filippatos G, Ferreira JP, Salsali A, Kaempfer C, Brueckmann C, Pocock SJ, Januzzi JL, Packer M. Uric acid and sodium-glucose cotransporter-2 inhibition with empagliflozin in heart failure with reduced ejection fraction: the EMPEROR-reduced trial. Eur Heart J. 2022;43:3435–46.
7. Kim KM, Henderson GN, Frye RF, et al. Simultaneous determination of uric acid metabolites allantoin, 6-aminouracil, and triuret in human urine using liquid chromatography–mass spectrometry. J Chromatogr B. 2009;877:65–70.
8. Bobulescu IA, Moe OW. Renal transport of uric acid: evolving concepts and uncertainties. Adv Chronic Kidney Dis. 2012;19:358–71.
9. Choi HK, Mount DB, Reginato AM. Pathogenesis of gout. Ann Intern Med. 2005;143:499.
10. Kotozaki Y, Satoh M, Tanno K, Ohmomo H, Otomo R, Tanaka F, Nasu T, Taguchi S, Kikuchi H, Kobayashi T, Shimizu A, Sakata K, Hitomi J, Sobue K, Sasaki M. Plasma xanthine oxidoreductase activity is associated with a high risk of cardiovascular disease in a general Japanese population. Int J Environ Res Public Health. 2021;18:1894.
11. Otaki Y, Watanabe T, Kinoshita D, et al. Association of plasma xanthine oxidoreductase activity with severity and clinical outcome in patients with chronic heart failure. Int J Cardiol. 2017;228:151–7.
12. Pacher P. Therapeutic effects of xanthine oxidase inhibitors: renaissance half a century after the discovery of allopurinol. Pharmacol Rev. 2006;58:87–114.
13. Landmesser U, Spiekermann S, Dikalov S, et al. Vascular oxidative stress and endothelial dysfunction in patients with chronic heart failure: role of xanthine-oxidase and extracellular superoxide dismutase. Circulation. 2002;106:3073–8.
14. Biscaglia S, Ceconi C, Malagù M, et al. Uric acid and coronary artery disease: an elusive link deserving further attention. Int J Cardiol. 2016;213:28–32.
15. Gersch C, Palii SP, Kim KM, et al. Inactivation of nitric oxide by uric acid. Nucleosides Nucleotides Nucleic Acids. 2008;27:967–78.
16. Zharikov S, Krotova K, Hu H, et al. Uric acid decreases NO production and increases arginase activity in cultured pulmonary artery endothelial cells. Am J Physiol Cell Physiol. 2008;295:C1183–90.
17. Dhanasekar C, Kalaiselvan S, Rasool M. Morin, a bioflavonoid suppresses monosodium urate crystal-induced inflammatory immune response in RAW 264.7 macrophages through the inhibition of inflammatory mediators, intracellular ROS levels and NF-κB activation. PloS One. 2015;10:e0145093.
18. Dick SA, Epelman S. Chronic heart failure and inflammation: what do we really know? Circ Res. 2016;119:159–76.
19. Janoudi A, Shamoun FE, Kalavakunta JK, Abela GS. Cholesterol crystal induced arterial inflammation and destabilization of atherosclerotic plaque. Eur Heart J. 2016;37:1959–67.
20. Borghi C, Desideri G. Urate-lowering drugs and prevention of cardiovascular disease: the emerging role of xanthine oxidase inhibition. Hypertension. 2016;67:496–8.

21. Mancia G, BombelliM FR, et al. Impact of different definitions of the metabolic syndrome on the prevalence of organ damage, cardiometabolic risk and cardiovascular events. J Hypertens. 2010;28:999–1006.
22. Alberti KGMM, Eckel RH, Grundy SM, et al. Harmonizing the metabolic syndrome: a joint interim statement of the international diabetes federation task force on epidemiology and prevention; National Heart, Lung, and Blood Institute; American Heart Association; World Heart Federation; International Atherosclerosis Society; and International Association for the Study of Obesity. Circulation. 2009;120:1640–5.
23. Ford ES. Risks for all-cause mortality, cardiovascular disease, and diabetes associated with the metabolic syndrome: a summary of the evidence. Diabetes Care. 2005;28:1769–78.
24. Han T, Lan L, Qu R, et al. Temporal relationship between hyperuricemia and insulin resistance and its impact on future risk of hypertension. Hypertension. 2017;70:703–11.
25. Grayson PC, Kim SY, La Valley M, Choi HK. Hyperuricemia and incident hypertension: a systematic review and meta-analysis: risk of incident hypertension associated with hyperuricemia. Arthritis Care Res. 2011;63:102–10.
26. Johnson RJ, Kang D-H, Feig D, et al. Is there a pathogenetic role for uric acid in hypertension and cardiovascular and renal disease? Hypertension. 2003;41:1183–90.
27. Menè P, Punzo G. Uric acid: bystander or culprit in hypertension and progressive renal disease? J Hypertens. 2008;26:2085–92.
28. Bellomo G, Venanzi S, Verdura C, et al. Association of uric acid with change in kidney function in healthy normotensive individuals. Am J Kidney Dis. 2010;56:264–72.
29. Hsu C, Iribarren C, McCulloch CE, et al. Risk factors for endstage renal disease: 25-year follow-up. Arch Intern Med. 2009;169:342.
30. Borghi C, Agnoletti D, Cicero AFG, Lurbe E, Virdis A. Uric acid and hypertension: a review of evidence and future perspectives for the management of cardiovascular risk. Hypertension. 2022;79:1927–36. https://doi.org/10.1161/HYPERTENSIONAHA.122.17956.
31. Gill D, Cameron AC, Burgess S, Li X, Doherty DJ, Karhunen V, Azmil H, Abdul-Rahim AH, Taylor-Rowan M, Zuber V, Tsao PS, Klarin D, VA Million Veteran Program, Evangelou E, Elliott P, Damrauer SM, Terence J, Quinn TJ, Abbas Dehghan A, Theodoratou E, Dawson J, Tzoulaki J. Urate, blood pressure, and cardiovascular disease: evidence from mendelian randomization and meta-analysis of clinical trials. Hypertension. 2021;77:383–92.
32. Hamaguchi S, Furumoto T, Tsuchihashi-Makaya M, et al. Hyperuricemia predicts adverse outcomes in patients with heart failure. Int J Cardiol. 2011;151:143–7.
33. Anker SD, Doehner W, Rauchhaus M, et al. Uric acid and survival in chronic heart failure: validation and application in metabolic, functional, and hemodynamic staging. Circulation. 2003;107:1991–7.
34. Kaufman M, Guglin M. Uric acid in heart failure: a biomarker or therapeutic target? Heart Fail Rev. 2013;18:177–86.
35. Chrysohoou C, Pitsavos C, Barbetseas J, et al. Serum uric acid levels correlate with left atrial function and systolic right ventricular function in patients with newly diagnosed heart failure: the Hellenic heart failure study: the Hellenic heart failure study. Congest Heart Fail. 2008;14:229–33.
36. Muiesan ML, Salvetti M, Virdis A, Masi S, Casiglia E, Tikhonoff BCM, Bombelli M, Cicero AFG, Cirillo M, Cirillo P, Desideri GB, D'Elia L, Ferri C, Galletti F, Gesualdo L, Giannattasio C, Iaccarino G, Mallamaci F, Maloberti A, Mazza A, Nazzaro P, Palatini P, Parati GF, Pontremoli R, Rattazzi M, Rivasi G, Tocci G, Ungar A, Verdecchia P, Viazzi F, Volpe M, Grassi G, Borghi C, from the Working Group on Uric Acid, Cardiovascular Risk of the Italian Society of Hypertension. Serum uric acid, predicts heart failure in a large Italian cohort: search for a cut-off value the URic acid right for heArt health study. J Hypertens. 2021;39:62–9.
37. Borghi C, Cosentino ER, Rinaldi ER, Cicero AFG. Uricaemia and ejection fraction in elderly heart failure outpatients. Eur J Clin Invest. 2014;44:573–7.
38. Amin A, Vakilian F, Maleki M. Serum uric acid levels correlate with filling pressures in systolic heart failure: filling pressures in systolic heart failure. Congest Heart Fail. 2011;17:79–83.

39. Vaduganathan M, Greene SJ, Ambrosy AP, et al. Relation of serum uric acid levels and outcomes among patients hospitalized for worsening heart failure with reduced ejection fraction (from the efficacy of vasopressin antagonism in heart failure outcome study with Tolvaptan trial). Am J Cardiol. 2014;114:1713–21.
40. Cicoira M, Zanolla L, Rossi A, et al. Elevated serum uric acid levels are associated with diastolic dysfunction in patients with dilated cardiomyopathy. Am Heart J. 2002;143:1107–11.
41. Filippatos GS, Ahmed MI, Gladden JD, et al. Hyperuricaemia, chronic kidney disease, and outcomes in heart failure: potential mechanistic insights from epidemiological data. Eur Heart J. 2011;32:712–20.
42. Shimizu T, Yoshihisa A, Kanno Y, et al. Relationship of hyperuricemia with mortality in heart failure patients with preserved ejection fraction. Am J Physiol Heart Circ Physiol. 2015;309:H1123–9.
43. Palazzuoli A, Ruocco G, Pellegrini M, et al. Prognostic significance of hyperuricemia in patients with acute heart failure. Am J Cardiol. 2016;117:1616–21.
44. Pascual-Figal DA, Hurtado-Martínez JA, Redondo B, et al. Hyperuricaemia and long-term outcome after hospital discharge in acute heart failure patients. Eur J Heart Fail. 2007;9:518–24.
45. Ben Salem C, Slim R, Fathallah N, Hmouda H. Drug induced hyperuricaemia and gout. Rheumatology (Oxford). 2017;56(5):679–88.
46. Maloberti A, Bombelli M, Facchetti R, Barbagallo CM, Bernardino B, Rosei EA, Casiglia E, Cicero AFG, Cirillo M, Cirillo P, Desideri G, D'elia L, Dell'Oro R, Ferri C, Galletti F, Giannattasio C, Loreto G, Iaccarino G, Lippa L, Mallamaci F, Masi S, Mazza A, Muiesan ML, Nazzaro P, Parati G, Palatini P, Pauletto P, Pontremoli R, Quarti-Trevano F, Rattazzi M, Rivasi G, Salvetti M, Tikhonoff V, Tocci G, Ungar A, Verdecchia P, Viazzi F, Volpe M, Virdis A, Grassi G, Borghi C, Working Group on Uric Acid, Cardiovascular Risk of the Italian Society of Hypertension (SIIA). Relationships between diuretic-related hyperuricemia and cardiovascular events: data from the URic acid right for heArt health study. J Hypertens. 2021;39(2):333–40.
47. Cicero AFG, Fogacci F, Kuwabara M, Borghi C. Therapeutic strategies for the treatment of chronic hyperuricemia: an evidence-based update. Medicina (Kaunas). 2021;57(1):58. https://doi.org/10.3390/medicina57010058.
48. Kario K, Nishizawa M, Kiuchi M, Kiyosue A, Tomita F, Ohtani H, Abe Y, Kuga H, Miyazaki S, Kasai T, Hongou M, Yasu T, Kuramochi J, Fukumoto Y, Hoshide S. Hisatome I comparative effects of topiroxostat and febuxostat on arterial properties in hypertensive patients with hyperuricemia. J Clin Hypertens (Greenwich). 2021;23(2):334–44.
49. Sakuma M, Toyoda S, Arikawa T, Koyabu Y, Kato S, Adachi T, Suwa H, Narita J, Anraku K, Ishimura K, Yamauchi F, Sato A, Inoue T, for Excited UA study Investigators. The effects of xanthine oxidase inhibitor in patients with chronic heart failure complicated with hyperuricemia: a prospective randomized controlled clinical trial of topiroxostat vs allopurinol—study protocol. Clin Exp Nephrol. 2018;22:1379–86.
50. Larsen KS, Pottegård A, Lindegaard HM, Hallas J. Effect of allopurinol on cardiovascular outcomes in hyperuricemic patients: a cohort study. Am J Med. 2016;129:299–306.
51. Agarwal V, Hans N, Messerli FH. Effect of allopurinol on blood pressure: a systematic review and meta-analysis. J Clin Hypertens. 2013;15:435–42.
52. Gunawardhana L, McLean L, Punzi HA, et al. Effect of Febuxostat on ambulatory blood pressure in subjects with hyperuricemia and hypertension: a phase 2 randomized placebo-controlled study. J Am Heart Assoc. 2017;6:e006683.
53. White WB, Saag KG, Becker MA, Borer JS, Gorelick PB, Whelton A, Hunt B, Castillo M, Gunawardhana L, CARES Investigators. Cardiovascular safety of Febuxostat or Allopurinol in patients with gout. N Engl J Med. 2018;378(13):1200–10.
54. Mackenzie IS, Ford I, Nuki G, Hallas J, Hawkey CJ, Webster J, Ralston SH, Walters M, Robertson M, De Caterina R, Findlay E, Perez-Ruiz F, McMurray JJV, MacDonald TM, FAST Study Group. Long-term cardiovascular safety of febuxostat compared with allopurinol in patients with gout (FAST): a multicentre, prospective, randomised, open-label, non-inferiority trial. Lancet. 2020;396(10264):1745–57.

55. Kim SC, Schneeweiss S, Choudhry N, et al. Effects of xanthine oxidase inhibitors on cardiovascular disease in patients with gout: a cohort study. Am J Med. 2015;128:653.e7–653.e16.
56. Gavin AD. Allopurinol reduces B-type natriuretic peptide concentrations and haemoglobin but does not alter exercise capacity in chronic heart failure. Heart. 2005;91:749–53.
57. Cingolani HE, Plastino JA, Escudero EM, et al. The effect of xanthine oxidase inhibition upon ejection fraction in heart failure patients: La Plata study. J Card Fail. 2006;12:491–8.
58. Ogino K, Kato M, Furuse Y, et al. Uric acid-lowering treatment with benzbromarone in patients with heart failure: a double-blind placebo-controlled crossover preliminary study. Circ Heart Fail. 2010;3:73–81.
59. Engberding N, Spiekermann S, Schaefer A, et al. Allopurinol attenuates left ventricular remodeling and dysfunction after experimental myocardial infarction: a new action for an old drug? Circulation. 2004;110:2175–9.
60. Xiao J, Deng SB, She Q, Li J, Kao GY, Wang JS, Ma Y. Allopurinol ameliorates cardiac function in non-hyperuricaemic patients with chronic heart failure. Eur Rev Med Pharmacol Sci. 2016;20:756–61.
61. Hare JM, Mangal B, Brown J, et al. Impact of oxypurinol in patients with symptomatic heart failure. J Am Coll Cardiol. 2008;51:2301–9.
62. Givertz MM, Anstrom KJ, Redfield MM, et al. Effects of xanthine oxidase inhibition in hyperuricemic heart failure patients: the xanthine oxidase inhibition for hyperuricemic heart failure patients (EXACT-HF) study. Circulation. 2015;131:1763–71.
63. Zhang M, Solomon DH, Desai RJ, et al. Assessment of cardiovascular risk in older patients with gout initiating febuxostat versus allopurinol: population-based cohort study. Circulation. 2018;138:1116–26.
64. Kojima S, Matsui K, Ogawa H, et al. Rationale, design, and baseline characteristics of a study to evaluate the effect of febuxostat in preventing cerebral, cardiovascular, and renal events in patients with hyperuricemia. J Cardiol. 2017;69:169–75.
65. Kimura K, Hosoya T, Uchida S, et al. Febuxostat therapy for patients with stage 3 CKD and asymptomatic hyperuricemia: a randomized trial. Am J Kidney Dis. 2018;72:798–810.
66. Yokota T, Fukushima A, Kinugawa S, et al. Randomized trial of effect of urate-lowering agent febuxostat in chronic heart failure patients with hyperuricemia (LEAF-CHF): study design. Int Heart J. 2018;59:976–82.
67. Chino Y, Samukawa Y, Sakai S, et al. SGLT2 inhibitor lowers serum uric acid through alteration of uric acid transport activity in renal tubule by increased glycosuria: uricosuric effect of SGLT2 inhibitor. Biopharm Drug Dispos. 2014;35:391–404.
68. Davies MJ, Trujillo A, Vijapurkar U, et al. Effect of canagliflozin on serum uric acid in patients with type 2 diabetes mellitus. Diabetes Obes Metab. 2015;17:426–9.
69. Neal B, Perkovic V, Mahaffey KW, et al. Canagliflozin and cardiovascular and renal events in type 2 diabetes. N Engl J Med. 2017;377:644–57.
70. Zinman B, Wanner C, Lachin JM, et al. Empagliflozin, cardiovascular outcomes, and mortality in type 2 diabetes. N Engl J Med. 2015;373:2117–28.
71. Inzucchi SE, Zinman B, Fitchett D, et al. How does empagliflozin reduce cardiovascular mortality? Insights from a mediation analysis of the EMPA-REG OUTCOME trial. Diabetes Care. 2018;41:356–63.
72. McDowell K, Welsh P, Docherty KF, Morrow DA, Jhund PS, de Boer RA, O'Meara E, Inzucchi SE, Køber L, Kosiborod MN, Martinez FA, Ponikowski P, Hammarstedt A, Langkilde AM, Sjöstrand M, Lindholm D, Solomon SD, Sattar N, Sabatine MS, McMurray JJV. Dapagliflozin reduces uric acid concentration, an independent predictor of adverse outcomes in DAPA-HF. Eur J Heart Fail. 2022;24(6):1066–76.
73. Anker SD, Butler J, Filippatos G, Ferreira JP, Bocchi E, Böhm M, Brunner-La Rocca HP, Choi DJ, Chopra V, Chuquiure-Valenzuela E, Giannetti N, Gomez-Mesa JE, Janssens S, Januzzi JL, Gonzalez-Juanatey JR, Merkely B, Nicholls SJ, Perrone SV, Piña IL, Ponikowski P, Senni M, Sim D, Spinar J, Squire I, Taddei S, Tsutsui H, Verma S, Vinereanu D, Zhang J, Carson P, CSP L, Marx N, Zeller C, Sattar N, Jamal W, Schnaidt S, Schnee JM, Brueckmann M, Pocock SJ,

Zannad F, Packer M, EMPEROR-Preserved Trial Investigators. Empagliflozin in heart failure with a preserved ejection fraction. N Engl J Med. 2021;385(16):1451–61.
74. McMurray JJV, Packer M, Desai AS, et al. Angiotensin–neprilysin inhibition versus enalapril in heart failure. N Engl J Med. 2014;371:993–1004.
75. Mogensen UM, Køber L, Jhund PS, et al. Sacubitril/valsartan reduces serum uric acid concentration, an independent predictor of adverse outcomes in PARADIGM-HF: sacubitril/valsartan, uric acid, and heart failure. Eur J Heart Fail. 2018;20:514–22.
76. Selvaraj S, Claggett BJ, Pfeffer MA, Desai AS, Mc Causland FR, McGrath MM, Anand IS, van Veldhuisen DJ, Kober L, Janssens S, Cleland JGF, Pieske B, Rouleau JL, Zile MR, Shi VC, Lefkowitz MP, McMurray JJV, Solomon SD. Serum uric acid, influence of sacubitril–valsartan, and cardiovascular outcomes in heart failure with preserved ejection fraction: PARAGON-HF. Eur J Heart Fail. 2020;22:2093–101.
77. Gheorghiade M, Greene SJ, Butler J, Filippatos G, Lam CSP, Maggioni AP, Ponikowski P, Shah SJ, Solomon SD, Kraigher-Krainer E, Samano ET, Müller K, Roessig L, Pieske B, for the SOCRATES-REDUCED Investigators and Coordinators. Effect of Vericiguat, a soluble guanylate cyclase stimulator, on natriuretic peptide levels in patients with worsening chronic heart failure and reduced ejection fraction. The SOCRATES-REDUCED randomized trial. JAMA. 2015;314(21):2251–62.

Myocardial Ischemia: A Link Between Hypertension and Heart Failure

11

Maria Dorobantu and Aura Vîjîiac

Hypertension remains one of the most important contributors to morbidity and mortality across the world [1], with an estimated 1.28 billion adults aged 30–79 years worldwide suffering from hypertension [2]. While almost half of the hypertensive patients are diagnosed and treated, only 21% of them have the disease adequately controlled [2], in accordance with current recommendations. This leads to more than 7.5 million deaths per year which can be attributed to hypertension [1], since the disease determines damage in other organs and systems, the most important targeted organs being the heart, the brain, and the kidneys.

Heart failure (HF) was believed to be the most common cause of death in hypertensive patients until the last few decades, particularly before antihypertensive therapy became widely available [3]. In the Framingham Heart Study, 91% of the individuals with HF had hypertension, while hypertensive males and females had a two and threefold risk, respectively, of developing HF, when compared to normotensive subjects [4].

In fact, the stages of hypertensive heart disease were categorized more than two decades ago into 4 ascending degrees [5, 6]: I—left ventricular (LV) diastolic dysfunction with no LV hypertrophy; II—LV diastolic dysfunction with LV hypertrophy; III—clinical signs and symptoms of HF with preserved ejection fraction (HFpEF); and IV—dilated cardiomyopathy with HF and reduced ejection fraction (HFrEF). Indeed, in most hypertensive patients, LV diastolic dysfunction is the first manifestation of cardiac disease, while the end stage of hypertensive heart disease (which occurs in long-standing, untreated hypertension) is a phenotype of dilated

M. Dorobantu (✉)
Carol Davila University of Medicine and Pharmacy, Bucharest, Romania

A. Vîjîiac
Carol Davila University of Medicine and Pharmacy, Bucharest, Romania

Emergency Clinical Hospital of Bucharest, Bucharest, Romania

M. Dorobantu et al. (eds.), *Hypertension and Heart Failure*, Updates in Hypertension and Cardiovascular Protection,
https://doi.org/10.1007/978-3-031-39315-0_11

cardiomyopathy with both systolic and diastolic dysfunctions [5]. Some authors argue that hypertensive HF is not merely a disorder of systole and/or diastole [7, 8]. Besides the impairment of LV function, chronotropic incompetence, microcirculatory dysfunction, lack of peripheral vasodilator reserve, and increased arterial stiffness play a role in the development of HF in hypertensive patients [7, 9, 10].

11.1 The Link Between Hypertension and Myocardial Ischemia

Whether hypertension can lead to systolic HF in the absence of myocardial infarction has long been a matter of debate [11]. Hypertension is a well-established risk factor for coronary artery disease, which in turn puts the patient at risk for developing HF [3]. The fact that atherosclerosis does not develop in vessels with low blood pressure (such as the veins or pulmonary arteries) highlights the critical role that hypertension plays in the atherosclerotic continuum [12]. On the one hand, 80% of the patients with ischemic heart disease are concomitantly hypertensive [13]. On the other hand, ischemic heart disease, defined as an impaired blood supply to the myocardium due to atherosclerosis of the epicardial coronary arteries, is the most frequent target-organ damage encountered in hypertensive patients [14].

While myocardial ischemia and hypertension are distinct clinical entities, they share a pathophysiological link which can be summarized in two main pathways [15]: atherosclerosis of the major epicardial coronary arteries and LV hypertrophy leading to abnormal coronary artery remodeling and microcirculatory dysfunction. Both atherosclerotic coronary artery disease and the endothelial, hemodynamic, and neurohumoral changes will lead to an imbalance between oxygen supply and demand, which will determine various clinical manifestations, ranging from silent ischemia to acute coronary syndromes.

The fact that hypertension is one of the most important risk factors for coronary artery disease has been known for half a century, since data from the Framingham Heart Study proved the independent predictive role of hypertension, dyslipidemia, diabetes, and smoking [16]—still accepted nowadays as the most important modifiable determinants of cardiovascular risk [17]. The triad of oxidative stress, endothelial dysfunction, and inflammation represent the foundation upon which the atherosclerotic process is built [18], and hypertension represents an important link between these three pathophysiological processes (Fig. 11.1). The increased arterial sheer stress seen in hypertensive patients activates endothelial mechanoreceptors, which will trigger a cascade of intracellular pathways leading to increased production of reactive oxygen species (ROS) and impairment in the synthesis and release of nitric oxide (NO) [18].

The formation of atherosclerotic plaques usually requires increased levels of plasma lipoproteins as well. The mechanical stress, together with NO dysregulation and ROS, will determine reduced vascular compliance and imbalanced tone regulation, as well as structural changes in the microvasculature [19]. The damaged endothelium has an increased permeability, which allows the oxidized LDL-cholesterol

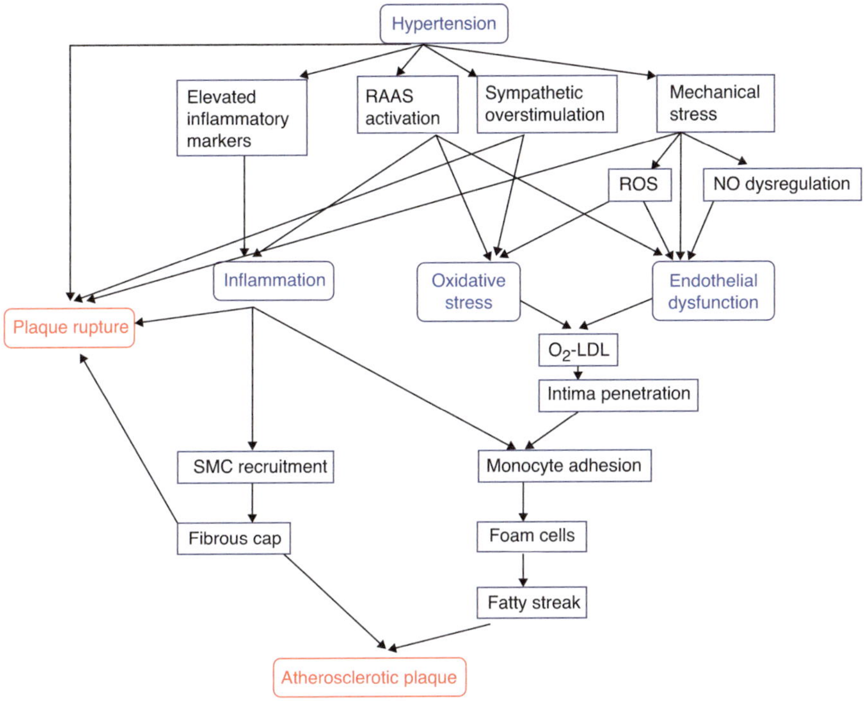

Fig. 11.1 Pathogenesis of atherosclerosis in hypertensive patients. *O_2-LDL* oxidized LDL cholesterol particles, *NO* nitric oxide, *RAAS* renin angiotensin aldosterone system

particles to infiltrate the intima and initiate an inflammatory process in the vascular wall [20]. The expression of leucocyte adhesion molecules attracts monocytes at the level of vascular injury; after ingestion of oxidized LDL molecules, monocytes will transform into foam cells and form a fatty streak [21]. Afterward, smooth muscle cells will migrate from the media to the intima of the vessel and produce collagen and elastin, leading to the formation of the fibrous cap which surrounds the atherosclerotic plaque [18].

Inflammation is both a local vascular process and a systemic one, associated with increased circulating levels of inflammatory markers such as acute phase reactants, adhesion molecules, and a wide range of cytokines [18]. Various studies suggested that hypertension was a chronic inflammatory process, since high blood pressure values were independently correlated with high circulating levels of C-reactive protein [22], interleukin-6 [23], interleukin-1β [24], and tumoral necrosis factor-α [25]. The inflammatory process associated with high blood pressure involves increased mechanical stress of the vascular wall, as well as neurohumoral factors such as the activation of the renin–angiotensin–aldosterone system. Angiotensin II is not just a potent vasoconstrictor but also has a profibrotic and proinflammatory role [26]. The relationship between hypertension and inflammation is bidirectional: on one side, hypertension is partly an inflammatory condition; on the other side, inflammation plays a role in the development of hypertension [18].

While atherogenesis is widely seen as a diffuse inflammatory process, inflammation also plays a key role in atherothrombosis. Atherosclerotic plaques which are morphologically vulnerable (thin-walled and with rich lipid core) may suffer erosion or rupture with subsequent thrombosis, leading to acute coronary syndromes [27]. Plaque rupture is triggered by an acute inflammatory process which leads to cap weakening and smooth muscle cell apoptosis [18, 28]. It is believed that the increased mechanical stress on the large vessels due to high blood pressure may contribute to plaque vulnerability by augmenting the inflammatory response at the level of the atherosclerotic plaques [29].

Even in the absence of epicardial coronary artery disease, hypertensive patients may develop myocardial ischemia through other mechanisms. High blood pressure leads to increased arterial stiffness, which is an established risk factor for adverse cardiovascular events [13]. Stiffening of the large arteries determines isolated systolic hypertension (which increases the myocardial oxygen demand) with augmentation of pulse pressure (which diminishes myocardial oxygen supply by decreasing the diastolic coronary pressure gradient). Consequently, increased arterial stiffness can induce myocardial ischemia in the absence of atherosclerosis.

Hypertension will also lead to an increase in LV afterload, which will determine changes in the pressure and flow generated by the ventricle [30], as well as changes in the ventricular structure and function. Chronic pressure overload will expose the LV to constant mechanical stress, which will lead to LV remodeling. Besides chronic hemodynamic overload, non-hemodynamic parameters also contribute to ventricular remodeling. It is well established that genetic factors, ethnicity, obesity, salt intake, activation of neurohormonal pathways (such as the renin–angiotensin system and the sympathetic nervous system) indirectly influence the response of the LV to chronic hypertension [31, 32].

The remodeling process includes hypertrophic growth of cardiomyocytes, increased rate of cardiomyocyte apoptosis, alteration of the extracellular matrix with excessive collagen deposition in the interstitium, and fibroblast proliferation, all these eventually leading to myocardial fibrosis and LV hypertrophy [31]. The increased LV mass leads to an increased oxygen demand, but coronary perfusion cannot adapt appropriately. There is hyperplasia of the intramyocardial arteries, with important medial arteriolar hypertrophy, leading to an increased wall-to-lumen ratio; moreover, the number of capillaries does not increase concurrently with the myocyte hypertrophy, resulting in a decreased relative capillary density [33]. There is also coronary arteriolar constriction due to perivascular fibrosis, hypertrophy of vascular smooth muscle cells, and increased deposition of connective tissue [34], a phenomenon generally described as vascular remodeling. All these structural modifications will determine an impaired diastolic coronary blood flow, with reduced coronary flow reserve and increased coronary microvascular resistance [33].

However, microcirculatory dysfunction in hypertensive patients is not determined just by structural changes but it is also the result of functional and extravascular mechanisms [35] (Fig. 11.2). Autoregulation of coronary blood flow is mainly mediated by arterioles less than 150 μm diameter [36]. Coronary resistance depends on the caliber of these small vessels and on their extravascular compression due to

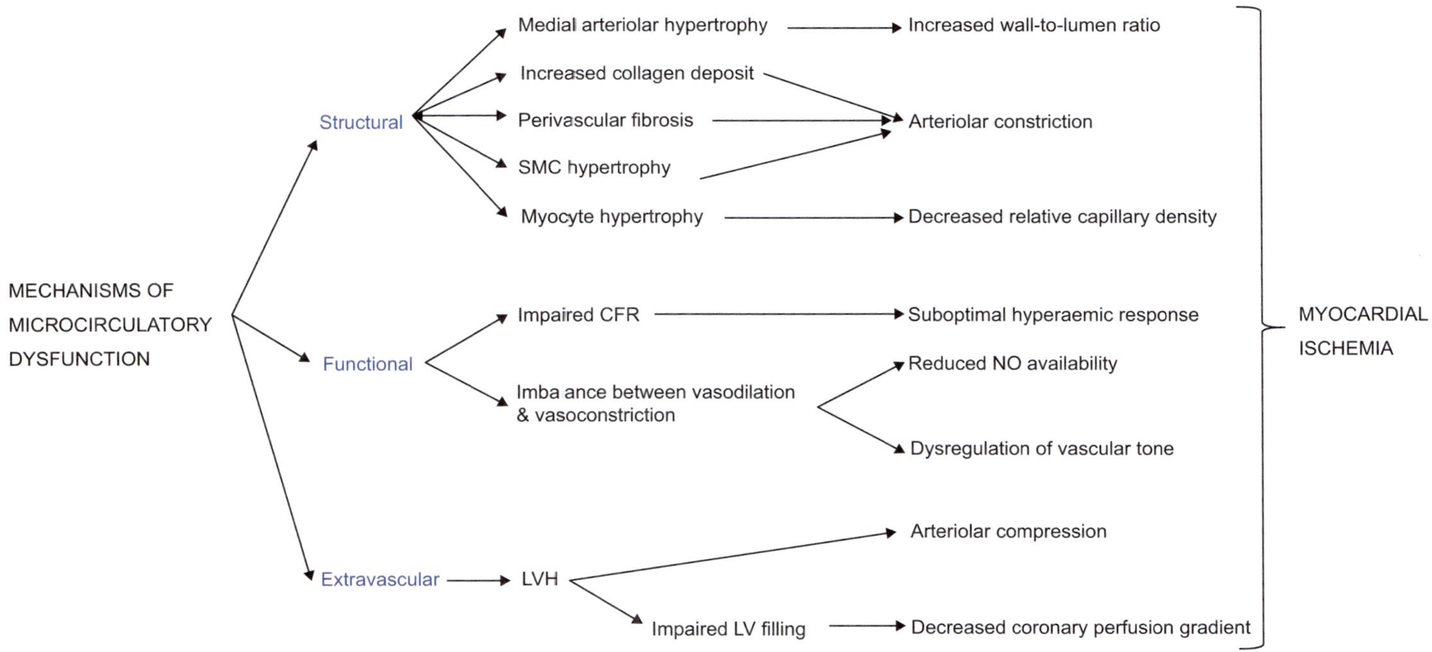

Fig. 11.2 Mechanisms of microcirculatory dysfunction in hypertensive patients. *CFR* coronary flow reserve, *LV* left ventricle, *LVH* left ventricular hypertrophy, *NO* nitric oxide, *SMC* smooth muscle cell

LV hypertrophy [35]. The increased intramyocardial systolic pressure encountered in LV hypertrophy has deleterious effects on coronary blood flow, which are more pronounced in the subendocardial than in the subepicardial layers [37].

The main functional changes in the microcirculation of the hypertrophied hearts are represented by impaired coronary flow reserve with dysregulation of myocardial blood flow, together with an imbalance between vasodilating and vasoconstrictive factors. In the hypertrophied LV, the myocardial blood flow is already increased at rest to meet the higher oxygen demand [35], leading to a reduced functional vasodilator capacity [38]. If the myocardial demand further increases, the suboptimal hyperemic response can determine myocardial ischemia, even in the absence of coronary artery stenoses.

The endothelium is an active "organ" regulating vascular tone, cell growth, being at the same time the interplay between the vessel wall and blood cells. The endothelial dysfunction encountered in hypertension refers to an impaired endothelium-dependent vasodilation due to reduced levels of NO [19, 39]. This is explained by an increased turnover of endothelial cells in vessels exposed to chronically elevated blood pressure, with regenerated endothelium being unable to adequately produce endothelium-derived relaxing factors [40]. The reduced availability of NO can increase the activity of endothelin, which has powerful vasoconstrictor and mitogenic effects [41], resulting in a dysregulation of vascular tone with impaired vasodilation and enhanced vasoconstriction [19]. Since NO, besides being a potent vasodilator, also inhibits smooth muscle cell proliferation, inflammation, platelet activation, and leukocyte adhesion [41], its reduced availability will lead to an increased risk of atherosclerotic vascular damage. This made some authors regard the endothelium as the earliest target "organ" to elicit changes of hypertensive damage [19] and thus an important target for tailored therapies.

11.2 The Link Between Myocardial Ischemia and Heart Failure

Heart failure remains a major public health concern and the most common cause of hospitalization in the elderly [42]. While the prognosis of cardiac conditions leading to HF (such as acute coronary syndromes, arrhythmias, hypertension, and valvular heart disease) has consistently improved during the last decades, the prevalence of HF is constantly growing [43]. Nowadays, it is usually claimed that ischemic heart disease is the major underlying cause of HF [3, 43], with nearly 70% of all cases of HF being attributable to myocardial ischemia [44]. The aging of the population, which leads to an increased incidence of myocyte apoptosis, together with the sustained increase in survival due to acute myocardial infarction (MI), are the two most likely explanations for the estimated 46% increase in HF prevalence by 2030 [43, 45]. HF caused by MI had a constantly increasing prevalence over the past decades [46], contrary to a decline in HF attributable to hypertension [44]. However, previous research showed that MI in hypertensive patients leads to more adverse outcome than in normotensive patients [3] and that LV dilation and prevalence of HF after acute MI is greater in hypertensive patients [47]. One probable explanation resides in the neurohormonal changes induced by chronically elevated blood pressure (such as activation of renin–angiotensin–aldosterone system), which predispose to adverse post-MI ventricular remodeling and leave the patient more susceptible to HF development after an acute MI [48].

A proper oxygen supply, consistent with the metabolic demands of the myocardium, is vital for myocyte viability and function. Moreover, oxygen is one of the main determinants of myocardial gene expression [49], while it is also needed for NO synthesis (which is one of the major regulators of vascular tone). Oxygen can be both indispensable and detrimental for cardiac function, since it also plays an important role in the synthesis of reactive oxygen species, which are involved in microcirculatory dysfunction, and which can induce irreversible cellular damage [49]. Consequently, cardiomyocyte hypoxia will determine cell malfunction, with impairment of ventricular performance. In acute myocardial ischemia, hypoxia occurs due to a sudden decrease in coronary circulation (usually secondary to an acute plaque rupture with thrombosis), which triggers acute cardiomyocyte injury [49, 50]. In chronic coronary syndromes, there is a progressive decrease in coronary circulation with an imbalance between oxygen demand and supply, which will lead to a reduced pumping performance of the myocardium, inadequate for the rising metabolic demands [49]. This contractile dysfunction is usually transient and reversible if adequate coronary flow is restored, a condition referred to as "hibernating myocardium." Given that ischemic HF is potentially reversible after myocardial revascularization, an important target in ischemic patients is to decide who will benefit from restoration of coronary perfusion.

Prolonged or recurrent ischemia will determine maladaptive cardiomyocyte remodeling with accumulation of extracellular matrix, which will usually lead to LV dilation and impairment of systolic performance [50, 51]. The event linking myocardial ischemia with the development of HF is usually an acute plaque rupture with thrombotic occlusion of a coronary artery [45], clinically manifested as an acute MI. The acute ischemic injury leads to cardiomyocyte apoptosis, with myocardial stunning and necrosis, endothelial dysfunction, myocardial inflammation, and subendocardial fibrosis [45]. Reperfusion usually determines a second wave of myocardial injury, caused by reactive oxygen species or embolization of microthrombi, which will perpetrate the microvascular dysfunction [52, 53]. The scar formation will activate neurohormonal pathways (such as renin–angiotensin–aldosterone and sympathetic overdrive), which will eventually lead to ventricular maladaptive remodeling, with dysfunction that sometimes extends to the surrounding non-infarcted myocardium [50]. The extent of LV remodeling depends on the size of the infarcted territory, on the timing and success of reperfusion, on pre-existent LV hypertrophy/fibrosis and on pre-existent endothelial dysfunction. A remodeled LV has an altered geometry, with wall thinning, further cardiomyocyte loss [52], and segmental dysfunction, which all lead to impaired ventricular performance and HFrEF.

The development of HF after a MI is directly related to several risk factors: the extent of ventricular remodeling and microvascular obstruction [52], the development of mitral regurgitation [45], genetic predictors such as microRNAs [54], magnitude of natriuretic peptides elevation [55], novel biomarkers of fibrosis such as galectin-3 [56], suppressor of tumorigenesis (ST2) [57], and various clinical risk factors. The latter category comprises elements such as age [58], prior MI [59], multivessel disease [60], prior hypertension [47], diabetes mellitus [61], and chronic kidney disease [62].

While HFrEF in ischemic heart disease appears because of scar tissue formation with subsequent LV remodeling, patients with myocardial ischemia can also develop HFpEF, due to isolated diastolic dysfunction. HFpEF was traditionally considered to develop in conditions leading to an increased LV afterload [63]. In hypertensive patients, LV hypertrophy represents an initial adaptive mechanism aimed to reduce wall stress and preserve cardiac output [1]. However, LV remodeling has deleterious effects over time, as LV fibrosis and stiffening will lead to impaired ventricular filling and diastolic dysfunction.

Echocardiographic studies showed that the initial LV functional changes in hypertensive patients lead to diastolic dysfunction and impaired longitudinal systolic function, while compensatory circumferential shortening and LV rotation maintain a normal ejection fraction [7, 64]. However, as the LV remodeling progresses over time, all the three myocardial layers and all the components of LV contraction become impaired and the ventricle begins to dilate, with subsequent decline in global systolic performance [7, 65].

Myocardial ischemia is a self-propagating process: during ischemia, stiffened cardiomyocytes have an impaired relaxation with ensuing elevation of LV filling pressures and wall stress, which in turn will increase the oxygen demand and further aggravate the ischemia [45, 66]. Speckle-tracking echocardiography showed that myocardial reperfusion after complete coronary occlusion is associated with transient diastolic stunning lasting for 24 h, even in myocardial segments with complete recovery of systolic function [67]. Myocardial ischemia impedes the normal calcium exchange managed by the myocyte sarcoplasmic reticulum, which will alter the calcium homeostasis and determine an impaired ventricular relaxation [68]. In the ischemic cascade, diastolic dysfunction appears before wall motion abnormalities, electrocardiographic changes or clinical symptoms, the alteration of ventricular distensibility being an early phenomenon in the setting of an acute MI [69]. Following scar formation, the degree of diastolic dysfunction is mainly determined by the extent of myocardial fibrosis and ventricular remodeling [66]. Extracardiac conditions such as aging, metabolic syndrome, obesity, and chronic kidney disease have been shown to perpetrate the LV fibrosis, dysfunction and remodeling through mechanisms involving systemic inflammation and endothelial dysfunction [45].

The cardiovascular continuum linking hypertension, myocardial ischemia, and heart failure is represented in Fig. 11.3.

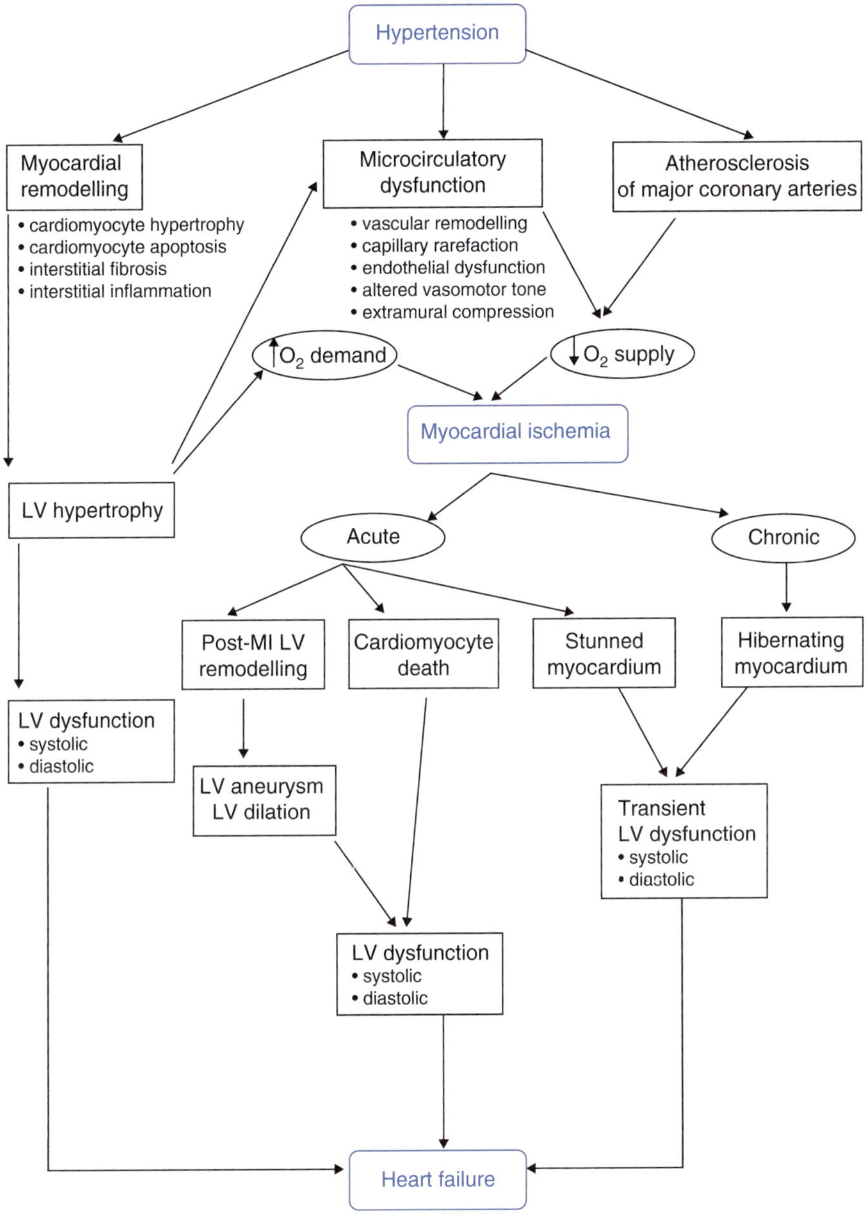

Fig. 11.3 The pathophysiological continuum of hypertension, myocardial ischemia, and heart failure. *LV* left ventricle, *MI* myocardial infarction

11.3 Conclusions

The controversial supremacy of hypertension or coronary artery disease as the only main determinant of HF needs to be readjusted. From a pathophysiological standpoint, hypertension, myocardial ischemia, and HF are closely intertwined clinical entities. While diastolic HF is mainly driven by uncontrolled hypertension, in systolic HF, coronary artery disease plays an important role [3]. This highlights the importance of a proper selection of patients who are eligible for myocardial revascularization and of a proper timing of intervention, besides the standard management of HF (regardless of etiology). However, since hypertension is a well-established risk factor for coronary atherosclerosis, it is equally important to detect and treat hypertensive patients as a means of preventing both atherogenesis and cardiac target organ damage. While attention shifted during the last decades to the discovery of innovative therapeutic approaches for HF, the importance of hypertension awareness, treatment, and adequate control is currently understated. Interrupting the continuum of hypertension—ischemia—HF would lead to a decrease in morbidity and mortality for these patients, subsequently reducing the global burden of cardiovascular disease.

References

1. Oh GC, Cho HJ. Blood pressure and heart failure. Clin Hypertens. 2020;26:1.
2. https://www.who.int/news-room/fact-sheets/detail/hypertension. Accessed 6 Mar 2022.
3. Kazzam E, Ghurbana BA, Obineche EN, Nicholls MG. Hypertension—still an important cause of heart failure? J Hum Hypertens. 2005;19(4):267–75.
4. Levy D, Larson MG, Vasan RS, Kannel WB, Ho KK. The progression from hypertension to congestive heart failure. JAMA. 1996;275:1557–62.
5. Messerli FH, Rimoldi SF, Bangalore S. The transition from hypertension to heart failure: contemporary update. JACC Heart Fail. 2017;5:543–51.
6. Iriarte M, Murga N, Sagastagoitia D, Morillas M, Boveda J, Molinero E, Etxebeste J, Salcedo A, Rodriguez E, Ormaetxe JM. Classification of hypertensive cardiomyopathy. Eur Heart J. 1993;14 Suppl J:95–101.
7. Yip GW, Fung JWH, Tan YT, Sanderson JE. Hypertension and heart failure: a dysfunction of systole, diastole or both? J Hum Hypertens. 2009;23:295–306.
8. Burkhoff D, Maurer SM, Packer M. Heart failure with a normal ejection fraction, is it really a disorder of diastolic function? Circulation. 2003;107:656–8.
9. Mottram PM, Haluska BA, Leano R, Carlier S, Case C, Marwick T. Relation of arterial stiffness to diastolic dysfunction in hypertensive heart disease. Heart. 2005;91:1551–6.
10. Borlaug BA, Melenovsky V, Russell SD, Kessler K, Pacak K, Becker LC, Kass DA. Impaired chronotropic and vasodilator reserves limit exercise capacity in patients with heart failure and a preserved ejection fraction. Circulation. 2006;114:2138–47.
11. Drazner MH. The progression of hypertensive heart disease. Circulation. 2011;123:327–34.
12. Spinar J. Hypertension and ischemic heart disease. Cor Vasa. 2012;54:E433–8.
13. Dorobantu M, Onciul S, Tautu OF, Cenko E. Hypertension and ischemic heart disease in women. Curr Pharm Des. 2016;22:3885–92.
14. Chobanian AV, Bakris GL, Black HR, Cushman WC, Green LA, Izzo JL Jr, Jones DW, Meterson BJ, Oparil S, Wright JT Jr, Roccella EJ. The seventh report of the joint national

committee on prevention, detection, evaluation, and treatment of high blood pressure: the JNC 7 report. JAMA. 2003;289:2560–72.
15. Olafiranye O, Zizi F, Brimah P, Jean-louis G, Makaryus AN, McFarlane S, Ogedegbe G. Management of hypertension among patients with coronary heart disease. Int J Hypertens. 2011;2011:653903. https://doi.org/10.4061/2011/653903.
16. Kannel WB, Dawber TR, Kagan A, Revotskie N, Stokes JIII. Factors of risk in the development of coronary heart disease—six-year follow-up experience. The Framingham Study. Ann Intern Med. 1961;55:33–50.
17. Yusuf S, Hawken S, Ounpuu S, Dans T, Avezum A, Lanas F, McQueen M, BudajA PP, Varigos J, Lisheng L. Effect of potentially modifiable risk factors associated with myocardial infarction in 52 countries (the INTERHEART study): case-control study. Lancet. 2004;364:937–52.
18. Cachofeiro V, Miana M, de las Heras N, Martin-Fernandez B, Ballesteros S, Balfagon G, Lahera V. Inflammation: a link between hypertension and atherosclerosis. Curr Hypertens Rev. 2009;5:40–8.
19. Bleakley C, Hamilton PK, Pumb R, Harbinson M, McVeigh GE. Endothelial function in hypertension: victim or culprit? J Clin Hypertens. 2015;17(8):651–4.
20. Hansson GK. Inflammation, atherosclerosis, and coronary artery disease. N Engl J Med. 2005;352:1685–95.
21. Libby P. Inflammation in atherosclerosis. Nature. 2002;420:868–74.
22. Bautista LE, Lopez-Jaramillo P, Vera LM, Casas JP, Otero AP, Guaracao AI. Is C-reactive protein an independent risk factor for essential hypertension? J Hypertens. 2001;19:857–61.
23. Bautista LE, Vera LM, Arenas IA, Gamarra G. Independent association between inflammatory markers (C-reactive protein, interleukin 6 and TNF alpha) and essential hypertension. J Hum Hypertens. 2005;19:149–54.
24. Peeters ACTM, Netea MG, Janssen MCH, Kullberg BJ, Van der Meer JWM, Thien T. Proinflammatory cytokines in patients with essential hypertension. Eur J Clin Invest. 2001;31:31–6.
25. Boos CJ, Lip GYH. Is hypertension an inflammatory process? Curr Pharm Des. 2006;12:1623–35.
26. Phillips MI, Kagiyama S. Angiotensin II as a pro-inflammatory mediator. Curr Opin Investig Drugs. 2002;3:569–77.
27. Agbor-Etang BB, Setaro JF. Management of hypertension in patients with ischemic heart disease. Curr Cardiol Rep. 2015;17:119.
28. Rodriguez JA, Orbe J, Paramo JA. Metalloproteases, vascular remodelling, and atherothrombotic syndromes. Rev Esp Cardiol. 2007;60:959–67.
29. Dirksen MT, van der Wal AC, van den Berg FM, van der Loos CM, Becker AE. Distribution of inflammatory cells in atherosclerotic plaques relates to the direction of flow. Circulation. 1998;98:2000–3.
30. Chirinos JA, Segers P. Noninvasive evaluation of left ventricular afterload: part 1: pressure and flow measurements and basic principles of wave conduction and reflection. Hypertension. 2010;56(4):555–62.
31. Nadruz W. Myocardial remodeling in hypertension. J Hum Hypertens. 2015;29:1–6.
32. du Cailar G, Pasquie JL, Ribstein J, Mimran A. Left ventricular adaptation to hypertension and plasma renin activity. J Hum Hypertens. 2000;14:181–8.
33. Stanton T, Dunn FG. Hypertension, left ventricular hypertrophy and myocardial ischemia. Med Clin North Am. 2017;101(1):29–41.
34. Chobanian AV. Vascular effects of systemic hypertension. Am J Cardiol. 1992;69:3E–7E.
35. Camici PG, Olivotto I, Rimoldi OE. The coronary circulation and blood flow in left ventricular hypertrophy. J Mol Cell Cardiol. 2012;52(4):857–64.
36. Bassenge E, Heusch G. Endothelial and neuro-humoral control of coronary blood flow in health and disease. Rev Physiol Biochem Pharmacol. 1990;116:77–165.
37. Duncker DJ, Bache RJ. Regulation of coronary blood flow during exercise. Physiol Rev. 2008;88(3):1009–86.

38. Hamasaki S, Al Suwaidi J, Higano ST, Miyauchi K, Holmes DR Jr, Lerman A. Attenuated coronary flow reserve and vascular remodeling in patients with hypertension and left ventricular hypertrophy. J Am Coll Cardiol. 2000;35:1654–60.
39. Feihl F, Liaudet L, Levy BI, Waeber B. Hypertension and microvascular remodelling. Cardiovasc Res. 2008;78:274–85.
40. Tang EH, Vanhoutte PM. Endothelial dysfunction: a strategic target in the treatment of hypertension? Pflugers Arch. 2010;459:995–1004.
41. Taddei S, Virdis A, Ghiadoni L, Sudano I, Salvetti A. Endothelial dysfunction in hypertension. J Cardiovasc Pharmacol. 2001;38:S11–4.
42. Rethy L, McCabe M, Pool LR, Vu THT, Kershaw KN, Yancy C, Vupputuri S, Feinglass J, Khan SS. Contemporary rates of hospitalization for heart failure in young and middle-aged adults in a diverse US state. Circ Cardiovasc Qual Outcomes. 2020;13(11):e007014.
43. Braunwald E. Heart failure. JACC Heart Fail. 2013;1:1–20.
44. Lloyd-Jones DM, Larson MG, Ep L, Beiser A, D'Agostino RB, Kannel WB, Murabito JM, Vasan RS, Benjamin EJ, Levy D. Lifetime risk for developing congestive heart failure: the Framingham Heart Study. Circulation. 2002;106(24):3068–72.
45. Elgendy IY, Mahtta D, Pepine CJ. Medical therapy for heart failure caused by ischemic heart disease. Circ Res. 2019;124(11):1520–35.
46. Gheorghiade M, Sopko G, De Luca L, Velasquez EJ, Parker JD, Binkley PF, Sadowski Z, Golba KS, Prior DL, Rouleau JL, Bonow RO. Navigating the crossroads of coronary artery disease and heart failure. Circulation. 2006;114:1202–13.
47. Kenchaiah S, Pfeffer MA, Sutton MSJ, Plappert T, Rouleau JL, Lamas GA, Sasson Z, Parker JO, Geltman EM, Solomon SD. Effect of antecedent systemic hypertension on subsequent left ventricular dilation after acute myocardial infarction (from the survival and ventricular enlargement trial). Am J Cardiol. 2004;94(1):1–8.
48. Richards AM, Nicholls MG, Troughton RW, Lainchbury JG, Elliott J, Framptom C, Espiner EA, Crozier IG, Yandle TG, Turner J. Antecedente hypertension and heart failure after myocardial infarction. J Am Coll Cardiol. 2002;39(7):1182–8.
49. Albakri A. Ischemic heart failure: a review of clinical status and meta-analysis of diagnosis and clinical management methods. Clin Med Invest. 2018;3(4):1–15.
50. Libby P, Theroux P. Pathophysiology of coronary artery disease. Circulation. 2005;111(25):3481–8.
51. Sutton MG, Sharpe N. Left ventricular remodelling after myocardial infarction: pathophysiology and therapy. Circulation. 2000;101:2981–8.
52. Pagliaro BR, Cannata F, Stefanini GG, Bolognese L. Myocardial ischemia and coronary disease in heart failure. Heart Fail Rev. 2020;25(1):53–65.
53. Jenca D, Melenovsky V, Stehlik J, Stanek V, Kettner J, Kautzner J, Adamkova V, Wohlfahrt P. Heart failure after myocardial infarction: incidence and predictors. ESC Heart Fail. 2021;8(1):222–37.
54. Niccoli G, Burzotta F, Galiuto L, Crea F. Myocardial no-reflow in humans. J Am Coll Cardiol. 2009;54:281–92.
55. Shah RV, Rong J, Larson MG, Yeri A, Ziegler O, Tanriverdi K, Murthy V, Liu X, Xiao C, Pico AR, Huan T, Levy D, Lewis GD, Rosenzweig A, Vasan RS, Das S, Freedman JE. Associations of circulating extracellular RNAs with myocardial remodelling and heart failure. JAMA Cardiol. 2018;3:871–6.
56. Carvalho LSF, Bogniotti LAC, de Almeida OLR, e Silva JC, Nadruz W, Coelho OR, Sposito AC. Change of BNP between admission and discharge after ST-elevation myocardial infarction (Killip I) improves risk prediction of heart failure, death, and recurrent myocardial infarction compared to single isolated measurement in addition to the GRACE score. Eur Heart J Acute Cardiovasc Care. 2019;8;643–51
57. Asleh R, Enriquez-Sarano M, Jaffe AS, Manemann SM, Weston SA, Jiang R, Roger VL. Galectin-3 levels and outcomes after myocardial infarction: a population-based study. J Am Coll Cardiol. 2019;73:2286–95.

58. Jenkins WS, Roger VL, Jaffe AS, Weston SA, Abou Ezzeddine OF, Jiang R, Manemann SM, Enriquez-Sarano M. Prognostic value of soluble ST2 after myocardial infarction: a community perspective. Am J Med. 2017;130:1112.e9–e15.
59. Sulo G, Igland J, Vollset SE, Nygard O, Ebbing M, Sulo E, Egeland GM, Tell GS. Heart failure complicating acute myocardial infarction; burden and timing of occurrence: a nation-wide analysis including 86771 patients from the cardiovascular disease in Norway (CVDNOR) project. J Am Heart Assoc. 2016;5:e002667.
60. Santoro GM, Carrabba N, Migliorini A, Parodi G, Valenti R. Acute heart failure in patients with acute myocardial infarction treated with primary percutaneous coronary intervention. Eur J Heart Fail. 2008;10:780–5.
61. Sorajja P, Gersh BJ, Cox DA, McLaughlin MG, Zimetbaum P, Costantini C, Stuckey T, Tcheng JE, Mehran R, Lansky AJ, Grines CL, Stone GW. Impact of multivessel disease on reperfusion success and clinical outcomes in patients undergoing primary percutaneous coronary intervention for acute myocardial infarction. Eur Heart J. 2007;28:1709–16.
62. Melchior T, Rask-Madsen C, Torp-Pedersen C, Hildebrandt P, Kober L, Jensen G. The impact of heart failure on prognosis of diabetic and non-diabetic patients with myocardial infarction: a 15-year follow-up study. Eur J Heart Fail. 2001;3:83–90.
63. Wellings J, Kostis JB, Sargsyan D, Cabrera J, Kostis WJ. Risk factors and trends in incidence of heart failure following acute myocardial infarction. Am J Cardiol. 2018;122:1–5.
64. Wang J, Khoury DS, Yue Y, Torre-Amione G, Nagueh SH. Preserved left ventricular twist and circumferential deformation, but depressed longitudinal and radial deformation in patients with diastolic heart failure. Eur Heart J. 2008;29:1283–9.
65. Sanderson JE. Heart failure with a normal ejection fraction. Heart. 2007;93:155–8.
66. Redfield MM. Heart failure with preserved ejection fraction. N Engl J Med. 2016;375:1868–77.
67. Zile MR, Brutsaert DL. New concepts in diastolic dysfunction and diastolic heart failure: part II: causal mechanisms and treatment. Circulation. 2002;105:1503–8.
68. Vanhecke TE, Kim R, Raheem SZ, McCullough PA. Myocardial ischemia in patients with diastolic dysfunction and heart failure. Curr Cardiol Rep. 2010;12:216–22.
69. Ishii K, Suyama T, Imai M, Maenaka M, Yamanaka A, Makino Y, Seino Y, Shimada K, Yoshikawa J. Abnormal regional left ventricular systolic and diastolic function in patients with coronary artery disease undergoing percutaneous coronary intervention: clinical significance of post-ischemic diastolic stunning. J Am Coll Cardiol. 2009;54(17):1589–97.

Role of Endothelial Dysfunction in the Progression from Hypertension to Heart Failure

12

Alessandro Mengozzi, Stefano Taddei, and Agostino Virdis

12.1 Endothelial Function

The endothelium is the inner layer of the vessel in direct contact with the blood. Thus, it is involved in the dialogue between the circulating factors and the vascular wall. However, defining endothelium as a barrier is reductive. Indeed, the most distal branches of the microvasculature, namely small resistance arteries (100–300 μm) and capillaries (7–100 μm), are the mediators of the dialogue between the bloodstream and the tissues, controlling critical actions needed to provide adequate oxygen and nutrients supply [1]. Being the microvasculature composed mainly of endothelial cells, it is clear how the endothelium role is central in regulating microvascular homeostasis and, consequently, the tissues' health.

The endothelium can respond to several stimuli, such as mechanical (e.g., shear stress) and chemical (e.g., acetylcholine and bradykinin) ones. Endothelial cells are endowed with paracrine and juxtracrine capacities: they can modulate the vessel tone, adapting the fluxes of gases (oxygen and carbon dioxide) and nutrients based on the tissue demands and the whole-body substrates' availability. Its several mediators also participate in different steps of the coagulation process and platelet

A. Mengozzi
Department of Clinical and Experimental Medicine, University of Pisa, Pisa, Italy

Health Science Interdisciplinary Center, Sant'Anna School of Advanced Studies, Pisa, Italy

Center for Translational and Experimental Cardiology (CTEC), Department of Cardiology, Zurich University Hospital and University of Zürich, Schlieren, Switzerland
e-mail: alessandro.mengozzi@unipi.it

S. Taddei · A. Virdis (✉)
Department of Clinical and Experimental Medicine, University of Pisa, Pisa, Italy
e-mail: stefano.taddei@unipi.it; agostino.virdis@unipi.it

M. Dorobantu et al. (eds.), *Hypertension and Heart Failure*, Updates in Hypertension and Cardiovascular Protection,
https://doi.org/10.1007/978-3-031-39315-0_12

aggregation. Latest but not least, they are involved in multiple stages of the inflammatory response, such as cellular proliferation and leukocyte adhesion [2].

Nitric oxide (NO) is the most important of the endothelium-derived factors. It can induce the relaxation of the underlying vascular smooth muscle cells and, acting as a second messenger molecule, activate endothelial-specific molecular pathways involved in all the aforementioned functions. NO is produced by the NO synthase, which is present in humans in several isoforms. Two are constitutive: endothelial NO synthase (eNOS) and neuronal NO synthase (nNOS), and another one is inducible (iNOS). eNOS has a heme group able to bind oxygen that used as a substrate together with L-arginine. eNOS catalyze their transformation to L-citrulline and NO. The cofactor tetrahydrobiopterin (BH_4) keeps eNOS in a dimeric form, granting the efficiency of the reaction. When BH_4 is absent, eNOS shifts to monomeric and becomes uncoupled. Uncoupled eNOS synthesizes superoxide anion instead of NO, unbalancing the oxidative tone toward a pro-oxidant state and perturbating the microvascular homeostasis. Asymmetric dimethylarginine (ADMA) and NG-monomethyl L-arginine (L-NMMA) are endogenous eNOS uncouplers, impairing eNOS and thus endothelial cells [3]. NO synthesis is constitutive; however, it can increase as a compensatory mechanism when the vessel is solicited by mechanical or chemical stimuli, as described above. Given the crucial role of NO in controlling many different endothelial processes, we identify endothelial function as the capacity of the endothelium to match the demanding stimulus with adequate NO production. When this is not possible, we enter an impaired condition, such as the one present in many metabolic diseases. NO-reduced bioavailability is considered a hallmark of endothelial dysfunction [4].

The circulatory system, in particular, the endothelium is one of the first, if not the first, targets of the cardiometabolic risk factors [5]. Therefore, it is not surprising that endothelial dysfunction characterizes cardiometabolic diseases since its earliest steps. Similarly, endothelial dysfunction is also a marker of accelerated (i.e., unhealthy) aging [6]. In general, endothelial dysfunction describes a very early subclinical damage: the first stage of a transition from a healthy to a pathologic phenotype. This, if not interrupted, is deemed to evolve into chronic and irreversible damage.

12.2 The Cardiometabolic Disease: Focus on the Microcirculatory System

According to the classical dogma, microvascular damage was considered consequential to alteration of central hemodynamics. This concept was also shared by the metabolic disease, where microvascular remodeling was thought to ensue systemic metabolic alterations [7]. The measure of large vessels' arterial stiffness (e.g., pulse wave velocity) has also been identified as a marker of arterial aging and thus of an impaired cardiometabolic state. However, epidemiologic studies [8] and GWAS [9] failed to recognize apparent determinants of macrovascular aging. Recent observations have thus led to the hypothesis of a paradigm change [10]: microvascular impairment precedes macrovascular damage. This is true, at least in some cardiometabolic phenotypes. The metabolic changes suffered by the microvasculature lead to an epigenetic signature that persists after the remotion of the *noxa*, as for the phenomenon of hyperglycemic memory [11]. This subclinical and irreversible damage exposes the microvessel to further deterioration. Also, small and large vessels dialogue and develop a vicious cyclical relationship: perturbation of microcirculatory homeostasis leads to macrovascular damage, further stressing the former.

Arterial hypertension is a state in which microvascular endothelial dysfunction is associated with systemic pressure overload. It is acknowledged that a pressure overload induces concentric LV hypertrophy [12]. Hypertensive heart disease is generally described as a multistep process [12]: the first stage is detecting a diastolic dysfunction, followed by apparent concentric hypertrophy that, when decompensated, leads to heart failure with preserved ejection fraction (HFpEF). If the macrovascular damage is perpetuated, the remodeling turns to left ventricle dilation with the loss of the contractile efficiency and the onset of heart failure with reduced ejection fraction (HFrEF). However, the hemodynamic alteration implicit in arterial hypertension is not the only one responsible for this evolution. Instead, it plays a controversial role. This is particularly evident in hypertensive patients, where intensive blood pressure control slows but does not halt the progression toward HFpEF [13, 14]. As pointed out by Paulus et al. about ten years ago [15], the underlying endothelial dysfunction is probably the main driver of this pathologic evolution. Hypertension is associated with accelerated senescence [16, 17] of the microvasculature, which is often neglected but plays a pivotal role in driving the deterioration of the phenotype [18]. The microvascular impairment, both systemic [19] and coronaric [20], hinders the endothelium's protective ability to regulate the physiological vascular remodeling. The accelerated microvascular aging, in turn, ultimately leads to overt heart failure.

12.3 Determinants of Endothelial Dysfunction in Hypertension and Their Role in the Progression to Heart Failure

The local increase in reactive oxygen species production is the final effector of the several pathways involved in the microvascular endothelial dysfunction [21]. Although several cellular factors contribute to a pro-oxidant environment (e.g., xanthine oxidase, cyclooxygenase, and cytochrome P-450 oxygenase), two are the major responsible: NAPDH-oxidases (in particular, the isoforms 2 and 4) and mitochondria. The increased ROS levels reduce NO bioavailability by reacting with NO. This, in turn, leads to the peroxynitrite generation, which elevates the pro-oxidant tone. Also, ROS uncouples eNOS, which causes a decrease in NO levels. Uncoupled eNOS itself is a source of ROS. This vicious cycle, present in several cardiometabolic diseases, ultimately concurs to a severely reduced NO bioavailability, a hallmark of endothelial dysfunction that further fuels the systemic low-grade proinflammatory state [22, 23]. As endothelial dysfunction is crucial in the progression from hypertension to heart failure (particularly HFpEF), several observations have been conducted to explore its main determinants.

12.3.1 Microvascular Inflammation

Arterial hypertension is characterized by a systemic low-grade proinflammatory state [24]. Salt-rich diet, angiotensin II, and aldosterone are independent propellants for the innate immune response [25]. Tissue sodium accumulation in salt-sensitive individuals promotes macrophage infiltration that further supports vascular inflammation. Angiotensin II induces hyperactivation of toll-like receptor (TLR)-4 [26] and the formation of neutrophil extracellular trap formation (NETosis) [27].

The abnormal, persistent activation of the immune-inflammatory response leads to increased oxidative stress that deteriorates the vascular wall, reduces NO availability, and fuels the vicious cycle leading to cardiometabolic derangement. Inflammatory cytokines (i.e., TNF-α, IL-2, IL-4, IL-5, IL-6, IL-10, IL-17a, and IFN-γ) are higher in hypertensive patients. Conversely, transforming growth factor (TGF)-β, platelet-derived growth factor, and insulin-like growth factor are reduced. This environment unbalances the microvascular remodeling toward a pathologic evolution [4]. The narrowing of the lumen and stiffening of the arteriolar wall lead to increased peripheral vascular resistance and reduced oxygen supply. Neoangiogenetic pathways activate to compensate for the impaired nourishment [28], but if the *stimuli* are not removed, this compensatory response per se becomes detrimental. Ultimately, this ubiquitous (subcutaneous, muscular, visceral, and coronary districts are all involved [15, 17]) microvascular damage induces the differentiation of the endothelial cells into a proinflammatory phenotype, which directly fosters the systemic low-grade inflammatory response [29, 30].

In the progression from hypertension to HFpEF, inflammatory markers such as soluble interleukin-1 receptor-like 1 (IL1R-L1), C-reactive protein (CRP), and growth differentiation factor-15 (GDF-15) [10] are found elevated in the plasma. At the myocardium level, the dysfunctional endothelium expresses adhesion molecules such as vascular cell adhesion molecule (VCAM) and E-selectin. It also secretes inflammatory cytokines and chemokines, which, in turn, recruit inflammatory cells. T- and B-lymphocytes and monocytes are attracted from the intravascular district to the interstitial space, the latter turning into macrophages. The final step of this process is the fibroblast activation, the consequent fibrosis, and the resulting increase in myocardial diastolic stiffness. It should be noted the crucial role of microvascular endothelial dysfunction in this process, as the increased hemodynamic overload is not sufficient to induce the increased myocardial stiffness.

The number of randomized controlled trials on anti-inflammatory molecules administered to people with cardiovascular disease has been growing in the last decade. Though some reported negative results, the most recent of them showed a good effect on cardiovascular risk reduction. These trials, however, have been conducted on people with a more deranged cardiovascular phenotype than the one discussed in this chapter. As microvascular inflammation is now an acknowledged hallmark of hypertension, anti-inflammatory therapies have been recently proposed as new drugs for the optimal control of arterial hypertension [31].

12.3.2 Cyclooxygenase-2

Cyclooxygenase-2 (COX-2) is another source of reactive oxygen species (ROS) in the peripheral microcirculation of hypertensive individuals [32]. This mechanism appears to be the derangement of a compensatory response, as it is not detected in healthy individuals [32, 33]. The first observations described an overexpression of the molecule at the endothelial level in the microcirculation of patients with hypertension. However, subsequent studies also hypothesized a role for the vascular smooth cell cyclooxygenase [34]. ROS per se increases the expression of COX-2 and its protein levels in both endothelial and vascular smooth cells. This effect has been suggested for both NAPDH-oxidase and mitochondria-derived ROS. In hypertension, many pathways of microvascular damage are mediated by a ROS-driven COX-2 hyperexpression. ROS and COX-2 generate a vicious cycle that produces microvascular damage by markedly unbalancing the oxidative tone toward a pro-oxidant state [35].

In vitro [36, 37] and ex vivo [32] evidence have shown promising therapeutic effects of selective COX-2 inhibition. However, the effect is controversial in humans, especially in patients suffering from cardiometabolic disease. In patients with coronary artery disease (CAD), no strong impact on endothelial dysfunction is observed after selective COX-2 inhibition with rofecoxib [38, 39]. Moreover, they have been associated with an increased risk of thrombotic cardiovascular events [40]. This appears to be particularly evident in older patients with previous cardiovascular disease [40, 41]. However, recent observations have been trying to repurpose these compounds [42]. It should be noted that the majority of these trials address a more deranged cardiovascular phenotype than the one in which the endothelial dysfunction is most involved (e.g., the microvascular cardiometabolic disease). This might be a primary reason to explain these discrepancies.

12.3.3 Mineralocorticoid Receptor

The endothelial mineralocorticoid receptor (EC-MR) is a contributor to microvascular dysfunction. While early observations reported conflicting results, it is now acknowledged that EC-MR has a different role in healthy patients compared to patients with cardiometabolic disease. In the former, it contributes to normal vasodilation, while in the latter it is a direct mediator of microvascular endothelial dysfunction [43]. Aberrant EC-MR activation leads to NADPH-oxidase-4 gene induction by a pathway involving small GTPase Rac1. This leads to an increased ROS generation and thus to an unbalance toward a pro-oxidant state [26]. Indeed, in patients with primary hyperaldosteronism, spironolactone improves endothelial function [44] by increasing the superoxide dismutase and catalase expression and reducing NADPH-oxidase-derived ROS [44, 45]. Though there are only a few in vivo observations, EC-MRs seem to play a different and sex-specific role in the deterioration of the microvascular homeostasis, involving potassium channels and endothelial-derived hyperpolarization [46]. In female mice, EC-MR impairs microvascular dysfunction by reducing the expression of estrogen-receptor α [47]. However, further studies are needed to assess these sex-specific mechanisms in humans.

12.3.4 Sphingolipid Metabolism and Sortilin

Although the most relevant determinants have been discussed above, the cardiometabolic *spectrum* is full of potential detrimental mediators of microvascular dysfunction. In particular, the sphingolipid metabolism is attracting attention as several observations have found its involvement in cardiometabolic disease [48, 49]. Sphingosine-1-phosphate (S1P) is a bioactive lysophospholipid regulator of several cellular processes and is involved in the pathogenesis of hypertension. Sortilin, a member of the vacuolar protein sorting 10 (VPS10P) receptor's family, has been recently reported to induce this S1P-mediated microvascular dysfunction in patients with subclinical or clinical arterial hypertension [50]. In detail, sortilin increases acid sphingomyelinase activity, leading to hyperproduction of S1P. The consequent hyperactivation of the NADPH-oxidase-2 at the cellular level boosts a pro-oxidant endothelial cell phenotype leading to the progressive loss of microvascular homeostasis and depicting the progression toward uncontrolled hypertension [50].

12.4 The Progression to HFpEF

The recent acknowledgement of the existence of HFpEF, an inflammatory-metabolic phenotype of heart failure [51], has further stressed the tight connection between hypertension and heart failure. Nowadays, HFpEF accounts for 50% of heart failure cases, and its prevalence relative to HfrEF continues to rise [10]. Arterial hypertension is the most prevalent comorbidity in HFpEF [52]. As described in the previous paragraphs, there is an apparent *continuum* between hypertension and HFpEF driven by microvascular endothelial dysfunction [15]. Microvascular endothelial function is essential to maintain adequate local homeostasis. Focusing on the heart, NO diffuses into cardiomyocytes from the coronary microcirculation and contributes directly to the modulation of cardiac functions. An impaired NO availability translates into altered activation of the soluble guanylate cyclase-cGMP-protein kinase G pathway. This signaling is crucial for the cardiomyocytes' health as its impairment leads to progressive cardiac remodeling. The myocardial relaxation became slower as the cardiomyocytes increase their stiffness. This process hesitates in cardiac interstitial fibrosis and cardiac chambers' hypetrofy, which are acknowledged features of HFpEF.

In turn, the comorbidities observed in HFpEF are worsened due to the systemic inflammation HFpEF-driven [53]. Systemic and coronary microvascular endothelium, together with the epicardial adipose tissue, became themselves promoters of the low-grade inflammatory state and the consequent increase of ROS production both at the coronary and peripheral microcirculation. Additionally, neurohormonal upregulation in HFpEF leads to increased oxidative stress and upregulation of collagen synthesis. HFpEF, in the end, worsens systemic endothelial dysfunction leading to progressive heart failure [54].

From the clinical point of view, past and ongoing studies are trying to evaluate microvascular endothelial dysfunction to identify a phenotype at greater risk of progression toward HFpEF [55, 56]. However, the structural assessment of the microcirculation and its endothelial function still present significant flaws. For instance, its gold standard technique is the wire or pressure micromyography [57], which relies on an invasive approach (fat tissue biopsies) and an operator dependency, highly limiting its diffusion in the routinary clinical practice [1]. Some noninvasive methods, like flow-mediated dilation, fail to explore the microvasculature district [58]. However, a growing number of techniques is becoming able to test endothelial function in humans. Peripheral arterial tonometry measures the reactive hyperemia index (RHI), which is related to coronary microvascular dysfunction, is reduced in people with HFpEF, and defines an HFpEF subgroup at high cardiovascular risk [59]. Exercise intolerance, another relevant feature of HFpEF, is related to endothelial dysfunction and can be measured with cardiopulmonary exercise testing with the assessment of VO_2 max, which shows a strong mortality prognostic power in HFpEF [60, 61]. Retinal endothelial dysfunction, measured through flicker light-induced arteriolar dilation, is another reliable and rather feasible noninvasive index showing promising prognostic potential [62]. However, no studies have explored retinal microcirculatory in people with HFpEF yet.

12.5 Conclusions

Endothelial dysfunction is a shared feature of the disease comprised in the cardiometabolic spectrum as well as a crucial stage in the transition from hypertension to heart failure (Fig. 12.1). Still, the mechanisms underlying its onset are to be grasped entirely. Further studies are required to comprehend this early step of the phenotype derangement, providing insight and thus novel specific biomarkers and potential new therapeutic targets. Together with noninvasive techniques, these need to be identified and implemented in the clinical routine to adequately explore endothelial dysfunction in humans and efficiently address it with therapeutic strategies.

Acknowledgments Figure 12.1 is created using biorender.com

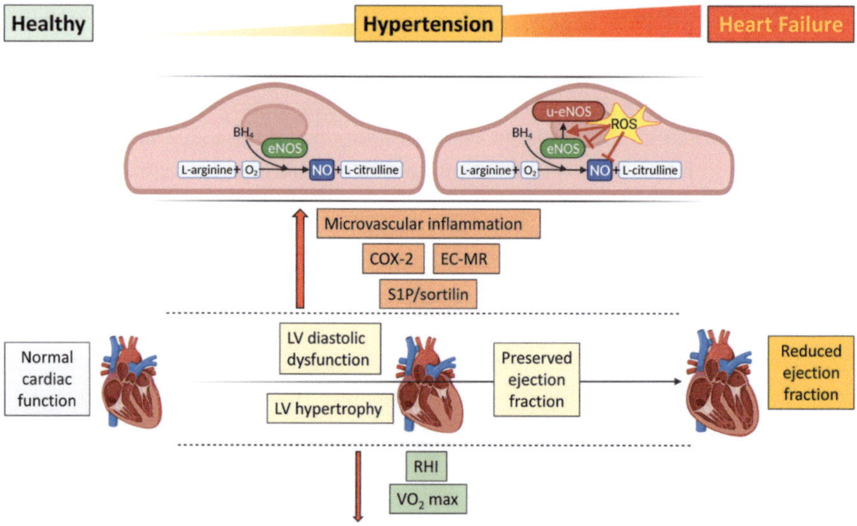

Fig. 12.1 Endothelial dysfunction plays a crucial role in the transition from healthy to hypertension and then heart failure. In patient with arterial hypertension, the low-grade inflammatory state, the overexpression of COX-2, the hyperactivation of the mineralocorticoid receptors at endothelial level, and the detrimental role of sortilin through the hyperactivation of the sphingosine-1-phosphate pathway all perturbate the microvascular homeostasis by increasing ROS production and inducing or worsening endothelial dysfunction. Microvascular endothelial dysfunction, in turn, damages the cardiomyocyte's health and leads to progressive pathologic cardiovascular remodeling. Though still few, some clinical indices, such the RHI and the VO_2 max, are able to appreciate the onset and the worsening of the deterioration of microvascular dysfunction. *COX-2* cyclo-oxygenase-2, *EC-MR* endothelial cell mineralocorticoid receptor, *eNOS* endothelial nitric oxide synthase, *LV* left ventricular, *NO* nitric oxide, *RHI* reactive hyperemia index, *ROS* reactive oxygen species, *u-eNOS* uncoupled eNOS

References

1. Rizzoni D, Mengozzi A, Masi S, Agabiti Rosei C, De Ciuceis C, Virdis A. New noninvasive methods to evaluate microvascular structure and function. Hypertension. 2022;79:874–86.
2. Mengozzi A, Pugliese NR, Taddei S, Masi S, Virdis A. Microvascular inflammation and cardiovascular prevention: the role of microcirculation as earlier determinant of cardiovascular risk. High Blood Press Cardiovasc Prev. 2021;29:41–8.
3. Vanhoutte PM, Shimokawa H, Feletou M, Tang EH. Endothelial dysfunction and vascular disease—a 30th anniversary update. Acta Physiol. 2017;219:22–96.
4. Mengozzi A, Pugliese NR, Chiriaco M, Masi S, Virdis A, Taddei S. Microvascular ageing links metabolic disease to age-related disorders: the role of oxidative stress and inflammation in promoting microvascular dysfunction. J Cardiovasc Pharmacol. 2021;78:S78–87.
5. Cooke J. Endotheliopathy of obesity. Circulation. 2020;142:380–3.
6. Masi S, Colucci R, Duranti E, Nannipieri M, Anselmino M, Ippolito C, et al. Aging modulates the influence of arginase on endothelial dysfunction in obesity. Arterioscler Thromb Vasc Biol. 2018;38:2474–83.
7. Barrett EJ, Liu Z, Khamaisi M, King GL, Klein R, Klein BEK, et al. Diabetic microvascular disease: an Endocrine Society scientific statement. J Clin Endocrinol Metab. 2017;102:4343–410.

8. Kim M, Kim M, Yoo HJ, Lee SY, Lee SH, Lee JH. Age-specific determinants of pulse wave velocity among metabolic syndrome components, inflammatory markers, and oxidative stress. J Atheroscler Thromb. 2018;25:178–85.
9. Park S, Lee J, Kim B, Lee S, Chang H, Choi D, et al. Lack of association between arterial stiffness and genetic variants by genome-wide association scan. Blood Press. 2015;24:258–61.
10. Shah SJ, Kitzman DW, Borlaug BA, van Heerebeek L, Zile MR, Kass DA, et al. Phenotype-specific treatment of heart failure with preserved ejection fraction: a multiorgan roadmap. Circulation. 2016;134:73–90.
11. Paneni F, Volpe M, Lüscher TF, Cosentino F. SIRT1, p66(Shc), and Set7/9 in vascular hyperglycemic memory: bringing all the strands together. Diabetes. 2013;62:1800–7.
12. Messerli FH, Rimoldi SF, Bangalore S. The transition from hypertension to heart failure: contemporary update. JACC Heart Fail. 2017;5:543–51.
13. Fan L, Pan JA, Lin H, Wang CQ, Zhang JF, Gu J. Optimal management of blood glucose, blood pressure and atrial fibrillation to reduce the risk of heart failure with preserved ejection fraction. Intern Med J. 2022;52:301–9.
14. Myhre PL, Selvaraj S, Solomon SD. Management of hypertension in heart failure with preserved ejection fraction: is there a blood pressure goal? Curr Opin Cardiol. 2021;36:413–9.
15. Paulus WJ, Tschöpe C. A novel paradigm for heart failure with preserved ejection fraction: comorbidities drive myocardial dysfunction and remodeling through coronary microvascular endothelial inflammation. J Am Coll Cardiol. 2013;62:263–71.
16. Taddei S, Virdis A, Mattei P, Ghiadoni L, Fasolo CB, Sudano I, et al. Hypertension causes premature aging of endothelial function in humans. Hypertension. 1997;29:736.
17. Taddei S, Virdis A, Ghiadoni L, Salvetti G, Bernini G, Magagna A, et al. Age-related reduction of NO availability and oxidative stress in humans. Hypertension. 2001;38:274–9.
18. Taddei S, Virdis A, Mattei P, Ghiadoni L, Gennari A, Fasolo CB, et al. Aging and endothelial function in normotensive subjects and patients with essential hypertension. Circulation. 1995;91:1981–7.
19. Weerts J, Mourmans SGJ, Barandiarán Aizpurua A, Schroen BLM, Knackstedt C, Eringa E, et al. The role of systemic microvascular dysfunction in heart failure with preserved ejection fraction. Biomolecules. 2022;12:278.
20. Carrick D, Haig C, Maznyczka AM, Carberry J, Mangion K, Ahmed N, et al. Hypertension, microvascular pathology, and prognosis after an acute myocardial infarction. Hypertension. 2018;72:720–30.
21. Taddei S, Virdis A, Ghiadoni L, Magagna A, Salvetti A. Vitamin C improves endothelium-dependent vasodilation by restoring nitric oxide activity in essential hypertension. Circulation. 1998;97:2222–9.
22. Virdis A, Duranti E, Colucci R, Ippolito C, Tirotta E, Lorenzini G, et al. Ghrelin restores nitric oxide availability in resistance circulation of essential hypertensive patients: role of NAD(P)H oxidase. Eur Heart J. 2015;36:3023–30.
23. Virdis A. Endothelial dysfunction in obesity: role of inflammation. High Blood Press Cardiovasc Prev. 2016;23:83–5.
24. Lam CSP, Voors AA, de Boer RA, Solomon SD, van Veldhuisen DJ. Heart failure with preserved ejection fraction: from mechanisms to therapies. Eur Heart J. 2018;39:2780–92.
25. Lu X, Crowley SD. Inflammation in salt-sensitive hypertension and renal damage. Curr Hypertens Rep. 2018;20:103.
26. Guzik TJ, Touyz RM. Oxidative stress, inflammation, and vascular aging in hypertension. Hypertension. 2017;70:660–7.
27. Targoński R, Sadowski J, Price S, Targoński R. Sodium-induced inflammation-an invisible player in resistant hypertension. Hypertens Res. 2020;43:629–33.
28. Saltiel AR, Olefsky JM. Inflammatory mechanisms linking obesity and metabolic disease. J Clin Invest. 2017;127:1–4.
29. Virdis A, Santini F, Colucci R, Duranti E, Salvetti G, Rugani I, et al. Vascular generation of tumor necrosis factor-alpha reduces nitric oxide availability in small arteries from visceral fat of obese patients. J Am Coll Cardiol. 2011;58:238–47.

30. Virdis A, Duranti E, Rossi C, Dell'Agnello U, Santini E, Anselmino M, et al. Tumour necrosis factor-alpha participates on the endothelin-1/nitric oxide imbalance in small arteries from obese patients: role of perivascular adipose tissue. Eur Heart J. 2015;36:784–94.
31. Madhur MS, Elijovich F, Alexander MR, Pitzer A, Ishimwe J, Van Beusecum JP, et al. Hypertension: do inflammation and immunity hold the key to solving this epidemic? Circ Res. 2021;128:908–33.
32. Virdis A, Bacca A, Colucci R, Duranti E, Fornai M, Materazzi G, et al. Endothelial dysfunction in small arteries of essential hypertensive patients: role of cyclooxygenase-2 in oxidative stress generation. Hypertension. 2013;62:337–44.
33. Verma S, Raj SR, Shewchuk L, Mather KJ, Anderson TJ. Cyclooxygenase-2 blockade does not impair endothelial vasodilator function in healthy volunteers: randomized evaluation of rofecoxib versus naproxen on endothelium-dependent vasodilatation. Circulation. 2001;104:2879–82.
34. Muñoz M, Sánchez A, Pilar Martínez M, Benedito S, López-Oliva ME, García-Sacristán A, et al. COX-2 is involved in vascular oxidative stress and endothelial dysfunction of renal interlobar arteries from obese Zucker rats. Free Radic Biol Med. 2015;84:77–90.
35. Hernanz R, Briones AM, Salaices M, Alonso MJ. New roles for old pathways? A circuitous relationship between reactive oxygen species and cyclo-oxygenase in hypertension. Clin Sci (Lond). 2014;126:111–21.
36. de Oliveira HT, Couto GK, Davel AP, Xavier FE, Rossoni LV. Chronic cyclooxygenase-2 inhibition prevents the worsening of hypertension and endothelial dysfunction induced by ouabain in resistance arteries of spontaneously hypertensive rats. Vascul Pharmacol. 2021;139:106880.
37. Virdis A, Colucci R, Fornai M, Blandizzi C, Duranti E, Pinto S, et al. Cyclooxygenase-2 inhibition improves vascular endothelial dysfunction in a rat model of endotoxic shock: role of inducible nitric-oxide synthase and oxidative stress. J Pharmacol Exp Ther. 2005;312:945–053.
38. Title LM, Giddens K, McInerney MM, McQueen MJ, Nassar BA. Effect of cyclooxygenase-2 inhibition with rofecoxib on endothelial dysfunction and inflammatory markers in patients with coronary artery disease. J Am Coll Cardiol. 2003;42:1747–53.
39. Bogaty P, Brophy JM, Noel M, Boyer L, Simard S, Bertrand F, et al. Impact of prolonged cyclooxygenase-2 inhibition on inflammatory markers and endothelial function in patients with ischemic heart disease and raised C-reactive protein: a randomized placebo-controlled study. Circulation. 2004;110:934–9.
40. Schjerning AM, McGettigan P, Gislason G. Cardiovascular effects and safety of (non-aspirin) NSAIDs. Nat Rev Cardiol. 2020;17:574–84.
41. Lu J, Wang L, Bennamoun M, Ward I, An S, Sohel F, et al. Machine learning risk prediction model for acute coronary syndrome and death from use of non-steroidal anti-inflammatory drugs in administrative data. Sci Rep. 2021;11:18314.
42. Chen W, Zhong Y, Feng N, Guo Z, Wang S, Xing D. New horizons in the roles and associations of COX-2 and novel natural inhibitors in cardiovascular diseases. Mol Med. 2021;27:123.
43. Davel AP, Anwar IJ, Jaffe IZ. The endothelial mineralocorticoid receptor: mediator of the switch from vascular health to disease. Curr Opin Nephrol Hypertens. 2017;26:97–104.
44. Chen L, Ding ML, Wu F, He W, Li J, Zhang XY, et al. Impaired endothelial repair capacity of early endothelial progenitor cells in hypertensive patients with primary hyperaldosteronemia: role of 5,6,7,8-tetrahydrobiopterin oxidation and endothelial nitric oxide synthase uncoupling. Hypertension. 2016;67:430–9.
45. Silva MA, Bruder-Nascimento T, Cau SB, Lopes RA, Mestriner FL, Fais RS, et al. Spironolactone treatment attenuates vascular dysfunction in type 2 diabetic mice by decreasing oxidative stress and restoring NO/GC signaling. Front Physiol. 2015;6:269.
46. Davel AP, Lu Q, Moss ME, Rao S, Anwar IJ, DuPont JJ, et al. Sex-specific mechanisms of resistance vessel endothelial dysfunction induced by cardiometabolic risk factors. J Am Heart Assoc. 2018;7:e007675
47. Biwer LA, Carvajal BV, Lu Q, Man JJ, Jaffe IZ. Mineralocorticoid and estrogen receptors in endothelial cells coordinately regulate microvascular function in obese female mice. Hypertension. 2021;77:2117–26.

48. Józefczuk E, Nosalski R, Saju B, Crespo E, Szczepaniak P, Guzik TJ, et al. Cardiovascular effects of pharmacological targeting of sphingosine kinase 1. Hypertension. 2020;75:383–92.
49. Jozefczuk E, Guzik TJ, Siedlinski M. Significance of sphingosine-1-phosphate in cardiovascular physiology and pathology. Pharmacol Res. 2020;156:104793.
50. Di Pietro P, Carrizzo A, Sommella E, Oliveti M, Iacoviello L, Di Castelnuovo A, et al. Targeting the ASMase/S1P pathway protects from sortilin-evoked vascular damage in hypertension. J Clin Invest. 2022;132:e146343.
51. Packer M, Lam CSP, Lund LH, Maurer MS, Borlaug BA. Characterization of the inflammatory-metabolic phenotype of heart failure with a preserved ejection fraction: a hypothesis to explain influence of sex on the evolution and potential treatment of the disease. Eur J Heart Fail. 2020;22:1551–67.
52. Cohen JB, Schrauben SJ, Zhao L, Basso MD, Cvijic ME, Li Z, et al. Clinical phenogroups in heart failure with preserved ejection fraction: detailed phenotypes, prognosis, and response to spironolactone. JACC Heart Fail. 2020;8:172–84.
53. Pugliese NR, Paneni F, Mazzola M, De Biase N, Del Punta L, Gargani L, et al. Impact of epicardial adipose tissue on cardiovascular hemodynamics, metabolic profile and prognosis in heart failure. Eur J Heart Fail. 2021;23:1858–71.
54. D'Amario D, Migliaro S, Borovac JA, Restivo A, Vergallo R, Galli M, et al. Microvascular dysfunction in heart failure with preserved ejection fraction. Front Physiol. 2019;10:1347.
55. Bayes-Genis A, Bisbal F, Núñez J, Santas E, Lupón J, Rossignol P, et al. Transitioning from preclinical to clinical heart failure with preserved ejection fraction: a mechanistic approach. J Clin Med. 2020;9:1110.
56. Ekström M, Hellman A, Hasselström J, Hage C, Kahan T, Ugander M, et al. The transition from hypertension to hypertensive heart disease and heart failure: the PREFERS Hypertension study. ESC Heart Fail. 2020;7:737–46.
57. Masi S, Rizzoni D, Taddei S, Widmer RJ, Montezano AC, Lüscher TF, et al. Assessment and pathophysiology of microvascular disease: recent progress and clinical implications. Eur Heart J. 2021;42:2590–604.
58. Flammer AJ, Anderson T, Celermajer DS, Creager MA, Deanfield J, Ganz P, et al. The assessment of endothelial function: from research into clinical practice. Circulation. 2012;126:753–67.
59. Shah SJ, Lam CSP, Svedlund S, Saraste A, Hage C, Tan RS, et al. Prevalence and correlates of coronary microvascular dysfunction in heart failure with preserved ejection fraction: PROMIS-HFpEF. Eur Heart J. 2018;39:3439–50.
60. Pugliese NR, De Biase N, Conte L, Gargani L, Mazzola M, Fabiani I, et al. Cardiac reserve and exercise capacity: insights from combined cardiopulmonary and exercise echocardiography stress testing. J Am Soc Echocardiogr. 2021;34:38–50.
61. Shafiq A, Brawner CA, Aldred HA, Lewis B, Williams CT, Tita C, et al. Prognostic value of cardiopulmonary exercise testing in heart failure with preserved ejection fraction. The Henry Ford HospITal CardioPulmonary EXercise Testing (FIT-CPX) project. Am Heart J. 2016;174:167–72.
62. Theuerle JD, Al-Fiadh AH, Amirul Islam FM, Patel SK, Burrell LM, Wong TY, et al. Impaired retinal microvascular function predicts long-term adverse events in patients with cardiovascular disease. Cardiovasc Res. 2021;117:1949–57.

Panendothelitis Due to the SARS COV 2 Infection: Consequences on Hypertension and Heart Failure

13

Victor Voicu

13.1 Differential Specific Characteristics of SARS CoV 2 and Pathogenic Consequences

13.1.1 Proximal Origin, Differential Genomic Characteristics

SARS-Cov-2 is the seventh known coronavirus to infect the human species[3]. It is impossible to say that the virus is the result of natural evolution and selection or is actually the result of multiple passages (in vitro or in vivo). SARS-CoV-2 has a genomic sequence that is 89% identical to SARS-Cov and 50% identical to MERS. Genomic features can only partially explain the transmissibility of SARS-Cov-2 in humans, as per the authors cited above [1]. It is stated in the before-mentioned article that the evidence showing that it is not a deliberately manipulated virus is practically impossible to confirm, as it is impossible to disprove the theories regarding the origin of the virus in question. The acquisition of both the polybasic cleavage site and the prediction of O-linked glycans is arguments against scenarios based on the current scientific culture.

The authors believe that the genesis of a polybasic cleavage site would have required repeated passages in cell or animal cultures that have a similar human ACE2 receptor, but this type of work has not been previously described. Furthermore, trying to predict the existence of O-linked glycans in cell cultures is quite impossible.

Upon careful analysis of SARS-Cov-2, we can agree that it meets all the criteria described for a biological weapon [2]. The SARS-Cov-2 virus has an affinity for the human host receptor, namely the conversion enzyme 2 (ACE2), about 10 to 20 times higher than the SARS-CoV, which is a key feature [3–5].

V. Voicu (✉)
Department of Pharmacology, Toxicology and Clinical Psychopharmacology, Faculty of Medicine, Carol Davila University of Medicine and Pharmacy, Bucharest, Romania

M. Dorobantu et al. (eds.), *Hypertension and Heart Failure*, Updates in Hypertension and Cardiovascular Protection,
https://doi.org/10.1007/978-3-031-39315-0_13

The conversion enzyme 2 (ACE2) is the receptor for SARS-Cov-2 that facilitates its entry into the cell with the co-participation of the serine protease TMPRSS2 (kallikrein 13) located on the cell surface [6].

It is already clearly defined that the princeps aggression takes place in the airways, especially in the lungs. The ACE2 receptor enzyme is expressed in the type II alveolar epithelium, more specifically in the secretory, ciliated, and basal glands, with a high expression in the nasal cavity and the hair cells. The authors comment that although the lung is the main target, just about 2% of the other lung tissues have ACE2 present. Instead, it is mentioned that ACE2 is present extensively in the small intestine, gallbladder, kidneys, testicles, thyroid, adipose tissue, myocardium, vagina, breast, ovary, and pancreas. Implicitly, it is admitted that the wide tissue distribution of ACE2 represents the structural functional background of the consequences of the interaction with SARS-CoV-2 and therefore generates multiple organ dysfunctions in certain patients [7]. The three characteristic acquisitions of the binding domain are: high affinity for ACE2, activation of the binding domain spike protein by an endogenous human protease (furin), and the masking of spike protein by an O-glycan group (also known as glycan shield) which thus protects it from host antibodies.

The identification of the 7 coronaviruses belonging to the family that contaminates various animals and humans was an important step.

People were surprised that an epidemic emergency caused by a new coronavirus associated with severe acute respiratory syndrome in 2003. Historically, the human coronavirus, CoV (CoV strains 229 E and OC43), generates mild, low-intensity infections associated with common symptoms, with multiple reinfections, even with the same strain. Coronaviruses have been recognized in veterinary medicine to cause enteric or respiratory infections, fatal to animals. The new strains of the new Covid variant can cause fatal disease in the naive animal population.

Recent data suggest that SARS-CoV has a wide range of non-human hosts. Recently, a phylogenetic analysis suggests that SARS-CoV may have evolved from a past recombination between S genes belonging to mammals (group 1) and those belonging to an avian group, making it a mosaic. It has been widely suggested that the CoV virus can evolve in the host population, obtaining new tissue tropisms or virulence through mutations or recombination, and that this type of event can also occur if the SARS-Cov virus keeps persisting in the human population [8].

These characteristics of the SARS-Cov-2 binding domain are compared to a secret weapon giving, in fact, the virus the determinants of a biological weapon [2, 9].

13.1.2 SARS-CoV-2 Virus Cellular Receptor

The receptor for SARS-CoV-2 is common to that previously identified in SARS-CoV, with the only difference being that the new coronavirus has an affinity 10 to 20 times higher than the membrane receptor of human host cells [10, 11]. SARS-CoV-2 infection, just like the SARS-CoV-1 infection, generates a reduction in the concentration of ACE2, following the interaction with this enzyme, as a result of the internalization of the complex formed with the virus. The generally known pulmonary and visual protective effect of ACE2 is thus severely disturbed, a process correlated with the severity of ARDS [12] (acute respiratory distress syndrome). The aggravation of tissue damage by increasing the concentration of angiotensin II, following a decrease in the presence of ACE2 and a reducement in the expression of this enzyme, suggests, at the same time, that this process has a compensatory, protective role, reducing the intracellular virus and its dispersion in the body.

13.1.3 Pathogenesis

Covid-19 is a global disease (pandemic) that has shocked the world with its morbidity and mortality, unprecedented in modern medicine [13, 14]. The cited authors point out that the pathogenicity and virulence of SARS-Cov-2 allows this virus to be considered a virtual, natural, perfect biological weapon. Another type of consideration is avoided, currently without concrete evidence, but only speculative, lacking another type of scientific explanation [13, 14].

It is considered that Covid-19 disease is not a clearly defined pathogenic entity, but a gradually evolving pathology, defining a series of stages, characterized by different biological and molecular mechanisms. Thus, 5 different phases are identified: incubation, respiratory damage, proinflammatory phase, prothrombotic phase, remission, or death.

In the initial stage of the disease, the virus directly triggers a series of cytopathic effects, followed by phases generated by the host reactions, all of them initiated, sustained, and amplified by the immune system, the complement, and the hemostatic system. These host reactions, including the cytokine storm, can be considered "almost suicidal" [13, 14].

13.1.4 Pathogenesis Associated with the Five Phases

During the incubation phase, which usually lasts 2–11 days, with an average of 6 days, the patient is contagious 1–3 days before the onset of symptoms. People with asymptomatic or mild symptoms (asymptomatic) might add up to over 50% of the cases. This pre-symptomatic phase is considered the most critical because along with the asymptomatic and slightly symptomatic forms, the viral load is comparable to that of patients with obvious disease. The relevance of viral transmission in this first phase is clear, with evidence being shown that 50–80% of new cases can be attributed to transmission in this category of patients. The relatively long incubation period and the high proportion of asymptomatic and with rough symptoms, together, explain the pandemic aspect and the implicit risk of contamination, with all the prophylaxis measures established on a public level.

Among the significant differences that distinguish SARS-CoV-2 virus from SARS-CoV is the ability of the former to intensely infect the upper respiratory tract (nasopharynx, oropharynx) due to a high affinity for the ACE2 receptor (which is highly expressed in those tissues). The SARS-CoV-2 virus can be easily detected in the upper respiratory tract during the prodromal period, which explains its rapid and highly effective transmissibility compared to SARS-CoV [15].

Phase 2, which involves the pathogenic process of airway involvement, includes about 30% of all patients with Covid-19, with typical symptoms of the disease. Respectively, they involve: fever (85% of all patients), dyspnea (70–80%), cough (40–80%), olfactory and gustatory dysfunction (50%), myalgia (30%), and gastrointestinal symptoms (10–30%).

In the respiratory tract, a series of typical symptoms might occur, such as: pharyngeal pain, nasal congestion, and rhinorrhea—all of them present in a small percentage of studied patients (about 20%). Dyspnea has a high prevalence, reflecting the progress of the respiratory involvement of the disease, usually with a tomographic equivalent, the image that is nowadays commonly referred to as "ground glass opacity," with or without pleural thickening.

This phase presents itself as a "silent hypoxemia," especially in the elderly population, as it is characterized by hypoxemia and respiratory failure without actual symptoms of respiratory distress.

The third phase, which occurs in about 15% of all patients infected with SARS-CoV-2, is considered the most challenging and pathogenically active [13, 14]. As we have mentioned above, this phase is a consequence of the direct cytopathic effects of the virus multiplying in the lung. This is a pro-inflammatory phase characterized by an exaggerated reaction of the host to the pathogen, either locally (pulmonary) or systemically, respectively suggesting an intense systemic inflammatory reaction (severe inflammatory response syndrome).

The partially simplified explanation is that the viral aggression activates the monocyte-macrophage line triggering the activation of the synthesis and secretion of proinflammatory cytokines (iL-6, iL-2, iL-7, TNF-α, and others). The consequences of the interaction of SARS-CoV-2 with ACE2 involve a clear increase in angiotensin II, with a decrease in its activity, generating vasoconstriction, proinflammatory reaction, oxidative stress, and fibrotic damage.

The virus damages T lymphocytes resulting in lymphopenia, while also producing cytopathic effects with disruptive consequences in cytokines. Consequently, lymphopenia, immunological disorders, cytokine storm with local lesions, and multiple organ damage occur.

It is emphasized that the cytopathic signs can be through direct viral effects on organs and tissues, mainly affecting the heart, kidney, liver, and intestine [16].

The fourth phase, which occurs in about 5% of those infected with SARS-CoV-2, is characterized by the onset of micro and macrovascular thrombosis, favored by the endothelial inflammatory process, which is also the primary focus of this paper. This thrombosis is favored by the inflammatory prothrombotic process at the venous and arterial level, with consequences mainly on pulmonary level, with extensive inflammatory infiltrates, microvascular thromboembolism, and complete destruction of the interalveolar septum. It is estimated that this pathogenic process must be controlled early in the course of the disease with anticoagulants and fibrinolytics [16]. About a third of Covid-19 patients who die are thought to have at least partially some evidence of pulmonary thrombosis as a cause of death.

In essence, Covid-19 disease is significantly and definitely associated with a pro-coagulating status characterized by micro and macrovascular thrombosis, which eventually results in lung damage and multiple organ failure. The pathogenic background is systemic endothelitis (i.e., panendothelitis) [17].

In the fifth phase, either death or remission occurs. Mortality in intensive care is 20–80%, with its variability being related to a multitude of factors. Lethality is a consequence of ARDS, pulmonary thrombosis, acute renal failure, acute heart damage, infections, and/or multiple organ failure [16].

We emphasize the conclusion of the quoted author, which essentially translates into the fact that the Covid-19 disease is a pathogenic storm produced by a perfect, possibly lethal, biological weapon, while also being virtually natural. For comparison, the general and specific characteristics of the before-mentioned biological weapon can be assessed in a recent synthesis [17].

13.2 ACE2, Physiological Functions, Tissue Distribution, and Functional Correlations

13.2.1 ACE2, the Receptor for SARS-CoV-2 Virus

The wide tissue distribution of ACE2, especially in the vascular endothelium and ultimately in the lung epithelial cells, largely explains the complex systemic pathogenesis (systemic endotheliitis), lung damage, and multiple organ pathology. A first sequence of the pathogenesis of Covid-19 disease is as follows:

- Respiratory viral invasion-interaction with ACE2—endothelial damage—activation of macrophages—triggering of the proinflammatory cascade (iL-6, iL-10, iL-2, TNT-α, VEGF, free radicals, and protease);
- Acute lung damage—respiratory failure [18].

Avoiding the immune response, a response that should occur after SARS-CoV-2 penetrates the host cells, thus allows cell infection by triggering viral multiplication unhindered. As evidenced, in some patients, a rapid evolution of the disease, the genesis of an acute respiratory distress syndrome (ARDS), a systemic inflammatory syndrome as well as multiple organ dysfunction syndrome may occur, with a lethality of about 10%.

The authors who analyzed this sequence of pathological events clearly point out that they are generated due to the presence of ACE2 in the lungs, heart, kidneys, liver, and other organs and tissues [19]. We reiterate the essential importance of the wide distribution of ACE2 in the vascular endothelium, with systemic consequences. As we will see, the endothelial expression of ACE has major consequences on this structure, causing endothelial dysfunction with thrombotic consequences, emboli, etc.

In terms of consequences, the interaction of ACE2 with SARS-CoV-2, there are controversies, in this case, regarding the process of attaching the virus to the ACE2 receptor. It is accepted that ACE2 expression does not differ significantly in patients with pre-existing comorbidities, except for smokers who have a high enzyme level compared to non-smokers [18].

There is another observation, namely that after infection with SARS-Cov-2, ACE2 reduces its activity either by reducing the genesis or by passing into plasma ("shedding").

The authors suggest that after binding of SARS-CoV-2 to ACE2, the latter, due to its complex role as a regulator of immune response, cytokine secretion, and post-infection viral replication, basically acts as a receptor and as a mediator of intracellular viral invasion.

We have also analyzed the major importance of the enzyme ACE2 in controlling the complex functions of the RAA system [20–22]. It is well known that the inhibitory substances of another enzyme, namely ACE (angiotensin converting enzyme I) or angiotensin II receptor blockers (AT1), are efficient antihypertensive agents.

It seems that these two types of antihypertensive drugs are able to indirectly increase the expression of ACE2. But the increase in ACE2 expression in conditions of SARS-CoV-2 infection facilitates the respective viral infection [23]. At an immediate direct analysis, the consequences seem undesirable and avoidable. Implicitly, a new idea arose, namely to suspend the antihypertensive therapy with conversion enzyme inhibitors and/or angiotensin II receptor 1 blockers in these particular situations. However, there are multicenter retrospective studies showing that antihypertensive treatment with conversion enzyme inhibitors and/or angiotensin receptor blockers has been associated with decreased mortality of any cause [23].

It should be noted that in a meta-analysis of over 1500 patients with Covid-19 in China, the prevalent comorbidities are high blood pressure (21%), diabetes (9.7%), and cardiovascular disease (8.4%). A similar study was conducted in Italy, highlighting the association between hypertension and mortality in intensive care services, with a prevalence of 49% hypertension, diabetes (17%), and cardiovascular disease, in terms of comorbidities [24].

We have analyzed various syntheses [20–22] targeting the complex functions of ACE2 in the context of the renin–angiotensin–aldosterone system, especially its wide tissue distribution and its protective functions on pulmonary, vascular, and cardiac levels. These complex protective effects are achieved, as it results from the cited synthesis, by decreasing and limiting the substrate of the other enzyme that controls the RAA system, respectively, ACE2 which degrades ang II converting it into angiotensin (1-7), with effects on MaR receptor (with an indirect effect as an antagonist of the stimulation of angiotensin II receptor 1) [23].

The aggression of the SARS-CoV-2 virus is, at least partially, the consequence of the interaction with ACE2, expressed in epithelial cells, endothelial cells of small vessels, allowing the virus to penetrate into the host cell. The processes controlled by ACE2, in the functional complex of the RAA system, are pulmonary and vascular protectors [20–22]. In various pathological conditions significant as severity, such as acute pulmonary aggression, pulmonary fibrosis, pulmonary hypertension, and ARDS, the other component, somewhat antagonistic, related to the functions of ACE and angiotensin II is functionally predominant, respectively, having a dysfunctional and lesional character, not being antagonized in the SARS-CoV 2 infection (Fig. 13.1).

ACE/ACE2 imbalance can be generated apart from genetic factors and pathological conditions, including (as per our case) SARS-CoV-2 aggression on the body, interaction with pulmonary and endothelial ACE2, decreased the presence of ACE2 in the tissue with loss of lung protective component, endothelial and tissue in general [20–22, 25–28].

The great variability of the clinical forms of Covid-19 can be explained by the functional status of each organism, especially regarding the immune system. This variability could be explained, from asymptomatic or asymptomatic forms to severe forms and critical life-threatening conditions. "There are no diseases, only sick people" is the old synthesis of the concept of personalized medicine or, as we more recently call it, precision medicine.

One cause of variability may be genetic polymorphism or genetic determinism of the ACE-A22 ratio, with a possible decrease in ACE2 activity, which predisposes to a vulnerability, to the genesis of acute respiratory distress syndrome, with high mortality.

It is noted that the ACE2 receptor binding region for SARS-CoV-2 is highly conserved in both animals and humans [29]. It should be noted that ACE2 is a receptor for SARS-CoV-2 in humans, dogs, cats, tigers, minks, and others, except for snakes.

ACE2 expression increases in human vessels, exposed to the pathological conditions of diabetes. Compared to the control lot, the expression of the enzyme also

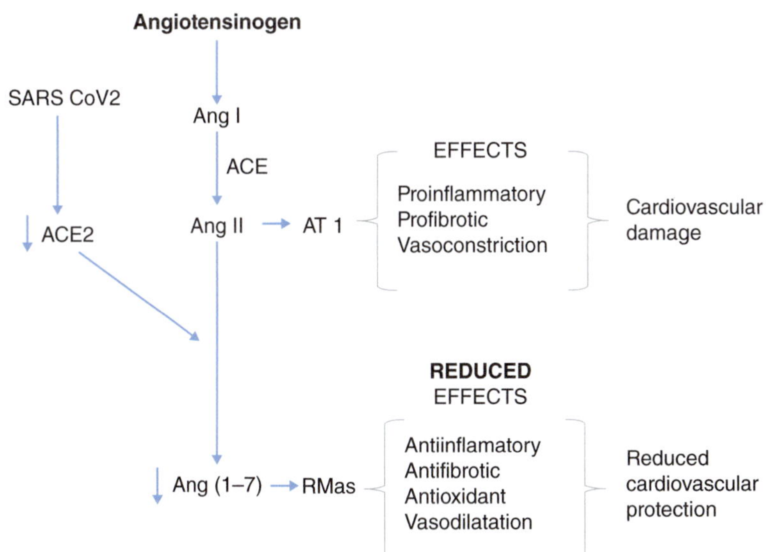

Fig. 13.1 The effects of the Sars Cov2 interaction with ACE2 and its consequences on RAA system

increases in conditions of cerebral ischemia, which explains the vulnerability to SARS-Cov-2 in these conditions [30].

In theory, the protective functional role of ACE2 in various pulmonary, vascular, and cardiovascular diseases is convincingly highlighted, which is mainly due to the dynamic balance of the two systems: ACE/Ang II/AT1 receptor versus ACE2/Ang (1-7)/Receiver Mas. ACE2 also has a functional role in the regulation of the quinine–kallikrein system which catalyzes the genesis of kallikrein, a type B2 receptor agonist and which, by hydrolysis, forms the ligand for the B1 receptor, des-Arg 9-bradykinin, and lys-des-Arg 9-bradykinin. Increasing the ACE2 levels is considered a protective factor for organs and tissues by reducing pulmonary vasoconstriction, remodeling, atherosclerosis, blood pressure, myocardial hypotrophy, fibrosis, and ventricular remodeling.

In contrast, decreased ACE2 levels is consistent with increased pulmonary vessel permeability, pulmonary edema, ARDS, atherosclerosis, hypertension, and pulmonary insufficiency. ACE2/Ang axis (1-7) Mas and inhibition of B1 receptor activation produce anti-inflammatory effects [31].

The problem with interpreting the increase of ACE2, in certain comorbidities, is that there is no possible correlation with the severity of Covid-19 disease. From systematic meta-analysis studies based on 2,065,805 Covid-19 treated hypertensive patients, the likelihood of death in patients treated with RAAS blockers was 35% lower than that in the untreated group.

An alternative interpretation for comorbidities, dysfunction, or absence of ACE2 leads to increased Ang II, hyperactivation of AT1 receptors with the production of inflammation, fibrosis, oxidative stress, vasoconstriction, and angioedema.

Increased Ang II levels in the plasma of Covid-19 patients with pneumonia compared to healthy control groups are correlated with viral load and lung damage. In those with comorbidities and RAAS dysfunction, AngS modulation by SARS-CoV-2 can reach critical values and ultimately produce a fatal inflammatory status [31].

13.2.2 ACE2 Receptor Polymorphism of SARS-CoV-2 and the Correlation with Clinical Forms of Covid-19

A recent study, linked to the broad spectrum of animal species that are potentially susceptible to SARS-CoV-2 infection, points out that a certain sequence of four amino acids in certain molecular regions of ACE2 is essential for interaction with SARS-CoV-2 binding domain spikes and, in particular, the presence of N90 makes susceptibility to the virus infection in question very likely [32].

Based on the "silico" studies, that are targeting a species that may be intermediate, capable of transmitting the virus to humans, hypothetically shows that species with amino acids K31 and K353 are more likely to interact with SARS-CoV-2 [33, 34]. The authors give relevant examples, among which the species *F. Catus*, which possesses in its structure the before-mentioned amino acids, while the resistant species express amino acids such as K31N and K353H.

Based on objective premises for a reasonable prediction, we have created a comprehensive list of species that are prone to infection, as well as a list of animals unfit to bind SARS-Cov-2 protein S1, on a cellular level, ACE2 [32].

The wide clinical variability of Covid-19 disease is due to variables, either known or often unknown, related to age, sex, host immune response, virus mutations, comorbidities, lifestyle, and genetic characteristics of the host [35, 36]. It is recognized that site-specific variations in the ACE2 gene may be considered the major genetic risk factor for Covid-19 [37]. Variants can affect ACE2 expression and potentially the function of the enzyme, contributing to the spread of the infection worldwide.

The problems generated by the wide individual variability of the response to the SARS-Cov-2 infection are reflected in the clinical forms, from asymptomatic to severe cirrhosis, and even potentially lethal effects. It is also significant that the basic reproductive number (RO) for SARS-CoV-2 was initially estimated at 2.68 resulting in an epidemic doubling at 6.4 days. Ro is the basic reproductive number, the number of secondary infections, when an indexed case is introduced in a population susceptible to infection. There are other estimates of Ro, up to 6.4. An important conclusion would be that Ro reflects the effectiveness of measures to prevent and control SARS-CoV-2 infection on a population level [38].

It is admitted that 33% of those infected with SARS-CoV-2 so far have not developed symptoms. On a large scale, the proportion of people who do not show symptoms but have tested positive for viral contamination is between 43 and 77%. However, patients with clinically asymptomatic infection may have objective pathological findings, such as "ground glass" images or atypical imaging abnormalities. Some asymptomatic patients may develop symptoms about 4–5 days after the positive RT-PCR test.

Symptomatic infection ranges from mild to critical. According to statistics from the CDC in China, out of 44,500 patients, the division based on the form of the disease is as it follows:

- 81% mild forms, with mild forms of pneumonia or without pneumonia.
- 14% severe illness (dyspnea, hypoxia, over 50% pulmonary involvement).
- 5% critical condition (respiratory failure, shock, multiple organ dysfunction).

In a study based on 1.3 million cases in the United States (CDC), at the end of May 2020, 14% were hospitalized, 2% were admitted to intensive care, and 5% died. Severe disease varies with age and comorbidities, a particularity that we have previously mentioned. Lethality increases exponentially with age, as shown by 27 studies on a community level, reaching 0.4% at 55 years, 1.4% at 65 years, 4.6% at 75 years, and 15% at 85 years old.

In hospitalized patients, for example, 2741 patients with Covid-19 hospitalized in New York City hospitals, out of 665 who needed ventilated, eventually 60 died. We are trying to analyze this wide variability based on the available molecular data, targeting the ACE2 receptor and the consequences of the interaction with the SARS-CoV-2 virus in human organs and tissues.

13.2.3 Tissue Expression of ACE2

Attempting to correlate ACE2 expression in different human tissues is of particular interest to clinical expression for both clinical forms and also to the specific type of pathological manifestations associated with damage to a particular organ or tissue [39].

In summary, the interaction of SARS-CoV-2 with ACE2, with an affinity of about 10–20 times higher than SARS-CoV-2, contributes to the high infectivity of the virus in question and the constant increase in the number of infected people [11]. The dissociation constant is much lower in SARS-CoV-2, allowing more intense binding at the ACE2 level than SARS-CoV-2, a process favoring the distribution of the high-density receptor enzyme in the oral cavity, nasal mucosa, lung, skin, lymph nodes, stomach, small intestine, colon, thymus, hematogenous marrow, kidney, spleen, and brain [19]. The authors rightly point out that this binding of the virus to ACE2, the cellular receptor, is the first essential step in the process of viral infection and at the same time a potential link that could be the target of pharmacotherapeutic means to block cell invasion [40]. SARS-CoV-2 binding at the ACE2 level reduces enzyme expression at the membrane level with consequences on the entire RAA system, modification of the ACE/ACE2 ratio, imbalance of the RAA system with functional predominance of ACE/AngII/RAT1, essentially predominates vasoconstriction and other described effects in context.

The kallikrein–kinin axis, a component that counterbalances the RAAS, has a cardiac and renal protective role [22]. Decreased ACE2 in the context of SARS-CoV-2 infection leads to an increase in bradykinin levels, which can activate the B2 receptor with consecutive vasoconstriction. It turns out that there could be, almost inevitably, a direct correlation between the expression of ACE2 in the tissues (both in the normal body, but especially in one with comorbidities) and the severity of Covid-19 disease [41].

One of the cited authors summarizes the data showing that ACE2 expression is elevated in cardiovascular disease, hypertension, lung disease and lung cancer, obesity, diabetes, male, old age, and conditions of use of R.A.A. system inhibitors [41]. All of these comorbidities are risk factors that increase the severity of the Covid-19 disease. We point out, once again, that all these comorbidities have one thing in common, namely the increase in ACE2 expression in the respective tissues, increasing the susceptibility to SARS-CoV-2 infection as well as the severity of the disease. Specifically, the phrase "increased ACE2 expression in tissues" has a quantitative reflection, the high risk target cells that contain more than 1% ACE2 expression [42]. This category includes the lower respiratory tract (2%), lung (>1%), heart (>7.5%), ileus (30%), esophagus (>1%), kidney (4%), and bladder (2.4%).

Interestingly, ACE2 in the nasal epithelium is correlated (proportionally) with age, thus explaining that young people have a low incidence of Covid-19.

Similarly, low ACE2 expression in both the nasal and bronchial epithelium in children compared to adults explains the low incidence of the disease in the first age group. A high expression of ACE2 has been identified in the mouth and tongue, explaining taste (ageuzia) and olfactory (anosmia) disorders.

Olfactory disorders are attributed to infection, not only in the olfactory receptor areas but also in the conduction pathways and the olfactory bulb. The expression of ACE in different areas of the brain is underlined in many studies, thus explaining the neurological symptoms. The viral load penetrates the brain, either hematologically, or through the cribriform plaque, penetrating the blood–brain barrier and reaching the ACE located in the brain tissue.

Although the viral invasion route is a respiratory one, there are alternative routes of viral contamination, namely the digestive tract, the fecal–oral route (which was suggested by highlighting the SARS-CoV-2 virus in fecal samples, although the nasopharyngeal samples were negative).

The quantitative expression of ACE2 explains the differences in susceptibility to Covid-19 disease of different ethnic groups, different sexes, and different ages.

Smoking increases 1.4 times the chance of developing severe forms of Covid-19 disease and 2.4 times the need for intensive care. The same authors mention that smoking increases the expression of ACE2 in the lungs and bronchi and nicotine stimulates the genesis of the enzyme in the brain, thus making these tissues more susceptible to SARS-CoV-2 infection.

A high-risk group of Covid-19 is that of pregnant women, with pregnancy increasing ACE2 expression in the uterus, kidneys, and placenta. Therefore, pregnant women with Covid-19 disease who end up being hospitalized, have a high risk of cesarean delivery, with a high risk of preterm birth.

It turns out that the many comorbidities mentioned are correlated with the high risk of facing severe forms of Covid-19 and high mortality.

The mechanisms of this increased susceptibility to SARS-CoV-2 infection and severe forms of Covid-19 disease, in the context of comorbidities, are mainly based on a chronic endothelial dysfunction, a background condition that promotes the genesis of SARS-CoV-2 endothelitis acceleration of vascular disease, preventing the recovery of ischemic disorders [22].

Following tissue damage, ACE2 is released from the plasma cell membrane and this circulating form can be used as a surrogate marker to predict mortality. This refers not only to Covid-19 but also to other conditions such as myocardial infarction, coronary heart disease, heart failure, arterial fibrillation, and aortic stenosis [43, 44]. SARS-CoV-2 interacts more frequently with ACE2 in men, as the level of this enzyme in plasma is usually higher in men than in women, but after a certain age becomes more common in women. This explains the differences in (higher) mortality in Covid-19 in men compared to women.

A problem of susceptibility and prognosis in Covid-19 is obesity which is associated with increased mortality. The risk factor for obesity increases sixfold, with a higher severity of Covid-19 disease, even after adjusting for age, gender, diabetes, hypertension, smoking, or dyslipidemia. People with diabetes and obesity express a high tissue level of ACE2 compared to normal weight diabetics [39]. It turns out that adipose tissue is a potential target for the SARS-CoV-2 virus and even a viral reservoir.

13.3 Vascular Endothelium and Its Functions as a Princeps Target in the Pathogenesis of Covid-19 Disease

13.3.1 Panendothelitis: The Pathogenic Basis Covid-19 Disease

It is admitted that viral aggression, on a cardiovascular level, is an extremely relevant prognostic factor regarding the respiratory disease in Covid-19. The results of the autopsies performed and published show aspects of great interest in understanding the pathogenesis of SARS-CoV-2. Thus, the published descriptions highlight evidence of direct endothelial damage and diffuse endothelial inflammation [45].

The authors emphasize the interaction of the virus with the ACE2 conversion enzyme, widely distributed in the pulmonary alveolar epithelium, in vascular endothelial cells, generating, lung damage and a wide endothelial (systemic) dysfunction associated with apoptosis. Endothelial dysfunction pathogenically underpins microvascular dysfunction, with altered vascular tone balance toward vasoconstriction and consequent ischemia, inflammation, edema, and a procoagulant status [46, 47]. Therefore, it is suggested that endothelitis in patients with Covid-19 induces several pathological alteration in multiple organs, due to the processes that it enhances induction of apoptosis and piroptosis and circulatory disorders [17, 48–52].

In essence, vascular damage is systemic (i.e., panendothelitis) and does not only appear in the lungs. Viral infection with SARS-CoV-2 causes cellular degeneration, necrosis, atrophy, focal hemorrhage, and infiltration of inflammatory cells into various organs such as the heart, blood vessels, liver, kidneys, spleen, brain, and skin [50, 53]. Although it is clear that panendothelitis (systemic endothelial dysfunction or systemic vasculitis) involves a systemic infection associated with the tissue distribution of ACE2, the latter phrase "systemic infection" (with SARS-CoV-2) is relevant in its anatomical form [17, 47, 51, 52].

The etiology and pathogenesis of respiratory failure and multiple organ failure in Covid-19 deaths are based on the endothelial dysfunction, thrombosis, and inflammation, with thrombosis being, implicitly, the consequence of the endothelial dysfunction [54].

13.3.2 Relevant Morphofunctional Characteristics of the Vascular Endothelium

Several morphofunctional features of the vascular endothelium show a great adaptation to functions, reflected by the well-known expression "division of labor" [55]. The cited author uses two relevant examples: the endothelium covering the segment of the postcapillary venules is "responsible" primarily for mediating leukocyte trafficking, while the arteriolar endothelium regulates the vasomotor tone. Coupled with other "evasive" characteristics, the endothelium shows a great morphological and functional heterogeneity, being considered to have an enormous behavioral (functional) repertoire. It seems that the heterogeneity of the endothelium has its immediate mechanism, of its specific properties, the response to the extracellular environment (reversible response) which can be differentiated, along with a number of other epigenetic properties [55].

That actually means that we are talking about a heterogeneity from an adaptive-evolutionary process of a cellular community that defines the endothelium. Regarding the functions (role) of the endothelium in a pathological context, phrases such as "endothelial cell activation" and "endothelial cell dysfunction" are used.

As we have previously stated, the endothelium has two "states," "off" and "on". Protective properties that in the "quiet" state are anticoagulant, antiadhesive, vasodilator properties, later become procoagulant, proadhesive, and vasoconstrictor during the active state. However, it is considered that vascular endothelial activation is not an "all or nothing" process commonly found in physiology. Moreover, correlated with its heterogeneity, the vascular endothelium has a broad spectrum of reactivity. The endothelium is usually active, receptive, and reactive to the environment (extracellular). The activated state can be adaptive or non-adaptive [55], an aspect of special interest. In essence, the endothelial dysfunction frequently invoked as a background of some phenomena of great interest in pathology is by definition maladaptive, especially regarding the vasodilating function.

The paradoxical vasoconstrictor effect of acetylcholine on atherosclerotic coronary arteries has been described since 1986 [56]. The vasodilating effect of acetylcholine is characteristic of the whole endothelium.

We evoke one of the first definitions of endothelial dysfunction, conceived as a non-adaptive change caused by pathophysiological stimuli, resulting in a local, acute, or chronic alteration and the interaction between cellular and macromolecular components in circulation with the blood vessel wall [57, 58].

Note that the activated endothelium is a state of an intact endothelium. The transition between proper functioning and dysfunction is not yet very clearly defined [59].

Treatment should reduce or prevent excessive response of the activated endothelium, this condition, within physiological limits, being characteristic of the functional endothelium [60].

In the context of endothelial dysfunction, the vasoconstrictive effect is generated by the direct action on the smooth muscles.

A concern of contemporary medicine is summed up by the question: is endothelial dysfunction reversible? Obviously, the approach aims at either a spontaneous reversibility or a follow-up to the treatment. The quoted author mentions the effects of improving endothelial function by hypocholesterolemic drugs administered for a period of about 6 months as well as antioxidant drugs (that promote the preservation and availability of NO, inactivated by free radicals).

Other classes of drugs that improve endothelial function are conversion enzyme inhibitors and calcium channel blockers. Vascular lesions increase in indirect correlation with vessel size, suggesting that in Covid-19 endothelitis targets small vessels and not the main coronary vessels. The presence of inflammatory neuropathy of the epicardial nerves is found as additional evidence that SARS-CoV-2 has an angio and neurotropic affinity, which contributes to the prevalence of cardiac complications such as myocardial injury and arrhythmias [61, 62].

Endothelial dysfunction is considered, in the context of Covid-19 disease, a major "player" of SARS-CoV-2 infection. Starting from the genesis of endothelial dysfunction, there are a number of consequences, namely the increased risk of thromboembolic systemic vasculitis, endothelial cell apoptosis, and systemic inflammation. The inflammatory process resulting from vascular endothelial activation and endothelial damage triggers the release of cytokines with prothrombotic consequences. It seems that the activation of coagulation with diffuse intravascular thrombosis (systemic endothelitis) in Covid-19 is rather the consequence of the systemic inflammatory process and not a specific effect of SARS-CoV-2 virus.

In essence, the sequence could be as follows: SARS-CoV-2 contaminates endothelial cells causing endothelial damage, perivascular inflammation, endothelial dysfunction, and altered microcirculation in various vascular areas causing serious complications such as thromboembolic disease with multiple organic involvement [63, 64]. However, it is considered, alternatively, that endothelial damage vasculopathy is the consequence of an indirect mechanism of endothelial damage [65].

Undoubtedly, the data presented and the sequence of arguments are of significant scientific interest.

In essence, the authors successfully present the two hypotheses and the scientific arguments in support of the two variants: the direct mechanism of genesis of epithelial injury and the indirect mechanism of genesis of endothelial damage and thrombosis in Covid-19. The indirect mechanism is based on the fact that vascular endothelial cells do not express the ACE2 receptor and are resistant to SARS-CoV-2 infection, vasculopathy being generated by indirect pathogenic mechanism. Attempts are made to justify, based on clinical and experimental data, on the pathogenesis, the complex polymorphic picture of Covid-19 disease, especially in later, complicated stages, that this entity is an endothelial disease [66]. The discussion remains open about the expression of ACE2 in the vascular endothelium. Recent work highlights the consistent presence of venous and arterial endothelial cells in all tissues studied [19, 67].

In the context of this approach, it is considered that a treatment strategy can be developed, as evidence accumulates to substantiate the guidance of therapeutic trials.

Resuming the discussion of the direct mechanism versus the indirect mechanism of endothelial damage by SARS-CoV-2, a pathogenic process dependent on the presence and density of ACE2, an important, yet not sufficient condition, requiring the colocalization of a serinprosthesis (TMPRSS2) [68].

The author cites immunoclinical research in which serine protease is not identified in the airway endothelial cells of deceased Covid-19 patients.

In addition, it is mentioned that spike proteins inside the endothelial membrane are not found in inflamed lung vessels. The cited evidence suggests that Covid-19 disease has a cycle of self-amplification of an excessive inflammatory process, preventing the antithrombotic properties of the vascular endothelium, synergistic with the action of circulating prothrombotic factors, leading to the formation of thrombi in capillaries, veins, and arteries. Excessive inflammation triggered by SARS-CoV-2 in non-endothelial cells can lead to endothelial damage and vascular thrombosis. Vasculopathy is triggered by the exaggerated immune response to viral infection of non-endothelial cells, causing indirect damage to vascular endothelial cells, released by prothrombotic factors, platelet hyperactivation, and thrombosis [65]. Pulmonary edema is a complication of vascular barrier alteration producing tissue edema, endothelitis, coagulation activation, and potential genesis of disseminated intravascular coagulation [69].

Vascular damage, with the described secondary processes, occurs through several mechanisms, namely the involvement of endothelial cells directly by the virus, with possible lysis and cell death. The second mechanism is the blocking of ACE2 activity after intracellular virus penetration. Decreased ACE2 activity allows the predominance of ang II effects, with known effects and activation of the kallikrein–bradykinin pathway.

13.3.3 The Endothelium and the Systemic Endothelial Dysfunction

The vascular endothelium by its anatomical-functional extent, by the size and complexity of its functions, is considered the largest endocrine organ, although we can agree that the name does not fully cover its functional expansion. This organ is amazing, especially when we understand the clinical significance of its abnormal condition: endothelial dysfunction. The vascular endothelium also known as "Achilles' heel" or "the corner stone" [70, 71] or "the major player" ("hardest challenger") [49, 63] of this organ is perhaps atypical for the foundations of our scientific thinking so far [72].

As we have pointed out in various chapters of this paper, the endothelium is not only the largest, most extensive organ but it also forms a continuous surface, a non-thrombogenic permeability barrier. This organ synthesizes and releases a wide range of active substances with vasoactive effects and more. Its alteration, called endothelial dysfunction, is probably the first step toward atherosclerosis with the complexity of the phenomena that define it [73].

Nitric oxide is released by vascular endothelial cells as a reaction to the stress (shearing) of the vascular wall at the pressure of blood flow. Nitric oxide (NO) is a free radical with a lifespan of a few seconds that can penetrate biological membranes, which comes from L-arginine and the cofactor tetrahydrobiopterin, a process catalyzed by NO synthase [74]. After diffusion, at the level of the vascular smooth muscle, it generates the increase of cyclic guanosine-monophosphate, with the consequent relaxation of the vascular smooth muscles.

NO and the enzyme that control its biosynthesis regulate vascular tone.

In the following paragraphs, we will list some of the vasoactive substances derived and released by the endothelium, essential for the functional integrity of this structure [74]. Functional balance between different vasoactive components is the key to endothelial homeostasis. We present in summary the vasoactive substances involved in endothelial homeostasis.

Vasodilators: nitric oxide (NO), prostacyclin, endothelium-derived vasodilators, adrenomedulin, and quinine.

Vasoconstrictors: angiotensin II, endothelin-1, and vasoconstrictor prostanoids.

Coagulants: factor V, heparan sulfate, protein C, protein S, thrombomodulin, tissue factor, and Willebrand factor.

Fibrinolytics: Plaminogen-activating inhibitor, tissue plasmonogen activator, and urokinase growth factors.

Inflammatory and immunomodulatory mediators: Cytokines (interleukins) and tumor necrosis factor-α.

Adhesion molecules: platelet intercellular cell adhesion molecules—endothelial cells, selectin, and vascular cell adhesion molecules.

Reactive oxygen species: H_2O_2, HO_2, and superoxide (O_2-).

Reactive species of nitrogen: nitrate (NO-2), nitrite (NO_{3-}), nitrogen dioxide (NO_2), and peroxynitrite (ONOO-).

Nitric oxide (NO) activates guanylate cyclase, generating the conversion of guanosine triphosphate to cyclic guanosine monophosphate and resulting in smooth muscle relaxation. It also produces antiplatelet, anti-inflammatory, antiproliferative, and anticellular effects, maintaining the integrity of the endothelium, a complex process in which other substances listed above participate. We are concerned about addressing the initiating mechanisms of endothelial dysfunction, but it does not seem to be a simple and unambiguous action. In essence, endothelial dysfunction is thought to create a blockage, prevent relaxation, endothelium-dependent vasodilation (vascular relaxation) in response to stimulation with acetylcholine, bradykinin, or increased wall stress, which causes dilation mediated by blood flow. But the description does not address an initiating factor or process but the consequences of endothelial dysfunction.

Risk factors that may favor endothelial dysfunction might be the following:

- Decreased bioavailability of NO which, starting with decreased availability of L-arginine, can be caused by many other causes, including oxidative stress.
- Hypercholesterolemia, causes endothelial cell activation, leukocyte recruitment and adhesion, platelet activation and adhesion to the vascular wall, initiation of an inflammatory process, a prothrombotic state, increased oxidative stress, all contributing to endothelial dysfunction.
- Hyperglycemia, diabetes that increases oxidative stress, following the activation of C-reactive protein, decreases the bioavailability of NO, increases platelet aggregation, increases endothelial apoptosis, etc.
- Low-density lipoproteins (LDLs), present in high concentrations in dyslipidemia, are captured in the vascular wall, initiating an oxidation process, generating, together with subendothelially migrated monocytes, the appearance of "foam cells." This process is considered to be one of the earliest in the onset of atherogenesis-atherosclerosis.
- HDL are, in turn, antioxidants that mediate the transport of cholesterol (reverse transport of cholesterol) from the periphery to the liver (transporter apolipoprotein

A-I) where, by biliary excretion, after prior metabolic processes, is eliminated in the intestine and excreted. HDL and paraoxonase have a strong antioxidant and protective activity for circulating cells and lipoproteins [46, 75–77].

The genesis of oxygen free radicals increases in the inflammatory process, in the repair processes and in the injury of various types. Free radicals are responsible for initiating and continuing pathological processes that activate thrombotic plaque and thrombotic response. The so-called risk factors, so called in a certain context of the discussion, are in fact either pathological entities or various aggressions, or lifestyle errors such as vascular oxidative stress, hypercholesterolemia, hypertension, diabetes, smoking, etc.[77]. In essence, it is accepted that, under conditions such as these, the genesis of free radicals of a certain magnitude exceeds the capacity of endogenous antioxidant systems, generating the so-called oxidative stress, an oxidative aggression targeting the enzyme systems, other proteins, lipids, membrane structures, nucleic acids which, in a certain association, generate endothelial dysfunction, implicitly the decrease of NO bioavailability. Hence, we believe the difficulty of defining and describing very targeted and chronologically the stages and initiating mechanisms of endothelial dysfunction.

Therefore, it comes as no surprise that given the interaction of the virus with the ACE2 enzyme, widely distributed in the endothelial tissue, viral particles are found in dead endothelial cells or are associated with dead inflammatory cells, highlighting a broad, diffuse inflammatory response to viral aggression [45]. The consequence of this viral aggression on the endothelium is permeability disorders, tissue destruction, and microcirculation. The mechanisms of viral aggression can be direct, by interaction with endothelial ACE2 with consecutive dysfunction at this level (outlined above) or indirect by the consequences generated, namely hypoxia, hyperinflammation, immune disorder, defining a severe form of Covid-19 [49]. These processes are secondary to the initial viral aggression.

The indirect mechanisms generating endothelial dysfunction in Covid-19, on the other hand, have been less studied. The lesions caused by acute hypoxia are caused, in the context of acute respiratory failure, by acute respiratory distress, in severe or critical forms, by Covid-19. Disease-associated hypoxemia causes a series of metabolic adaptations designed to maintain ATP production, with conversion of oxidative phosphorylation to anaerobic metabolism. This change in metabolism, in a hypoxic context, determines the genesis of free radicals, with the alteration of the redox balance, at the cellular level.

The reduction of ACE2, following viral aggression, leads to the accumulation of angiotensin II, the activation of the AT1 receptor, a context in which, in addition to vasoconstruction, increased vascular permeability, inflammation and pulmonary edema might also occur, all of them being linked to an increase in free radical genesis. The consequences of these processes are, in particular, the increase of angiotensin II and free radicals, the consequent conversion of NO to peroxynitrite (ONOO- free radical), which results in increased cell damage.

Furthermore, hypoxia is associated with cellular acidosis following the conversion of pyruvate to lactate, causing impairment of cellular functions. Phenomena following the processes previously described, caused by lysosomal damage and enzyme release and further cell damage, result in most of the ion pumps being severely affected by ATP reduction, with apoptosis and myocardial damage.

This first process, hypoxia, underlies the evolution to the stage of multiple organ failure. In essence, reducing the bioavailability of endothelial NO, resulting in the production of excess free radicals and redox imbalance, is the benchmark for understanding endothelial dysfunction and associated thrombotic events. Moreover, the disorder of NO deficiency signaling pathways creates a permissive cellular environment for viral penetration and replication [49]. In a broad analysis of the mechanisms of endothelial damage, as a support for systemic vasculopathy, panendothelitis, interesting arguments are made in favor of indirect mechanisms.

Interestingly, the hypothesis that endothelial lesions are directly generated by the Sars CoV 2 infection cannot be clearly extracted as a take home message from the studied data. It is not implicit, from the available data, that endothelial cells are susceptible to SARS-CoV-2 infection [65]. Very relevant are the data obtained based on cultures of human endothelial cells, in the presence of SARS-CoV-2 virus, proving that the virus cannot primarily infect human endothelial cells. The data show that human primary endothelial cells do not have ACE2 expressed in membranes, but the enzyme is found in the middle muscle tunic, in the smooth muscles, not in the endothelium. The authors point out in this context that primary endothelial cells do not possess ACE2 but possess the proteases necessary for the intracellular viral penetration process, implicitly implying that the known lesions, caused by endocrine disruption, are indirectly, independent of the existence of ACE2 at this level.

Endothelial damage in Covid-19 and activation of the response is due to ACE2 blockade and ARAA imbalance, loss of angiontensin protection (1-7), RMas, and the predominant direct effects of ang II on the endothelium.

Of course, there are conflicting data and approaches, which show that SARS-CoV-2 infects the vascular endothelium, including lung endothelial cells, both in vivo and in vitro, generating known pathogenic processes in which panendothelitis is the basis of intracellular penetration seen in the Covid-19 disease [21, 22, 46, 78–81].

We further argue in this paper that 58% of patients with severe and hospitalized Covid-19 are hypertensive, 25% have other cardiovascular comorbidities, and 44% have cardiac arrhythmias [82]. Reports of Covid-19 mortality in China have been reported, stating that only 11.8% of patients who died had no cardiovascular comorbidities. Specifically, 35% were hypertensive and 17% had a history of coronary heart disease.

The consequences of these comorbidities on the prognosis of Covid-19 in medium and severe forms generated, on the one hand, by endothelial dysfunction are already present and, on the other hand the internalization of ACE2 given by the virus, generates an increase in angiotensin II levels, with further alteration of the vascular system and heart with increased mortality.

The consequences of the decrease of ACE2 in the cerebral level are mentioned, with the decrease of the sensitivity of the baroreceptor reflex, avoiding the neurogenic component of the hypertension by intensifying the sympatho-adrenergic activity and the endocrine component [83].

13.3.4 Protection and Functional Restoration of the Endothelium in Covid-19

As previously mentioned, ACE2 is widely distributed within the vascular endothelium, both arterial and venous. It is found in the microvascularization of the pulmonary endothelial cells, in the myocardial capillaries, in the endothelium of the smooth muscles of the cerebral capillaries [19, 67].

By catalyzing the conversion of Angiotensin I and Angiotensin II to Angiotensin (1-7), ACE2 antagonizes vasoconstruction, proinflammatory, proapoptotic, mitogenic, metabolic effects of Angiotensin II, via the AT1 receptor. Through Angiotensin (1-7), ACE2 decreases the endothelial production of oxygen free radicals, the vascular adhesion molecule-1 (VCAM-1) decreases the adhesion of monocytes to endothelial cells and the maturation of endothelial cells in vascular structures [46]. As from previous analysis, ACE2 is a receptor for coronaviruses, including SARS-CoV-2, which interacts with the S proteins with the enzyme and which, after fusion with the host cell membrane, penetrates intracellularly. Thus, the entry of the virus into the cell is achieved by association with ACE2, which is thus diminished at the cell surface.

The internalization of ACE2 generates a change in the balance of vasoactive mediators with the increase of angiotensin II, with the onset of vasoconstruction, inflammation, and procoagulant status. SARS-CoV-2 has an affinity for endothelial cells, can infect them, and can multiply at this level [25]. Thus, it is admitted that the virus has a second target consisting of endothelium, vessels, and ubiquitous ACE2 at this level. Thus, they can be targeted for SARS-CoV-2, and under its pathogenic action, multiple organ pathology is generated.

Infection of endothelial cells by SARS-CoV-2, decrease or disappearance of ACE2 from the surface of the above-mentioned cells, under the action of the virus, confers a proinflammatory, proapoptatic and procoagulant phenotype to the cells in question, generating a vascular pathology in patients with Covid 19.

Furthermore, SARS-CoV modulates myocardial ACE2 expression in coronavirus patients. Specifically, SARS-CoV decreases enzyme production, myocardial inflammation, and myocardial dysfunction with adverse effects on patients with Covid [46].

Postmortem, morphopathological data show that in Covid-19, there is congestion in small vessels, tissues, myocardium, kidneys, small intestine, liver, and lungs; vascular microthrombi occur in various organs, especially in the lungs, capillary inflammation, and acute cerebral infarction [46].

Coagulopathy is already well identified in Covid-19, especially in the severe forms of this disease, characterized by microthrombi disseminated in the pulmonary vessels suggesting a vasculopathy characteristic of the disease in question [75].

The endothiopathy produced by SARS-CoV-2 generates and maintains coagulation, especially in the lungs, in microvascularization.

As we mentioned in this paper, one of the authors finds that the interaction of SARS-CoV-2 with ACE2 generates a multiple endothelial organ disease, vascular disease what we called panendothelitis [17, 21, 22, 51, 84, 85].

In the next chapter, we will try to synthesize the pharmacotherapeutic means that can prevent endothelial dysfunction, reduce the amplitude and dispersion of endothelial damage, and possibly contribute to the functional recovery of vascular endothelium.

13.3.5 Pharmacotherapy of Endothelial Dysfunction

Can this endothelial SARS-CoV-2-induced pathogenesis be prevented?

A number of clinical trials have been conducted in which various classes of drugs, such as anticoagulants, have been tested to prevent thrombosis and thromboembolism. 6, renin-angiotensin-aldosterone system inhibitors, etc.

Interestingly, the literature mentions that AT1 receptor blockers of angiotensin II attenuate the lung injury caused by angiotensin II following the blockade of the transducer and triggering of its effects.

The effects at SARS-Cov-2 level depend on pre-existing cormobidia, systemic inflammation, and viral load. Immune plasma, according to recent data, is not a consistently effective solution. Recombinant ACE2, in clinical trials, can block viremia by competitively neutralizing viremia versus ACE2 at the cellular level.

The use of medications to reduce or control IL-6 activity depends on the outcome of evolving clinical trials. IL-6 has complex pro and anti-inflammatory effects in different functional contexts.

IL-6 in high concentrations interacting with its soluble receptor damages endothelial cells, the interleukin in question being a marker of severity of Covid-19 disease. IL-6 and IL-6 receptor inhibitors are approved for the treatment of rheumatoid arthritis. IL-6 induces VEGF-A production which generates vasculopathy with increasing permeability and produces angiogenesis. In addition to this factor, angiopoietin 2, abnormally increased in Covid-19, is also the target for inhibiting the angiogenesis and vascular inflammatory process in Covid-19.

We note the relatively speculative assertion that patients with Covid-19 should be identified for the risk of vascular complications in Covid-19, to act quickly to reduce the viral load, by treatment with specific drugs for vascular protection (endothelial), validated by clinical trials [46].

13.3.6 The Effects of Antihypertensive Drugs on Endothelial Dysfunction and Inflammation in Covid-19

Of particular importance are the characteristics of the effects of different classes of antihypertensives on endothelial function (dysfunction) in conditions of associated comorbidities such as hypertension or, in our case, Covid-19 disease. It is well established that hypertension induces functional and morphological disorders of the blood vessels progressively generating endothelial dysfunction, vascular inflammation, and arterial stiffness, which in turn maintain hypertension [86, 87].

As we mentioned, in the chapters related to this paper, angiotensin II through its essential actions for cardio-circulatory homeostasis generates a series of effects controlled, however, by counter-regulatory links (Angiotensin (1-7) -Receptor Mas) and decreased bioavailability NO, increased free radical genesis, proinflammatory condition.

Conversion enzyme inhibitors decrease free radical production and proinflammatory state by implicitly reducing endothelial dysfunction and proinflammatory condition. In addition, conversion enzyme inhibitors, by reducing the degradation of bradykinin, facilitate the genesis and release of NO by promoting endothelin-dependent vasodilation [86].

Randomized clinical trials have shown that conversion enzyme inhibitors improve endothelial function, reduce vascular inflammation, and thus reduce the risk of cardiovascular events. Angiotensin II receptor blockers II also have beneficial effects on endothelial function and vascular inflammatory process.

Of course, in this context, there is the issue of identifying the evidence based effects of antihypertensives in the prevention and/or treatment of systemic endothelitis caused by SARS-Cov-2. Plasma levels of angiotensin II are markedly elevated in most cases of Covid-19 (90.2%) and even more pronounced in critically ill patients, compared to those in average condition [88, 89]. AT1 receptor blockers of angiotensin II implicitly block its proinflammatory effects, diverting the action of angiotensin II to the AT2 receptor with beneficial effects described in this paper. Conversion enzyme inhibitors have direct consequences by decreasing the genesis of angiotensin II in angiotensin[1].

13.3.6.1 The Inflammasome and Pharmacological Inhibitors

In a viral infection, innate immune system signaling and inflammation activation are major barriers to viral infection. The activation of the innate immune system must be controlled to a great extent, as excessive activation can lead to systemic inflammation and tissue damage, with damaging effects on the host [90].

After the recognition of viral RNA by infected cells (by PRRSx) in the upper respiratory tract, the production of proinflammatory cytokines is triggered, inducing an antiviral state [90].

As a prototype, conceptually, proinflammatory cytokines are widely benefited by triggering infiltration with immune cells to remove the viral agent. In this context, the disorder, the dysfunctional response of proinflammatory cytokines, triggers a systemic hyperinflammation, cytokine storm, with multi-organ lung tissue damage [90].

SARS-CoV-2 stimulates the activation of inflammation (NLRP3xx) and the activation of caspase-1. The triggered inflammatory response is mediated by the key inflammatory cytokines, IL-1β and IL-18, increased in plasma, a process correlated with the severity and mortality in Covid-19[1].

[1] PRRS: Pattern recognition receptors and induces the innate immune response.

NLRP3: Critical component of the innate immune system (intracellular sensor protein) which controls the secretion of proinflammatory interleukins in response to infections and cell damage.

A number of experimental data show that activation of NLRP3 inflammasome is usually vital for the antiviral immune response, with inflammation being activated in response to viral RNA, inhibiting viral replication and reducing mortality in experimental animals. In the context of the mentioned experiments (SARS-CoV-2 infection model) proinflammatory cytokines, TNF and IL-6 produce lung lesions and acute respiratory distress syndrome (ARDS).

Although the activation of the inflammasome is beneficial to the host, as we have previously pointed out, the excessive, aberrant activation is detrimental [90]. The molecular mechanism of this process and its role in the genesis of ARDS are not completely deciphered. In the described context, programmed death occurs, with nuances in the mechanism of production: apoptosis, pyroptosis, and necroptosis.

The proteins of various coronaviruses (SARS-CoV, MERS-CoV, and SARS-CoV-2) produce the activation of inflammation, with the consequent activation of caspase-1. Activated protease cleaves the pro-interleukins IL-1β and IL-18, initiating various signaling cascades generating apoptosis, pyroptosis, and necroptosis [91].

Caspases generally regulate the process of apoptosis, inflammation, and the immune response. Caspaza-6 promotes the activation of the programmed cell death program, namely pyroptosis, apoptosis, and necroptosis (PANoptosis).

Activation of the mechanisms of programmed cell death generates tissue damage and severe inflammatory processes.

As a result of this complex process of programmed death with tissue dysfunction, the serum lactate dehydrogenase (LDH) concentration increases markedly, being considered an important marker for predicting the early identification of the severity of lung damage in severe forms of Covid-19 [92].

The three processes of programmed death communicate and are co-regulated being called PANoptosis, i.e., controlled by a single programmed death program, called PANoptosomme [93].

The NLRP3 inflammasome, with a crucial role in innate immunity, consists of a binding domain, rich in leucine and a domain of pyrin (protein bound to inflammation, making a complex that promotes inflammation). By being activated by a wide variety of stimuli, inflammasome NLRP3 recruits an adapter protein, AUC (a protein associated with apoptosis containing a caspase recruitment domain) which recruits a pro-caspase and which, by cleavage, induces the maturation and secretion of inflammatory cytokines and pyroptosis [94, 95].

Aberrant activation of NLRP3 inflammation is possible in various conditions, such as diabetes, atherosclerosis, metabolic syndrome, cardiovascular disease, and neurodegenerative diseases. Implicitly, research and clinical validation of some NLRP3 inhibitors of inflammation are of particular interest [96].

The inflammasome therefore recognizes and reacts to a wide range of stimuli such as: protozoa, fungi, various viruses, various bacteria, endogenous factors, such as uric acid crystals, amyloid peptides, extracellular ATP, etc.

The first stage of inflammation signaling involves the initiation of transcription of this factor and pro-IL-1β. In the second stage of this process, the secretion of IL-1β and IL-18 is processed.

The problem of researching some pharmacological means begins with the identification of some targets, in the complex process of the signaling cascade of this essential factor in pathology.

Potential targets may be:

- Inhibition of inflammation activation, inhibition of caspase-1, neutralization of inflammatory cytokines, inhibition of the whole inflammation, blockade of ATP binding, blockage of ATPase, etc.

Examples of indirect inhibitors:

- Gliburid, antidiabetic (type 2), compound, with sulfonylureal structure, inhibits ATP-sensitive K + channels (KATP) in pancreatic β cells.

Gliburide produces an effective inhibition of caspase-1 and IL-1β, being active in vivo and in vitro in the process of activating NLRP3 inflammation.

Beta-hydroxybutyrate produces an effective decrease in IL-1β and IL-18 in human monocytes. It blocks the activation of the inflammasome under the action of free radicals, activated protein kinase AMP, glycolytic inhibition, and it can be anticipated as a potential preparation to fight chronic inflammation.

Direct inhibitors include:

- Tranilast, a metabolite analogue of tryptophan with inhibitory effects on inflammation. The compound is well tolerated in patients, with significant therapeutic and preventive effects in gout, clinically and experimentally and in type 2 diabetes.
- Oridonium, a diterpenoid, a compound of plant origin (Rabdosia rubescens), has anticancerous actions, suppresses angiogenesis, and induces apotosis. The compound inhibits NK-κB and MAPK activation inhibits the release of inflammatory proinflammatory cytokines. It has therapeutic effects in neuroinflammation, sepsis, and colitis.

There are also distinct preparations that block IL-1β, such as anti-IL-1β antibodies, canakinumab, recombinant IL-1 antagonist, anakinra, IL-1 receptor blocker, and rilonacept. Pharmacological inhibitors of inflammasome seem to be the best option in diseases with associated inflammation [96].

13.4 SARS-CoV-2 and Cardiovascular Comorbidities. Common Link: Panendothelitis (Systemic Endothelitis)

In principle, there is a consensus on the vulnerability of patients with pre-existing cardiovascular disease to SARS-CoV-2 aggression.

Of course, health systems are concerned with two biunivocal issues, namely the consequences of pre-existing comorbidities on the virulence and pathogenicity of SARS-Cov-2 and potential post-Covid-19 cardiovascular sequelae [17, 51, 52].

We also pointed out the broad symptomatic picture of Covid-19 disease, most likely correlated with the individual characteristics of patients affected by SARS-CoV-2 [9]. A systematic research of scientific open access articles published between January and August 2020, based on biopsies and autopsies performed on 603 cases of Covid-19 that meet the selection criteria, have shown a series of high scientific interest aspects regarding visceral and systemic lesions and potential causes of death [79] (Table 13.1).

The authors' synthesis shows that, on a pulmonary level, there is a diffuse alveolar lung damage (92%), superimposed on bronchopneumonia (27%), myocarditis (11.4%), and vasculitis, highlighted on the vessels of skin biopsies (25%). Endothelial cell damage is identified in the lungs (18%) and in the heart (4%). Lesions due to hypoxia/hypoperfusion were identified in the kidneys (53%), liver (48%), CNS (31%), and gastrointestinal tract (18%).

In the previously mentioned research study, the authors were concerned with the evaluation of the direct cytopathic effects of the virus at the level of organs and systems and of those mediated by the immuno-inflammatory mediated processes that generated microangiopathies and coagulation.

The marker research shows an increase in D-dimers (83.2%), procalcitonin (66.7%), systemic inflammatory markers, C-reactive protein, ferritin, IL-6 (identified as having increased values to 91%, 83%, and 74% of the patients studied).

The leading cause of death in 227 Covid-19 patients was respiratory failure (70.9%), followed by multiple organ failure (11%), cardiovascular disease (10.6%), the Covid-19 infection per se (7%), and pneumonia (6.6%).

As shown above, diffuse alveolar lung damage is the most present (92.1%) as an overlap of focal or diffuse bronchopneumonia lesions (27.4%). Myocarditis was present in 11.4% of the cases.

Table 13.1 Criteria for clinical severity (Covid-19 confirmed cases)

Mild disease
• Fever <38 °C, resolves without treatment
• With or without cough, without dyspnea
• No breathing difficulties, no chronic disease
Moderate disease
• Fever, respiratory symptoms
• Image suggestive for pneumonia
• Evidence for pneumonia
Severe disease
• Patient presents any of the following symptoms:
– Respiratory distress—severe respiratory disorders
– Respiratory rate >30/min
– Saturation with O_2, $SpO_2 \geq 93\%$ at rest
– $PaO_2/FiO_2 \leq 300$ mmHg
Critical disease
• Patient presents any of the following symptoms:
– Respiratory failure, requires mechanical ventilatory assistance
– Shock
– Multiple organ failure, requires assistance in ATI

Table adapted from an article published by Zi Yue Zu et al. (2020) [97]
FiO$_2$ fraction of inspired oxygen, *PaO*$_2$ partial pressure of oxygen

The most common endothelial injury was identified in the lungs (17.8%) and in the heart (3.8%). Thromboembolic disease was found in the liver (13.7%), heart (21.4%) and lung (13.7%).

Viral cytopathic effects were demonstrated in the myocardium and vascular endothelium, with myocarditis being present, as mentioned above, in 11% of cases.

The research describes about 4 mechanisms that cause heart damage by SARS-CoV-2, respectively: myocarditis, cytokine storm, coronary artery ischemia suggesting coronary artery disease, and increased thrombosis of small and large coronary arteries, in the absence of coronary artery disease [98].

In essence, the disease can be either asymptomatic or moderate with low symptoms, such as fever, dry cough, and fatigue. Another category of patients develops severe pre-pneumonia and possibly respiratory distress syndrome (ARDS) with potentially lethal effects [49]. Comorbidities are of great importance in their association with the Covid-19 disease both in terms of prevalence and increased mortality [99]. As an example, the Chinese Center for Disease Control and Prevention performed a study based on a number of 72,314 cases of Covid-19, which pointed out a general mortality of 2.3%. However, for those with pre-existing cardiovascular disease, there is a general mortality of 10.5%, which goes down to 7.3% for those with diabetes, 6.3% for those with pre-existing chronic respiratory diseases and 6% for those associating only hypertension [89, 100].

Cardiovascular complications, in severe forms of Covid-19, appear to be among the most serious, some of them even potentially fatal.

A meta-analysis based on 46,248 cases found that hypertension and diabetes are the most common comorbidities associated with severe forms of Covid-19, a context in which severe acute respiratory distress syndrome and heart damage are the most severe complications [49, 101].

If endothelial dysfunction is the most difficult challenge, a feature, and a background on which cardiovascular disease develops and progresses, it turns out that a new aggression, generated by SARS-CoV-2 infection, will aggravate that dysfunction and pre-existing associated diseases, such as would be hypertension, diabetes, heart failure, and obesity. Partial or accentuated loss of endothelial functions, such as vasodilation, hemostasis, vascular permeability, vascular tone, antiplatelet effects, recruitment of neutrophils, fibrinolysis, etc. are conditions that, in a pathological context, may be exacerbated by SARS-Cov-2 infection [17, 49, 51].

A direct and predictive relationship of endothelial dysfunction with hypertension is not currently confirmed. However, the pre-existing consequences of hypertension and the spread of endothelial dysfunction in Covid-19 are relevant.

Hypertension reduces NO availability, increases oxidative stress, by increasing reactive oxygen species, decreasing antioxidant capacity. This aspect is further developed in another sub-chapter. It seems beyond any doubt that the severity of Covid-19 disease is correlated with the pre-existence of cardiovascular disease.

Various cardiovascular diseases, especially hypertension, coronary heart disease and heart failure are associated with an unfavorable prognosis [102].

It is interesting to note that in a meta-analysis on 1576 infected patients, the prevalence of hypertension in 21.1% of patients, diabetes in 9.7%, and other cardiovascular diseases in 8.4% is mentioned [100].

In the context of Covid-19 disease, myocardial damage can be either direct or indirect and as a consequence these heart injuries are accompanied by high mortality. A recent meta-analysis shows that the proportion of heart damage is 22% and 28%, respectively, in patients hospitalized or with severe forms of Covid-19 disease [102].

In patients over the age of 60, the incidence of heart damage is 30%, twice as high as in those under 60 (15%). Another meta-analysis shows that the variation and spread of disease forms and the severity of Covid-19 are broad and multifactorial, correlated with patients' vulnerabilities [103].

An analysis of asymptomatic forms raises a number of questions, generating various hypotheses. The latter are correlated either with the host, in terms of susceptibility to infection, or with the virus, in terms of virulence [102].

The factors associated with the host, in terms of asymptomatic forms, include, for example: (1) possible insufficient expression of ACE2 in the nasopharynx or oral cavity, causing a viral load and thus reduced infection; (2) the ACE2 polymorphism treated by us in this chapter; (3) TMPRSS2 polymorphism, both functionally and structurally; (4) variability related to sex, women develop lighter and uncomplicated forms compared to men; (5) variations related to blood group, those with group "A" are exposed to a higher risk in the purchase with the group "0"; and (6) duration of antibody production, etc. [102].

The authors cited above consider that in the context of the impact between SARS-CoV-2 and the ACE2 receptor axes, a number of complex mechanisms are triggered in the heart and vascular tissues, either aggravating pre-existing comorbidities or causing the onset of new dysfunctions, at these levels, during the infection in question.

We emphasize that the invasion of the infectious agent and the wide tissue dissemination, correlated with the dispersion of ACE2, are the key to understanding the dysfunctions and multiple insufficiency of the vital organs. The basis of the mechanisms of multiple organ failure in SARS-CoV-2 infection includes endothelial damage, thromboinflammation, impaired immune response, and renin-angiotensin-aldosterone imbalance. In order to prevent profound impairment of vital organ functions, close, individual supervision of patients, post Covid-19, is required to prevent cardiovascular damage.

Hypertension, in the context of Covid-19 disease, has two possible connotations: hypertension, as a pre-existing comorbidity of contacting Covid-19 disease, and the consequences of this infection on the evolution of hypertension, as we set out to address, hypertension as a possible complication, a sequela of Covid-19.

The Japanese Society of Hypertension addresses these issues. There are a number of data showing the prevalence of hypertension in patients with Covid-19, although there are no relevant data to show that the risk of SARS-CoV-2 increases in the presence of hypertension in some patients [104].

The question is, however, whether the severity of Covid-19 disease is exacerbated by hypertension and whether it is a factor in aggravating the infectious disease.

A retrospective study in China of 487 Covid-19 patients found the prevalence of hypertension in 49 severe cases to 438 cases of moderate severity. Similarly, other studies have shown a significant prevalence of hypertension in patients with severe forms of Covid-19.

The authors also note that hypertensive patients with Covid-19 have other comorbidities, which are major factors in the severity of the infectious disease. These risk factors include the age of the patients.

For example, the authors note that in the age group 70–79 years and >80 years, the mortality was 6.8% and 14.8%, respectively, while the mortality generated in all age groups was 2.6%. Hypertension remains a marker of the severity of Covid-19 disease. The extent to which hypertension worsens or does not remain to be determined by other clinical trials.

In vascular physiology, the criteria for functioning and reactivity, i.e., proper functioning, is vasodilation, which, implicitly and axiomatically, increases blood flow allowing the initiation of function and an adequate biological response.

It follows that the gold standard of endothelial function is the vasodilator response of the vessels and the hyperemia (peripheral and coronary vessels) following the infusion of acetylcholine. It follows that vasodilation is the defining function of the functional whole endothelium and not vasoconstriction [105, 106].

Interestingly, a narrower definition of the endothelial dysfunction refers to the decrease of the vasodilative properties of the endothelium and to the changes of the endothelium's homeostasis. [106].

The authors consider that the relevant convergent forces, manifested in the context of endothelial dysfunction leading to atherosclerosis are those with an impact on vasomotor control and chronic inflammation, unresolved. The therapeutic effects on inflammation are more difficult to quantify compared to the quantification of vasodilation.

13.4.1 Cardiovascular Consequences of SARS-CoV-2 Interaction with the ACE2 Receptor When There Are Pre-Existing Comorbidities and/or Conditions That Are Favoring This Reaction

The interaction of the SARS-CoV-2 virus with the ACE2 receptor is of particular interest for the pathogenesis induced by the respective virus correlated with the role of ACE2 as part of the renin-angiotensin-aldosterone system [21, 22].

The enzyme ACE2, the SARS-CoV-2 receptor, has a number of correlations by function and expression in various tissues with comorbidities that influence the severity of Covid-19 disease. More explicitly, the level of ACE2 expression and associated molecules (TMPRSS2) are factors of great relevance for viral load and associated infection. Hence, a number of risk factors related to groups of patients who have conditions for severe Covid-19 disease. These include age, male gender, and underlying comorbidities such as obesity, hypertension, other cardiovascular diseases, diabetes, and chronic respiratory diseases [107].

The synthesized data show that ACE2 expression increases in cardiovascular diseases, chronic respiratory diseases, obesity, diabetes (pancreatic islets), in males and also increases with age.

It should be noted that ACE2 is highly expressed in pericytes and, in context, the heart damage in Covid-19 varies from 2% in patients who do not require intensive care to 59% in those in intensive care and who subsequently died. Contamination of the pericytes generates endothelial dysfunction with microcirculatory disorders [108].

Patients with heart failure, pre-existing with SARS-CoV-2 infection, have a high expression of ACE2 at the ventricular level, indicating that patients with this condition are more susceptible to viral infection and may develop new myocardial lesions with the risk of critical conditions [108].

Hypertension is a comorbidity associated with unfavorable results in Covid-19, a 2.5-fold increase in the risk of severe or fatal forms is assessed.

In addressing the consequences of viral aggression on the endothelium, with the implicit genesis of endothelial dysfunction which is the pathological background of pathogenic processes and/or pathological entities such as hypertension, diabetes, obesity, aggravating comorbidities of the evolution of Covid-19 disease. We emphasized their relevance for the prognosis of this disease [21, 22].

We acknowledge that a number of pathological entities or comorbid conditions such as hypertension, diabetes, obesity, and heart failure may pre-exist Covid-19 disease or, consequently, as a series of post-Covid-19 sequelae, included in the so-called post-Covid syndrome (long Covid-19 syndrome/disease) [21, 22].

Regarding cardiovascular lesions, developed in the context of Covid-19 disease, the expression of ACE2 is considered to be a major factor in the genesis of specific tissue lesions induced by SARS-CoV-2. ACE2 is thus evaluated as a component part of the RAA system and involved in the genesis of hypertension, diabetes, and heart failure [109].

The authors mention that the expression of ACE2 in the heart and coronary arteries is wider than in the lungs. In humans, ACE2 is highly expressed. On this basis, it is recognized that SARS-CoV-2 can directly affect various types of heart and cardiovascular cells including cardiomyocytes, endothelial cells and pericytes [109].

It is interesting to note that the expression of ACE2 in these tissues is not itself high enough to allow the virus to enter the cell, but viral replication and its release can be ways to facilitate the infection of host cells. Hence, the conclusion that SARS-CoV-2 can indirectly affect the cardiovascular system.

It should not be overlooked that SARS-CoV-2 has a very diverse tissue tropism, in addition to the lung, a number of organs and tissues are affected, of which we mention, consecutive clinical manifestations at the cardiovascular level, stroke, liver damage, renal dysfunction, gastrointestinal disorders, etc. We do not omit the importance of systemic hyperinflammation induced by viral infection of the lungs and other organs and systems.

As far as recent studies of Covid-19 infection have shown, cardiovascular damage is described in patients with myocardial infarction, acute coronary syndrome, heart disease, venous thromboembolism, arrhythmia, and cardiogenic shock, all of which are associated with increased mortality. The high prevalence of comorbidities, the pre-existing cardiovascular diseases in this case hypertension and other cardiovascular diseases as well as diabetes due to the known cardiovascular complications, mark the evolution of Covid-19 [54, 109, 110].

The highest prevalence remains hypertension as shown by the vast majority of studies (58% of patients admitted to intensive care, for example) who have a high mortality in the context of Covid-19 pathology due to cardiovascular comorbidities [109].

Heart failure is a common complication of Covid-19 disease with a 24% incidence of all patients and 49% of deaths.

It is well known today that most patients infected with SARS-Cov-2 develop a mild to moderate form of the disease that resolves clinically in about 6–10 days, but about 20% of them develop a severe form of bilateral atypical interstitial pneumonia with respiratory distress syndrome and eventually death. Massive endothelial damage, coagulopathy, and associated thrombosis lead to systemic microangiopathy and thromboembolism, complications that can lead to multiple organ failure and potential death [54]. Renal complications in this context are associated with an increased risk in severe patients.

Conclusive evidence highlights the loss of vascular barrier integrity and the development of a procoagulant endothelium that allows the onset and spread of respiratory distress syndrome following pulmonary endothelitis and inflammatory infiltrate at this level.

As we have presented in this text, the alteration of vascular homeostasis by SARS-Cov-2 is due to the interaction of the virus with ACE2 with a large distribution at the endothelial level.

Microangiopathy and vascular thrombosis are consecutive activation of complement, an integral component of the innate immune system. In the context of prolonged SARS-Cov-2 infection, restricted complement activation results in endothelial damage, thrombosis, and intravascular coagulation in the terminal phase [54].

In essence, the cited authors point out that the available evidence shows that SARS-Cov-2 induces immune dysfunction, extensive endothelial injury, coagulopathy, and systemic microangiopathy.

Venous thromboembolism, arterial thrombosis, and microangiopathy through their complex interactions generate thromboinflammation in Covid-19. It should be noted that coagulopathy, thrombocytopathy and endothiopathy associated with cardiovascular risk factors (diabetes, obesity, old age) aggravate thrombosis with increasing mortality [110].

Endotheliopathy or endothelial dysfunction are defining features of Covid-19, demonstrated by electronic microscopy by the presence of endothelial cell damage and its apoptosis.

A number of biomarkers of endothelial dysfunction, present in high concentrations, in patients with Covid-19, respectively thrombomodulin, angiopactin 2, with prognostic value, are associated with severe forms of disease and increased mortality [111].

The mechanisms associated with the interaction of SARS-Cov-2 with ACE2 and the disorder of the protective component of the renin-angiotensin-aldosterone system have been presented in this chapter.

The fatal combination of thrombocytopathy and endotheliopathy with platelet and endothelial activation increases the lethal risk of Covid-19.

In the context of the coexistence of cardiovascular risk factors (hypertension, diabetes, lung disease, old age, and obesity), their association with platelet and endothelial dysfunction is described.

Thrombocytopathy and endotheliopathy, which in turn are associated with inflammation and coagulopathy, exacerbate SARS-Cov-2 infection with a risk of thrombosis and pulmonary embolism in other organs such as the heart, kidneys, and liver, suggesting a Covid-19 disease thrombotic microangiopathies in those patients [110].

A number of markers of vascular endothelial damage have been described, highlighting the prothrombotic phenotype, coagulopathy with increased risk of vascular thrombosis [112].

SARS-Cov-2 alters cardiac vasculature generating endothelitis of small epicardial and intramiocardial vessels revealing an important parogenic mechanism of myocardial damage in Covid-19. Data were obtained by cardiac autopsy on a number of 6 patients who died of multiple organ failure and ARDS [62].

13.4.2 Myocarditis or the Direct Viral Infection of the Myocardium with Variable Evidence of Inflammation in Covid-19 Disease

Some forms of viral infection are not accompanied by the typical inflammatory infiltrate. Both forms of viral disease with and without obvious inflammation are called or considered myocarditis.

Activation of the immune system associated with viral infection cause myocarditis. Cardiac damage is achieved by the genesis of direct cytopathic effects in cardiac myocytes or by activating immune processes with inflammation and cellular infiltration. There is a variety of evidence of viral heart damage, including direct histological damage, by markers of heart damage [113].

It is recognized that viral heart damage may be the result of direct infection of myocytes or other types of myocardial cells such as fibroblasts, endothelial cells, pericytes, etc., lesions associated with in-hospital mortality [114].

A recent synthesis [115] of the association of coronavirus with cardiovascular disease, myocardial injury and arrhythmia shows that at the mentioned anatomical-clinical level, there is a various impairment in Covid-19 disease, myocardial damage being detected in 25% of hospitalized patients, lesions related to mortality.

The cited synthesis analyzes the mechanisms of myocardial injury mentioning the imbalance between oxygen demand and supply, direct myocardial invasion, inflammation, rupture of the coronary plaque, with acute myocardial infarction, microvascular thrombosis, and adrenergic stress.

Arrhythmias are considered a common manifestation in Covid-19 either as an initial manifestation, at the presentation of the patient (7.3%) or a later phenomenon (16.7%) [115]. In the genesis of arrhythmias occurring in the Covid-19 disease, several mechanisms are involved, namely the mentioned myocardial injury, with a frequency of 19.7–27.8%, which is at the same time correlated with a substantial increase in the incidence of arrhythmias.

Another mechanism that is frequently mentioned, namely cardiac electrical instability with prolonged QT interval, is also explained by hypokalemia or hypomagnesemia, especially in critically ill patients. It also mentions the side effects of some drug treatments including diuretics, antimalarials, used in context and azithromycin (which can prolong the QT interval). Hyperinflammatory conditions favor the appearance of arrhythmias, mentioning inflammatory cytokines that prolong the duration of ventricular action potential.

In addition, cytokines can indirectly generate arrhythmias through sympathoadrenergic hyperactivity that they induce with potential, life-threatening arrhythmias.

Hydro-electrolyte abnormalities following impaired renal function in Covid-19 may contribute to the genesis or worsening of Covid-19 arrhythmias.

In essence, arrhythmias are the direct or indirect consequence of viral aggression on the myocardium, namely:

- direct viral myocardial injury
- inflammation related to myocardial injury
- microvascular dysfunction
- microvascular thrombosis
- stress-induced cardiomyopathy
- imbalance of the necessary ratio/oxygen supply
- rupture of the coronary plate

Thromboembolic complications, in Covid-19, in the context of vascular lung disease (ARDS) are common, without being able to specify their exact incidence [115].

Covid-19 arterial or venous thromboembolism has a complex pathogenesis, with the synergistic participation of the following processes: inflammation, hypoxia, some side effects of therapeutic procedures, and even hospitalization. In essence, the synergistic factors are: excessive, uncontrollable inflammation, endothelial activation, and dysfunction of this structure, arterial, and venous stasis, hypoxia that define moderate and severe cases of Covid-19. Supraventricular tachyarrhythmias may increase the incidence of Covid-19 thromboembolism.

13.4.3 Prevalence of Hypertension and Other Conditions as Risk Factors for Covid-19 Disease

Research by the Japanese Hypertension Society shows that hypertension is prevalent in patients with Covid-19.

Thus, hypertension as a risk factor may be associated with age, which in turn is a key risk factor for increasing the severity and mortality of Covid-19. If we admit that a large part of Covid-19 patients with atherosclerotic disease have other risk factors including hypertension, obesity, diabetes, smoking, etc. which has aggravating clinical consequences.

It turns out that the co-participation in the genesis of severity and mortality of different risk factors can be assessed, without being able to separate the independent involvement of factors such as age or atherosclerosis [104].

For example, hypertension, diabetes, and dyslipidemia accelerate atherosclerosis, which in turn causes cardiovascular complications, increasing the severity and mortality of Covid-19.

Hypertension is associated, or develops, with endothelial dysfunction, especially in the elderly, and endothelial damage can lead to thromboembolism, a known complication in Covid-19.

The authors evoke, in the context of disasters, the term "disaster hypertension." Similarly, there is an increased risk of mental disorders (anxiety, depression) that can further impair blood pressure control [116, 117].

Adequate monitoring, in the context of Covid-19 disease, of hypertensive patients is a recommendation to be able to identify, in time, the onset of another cardiovascular disease [104]. The importance of heart failure, both as a risk factor, for worsening the course of Covid-19 disease and increasing mortality and as a possible consequence of myocardial injury following SARS-Cov-2 infection, is emphasized [118].

Several studies have shown cardiomyopathy in patients with Covid-19 and 33% of critically ill patients. It is estimated that heart failure is a common complication in Covid-19 disease that is due to either the deterioration of pre-existing heart dysfunction or new cardiomyopathy and myocarditis in the context of the disease.

A multicenter study found heart failure in 23% of patients, with a clear prevalence in non-survivors, compared to survivors, respectively, 52% versus 12%.

Heart failure, in the setting of Covid-19 disease, is caused by myocardial injury, systemic inflammation, pulmonary hypertension and acute respiratory distress syndrome, renal dysfunction, hydrosaline retention, and the imbalance between the necessary and myocardial oxygen supply.

There is a consensus in assessing pre-existing cardiovascular disease in the evolution of Covid-19, in this case, these conditions worsen the prognosis of viral disease.

Cardiovascular complications also revealed by biomarkers are myocarditis, heart failure and thromboembolism which are accentuated by delayed medical care.

Angyotensin II and the AT1 receptor activate mitogen-activated protein kinase (MAPK) while Angiotensin (1-7) represses this signaling. It should be noted that cardiovascular disease is associated with decreased ACE2 and activation of the damaging component of RAAS (Angiotensin II/R AT1). Inhibition of MAPK signaling suppresses SARS-Cov-2 replication.

In those with pre-existing cardiovascular disease, the increased severity of Covid-19 disease is partly due to the activation of the MAPK pathway (MAPK protein kinases are a family of serine-threonine protein kinases involved in the host's defense mechanism and in apoptosis), while Angiotensin (1-7) represses MAPK signaling.

The inhibition of the latter results in the suppression of the multiplication of the coronary virus. ACE2 in these cells decreases with age, and the severity of Covid-19 disease is inversely proportional to the tissue expression of ACE2 [118].

Comorbidities also show an inverse relationship of ACE2 expression with Covid-19 severity. The proinflammatory state evolves with the decrease of the

ACE2/ACE ratio, following the decrease of the ACE2 expression, as it is observed in the acute respiratory distress syndrome.

With age, there is an increase in inflammatory processes contributing to the phenomenon of aging and senescence.

Decreased ACE2 (down-regulation) following the interaction with SARS-Cov-2 could suggest a decrease in the impact of the virus on cells, limiting its entry into cells. But the intense affinity of SARS-Cov-2 (10–20 times more intense than SARS-Cov-1) is discussed, implicitly, this relative decrease in the receptor has no consequences on viral dispersion. Recent studies show that Covid-19 disease is more severe in the elderly and with comorbidities.

The clinical prevalence of comorbidities in those with Covid-19 disease is evident, in intensive care units this type of association between the disease and at least one state of comorbidity reaches from 68% to 72% [107].

Relevant data highlight the correlation between ACE2 expression, different comorbidities as risk factors, and the severity of Covid-19 disease [107].

All these comorbidities (cardiovascular disease, chronic respiratory disease, obesity, diabetes, age, male gender, and plus RAA system inhibitors) are characterized by increased tissue expression of ACE2, except for the latter category.

Noting that ACE2 has a low expression in cardiomyocytes, but is highly expressed in pericytes, which are associated with vascular endothelium, which, in the context of the disease, explains the prevalence of heart damage reaching 59% of patients in intensive care.

It is admitted that this pathway, of viral aggression of pericytes, produces endothelial dysfunction and microcirculatory disorders [108].

Of great interest for the perception of the pathogenic context is the fact that ACE2 expression is high in cardiovascular diseases, respectively in ischemic heart failure, dilated cardiomyopathy, pulmonary hypertension, and heart failure.

Patients with heart failure have high ACE2 expression in the ventricles and generally throughout the myocardium [108, 119, 120].

We can interpret, without risk of error, that this increase in ACE2 expression in cardiovascular diseases and especially in heart failure has a possible explanation, as an adaptive compensatory reaction, but not related to SARS-Cov-2.

The authors cite that the context shows that patients with heart failure are more susceptible to SARS-Cov-2 infection and may develop additional life-threatening heart damage [107].

The fact that ACE2 increases in virtually all conditions associated with risk factors, heart failure, obstructive pulmonary disease, lung cancer, correlated with age and hypertension, could argue our interpretation in the adaptive sense of a dysfunction and a specific pathology.

The increase in ACE2 in pathological conditions is a precondition that favors viral aggression, the process being favored by endothelial dysfunction as a background and other consequences of the risk factors mentioned in the medium and long term.

Viral aggression, on a vascular level, is clearly revealed by the fact that patients with severe Covid-19 disease have prolonged prothrombin time, increased dimer D, and some of them develop disseminated intravascular coagulopathy. Covid-19 associated coagulopathy is prothrombotic, with a high prevalence of venous thromboembolism, deep vein thrombosis, pulmonary embolism, aspects observed in patients in intensive care units. It is admitted that the decrease in ACE2 expression in Covid-19 acts as a transient molecular disease that causes disruption of the RAA system with tissue damage in the microcirculation environment, a fundamental aspect of the pathogenesis of Covid-19 [121].

As a result of these disorders there are exaggerated reactions of the innate immune system with microcirculatory immunothrombosis. Cell activation (neutrophils, platelets) and activation of cytokine-releasing complement can shape the cytokine storm with extensive tissue damage and thromboembolism.

We emphasize that this pathogenic context is possible through viral interaction with endothelial ACE, i.e., arranged in a large structure of the body, reflecting the microcirculatory system of the whole body.

It is considered, of course, that endothelial dysfunction is a major condition in the critically ill patient that conditions the development of multiple organ dysfunction syndrome, with potential death [121].

Apart from the outlined pathogenic picture of endothelial dysfunction, it is believed that this cardinal process is not fully understood in the context of Covid-19 disease. The authors note that in most cases of microvascular thrombosis, the diagnosis is usually made by clinical reasoning and sometimes confirmed histologically. It turns out that this extensive microvascular damage can occur much more frequently than is diagnosed, as there is no surrogate laboratory test.

Early identification is essential to initiate appropriate therapy. Analysis of the pathogenic structure of severe cases of Covid-19 (5%) shows that they develop severe acute respiratory distress syndrome, systemic inflammatory response syndrome, and multiple organ failure and dysfunction with high mortality.

It is estimated, in context, that respiratory failure associated with severe acute respiratory distress syndrome is the leading cause of death in Covid-19 disease. By default, the severity of Covid-19 disease is estimated to be correlated with plasma levels of IL-6, a C-reactive protein, as markers of proinflammatory status.

The distinctive perception of the functions of the systemic, local-tissue and intracellular renin-angiotensin-aldosterone system generates a more complex and relevant image of it. The classic concept is involved in the control and regulation of major functions such as hydroelectrolytic homeostasis and vascular tone in the vital organs. The action focuses on angiotensin II derived from angiotensin I, a process catalyzed by the ACE conversion enzyme in pulmonary vascular endothelial cells. Angiotensin II has major functions through AT1 and AT2 receptors.

The local or tissue RAA system exerts its effects on the heart, lung, brain, kidney, liver, and digestive tract having the functions of regulating the inflammatory process, vascular tone and permeability, apoptosis, growth, migration, and cell differentiation. Through the signaling and effector link, angiotensin II activates the AT1 receptor generating the release of proinflammatory cytokines, the production of free radicals, amplifying cell damage, and the inflammatory process.

Counteracting and balancing the recalled processes is achieved by stimulating RAT2 and especially by involving ACE2 which catalyzes the production of Angiotensin (1-7) with the consecutive activation of R Mas, an axis of great relevance with protective and anti-inflammatory properties.

In summary, the ACE2/Ang (1-7)/R Mas link antagonizes the ACE/Ang II/RAT1 link, both components of the renin-angiotensin-aldosterone system [21, 22].

Viral aggression, implicitly triggering, through the dysfunctional reaction of the host, the cytokine storm that affects the endothelial and epithelial cells in the lungs, amplifies the release of cytokines, with systemic inflammatory process, including panendothelitis.

Coexpression of ACE2 and serum transmembrane protease (TMPRSS2) in type I pneumococytes, lymphocytes, macrophages, smooth muscle, enterocytes, vascular smooth muscle, cardiomyocytes, hepatocytes, kidney cells, and neurons is an illustrative picture of systemic damage in Covid-19 and its clinical consequences [80].

13.4.4 Mechanisms of Cardiovascular Disorders in Covid-19 Disease

The approach of the pathogenic, molecular mechanisms, at cardiovascular level, identifies as initiating link the interaction of SARS-Cov-2 with the enzyme ACE2, arranged at the level of the cell membranes belonging to the vital organs of the human and animal organism. As a result of the interaction with SARS-Cov-2, ACE2 in the lung and cardiovascular structures decreases quantitatively, diminishing the protective link of the RAA system, by decreasing the genesis and the interaction with the Mas receptor of Ang (1-7) and implicitly, consecutively, decreases lung and cardiovascular protection [21, 22]. We note that ACE2 is elevated in diabetes, cardiovascular disease, including heart failure and ischemic cardiomyopathy, which could suggest a negative prognosis in the context of the association of these conditions with Covid-19 disease.

We emphasize that the binding of the virus to ACE2 and the decrease in its presence and activity cancel (at least partially) the lung and cardiovascular protection by the link controlled by this enzyme (through anti-inflammatory, antifibrotic, antioxidant, and vasodilation effects) favoring proinflammatory processes, lung and cardiovascular damage, effects controlled by Ang II, via the AT1 receptor.

Increasing the Ang level (1-7) by pharmacological methods, counteracts the pathological pulmonary and cardiovascular consequences induced by the virus, being a potential route of Covid-19 pharmacotherapy.

Understanding and addressing the distribution and significance of ACE2 in SARS-CoV-2 infection may provide several ways to prevent and treat Covid-19 disease (the severity of Covid-19 disease) is correlated with the expression of ACE2 in various organs and tissues or with differentiation according to the sex or age of the patient? [122]. The differentiation of the degree of damage to a particular tissue is not correlated with the particularities generated by the sex or age of the patient, suggesting that other factors are involved in the pathogenesis of Covid-19 disease. An argument for this approach is for example that although the density of ACE2 is similar in lung and liver damage is very different.

Cytokine storm, generated in the context of SARS-CoV-2 infection, is an exaggerated reaction to release proinflammatory cytokines (interferon-ɣ, TNF-α, IL-1, IL-6, IL-7, IL-10, etc.) with systemic inflammatory syndrome, hyperferinemia, hemodynamic instability, with the potential for development of acute respiratory distress syndrome and multiple organ failure with an unfavorable prognosis.

Of course, one of the questions generated, in the context described, concerns the mechanism and significance of the decrease in ACE2 after the interaction with SARS-CoV-2. Is this decrease in viral load dependent and consequent on the decrease in the membrane presence of ACE2? [60] Is this a transient decrease, recovering theoretically through neosynthesis, to the extent that the viral spread in the body does not further decrease ACE2?

Is the decrease in ACE2, more sustainable after the viral invasion, a more complex process than the simple decrease following the viral invasion?

Let's try and answer these and other consensual questions.

The aspects outlined above appear to us as indirect molecular mechanisms of vascular endothelial damage with already known consequences.

We are talking, of course, about a direct mechanism that is also defined by identifying alternative receptors that are arranged, with high expression, in the endothelial cells [123, 124]. These are neuropilin and CD209L, proposed as facilitators of SARS-CoV-2 cell entry. Neuropilin-1 (NRP1) potentiates SARS-CoV-2 infectivity and is abundantly expressed in the olfactory and respiratory epithelium of endothelial and epithelial cells [125].

Neuropilin-1 is quantitatively increased in vessels infected with SARS-CoV-2, being involved in the development of the cardiovascular system, angiogenesis, and the formation of neural circuits. Experimental data show that NRP1 can mediate the entry of SARS-CoV-2 into cells in general, including the CNS, thus increasing infectivity [126].

In SARS-Cov-2 infection, endothelitis, respectively vascular inflammation is highlighted by swelling of endothelial cells, their damage, and their separation from the basement membrane by subendothelial lymphocytes. Diffuse endothelial inflammation has been identified, in terms of accumulation of inflammatory cells, apoptosis and endothelial necrosis, in various organs such as heart, lung, kidney, liver, small intestine, and skin, with the presence of viral components in endothelial cells [45, 81].

It is considered that the widespread endothelial dysfunction, endothelitis, is the consequence of a direct action, by viral invasion of the endothelium, which triggers inflammation and the initiation of the innate immune response. The clinical consequences of these processes are microthrombotic complications, which can lead to deep vein thrombosis, pulmonary embolism and stroke [124].

Cardiac dysfunction in patients with Covid-19 has a weight of 20–50%, regardless of the severity of the disease, but is strongly associated with mortality [127].

A large proportion of patients with heart damage remain with long-term sequelae after Covid-19, given that the heart does not have innate regenerative mechanisms.

The cited paper analyzes the cytopathic effects induced by SARS-Cov-2, especially in cardiomyocytes, characterized by myofibrillar fragmentation in sarcomeric units and loss of DNA staining in the body of intact cells. This type of lesions with sarcomeric structural alterations is observed in patients who die of Covid-19.

SARS-Cov-2 infection causes disruption of the myofibrillar structure with dissolution of the cardiac contractile machinery that can lead to functional collapse [127].

It is noteworthy that recently a SARS-Cov-2 viral protein has been shown to induce endothelial activation and dysfunction with all known pathogenic consequences [128].

The SARS-Cov-2 genome encodes 29 viral proteins, of which 16 are non-structural proteins, 4 are structural proteins that include spike proteins, membrane proteins, nucleocapsid proteins, envelope proteins, and 9 accessory proteins. In the present case, it is emphasized that the nucleocapsid protein (N) produces endothelial activation, via receptor 2 (TLR2)/NF-KB and activated mitogen protein kinase (MAPK)[2].

The nucleocapillary of other viruses (SARS-Cov, MERS, H1N1, etc.) does not activate endothelial cells.

The N protein is abundant in the SARS-Cov-2 virus, being involved in the entry into the host cell binds to the viral RNA genome forming a ribonucleoprotein that facilitates its own replication, assembly of the virus and its release outside the cell. It is highlighted, in the cited paper, a fact of special significance, namely that the N protein acts as an associated pathogen, binding to TLR2, activating the signaling by NF-KB and MAPK [129].

As mentioned, SARS-Cov-2 N protein is the most abundant in SARS-Cov-2 and, after lysis of the infected cell, is widely distributed in the host organism, especially extracellular or circulating, being present in the circulation of positive patients in proportion of 76% [130].

The authors of the paper mention that the circulating N protein can probably amplify endothelial cells by generating thrombosis and multiple organ failure in patients with severe forms of Covid-19.

In addition, it is of great interest that statins, used early in the treatment of Covid-19, to prevent cytokine storm and acute respiratory distress syndrome, reduce the risk of death in these patients.

Among the known stands, only simvastatin has been shown to be an intense inhibitor of N protein activity, inducing endothelial activation.

The effects of simvastatin are known.

The authors [128] demonstrated that the SARS-Cov-2 N protein induces proinflammatory cytokines and chemokines in human peripheral blood monucleores, suggesting that N protein is involved in cytokine storms. A very relevant aspect is that only the N protein from SARS-Cov-2 has retained its property of inducing

[2] MAPK—Mitogen-activated protein kinase.

TLR2—Toll-type receiver 2, located membrane recognizes lipoproteins from pathogens (bacteria, viruses, fungi, etc.) and signals the production of proinflammatory cytokines.

endothelial activation, because the N protein from SARS-Cov, MERS and H1N1 influenza does not affect the vascular endothelium.[3]

The amino acid sequence of the nucleocapsid in SARS-Cov-2 and SARS-Cov has an identity of 91%, suggesting that the difference could contribute to the genesis of the mentioned effects, of great interest.

In addition, the effects of simvastatin are particularly striking, as a potential drug to prevent the pathogenesis of systemic panendothelitis, the basis of multiple organ failure [21, 22, 131].

Another viral protein is majorly involved in the pathogenesis of SARS-Cov-2.

Inflammation signaling leads to the secretion of cytokines, such as TNFα and IFNɣ, a process in which TLR2 receives SARS-Cov-2 infection by identifying protein E inducing cytokine release. Interestingly, the activation of TLR2 receptor signaling is independent of the entry into the cell of the virus in question and its replication.

The enveloping viral protein E is the one that is "detected" by TLR2, thus triggering the activation of the inflammatory process, following the production and release of cytokines.

It seems remarkable that the identification of TLR2 function in SARS-Cov-2 infection can be further argued by the fact that blocking signaling through said receptor reduces IL-6 release, with a significant increase in the survival of experimental animals (mice) infected with SARS-Cov-2. The authors cite that blocking the TLR2 receptor could provide protection against mortality through SARS-Cov-2 [129].

It is accepted that mortality in Covid-19 is caused by the inflammatory process, by the genesis of pulmonary endothelial barrier dysfunction and by left ventricular dysfunction that aggravates each other.

The association between age, hypertension, and the severity of Covid-19 infection can also be explained by the fact that hypertension is very closely associated with age. The most relevant predictor of severe disease and mortality is age, which can be explained by immunosuppression, chronic inflammatory processes, excessive signaling of the AT1 receptor and angiotensin II, with prothrombotic and profibrotic proinflammatory effects.

Endothelial injury and thrombosis are interconnected processes, endothelial damage, through viral aggression, activates the coagulant system.

The interaction of SARS-Cov-2 with ACE2 not only reduces the functions of the enzyme, as a major regulator of the RAA system, reducing the formation of Ang (1-7), but also induces the production of plasminogen-activating inhibitory factor (PAF-1), preventing fibrinolysis and thrombus dissolution. . The inflammatory process in Covid-19 is accompanied by disorders of the coagulation process and the fibrinolotic system, resulting in impairment of factors involved in thrombosis such

[3]MAPK—Mitogen-activated protein kinase.

TLR2—Toll-type receiver 2, located membrane recognizes lipoproteins from pathogens (bacteria, viruses, fungi, etc.) and signals the production of proinflammatory cytokines.

as inhibitory factor TF (tissue factor involved in coagulation, thrombin formation), antithrombin III, and C-reactive protein [131, 132].

It has recently been shown that viral binding protein S causes endothelial damage by altering mitochondrial function, decreasing NO synthase activity (eNOS), reducing NO availability, increasing oxidative stress, and decreasing ACE2 [133, 134].

The decrease in ACE2, at least in the lungs, is due to the interaction with the spike proteins of SARS-Cov-2 for at least two reasons. After interaction with the spike protein, ACE2 is internalized on the one hand or detached from the membrane, after cleavage by the protease, on the other hand. The increase in angiotensin II is due to a decrease in the presence and specific activity of ACE2.

The incidence of cardiac manifestations in patients with Covid-19 is high, processes being correlated with the disruption of ACE2 functions in context and by the onset of cytokine storm, following the imbalance between the response of type 1 and 2 helper cells, respiratory disorders, and hypoxemia, resulting in myocardial cell damage. In the study cited, patients with severe forms of Covid-19 had a proportion of 58% hypertension, 25% other heart disease and 44% arrhythmia. Mortality is 35% in those with a history of hypertension and 17% in those with coronary heart disease [135].

ACE2 expression is elevated not only in the lung, type II alveolar cells and lower respiratory tract, but also in the myocardium, esophageal vascular endothelium, proximal convoluted tube, renal and urinary bladder, which are implicitly organs with high risk for SARS-Cov-2 infection.

Decreased ACE2 expression causes damage or worsening of pre-existing heart damage.

By altering the protective linkage of the RAA system, respectively the decrease of Ang (1-7), produces the decrease of vasodilation, of the decrease of the blood pressure, the antiproliferative, anti-inflammatory, antifibrotic and antiapoptotic effects from the pulmonary alveolar level are reduced.

Abnormal increase in Ang II aggravates pre-existing cardiovascular disease. S proteins, separated from the viral structure, can damage endothelial cells by lowering ACE2 and consequently inhibiting mitochondrial function [133]. Since 2005, Kuba et al. demonstrated the crucial role of ACE2 in the genesis of lung damage in Covid-19. The authors cited above have shown that separate S protein can damage the endothelium, a process manifested by impaired mitochondrial function and NO synthase activity, with increased glycolysis.

The severity of Covid-19 disease is increased by the degradation of ACE2 by the SARS-Cov-2 virus, a process considered the main trigger of excessive inflammation in severe forms of the disease [118].

Observational studies and meta-analyzes show a decrease in ACE2 in the biological context, respectively, in the male sex, positively correlated with old age and comorbidity such as hypertension, cardiovascular disease, chronic obstructive pulmonary disease, diabetes, and conditions considered risk factors with increasing severity Covid-19.

In hypertension, for example, following the increase in Ang II, the activity of the AT1 receptor intensifies with the decrease of the presence of ACE2 and the disturbance of the ACE2/ACE balance, with the increase of blood pressure, ACE2 being markedly low. ACE2 is also dominant in hypoxia as well [122].

The association of heart failure and septic shock generates a high mortality, inevitably (7090%), versus 20% in patients without cardiovascular comorbidities [132, 136].

The background of the pathogenic mechanisms of Covid-19 disease is the widespread endothelial dysfunction—panendothelitis and immunothrombosis as microthrombotic complications, after the onset of coagulation cascade with the formation of intravascular clots, generating severe acute respiratory distress syndrome and other dysfunctions [137].

13.5 General Conclusions

The SARS-Cov-2 virus, a 30Kb nucleotide RNA coronavirus, is a variant resulting from SARS-CoV, identified in the 2003 epidemics in host tissues, human or many other species.

The affinity for ACE2 is 10–20 times higher than that of SARS-CoV-2, which gives the SARS-CoV-2 virus a higher rate of transmissibility, as measured by the Ro index. The membrane of the ACE2 enzyme has fundamental functions in the regulation and control of the renin-angiotensin-aldosterone system, targeting major functions such as blood pressure, hydro-electrolytic homeostasis, vascular reactivity. The wide distribution of ACE2 in human tissues and organs and not least in the vascular endothelium also explains the extent of systemic damage (panendothelitis) by the virus in question, but does not explain the individual differences in Covid-19 disease [82].

In addition, the widespread multi-organ damage in severe forms of Covid-19 disease is largely explained by systemic endothelial damage (panendothelitis) [46, 110].

The causes and mechanisms of individual receptivity (susceptibility), severity and prognosis of the disease are not yet elucidated. The relationship between the wide tissue distribution and the viral aggression is only partially explained, as there is no direct and relevant clinical relationship between the two components (the presence of the receptor enzyme and the specific viral effects).

Of course, it is relevant that the interaction of SARS-Cov-2 with membrane ACE2 disrupts the function of the enzyme, decreases its functional presence by endocytosis concomitant with the virus, compromises consecutively link ang (1-7)/Mas receptor with the onset of lung injury, the result of predominance of the effects of angiotensin II. They can be caused by hypoxia, multiple organ failure to which it contributes and direct and indirect viral damage to tissues including the cardiovascular system.

Cardiovascular comorbidities, age over 60, diabetes and obesity may contribute to vulnerability to the prognosis of Covid-19 disease. The paradox is generated by the fact that a large proportion of those infected are asymptomatic (about 45% at the onset of the pandemic) which does not correlate with the presence and tissue density of the ACE2 receptor.

Asymptomatic patients transmit the virus to other people in a significant proportion. Patients with severe disease account for about 20%, of those hospitalized develop viral sepsis and acute respiratory distress syndrome. About 80% of hospitalized patients and 60% of those admitted to intensive care units survive, but many of these survivors remain with long-term sequelae [21, 22, 138].

References

1. Andersen K, et al. The proximal origin of SARS-Cov-2. Nat Med. 2020;26:450–2.
2. McIntyre CR, et al. Converging and emerging threat to health security. Environ Syst Decis. 2018;38:198–207.
3. Imai Y, et al. Angiotensin-converting enzyme (ACE_2) in disease pathogenesis. Circ J. 2010;74:105–410.
4. Imai Y, et al. Angiotensin-converting enzyme 2 in acute respiratory distress syndrome. Cell Mol Life Sci. 2007;64:2006–12.
5. Imai Y, et al. Angiotensin-converting enzyme protects from severe acute lung failure. Nature. 2005;436:112–6. https://doi.org/10.1038/nature03712.
6. Matsuyama S, et al. Efficient activation of the severe acute respiratory syndrome coronavirus spike protein by transmembrane protease TMPRSS2. J Virol. 2010;84:12658–64.
7. Zhou F, et al. Clinical course and risk factors for mortality of adult inpatients with Covid-19 in Wuhan, China: a retrospective cohort study. Lancet. 2020;395:1054–62.
8. Saif LJ. Animal coronaviruses: lessons for SARS. Washington, DC: National Academic Press; 2004.
9. Voicu V, Cernescu C, Popescu I. Pandemia Covid-19 în Romania. In: Aspecte clinice şi epidemiologice. Bucureşti: Editura Academiei Române; 2020.
10. Hoffman M, et al. SARS-Cov-2 cell entry depends on ACE_2 and TMPRSS2 and is blocked by clinical proven protease inhibitor. Cell. 2020;181:271–280.e8.
11. Wrapp D, et al. Cryo-EM structure of the 2019-n Cov spike in the prefusion conformation. Science. 2020;367:1260–3.
12. Gibson P, et al. Covid-19 acute respiratory distress syndrome (ARDS) clinical features and differences from typical pre-Covid-19 ARDS. Med J Aust. 2020;213:54–56.e1. https://doi.org/10.5694/mja2.50674.
13. Lippi G, et al. Coronavirus disease 2019 (Covid-19) the portrait of perfect storm. Ann Transl Med. 2020;8:497.
14. Lippi G, et al. Covid-19: unravelling the clinical progression of nature's virtually perfect biological weapon. Ann Transl Med. 2020,8(11):693.
15. Harrison A, et al. Mechanisms of SARS-Cov-2 transmission and pathogenesis. Trends Immunol. 2020;41(12):1100–14.
16. Lippi G, et al. Coronavirus disease 2019 (Covid-19): the portrait of a perfect storm. Ann Transl Med. 2020;8(7):497. https://doi.org/10.21037/atm.2020.03.157.
17. Voicu V. SARS-Cov-2, ACE_2, panendothelitis and viral neuroinvasion. Systemic consequences. J Hypertens Res. 2021;7(1):4–11.
18. Jha NK, et al. Current understanding of novel Coronavirus: molecular pathogenesis diagnosis and treatment approaches. Immuno. 2021;1:30–66.
19. Hamming I, et al. Tissue distribution of ACE_2 protein, the functional receptor for SARS-coronavirus. A first step in understanding SARS pathogenesis. J Pathol. 2004;203:631–7.
20. Voicu V. Mecanisme patogenice în infecţia cu virusul SARS-Cov-2. SARS-Cov-2-receptor, penetrarea în celula umană, patogenie, mecanisme. In: Voicu V, Cernescu C, Popescu I, editors. Patogenia Covid-19 în Romania, aspecte clinice şi epidemiologice. Bucureşti: Editura Academiei Române; 2020.
21. Voicu V. The impact of SARS-Cov-2 infection on the renin-angiotensin-aldosterone system and its axes. Pathogenic consequences. J Hypertens Res. 2020;6(4):99–110.
22. Voicu V. SARS-Cov-2, ACE_2, panendothelitis and viral neuro-invasion. Systemic consequences. J Hypertens Res. 2021;7(1):4–12.
23. Saponaro F, et al. ACE_2 in the era of SARS-Cov-2: controversies and novel perspectives. Front Mol Biosci. 2020;7:588618.

24. Grasselli G, et al. Baseline characteristics and outcome of 1591 patients infected with SARS-Cov-2 admitted to ICUs of the Lombardi region, Italy. JAMA. 2020;323:1574–81.
25. Gheblawi M, et al. Angiotensin-converting enzyme 2: SARS-Cov-2 receptor and regulator of the renin-angiotensin system. Circ Res. 2020;126:1457–75.
26. Rajendran P, et al. The vascular endothelium and human disease. Int J Biol Sci. 2013;9:1057–69.
27. Lang M, et al. Pulmonary vascular manifestations of Covid-19 pneumonia. Radiol Cardiothorac Imaging. 2020;2(3):e200277.
28. Evans PC, et al. Endothelial injury and dysfunction in Covid-19. Cardiovasc Res. 2020;116(14):2177–84.
29. Hayashi T, et al. Highly conserved binding region of ACE_2 as a receptor for SARS-Cov-2 between human and animals. Vet Q. 2020;40(1):243–9.
30. Choi J-Y, et al. Altered Covid-19 receptor ACE_2 expression in a higher risk group for cerebrovascular disease and ischemic stroke. Biochem Biophys Res Commun. 2020;528:413–9.
31. Cuervo NZ, et al. ACE_2: evidence of role as entry receptor for SARS-Cov-2 and implications in comorbidities. Elife. 2020;9:e61390. https://doi.org/10.7554/eLife.61390.
32. Devaux C, et al. Can ACE_2 receptor polymorphism predict species susceptibility to SARS-Cov-2. Front Public Health. 2021;8:6087665.
33. Shang J, et al. Structural basis of receptor recognition by SARS-Cov-2. Nature. 2020;581:221–4.
34. Yan R, et al. Structural basis for the recognition of the SARS-Cov-2 by full-length human ACE_2. Science. 2020;367:1444–8.
35. Bakhshandeh B, et al. Variants in ACE_2 potential influences on virus infection and Covid-19 severity. Infect Genet Evol. 2021;90:104773.
36. Qui H, et al. Clinical and epidemiological feature of 36 children with coronavirus disease 2019 (Covid-19) in Zhejiang, China: on observational study. Lancet Infect Dis. 2020;20:689–96.
37. Benetti E, et al. ACE_2 gene variants may underlie interindividual variability and susceptibility to Covid-19 in the Italian population. Eur J Hum Genet. 2020;28:1602–14.
38. Re R, et al. Estimating the reproductive number Ro of SARS-Cov-2 in the United State and eight European countries and implication for vaccination. J Theor Biol. 2021;517:110621.
39. Beyerstedt S, et al. Covid-19: angiotensin-converting enzyme 2 (ACE_2) expression and tissue susceptibility to SARS-Cov-2 infection. Eur J Clin Microbiol Infect Dis. 2021;40:905–19. https://doi.org/10.1007/s10096-020-04138-6.
40. Nguyen HL, et al. Does SARS-Cov-2 bind to human ACE_2 more strongly than does SARS-Cov? J Phys Chem. 2020;124:7336–47.
41. Rodrigues R, et al. The impact of ACE_2 expression levels in patients with comorbidities on Covid-19 severity: a comprehensive review. Microorganisms. 2021;9:1692.
42. Zou X, et al. Single-cell RNA-Seq. data analysis on the receptor ACE_2 expression reveals the potential risk of different human organs vulnerable to 2019-nCov infection. Front Med. 2020;14:185–92.
43. Li Y, et al. Molecular mechanisms of sex bias differences in Covid-19 mortality. Crit Care. 2020;24:405.
44. Kornilov SA, et al. Plasma level of soluble ACE_2 are associated with sex, metabolic syndrome, and its biomarkers in a large cohort, pointing to a possible mechanisms for increased severity in Covid-19. Crit Care. 2020;24:452.
45. Varga Z. Endothelial cell infection and endothelitis in Covid-19. Lancet. 2020;395:1417–8.
46. Labo N, et al. Vasculopathy and coagulopathy associated with SARS-Cov-2 infection. Cell. 2020;9:1583.
47. Almashat SA. Vasculitis in Covid-19: a literature review. J Vasculitis. 2020;6:1. https://doi.org/10.37421/JOV.2020.6.129.
48. Gavrilaki E, et al. Endothelial dysfunction in Covid-19: lessons learned from coronaviruses. Curr Hypertens Rep. 2020;22:63.
49. Del Turco S, et al. Covid-19 and cardiovascular consequences: is the endothelial dysfunction the hardest challenge? Thromb Res. 2020;196:143–51.

50. Calabrese F, et al. Pulmonary pathology and Covid-19: lesson from autopsy. The experience of European pulmonary pathologists. Virchows Arch. 2020;477:359–72. https://doi.org/10.1007/s00428-020-02886-6.
51. Voicu V. Post-Covid-19 sequelae or persistent Covid-19 disease. J Hypertens Res. 2021;7(3):84–8.
52. Voicu V. Sechele post-Covid-19 sau Boala post-Covid-19 prelungită (Sindromul post-Covid-19). Bucureşti: Editura Academiei Române; 2021.
53. Jung F, et al. Covid-19 and the endothelium. Clin Hemorheol Microcirc. 2020;75:7–11.
54. Perico L, et al. Immunity, endothelial injury and complement-induced coagulopathy in Covid-19. Nat Rev Nephrol. 2020;17:46–64. https://doi.org/10.1038/s41581-020-00357-4.
55. Aird WC. Endothelium in health and disease. Pharmacol Rep. 2008;60:139–43.
56. Ludmer PL, et al. Paradoxical vasoconstriction induced by acetylcholine in atherosclerotic coronary arteries. N Engl J Med. 1986;315:1046–51.
57. Gimbrone MA Jr. Endothelial dysfunction and the pathogenesis of atherosclerosis. New York: Springer Verlag; 1980.
58. Cybulsky MI, et al. Endothelial expression of a mononuclear leucocyte adhesion molecule during atherosclerosis. Science. 1991;251:788–91.
59. Furchgott RF, et al. The obligatory role of endothelial cells in the relaxation of arterial smooth muscle by acetylcholine. Nature. 1980;288:373–6.
60. Celermajer D. Endothelial dysfunction: does it matter? Is it reversible? J Am Coll Cardiol. 1997;30:325–33.
61. Maccio U, et al. SARS-Cov-2 leads to a small vessel endotheliitis in the heart. Ebio Medicine. 2021;63:103182.
62. Maccio U, et al. SARS-Cov-2 leads to a small vessel endotheliitis in the heart. EBioMedicine. 2021;63:103182.
63. Huertas A, et al. Endothelial cell dysfunction: a major player in SARS-Cov-2 infection? Eur Respir J. 2020;54:2001634.
64. Huertas A, et al. Endothelial cell dysfunction: a major player in SARS-Cov-2 infection (Covid-19?). Eur Respir J. 2001;2020(56):634.
65. Nicosia R, et al. Covid-19 vasculopathy: mounting evidence for the indirect mechanism of endothelial injury. Am J Pathol. 2021;191(8):1374–84.
66. Libby P, Lüscher T. Covid-19 is, in the end, an endothelial disease. Eur Heart J. 2020;41:3038–44.
67. Hamming I, et al. The emerging role of ACE_2 in physiology and pathology. J Pathol. 2007;212:1–11.
68. Basta G. Direct or indirect endothelial damage? An unresolved question. EBioMedicine. 2021;64:103215.
69. Teuwen L-A, et al. Covid-19: the vasculature unleashed. Nat Rev Immunol. 2020;20:389–91.
70. Pons S, et al. The vascular endothelium: the cornerstone of organ dysfunction in severe SARS-Cov-2 infection. Crit Care. 2020;24(353):1–8.
71. Pons S, et al. The vascular endothelium: the cornerstone of organ dysfunction in severe SARS-Cov-2 infection. Crit Care. 2020;24:353.
72. Kaur S, et al. The enigma of endothelium in Covid-19. Front Physiol. 2020;11:989.
73. Sudano I, et al. Protection of endothelial function: targets for nutritional and pharmacological intervention. J Cardiovasc Pharmacol. 2006;47(Suppl 2):S-136–S150.
74. Su JB. Vascular endothelial dysfunction and pharmacological treatment. World J Cardiol. 2015;7(11):719–41.
75. O'Sullivan J, et al. Endothelial cells orchestrate Covid-19. Lancet Hematol. 2020;7(8):e553–5. https://doi.org/10.1016/s2352-3026(20)30215-5.
76. Bikdeli B, et al. Covid-19 and thrombotic or thromboembolic disease: implications for prevention, antithrombotic therapy and fellow up, JAAC State of the art review. J Am Coll Cardiol. 2020;75(23):2950–73.
77. Brunner H, et al. Endothelial function and dysfunction. Part II: association with cardiovascular risk factors and disease. A statement of the working group on endothelins and endothelial factors of European Society of Hypertension. J Hypertens. 2005;23:233–46.

78. Liu F, et al. SARS-Cov-2 infects endothelial cells in vivo and in vitro. Front Cell Infect Microbiol. 2021;11:701278.
79. Peiris S, et al. Pathological findings in organs and tissue of patients with Covid-19: a systematic review. PLoS One. 2021;16:e0250708. https://doi.org/10.1371/journal.pone.0250708.
80. Amraei R, et al. Covid-19, renin-angiotensin system and endothelial dysfunction. Cell. 2020;9:1652.
81. Ackermann M, et al. Pulmonary vascular endothelitis, thrombosis and angiogenesis in Covid-19. N Engl J Med. 2020;383:120–8.
82. Ashrof UM, et al. SARS-Cov-2, ACE2 expression, and systemic organ invasion. Physiol Genomics. 2021;53:51–60.
83. Xia H, et al. Brain angiotensin-converting enzyme type 2 shedding contributes to the development of neurogenic hypertension. Circ Res. 2013;113:1087–96.
84. Stein R, et al. From ACE_2 to Covid-19: a multiorgan endothelial disease (editorial). Int J Infect Dis. 2020;100:425. https://doi.org/10.1016/J.iJid.2020.08.083.
85. Leisman D, et al. Facing Covid-19 in the ICU: vascular dysfunction, thrombosis and dysregulated inflammation. Intensive Care Med. 2020;46:1105–68.
86. Gomes Silva IV, et al. Effects of different classes of antihypertensive drugs on endothelial function and inflammation. Int J Mol Sci. 2019;20(3458):9–15.
87. Cameron AC, et al. Drug treatment of hypertension: focus on vascular health. Drugs. 2016;76:1529–50.
88. Wu Z, et al. Elevation of plasma angiotensin II level is a potential pathogenesis for the critical ill Covid-19 patients version 2. Crit Care. 2020;24:290.
89. Wu Z, et al. Characteristics of and important lessons from coronavirus disease (Covid-19) outbreak in China: summary of a report of 72314 cases from Chinese Center for Disease Control and Prevention. JAMA. 2020;323(13):1239–42.
90. Le SJ, et al. Coronavirus: innate immunity, inflammasome activation, inflammatory cell death and cytokines. Trends Immunol. 2020;41:1083–99.
91. Bergsbaken T, et al. Pyroptosis: host cell death and inflammation. Nat Rev Microbiol. 2009;7:99–109.
92. Yan L, et al. An interpretable mortality prediction model for Covid-19 patients. Nat Mach Intell. 2020;2:283–8.
93. Christgen S, et al. Identification of the PANoptosome a molecular platform trigering pyroptosis, apoptosis and necroptosis (PANaptosis). Front Cell Infect Microbiol. 2020;10:237.
94. Heiling R, et al. Function and mechanism of the pyrin inflammasome. Eur J Immunol. 2018;18:230–8.
95. Schnappauf O, et al. The pyrin inflammasome in health and disease. Front Immunol. 2019;10:1745.
96. Zahid A, et al. Pharmacological inhibitors of the NIRP3 inflammasome. Front Immunol. 2019;10:2538.
97. Zu ZY, et al. Coronavirus disease 2019 (Covid-19): a perspective from China. Radiology. 2020;296:E15–25.
98. Knowlton K. Pathogenesis of SARS-Cov-2 induced cardiac injury from perspective of the virus. J MolCellCardiol. 2020;147:12–7.
99. Tomasoni D, et al. Covid-19 and heart failure: from infection to inflammation and angiotensin II stimulation, searching for evidence from a new disease. Eur J Heart Fail. 22:957–66.
100. Yang J, et al. Prevalence of comorbidities and its effects in patients infected with SARS-Cov-2: a systematic review and meta-analysis. Int J Infect Dis. 2020;94:91–5.
101. Hu Y, et al. Prevalence and severity of coronavirus disease 2019 (Covid-19): a systematic review and meta-analysis. J Clin Virol. 2020;127:104371.
102. Veluswami P, et al. The SARS-Cov-2/receptor axis in heart and blood vessels: a crisp update on Covid-19 disease with cardiovascular complications. Viruses. 2021;13:1346.
103. Fu L, et al. Prevalence and impact of cardiac injury on Covid-19. A systematic review and meta-analysis. Clin Cardiol. 2021;44:276–83.

104. Shibata S, et al. Hypertension and related diseases in the era of Covid-19: a report from the Japanese Society of hypertension task force on Covid-19. Hypertens Res. 2020;43:1028–46.
105. Anderson TJ, et al. Close relation of endothelial function in the human coronary and peripheral circulation. J Am Coll Cardiol. 1995;26:1235–41.
106. Little PJ, et al. Endothelial dysfunction and cardiovascular disease: history and analysis of the clinical utility of the relationship. Biomedicine. 2021;9:699.
107. Rodriguez R, et al. The impact of ACE_2 expression levels in patients with comorbidities on Covid-19 severity: a comprehensive review. Microorganisms. 2021;9:1692.
108. Liang C, et al. The ACE_2 expression in human heart indicates new potential mechanisms of heart injury among patients infected with SARS-Cov-2. Cardiovasc Res. 2020;116:1097–100.
109. Nishiga M, et al. Covid-19 and cardiovascular disease: from basic mechanisms to clinical perspectives. Nat Rev Cardiol. 2020;17:543–58.
110. Gu SX, et al. Thrombocytopathy and endotheliopathy: crucial contributors to Covid-19 thromboinflammation. Nat Rev Cardiol. 2021;18:194–209.
111. Smadja D, et al. Angiopoietin-2 as a marker of endothelial activation is a good predictor factor Covid-19 patients. Angiogenesis. 2020;23:611–20. https://doi.org/10.1007/s10456-020-09730-0.
112. Dupont A, et al. Vascular endothelial damage in the pathogenesis of organ injury in severe Covid-19. Arterioscler Thromb Vasc Biol. 2021;41:1760–73.
113. Calvo-Fernandez A, et al. Markers of myocardial injury in the prediction of short Covid-19 prognosis. Rev Esp Cardiol (Engl Ed). 2021;74(7):576–83.
114. Shi S, et al. Association of cardiac injury with mortality in hospitalized patients with Covid-19, in Wuhan, China. JAMA Cardiol. 2020;5(7):802–10.
115. Giustino G, et al. Coronavirus and cardiovascular disease, myocardial injury and arrhythmia: JACC focus seminar. J Am Coll Cardiol. 2020;76(17):2011–23.
116. Narita K, et al. Disaster hypertension and cardiovascular events in disaster and Covid-19 pandemic. J Clin Hypertens. 2021;23:575–83.
117. Narita K, et al. Time course of disaster related cardiovascular disease and blood pressure elevation. Hypertens Res. 2021;44:1534–9.
118. Bastola U. Is Covid-19 severity associated with ACE_2 degradation. Front Drug Discov. 2021. https://arxiv.org.
119. Zisman LS, et al. Increased angiotensin (1-7) forming activity in human heart ventricles. Evidence for upregulation of angiotensin-converting enzyme homologue ACE_2. Circulation. 2003;108:1707–12.
120. Goulter AB, et al. ACE_2 gene expression is up-regulated in human heart failing. BMC Med. 2004;2:19.
121. Ramos SG, et al. ACE_2 down-regulation may act as a transient molecular disease causing RAAS dysregulation and tissue damage in the microcirculatory environment among Covid-19 patients. Am J Pathol. 2021;191:1154–64.
122. Saba L, et al. Molecular pathway triggered by Covid-19 in different organs: ACE_2 receptor expressing, cells under attack? A review. Eur Rev Med Pharmacol Sci. 2020;24(23):12609–22.
123. Jeffers SA, et al. CD209L (L-SIGN) is a receptor for severe acute respiratory syndrome coronavirus. PNAS. 2004;101:15748–53.
124. Giordo R, et al. SARS-Cov-2 and endothelial cell interaction in Covid-19: molecular perspectives. Vasc Biol. 2021;3:R15–23. https://doi.org/10.1530/VB-20-0017.
125. Daly JL, et al. Neuropilin-1 is a host factor for SARS-Cov-2 infection. Science. 2020;370:861–5.
126. Cantuti-Castelvetri L, et al. Neuropilin-1 facilitates SARS-Cov-2 cell entry and provides a possible pathway into the central nervous system. bioRxiv. 2020. https://doi.org/10.1101/2020.06.07.137802.
127. Perez-Bermejo JA, et al. SARS-Cov-2 infection of human iPSC-derived cardiac cell reflects cytopathic features in hearts of patients with Covid-19. Sci Transl Med Res. 2021;13:eabf7872.

128. Qian Y, et al. Direct activation of endothelial cells by SARS-Cov-2 nucleocapsid protein is blocked by simvastatin. J Virol. 2021;95(23):e01396–21.
129. Zheng M, et al. TLR_2, senses the SARS-Cov-2 envelope protein to produce inflammatory cytokines. Nat Immunol. 2021;22:829–33.
130. Li T, et al. Serum SARS-Cov-2 nucleocapsid protein: a sensitivity and specificity early diagnostic marker for SARS-Cov-2 infection. Front Cell Infect Microb. 2020;10:470. https://doi.org/10.3389/fcimb.2020.00470.
131. Cenko E, et al. Cardiovascular disease and Covid-19: a consensus paper from ESC Working group on coronary pathophysiology & circulation, ESC Working group on thrombosis and the association for acute cardiovascular care, in collaboration with European Heart Rhythm Association (EHRA). Cardiovasc Res. 2021;117:2705–29.
132. Ruan Q, et al. Clinical predictors of mortality due to Covid-19 based on analysis of data of 150 patients from Wuhan, China. Intensive Care Med. 2020;46(5):846–8.
133. Lei Y, et al. SARS-Cov-2 spike protein impairs endothelial function via downregulation of ACE_2. Circ Res. 2021;128:1323–6.
134. Mackman N. Role of tissue factor in hemostasis, thrombosis and vascular development. Atheroscler Thromb Vasc Biol. 2004;24:1015–22.
135. Zheng Y-Y, et al. Covid-19 and the cardiovascular system. Nat Rev. 2020;17:259–60.
136. Bader F, et al. Heart failure and Covid-19. Heart Fail Rev. 2020;26:1–10. https://doi.org/10.1007/s10741-020-10008-2.
137. Bonaventura A, et al. Endothelial disfunction and immunothrombosis as key pathogenic mechanisms in Covid-19. Nat Rev. 2021;21:319–28.
138. Prescott H, et al. Recovery from severe Covid-19: Leveraging the lessons survival from sepsis. JAMA. 2020;324(8):739.
139. Yan LM, et al. SARS-Cov-2 is an unrestricted bioweapon: a truth revealed through uncovering a large-scale, organised scientific fraud. https://doi.org/10.5281/zenoda.4073131.svg.
140. Jain R, et al. SARS-Cov-2 pandemic: a critical review on novel coronavirus pathogenesis, clinical diagnosis and treatment. J Drug Deliv Therap. 2020;10(3):241–52.
141. Ben H, et al. Characteristics of SARS-Cov-2 and Covid-19. Nat Rev Microbiol. 2020;19:141–54. https://doi.org/10.1038/s41579-020-00459-7.
142. Tang D, et al. The hallmarks of Covid-19 disease. PLoS Pathog. 2020;16:e1008536. https://doi.org/10.1371/journal.ppat.1008536.
143. Kai H, et al. Interaction of coronavirus with ACE_2, angiotensin II and RAS inhibitors-lessons from available evidence and insight into Covid-19. Hypertens Res. 2020;43:648–54. https://doi.org/10.1038/s41440-020-0455-8.
144. Bourgonje A, et al. Angiotensin-converting enzyme 2 (ACE_2), SARS-Cov-2 and the pathophysiology of coronavirus disease 2019 (Covid-19). J Pathol. 2020;251:228–48.
145. Glowacka I, et al. Evidence that TMPRSS2 activates the severe acute respiratory syndrome coronavirus spike protein for membrane fusion and reduce, viral control by the humoral immune response. J Virol. 2011;85:4122–34.
146. Synowiec A, et al. Severe acute respiratory syndrome coronavirus 2 (SARS-Cov-2): a systemic infection. Clin Microbiol Rev. 2021;34(2):e00133–20.
147. Pinto P, et al. ACE_2 expression is increased in the lungs of patients with comorbidities associated with severe Covid-19. J Infect Dis. 2020;222:556. https://doi.org/10.1093/infectis/jiaa.
148. Khayat AS, et al. ACE_2 polymorphisms as potential players in Covid-19 outcome. PLoS One. 2020;15:e0243887. https://doi.org/10.1371/journal.pone.0243887.
149. Yuen K-S, et al. SARS-Cov-2 and Covid-19: the most important research questions. Cell Biosci. 2020;10:40. https://doi.org/10.1186/s13578-020-00404-4.
150. Wang J, et al. Molecular simulation of SARS-Cov-2 spike protein binding to pangolin ACE_2 or human ACE_2 natural variants reveals altered susceptibility to infection. J Gen Virol. 2020;101:921–4.
151. Sivaraman H, et al. Structural basis of SARS-Cov-2 and SARS-Cov receptor binding and small molecule blockers as potential therapeutics. Annu Rev Pharmacol Toxicol. 2021;71:465–93.

152. Sims J, et al. Characterization of the cytokine storm—reflects hyperinflammatory endothelial dysfunction in Covid-19. J Allergy Clin Immunol. 2021;147:107–11. https://doi.org/10.1016/J.Jaci.2020.08.031.
153. Aird WC. Mechanisms of endothelial cell heterogeneity in health and disease. Circ Res. 2006;98:159–62.
154. Celermajep D, et al. Endothelial dysfunction: does it matter? It is reversible? J Am Coll Cardiol. 1997;30(2):325–33.
155. Gladka M, et al. The endothelium as Achilles' heel in Covid-19 patients. Cardiovasc Res. 2020;116(14):e195–7.
156. Thomas PG, et al. The intracellular sensor NLRP3 mediates key innate and healing responses in influenza a virus via the regulation of caspase-1. Immunity. 2009;30:566–75.
157. Nieto-Torres JL, et al. Severe acute respiratory syndrome coronavirus envelope protein ion channel activity pronates virus fitness and pathogenesis. PLoS Pathog. 2014;10:e1004077.
158. Vardecchia P, et al. The pivotol link between ACE_2 deficiency and SARS-Cov-2 infection. Eur J Intern Med. 2020;76:14–20.
159. Pasparakis M, et al. Necroptosis and its role in inflammation. Nature. 2015;517:311–20.
160. Zheng M, et al. Caspase-6 is a key regulator of innate immunity, inflammasome activation and host defense. Cell. 2020;181:674–687.e13.
161. Patak EM. Convalescent plasma is ineffective for Covid-19. BMJ. 2020;371:m4072.
162. Widlansky M, et al. The clinical implications of endothelial dysfunctions. J Am Coll Cardiol. 2003;42(7):1149–60.
163. Al-Farabi MJ, et al. Biomarkers of endothelial dysfunction and outcomes in coronavirus disease 2019 (Covid-19) patients: a systematic review and meta-analysis. Microvasc Res. 2021;138:104224.
164. Banon-Gonzales R, et al. Autopsies of suspected SARS-Cov-2 cases. Spanish J Legal Med. 2020;46(3):93–100.
165. Umbrajkar S, et al. Cardiovascular health and disease in the context of Covid-19. Cardiol Res. 2021;12(2):67–79.
166. Farshidfar F. Cardiovascular complication of Covid-19. JCI Insight. 2021;6(13):e148980.
167. Manganaro R, et al. Endothelial dysfunction and coronary artery disease: new insight from reactive hyperemia test. Vessel Plus. 2021;1(5):37.
168. van Eijk LE, et al. Covid-19: immunopathology, pathophysiological mechanism and treatment option. J Pathol. 2021;254:307–31.
169. Amraei R, et al. CD209L/L-SIGN and CD209/DC-SIGN act as receptor for SARS-Cov-2 and are differentially expressed in lung and kidney epithelial and endothelial cells. bioRxiv. 2020. https://doi.org/10.1101/2020.06.22.165803
170. Chen L, et al. The ACE_2 expression in human heart indicates new potential mechanism of heart injury among patients infected with SARS-Cov-2. Cardiovasc Res. 2020;116:1097–110.
171. Xudong X, et al. Age-and gender related difference of ACE_2 expression in rat lung. Life Sci. 2006;78:2166–71.
172. McFadyen JD, et al. The emerging threat of (micro) thrombosis in Covid-19 and its therapeutic implications. Circ Res. 2020;127:571–87.
173. Mehra MR, et al. Cardiovascular disease, drug therapy and mortality in Covid-19. N Engl J Med. 2020;382(25):e102.
174. Voicu V. Post-Covid-19 sequellae or persistent Covid-19 disease. J Hypertens. 2021;7(3):84–8.

Pathophysiology of Hypertensive Heart Disease

14

Cesare Cuspidi, Marijana Tadic, Carla Sala, and Guido Grassi

14.1 Introduction

Numerous alterations of cardiac structure and function have been consistently reported in patients with sustained hypertension, including left ventricular (LV) concentric remodeling, concentric and eccentric LV hypertrophy (LVH), left atrial dilatation, aortic enlargement, systolic and diastolic LV dysfunction, reduced coronary reserve, and increased risk of supraventricular and ventricular arrhythmias (Fig. 14.1) [1, 2]. Early alterations in LV structure and function have been detected in patients with high-normal blood pressure (BP), mild hypertension, and isolated office hypertension [3, 4].

Among the manifestations of hypertension-mediated cardiac damage, most attention has been paid to LVH, due to its high prevalence both in the general hypertensive population and in subgroups with comorbidities; LVH unfavorably affects systolic and diastolic function, which in turn carry an increased risk of heart failure (HF), coronary disease, ictus, and cardiovascular mortality [5–7].

Among the causes of hospitalization for cardiovascular disease in developed countries, chronic HF is by far the most frequent one preceding cerebrovascular and coronary heart disease [8]. Increased age, prevalence of hypertension, LVH, obesity, and diabetes are the factors closely related to HF risk in the general population worldwide [9].

C. Cuspidi (✉) · G. Grassi (✉)
University of Milano-Bicocca, Milano, Italy
e-mail: cesare.cuspidi@unimib.it; guido.grassi@unimib.it

M. Tadic
Klinik für Innere Medizin II, Universitätsklinikum, Ulm, Germany

C. Sala
Department of Clinical Sciences and Community Health, University of Milano and Fondazione IRCCS Ospedale Maggiore Policlinico, Milano, Italy

M. Dorobantu et al. (eds.), *Hypertension and Heart Failure*, Updates in Hypertension and Cardiovascular Protection,
https://doi.org/10.1007/978-3-031-39315-0_14

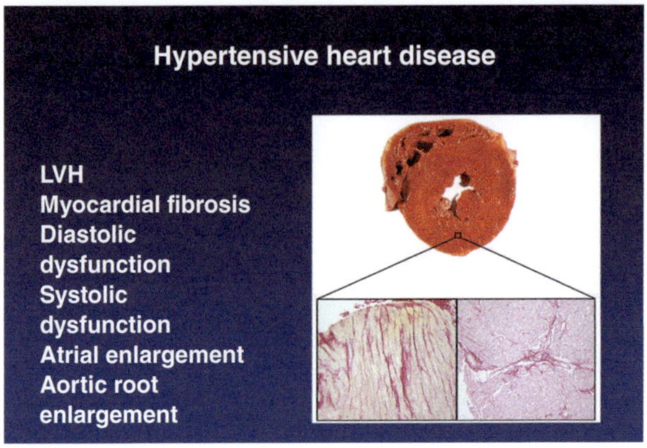

Fig. 14.1 Abnormal structural and functional phenotypes in hypertensive heart disease

In this chapter, the pathophysiological mechanisms underlying development and progression of HHD will be discussed. Particular attention will be paid to LVH and its subtypes, diastolic and systolic dysfunction, as assessed by conventional and new echocardiographic techniques, all of which will be described in separate sections.

14.2 Left Ventricular Hypertrophy

LVH is the key feature of hypertension-mediated cardiac organ damage due to its detrimental clinical and prognostic implications. Numerous prospective studies have shown that LVH has a strong, independent adverse prognostic significance. Therefore, identification of this phenotype is a fundamental step in the evaluation of hypertensive patients. LVH can be regarded as an integrated marker of cardiovascular risk, reflecting cardiac effects of hemodynamic and non-hemodynamic factors operating in hypertension [10].

The rise in LV wall stress induced by the increased intraventricular systolic pressure is a powerful stimulus for the development of LVH, which, in turn, tends to reduce wall stress. According to Laplace's law, indeed, LV wall stress is directly related to intraventricular pressure, to radius and inversely related to wall thickness ($P = 2T\ d/r$). The thickening of cardiac wall is the result of an increased number of sarcomeres arranged in parallel in concentric LVH and longitudinally in eccentric LVH, which initially develops at interventricular septum.

Hypertensive LVH has been reported to be associated with an over-expression of fetal isoforms of contractile proteins such as β-myosin heavy chain, β-troponin, and skeletal α-actin. Thus, LVH may represent a regression to early stages of myocardial development rather than an overproduction of normal contractile proteins.

Both postmortem and endomyocardial biopsies have consistently shown that along with increased LV mass, the collagen fraction of myocardium is increased in hypertensive subjects compared to normotensive counterparts. Endocardial and

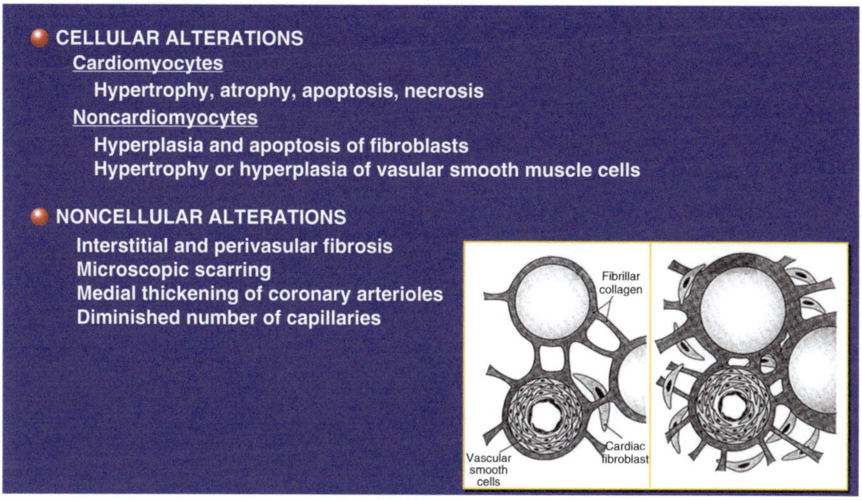

Fig. 14.2 Hypertensive left ventricular hypertrophy as a consequence of cellular and noncellular alterations

interstitial fibrosis as well as focal scars have been documented in LVH of mild degree even in the early phases of human hypertension [11].

In hypertensive LVH, a disproportionate accumulation of cells other than cardiomyocytes, such as fibroblasts, accounts for the development of myocardial fibrosis [12]. These adaptive changes occurring in LV texture initially preserve cardiac function from BP overload, but in the long-term chronic promotes a progressive increase in LV collagen leading to heart failure (Fig. 14.2) [13].

The mechanisms underlying the alterations in LV structure and geometry are incompletely elucidated. BP load is considered the major trigger; this parameter, however, even when accurately defined by out-of-office measurements (i.e., home or ambulatory BP) accounts for only 30% of the observed variance of LV mass. Accumulating evidence supports the view that growth factors, cytokines, and neurohormones (i.e., angiotensin II, endothelin I, catecholamines, aldosterone, and insulin-like growth factor), combined with increased aortic stiffness, myocardial ischemia due to impaired coronary microcirculation, excessive salt intake, and ethnic/genetic predisposition, overall concur to trigger and worsen subclinical cardiac damage (Fig. 14.3).

Several lines of evidence support a role for circulating and local renin–angiotensin–aldosterone system (RAAS) in the development of myocardial hypertrophy and fibrosis, independently of loading conditions [14]. Binding of angiotensin II to angiotensin type 1 (AT1) receptors initiates a signal transduction cascade resulting in myocytes hypertrophy and enhanced collagen synthesis by cardiac fibroblasts. The activity of angiotensin converting enzyme (ACE) in degradation of bradykinin and inactivation of nitric oxide may also be relevant. Furthermore, aldosterone has

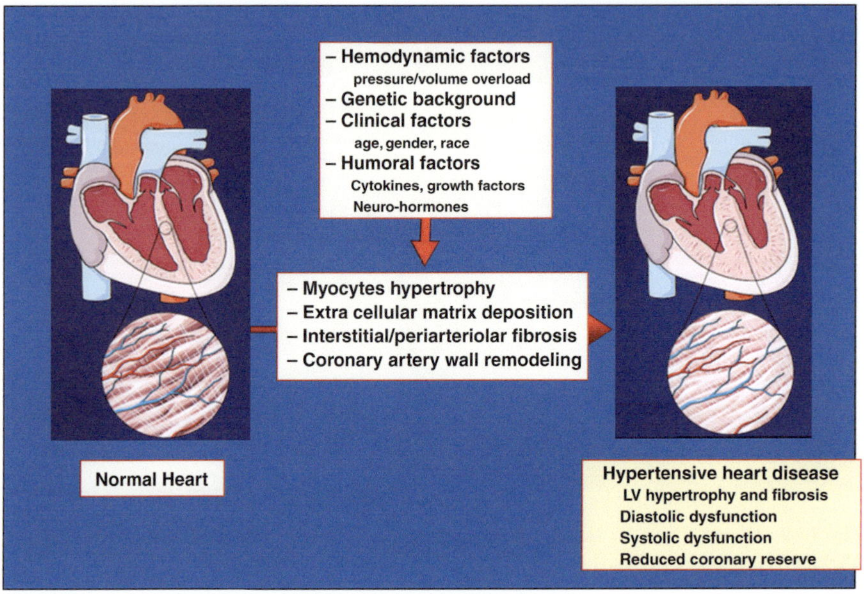

Fig. 14.3 Pathogenesis of hypertensive heart disease

been consistently shown to increase myocardial mass and promote fibroblast proliferation, collagen deposition, oxidative stress, and inflammation.

Convincing evidence of the independent influence of aldosterone excess on cardiac structure has been obtained in patients with primary aldosteronism, in whom higher LV mass index, left atrial (LA) volumes, and prevalence rates of LVH compared with essential hypertensives have been documented. In addition, LVH regression after surgical correction of primary aldosteronism strongly supports a direct, causal role of aldosterone in LVH development [15]. Finally, genetic variations in RAAS components have been suggested to impact on the degree of cardiac cell growth during physiological or pathological states.

The hypothesis that sympathetic influences exert cardiotrophic effects and enhance myocardial hypertrophy has been supported by several experimental studies. In animal models, the rise in LV wall thickness and mass was paralleled by enhanced cardiac sympathetic drive. The adverse influence of sympathetic nervous system (SNS) probably involves more complex mechanisms than activation of α- and β-adrenergic receptors on cardiac myocytes and fibroblasts; the inflammatory response driven by sympathetic over-activity, indeed, has been demonstrated to be an important component of hypertension-induced cardiac remodeling. Available evidence in favor of a cause–effect relationship between adrenergic factors and cardiac hypertrophy is more controversial in humans; an association between LV mass and sympathetic nerve traffic in the peripheral muscle circulation in essential hypertensives has been found by some studies but not by others [16, 17]. New insights supporting a SNS role in the pathogenesis of LVH may indirectly derive from renal denervation studies reporting that a reduction of renal sympathetic afferent and efferent activity may facilitate LVH regression [18].

A causal role for mast cells in regulating myocardial cytokines, macrophage recruitment, and development of fibrosis in the hypertensive heart has been shown in spontaneously hypertensive rats.

Overweight, obesity, and hypertension are closely linked phenotypes sharing a variety of cardiac and hemodynamic alterations. Adiposity excess affects the heart by causing hemodynamic, metabolic, and inflammatory alterations leading to epicardial and intramural fat accumulation, development of eccentric or, more frequently, concentric LVH [19]. Pathophysiological mechanisms advocated to explain the link between obesity and LVH in both normotensive and hypertensive individuals include: (1) increased intravascular volume; (2) activation of sympathetic tone and RAAS; (3) abnormal release of myocardial growth substances from abdominal and cardiac adipose tissue; and (4) metabolic alterations contributing to cardiac load by increasing arterial stiffness and peripheral vascular resistances.

This last mechanism also plays a pivotal role in different hypertensive settings such as elderly, diabetes and metabolic syndrome. The progressive stiffening of aorta and large arteries in response to multiple risk factors and aging increases the amplitude and velocity of reflected pulse waves generated by resistance arterioles. These arterioles, indeed, tend to adapt to chronic arterial pressure increments through smooth muscle hypertrophy: this process, on one hand, tends to reduce end-arteriolar wall tension, but on the other hand, further increases large artery pressure and pulse wave velocity. As a result, the reflected waves, which normally reaches central aorta after aortic valve closure, increase its velocity enough to return to the proximal aorta in late systole. This phenomenon leads to a further increase in the after load and contributes to LVH progression/development [20].

A large amount of data provided by animal and human studies suggest a link between salt intake and LV growth. Saline overload has been consistently reported to increase LV mass; conversely, dietary salt restriction has been shown to regress cardiac hypertrophy in hypertensive rats and in salt-sensitive hypertensive subjects. Although the mechanisms of cellular growth induced by sodium overload are incompletely understood, SNS activation is believed to play a role in this process. Salt intake also stimulates myocardial growth via hemodynamic mechanisms such as BP and volume overload [21] and stimulates phospholipase C-activity mediated by platelet-derived growth factors.

Sleep disordered breathing, in particular obstructive sleep apnea syndrome (OSA), has been reported to unfavorably affect cardiac structure and function. The increased likelihood of developing LVH in OSA patients has been attributed to the synergistic effect of hypoxemia, RAAS activation, catecholamine excess, nocturnal BP elevation (associated to non-dipping and reverse dipping pattern), and intrathoracic pressure swings affecting pre- and after-load as well as LV transmural pressures [22].

Genetic factors related to essential hypertension also influence myocardial growth response to stimuli [23]. Clinical observations support the view that genetic predisposition to myocardial growth contributes up to 60% of LVH risk in essential hypertensives. In particular, higher LV mass indexes and LVH prevalence rates have been reported in normotensive relatives of hypertensive patients as compared to age-matched counterparts without family history of hypertension. Moreover, LVH has been shown to develop even in hypertensive patients on effective antihypertensive medications, to occur more frequently and severely in black than in white hypertensive patients. Several candidate genes responsible for LVH have been studied in different ethnic populations: the results of these studies, however, remain inconsistent.

14.3 Abnormal Left Ventricular Geometric Patterns

In the early 90s, Ganau and coworkers proposed a comprehensive classification of LV geometry based on four echocardiographic patterns: normal LV geometry (normal LV mass and relative wall thickness, RWT), concentric remodeling (normal LV mass and increased RWT), eccentric hypertrophy (increased LV mass and normal RWT), and concentric hypertrophy (increased LV mass and RWT) [24]. In the last three decades, studies carried out in several settings (i.e., general population samples, hypertensive cohorts, patients with ischemic heart disease, type 2 diabetes, and chronic heart failure) have shown that these patterns differ from each other in many respects such as LV systolic/diastolic function, LA size and function, systemic hemodynamic, office and ambulatory BP levels, macro- and micro-vascular organ damage.

In hypertensive patients with concentric LVH, BP is characterized by increased peripheral resistances, cardiac output being slightly above normal average, whereas in those with eccentric hypertrophy, cardiac output is generally increased with minimal or no elevations of peripheral resistances. Patients with concentric LV remodeling have normal or slightly reduced cardiac output and increased peripheral resistances. A progressive increase of LV preload and plasma volume has been described from LV concentric remodeling, to concentric and eccentric LVH.

Noninvasive ultrasound imaging techniques have allowed to accurately study human cardiac and vascular structure as well as the impact of altered arterial properties on ventricular adaptation. Numerous studies have documented a tight association between abnormal LV geometric patterns and subclinical arterial disease. In 1074 mild to moderate untreated hypertensive patients, we found that LV mass index was positively related to systemic atherosclerosis detected by carotid ultrasound, independently of age, gender, BP, and conventional risk factors [25]. A stepwise increase in carotid intima-media thickness occurred from patients with normal LV mass and geometry to patients with LV concentric remodeling, eccentric hypertrophy, and concentric LVH. Concentric remodeling and concentric hypertrophy were the cardiac phenotypes associated with higher carotid relative wall thickness compared to normal LV geometry and eccentric LVH.

The classification by Ganau et al. [24] has been criticized for not taking into account the absolute values of LV wall thickness and internal dimensions, considering only their ratio. Investigators of the Dallas Heart Study, a population-based sample of 2803 residents of Dallas County undergone to cardiac magnetic resonance imaging, have proposed an updated LVH classification based on four subtypes: eccentric non-dilated and dilated LVH; concentric non-dilated and dilated LVH [26]. In that study participants with eccentric non-dilated LVH, at variance from other LVH subtypes, showed similar levels of LV function and biomarkers of cardiac stress (i.e., brain natriuretic peptide) as their counterparts without LVH. Thereafter, Bang and co-workers [27] investigating the prognostic value of such LVH sub-classification in 939 participants of Losartan Intervention for Endpoint Reduction (LIFE) echocardiographic sub-study documented that, compared to patients with normal LV mass, those with eccentric dilated and concentric both non-dilated and dilated LVH had an increased risk of cardiovascular and all-cause mortality; this was not the case for patients with eccentric non-dilated LVH. Both reports documented a marginal impact of eccentric non-dilated LVH on ventricular function and cardiovascular prognosis and supported the view that the new classification of LV geometry may improve the clinical/prognostic evaluation of patients with increased LV mass.

In the PAMELA study, individuals with LV geometric abnormalities, defined according to Dallas classification, were older, had higher body mass index (BMI), office and ambulatory BPs, glucose and total cholesterol levels, and lower HDL-cholesterol compared to subjects with normal LV geometry [28]. Significant differences were also found among groups with abnormal LV geometry: subjects with concentric and eccentric dilated LVH exhibited higher BMI and BP levels than concentric remodeling. As for the prognostic value, however, only concentric LVH persisted to be an independent predictor of cardiovascular outcomes after adjustment for baseline differences in LV mass index. Finally, a large study, including 8848 hypertensive patients from the Campania Salute Network, suggested that the excess of cardiovascular risk related to eccentric dilated LVH, concentric non-dilated LVH, and concentric dilated LVH was mostly dependent on the magnitude of LV mass index [29].

14.4 Diastolic Dysfunction

LV diastolic function is a complex dynamic process schematically divided into four phases, starting from aortic valve closure. The first diastolic phase refers to isovolume relaxation, which does not contribute to ventricular filling. Second early, rapid filling phase provides about 60–85% of LV ventricular filling. Third slow filling phase, defined as diastasis, contributes only 5% of total filling. The final atrial booster phase normally accounts for the remaining 5–30%, its contribution increasing with aging.

Diastolic LV relaxation and filling is an active process requiring Ca^{2+} reuptake from cytosol into sarcoplasmic reticulum by the calcium-handling protein Ca^{2+} ATPase in the presence of ATP [30]. LV diastolic properties are primarily affected by pathological conditions impairing the active process of calcium reuptake such as myocardial hypertrophy and myocardial ischemia. In hypertensive hearts, multiple

Table 14.1 Determinants of left ventricular diastolic properties

Intrinsic factors
Completeness of ventricular relaxation (influenced by myocardial perfusion)
Passive elastic properties of the ventricle (compliance), determined by the thickness and composition of the ventricular walls (amount of myocardial fibrosis)
Extrinsic factors
Pressor overload
Volume overload
Strength of atrial contraction
Pericardial properties

factors including increased BP itself, structural LV changes (i.e., increased collagen matrix, disorganization of collagen fibers, and abnormal collagen type I/III ratio), and impaired coronary microcirculation all contribute to impair LV relaxation and to increase wall stiffness (Table 14.1). Prevalence and severity of diastolic dysfunction have been reported to be related to degree and type of cardiac hypertrophy, although diastolic dysfunction may also occur in the absence of LVH; a large fraction of hypertensive patients without LVH, indeed, fulfill echocardiographic diagnostic criteria of altered LV relaxation and filling. In animal models, diastolic dysfunction has been shown to precede LVH development. Multiple factors such as insulin resistance, sleep apnea syndrome, obesity, and diabetes mellitus have been documented to contribute to diastolic dysfunction in the absence of LVH [31].

In the London Life Sciences Prospective Population (LOLIPOP) study including 1074 hypertensive patients without cardiovascular disease, LV diastolic dysfunction was related to degree of LV concentric remodeling, determined by relative wall thickness, and to LV geometric pattern [32]. Compared with patients with normal LV geometry or non-hypertrophic concentric remodeling, those with LVH exhibited a worse diastolic function and higher LV filling pressures. Investigators of the Assessment of Prevalence Observational Study of Diastolic Dysfunction (APROS diadys) project, a cross-sectional observational study on elderly hypertensives without systolic dysfunction ($n = 2545$) aimed to assess the prevalence of diastolic dysfunction in relation to demographic and clinical characteristics, found that 1 gram increase in LV mass increased the risk of diastolic dysfunction by 1.3% [33]. Prevalence of diastolic dysfunction progressively increased from normal geometry to LV concentric remodeling, eccentric, and concentric LVH.

The prevalence of diastolic dysfunction increases when hypertension is associated with overweight and obesity; the link between isolated obesity and diastolic dysfunction, however, remains controversial [34]. Also controversial is the association between pre-hypertension or mild hypertension with cardiac hypertrophy and diastolic dysfunction, when these conditions are diagnosed by conventional echocardiographic parameters. Recently developed echocardiographic techniques, such as three-dimensional echocardiography and speckle tracking imaging, provide more detailed insights into LV systolic and diastolic mechanics in three spatial directions [2]. These innovative techniques have been reported to accurately detect early LV functional alterations and to identify subtle LV mechanical abnormalities even in subjects with normal cardiac function by routine ultrasound examinations [35]. It has been recently reported that in patients with pre-hypertension and white coat hypertension, the likelihood of impaired LV mechanics (longitudinal, circumferential, and radial) is higher than in counterparts with normal or optimal BP values. In conclusion, it is likely that data on prevalence and correlates of diastolic dysfunction in hypertension without apparent cardiac organ damage based on traditional echocardiographic and Doppler parameters, including tissue Doppler imaging, substantially underestimate the real, adverse impact of this condition on the heart.

The view of isolated diastolic dysfunction in hypertensive heart disease is challenged by increasing evidence supporting an association between altered LV ventricular mechanics and deterioration of diastolic function. In patients with moderately severe hypertension, a prolonged isovolumetric relaxation time, an index of impaired early diastolic LV relaxation, has been found to be associated with a worsen LV systolic myocardial function, as assessed by mid-wall shortening and stress-corrected mid-wall shortening.

The increase in LA size may be seen as a mechanism counterbalancing the impaired LV compliance and LV diastolic dysfunction in hypertrophied ventricles [36]. Enlargement of LA in hypertensive heart disease is regarded as a reliable marker of chronically elevated LV filling pressure and diastolic dysfunction in the absence of mitral valve diseases.

14.5 Systolic Dysfunction

Altered contractility in LVH is related to structural and functional abnormalities involving extracellular matrix and fibrous tissue, vasculature as well as cardiomyocytes themselves [37]. In the large majority of hypertensive patients without frank LV mass increases, LV systolic performance as traditionally assessed by LV ejection fraction (LVEF) at rest is preserved or even mildly increased.

Unfortunately, LVEF, which has been regarded as the gold standard of LV systolic function for decades, does not provide a direct measure of myocardial fiber shortening, as circumferential fibers responsible for LV shortening in the short axis are located in the mid-portion of LV wall, between two longitudinal shells promoting long axis shortening and twisting. In addition, two-dimensional echocardiographic evaluation of LVEF is faced with multiple limitations: foreshortening, geometry assumption, moderate accuracy, interobserver and intraobserver reproducibility.

Evaluation of LV systolic function as fractional shortening at mid-wall, as opposed to endocardium level, has been proposed as a more reliable method than LVEF for assessing LV performance in the presence of LVH [38]. At the endocardium level, indeed, LV systolic function has been shown to be preserved or even increased in most uncomplicated hypertensives, thus, loosing any independent predictive value. On the contrary, subnormal LV function as assessed by mid-wall shortening has been associated with unhealthy factors at any BP level and with a marked increase in cardiac morbidity and mortality, independent of age and LV mass.

Numerous studies in hypertensive patients have reported that systolic dysfunction assessed by impaired LV mid-wall function was three- to fourfold more prevalent than systolic dysfunction assessed by conventional LV endocardial mechanics (i.e. LVEF).

In the last decade, LV deformation during systole has been quantified in multiple planes using speckle tracking echocardiography or magnetic resonance imaging tissue tagging. Planes of deformation have been defined in relation to myocardial fiber orientation, including longitudinal, radial, and circumferential shortening (strain). Among available strain parameters, LV global longitudinal strain (GLS) appears to have the best clinical relevance, as good predictor of cardiovascular and total morbidity and mortality in members of the general population and hypertensive cohorts [39–41]. In hypertensive patients, abnormalities of this measure of LV longitudinal contractile ability have been reported even in the absence of structural and functional LV alterations as assessed by conventional echocardiography [42].

Thus, new echocardiographic techniques may significantly improve early diagnosis of functional alterations in hypertensive heart disease and identify the high-risk individuals to be appropriately treated. In this regard, a recent meta-analysis of our group showed that GLS assessment unmasked systolic dysfunction undetected by conventional ejection fraction in essential hypertensive patients [43]. The clinical implication of this observation is that this parameter should be incorporated into diagnostic work-up of hypertensive patients.

14.6 Future Directions

Inadequate prevention strategies such as late identification of subclinical hypertensive cardiac damage and suboptimal treatment of hypertension are among the most important factors responsible for advanced cardiac remodeling associated with marked functional alterations ultimately leading to chronic HF. Considering that over the past 30 years, cardiac imaging, based primarily on echocardiography, provided a major contribution to understand the pathophysiology of hypertensive heart disease, the increasing application of new imaging modalities (i.e., 3D echocardiography, 2D and 3D STE) in clinical practice, may contribute to early identification of hypertensive organ damage (in the potentially reversible phase), improve cardiovascular risk stratification, and optimize therapeutic interventions (Fig. 14.2). Finally, multimodal imaging (i.e., echocardiography, computed tomography, and resonance magnetic imaging) more widely implemented in hypertensive patients will provide relevant clinical information on the presence of fibrosis, inflammation, microvascular changes, and characteristics of myocardial hypertrophy [44].

Sources of Funding None.

Disclosure No conflict of interest.

References

1. Perrone-Filardi P, Coca A, Galderisi M, et al. Non-invasive cardiovascular imaging for evaluating subclinical target organ damage in hypertensive patients: a consensus paper from the European Association of Cardiovascular Imaging (EACVI), the European Society of Cardiology Council on Hypertension, and the European Society of Hypertension (ESH). Eur Heart J Cardiovasc Imaging. 2017;18:945–60.
2. Tadic M, Cuspidi C, Bombelli M, et al. Hypertensive heart disease beyond left ventricular hypertrophy: are we ready for echocardiographic strain evaluation in everyday clinical practice? J Hypertens. 2018;36:744–53.
3. Cuspidi C, Facchetti R, Bombelli M, et al. High normal blood pressure and left ventricular hypertrophy echocardiographic findings from the PAMELA population. Hypertension. 2019;73:612–9.
4. Mancia G, Facchetti R, Bombelli M, et al. White-coat hypertension: pathophysiological and clinical aspects: excellence award for hypertension research 2020. Hypertension. 2021;78:1677–88.
5. Cuspidi C, Sala C, Negri F, et al. Prevalence of left-ventricular hypertrophy in hypertension: an updated review of echocardiographic studies. J Hum Hypertens. 2012;26:343–9.
6. Pedersen LR, Kristensen AM, Petersen SS, et al. Prognostic implications of left ventricular hypertrophy diagnosed on electrocardiogram vs echocardiography. J Clin Hypertens. 2020;22:1647–58.
7. Vasan RS, Song RJ, Xanthakis V, et al. Hypertension-mediated organ damage prevalence, correlates, and prognosis in the community. Hypertension. 2022;79:505–15.
8. Salah HM, Minhas AMK, Khan MS, et al. Trends and characteristics of hospitalizations for heart failure in the United States from 2004 to 2018. ESC Heart Fail. 2022;9(2):947–52.
9. Faulkner JL. Obesity-associated cardiovascular risk in women: hypertension and heart failure. Clin Sci. 2021;135:1523–44.
10. Joseph G, Marott JL, Biering-Sørensen T, et al. level of physical activity, left ventricular mass, hypertension, and prognosis. Hypertension. 2020;75:693–701.
11. Rossi MA. Pathologic fibrosis and connective tissue matrix in left ventricular hypertrophy due to chronic arterial hypertension in humans. J Hypertens. 1998;16:1031–41.
12. Stacey RB, Hundley WG. Integrating measures of myocardial fibrosis in the transition from hypertensive heart disease to heart failure. Curr Hypertens Rep. 2021;23(4):22.
13. López B, Ravassa S, Moreno MU, et al. Diffuse myocardial fibrosis: mechanisms, diagnosis and therapeutic approaches. Nat Rev Cardiol. 2021;18:479–98.
14. Liu Y, Lin Y, Zhang MM, Li XH, et al. The relationship of plasma renin, angiotensin, and aldosterone levels to blood pressure variability and target organ damage in children with essential hypertension. BMC Cardiovasc Disord. 2020;20(1):296.
15. Cuspidi C, Tadic M, Sala C, et al. Regression of left ventricular hypertrophy in primary aldosteronism after adrenalectomy: a meta-analysis of echocardiographic studies. J Hypertens. 2021;39:775–83.
16. Grassi G, Seravalle G, Mancia G. Sympathetic activation in cardiovascular disease: evidence, clinical impact and therapeutic implications. Eur J Clin Investig. 2015;45:1367–75.
17. Burns J, Sivananthan MU, Ball SG, et al. Relationship between central sympathetic drive and magnetic resonance imaging-determined left ventricular mass in essential hypertension. Circulation. 2007;115:1999–2005.
18. Kordalis A, Tsiachris D, Pietri P, et al. Regression of organ damage following renal denervation in resistant hypertension: a meta-analysis. J Hypertens. 2018;36:1614–21.
19. da Silva-deAbreu A, Alhafez BA, Lavie CJ, et al. Interactions of hypertension, obesity, left ventricular hypertrophy, and heart failure. Curr Opin Cardiol. 2021;36:453–60.
20. Chirinos JA, Segers P, Hughes T, et al. Large-artery stiffness in health and disease: JACC state-of-the-art review. Am Coll Cardiol. 2019;74:1237–63.

21. Haring B, Wang W, Lee ET, et al. Effect of dietary sodium and potassium intake on left ventricular diastolic function and mass in adults≤40 years (from the Strong Heart Study). Am J Cardiol. 2015;115:1244–8.
22. Cuspidi C, Tadic M, Gherbesi E, et al. Targeting subclinical organ damage in obstructive sleep apnea: a narrative review. J Hum Hypertens. 2021;35:26–36.
23. Adeoye AM, Ovbiagele B, Kolo A, et al. SIREN Team as part of H3Africa consortium exploring overlaps between the genomic and environmental determinants of LVH and stroke: a multicenter study in West Africa. Glob Heart. 2017;12:107–13.
24. Ganau A, Devereux RB, Roman MJ, et al. Patterns of left ventricular hypertrophy and geometric remodeling in essential hypertension. J Am Coll Cardiol. 1992;19:1550–8.
25. Cuspidi C, Mancia G, Ambrosioni E, et al. Left ventricular and carotid structure in untreated, uncomplicated essential hypertension: results from the Assessment Prognostic Risk Observational Survey (APROS). J Hum Hypertens. 2004;18:891–6.
26. Khouri MG, Peshock RM, Ayers CR, et al. A 4-tiered classification of left ventricular hypertrophy based on left ventricular geometry: the Dallas Heart Study. Circ Cardiovasc Imaging. 2010;3:164–71.
27. Bang CN, Gerdts E, Aurigemma GP, et al. Four-group classification of left ventricular hypertrophy based on ventricular concentricity and dilatation identifies a low-risk subset of eccentric hypertrophy in hypertensive patients. Circ Cardiovasc Imaging. 2014;7:422–9.
28. Cuspidi C, Facchetti R, Bombelli M, et al. Risk of mortality in relation to an updated classification of left ventricular geometric abnormalities in a general population: the Pamela study. J Hypertens. 2015;33:2133–40.
29. de Simone G, Izzo R, Aurigemma GP, et al. Cardiovascular risk in relation to a new classification of hypertensive left ventricular geometric abnormalities. J Hypertens. 2015;33:745–54.
30. Voorhees AP, Han HC. Biomechanics of cardiac function. Compr Physiol. 2016;5:1623–44.
31. Wenzel JP, Kellen RB, Magnussen C, et al. Diastolic dysfunction in individuals with and without heart failure with preserved ejection fraction. Clin Res Cardiol. 2021;111(4):416–27.
32. Chahal NS, Lim TK, Jain P, et al. New insights into the relationship of left ventricular geometry and left ventricular mass with cardiac function: a population study of hypertensive subjects. Eur Heart J. 2010;31:588–94.
33. Zanchetti A, Cuspidi C, Comarella L, et al. Left ventricular diastolic dysfunction in elderly hypertensives: results of the APROS-diadys study. J Hypertens. 2007;25:2158–67.
34. Canepa M, Strait JB, Milaneschi Y, et al. The relationship between visceral adiposity and left ventricular diastolic function: results from the Baltimore Longitudinal Study of Aging. Nutr Metab Cardiovasc Dis. 2013;23:1263–70.
35. Skaarup KG, Lassen MCH, Marott JL, et al. Diastolic function assessed with speckle tracking over a decade and its prognostic value: The Copenhagen City Heart Study. Echocardiography. 2021;38:964–73.
36. Airale L, Paini A, Ianniello E, et al. Left atrial volume indexed for height(2) is a new sensitive marker for subclinical cardiac organ damage in female hypertensive patients. Hypertens Res. 2021;44:692–9.
37. Pichler G, Redon J, Martínez F, et al. Cardiac magnetic resonance-derived fibrosis, strain and molecular biomarkers of fibrosis in hypertensive heart disease. J Hypertens. 2020;38:2036–42.
38. Burocchi S, Gori M, Cioffi G, et al. Risk stratifying asymptomatic left ventricular systolic dysfunction in the community: beyond left ventricular ejection fraction. Eur Heart J Cardiovasc Imaging. 2020;21:1405–11.
39. Biering-Sørensen T, Biering-Sørensen SR, Olsen FJ, et al. Global longitudinal strain by echocardiography predicts long-term risk of cardiovascular morbidity and mortality in a low-risk general population: the Copenhagen City Heart Study. Circ Cardiovasc Imaging. 2017;10:e005521.
40. Stylidis M, Leon DA, Rösner A, et al. Global myocardial longitudinal strain in a general population-associations with blood pressure and subclinical heart failure: The Tromso Study. Int J Cardiovasc Imaging. 2020;36:459–70.

41. Modin D, Biering-Sørensen SR, Mogelvang R, et al. Prognostic value of echocardiography in hypertensive versus nonhypertensive participants from the general population. Hypertension. 2018;71:742–51.
42. Tadic M, Cuspidi C, Majstorovic A, et al. The relationship between left ventricular deformation and different geometric patterns according to the updated classification: findings from the hypertensive population. J Hypertens. 2015;33:1954–61.
43. Tadic M, Sala C, Carugo S, et al. Myocardial strain in hypertension: a meta-analysis of two-dimensional speckle tracking echocardiographic studies. J Hypertens. 2021;39:2103–12.
44. Tadic M, Cuspidi C, Plein S, et al. Comprehensive assessment of hypertensive heart disease: cardiac magnetic resonance in focus. Heart Fail Rev. 2021;26:1383–90.

Part III

New Perspectives on Diagnosis of Heart Failure in Hypertensive Patients

Left Ventricular Diastolic Dysfunction

15

Vincenzo Castiglione, Francesco Gentile, Alberto Aimo, and Michele Emdin

Abbreviations

ATP	Adenosine triphosphate
CMR	Cardiac magnetic resonance
dPAP	Diastolic pulmonary artery pressure
EDV	End-diastolic volume
GLS	Global longitudinal strain
HF	Heart failure
HFpEF	Heart failure with preserved ejection fraction
HFrEF	Heart failure with preserved ejection fraction
HR	Heart rate
IVRT	Isovolumetric relaxation time
LA	Left atrium
LAVi	Left atrial volume index
LIFE	Losartan intervention for end point reduction in hypertension
LV	Left ventricle
LVEDP	Left ventricular end-diastolic pressure
LVEF	Left ventricular ejection fraction
LVH	Left ventricular hypertrophy
mPAP	Mean pulmonary artery pressure
PCWP	Pulmonary capillary wedge pressure
PFR	Peak filling rate
PV	Pulmonary vein

V. Castiglione · F. Gentile · A. Aimo · M. Emdin (✉)
Health Science Interdisciplinary Research Center, Scuola Superiore Sant'Anna, Pisa, Italy

Fondazione Toscana Gabriele Monasterio, Pisa, Italy
e-mail: vcastiglione@ftgm.it; aimoalb@ftgm.it; m.emdin@santannapisa.it; emdin@ftgm.it

M. Dorobantu et al. (eds.), *Hypertension and Heart Failure*, Updates in Hypertension and Cardiovascular Protection,
https://doi.org/10.1007/978-3-031-39315-0_15

SERCA	Sarco/endoplasmic reticulum calcium ATPase
sPAP	Systolic pulmonary artery pressure
SR	Strain rate
STE	Speckle-tracking echocardiography
TPFR	Time to peak filling rate
TR	Tricuspid regurgitation
TTE	Transthoracic echocardiography

15.1 Introduction

"*Diastolic dysfunction*" refers to a disturbance in ventricular relaxation, distensibility, or filling, regardless of whether left ventricular (LV) ejection fraction (LVEF) is normal or depressed and whether the patient is symptomatic or asymptomatic [1]. A patient with preserved LVEF (≥50%), symptoms and/or signs of heart failure (HF), and "objective evidence of cardiac structural and/or functional abnormalities consistent with the presence of LV diastolic dysfunction/raised LV filling pressures, including raised natriuretic peptides" can be diagnosed with HF with preserved EF (HFpEF) [2].

Hypertension is a major risk factor for HFpEF [3]. Messerli et al. proposed the following stages in the evolution of hypertensive heart disease [4]:

degree I, hypertension without LV hypertrophy (LVH);
degree II, asymptomatic hypertension with LVH;
degree III, symptomatic HFpEF;
degree IV, symptomatic HF with reduced EF (HFrEF).

This cascade starts with a chronically increased afterload due to high blood pressure. Concentric hypertrophy represents, at least initially, an attempt by the LV to maintain wall stress in response to elevated systemic pressures, in agreement with Laplace's law. Hypertension may also promote cardiac remodeling by eliciting systemic inflammation, either as a direct effect or because of the prevalent comorbidities, such as diabetes or obesity.

The link between hypertension and the development of LVH is well established [5, 6]. In major clinical trials, the prevalence of LVH among hypertensive patients ranged from 23% to 48% [7]. Hypertension appears to cause LVH in a "dose-dependent" fashion, and even prehypertensive patients are at risk for LV remodeling [6, 8, 9]. Although initially an adaptive response, LVH ultimately becomes pathologic. There are clear differences in the pattern of hypertrophy in patients with LVH owing to hypertension, compared with individuals with physiologic hypertrophy, such as athletes. Indeed, it has been reported that patients with hypertensive hypertrophy have significantly lower global longitudinal strain (GLS) values and greater degrees of diastolic dysfunction than age-matched competitive athletes despite lower mean LV mass [10]. In addition, LV mass inversely correlates with cardiac index in hypertensive patients [7].

LVH development is accompanied by abnormalities in myocyte relaxation and progressive extracellular fibrosis, which ultimately cause diastolic dysfunction. LV diastolic dysfunction leads to an elevation in LV and left atrial (LA) filling pressures, causing the development of dyspnea symptom and the transition to overt HF, whose risk increases dramatically once LVH develops [6, 11]. This relationship appears to be "dose-dependent." Indeed, each 1% increase in echocardiographic LV mass above the normal range has been associated with a 1% increase in incidence of HF after controlling for other risk factors, such as prior myocardial infarction [11]. Elevation of natriuretic peptides and troponins may identify patients with LVH at higher risk of progression to HF and worse outcome [12, 13]. Once clinical HFpEF has manifested, there is minimal hope for reversing the disease process. Thus, preventing the development of diastolic dysfunction or blocking its progression are important treatment goals in hypertensive patients. This chapter will provide an overview of diastolic dysfunction in hypertension from its mechanisms to its clinical implications.

15.2 Pathophysiology of Diastolic Dysfunction

Diastolic dysfunction may derive from alterations in one or more of the following determinants of diastole: active relaxation, elastic properties of the LV, heart rate (HR), ventricular load, and uniformity of ventricular relaxation [14].

15.2.1 Active Relaxation

Myocardial inactivation is caused by calcium extrusion from the cytosol and cross-bridge detachment. The determinants of myocardial inactivation are then mechanisms maintaining calcium homeostasis and regulators of cross-bridge cycling. A lower concentration or activity of the sarco/endoplasmic reticulum calcium ATPase (SERCA) can slow the removal of calcium from the cytosol. Increased levels or activity of the SERCA inhibitor phospholamban can also impair relaxation. Increased cyclic adenosine monophosphate, resulting from beta-adrenergic stimulation or inhibition of cardiac phosphodiesterase, promotes the phosphorylation of phospholamban to remove its inhibitory effect on SERCA, thus improving myocardial relaxation. Pathological LVH secondary to hypertension or to aortic stenosis results in decreased SERCA and increased phospholamban, leading to impaired relaxation. Similar changes are seen in the myocardium of patients with hypertrophic or dilated cardiomyopathy. SERCA expression also decreases with age, in parallel with diastolic dysfunction worsening [15].

15.2.2 Elastic Properties

Diastolic relaxation is influenced by the elastic properties of the myocardium, specifically the cytoskeleton and the extracellular matrix [16]. Most of the elastic force of cardiomyocytes is thought to reside in the macromolecule titin, whose mechanical properties differ across titin isoforms and the phosphorylation status [17]. The contribution of microtubules and intermediate filaments appears much less important at normal sarcomere lengths [18]. Modifications of myofilament calcium sensitivity includes changes associated with protein kinase A- or protein kinase G-mediated phosphorylation of myosin light chain 2 and troponin I. In this setting, nitric oxide and cyclic guanosine monophosphate increase resting diastolic cell length as a result of protein kinase G-mediated phosphorylation of myofilaments. Furthermore, myocardial stiffness is modulated by load and endothelin-1: load elevations acutely increase LV stiffness, while endothelin-1 acutely decreases myocardial stiffness [17]. As for the extracellular matrix, fibrillar collagen seems the most important contributor to diastolic dysfunction. Its amount, distribution, degree of cross-linking, and ratio of type I to type III collagen are often altered in disorders leading to diastolic dysfunction, such as hypertension [16].

15.2.3 Heart Rate

A high HR increases myocardial oxygen demand and reduces the time for coronary perfusion. This can lead to diastolic dysfunction even in the absence of coronary artery disease, especially in hypertrophic hearts. Furthermore, a high HR can shorten diastole to an extent that prevents complete relaxation between beats, resulting in increased filling pressures. This may occur at lower HR in failing hearts than in normal hearts [16].

15.2.4 Ventricular Load

In the normal heart, a mild to moderate afterload elevation imposed early during the cardiac cycle will delay the onset and accelerate the rate of pressure fall. A severe afterload elevation or an afterload increase occurring late during the cardiac cycle will induce a premature onset and a pronounced slowing of the pressure fall. This slowing may lead to an incomplete relaxation and then to elevated filling pressures, particularly when the preload is elevated as well [19]. In acute hypertensive crises, high blood pressure values represent a heavy afterload to the LV and may cause diastolic dysfunction and pulmonary congestion [20].

15.2.5 Uniformity of Ventricular Relaxation

During isovolumetric relaxation, extension of one ventricular segment is accompanied by post-systolic shortening of another segment. The ventricular volume remains unchanged, but the ventricular shape changes. Loss of synchronicity in this process reduces the rate of ventricular pressure fall and may contribute to diastolic dysfunction, especially in patients with coronary artery disease or intraventricular conduction disturbances [14, 21].

15.3 Factors Leading to Diastolic Dysfunction in Hypertension

The response of the myocardium to chronic-increased stretch is hypertrophy. Schematically, an increased preload (volume overload) leads to eccentric hypertrophy, while an increased afterload (pressure overload) promotes concentric hypertrophy. Early in the evolution of concentric hypertrophy, myocardial wall thickening allows ventricular function and ejection fraction to keep pace with increased afterload. The thickened wall is intrinsically less efficient, and the force of contraction per unit of muscle mass declines in parallel with the progression of LVH. A rise in LV end-diastolic pressure (LVEDP), the hallmark of diastolic dysfunction, eventually develops [22].

The afterload may be conceived as the integrated sum of three components: systemic vascular resistance (which is increased by atherosclerosis and media-intima calcification of aortic wall and is also intrinsically higher in smaller vessels, as typically occurs in women) [23], ventricular–vascular coupling during early systole (which depends on cardiac emptying and aortic stiffness), and late systolic pressure augmentation (produced by the incident pressure wave and reflected pressure wave) [22]. The response to chronically increased afterload is modulated by neurohormones such as norepinephrine, epinephrine, angiotensin II, aldosterone, and paracrine substances such as endothelin, which exert direct growth-promoting influences on cardiomyocytes [24, 25]. Although hypertrophy is believed to be mediated largely by AT1 receptors, AT2 receptor stimulation might also contribute to hypertrophy [26]. The beneficial effects of angiotensin-converting enzyme inhibitors/angiotensin receptor blockers on LVH are partially related on the inhibition of angiotensin II stimulation, in addition to their effects on preload and afterload [27]. Angiotensin II and aldosterone are also powerful profibrotic stimuli, thus contributing to ventricular stiffness and diastolic dysfunction [28, 29].

A genetic component in response to chronically elevated blood pressure is suggested by the finding that diastolic abnormalities occur in the offspring of individuals with LVH [30] and by the considerable interindividual variation in the hypertrophic response to similar blood pressure values. LVH is considered a complex genetic disease, likely representing an interaction of several genes with the environment. The heritability of LV mass, measured as a quantitative trait, ranges from 30% to 70% in different populations [31] and is higher in African Americans [32]. Genetic susceptibility to LVH has been attributed to a variety of genes encoding structural proteins of the LV; cell transduction signals, hormones, growth factors, calcium homeostasis, substrate metabolism, and blood pressure are likely candidates for the development of common forms of LVH [32, 33].

In summary, LVH and the resulting diastolic dysfunction in patients with hypertension mainly derive from a chronically increased afterload but also depend on the degree of neurohormonal activation (most notably of the renin–angiotensin–aldosterone axis) and on genetic factors.

15.4 Assessment of Left Ventricular Diastolic Dysfunction

15.4.1 Invasive Assessment of Left Ventricular Diastolic Dysfunction

Cardiac catheterization represents the gold standard method to assess LV pressures and to diagnose diastolic dysfunction [34]. Traditionally, the direct measures of LVEDP have been used to identify diastolic dysfunction in the experimental setting [35]. However, left chambers' pressures are rarely directly evaluated in the clinical practice. The measure of the pulmonary capillary wedge pressure (PCWP) is hence widely accepted as a valuable surrogate, although it estimates LA more than LV filling pressures [35]. In resting conditions, LVEDP and PCWP are considered altered for values ≥16 and ≥15 mmHg, respectively [36], although PCWP values >12 mmHg retain prognostic significance as well [36, 37].

In patients with effort-related symptoms (most commonly dyspnea) suggestive of HFpEF, but with normal diastolic function at rest, a diastolic stress test is recommended [36, 38, 39]. During exercise, an abnormal (or disproportionate) increase in LVEDP and PCWP, consequent to the inability of the failing heart to adapt to the changes in preload, afterload, and HR, may indeed favor lung congestion and symptoms occurrence [39]. An invasive stress test, by using supine cycling during cardiac catheterization, may therefore unveil the impact of exercise on LV pressures and conclude the diagnostic work-up for HFpEF. In this setting, reaching PCWP values ≥25 mmHg may accurately imply a cardiac origin of symptoms [40].

Although LVEDP and PCWP have been extensively utilized to assess diastolic dysfunction, they are exquisitely load-dependent and do not mirror the active phase of diastole [41]. Therefore, they may not be reliable in some specific conditions such as hypovolemia, pericardial effusion, or during either invasive or noninvasive ventilation. In these cases, the slope of the LV end-diastolic pressure–volume relationship, which is shifted upward and leftward in patients with HFpEF compared with healthy individuals, may be used to assess LV compliance. However, this measure should be assessed in different loading conditions, and it is affected by extra-LV factors (such as pericardial, right ventricular, and respiratory forces) [42, 43]. Finally, various parameters have been tested to evaluate LV active relaxation, but their use is almost entirely confined to the experimental setting [39, 41]. Among these, the maximal rate of LV pressure reduction (dP/dt_{min}) reflects active relaxation, but it is influenced by HR, preload, and afterload [44]. The time constant τ (*tau*), estimating the time necessary for LV pressure recovery during isovolumetric relaxation, is a more specific index of LV relaxation, being less affected by hemodynamic conditions [44, 45]. Nonetheless, the necessity of complex mathematical calculations and sophisticate instrumentation limit its clinical use.

Thanks to the advances and widespread availability of noninvasive tests, cardiac catheterization is rarely necessary to evaluate diastolic function in the clinical practice. However, the direct measure of LV pressures—either at rest or during exercise—may still be useful in case of inconclusive or discordant findings, or when a differential diagnosis with other conditions (e.g., pre-capillary pulmonary hypertension) is required [39].

15.4.2 Noninvasive Imaging Modalities for the Evaluation of Left Ventricular Diastolic Dysfunction

15.4.2.1 Standard Echocardiography

Transthoracic echocardiography (TTE) in resting conditions is the most used test to assess LV diastolic function in the clinical practice. Although various parameters have been proposed to identify and classify diastolic dysfunction and to estimate LV filling pressures at TTE, each measure holds some limitations or is less accurate in specific conditions. Therefore, the current guidelines suggest the use of a multiparametric and patient-tailored approach [39, 46].

Transmitral Flow Velocities

In the last century, landmark experimental studies showed that transmitral diastolic flow velocities, assessed though pulsed-wave Doppler, were related to LV compliance and relaxation capacities [45, 47]. In particular, E wave velocity corresponds to the early passive phase of LV filling (i.e., consequent to LV relaxation) and A wave velocity to the late active (i.e., consequent to LA systole) phase [47, 48]. Notably, while A wave velocity depends on LA contractility, E wave velocity is mainly determined by the LA/LV pressure gradient; hence, it is influenced by both LA pressures and LV relaxation (expressed by the time constant τ) [49]. The evaluation of these velocities and their ratio is therefore routinely performed to estimate diastolic function.

While E/A is usually >1 in euvolemic healthy individuals, delayed relaxation is associated with an inversion of this ratio. This pattern may be physiologically observed with aging not only because of reduced LV compliance and slower relaxation but also in case of hypertension, myocardial ischemia, or cardiomyopathies [50, 51]. As the disease progresses, the increase in LA pressure may heighten E wave velocity, leading to the so-called "pseudonormalization" of E/A ratio up to values >2 [52], corresponding to the ominous "restrictive" pattern [53]. Notably, the normal and the pseudonormal patterns may be distinguished by varying loading conditions (e.g., through Valsalva maneuver), unveiling impaired LV relaxation in case of a pseudonormal filling [54]. Although the evaluation of E/A is widely used to assess diastolic function, it may be affected by HR, while rhythm and conduction disorders may hamper wave visualization and interpretation [46, 48].

The evaluation of diastolic function by transmitral flow velocities can be complemented with other parameters. The deceleration time of E wave is inversely proportional to LV compliance, and values lower than 125 or 150 ms have been associated with increased LV stiffness and worse outcome [48, 55]. Some patients may show an intermediate wave between E and A waves, i.e., the L-wave, which has been associated with severely increased filling pressure and risk of incident HF [56]. Finally, isovolumetric relaxation time (IVRT) can be directly measured as the interval between aortic valve closure and mitral valve opening. While healthy individuals show an IVRT ≤70 ms, longer durations have been associated with impaired LV relaxation and abnormal filling pressures [46, 57].

Mitral Annulus Deformation Imaging

The evaluation of LV longitudinal motion helps assess diastolic function and can be altered in case of increased filling pressure (e.g., hypertension, obesity, and cardiomyopathies), with clinical and prognostic implications [58]. The use of tissue Doppler imaging allows a load-independent evaluation of LV shortening during TTE. Three waves can be described: the positive s' wave (representing systolic shortening), and the negative e' and a' waves (representing the early and the late phases of LV longitudinal relaxation, respectively) [58].

Physiologically, the lateral mitral annulus shows larger oscillations than the septal region; therefore, lateral e' wave is normally higher than septal *e′* wave. Accordingly, lateral and septal *e′* wave velocities <7 and <10 cm/s have been associated with impaired LV relaxation [39, 46]. Since *e′* wave velocity is inversely proportional to the time constant τ, the use of *E*/*e′* ratio has been proposed to estimate LV filling pressures [59]. Whereas healthy individuals usually show a ratio <8, lateral *E*/*e′* >12, septal *E*/*e′* >15, or average *E*/*e′* >14 have been consistently associated with a high probability of increased filling pressures [39, 46, 60].

Although largely utilized in the clinical practice, also *E*/*e′*—similar to E/A—is age-specific and it is weakly related to invasively measured LVEDP, particularly in case of low or intermediate values [61]. Moreover, *e′* waves may be difficultly evaluated in case of rhythm and conduction disorders, whereas other condition such as mitral annulus calcification, prosthetic mitral valves, severe mitral stenosis or regurgitation, LV regional wall abnormalities, right ventricular dysfunction, and pericardial disease may affect the diagnostic and prognostic yields of these measures.

Left Atrial Dimensions and Filling

In the absence of atrial fibrillation, the structural remodeling of LA may be a direct consequence of increased filling pressure. Therefore, the evaluation of LA dimension—optimally expressed as the maximal volume indexed to body surface area (LAVi)—is often considered in the echocardiographic assessment of LV diastolic function [62]. Of note, among HFpEF patients, enlarged LA (i.e., with a LAVi >34 mL/m^2) has been associated with elevated LV filling pressure and worse outcome [63, 64]. However, the sole evaluation of LAVi may be less sensitive in the general population, and values >34 mL/m^2 may be detected in >10% individuals in the absence of other abnormalities [65]. To overcome this inaccuracy, growing evidence has outlined the diagnostic and prognostic significance of LA strain [66]. Indeed, as detailed below in this chapter, the evaluation of these parameters may detect diastolic dysfunction at an earlier stage and refine risk stratification in patients with HFpEF [39, 66].

Finally, the evaluation of LA filling, assessed through the sampling of pulmonary vein (PV) flows, can also provide insight on diastolic function. Normally, PV flow is characterized by a dominant S-wave and a recessive D-wave, with a positive S/D ratio, which can be amplified in case of mild diastolic dysfunction [67]. At more advanced disease stages, most of LA filling occurs in the early stage of diastole because of increased filling pressures and the S/D ratio is often inverted [67, 68]. Furthermore, the increase in LA pressure may promote an inverse flow toward PV during atrial contraction, depicted by a larger and prolonged negative A_r-wave [68]. Nevertheless, the poor specificity and the possible difficulties in obtaining an accurate visualization of PV flow in the absence of an optimal acoustic window hamper the use of these measures in the clinical practice.

Estimation of Pulmonary Artery Pressures

In the absence of noncardiac causes of pulmonary hypertension, the evaluation of pulmonary artery pressures may corroborate the suspect of elevated LV filling pressures. TTE allows to estimate systolic pulmonary artery pressure (sPAP) as the sum of systolic trans-tricuspid pressure gradient and right atrial pressure, which is indirectly derived from the diameter and respiratory collapsibility of the inferior vena cava [69]. The visualization of a tricuspid regurgitation (TR) jet is therefore necessary, and a TR velocity >2.8 m/s is usually a synonym of pulmonary hypertension [70]. Similarly, diastolic pulmonary artery pressure (dPAP) can be estimated by measuring the end-diastolic velocity of pulmonary regurgitation [71], while mean pulmonary artery pressure (mPAP) can be calculated as 1/3 (sPAP) + 2/3 (dPAP) [70]. Notably, integrating such measure of mPAP with other parameters, routinely collected during TTE, allowed an accurate estimation of PCWP and of pulmonary vascular resistances in a large cohort of patients ($n = 795$), which also underwent cardiac catheterization [72].

Other Measures

Several other TTE parameters have been proposed to estimate diastolic function. Among these, the evaluation of mitral-to-apical flow propagation, assessed with color M-mode Doppler, has been proposed as a measure of LV apical suction. However, other factors beyond LV filling pressures may affect this parameter, limiting its clinical use [73]. While the duration of trans-mitral A wave is often reduced in the presence of diastolic dysfunction, that of pulmonary vein reverse A wave (Ar) may be prolonged. Accordingly, it has been reported that a larger Ar-A difference (i.e., >30 ms) corresponds to LV diastolic dysfunction. However, this parameter may only be assessed during sinus rhythm and normal atrioventricular conduction [74]. Finally, the evaluation of the time interval $T_{E\text{-}e'}$ has been proposed to estimate impaired LV relaxation. Notably, such parameter may prove value in case of mitral valve disease or to discriminate restrictive from constrictive conditions [46].

Multiparametric Algorithm

The assessment and the grading of diastolic dysfunction, as well as the estimation of LV filling pressures, hold prognostic significance and may guide treatment in HF patients independently of LVEF [75]. However, as discussed above, no single TTE measure can per se provide an accurate estimation of diastolic function. Therefore, the use of a multiparametric algorithm has been proposed (Fig. 15.1) [46]. In patients with no structural abnormalities, the presence of four parameters should be assessed: average E/e' >14, septal e' velocity <7 cm/s or lateral e' velocity <10 cm/s, TR velocity >2.8 m/s, and LAVi >34 mL/m^2. When either <2 or >2 of these

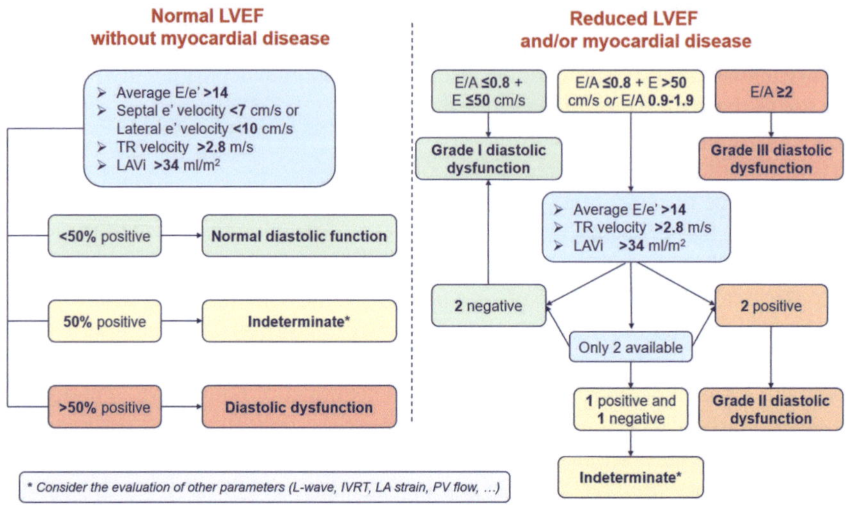

Fig. 15.1 Multiparametric algorithm for diastolic function assessment in patients with either normal or reduced left ventricular ejection fraction (LVEF). According to the latest international recommendations, a multiparametric approach should be adopted to estimate diastolic function. Although the evaluation of trans-mitral flow and mitral annulus velocities, tricuspid regurgitation (TR) max velocity, and LA volume index (LAVi) is often sufficient to estimate diastolic function and filling pressures, the use of further parameters may be necessary in selected cases. *IVRT* isovolumetric relaxation time, *PV* pulmonary veins

parameters are identified, diastolic function should be considered either normal or abnormal, respectively, while it remains undetermined in the other cases [46].
The evaluation of mitral flow velocities constitutes the first step to assess diastolic dysfunction in patients with either suspected or known HF. Whereas an $E/A \leq 0.8$ with an E wave velocity ≤50 cm/s is highly specific of normal filling pressures and indicates grade I diastolic dysfunction, an $E/A \geq 2$ often corresponds to increased filling pressures and indicates grade III diastolic dysfunction [46]. Otherwise, the presence of further parameters should be tested, namely average E/e' >14, TR velocity >2.8 m/s, and LAVi >34 mL/m^2. If ≥2 of these criteria are met, filling pressures are usually elevated and a grade II diastolic dysfunction is attributed. Conversely, if ≥2 of these measures do not reach the proposed cut-offs, filling pressures are usually normal and grade I diastolic dysfunction is attributed. If uncertainty persists, other measures (e.g., PV flow) should be evaluated [46].

Although broadly validated [76], recent studies suggest that the diagnostic accuracy of such algorithm may be lower in patients with suspected HFpEF [76, 77]. The introduction of novel parameters, such as LA strain, may thus prove value in this context [78]. Moreover, considering that many patients with suspected HFpEF show increased filling pressure only during exercise [40], the implementation of stress diastolic testing may be of help [39].

15.4.2.2 Speckle-Tracking Echocardiography

Speckle-tracking echocardiography (STE) analyses the motion of tissues in the heart by using the naturally occurring speckle pattern in the myocardium. By the analysis of speckle motion in 2-dimensional images, this technique allows a non-Doppler, angle-independent analysis of myocardial deformation, with the possibility to quantify thickening, shortening, and rotation dynamics of cardiac function. STE has mainly been employed to derive several indices of systolic function, the most validated being GLS, which represents the mean of the shortening (strain) of all myocardial segments along the longitudinal axis of the heart [79, 80]. GLS is a sensitive marker of systolic dysfunction; indeed, reduced GLS can be detected even in patients with hypertension and normal ejection fraction [81]. Nonetheless, STE can also be used to assess diastolic function. Strain rate (SR) measured during the isovolumetric relaxation period (SR_{IVR}) or the early diastolic filling (SR_E) has been demonstrated as reliable indices of myocardial relaxation. In particular, SR_{IVR} has shown a strong correlation with the time constant τ [82]. By combining these indices with the mitral E wave velocity, new parameters (E/SR_{IVR} and E/SR_E) have been obtained, able to accurately estimate invasively measured LV filling pressures with better precision than indices derived from tissue Doppler imaging (i.e., E/e') [82–85]. LV untwisting rate is another index of LV relaxation derived from the analysis of LV torsion mechanics by STE. LV untwisting rate appears to correlate well with both τ and invasively measured LV pressure gradients but can be influenced by LV recoil and thus by LV systolic function [86, 87]. The timing to peak LV untwisting rate can also inform diastolic function, since an impaired relaxation can often lead to a delayed peak of LV untwisting rate [86]. STE analysis also allows to investigate LA functions of reservoir and active contraction: peak LA longitudinal strain is measured during ventricular systole, while peak LA contraction strain is measured during atrial contraction [88]. LA strain indices inversely correlate with LVEDP, hence can be used as surrogate markers of diastolic dysfunction [89]. However, although STE provides several reliable indices of diastolic dysfunction, their use in the clinical practice is limited by the fact that STE analysis is time-consuming and requires dedicated software, with significant variability among vendors [79].

15.4.2.3 Stress Echocardiography

Patients with mild diastolic dysfunction can have similar cardiac output and filling pressures at rest compared to heathy individuals. Exercise test can unmask reduced diastolic capacity. Indeed, in heathy individuals, stroke volume increases during exercise without elevation of LV filling pressures thanks to augmented myocardial relaxation. At echocardiography, this determines a parallel increase in mitral E wave and mitral annular e' velocities, resulting in preserved E/e' ratio with exercise. The impairment of myocardial relaxation is the first manifestation of diastolic dysfunction; hence, patients with diastolic dysfunction can increase stroke volume during exercise only through an elevation of filling pressures. At echocardiography, these patients show a slower e' velocity increase with exercise, resulting in higher E/e' ratio [38, 46]. This constitutes the pathophysiological rationale for diastolic stress test, which can be performed using either a supine bicycle or a treadmill; the former being preferred because it allows the acquisition of Doppler recordings (i.e., mitral inflow velocities, mitral annular velocities, and TR velocity) throughout the test. The ideal target population of diastolic stress test is neither patients with completely normal hearts nor those with abnormal findings at baseline already consistent with elevated LV filling pressures, but rather those patients with grade I diastolic dysfunction usually complaining of unexplained exertional dyspnea [38, 46]. Notably, patients with hypertension, LVH and normal EF often fall in this category [90]. The test is considered indicative of diastolic dysfunction if all the following criteria are met: (1) average E/e' ratio >14 or septal E/e' ratio >15 with exercise; (2) peak TR velocity >2.8 m/s with exercise; and (3) septal e' velocity <7 cm/s (or lateral e' velocity <10 cm/s) at baseline. An increase in TR velocity during exercise alone is not sufficient for a positive test, since it can also occur in healthy subjects as a result of increased pulmonary blood flow and not only as a consequence of elevated LV filling pressures. When the above criteria are not met, the test is considered indetermined and invasive hemodynamic assessment is warranted [38, 46].

15.4.2.4 Nuclear Cardiology

Nuclear imaging can be used for diastolic function evaluation, albeit it only allows the assessment of volumetric parameters, such as peak filling rate (PFR) and time to peak filling rate (TPFR). Multi-gated-radionuclide angiography or gated single-photon emission computerized tomography provides LV time–volume curves, from which quantitative parameters of diastolic function can be derived [91, 92]. PFR corresponds to the point of maximum positive slope of the time–volume curve, and normal values usually are >2.5 end-diastolic volume (EDV)/s. PFR is influenced by age, HR, EDV, end-systolic volume, and LVEF, which should be considered when comparing PFR values between different patients [91, 92]. Normalization of PFR to the peak ejection rate provides a HR-independent index of diastolic filling [93]. TPFR is the time interval between end-systole and PFR, which is normally <180 ms. Patients with diastolic dysfunction show a decrease in PFR and a prolongation of TPFR [91, 92].

15.4.2.5 Cardiac Magnetic Resonance

Cardiac magnetic resonance (CMR) represents the gold standard imaging modality for the assessment of atrial and ventricular dimensions and for the quantification of the ejection fraction. Consequently, CMR allows to accurately characterize the structural remodeling underlying diastolic dysfunction in patients with hypertensive heart disease and HFpEF [94, 95]. CMR is superior to echocardiography in the evaluation of LVH and quantification of LV mass since it does not rely on geometric assumptions. LV mass is calculated by multiplying the specific gravity of the myocardium (1.05 g/ml) by the LV myocardial volume, which is measured as the area between endocardial and epicardial borders in contiguous short-axis cine images of the entire LV multiplied by the interslice distance [96, 97]. Together with LVH, LA dilation is a hallmark of structural remodeling related to diastolic dysfunction.

CMR-derived LA volumes are measured similarly to echocardiography, by using the area–length method or Simpson's biplane method in the 2-chamber and 4-chamber views. Nonetheless, LA volumes assessed through CMR are usually larger than echocardiography-derived ones because of better alignment and visualization of the LA [98, 99].

CMR also allows to noninvasively catheterize the tissue alterations underlying cardiac structural remodeling. CMR mapping techniques measure myocardial tissue relaxation times (T1, longitudinal relaxation time; T2, transversal relaxation time), which have been extensively established as quantifiable markers of interstitial fibrosis and myocardial edema. Pre- and post-contrast T1, along with blood hematocrit, allow a direct measure of the myocardial extracellular volume, which is directly related to the expansion of the cardiac extracellular matrix, a major contributor to the impairment of cardiac relaxation and stiffness. These techniques have become unvaluable tools to differentiate the underlying pathologies in patients with HF and prevalent diastolic dysfunction [100].

Direct diastolic function assessment by CMR is also feasible, albeit it has not yet entered the clinical practice because of restricted availability of this imaging modality as well as slow image acquisition and data processing. Cine-CMR acquisitions provide LV time–volume curves, from which PFR and TPFR can be obtained in a similar way to nuclear imaging. These parameters show a good correlation with the echocardiographic evaluation of diastolic dysfunction [101]. LV time–volume curves also allow to derive the "diastolic volume recovery," i.e., the percentage of diastole necessary for the recovery of 80% of the stroke volume, which is another sensitive index of diastolic dysfunction [101].

Phase-contrast CMR imaging enables quantitative assessment of flow through an orifice, allowing to obtain measurements of the transmitral (E and A waves) and pulmonary flows. These parameters show a good correlation with the counterparts calculated by echocardiography, although the velocities measured with CMR are systematically lower, mainly because the CMR acquisition has a lower temporal resolution than pulsed-wave Doppler (30–40 vs. <10 ms) [102]. Similar to tissue Doppler imaging, phase-contrast CMR imaging can also be used to calculate mitral annular septal and lateral velocities, with good correlation between the 2 imaging modalities [103]. Notably, in a study on hypertensive patients, CMR-derived E/e' also showed a strong correlation with invasively measured PCWP [104].

Finally, CMR can also be used for the assessment of myocardial strain. Myocardial tagging is the most studied technique for myocardial strain analysis and consists in labeling the myocardium in a grid fashion through spatial modulation of the magnetization and then following these tags over time, thus providing measurements of myocardial relaxation and deformation [105]. More recently, feature-tracking software has been implemented in CMR imaging for endocardial border tracking in a similar way to STE, thus allowing to assess both LV and LA strain. CMR-derived strain parameters show a close correlation with those obtained by echocardiography [106].

15.5 Clinical Implications of Left Ventricular Diastolic Dysfunction in Hypertension

Diastolic dysfunction is a strong predictor of new onset HF in hypertensive individuals, regardless of blood pressure levels and LV mass [107, 108]. Diastolic dysfunction is also predictive of cardiovascular events [109–111], possibly because the severity of diastolic dysfunction reflects the degree of organ damage from hypertension [112]. It has been consistently demonstrated that lowering blood pressure improves diastolic dysfunction independently of the antihypertensive drug used [107, 113–116]. Optimal management of comorbidities and reduction in dietary sodium intake may further contribute to improve diastolic function in hypertensive patients [117, 118]. However, it is still uncertain if relieving diastolic dysfunction translates into better quality of life and outcomes. In the Losartan Intervention for End Point Reduction in Hypertension (LIFE) study, antihypertensive treatment improved LV diastolic function, but such improvements were not associated with reductions in the composite endpoint of cardiovascular mortality, myocardial infarction, or stroke [107]. Similarly, a survival benefit from antihypertensive drugs did not emerge in subgroup analyses of HFpEF trials [119].

15.6 Conclusions

Hypertension imposes an increased afterload on the LV, causing progressive cardiac remodeling, whose hallmarks are LVH and diastolic dysfunction. Diastolic dysfunction is associated with an elevation of LA and LV filling pressures, which promote the development of symptoms like exertional dyspnea and, consequently, the transition to clinical HF, at first in the form of HFpEF. Diastolic dysfunction is also predictive of cardiovascular events, possibly because the severity of diastolic dysfunction reflects the degree of organ damage from hypertension. Therefore, the assessment of diastolic function is essential to characterize cardiac damage caused by hypertension. Invasive measurement of cardiac filling pressure through cardiac catheterization is the gold standard for the study of diastolic function but is rarely performed. Indeed, Doppler echocardiography provides noninvasive reliable indices of diastolic function that can be combined into validated multiparametric scores for the classification diastolic dysfunction. Moreover, stress echocardiography can be used to unveil even subtle diastolic abnormalities. There are also other imaging techniques, such as nuclear cardiology and CMR, which can provide additional indices of diastolic dysfunction, but their role in the routine clinical practice is still limited by accessibility and slow processing time. Although great progress has been made in the ability to diagnose diastolic dysfunction in hypertensive patients, it is still unclear whether diastolic improvement following the use of antihypertensive drugs results in a better prognosis.

References

1. Leite-Moreira AF. Current perspectives in diastolic dysfunction and diastolic heart failure. Heart Br Card Soc. 2006;92(5):712–8. https://doi.org/10.1136/hrt.2005.062950.
2. McDonagh TA, et al. 2021 ESC guidelines for the diagnosis and treatment of acute and chronic heart failure. Eur Heart J. 2021;42(36):3599–726. https://doi.org/10.1093/eurheartj/ehab368.
3. Kasiakogias A, et al. Hypertension and heart failure with preserved ejection fraction: position paper by the European Society of Hypertension. J Hypertens. 2021;39(8):1522–45. https://doi.org/10.1097/HJH.0000000000002910.
4. Messerli FH, Rimoldi SF, Bangalore S. The transition from hypertension to heart failure: contemporary update. JACC Heart Fail. 2017;5(8):543–51. https://doi.org/10.1016/j.jchf.2017.04.012.
5. Kannel WB, Gordon T, Offutt D. Left ventricular hypertrophy by electrocardiogram. Prevalence, incidence, and mortality in the Framingham study. Ann Intern Med. 1969;71(1):89–105. https://doi.org/10.7326/0003-4819-71-1-89.
6. Heckbert SR, et al. Traditional cardiovascular risk factors in relation to left ventricular mass, volume, and systolic function by cardiac magnetic resonance imaging: the Multiethnic Study of Atherosclerosis. J Am Coll Cardiol. 2006;48(11):2285–92. https://doi.org/10.1016/j.jacc.2006.03.072.
7. Devereux RB, Pickering TG, Alderman MH, Chien S, Borer JS, Laragh JH. Left ventricular hypertrophy in hypertension. Prevalence and relationship to pathophysiologic variables. Hypertension. 1987;9(2 Pt 2):53–60. https://doi.org/10.1161/01.hyp.9.2_pt_2.ii53.
8. Parati G, Pomidossi G, Albini F, Malaspina D, Mancia G. Relationship of 24-hour blood pressure mean and variability to severity of target-organ damage in hypertension. J Hypertens. 1987;5(1):93–8. https://doi.org/10.1097/00004872-198702000-00013.
9. Santos ABS, et al. Prehypertension is associated with abnormalities of cardiac structure and function in the atherosclerosis risk in communities study. Am J Hypertens. 2016;29(5):568–74. https://doi.org/10.1093/ajh/hpv156.
10. Galderisi M, et al. Differences of myocardial systolic deformation and correlates of diastolic function in competitive rowers and young hypertensives: a speckle-tracking echocardiography study. J Am Soc Echocardiogr. 2010;23(11):1190–8. https://doi.org/10.1016/j.echo.2010.07.010.
11. de Simone G, Gottdiener JS, Chinali M, Maurer MS. Left ventricular mass predicts heart failure not related to previous myocardial infarction: the Cardiovascular Health Study. Eur Heart J. 2008;29(6):741–7. https://doi.org/10.1093/eurheartj/ehm605.
12. Neeland IJ, et al. Biomarkers of chronic cardiac injury and hemodynamic stress identify a malignant phenotype of left ventricular hypertrophy in the general population. J Am Coll Cardiol. 2013;61(2):187–95. https://doi.org/10.1016/j.jacc.2012.10.012.
13. Seliger SL, et al. Older adults, "malignant" left ventricular hypertrophy, and associated cardiac-specific biomarker phenotypes to identify the differential risk of new-onset reduced versus preserved ejection fraction heart failure: CHS (Cardiovascular Health Study). JACC Heart Fail. 2015;3(6):445–55. https://doi.org/10.1016/j.jchf.2014.12.018.
14. Brutsaert DL, Sys SU. Relaxation and diastole of the heart. Physiol Rev. 1989;69(4):1228–315. https://doi.org/10.1152/physrev.1989.69.4.1228.
15. Angeja BG, Grossman W. Evaluation and management of diastolic heart failure. Circulation. 2003;107(5):659–63. https://doi.org/10.1161/01.cir.0000053948.10914.49.
16. Zile MR, Brutsaert DL. New concepts in diastolic dysfunction and diastolic heart failure: part II: causal mechanisms and treatment. Circulation. 2002;105(12):1503–8. https://doi.org/10.1161/hc1202.105290.
17. Leite-Moreira AF, Bras-Silva C, Pedrosa CA, Rocha-Sousa AA. ET-1 increases distensibility of acutely loaded myocardium: a novel ETA and Na+/H+ exchanger-mediated

effect. Am J Physiol Heart Circ Physiol. 2003;284(4):1332–9. https://doi.org/10.1152/ajpheart.00715.2002.
18. Kass DA, Bronzwaer JGF, Paulus WJ. What mechanisms underlie diastolic dysfunction in heart failure? Circ Res. 2004;94(12):1533–42. https://doi.org/10.1161/01.RES.0000129254.25507.d6.
19. Leite-Moreira AF, Correia-Pinto J. Load as an acute determinant of end-diastolic pressure-volume relation. Am J Physiol Heart Circ Physiol. 2001;280(1):51–9. https://doi.org/10.1152/ajpheart.2001.280.1.H51.
20. Gandhi SK, et al. The pathogenesis of acute pulmonary edema associated with hypertension. N Engl J Med. 2001;344(1):17–22. https://doi.org/10.1056/NEJM200101043440103.
21. Leite-Moreira AF, Gillebert TC. The physiology of left ventricular pressure fall. Rev Port Cardiol. 2000;19(10):1015–21.
22. Izzo JL, Gradman AH. Mechanisms and management of hypertensive heart disease: from left ventricular hypertrophy to heart failure. Med Clin North Am. 2004;88(5):1257–71. https://doi.org/10.1016/j.mcna.2004.06.002.
23. Mitchell GF, et al. Determinants of elevated pulse pressure in middle-aged and older subjects with uncomplicated systolic hypertension: the role of proximal aortic diameter and the aortic pressure-flow relationship. Circulation. 2003;108(13):1592–8. https://doi.org/10.1161/01.CIR.0000093435.04334.1F.
24. Laks MM, Morady F. Norepinephrine--the myocardial hypertrophy hormone? Am Heart J. 1976;91(5):674–5. https://doi.org/10.1016/s0002-8703(76)80156-1.
25. Waeber B, Brunner HR. Cardiovascular hypertrophy: role of angiotensin II and bradykinin. J Cardiovasc Pharmacol. 1996;27(2):36–40. https://doi.org/10.1097/00005344-199600002-00008.
26. Ichihara S, Senbonmatsu T, Price E, Ichiki T, Gaffney FA, Inagami T. Angiotensin II type 2 receptor is essential for left ventricular hypertrophy and cardiac fibrosis in chronic angiotensin II-induced hypertension. Circulation. 2001;104(3):346–51. https://doi.org/10.1161/01.cir.104.3.346.
27. Schmieder RE, Schlaich MP. Comparison of therapeutic studies on regression of left ventricular hypertrophy. Adv Exp Med Biol. 1997;432:191–8. https://doi.org/10.1007/978-1-4615-5385-4_21.
28. Sun Y, Weber KT. Cardiac remodelling by fibrous tissue: role of local factors and circulating hormones. Ann Med. 1998;30(Suppl 1):3–8.
29. Weber KT. Metabolic responses of extracellular matrix in tissue repair. Ann Med. 1997;29(4):333–8. https://doi.org/10.3109/07853899708999357.
30. Galderisi M, Celentano A, Garofalo M, Tammaro P, Crivaro M, de Divitiis O. Parental left ventricular hypertrophy predicts Doppler-derived diastolic filling in young offspring of hypertensive subjects. J Hypertens. 1993;11(5):S76–7.
31. Arnett DK, de Fuentes L, Broeckel U. Genes for left ventricular hypertrophy. Curr Hypertens Rep. 2004;6(1):36–41. https://doi.org/10.1007/s11906-004-0009-5.
32. Arnett DK, et al. Sibling correlation of left ventricular mass and geometry in hypertensive African Americans and whites: the HyperGEN study. Hypertension Genetic Epidemiology Network. Am J Hypertens. 2001;14(12):1226–30. https://doi.org/10.1016/s0895-7061(01)02200-2.
33. Arnett DK, et al. Linkage of left ventricular contractility to chromosome 11 in humans: The HyperGEN Study. Hypertension. 2001;38(4):767–72. https://doi.org/10.1161/hy1001.092650.
34. Balaney B, et al. Invasive validation of the echocardiographic assessment of left ventricular filling pressures using the 2016 diastolic guidelines: head-to-head comparison with the 2009 guidelines. J Am Soc Echocardiogr. 2018;31(1):79–88. https://doi.org/10.1016/J.ECHO.2017.09.002.
35. Kass DA. Assessment of diastolic dysfunction. Invasive modalities. Cardiol Clin. 2000;18(3):571–86. https://doi.org/10.1016/S0733-8651(05)70162-4.

36. Paulus WJ, et al. How to diagnose diastolic heart failure: a consensus statement on the diagnosis of heart failure with normal left ventricular ejection fraction by the Heart Failure and Echocardiography Associations of the European Society of Cardiology. Eur Heart J. 2007;28(20):2539–50. https://doi.org/10.1093/EURHEARTJ/EHM037.
37. Dorfs S, et al. Pulmonary capillary wedge pressure during exercise and long-term mortality in patients with suspected heart failure with preserved ejection fraction. Eur Heart J. 2014;35(44):3103–12. https://doi.org/10.1093/EURHEARTJ/EHU315.
38. Ha JW, Andersen OS, Smiseth OA. Diastolic stress test: invasive and noninvasive testing. JACC Cardiovasc Imaging. 2020;13(1):272–82. https://doi.org/10.1016/J.JCMG.2019.01.037.
39. Smiseth OA, et al. Multimodality imaging in patients with heart failure and preserved ejection fraction: an expert consensus document of the European Association of Cardiovascular Imaging. Eur Heart J Cardiovasc Imaging. 2022;23(2):e34–61. https://doi.org/10.1093/ehjci/jeab154.
40. Borlaug BA, Nishimura RA, Sorajja P, Lam CSP, Redfield MM. Exercise hemodynamics enhance diagnosis of early heart failure with preserved ejection fraction. Circ Heart Fail. 2010;3(5):588–95. https://doi.org/10.1161/CIRCHEARTFAILURE.109.930701.
41. Ogilvie LM, et al. Hemodynamic assessment of diastolic function for experimental models. Am J Physiol Heart Circ Physiol. 2020;318(5):H1139–58. https://doi.org/10.1152/AJPHEART.00705.2019.
42. Pak PH, Maughan WL, Baughman KL, Kass DA. Marked discordance between dynamic and passive diastolic pressure-volume relations in idiopathic hypertrophic cardiomyopathy. Circulation. 1996;94(1):52–60. https://doi.org/10.1161/01.CIR.94.1.52.
43. Dauterman K, et al. Contribution of external forces to left ventricular diastolic pressure. Implications for the clinical use of the Starling law. Ann Intern Med. 1995;122(10):737–42. https://doi.org/10.7326/0003-4819-122-10-199505150-00001.
44. Hirota Y. A clinical study of left ventricular relaxation. Circulation. 1980;62(4):756–63. https://doi.org/10.1161/01.CIR.62.4.756.
45. Weiss JL, Frederiksen JW, Weisfeldt ML. Hemodynamic determinants of the time-course of fall in canine left ventricular pressure. J Clin Invest. 1976;58(3):751. https://doi.org/10.1172/JCI108522.
46. Nagueh SF, et al. Recommendations for the evaluation of left ventricular diastolic function by echocardiography: an update from the American Society of Echocardiography and the European Association of Cardiovascular Imaging. J Am Soc Echocardiogr. 2016;29(4):277–314. https://doi.org/10.1016/J.ECHO.2016.01.011.
47. Thomas JD, Weyman AE. Echocardiographic Doppler evaluation of left ventricular diastolic function: physics and physiology. Circulation. 1991;84(3):977–90. https://doi.org/10.1161/01.CIR.84.3.977.
48. Mitter SS, Shah SJ, Thomas JD. A test in context: E/A and E/e′ to assess diastolic dysfunction and LV filling pressure. J Am Coll Cardiol. 2017;69(11):1451–64. https://doi.org/10.1016/J.JACC.2016.12.037.
49. Choong CY, Abascal VM, Thomas JD, Luis Guerrero J, McGlew S, Weyman EAE. Combined influence of ventricular loading and relaxation on the transmitral flow velocity profile in dogs measured by Doppler echocardiography. Circulation. 1988;78(3):672–83. https://doi.org/10.1161/01.CIR.78.3.672.
50. Bryg RJ, Williams GA, Labovitz AJ. Effect of aging on left ventricular diastolic filling in normal subjects. Am J Cardiol. 1987;59(9):971–4. https://doi.org/10.1016/0002-9149(87)91136-2.
51. Bryg RJ, Pearson AC, Williams GA, Labovitz AJ. Left ventricular systolic and diastolic flow abnormalities determined by doppler echocardiography in obstructive hypertrophic cardiomyopathy. Am J Cardiol. 1987;59(9):925–31. https://doi.org/10.1016/0002-9149(87)91127-1.
52. Whalley GA, et al. Pseudonormal mitral filling pattern predicts hospital re-admission in patients with congestive heart failure. J Am Coll Cardiol. 2002;39(11):1787–95. https://doi.org/10.1016/S0735-1097(02)01868-5.
53. Nijland F, Kamp O, Karreman AJP, Van Eenige MJ, Visser CA. Prognostic implications of restrictive left ventricular filling in acute myocardial infarction: a Serial Doppler

Echocardiographic Study. J Am Coll Cardiol. 1997;30(7):1618–24. https://doi.org/10.1016/S0735-1097(97)00369-0.
54. Wijbenga AAM, et al. Potentials and limitations of the valsalva maneuver as a method of differentiating between normal and pseudonormal left ventricular filling patterns. Am J Cardiol. 1999;84(1):76–81. https://doi.org/10.1016/S0002-9149(99)00195-2.
55. Giannuzzi P, et al. Independent and incremental prognostic value of doppler-derived mitral deceleration time of early filling in both symptomatic and asymptomatic patients with left ventricular dysfunction. J Am Coll Cardiol. 1996;28(2):383–90. https://doi.org/10.1016/0735-1097(96)00163-5.
56. Lam CSP, Han L, Ha JW, Oh JK, Ling LH. The mitral L wave: a marker of pseudonormal filling and predictor of heart failure in patients with left ventricular hypertrophy. J Am Soc Echocardiogr. 2005;18(4):336–41. https://doi.org/10.1016/J.ECHO.2004.10.019.
57. Nagueh SF, Kopelen HA, Quiñones MA. Assessment of left ventricular filling pressures by Doppler in the presence of atrial fibrillation. Circulation. 1996;94(9):2138–45. https://doi.org/10.1161/01.CIR.94.9.2138.
58. Opdahl A, et al. Determinants of left ventricular early-diastolic lengthening velocity independent contributions from left ventricular relaxation, restoring forces, and lengthening load. Circulation. 2009;119(19):2578–86. https://doi.org/10.1161/CIRCULATIONAHA.108.791681.
59. Nagueh SF, Sun H, Kopelen HA, Middleton KJ, Khoury DS. Hemodynamic determinants of the mitral annulus diastolic velocities by tissue Doppler. J Am Coll Cardiol. 2001;37(1):278–85. https://doi.org/10.1016/S0735-1097(00)01056-1.
60. Nagueh SF, et al. Echocardiographic evaluation of hemodynamics in patients with decompensated systolic heart failure. Circ Cardiovasc Imaging. 2011;4(3):220–7. https://doi.org/10.1161/CIRCIMAGING.111.963496.
61. Sharifov OF, Schiros CG, Aban I, Denney TS, Gupta H. Diagnostic accuracy of tissue Doppler index E/e' for evaluating left ventricular filling pressure and diastolic dysfunction/heart failure with preserved ejection fraction: a systematic review and meta-analysis. J Am Heart Assoc. 2016;5:1. https://doi.org/10.1161/JAHA.115.002530.
62. Lester SJ, Ryan EW, Schiller NB, Foster E. Best method in clinical practice and in research studies to determine left atrial size. Am J Cardiol. 1999;84(7):829–32. https://doi.org/10.1016/S0002-9149(99)00446-4.
63. Lim TK, Ashrafian H, Dwivedi G, Collinson PO, Senior R. Increased left atrial volume index is an independent predictor of raised serum natriuretic peptide in patients with suspected heart failure but normal left ventricular ejection fraction: implication for diagnosis of diastolic heart failure. Eur J Heart Fail. 2006;8(1):38–45. https://doi.org/10.1016/J.EJHEART.2005.05.008.
64. Beinart R, et al. Long-term prognostic significance of left atrial volume in acute myocardial infarction. J Am Coll Cardiol. 2004;44(2):327–34. https://doi.org/10.1016/J.JACC.2004.03.062.
65. Caballero L, et al. Echocardiographic reference ranges for normal cardiac Doppler data: results from the NORRE Study. Eur Heart J Cardiovasc Imaging. 2015;16(9):1031–41. https://doi.org/10.1093/EHJCI/JEV083.
66. Morris DA, et al. Potential usefulness and clinical relevance of adding left atrial strain to left atrial volume index in the detection of left ventricular diastolic dysfunction. JACC Cardiovasc Imaging. 2018;11(10):1405–15. https://doi.org/10.1016/J.JCMG.2017.07.029.
67. Buffle E, et al. Added value of pulmonary venous flow Doppler assessment in patients with preserved ejection fraction and its contribution to the diastolic grading paradigm. Eur Heart J Cardiovasc Imaging. 2015;16(11):1191–7. https://doi.org/10.1093/EHJCI/JEV126.
68. Fadel BM, et al. Spectral Doppler interrogation of the pulmonary veins for the diagnosis of cardiac disorders: a comprehensive review. J Am Soc Echocardiogr. 2021;34(3):223–36. https://doi.org/10.1016/J.ECHO.2020.09.012.
69. Parasuraman S, et al. Assessment of pulmonary artery pressure by echocardiography—a comprehensive review. Int J Cardiol Heart Vasc. 2016;12:45. https://doi.org/10.1016/J.IJCHA.2016.05.011.

70. Galiè N, et al. 2015 ESC/ERS guidelines for the diagnosis and treatment of pulmonary hypertension. Eur Respir J. 2015;46(4):903–75. https://doi.org/10.1183/13993003.01032-2015.
71. Masuyama T, Kodama K, Kitabatake A, Sato H, Nanto S, Inoue M. Continuous-wave Doppler echocardiographic detection of pulmonary regurgitation and its application to noninvasive estimation of pulmonary artery pressure. Circulation. 1986;74(3):484–92. https://doi.org/10.1161/01.CIR.74.3.484.
72. Chubuchny V, et al. A novel echocardiographic method for estimation of pulmonary artery wedge pressure and pulmonary vascular resistance. ESC Heart Fail. 2021;8(2):1216–29. https://doi.org/10.1002/ehf2.13183.
73. Barbier P, Grimaldi A, Alimento M, Berna G, Guazzi MD. Echocardiographic determinants of mitral early flow propagation velocity. Am J Cardiol. 2002;90(6):613–9. https://doi.org/10.1016/S0002-9149(02)02565-1.
74. Nakatani S, Yoshitomi H, Wada K, Beppu S, Nagata S, Miyatake K. Noninvasive estimation of left ventricular end-diastolic pressure using transthoracic Doppler-determined pulmonary venous atrial flow reversal. Am J Cardiol. 1994;73(13):1017–8. https://doi.org/10.1016/0002-9149(94)90162-7.
75. Rohde LE, Palombini DV, Polanczyk CA, Goldraich LA, Clausell N. A hemodynamically oriented echocardiography-based strategy in the treatment of congestive heart failure. J Card Fail. 2007;13(8):618–25. https://doi.org/10.1016/J.CARDFAIL.2007.05.003.
76. Prasad SB, Holland DJ, Atherton JJ, Whalley EG. New diastology guidelines: evolution, validation and impact on clinical practice. Heart Lung Circ. 2019;28(9):1411–20. https://doi.org/10.1016/j.hlc.2019.03.013.
77. van de Bovenkamp AA, et al. Validation of the 2016 ASE/EACVI guideline for diastolic dysfunction in patients with unexplained dyspnea and a preserved left ventricular ejection fraction. J Am Heart Assoc. 2021;10(18):21165. https://doi.org/10.1161/JAHA.121.021165.
78. Smiseth OA, Baron T, Marino PN, Marwick TH, Flachskampf FA. Imaging of the left atrium: pathophysiology insights and clinical utility. Eur Heart J Cardiovasc Imaging. 2021;23(1):2–13. https://doi.org/10.1093/EHJCI/JEAB191.
79. Voigt J-U, et al. Definitions for a common standard for 2D speckle tracking echocardiography: consensus document of the EACVI/ASE/Industry Task Force to standardize deformation imaging. Eur Heart J Cardiovasc Imaging. 2015;16(1):1–11. https://doi.org/10.1093/ehjci/jeu184.
80. Lang RM, et al. Multimodality imaging of diseases of the thoracic aorta in adults: from the American Society. Eur Heart J Cardiovasc Imaging. 2015;28(2):119–82. https://doi.org/10.1093/ehjci/jev014.
81. Bendiab NST, et al. Factors associated with global longitudinal strain decline in hypertensive patients with normal left ventricular ejection fraction. Eur J Prev Cardiol. 2017;24(14):1463–72. https://doi.org/10.1177/2047487317721644.
82. Wang J, Khoury DS, Thohan V, Torre-Amione G, Nagueh SF. Global diastolic strain rate for the assessment of left ventricular relaxation and filling pressures. Circulation. 2007;115(11):1376–83. https://doi.org/10.1161/CIRCULATIONAHA.106.662882.
83. Dokainish H, Sengupta R, Pillai M, Bobek J, Lakkis N. Usefulness of new diastolic strain and strain rate indexes for the estimation of left ventricular filling pressure. Am J Cardiol. 2008;101(10):1504–9. https://doi.org/10.1016/j.amjcard.2008.01.037.
84. Meluzin J, et al. Estimation of left ventricular filling pressures by speckle tracking echocardiography in patients with idiopathic dilated cardiomyopathy. Eur J Echocardiogr. 2011;12(1):11–8. https://doi.org/10.1093/ejechocard/jeq088.
85. Kimura K, et al. Speckle tracking global strain rate E/E′ predicts LV filling pressure more accurately than traditional tissue Doppler E/E′. Echocardiography. 2012;29(4):404–10. https://doi.org/10.1111/j.1540-8175.2011.01587.x.
86. Notomi Y, et al. Ventricular untwisting: a temporal link between left ventricular relaxation and suction. Am J Physiol Heart Circ Physiol. 2008;294(1):H505–13. https://doi.org/10.1152/ajpheart.00975.2007.

87. Wang J, Khoury DS, Yue Y, Torre-Amione G, Nagueh SF. Left ventricular untwisting rate by speckle tracking echocardiography. Circulation. 2007;116(22):2580–6. https://doi.org/10.1161/CIRCULATIONAHA.107.706770.
88. Cameli M, et al. Multicentric atrial strain comparison between two different modalities: MASCOT HIT study. Diagn Basel Switz. 2020;10(11):E946. https://doi.org/10.3390/diagnostics10110946.
89. Wakami K, et al. Correlation between left ventricular end-diastolic pressure and peak left atrial wall strain during left ventricular systole. J Am Soc Echocardiogr. 2009;22(7):847–51. https://doi.org/10.1016/j.echo.2009.04.026.
90. Kosmala W, Jellis CL, Marwick TH. Exercise limitation associated with asymptomatic left ventricular impairment: analogy with stage B heart failure. J Am Coll Cardiol. 2015;65(3):257–66. https://doi.org/10.1016/j.jacc.2014.10.044.
91. Bonow RO. Radionuclide angiographic evaluation of left ventricular diastolic function. Circulation. 1991;84(3):208–15.
92. Akincioglu C, et al. Assessment of diastolic function using 16-frame 99mTc-sestamibi gated myocardial perfusion SPECT: normal values. J Nucl Med. 2005;46(7):1102–8.
93. Lee KJ, et al. Normalised radionuclide measures of left ventricular diastolic function. Eur J Nucl Med. 1989;15(3):123–7. https://doi.org/10.1007/BF00254623.
94. Chamsi PMA, Zhan Y, Debs D, Shah DJ. CMR in the evaluation of diastolic dysfunction and phenotyping of HFpEF. JACC Cardiovasc Imaging. 2020;13:283–96. https://doi.org/10.1016/j.jcmg.2019.02.031.
95. Barison A, Aimo A, Todiere G, Grigoratos C, Aquaro GD, Emdin M. Cardiovascular magnetic resonance for the diagnosis and management of heart failure with preserved ejection fraction. Heart Fail Rev. 2022;27(1):191–205. https://doi.org/10.1007/s10741-020-09998-w.
96. Armstrong AC, Gidding S, Gjesdal O, Wu C, Bluemke DA, Lima JAC. LV mass assessed by echocardiography and CMR, cardiovascular outcomes, and medical practice. JACC Cardiovasc Imaging. 2012;5(8):837–48. https://doi.org/10.1016/j.jcmg.2012.06.003.
97. Bacharova L, et al. Determinants of discrepancies in detection and comparison of the prognostic significance of left ventricular hypertrophy by electrocardiogram and cardiac magnetic resonance imaging. Am J Cardiol. 2015;115(4):515–22. https://doi.org/10.1016/j.amjcard.2014.11.037.
98. Kawel-Boehm N, et al. Normal values for cardiovascular magnetic resonance in adults and children. J Cardiovasc Magn Reson. 2015;17(1):29. https://doi.org/10.1186/s12968-015-0111-7.
99. Maceira AM, Cosín-Sales J, Roughton M, Prasad SK, Pennell DJ. Reference left atrial dimensions and volumes by steady state free precession cardiovascular magnetic resonance. J Cardiovasc Magn Reson. 2010;12(1):65. https://doi.org/10.1186/1532-429X-12-65.
100. Baggiano A, et al. Role of CMR mapping techniques in cardiac hypertrophic phenotype. Diagnostics. 2020;10(10):770. https://doi.org/10.3390/diagnostics10100770.
101. Kawaji K, et al. Automated segmentation of routine clinical cardiac magnetic resonance imaging for assessment of left ventricular diastolic dysfunction. Circ Cardiovasc Imaging. 2009;2(6):476–84. https://doi.org/10.1161/CIRCIMAGING.109.879304.
102. Rathi VK, et al. Routine evaluation of left ventricular diastolic function by cardiovascular magnetic resonance: a practical approach. J Cardiovasc Magn Reson. 2008;10(1):36. https://doi.org/10.1186/1532-429X-10-36.
103. Buss SJ, et al. Classification of diastolic function with phase-contrast cardiac magnetic resonance imaging: validation with echocardiography and age-related reference values. Clin Res Cardiol. 2014;103(6):441–50. https://doi.org/10.1007/s00392-014-0669-3.
104. Paelinck BP, et al. Feasibility of tissue magnetic resonance imaging: a pilot study in comparison with tissue Doppler imaging and invasive measurement. J Am Coll Cardiol. 2005;45(7):1109–16. https://doi.org/10.1016/j.jacc.2004.12.051.
105. Yoneyama K, Venkatesh BA, Bluemke DA, McClelland RL, Lima JAC. Cardiovascular magnetic resonance in an adult human population: serial observations from the multi-ethnic study of atherosclerosis. J Cardiovasc Magn Reson. 2017;19(1):52. https://doi.org/10.1186/s12968-017-0367-1.

106. Onishi T, et al. Global longitudinal strain and global circumferential strain by speckle-tracking echocardiography and feature-tracking cardiac magnetic resonance imaging: comparison with left ventricular ejection fraction. J Am Soc Echocardiogr. 2015;28(5):587–96. https://doi.org/10.1016/j.echo.2014.11.018.
107. Wachtell K, et al. Prognostic significance of left ventricular diastolic dysfunction in patients with left ventricular hypertrophy and systemic hypertension (the LIFE Study). Am J Cardiol. 2010;106(7):999–1005. https://doi.org/10.1016/j.amjcard.2010.05.032.
108. Peterson GE, et al. Relationship of left ventricular hypertrophy and diastolic function with cardiovascular and renal outcomes in African Americans with hypertensive chronic kidney disease. Hypertension. 2013;62(3):518–25. https://doi.org/10.1161/HYPERTENSIONAHA.111.00904.
109. Schillaci G, et al. Prognostic significance of left ventricular diastolic dysfunction in essential hypertension. J Am Coll Cardiol. 2002;39(12):2005–11. https://doi.org/10.1016/s0735-1097(02)01896-x.
110. Wang M, et al. Tissue Doppler imaging provides incremental prognostic value in patients with systemic hypertension and left ventricular hypertrophy. J Hypertens. 2005;23(1):183–91. https://doi.org/10.1097/00004872-200501000-00029.
111. Sharp ASP, et al. Tissue Doppler E/E' ratio is a powerful predictor of primary cardiac events in a hypertensive population: an ASCOT substudy. Eur Heart J. 2010;31(6):747–52. https://doi.org/10.1093/eurheartj/ehp498.
112. Nadruz W, Shah AM, Solomon SD. Diastolic dysfunction and hypertension. Med Clin North Am. 2017;101(1):7–17. https://doi.org/10.1016/j.mcna.2016.08.013.
113. Terpstra WF, et al. Long-term effects of amlodipine and lisinopril on left ventricular mass and diastolic function in elderly, previously untreated hypertensive patients: the ELVERA trial. J Hypertens. 2001;19(2):303–9. https://doi.org/10.1097/00004872-200102000-00018.
114. Müller-Brunotte R, Kahan T, Malmqvist K, Ring M, Edner M. Tissue velocity echocardiography shows early improvement in diastolic function with irbesartan and atenolol therapy in patients with hypertensive left ventricular hypertrophy. Results form the Swedish Irbesartan Left Ventricular Hypertrophy Investigation vs Atenolol (SILVHIA). Am J Hypertens. 2006;19(9):927–36. https://doi.org/10.1016/j.amjhyper.2006.02.009.
115. Solomon SD, et al. Effect of intensive versus standard blood pressure lowering on diastolic function in patients with uncontrolled hypertension and diastolic dysfunction. Hypertension. 2010;55(2):241–8. https://doi.org/10.1161/HYPERTENSIONAHA.109.138529.
116. Tapp RJ, et al. Differential effects of antihypertensive treatment on left ventricular diastolic function: an ASCOT (Anglo-Scandinavian Cardiac Outcomes Trial) substudy. J Am Coll Cardiol. 2010;55(17):1875–81. https://doi.org/10.1016/j.jacc.2009.11.084.
117. Hummel SL, et al. Low-sodium DASH diet improves diastolic function and ventricular-arterial coupling in hypertensive heart failure with preserved ejection fraction. Circ Heart Fail. 2013;6(6):1165–71. https://doi.org/10.1161/CIRCHEARTFAILURE.113.000481.
118. Carluccio E, et al. Effect of revascularizing viable myocardium on left ventricular diastolic function in patients with ischaemic cardiomyopathy. Eur Heart J. 2009;30(12):1501–9. https://doi.org/10.1093/eurheartj/ehp125.
119. Shah SJ, et al. Phenotype-specific treatment of heart failure with preserved ejection fraction: a multiorgan roadmap. Circulation. 2016;134(1):73–90. https://doi.org/10.1161/CIRCULATIONAHA.116.021884.

New Perspectives on Cardiac Magnetic Resonance Imaging in Hypertensive Patients with Heart Failure

16

Sebastian Onciul

Hypertensive heart disease (HHD) refers to structural and functional alterations of the heart in the context of arterial hypertension. Depending on the severity and duration of hypertension, HHD can be characterized by left ventricular hypertrophy (LVH), diastolic and eventually systolic LV dysfunction, diffuse myocardial fibrosis, and left atrial (LA) dilation and dysfunction [1].

Cardiovascular imaging aims to identify these alterations in order to modulate hypertension treatment and improve the prognostic value of hypertensive patients. Echocardiography is the first-line imaging modality, offering, in most cases, accurate information regarding LVH, diastolic dysfunction, and LA anatomy and function.

Cardiovascular magnetic resonance (CMR) is a unique imaging modality because it offers high-resolution imaging independently of the body habitus; it is the gold standard for measurement of the cardiac chambers' dimensions and function, and it allows for myocardial perfusion and tissue characterization—all in one single examination.

LV myocardial mass assessment by CMR is based on direct contouring of the LV epicardium and endocardium in serial short-axis slices. As it does not imply any geometrical assumptions, it is a highly accurate noninvasive measure of myocardial mass. An increased myocardial mass is associated with worse outcomes.

Furthermore, CMR parametric mapping is a powerful tool for myocardial tissue characterization, currently available in most CMR departments worldwide. T1 mapping techniques allow for identification and quantification of diffuse myocardial fibrosis, a pathological finding in patients with long-standing hypertension and LVH.

S. Onciul (✉)
Carol Davila University of Medicine and Pharmacy, Bucharest, Romania

Cardiology Department, Clinical Emergency Hospital, Bucharest, Romania
e-mail: sebastian.onciul@gmail.ro

M. Dorobantu et al. (eds.), *Hypertension and Heart Failure*, Updates in Hypertension and Cardiovascular Protection,
https://doi.org/10.1007/978-3-031-39315-0_16

16.1 Imaging of Myocardial Fibrosis

Myocardial fibrosis is a complication of long-standing arterial hypertension and is the pathophysiological link between high blood pressure, diastolic and systolic dysfunction, and clinical overt heart failure. Moreover, myocardial fibrosis represents the arrhythmogenic substrate, which explains the higher-than-normal incidence of ventricular arrhythmias in hypertensive patients.

Myocardial fibrosis can be accurately identified on endomyocardial biopsy and noninvasively by CMR. Histological studies of hypertensive hearts have shown that the amount of interstitial collagen accumulation is proportionate to the degree of myocyte hypertrophy and heart weight [2].

Traditionally, late gadolinium enhancement (LGE) imaging was used to detect and quantify myocardial fibrosis. As gadolinium is an extracellular contrast agent, it is mainly distributed in tissues with expanded extracellular space such as in myocardial focal fibrosis. Half of the hypertensive patients with LVH have focal myocardial fibrosis as identified on LGE imaging, and the amount of fibrosis is proportional to the degree of LV remodeling [3].

However, there is no specific LGE pattern in HHD. More frequently, focal nonsubendocardial areas of myocardial fibrosis may be encountered in the basal inferolateral and inferior LV walls (Figs. 16.1 and 16.2). Nonspecific focal fibrosis at the right ventricular insertion points can also be identified (Fig. 16.3). There is a phenotypic LGE imaging overlap even between HHD and athletic heart, with nonspecific focal myocardial fibrosis described in highly competitive athletes [4]. As in HHD, the localization of myocardial fibrosis is also in the basal inferior and lateral walls. In athletes, focal fibrosis seems to correlate with the peak exercise systolic blood pressure and race distances [4]. Therefore, the presence of LGE cannot accurately differentiate between HHD and athletic heart, and high blood pressure during exercise might be the physiological link between strenuous exercise and development of myocardial fibrosis in highly trained athletes.

Due to its inherent qualitative nature, LGE imaging can only detect and quantify scars and focal areas of myocardial fibrosis, but it cannot detect diffuse interstitial myocardial fibrosis. However, the latter can be identified and quantified using T1 mapping and extracellular volume (ECV) computation.

T1 mapping refers to pixelwise illustrations of absolute T1 relaxation times on a map (Figs. 16.1, 16.2, and 16.3) [5]. As such, having the potential to detect diffuse myocardial fibrosis, T1 mapping allows direct T1 quantification. Estimation of the extracellular volume (ECV) requires measurement of myocardial and blood T1 before and after administration of contrast agents as well as the patient's hematocrit value [5]. Apart from amyloid, an increased ECV is most often explained by excessive collagen deposition and is thus a highly accurate measure of diffuse interstitial myocardial fibrosis [5].

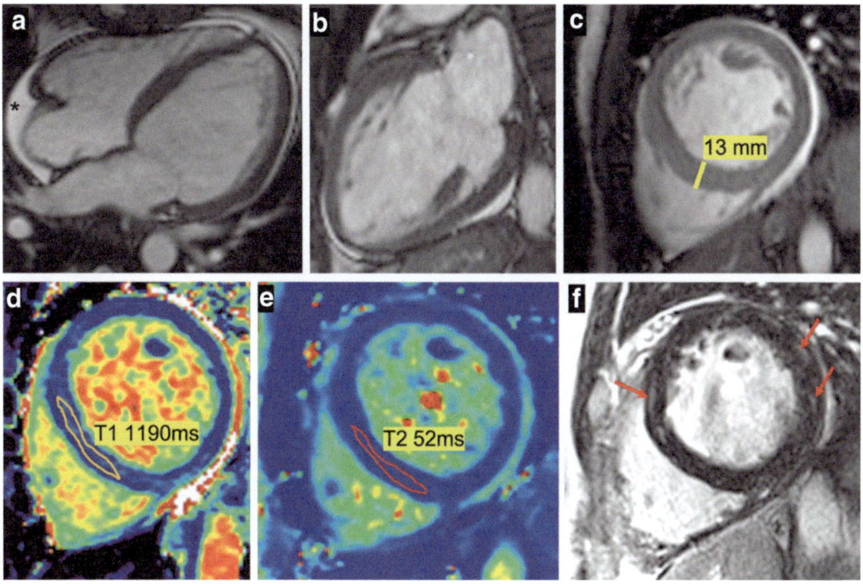

Fig. 16.1 Contrast-enhanced cardiovascular magnetic resonance in a 30-year-old hypertensive male with chronic kidney disease on hemodialysis. (**a–c**) Balanced-SSFP cine images diastolic frames in 4, 2 chambers and short axis, respectively. There is mild-to-moderate left ventricular (LV) hypertrophy with an interventricular septum (IVS) thickness of 13 mm and an indexed myocardial mass of 98 g/m^2. The LV is dilated (151 ml/m^2) with a mildly decreased LV ejection fraction of 47%. The right ventricular ejection fraction was normal. A small pericardial effusion is seen (the small asterisk). (**d**) T1 mapping shows increased native T1 values, reaching 1190 ms on the IVS. The calculated extracellular volume fraction (ECV) was 30%. (**e**) T2 mapping shows mildly increased T2 values (with a normal reference value ≤50 ms), suggesting a mild degree of myocardial edema. The increased T1 values are explained by both interstitial diffuse myocardial fibrosis and myocardial water content. (**f**) Late gadolinium enhancement (LGE) imaging shows small, linear areas of hyperenhancement with a nonischemic pattern, suggestive of focal replacement fibrosis. The overall CMR appearance is not suggestive of cardiac amyloidosis. A complete genetic testing for cardiomyopathies was negative

Kuruvilla et al. showed that hypertensive patients with LVH have higher ECV values than do those without LVH [6]. The work by Treibel et al. confirmed that a significant increase in the ECV can be detected only in LVH hypertensive patients and not in hypertensive patients with normal LV wall thickness [7]. An electrocardiogram (ECG)–CMR correlation study showed that patients with an ECG pattern of LV strain have higher indexed LV mass and more diffuse interstitial fibrosis (quantified by ECV) compared to hypertensive subjects without ECG strain [8]. Thus, LVH is associated with increased diffuse myocardial fibrosis, which can be noninvasively quantified by T1 mapping techniques.

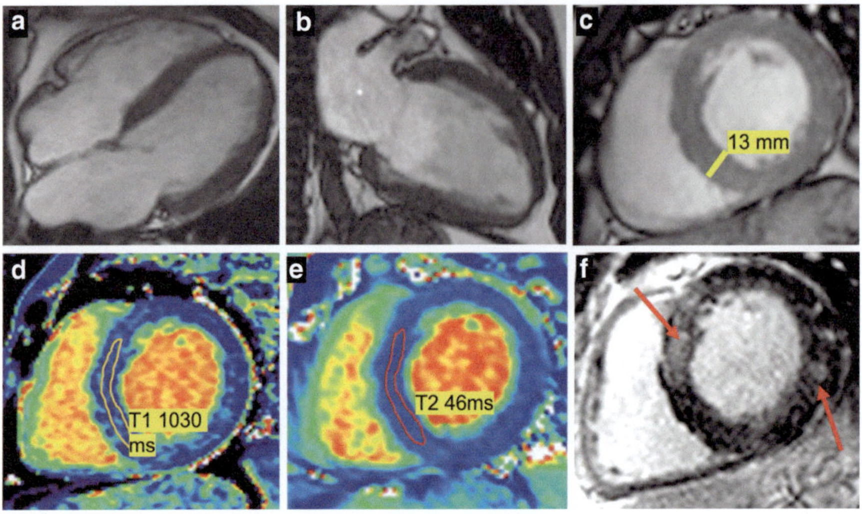

Fig. 16.2 Contrast-enhanced cardiovascular magnetic resonance in a 45-year-old male with secondary arterial hypertension due to primary hyperaldosteronism. (**a–c**) Balanced SSFP cine images diastolic frames in 4, 2 chambers and short axis, respectively. There is mild left ventricular (LV) hypertrophy with an interventricular septum thickness (IVS) of 13 mm. Both LV and the right ventricle have normal dimensions and ejection fraction. (**d**) T1 mapping shows mildly increased T1 values, reaching 1030 ms on the IVS. (**e**) T2 mapping shows normal T2 values, excluding myocardial edema. (**f**) Late gadolinium enhancement (LGE) shows patchy areas of mid-myocardial hyperenhancement in the basal IVS and lateral wall (red arrows), suggestive of nonischemic replacement myocardial fibrosis. The overall CMR appearance is suggestive of hypertensive heart disease

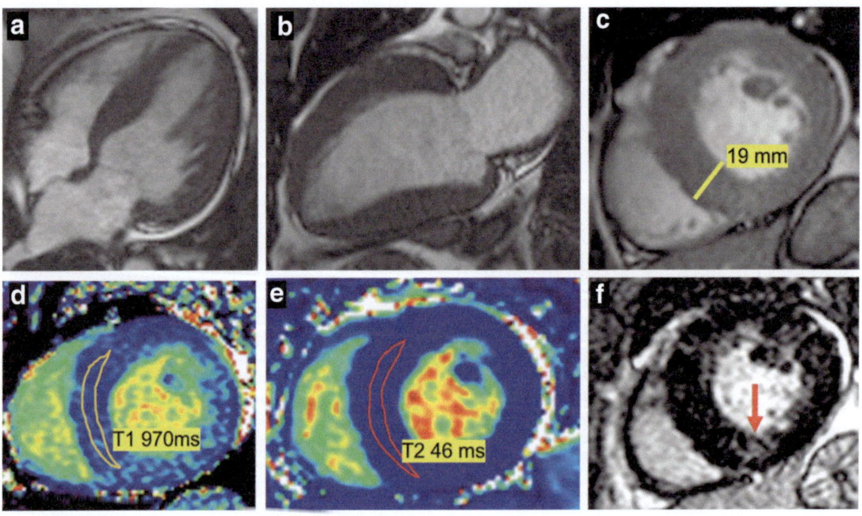

16.2 The Role of CMR in the Assessment of LV Diastolic Function

Alteration in LV diastolic function is common in hypertensive patients even in the absence of LVH. Hypertensive patients with diastolic dysfunction have an increased cardiovascular risk, independently of LV mass and ambulatory BP [9].

Undoubtedly, echocardiography is the mainstay of noninvasive imaging assessment of LV diastolic function and filling pressures. It is a simple, noninvasive, and easily repeatable method of assessing LV diastolic filling. Moreover, the feasibility of assessing diastolic function during physical exercise is one of the main advantages of echocardiography. On the other hand, assessment of diastolic function by CMR is cumbersome, time-consuming, and is thus not implemented in everyday clinical practice.

Assessment of diastolic function by echocardiography implies measurement of several Doppler-derived parameters such as mitral E and A velocities, septal E velocity, and tricuspid regurgitation velocity. All these parameters can be measured during a CMR examination by phase-contrast imaging, a well-defined imaging technique for flow measurement. Studies have shown that the above flow parameters measured by phase-contrast imaging correlate well with echocardiography; however, the acquisition is time-consuming and the lower temporal resolution of CMR systematically underestimates the velocities compared to echocardiography [10].

In recent years, efforts have been made to add value to CMR by searching for imaging parameters of diastolic function. Currently, there are several CMR parameters that assess LV diastolic dysfunction, highly similar to echocardiography: LV mass and hypertrophy, LA size and function, mitral inflow and pulmonary venous velocity profiles, and myocardial deformation imaging with strain [10].

A recent study has enrolled patients with suspected HF who underwent right heart catheterization (RHC), CMR, and transthoracic echocardiography (TTE)

Fig. 16.3 Contrast-enhanced cardiovascular magnetic resonance (CMR) in a 42-year-old patient with severe arterial hypertension (mean blood pressure 200/100 mmHg on ambulatory blood pressure monitoring). His brain MRI showed changes compatible with a diagnosis of hypertensive angiopathy. (**a–c**) Balanced-SSFP diastolic frames of cine images in 4, 2 chambers and short axis, respectively. There is severe left ventricle (LV) hypertrophy with a maximal thickness of the interventricular septum (IVS) of 19 mm. The left ventricle is dilated (120 ml/m^2) with a normal ejection fraction (EF 58%). (**d**) T1 mapping shows normal native T1 values of 970 ms at the level of the IVS, suggesting the absence of diffuse interstitial myocardial fibrosis. (**e**) T2 mapping shows normal T2 values, suggesting the absence of myocardial edema. (**f**) Late gadolinium enhancement (LGE) imaging shows focal hyperenhancement at the inferior insertion point, suggestive of nonspecific myocardial fibrosis. In the context of coexistence of other target organ lesions (cerebral hypertensive angiopathy), the CMR phenotype is compatible with a diagnosis of hypertensive heart disease

within 24 h of each other [11]. Garg et al. showed that only LV mass and left atrial volume measured by CMR are associated with LV filling pressures. They elegantly demonstrated that CMR-modeled pulmonary capillary wedge pressure (PCWP) is superior to both TTE, in classifying patients as normal, and raised filling pressures [11]. Furthermore, they showed that CMR-modeled PCWP is associated with an increased risk of death.

LV time–volume filling curves can be obtained from LV cine volumetric images, which are acquired during a standard CMR examination. Parameters such as peak filling rates, time to peak filling, and diastolic volume recovery can be obtained, but these are time-consuming and have not yet been used in clinical practice [10].

16.3 The Added Value of CMR-Derived Myocardial Strain Imaging

Fast strain-encoded (SENC) imaging is an advanced CMR method used for quantification of strain at free breathing and with high reproducibility. Using out-of-plane phase encoding gradients along the slice-selected direction, global longitudinal and circumferential strain can be calculated [12]. Giusca et al. showed that multiparametric CMR combining LGE pattern and SENC-derived myocardial deformation offers high precision for the differentiation between hypertrophic cardiomyopathy (HCM) and HHD and HCM and amyloidosis. In the abovementioned study, the athletes and healthy controls had similar myocardial strain values, as opposed to HHD and HCM patients, who showed worse strain values. This observation may be useful in differentiating HHD from athlete heart.

Cardiovascular magnetic resonance myocardial feature tracking (CMR-FT) has been a recent advanced imaging modality employed for the assessment of measures of myocardial deformation such as strain, torsion, and dyssynchrony [13]. The main advantage of CMR-FT is that no special acquisition is required, but, instead, the analysis can be performed on routinely acquired ECG-triggered steady-state free precession (SSFP) CMR images [14].

Feature tracking-derived myocardial deformation parameters can identify subclinical functional alterations in asymptomatic hypertensive patients. Compared with controls, the LV peak diastolic strain rate in the radial, circumferential, and longitudinal directions and the LA reservoir and conduit function are significantly impaired in Hypertension (HTN) patients regardless of LVH [15].

CMR-FT-derived global longitudinal strain is significantly lower in HCM than in HHD [16]. Therefore, CMR-derived myocardial strain may be incorporated into future multiparametric scores that will be able to accurately differentiate between HHD and other etiologies of LVH.

Assessment of left atrial deformation by CMR in hypertensive patients is also feasible. One study showed that in diabetic patients with hypertension, the LA strain is ameliorated compared with that in diabetic normotensive patients. This may be explained through a compensatory LA mechanism in the context of increased LV stiffness. Furthermore, diuretic treatment was significantly related to greater LA global strain. However, as the authors mention, whether diuretics can prevent, delay, or reverse the impairment of LA strain requires further study [17].

16.4 Differentiation Among Various LVH Causes

One of the main indications for performing CMR in a hypertensive patient is to differentiate among other possible etiologies of LVH. Accurate distinction among HHD, hypertrophic cardiomyopathy (HCM), athlete heart, or cardiac amyloidosis may be difficult, and, consequently, it represents a hot topic for research. As there is no single CMR parameter able to accurately differentiate among these entities, recent studies have focused on the multiparametric approach of the CMR, including newly developed magnetic resonance imaging (MRI) sequences [12].

Research efforts were made to identify quantitative CMR parameters for defining the different etiologies of LVH both at 1.5 T and 3 T [18]. Using native T1 cutoff values for differentiating between HHD, HCM, and amyloidosis is appealing. However, as both sarcomeric HCM and HHD share the common pathophysiological pathway of diffuse interstitial myocardial fibrosis, one would expect similar native T1 values, thus making a clear distinction between HHD and HCM difficult. Nevertheless, it was demonstrated that native T1 and extracellular volume fraction are significantly higher in HCM patients compared with patients with hypertension, including in a subgroup of hypertensive patients with an LV wall thickness >15 mm [19].

"Cardiac amyloidosis (CA)" is an infiltrative cardiomyopathy characterized by extracellular accumulation of amyloid molecules, which results in expansion of the myocardial interstitial space and LVH. Among all LVH etiologies, cardiac amyloidosis has probably the most distinctive CMR imaging phenotype. Conventional CMR imaging using LGE shows a typical pattern of a diffuse, mostly subendocardial to transmural pattern of hyperenhancement that predominantly affects all basal to mid-LV segments [20].

Similarly, T1 mapping has unique characteristics, which can be easily differentiated from other etiologies of LVH. Because of the severe interstitial expansion, the native T1 values are characteristically extremely high in CA. Baggiano et al. propose the use of native T1 cutoff values to exclude or confirm CA: a native T1 <1036 ms was associated with a 98% negative predictive value for CA, whereas a native T1 >1164 ms was associated with a 98% positive predictive value for CA [21].

"Uremic cardiomyopathy" may be regarded as a subtype of HHD characterized by a unique pathophysiology, and, consequently, specific myocardial alterations would be expected in these cases. Patients with chronic kidney disease develop LVH, myocardial fibrosis, and a degree of myocardial edema that could be identified and quantified by T1 and T2 mapping techniques.

T1 mapping can quantify diffuse interstitial myocardial fibrosis. However, T1 time is a relatively nonspecific measure, as it is increased by both myocardial fibrosis and edema. On the other hand, high T2 values are characteristic of myocardial edema. Therefore, using both T1 and T2 mapping, one can accurately characterize the myocardium with respect to the amount of fibrosis and myocardial water content.

A degree of myocardial edema was demonstrated by CMR in uremic cardiomyopathy [22, 23]. Myocardial edema quantification by T2 mapping has identified higher T2 values in uremic cardiomyopathy as compared to HCM, suggesting higher myocardial water content in CKD associated LVH. Moreover, a reduction of T2 values (a decrease in myocardial water content) was noted after volume removal during hemodialysis [22, 24]. Native T2 was the strongest independent myocardial measure discriminating the CKD cohort from HCM patients (area under the curve (AUC) 0.891; $p<0.001$) [22]. In conclusion, T1 and T2 mapping should be used as complementary techniques for characterization of diffuse interstitial myocardial fibrosis and myocardial edema.

References

1. Tadic M, Cuspidi C, Plein S, Milivojevic IG, Wang DW, Grassi G, et al. Comprehensive assessment of hypertensive heart disease: cardiac magnetic resonance in focus. Heart Fail Rev. 2020;26(6):1383–90.
2. Rossi MA. Pathologic fibrosis and connective tissue matrix in left ventricular hypertrophy due to chronic arterial hypertension in humans. J Hypertens. 1998;16(7):1031. Available from https://journals.lww.com/jhypertension/Fulltext/1998/16070/Pathologic_fibrosis_and_connective_tissue_matrix.18.aspx.
3. Rudolph A, Abdel-Aty H, Bohl S, Boyé P, Zagrosek A, Dietz R, et al. Noninvasive detection of fibrosis applying contrast-enhanced cardiac magnetic resonance in different forms of left ventricular hypertrophy relation to remodeling. J Am Coll Cardiol. 2009;53:284.
4. Tahir E, Starekova J, Muellerleile K, von Stritzky A, Münch J, Avanesov M, et al. Myocardial fibrosis in competitive triathletes detected by contrast-enhanced CMR correlates with exercise-induced hypertension and competition history. JACC Cardiovasc Imaging. 2018;11(9):1260–70. Available from https://www.sciencedirect.com/science/article/pii/S1936878X17309816.
5. Haaf P, Garg P, Messroghli DR, Broadbent DA, Greenwood JP, Plein S. Cardiac T1 mapping and extracellular volume (ECV) in clinical practice: a comprehensive review. J Cardiovasc Magn Reson. 2016;18(1):89. https://doi.org/10.1186/s12968-016-0308-4.
6. Kuruvilla S, Janardhanan R, Antkowiak P, Keeley EC, Adenaw N, Brooks J, et al. Increased extracellular volume and altered mechanics are associated with LVH in hypertensive heart disease, not hypertension alone. JACC Cardiovasc Imaging. 2015;8(2):172–80.
7. Treibel TA, Zemrak F, Sado DM, Banypersad SM, White SK, Maestrini V, et al. Extracellular volume quantification in isolated hypertension - changes at the detectable limits? J Cardiovasc Magn Reson. 2015;17(1):74. Available from: http://www.ncbi.nlm.nih.gov/pmc/articles/PMC4534050/.
8. Rodrigues JCL, Amadu AM, Ghosh Dastidar A, McIntyre B, Szantho GV, Lyen S, et al. ECG strain pattern in hypertension is associated with myocardial cellular expansion and diffuse interstitial fibrosis: a multi-parametric cardiac magnetic resonance study. Eur Heart J Cardiovasc Imaging. 2017;18(4):441–50. https://doi.org/10.1093/ehjci/jew117.
9. Schillaci G, Pasqualini L, Verdecchia P, Vaudo G, Marchesi S, Porcellati C, et al. Prognostic significance of left ventricular diastolic dysfunction in essential hypertension. J Am Coll Cardiol. 2002;39(12):2005–11. Available from https://www.sciencedirect.com/science/article/pii/S073510970201896X.
10. Chamsi-Pasha MA, Zhan Y, Debs D, Shah DJ. CMR in the evaluation of diastolic dysfunction and phenotyping of HFpEF: current role and future perspectives. JACC Cardiovasc

Imaging. 2020;13(1):283–96. Available from https://www.sciencedirect.com/science/article/pii/S1936878X19303730.
11. Garg P, Gosling R, Swoboda P, Jones R, Rothman A, Wild JM, et al. Cardiac magnetic resonance identifies raised left ventricular filling pressure: prognostic implications. Eur Heart J. 2022;43(26):2511–22. https://doi.org/10.1093/eurheartj/ehac207.
12. Giusca S, Steen H, Montenbruck M, Patel AR, Pieske B, Erley J, et al. Multi-parametric assessment of left ventricular hypertrophy using late gadolinium enhancement, T1 mapping and strain-encoded cardiovascular magnetic resonance. J Cardiovasc Magn Reson. 2021;23(1):92. https://doi.org/10.1186/s12968-021-00775-8.
13. Schuster A, Hor KN, Kowallick JT, Beerbaum P, Kutty S. Cardiovascular magnetic resonance myocardial feature tracking. Circ Cardiovasc Imaging. 2016;9(4):e004077. https://doi.org/10.1161/CIRCIMAGING.115.004077.
14. Lange T, Schuster A. Quantification of myocardial deformation applying CMR-feature-tracking—all about the left ventricle? Curr Heart Fail Rep. 2021;18(4):225–39. https://doi.org/10.1007/s11897-021-00515-0.
15. Song Y, Li L, Chen X, Shao X, Lu M, Cheng J, et al. Early left ventricular diastolic dysfunction and abnormal left ventricular-left atrial coupling in asymptomatic patients with hypertension: a cardiovascular magnetic resonance feature tracking study. J Thorac Imaging. 2022;37(1):26. Available from https://journals.lww.com/thoracicimaging/Fulltext/2022/01000/Early_Left_Ventricular_Diastolic_Dysfunction_and.4.aspx.
16. Neisius U, Myerson L, Fahmy AS, Nakamori S, El-Rewaidy H, Joshi G, et al. Cardiovascular magnetic resonance feature tracking strain analysis for discrimination between hypertensive heart disease and hypertrophic cardiomyopathy. PLoS One. 2019;14(8):e0221061. https://doi.org/10.1371/journal.pone.0221061.
17. Shao G, Cao Y, Cui Y, Han X, Liu J, Li Y, et al. Early detection of left atrial and bi-ventricular myocardial strain abnormalities by MRI feature tracking in normotensive or hypertensive T2DM patients with preserved LV function. BMC Cardiovasc Disord. 2020;20(1):196. https://doi.org/10.1186/s12872-020-01469-2.
18. Lavall D, Vosshage NH, Stoebe S, Denecke T, Hagendorff A, Laufs U. Native T1 mapping for differential diagnosis of left ventricular hypertrophy. Eur Heart J Cardiovasc Imaging. 2021;22(2):54. https://doi.org/10.1093/ehjci/jeab090.054.
19. Hinojar R, Varma N, Child N, Goodman B, Jabbour A, Yu C-Y, et al. T1 mapping in discrimination of hypertrophic phenotypes: hypertensive heart disease and hypertrophic cardiomyopathy. Circ Cardiovasc Imaging. 2015;8(12):e003285. https://doi.org/10.1161/CIRCIMAGING.115.003285.
20. Yilmaz A. The "Native T1 versus extracellular volume fraction paradox" in cardiac amyloidosis: answer to the million-dollar question? JACC Cardiovasc Imaging. 2019;12(5):820–2. Available from https://www.sciencedirect.com/science/article/pii/S1936878X18309264.
21. Andrea B, Michele B, Ana M-N, Tushar K, Aviva P, Tamer R, et al. Noncontrast magnetic resonance for the diagnosis of cardiac amyloidosis. JACC Cardiovasc Imaging. 2020;13(1):69–80. https://doi.org/10.1016/j.jcmg.2019.03.026.
22. Arcari L, Hinojar R, Carr-White G, Zainal H, Zhou H, Vasques M, et al. Excess of myocardial water and fibrosis define myocardial hypertrophy in uremic but not in hypertrophic cardiomyopathy - TrueTypeCKD study. Eur Heart J Cardiovasc Imaging. 2019;20(2):jez111.002. https://doi.org/10.1093/ehjci/jez111.002.
23. Arcari L, Hinojar R, Engel J, Freiwald T, Platschek S, Zainal H, et al. Native T1 and T2 provide distinctive signatures in hypertrophic cardiac conditions – comparison of uremic, hypertensive and hypertrophic cardiomyopathy. Int J Cardiol. 2020;306:102–8. Available from: https://www.sciencedirect.com/science/article/pii/S0167527319347539.
24. Kotecha T, Martinez-Naharro A, Yoowannakul S, Lambe T, Rezk T, Knight DS, et al. Acute changes in cardiac structural and tissue characterisation parameters following haemodialysis measured using cardiovascular magnetic resonance. Sci Rep. 2019;9(1):1388. https://doi.org/10.1038/s41598-018-37845-4.

Cardiac Biomarkers in the Progression to Heart Failure in Hypertension

17

Nicholas Cauwenberghs

17.1 Heart Failure and Hypertension

Heart failure (HF) still poses a major burden on societal health and healthcare. The HF epidemic continues to grow due to aging population and rising prevalence of cardiometabolic risk factors and comorbidities [1]. Hypertension is the major modifiable risk factor for HF [2]. Hypertension promotes the initiation and progression of HF by mediating excessive neurohormonal activation, chronic low-grade inflammation and myocardial stretch and injury, and extracellular matrix remodeling in the heart [3]. In response, the heart gradually remodels and its function progressively worsens for years until the heart decompensates and HF symptoms present [4]. As this gradual process of cardiac remodeling and dysfunction already starts long before the onset of HF symptoms, we may detect HF in a subclinical stage in order to proactively modulate its risk factors and prevent the transition to overt disease [5]. To date, however, it remains challenging to diagnose overt HF, let alone to detect HF in a subclinical phase. Biomarkers may improve our understanding of the complex molecular machinery behind HF pathogenesis and pathophysiology and facilitate the detection of patients progressing toward overt HF. As such, biomarkers may cost-effectively supplement current strategies for early detection and timely management of HF to tackle the pressing HF epidemic.

N. Cauwenberghs (✉)
Research Unit Hypertension and Cardiovascular Epidemiology, Department of Cardiovascular Sciences, University of Leuven, Leuven, Belgium
e-mail: nicholas.cauwenberghs@kuleuven.be

M. Dorobantu et al. (eds.), *Hypertension and Heart Failure*, Updates in Hypertension and Cardiovascular Protection,
https://doi.org/10.1007/978-3-031-39315-0_17

17.2 The Ideal HF Biomarker

The World Health Organization defines a biological marker (or biomarker in short) as "any substance, structure or process that can be measured in the body or its products and influence or predict the incidence of outcome or disease" [6]. The ideal biomarker should be measurable in a quick, precise, and accurate way at a reasonable cost using a widely available assay. Biological variation and reference limits of the biomarker should be known, and pre-analytical, analytical, and post-analytical sources of error should be identified [7]. In addition, a HF biomarker should characterize an important pathological pathway involved in the initiation and progression of HF and provide information that cannot be obtained through routine examination and laboratory investigation. Evidently, a HF biomarker should enable adequate prediction, diagnosis, staging, prognosis, and/or management of this condition. Most likely, we will never find a single HF biomarker that fulfills all these clinical needs on its own. Instead, we may need to search for the *ideal combination of biomarkers* that captures the full spectrum of HF pathophysiology and that adds essential diagnostic and prognostic information to the clinical decision-making process. Here, we focus on *blood-borne or circulating biomarkers* for detection of subclinical heart maladaptation. Other HF-related biomarkers, including imaging and (epi)genetic markers, have been described elsewhere in this book.

Figure 17.1 outlines the established and promising HF biomarkers associated with hypertension-mediated mechanisms underlying HF pathogenesis [3]. In this chapter, we predominantly discuss cardiac-specific biomarkers, namely biomarkers of myocardial stretch, injury, and remodeling. To date, it has not been sufficiently demonstrated that circulating cardiac biomarkers could cost-effectively supplement conventional screening tools such as routine biochemistry and imaging modalities if used for HF risk assessment in patients with hypertension. Therefore, both American and European guidelines on the management of hypertension do not recommend the routine measurement of circulating biomarkers of subclinical heart deterioration to evaluate hypertension-mediated organ damage (HMOD) [8, 9].

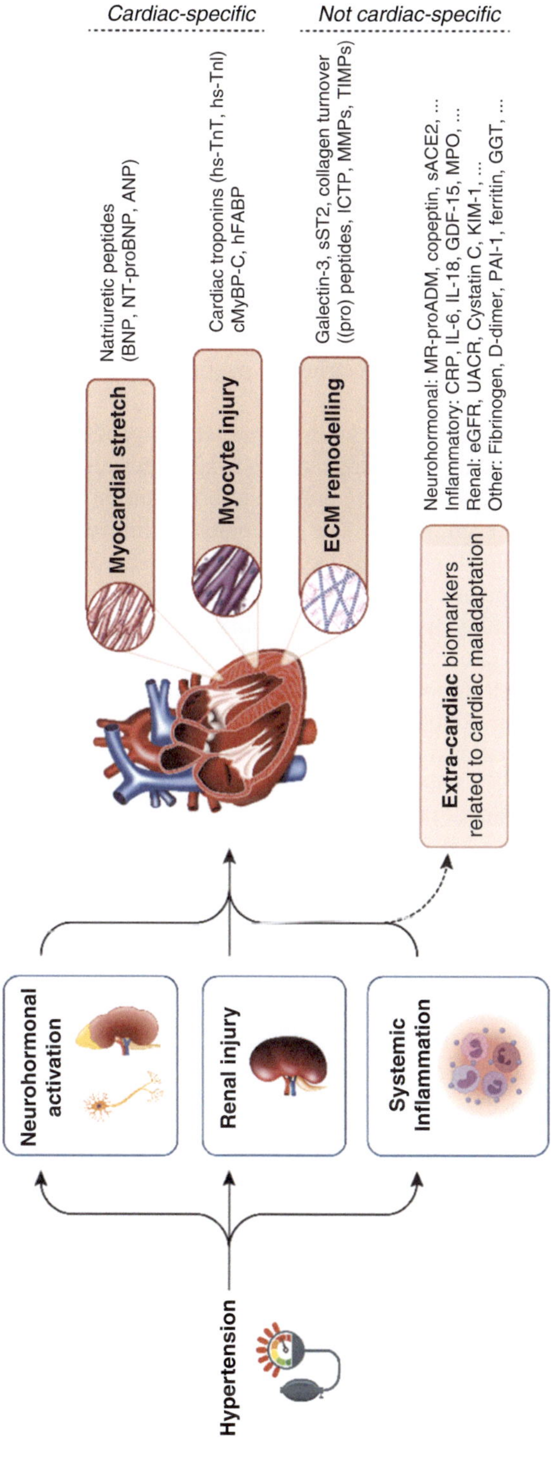

Fig. 17.1 Circulating biomarkers reflecting hypertension-mediated mechanisms underlying heart failure (HF) pathogenesis. *ANP* atrial natriuretic peptide, *BNP* brain natriuretic peptide, *cMyBP-C* cardiac myosin binding protein-C, *eGFR* estimated glomerular filtration rate, *GDF-15* growth/differentiation factor 15, *GGT* gamma-glutamyl transferase, *hFABP* heart-type fatty acid binding protein, *hs-TnI and hs-TnT* high-sensitivity troponin I and T, *ICTP* C-terminal telopeptide of type I collagen, *IL* interleukin, *KIM-1* kidney injury molecule 1, *MMPs* matrix metalloproteinases, *MPO* myeloperoxidase, *MR-proADM* mid-regional proadrenomedullin, *PAI-1* plasminogen activator inhibitor-1, *sACE2* soluble angiotensin-converting enzyme 2, *sST2* soluble suppression of tumorigenesis-2, *TIMPs* tissue inhibitors of metalloproteinases, *UACR* urine albumin-to-creatinine ratio

17.3 Markers of Myocardial Stretch: The Natriuretic Peptides

When excessively stretched, cardiomyocytes actively release natriuretic peptides that activate compensatory mechanisms, including blood pressure-lowering mechanisms such as natriuresis and vasodilation [10]. Imbalances in the natriuretic peptide system have been strongly linked to the development of both hypertension and HF [11]. In longstanding hypertension, the natriuretic peptide system is pushed to its limits and no longer exerts its cardioprotective effects adequately, thus failing to normalize cardiac wall stress [10]. The fact that natriuretic peptides are cardiac-specific hormones actively released upon pathologic cardiac stress makes them unrivaled as circulating biomarkers for HF prediction, diagnosis, and management [11]. Chapter 19 discusses the value of natriuretic peptides for evaluation of HF progression in hypertension in more depth.

Cardiac natriuretic peptides and their amino terminal pro-peptides are widely used as HF biomarkers in the daily cardiology clinic [5, 12]. They are the gold standard for the diagnosis of symptomatic HF, provide information on HF prognosis, steer HF therapies, and are used as surrogate endpoints in clinical trials [5, 12]. Community-based studies also highlighted the role of natriuretic peptides for HF risk stratification in asymptomatic individuals, as measurement of natriuretic peptides was found to predict future HF and detect people at highest risk of subclinical heart maladaptation. In a meta-analysis on the data of 95,617 individuals and more than 2000 HF events, natriuretic peptide levels significantly improved the prediction of new-onset HF beyond conventional risk factors [13]. Moreover, natriuretic peptide levels independently predicted both incident HF with preserved ejection fraction (HFpEF) and incident HF with reduced ejection fraction (HFrEF) in 22,756 community-dwelling individuals [14]. In a cohort of 3842 individuals, 86% of them having hypertension, N-terminal pro–B-type natriuretic peptide (NT-proBNP) alone predicted a 5-year incidence of HF with reasonable sensitivity and specificity (area under the receiver operating characteristic curve = 0.79) [15].

Natriuretic peptides were also found to detect subclinical cardiac maladaptation in the community. For instance, in 727 community-dwelling subjects, NT-proBNP levels were associated with echocardiographic profiles of subclinical heart maladaptation, including left ventricular (LV) remodeling and diastolic dysfunction [16]. Screening for natriuretic peptides may even complement cardiac imaging for HF risk stratification. Indeed, in the Dallas Heart Study, the risk of future HF or cardiovascular (CV) death increased substantially if elevated NT-proBNP levels and LVH were concomitantly present [17]. Studies remain warranted to derive sex- and race-specific cutoffs for NT-proBNP for use in at-risk individuals [18].

Given the current evidence, the 2022 HF guidelines endorsed by the American College of Cardiology (ACC) and the American Heart Association (AHA) consider brain natriuretic peptide (BNP)- or NT-proBNP-based screening useful in patients at risk of developing HF, including patients with hypertension, to prevent the development of LV dysfunction or new-onset HF (Class 2A recommendation, B-R level of evidence) [5]. There is also consensus between American, European, and Japanese HF societies that elevated natriuretic peptide levels substantiate the presence of stage B HF ("pre-HF"), a HF precursor stage, in patients with hypertension or other HF risk factors [5, 19]. By contrast, American and European hypertension guidelines do not recommend routine assessment of natriuretic peptides for evaluation of HMOD [8, 9].

17.4 Markers of Myocardial Injury

Cardiac troponins I and *T. Cardiac troponins* are proteins regulating actin-myosin interaction inside the myocardial contractile apparatus. Troponins are passively released with normal cardiomyocyte turnover or due to pathological cardiomyocyte death and increased cardiomyocyte wall permeability [20]. Solely expressed in cardiomyocytes, the cardiac isoforms of troponin I and troponin T are the biomarkers of choice for biochemical detection of myofibrillar injury. Moreover, these cardiac-specific troponins are, like natriuretic peptides, the gold standard for the diagnosis and prognosis of several cardiac diseases, including symptomatic HF [5, 12].

Nowadays, highly sensitive assays enable cardiac troponin detection in almost any individual [21]. This has allowed researchers to study cardiac troponins as a prognostic biomarker in patients with non-acute heart disease and in the community. Some studies linked discrete elevations in cardiac troponin levels to the development of hypertension [21, 22]. For instance, in 6516 ARIC study participants free of hypertension and cardiovascular disease at baseline, high-sensitivity cardiac troponin T (hs-cTnT) was associated with incident hypertension [22]. Population studies also consistently identified hs-cTnT as an independent predictor of incident HF [14, 23, 24]. A meta-analysis on 16 prospective studies revealed that individuals in the highest high-sensitivity troponin I (hs-cTnI) or hs-cTnT tertile have a two-fold greater risk of developing HF in the future than those in the lowest tertile [25].

Numerous studies in the community [16, 23, 26, 27] and in patients with hypertension [21, 22, 28, 29] found that measurement of high-sensitivity cardiac troponins was also useful for identifying individuals with subclinical heart maladaptation, especially LVH. For instance, in 592 community-dwelling individuals, elevated hs-cTnT levels at baseline were associated with a greater risk of developing or maintaining LVH over time, independent of CV risk factors, such as hypertension [27]. As such, discrete but persistent elevations in cardiac troponins may mark individuals whose heart is undergoing adverse remodeling and who are thus in the process of developing clinical HF. Like markers of myocardial stretch, biomarkers of myocardial injury may complement imaging modalities in HF risk assessment [17, 30]. In the Jackson Heart Study (n = 3987), for instance, participants with concomitant LVH and abnormally high hs-cTnI were at highest risk of developing overt HF [30].

In line, HF societies emphasize that persistently elevated cardiac troponins mark the presence of stage B HF ("pre-HF") in patients with hypertension, unless other conditions are present that may explain the rise in troponins [5, 19]. However, HF guidelines do not yet specify cutoffs for cardiac troponins to discern HF stage B. In the "Biomarker for Cardiovascular Risk Assessment in Europe" consortium pooling individual-level data from 4 prospective population-based cohorts (n = 48,455 individuals), the optimal hs-cTnI cutoffs for selecting individuals at risk of new-onset HF were 2.6 ng/l for women and 4.2 ng/l for men [24]. To date, however, research initiatives should validate sex- and race-specific thresholds and their complementarity with current screening tools to detect HF progression before routine assessment of cardiac troponins in the daily hypertension clinic can be recommended. In extent, the clinical value of cardiac troponins for monitoring blood pressure (BP) therapy remains scarcely explored. In the Randomized TEAMSTA Protect I Trial, a reduction in blood pressure following 6 months of antihypertensive treatment was associated with a reduction in Hs-cTnI but not in hs-cTnT [31]. It remains to be studied if and to what extent cardiac troponins could provide additional information beyond conventional tools to monitor changes in cardiovascular risk upon BP management.

Cardiac Myosin-Binding Protein C (cMyBP-C) As a constituent of thick sarcomere filaments, cMyBP-C modulates cardiomyocyte contractility by regulating actin-myosin cross-bridging [32]. Like cardiac troponins, cMyBP-C gets released upon myofibrillar injury. Circulating cMyBP-C has predominantly been studied as a marker of acute cardiac injury [33]. During myocardial ischemia, a N-terminal fragment of ~40 kDa from cMyBP-C is cut and released into the circulation to protect cMyBP-C from proteolysis and thus to help preserve cardiac performance. Recent studies suggest that cMyBP-C may outperform cardiac troponins for early detection of myocardial infarction upon hospital admission [34]. As a cardiac-specific and highly sensitive marker, circulating cMyBP-C is suggested as a candidate biomarker for evaluation of subclinical cardiovascular disease [35]. So far, however, only one small study on 158 individuals has explored the usefulness of circulating cMyBP-C as a predictor of cardiovascular events [36]. Another study on 161 patients found an association between serum cMyBP-C and myocardial hypertrophy and fibrosis and the risk of all-cause mortality in aortic stenosis [37]. Overall, much more research is needed to determine if cMyBP-C measurements are useful for HF risk stratification.

Heart-Type Fatty Acid Binding Protein (hFABP) hFABP is a cytoplasmic protein in cardiomyocytes that supplies mitochondria with long-chain, polyunsaturated fatty acids [38]. Like cardiac troponins and cMyBP-C, hFABP is passively released by cardiomyocytes upon membrane injury and is thus classified as a marker of myocardial injury and cardiac cell death [39]. The diagnostic and prognostic value of hFABP in the setting of acute coronary syndromes has been reviewed elsewhere [40].

Like cardiac troponins, discrete elevations in hFABP could mark long-term, ongoing myocardial injury and may thus be useful for HF risk stratification. Indeed, previous reports suggest that leakage of hFABP starts an adverse vicious cycle that accelerates HF progression, where extracellular hFABP worsens cardiac contraction by reducing intracellular Ca^{2+}, which in turn leads to more cardiac damage with more hFABP release [41]. hFABP levels were increased in both HFpEF and HFrEF, and higher hFABP was prognostic for a worse HF outcome [42, 43]. Of note, hFABP was reportedly more sensitive than standard troponins for detection of latent myocardial damage in patients with congestive HF [44]. Moreover, hFABP may also complement natriuretic peptides, as the combined measurements of hFABP and natriuretic peptides appeared to better forecast adverse events in patients with overt HF [45, 46].

Thanks to highly sensitive assays, circulating hFABP is detectable in nearly everybody nowadays [47], paving the way for studies on hFABP in primary care. In community-dwelling people, circulating hFABP originates from normal cardiomyocyte turnover or latent myocardial damage inflicted by CV risk factors, such as hypertension. In population studies, hFABP levels were found to be associated with hypertension [47, 48] and a future risk of cardiovascular death [48]. In 46 patients with newly diagnosed hypertension, hFABP levels were increased as compared to an age-matched control group [49]. Furthermore, hFABP levels predicted new-onset CV disease beyond conventional risk factors in 192 patients with essential hypertension [50]. Although more research is warranted to elucidate the role of hFABP in HF progression, current literature indicates hFABP as a promising biomarker for evaluation of cardiac damage in hypertension.

17.5 Markers of Myocardial Remodeling and Fibrosis

If afterload is chronically elevated, the ventricular wall remodels in an attempt to accommodate the increased load and normalize wall stress [4]. If chronic, myocardial wall injury activates tissue-repairing processes that disturb the myocardial collagen network (e.g., fibrosis) and contribute to early LV remodeling in the long run [16]. It should be noted that fibrosis is not a cardiac-specific phenomenon but a ubiquitous process in the human body. As such, circulating fibrosis markers originate from multiple tissues and different cell types throughout the human body, often impeding their clinical value as HF markers. The 2017 ACC/AHA HF guidelines did recommend measurement of cardiac fibrosis markers, particularly galectin-3 and Soluble suppression of tumorigenicity-2 (sST2), for prognostic stratification in patients with symptomatic HF (class IIb recommendation) [51]. Nowadays, however, both American and European HF guidelines do not endorse measurement of cardiac fibrosis markers, neither for HF prognosis in patients nor for risk stratification in asymptomatic individuals [5, 12]. Fibrosis markers may perhaps be informative for HF risk stratification when part of a multi-biomarker panel, but such panels remain under construction.

Galectin-3 Galectin-3 is a ubiquitous protein linked to the pathogenesis of various inflammatory and/or fibrotic disorders, including the development of various heart diseases, renal fibrosis, chronic obstructive pulmonary disease, and different types of cancers [52]. Although galectin-3 is not cardiac-specific, rat models for hypertension and HF highlighted galectin-3 as a marker for hypertrophic hearts prone to developing HF [53]. In rats, secretion of galectin-3 by activated macrophages was found to mediate ventricular remodeling and dysfunction by inducing inflammation, fibroblast proliferation, and excessive collagen deposition in the heart [53, 54].

The added prognostic value of plasma galectin-3 in overt HF remains contested. In chronic HF, plasma galectin-3 was found elevated [55] and correlated with echocardiographic indexes of LV volume and function [56, 57]. By contrast, galectin-3 did not correlate with echocardiographic parameters of myocardial structure and function in 132 patients with HFrEF [58]. Moreover, galectin-3 was not associated with future HF-related events when adjusting for NT-proBNP [59]. Yet, data from the Deventer-Alkmaar Heart Failure study ($n = 232$), the Controlled Rosuvastatin Multinational Trial in Heart Failure ($n = 1329$), and the Comparison of Outcomes and Access to Care for Heart Failure trial ($n = 324$) studies suggest that galectin-3 may be useful in predicting future events in HF patients when measured serially [60–62].

It also remains questioned if galectin-3 measurements are useful to detect subclinical heart maladaptation or predict new-onset HF in at-risk individuals. In the Framingham Heart Study Offspring cohort ($n = 2477$), galectin-3 increased independently stronger over time in participants who were hypertensive at baseline [63]. Moreover, in 107 patients, plasma galectin-3 was significantly higher in hypertensive patients with LVH as compared to those without [64]. In 3353 participants of the Framingham Offspring cohort, higher galectin-3 was associated with higher LV mass in age- and sex-adjusted analyses and with increased risk of incident HF (adjusted hazard ratio (HR): 1.23 per 1 standard deviation (SD) increase in log galectin-3; $P = 0.02$) [65]. Yet, adding galectin-3 to a clinical risk model led to negligible changes in C-statistics and only a small net reclassification improvement [65]. In the same cohort ($n = 2477$), the change in galectin-3 over time predicted future HF in the community (adjusted HR per 1 SD increase: 1.39, 1.13 to 1.71). By contrast, De Boer et al. did not find an association between galectin-3 and incident HF in a meta-analysis of 95,617 individuals after adjustment for potential confounders [14]. When pooling individual-level data from the same four cohorts ($n = 22{,}756$), higher galectin-3 did predict incident HF in women after adjustment for clinical covariables, but this association was lost after additional adjustment for natriuretic peptides [66].

To date, uncertainty about the tissue of origin hampers routine evaluation of galectin-3 in HF prediction, diagnosis, and prognosis. Indeed, the ubiquitous expression of galectin-3 by different cell types across different tissues makes it difficult to discern the primary sources of high circulating galectin-3 levels. Hence, increased plasma galectin-3 does not necessarily correspond to increased galectin-3 production in the heart. For instance, in 115 patients with cardiomyopathy, plasma galectin-3 levels did not correlate with myocardial galectin-3 levels as measured in endomyocardial biopsies nor with histologically assessed cardiac fibrosis [67].

Soluble Suppression of Tumorigenicity-2 (sST2) sST2 acts as a decoy receptor for interleukin-33 (IL-33), a cytokine secreted in response to tissue injury [68]. Under normal conditions, IL-33 binds to the transmembrane isoform of the soluble suppression of tumorigenicity-2 protein (ST2 ligand or ST2L) to exert anti-apoptotic, anti-fibrotic, and anti-hypertrophic effects that mitigate cardiac hypertrophy and fibrosis [68]. In response to hemodynamic overload and profibrotic stimuli, however, endothelial cells, cardiomyocytes, and cardiac fibroblasts shed the soluble ST2 isoform, sST2, which obstructs the cardioprotective effects of the IL-33/ST2L axis by neutralizing circulating IL-33 [69]. Nowadays, sST2 can be assessed accurately at a reasonable cost, boosting research interest, for this protein.

Numerous studies consistently demonstrated sST2 as a strong prognosticator of HF morbidity and mortality in patients with acute [70] and chronic HF [71], even adding prognostic information beyond natriuretic peptides and hs-cTnT [72, 73]. In an ambulatory HF cohort of 876 patients, sST2 was found superior to galectin-3 for prediction of 5-year all-cause and CV mortality and HF hospitalization [74]. To date, however, current HF guidelines do not recommend sST2 measurements for prognostic stratification of patients with symptomatic HF [5, 12].

In addition, disturbances in the IL-33/ST2 axis appear to be involved in the development of hypertension, suggesting a diagnostic and prognostic value for sST2 in hypertension. For instance, serum levels of sST2 were markedly higher and ST2L levels significantly lower in patients with essential hypertension than in healthy controls [75]. In community-based studies, higher serum sST2 was associated with higher systolic blood pressure and hypertension [76, 77]. Moreover, in patients with hypertension, plasma sST2 allowed discrimination of patients with LVH from LVH-free subjects [78, 79]. However, in the FINRISK97 cohort of 8,444 participants, sST2 failed to independently predict the long-term incidence of cardiovascular events, including HF [80]. In 3915 elderly people from the Cardiovascular Health Study, Parikh RV et al. found an association between higher risk of incident HF and higher sST2, but this association lost statistical significance after adjustment for NT-proBNP and hsTnT despite strong associations with risk factors and underlying heart disease [81]. In line, sST2 levels did not independently predict incident HFrEF nor HFpEF in a recent meta-analysis on population data [14], likely due to the multi-organ production of sST2. As sST2 is not cardiac-specific, elevated sST2 does not necessarily mirror cardiac production and remodeling. Moreover, the IL-33/ST2L axis is linked to the development of non-cardiac diseases as elevations in sST2 were observed in breast and gastric cancer, nephropathy, and liver failure [68].

Collagen Propeptides, Matrix Metalloproteinases (MMPs), and Tissue Inhibitors of Metalloproteinases (TIMPs) Whatever the etiology, HF is characterized by diffuse deposition and extensive crosslinking of collagen fibers in the extracellular matrix space [82]. Multiple circulating markers of collagen type I and III turnover have been investigated for non-invasive detection of cardiac fibrosis. Most research focused on collagen-derived peptides, particularly (1) the N-terminal (PINP) and C-terminal propeptide of collagen type I (PICP), a marker of collagen I synthesis; (2) N-terminal propeptide of collagen type III (PIIINP), a marker of collagen III synthesis; and (3) the C-terminal telopeptide of collagen type I (ICTP), reflecting collagen type I degradation [83]. As key modulators of collagen synthesis and degradation, enzymes that degrade (MMPs) or protect collagen fibers (TIMPs) have also been thoroughly investigated. As summarized elsewhere, clinical studies have observed abnormal levels of fibrosis biomarkers (reflecting disbalanced collagen turnover) in patients with symptomatic HF, including hypertensive heart disease [83]. Circulating biomarkers of collagen homeostasis may thus be important prognosticators to forecast clinical outcomes in HF patients.

Because collagen metabolism already gets perturbed early in HF development, it may be clinically meaningful to monitor disturbances in the quality and quantity of myocardial collagen in patients with hypertension to identify individuals at high risk of HF. Previous studies found associations between disbalanced collagen homeostasis and hypertension. In hypertensive patients for instance, higher BP load was related to higher PICP/ICTP ratio (i.e., higher net collagen I synthesis) [84]. In line, collagen I degradation was depressed in patients with essential hypertension but normalized after treatment with lisinopril [85]. In 100 subjects with never-treated hypertension, plasma PICP, CITP, and TIMP-1 were significantly higher as compared to 50 normotensive controls [86]. Moreover, in the same study, the rise in TIMP-1 detected diastolic dysfunction with a specificity of 97% and a positive predictive value of 96% [86]. Also, in patients with hypertension, increased collagen III synthesis may hallmark progressive deterioration in cardiac function, whereas changes in collagen I turnover may promote cardiac dysfunction in more advanced hypertensive heart disease [87]. In patients with never-treated hypertension, serum collagen type III synthesis (PIIINP) was higher if LV hypertrophy (LVH) was concomitantly present [88].

In line, cross-sectional and longitudinal observations in the general population found that circulating biomarkers of cardiac collagen composition may improve the identification of subjects at risk of subclinical cardiac maladaptation, i.e., those who are progressing toward overt HF [27]. In the Flemish Study on Environment, Genes and Health Outcomes (FLEMENGHO) with 782 participants, LV diastolic dysfunction correlated with higher collagen I (higher CITP, TIMP-1) and collagen III deposition (higher PIIINP) [89]. In 592 FLEMENGHO participants, LV longitudinal strain decreased stronger during a 5-year follow-up period if collagen I production and degradation was higher at baseline [27]. In the same cohort, the *E*/*e′* ratio rose stronger with a lower MMP-1 and MMP-1/TIMP-1 ratio at baseline. In the multi-ethnic study of atherosclerosis with 3187 participants, circulating CITP and PIIINP independently predicted incident HFpEF, but not HFrEF, during a follow-up period of 13 years [90]. By contrast, PINP levels were not predictive of incident CV events during a follow-up period of 9 years in 1324 asymptomatic individuals beyond traditional risk factors [91].

Taken together, measurement of fibrosis markers could be a useful tool to non-invasively assess myocardial remodeling in hypertension, but several gaps remain to be addressed before non-invasive fibrosis biomarkers may find their way to the routine hypertension clinic. First, longitudinal data from large-scale cohorts of asymptomatic individuals remain scarce [27]. Second, cutoff points remain undefined, which is for some collagen turnover markers hampered by differences between assays used for detection (e.g., radioimmunoassay versus enzyme-linked immunosorbent assays). Third, it remains to be investigated if and to what extent fibrosis markers have additional value beyond established cardiovascular markers (including imaging features and natriuretic peptides) for evaluation of HMOD in asymptomatic patients and, if so, how they can be cost-efficiently integrated into the current clinical workflow. Last but foremost, although correlating with histologically assessed myocardial collagen [92], collagen-derived peptides are not cardiac-specific. Indeed, collagen-derived peptides can also be produced by other organs, such as liver, kidneys, and lungs, and impaired collagen I and III turnover can also reflect extracardiac pathologies [83]. We may need to identify molecules specific for myocardial fibrosis and develop multi-dimensional tools to percolate the most useful information from the complex network of (extra)cardiac fibrosis markers for evaluation of HMOD.

17.6 Extracardiac Biomarkers of HF Progression

Extracardiac biomarkers may complement HF risk prediction strategies in hypertension because they mirror pathological pathways mediated by hypertension that expedite HF progression. For instance, hypertension incites excessive neurohormonal activation, including increased secretion of natriuretic peptides and overactivation of the sympathetic nervous system (SNS) and the renin-angiotensin-aldosterone system (RAAS) [93]. Previous studies identified mid-regional propeptide (MR-proADM), a precursor of adrenomedullin, as a promising marker of SNS overactivation that predicts HF [94]. Markers of RAAS overactivation have also been linked to HF pathogenesis in hypertension, such as copeptin, a vasopressin precursor peptide [95], and soluble angiotensin-converting enzyme 2 (sACE2) [96].

Hypertension also promotes chronic low-grade inflammation and oxidative stress, two hallmarks of HFpEF pathogenesis [97]. Among others, the following biomarkers of systemic inflammation and oxidative stress were found elevated in hypertension and related to subclinical cardiac remodeling and/or dysfunction: (1) C-reactive protein (CRP) [98, 99], (2) IL-6 and IL-18 [100–102], (3) growth differentiation factor-15 (GDF-15) [103, 104], and (4) myeloperoxidase (MPO) [105]. Taken together, patients with hypertension and evidence of chronic low-grade inflammation seem at high risk of HF, but more research is needed to translate these observations into clinical practice.

Other extracardiac biomarkers may be useful for HF risk assessment in hypertension because they reflect comorbidities that frequently coincide with hypertension and that accelerate HF progression. [106]. First and foremost, associations between HF development and circulating markers of cardiometabolic disturbances have been described. For instance, in the FLEMENGHO cohort, insulin resistance at baseline was strongly associated with subclinical worsening of LV diastolic function during a follow-up period of 5 years [107]. Markers of renal function and injury are associated with progressive cardiac damage as well, including in patients with hypertension [108]. Among others, renal markers linked to HF progression include standard measures such as estimated glomerular filtration rate [109] and the urine albumin-to-creatinine ratio (UACR for microalbuminuria) [14, 110] and more novel markers such as cystatin C [14] and kidney injury molecule 1 [96]. The development of HF in both the general population [14, 100] and hypertensive patients [111] has also consistently been linked to a prothrombotic state as marked by alterations in circulating fibrinogen, D-dimer, or plasminogen activator inhibitor-1 (PAI-1). Furthermore, markers of iron deficiency and liver damage (e.g., gamma-glutamyl transferase) correlated with subclinical heart dysfunction [100] and independently predicted incident HF in the community [112, 113].

17.7 The Future: Integrative Bioprofiling for Personalized Risk Assessment

To evaluate HF risk in hypertension, a combination of biomarkers may be more informative than individual biomarkers [114]. Multi-marker models may be more specific and more sensitive than single-marker interpretations as they could merge the most informative biomarkers of different pathways linked to HF pathogenesis. Several studies already tested multi-marker scores for risk prediction of incident HF in the general population [115, 116], but the biomarkers included were arbitrarily selected and analyses were limited to traditional regression models that do not account for biomarker interrelationships (Fig. 17.2a). Instead, HF biomarkers should be identified, validated, and integrated into multi-marker models in a systematic and unbiased way to exploit their full value.

Advancements in proteomic profiling continue to facilitate the search for novel HF biomarkers and construct multi-marker scores for HF prediction in a systematic and unbiased way (Fig. 17.2b). Indeed, improvements in laboratory technology continuously boost the scale and precision of biomarker assays. Currently, high-throughput proteomic panels can detect thousands of proteins simultaneously with high sensitivity and precision using a minimal amount of sample [117]. Moreover, advanced analytical approaches such as machine learning algorithms continuously improve the capability to deal with the complex, multi-dimensional data retrieved by high-throughput assays.

High-throughput proteomic profiling has already identified circulating biomarkers useful for prediction of incident HF [118, 119] or detection of subclinical heart maladaptation in at-risk individuals [96, 120]. For instance, in the Malmö Diet and Cancer Study, a case-control study, nt-proBNP, troponin T, CRP, sST2, and tumor necrosis factor alpha (TNFα) were higher in 185 subjects who later on developed HF as compared to a random subset of 583 controls from the same cohort [119]. Of the 1305 proteins measured, 16 allowed discrimination between cases and controls independent of age and sex [119]. A case-control study (the Heart "OMics" in AGEing Study) found 38 proteins associated with HF incidence, of which 8 survived stepwise regression with traditional HF risk factors [121]. Proteomic analysis in an elderly cohort without HF demonstrated that 9 out of 92 proteins, including GDF-15 and sST2, improved prediction of incident HF when added separately to a Cox proportional hazards regression model with established risk factors [118]. The aforementioned studies, however, identified HF-related proteins by comparing individual

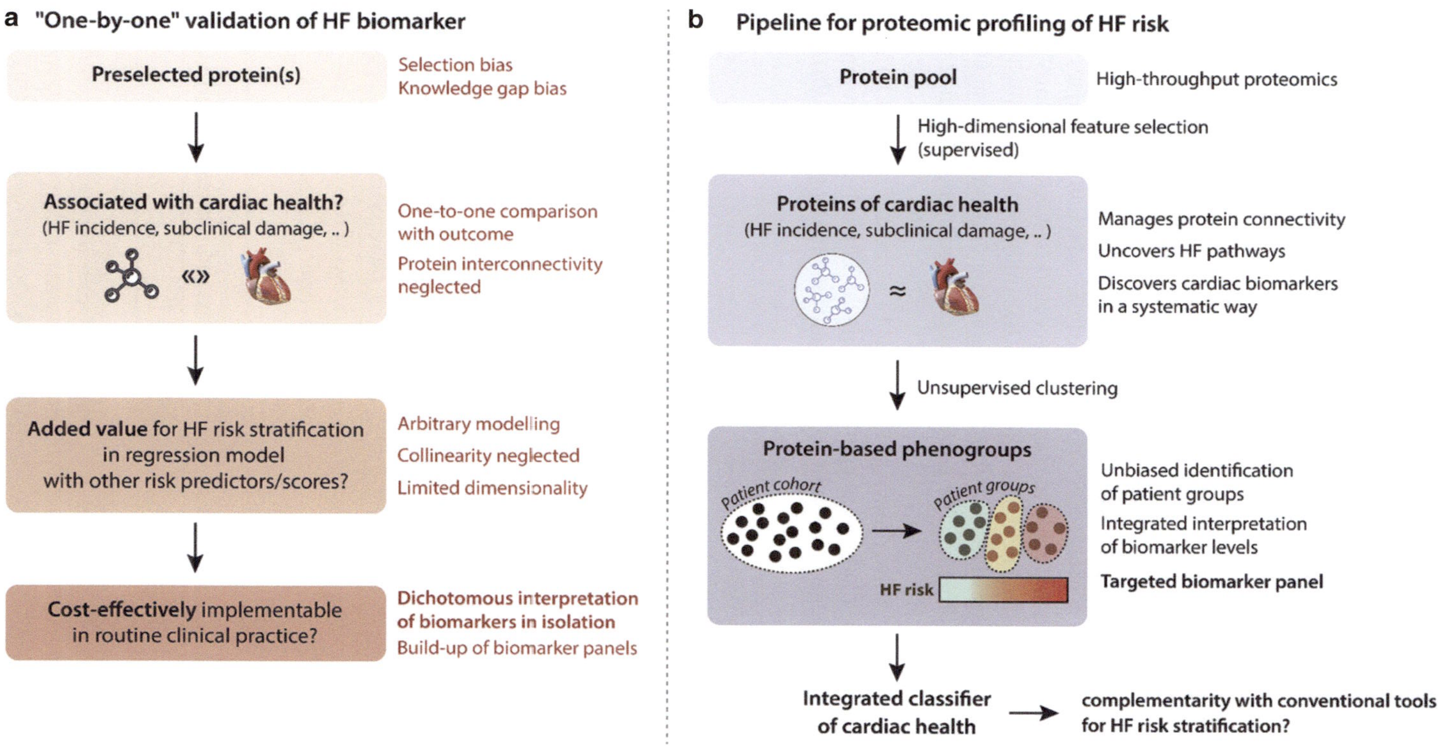

Fig. 17.2 Current workflow (**a**) and innovative proteomic pipeline (**b**) for discovery and validation of circulating HF biomarkers. (**a**) At present, a potential HF biomarker is preselected and validated to clinical parameters of cardiac health (e.g., HF incidence, evidence of cardiac damage), after which its added value for HF risk stratification is tested in traditional regression models. If cost-effective, the HF biomarker could be implemented in routine clinical practice and dichotomously interpreted (normal or abnormal level). (**b**) High-throughput proteomics and advanced analytics facilitate the systematic and unbiased search for HF-related protein and integrate them in a protein-based classifier of cardiac health. If cost-effective, targeted biomarker panels with integrative interpretation of the biomarker levels could complement conventional HF risk assessment tools

protein levels one by one to the outcome (e.g., new-onset HF), thereby neglecting protein interconnectivity, whereas advanced analytical approaches are available that can identify the most informative markers from the cloud of protein data while considering the vast network of protein interactions.

To date, there is still a need for better protein-based characterization and detection of asymptomatic cardiac remodeling and dysfunction. Indeed, as extensively illustrated throughout this chapter, studies on subclinical heart maladaptation often investigate only a limited and biased selection of proteins at a time while ignoring protein interconnectivity. However, a study on 575 community-dwelling individuals recently proposed a pipeline to extract pathologically relevant proteins from high-throughput proteomic data and integrate them into a protein-based classifier of cardiac health [96]. From a panel of 87 proteins, feature selection techniques first identified 13 proteins (including kidney injury molecule-1, sACE2, galectin-9, and IL-6) as important for detection of subclinical echocardiographic abnormalities. Subsequently, a clustering algorithm divided the cohort into distinct phenogroups solely based on the 13 proteins. The protein-based phenogrouping identified individuals at high risk of cardiac remodeling and dysfunction [96]. As such, the proposed workflow for integrative protein profiling may be a first step toward a targeted protein-based screening platform integrable within the clinical decision-making process for cardiac risk stratification in asymptomatic individuals (Fig. 17.2b).

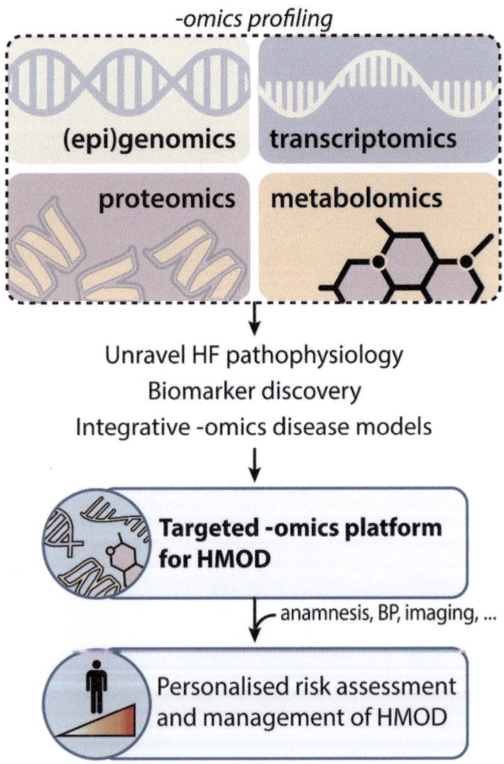

Fig. 17.3 Targeted -omics profiling for detection and management of hypertension-mediated organ damage (HMOD). -Omics profiling may further unravel the molecular pathways that mediate HF progression in hypertension and lead to integrative -omics models of diseases, such as heart failure (HF). In the future, a targeted -omics platform may complement conventional clinical and imaging profiles and enable personalized detection and management of subclinical heart maladaptation or other HMOD

Even one step further, the integration of proteomic data with clinical and imaging profiles as well as with other-omics data, such as genomics, epigenomics, transcriptomics, and metabolomics, may pave the way toward even more personalized risk assessment (Fig. 17.3) [122]. Systems biology may further unravel the molecular pathways and regulatory networks mediating HF progression in hypertension and construct integrative-omics models that can identify patient clusters, stratify HF risk, and predict patient outcome. Advanced bioinformatics, including artificial intelligence approaches, are being developed to match the different biological inputs from the different molecular platforms while dealing with the complex interrelationships within the highly dimensional data [123]. To date, however, the field of multi-omics in hypertension and HF is in its infancy. Much more research is needed to interpret the complex multi-omics puzzle of biological processes in hypertension, particularly if intended for personalized BP management and risk stratification for HF or other HMOD in clinic [123].

17.8 Conclusions

The ever-growing burden of HF on our society highlights the pressing need to better understand HF development in order to improve screening, prevention, and treatment of this life-threatening condition. Circulating biomarkers can help unravel the molecular mechanisms behind HF pathogenesis, identify individuals at highest risk of developing overt HF, and monitor the effects of preventative therapies intended to block HF progression. Numerous HF biomarkers have been identified. In particular, biomarkers of myocardial stretch (natriuretic peptides) and injury (cardiac troponins) were found to detect subclinical heart dysfunction suggestive of (sub) clinical HF and are therefore considered gold standards for HF risk stratification, diagnosis, and prognosis. Most other circulating markers of HF are not cardiac-specific, reflecting rather systemic pathological processes such as fibrosis and chronic low-grade inflammation and, as a consequence, their clinical utility to evaluate HF progression in subclinical stages remains debated. To date, American and European guidelines on the management of hypertension do not recommend the routine measurement of any circulating biomarker of subclinical heart deterioration to evaluate HMOD. In the future, multi-dimensional and integrative biomarker profiling may complement HF risk-stratification strategies in hypertension. Randomized clinical trials are warranted to corroborate the clinical utility, added value, and cost-effectiveness of biomarker-guided detection and management of subclinical heart maladaptation in hypertensive patients beyond conventional screening tools.

References

1. Seferović PM, Vardas P, Jankowska EA, Maggioni AP, Timmis A, Milinković I, et al. The Heart Failure Association Atlas: heart failure epidemiology and management statistics 2019. Eur J Heart Fail. 2021;23(6):906–14.
2. Levy D, Larson MG, Vasan RS, Kannel WB, Ho KK. The progression from hypertension to congestive heart failure. JAMA. 1996;275(20):1557–62.

3. Braunwald E. Heart failure. JACC Heart Fail. 2013;1(1):1–20.
4. Cauwenberghs N, Knez J, D'hooge J, Thijs L, Yang W-Y, Wei F-F, et al. Longitudinal changes in LV structure and diastolic function in relation to arterial properties in general population. JACC Cardiovasc Imaging. 2017;10(11):1307–16.
5. Heidenreich PA, Bozkurt B, Aguilar D, Allen LA, Byun JJ, Colvin MM, et al. 2022 AHA/ACC/HFSA guideline for the management of heart failure: a report of the American College of Cardiology/American Heart Association Joint Committee on clinical practice guidelines. Circulation. 2022;79(17):e263–421.
6. World Health Organization. Biomarkers in risk assessment: validity and validation. Geneva: World Health Organization; 2001. p. 235–6, 222.
7. Morrow DA, de Lemos JA. Benchmarks for the assessment of novel cardiovascular biomarkers. Circulation. 2007;115(8):949–52.
8. Whelton PK, Carey RM, Aronow WS, Casey DEJ, Collins KJ, et al. 2017 ACC/AHA/AAPA/ABC/ACPM/AGS/APhA/ASH/ASPC/NMA/PCNA guideline for the prevention, detection, evaluation, and management of high blood pressure in adults: a report of the American College of Cardiology/American Heart Association Task Force on clinical Pr. Circulation. 2018;138(17):484–594.
9. Williams B, Mancia G, Spiering W, Agabiti Rosei E, Azizi M, Burnier M, et al. 2018 ESC/ESH guidelines for the management of arterial hypertension. Eur Heart J. 2018;39(33):3021–104.
10. Volpe M, Carnovali M, Mastromarino V. The natriuretic peptides system in the pathophysiology of heart failure: from molecular basis to treatment. Clin Sci. 2015;130(2):57–77.
11. Maisel AS, Duran JM, Wettersten N. Natriuretic peptides in heart failure. Heart Fail Clin. 2018;14(1):13–25.
12. McDonagh TA, Metra M, Adamo M, Gardner RS, Baumbach A, Böhm M, et al. 2021 ESC guidelines for the diagnosis and treatment of acute and chronic heart failure. Eur Heart J. 2021;42(36):3599–726.
13. Willeit P, Kaptoge S, Welsh P, Butterworth AS, Chowdhury R, Spackman SA, et al. Natriuretic peptides and integrated risk assessment for cardiovascular disease: an individual-participant-data meta-analysis. Lancet Diabetes Endocrinol. 2016;4(10):840–9.
14. de Boer RA, Nayor M, deFilippi CR, Enserro D, Bhambhani V, Kizer JR, et al. Association of cardiovascular biomarkers with incident heart failure with preserved and reduced ejection fraction. JAMA Cardiol. 2018;3(3):215–24.
15. Campbell DJ, Gong FF, Jelinek MV, Castro JM, Coller JM, McGrady M, et al. Prediction of incident heart failure by serum amino-terminal pro-B-type natriuretic peptide level in a community-based cohort. Eur J Heart Fail. 2019;21(4):449–59.
16. Ravassa S, Kuznetsova T, Varo N, Thijs L, Delles C, Dominiczak A, et al. Biomarkers of cardiomyocyte injury and stress identify left atrial and left ventricular remodelling and dysfunction: a population-based study. Int J Cardiol. 2015;185:177–85.
17. Neeland IJ, Drazner MH, Berry JD, Ayers CR, DeFilippi C, Seliger SL, et al. Biomarkers of chronic cardiac injury and hemodynamic stress identify a malignant phenotype of left ventricular hypertrophy in the general population. J Am Coll Cardiol. 2013;61(2):187–95.
18. Myhre PL, Claggett B, Yu B, Skali H, Solomon SD, Røsjø H, et al. Sex and race differences in N-terminal pro–B-type natriuretic peptide concentration and absolute risk of heart failure in the community. JAMA Cardiol. 2022;7(6):623–31.
19. Bozkurt B, Coats AJS, Tsutsui H, Abdelhamid CM, Adamopoulos S, Albert N, et al. Universal definition and classification of heart failure: a report of the Heart Failure Society of America, Heart Failure Association of the European Society of Cardiology, Japanese Heart Failure Society and Writing Committee of the Universal Definition o. Eur J Heart Fail. 2021;23(3):352–80.
20. White HD. Pathobiology of troponin elevations. J Am Coll Cardiol. 2011;57(24):2406–8.
21. Aeschbacher S, Schoen T, Bossard M, van der Lely S, Glättli K, Todd J, et al. Relationship between high-sensitivity cardiac troponin i and blood pressure among young and healthy adults. Am J Hypertens. 2015;28(6):789–96.
22. McEvoy JW, Chen Y, Nambi V, Ballantyne CM, Sharrett AR, Appel LJ, et al. High-sensitivity cardiac troponin T and risk of hypertension. Circulation. 2015;132(9):825–33.

23. Seliger SL, Hong SN, Christenson RH, Kronmal R, Daniels LB, Lima JAC, et al. High-sensitive cardiac troponin T as an early biochemical signature for clinical and subclinical heart failure: MESA (multi-ethnic study of atherosclerosis). Circulation. 2017;135(16):1494–505.
24. Yan I, Börschel CS, Neumann JT, Sprünker NA, Makarova N, Kontto J, et al. High-sensitivity cardiac troponin I levels and prediction of heart failure. JACC Heart Fail. 2020;8(5):401–11.
25. Evans JDW, Dobbin SJH, Pettit SJ, Di Angelantonio E, Willeit P. High-sensitivity cardiac troponin and new-onset heart failure. JACC Heart Fail. 2018;6(3):187–97.
26. de Lemos JA, Drazner MH, Omland T, Ayers CR, Khera A, Rohatgi A, et al. Association of troponin T detected with a highly sensitive assay and cardiac structure and mortality risk in the general population. JAMA. 2010;304(22):2503–12.
27. Cauwenberghs N, Ravassa S, Thijs L, Haddad F, Yang W-Y, Wei F-F, et al. Circulating biomarkers predicting longitudinal changes in left ventricular structure and function in a general population. J Am Heart Assoc. 2019;8(2):e010430.
28. Siciliano M, Mettimano M, Dondolini-Poli A, Ballarin S, Migneco A, Annese R, et al. Troponin I serum concentration: a new marker of left ventricular hypertrophy in patients with essential hypertension. Ital Heart J. 2000;1(8):532–5.
29. Sato Y, Yamamoto E, Sawa T, Toda K, Hara T, Iwasaki T, et al. High-sensitivity cardiac troponin T in essential hypertension. J Cardiol. 2011;58(3):226–31.
30. Pandey A, Keshvani N, Ayers C, Correa A, Drazner MH, Lewis A, et al. Association of cardiac injury and malignant left ventricular hypertrophy with risk of heart failure in African Americans. JAMA Cardiol. 2019;4(1):51.
31. Jagodzinski A, Neumann JT, Ojeda F, Sörensen NA, Wild P, Münzel T, et al. Cardiovascular biomarkers in hypertensive patients with medical treatment—results from the randomized TEAMSTA protect I trial. Clin Chem. 2017;63(12):1877–85.
32. Ackermann MA, Kontrogianni-Konstantopoulos A. Myosin binding protein-C: a regulator of actomyosin interaction in striated muscle. J Biomed Biotechnol. 2011;2011:1–9.
33. Govindan S, Kuster DW, Lin B, Kahn DJ, Jeske WP, Walenga JM, et al. Increase in cardiac myosin binding protein-C plasma levels is a sensitive and cardiac-specific biomarker of myocardial infarction. Am J Cardiovasc Dis. 2013;3(2):60–70.
34. Kaier TE, Twerenbold R, Puelacher C, Marjot J, Imambaccus N, Boeddinghaus J, et al. Direct comparison of cardiac myosin-binding protein C with cardiac troponins for the early diagnosis of acute myocardial infarction. Circulation. 2017;136(16):1495–508.
35. Shrivastava A, Haase T, Zeller T, Schulte C. Biomarkers for heart failure prognosis: proteins, genetic scores and non-coding RNAs. Front Cardiovasc Med. 2020;7:601364.
36. Tong CW, Dusio GF, Govindan S, Johnson DW, Kidwell DT, De La Rosa LM, et al. Usefulness of released cardiac myosin binding protein-C as a predictor of cardiovascular events. Am J Cardiol. 2017;120(9):1501–7.
37. Anand A, Chin C, Shah ASV, Kwiecinski J, Vesey A, Cowell J, et al. Cardiac myosin-binding protein C is a novel marker of myocardial injury and fibrosis in aortic stenosis. Heart. 2018;104(13):1101–8.
38. Otaki Y, Watanabe T, Kubota I. Heart-type fatty acid-binding protein in cardiovascular disease: a systemic review. Clin Chim Acta. 2017;474:44–53.
39. Ishii J, Ozaki Y, Lu J, Kitagawa F, Kuno T, Nakano T, et al. Prognostic value of serum concentration of heart-type fatty acid–binding protein relative to cardiac troponin T on admission in the early hours of acute coronary syndrome. Clin Chem. 2005;51(8):1397–404.
40. Goel H, Melot J, Krinock MD, Kumar A, Nadar SK, Lip GYH. Heart-type fatty acid-binding protein: an overlooked cardiac biomarker. Ann Med. 2020;52(8):444–61.
41. Lamounier-Zepter V, Look C, Alvarez J, Christ T, Ravens U, Schunck W-H, et al. Adipocyte fatty acid-binding protein suppresses cardiomyocyte contraction. Circ Res. 2009;105(4):326–34.
42. Arimoto T, Takeishi Y, Shiga R, Fukui A, Tachibana H, Nozaki N, et al. Prognostic value of elevated circulating heart-type fatty acid binding protein in patients with congestive heart failure. J Card Fail. 2005;11(1):56–60.
43. Gruson D, Adamantidou C, Ahn SA, Rousseau MF. Heart-type fatty acid binding protein is related to severity and established cardiac biomarkers of heart failure. Adv Lab Med. 2021;2(4):541–4.

44. Niizeki T, Takeishi Y, Arimoto T, Takabatake N, Nozaki N, Hirono O, et al. Heart-type fatty acid-binding protein is more sensitive than troponin T to detect the ongoing myocardial damage in chronic heart failure patients. J Card Fail. 2007;13(2):120–7.
45. Ishino M, Takeishi Y, Niizeki T, Watanabe T, Nitobe J, Miyamoto T, et al. Risk stratification of chronic heart failure patients by multiple biomarkers implications of BNP, H-FABP, and PTX3. Circ J. 2008;72(11):1800–5.
46. Niizeki T, Takeishi Y, Arimoto T, Takahashi T, Okuyama H, Takabatake N, et al. Combination of heart-type fatty acid binding protein and brain natriuretic peptide can reliably risk stratify patients hospitalized for chronic heart failure. Circ J. 2005;69(8):922–7.
47. Niizeki T, Takeishi Y, Takabatake N, Shibata Y, Konta T, Kato T, et al. Circulating levels of heart-type fatty acid-binding protein in a general Japanese population effects of age, gender and physiologic characteristics. Circ J. 2007;71(9):1452–7.
48. Otaki Y, Watanabe T, Takahashi H, Hirayama A, Narumi T, Kadowaki S, et al. Association of heart-type fatty acid-binding protein with cardiovascular risk factors and all-cause mortality in the general population: The Takahata Study. PLoS One. 2014;9(5):e94834.
49. Gedikli O, Ozturk S, Yilmaz H, Baykan M, Kiris A, Durmus I, et al. Relationship between arterial stiffness and myocardial damage in patients with newly diagnosed essential hypertension. Am J Hypertens. 2008;21(9):989–93.
50. Setsuta K, Seino Y, Mizuno K. Heart-type fatty acid-binding protein is a novel prognostic marker in patients with essential hypertension. Int J Cardiol. 2014;176(3):1323–5.
51. Yancy CW, Jessup M, Bozkurt B, Butler J, Casey DEJ, Colvin MM, et al. 2017 ACC/AHA/HFSA focused update of the 2013 ACCF/AHA guideline for the management of heart failure: a report of the American College of Cardiology/American Heart Association Task Force on clinical practice guidelines and the heart failure Society of America. J Am Coll Cardiol. 2017;70(6):776–803.
52. Dong R, Zhang M, Hu Q, Zheng S, Soh A, Zheng Y, et al. Galectin-3 as a novel biomarker for disease diagnosis and a target for therapy (review). Int J Mol Med. 2017;41:599.
53. Sharma UC, Pokharel S, van Brakel TJ, van Berlo JH, Cleutjens JPM, Schroen B, et al. Galectin-3 marks activated macrophages in failure-prone hypertrophied hearts and contributes to cardiac dysfunction. Circulation. 2004;110(19):3121–8.
54. Calvier L, Martinez-Martinez E, Miana M, Cachofeiro V, Rousseau E, Sádaba JR, et al. The impact of galectin-3 inhibition on aldosterone-induced cardiac and renal injuries. JACC Heart Fail. 2015;3(1):59–67.
55. Filipe MD, Meijers WC, Rogier van der Velde A, de Boer RA. Galectin-3 and heart failure: prognosis, prediction & clinical utility. Clin Chim Acta. 2015;443:48–56.
56. Lok DJ, Lok SI, Bruggink-André de la Porte PW, Badings E, Lipsic E, van Wijngaarden J, et al. Galectin-3 is an independent marker for ventricular remodeling and mortality in patients with chronic heart failure. Clin Res Cardiol. 2013;102(2):103–10.
57. Michalski B, Trzciński P, Kupczyńska K, Miśkowiec D, Pęczek Ł, Nawrot B, et al. The differences in the relationship between diastolic dysfunction, selected biomarkers and collagen turn-over in heart failure patients with preserved and reduced ejection fraction. Cardiol J. 2017;24(1):35–42.
58. Stoltze Gaborit F, Bosselmann H, Kistorp C, Iversen K, Kumler T, Gustafsson F, et al. Galectin 3: association to neurohumoral activity, echocardiographic parameters and renal function in outpatients with heart failure. BMC Cardiovasc Disord. 2016;16(1):117.
59. Felker GM, Fiuzat M, Shaw LK, Clare R, Whellan DJ, Bettari L, et al. Galectin-3 in ambulatory patients with heart failure. Circ Heart Fail. 2012;5(1):72–8.
60. van der Velde AR, Gullestad L, Ueland T, Aukrust P, Guo Y, Adourian A, et al. Prognostic value of changes in galectin-3 levels over time in patients with heart failure. Circ Heart Fail. 2013;6(2):219–26.
61. Lok DJA, Van Der Meer P, de la Porte PWB-A, Lipsic E, Van Wijngaarden J, Hillege HL, et al. Prognostic value of galectin-3, a novel marker of fibrosis, in patients with chronic heart failure: data from the DEAL-HF study. Clin Res Cardiol. 2010;99(5):323–8.

62. Gullestad L, Ueland T, Kjekshus J, Nymo SH, Hulthe J, Muntendam P, et al. The predictive value of galectin-3 for mortality and cardiovascular events in the controlled rosuvastatin multinational trial in heart failure (CORONA). Am Heart J. 2012;164(6):878–83.
63. Ghorbani A, Bhambhani V, Christenson RH, Meijers WC, de Boer RA, Levy D, et al. Longitudinal change in galectin-3 and incident cardiovascular outcomes. J Am Coll Cardiol. 2018;72(25):3246–54.
64. Yao Y, Shen D, Chen R, Ying C, Wang C, Guo J, et al. Galectin-3 predicts left ventricular remodeling of hypertension. J Clin Hypertens. 2016;18(6):506–11.
65. Ho JE, Liu C, Lyass A, Courchesne P, Pencina MJ, Vasan RS, et al. Galectin-3, a marker of cardiac fibrosis, predicts incident heart failure in the community. J Am Coll Cardiol. 2012;60(14):1249–56.
66. Suthahar N, Lau ES, Blaha MJ, Paniagua SM, Larson MG, Psaty BM, et al. Sex-specific associations of cardiovascular risk factors and biomarkers with incident heart failure. J Am Coll Cardiol. 2020;76(12):1455–65.
67. Besler C, Lang D, Urban D, Rommel K-P, von Roeder M, Fengler K, et al. Plasma and cardiac galectin-3 in patients with heart failure reflects both inflammation and fibrosis. Circ Heart Fail. 2017;10(3):e003804.
68. Pascual-Figal DA, Januzzi JL. The biology of ST2: The International ST2 consensus panel. Am J Cardiol. 2015;115(7):3B–7B.
69. Aimo A, Januzzi JL, Bayes-Genis A, Vergaro G, Sciarrone P, Passino C, et al. Clinical and prognostic significance of sST2 in heart failure. J Am Coll Cardiol. 2019;74(17):2193–203.
70. Wang Z, Pan X, Xu H, Wu Y, Jia X, Fang Y, et al. Serum soluble ST2 Is a valuable prognostic biomarker in patients with acute heart failure. Front Cardiovasc Med. 2022;9:812654.
71. Aimo A, Vergaro G, Ripoli A, Bayes-Genis A, Pascual Figal DA, de Boer RA, et al. Meta-analysis of soluble suppression of tumorigenicity-2 and prognosis in acute heart failure. JACC Heart Fail. 2017;5(4):287–96.
72. van Vark LC, Lesman-Leegte I, Baart SJ, Postmus D, Pinto YM, Orsel JG, et al. Prognostic value of serial ST2 measurements in patients with acute heart failure. J Am Coll Cardiol. 2017;70(19):2378–88.
73. Boisot S, Beede J, Isakson S, Chiu A, Clopton P, Januzzi J, et al. Serial sampling of ST2 predicts 90-day mortality following destabilized heart failure. J Card Fail. 2008;14(9):732–8.
74. Bayes-Genis A, de Antonio M, Vila J, Peñafiel J, Galán A, Barallat J, et al. Head-to-head comparison of 2 myocardial fibrosis biomarkers for long-term heart failure risk stratification. J Am Coll Cardiol. 2014;63(2):158–66.
75. Yin X, Cao H, Wei Y, Li H-H. Alteration of the IL-33-sST2 pathway in hypertensive patients and a mouse model. Hypertens Res. 2019;42(11):1664–71.
76. Coglianese EE, Larson MG, Vasan RS, Ho JE, Ghorbani A, McCabe EL, et al. Distribution and clinical correlates of the interleukin receptor family member soluble ST2 in the Framingham Heart Study. Clin Chem. 2012;58(12):1673–81.
77. Ho JE, Larson MG, Ghorbani A, Cheng S, Vasan RS, Wang TJ, et al. Soluble ST2 predicts elevated SBP in the community. J Hypertens. 2013;31(7):1431–6.
78. Ojji DB, Opie LH, Lecour S, Lacerda L, Adeyemi OM, Sliwa K. The effect of left ventricular remodelling on soluble ST2 in a cohort of hypertensive subjects. J Hum Hypertens. 2014;28(7):432–7.
79. Ojji DB, Opie LH, Lecour S, Lacerda L, Adeyemi O, Sliwa K. Relationship between left ventricular geometry and soluble ST2 in a cohort of hypertensive patients. J Clin Hypertens. 2013;15(12):899–904.
80. Hughes MF, Appelbaum S, Havulinna AS, Jagodzinski A, Zeller T, Kee F, et al. ST2 may not be a useful predictor for incident cardiovascular events, heart failure and mortality. Heart. 2014;100(21):1715–21.
81. Parikh RH, Seliger SL, Christenson R, Gottdiener JS, Psaty BM, DeFilippi CR. Soluble ST2 for prediction of heart failure and cardiovascular death in an elderly, community-dwelling population. J Am Heart Assoc. 2016;5(8):e003188.

82. Liu T, Song D, Dong J, Zhu P, Liu J, Liu W, et al. Current understanding of the pathophysiology of myocardial fibrosis and its quantitative assessment in heart failure. Front Physiol. 2017;8:238.
83. Nikolov A, Popovski N. Extracellular matrix in heart disease: focus on circulating collagen type I and III derived peptides as biomarkers of myocardial fibrosis and their potential in the prognosis of heart failure: a concise review. Meta. 2022;12(4):297.
84. Olsen MH, Christensen MK, Wachtell K, Tuxen C, Fossum E, Bang LE, et al. Markers of collagen synthesis is related to blood pressure and vascular hypertrophy: a LIFE substudy. J Hum Hypertens. 2005;19(4):301–7.
85. Laviades C, Varo N, Fernández J, Mayor G, Gil MJ, Monreal I, et al. Abnormalities of the extracellular degradation of collagen type I in essential hypertension. Circulation. 1998;98(6):535–40.
86. Lindsay MM, Maxwell P, Dunn FG. TIMP-1: a marker of left ventricular diastolic dysfunction and fibrosis in hypertension. Hypertension. 2002;40(2):136–41.
87. Plaksej R, Kosmala W, Frantz S, Herrmann S, Niemann M, Störk S, et al. Relation of circulating markers of fibrosis and progression of left and right ventricular dysfunction in hypertensive patients with heart failure. J Hypertens. 2009;27(12):2483–91.
88. dos Santos MC, Serejo F, Alcântara P, Ramalhinho V, Braz NJ. Procollagen type III amino terminal peptide and myocardial fibrosis: a study in hypertensive patients with and without left ventricular hypertrophy. Rev Port Cardiol. 2015;34(5):309–14.
89. Zhang Z-Y, Ravassa S, Yang W-Y, Petit T, Pejchinovski M, Zürbig P, et al. Diastolic left ventricular function in relation to urinary and serum collagen biomarkers in a general population. PLoS One. 2016;11(12):e0167582.
90. Duprez DA, Gross MD, Kizer JR, Ix JH, Hundley WG, Jacobs DR. Predictive value of collagen biomarkers for heart failure with and without preserved ejection fraction: MESA (multiethnic study of atherosclerosis). J Am Heart Assoc. 2018;7(5):e007885.
91. Pareek M, Bhatt DL, Vaduganathan M, Biering-Sørensen T, Qamar A, Diederichsen AC, et al. Single and multiple cardiovascular biomarkers in subjects without a previous cardiovascular event. Eur J Prev Cardiol. 2017;24(15):1648–59.
92. Querejeta R, Varo N, López B, Larman M, Artiñano E, Etayo JC, et al. Serum carboxyterminal propeptide of procollagen type I is a marker of myocardial fibrosis in hypertensive heart disease. Circulation. 2000;101(14):1729–35.
93. Packer M. The neurohormonal hypothesis: a theory to explain the mechanism of disease progression in heart failure. J Am Coll Cardiol. 1992;20(1):248–54.
94. Funke-Kaiser A, Havulinna AS, Zeller T, Appelbaum S, Jousilahti P, Vartiainen E, et al. Predictive value of midregional pro-adrenomedullin compared to natriuretic peptides for incident cardiovascular disease and heart failure in the population-based FINRISK 1997 cohort. Ann Med. 2014;46(3):155–62.
95. Schill F, Timpka S, Nilsson PM, Melander O, Enhörning S. Copeptin as a predictive marker of incident heart failure. ESC Heart Fail. 2021;8(4):3180–8.
96. Cauwenberghs N, Sabovčik F, Magnus A, Haddad F, Kuznetsova T. Proteomic profiling for detection of early-stage heart failure in the community. ESC Heart Fail. 2021;8:2928.
97. Wu J, Dong E, Zhang Y, Xiao H. The role of the inflammasome in heart failure. Front Physiol. 2021;12:709703.
98. Kardys I, Knetsch AM, Bleumink GS, Deckers JW, Hofman A, Stricker BHC, et al. C-reactive protein and risk of heart failure. The Rotterdam Study. Am Heart J. 2006;152(3):514–20.
99. Anand IS, Latini R, Florea VG, Kuskowski MA, Rector T, Masson S, et al. C-reactive protein in heart failure. Circulation. 2005;112(10):1428–34.
100. Cauwenberghs N, Sabovčik F, Vandenabeele E, Kobayashi Y, Haddad F, Budts W, et al. Subclinical heart dysfunction in relation to metabolic and inflammatory markers: a community-based study. Am J Hypertens. 2020;34:46–55.
101. Chia YC, Kieneker LM, van Hassel G, Binnenmars SH, Nolte IM, van Zanden JJ, et al. Interleukin 6 and development of heart failure with preserved ejection fraction in the general population. J Am Heart Assoc. 2021;10(11):e018549.

102. Kuznetsova T, Haddad F, Knez J, Rosenberg-Hasson Y, Sung J, Cauwenberghs N, et al. Cytokines profile in hypertensive patients with left ventricular remodeling and dysfunction. J Am Soc Hypertens. 2015;9(12):975–84.
103. Xue H, Fu Z, Chen Y, Xing Y, Liu J, Zhu H, et al. The association of growth differentiation factor-15 with left ventricular hypertrophy in hypertensive patients. PLoS One. 2012;7(10):e46534.
104. Wesseling M, Poel JHC, Jager SCA. Growth differentiation factor 15 in adverse cardiac remodelling: from biomarker to causal player. ESC Heart Fail. 2020;7(4):1488–501.
105. Ndrepepa G. Myeloperoxidase – a bridge linking inflammation and oxidative stress with cardiovascular disease. Clin Chim Acta. 2019;493:36–51.
106. von Bibra H, Paulus W, St. John Sutton M. Cardiometabolic syndrome and increased risk of heart failure. Curr Heart Fail Rep. 2016;13(5):219–29.
107. Cauwenberghs N, Knez J, Thijs L, Haddad F, Vanassche T, Yang W-Y, et al. Relation of insulin resistance to longitudinal changes in left ventricular structure and function in a general population. J Am Heart Assoc. 2018;7(7):e008315.
108. Han X, Zhang S, Chen Z, Adhikari BK, Zhang Y, Zhang J, et al. Cardiac biomarkers of heart failure in chronic kidney disease. Clin Chim Acta. 2020;510:298–310.
109. Kottgen A, Russell SD, Loehr LR, Crainiceanu CM, Rosamond WD, Chang PP, et al. Reduced kidney function as a risk factor for incident heart failure: the atherosclerosis risk in communities (ARIC) study. J Am Soc Nephrol. 2007;18(4):1307–15.
110. Mettimano M, Specchia ML, Migneco A, Savi L. Microalbuminuria as a marker of cardiac damage in essential hypertension. Eur Rev Med Pharmacol Sci. 2001;5(1):31–6.
111. Catena C, Colussi G, Fedrizzi S, Sechi LA. Association of a prothrombotic state with left-ventricular diastolic dysfunction in hypertension. J Hypertens. 2013;31(10):2077–84.
112. Dhingra R, Gona P, Wang TJ, Fox CS, D'Agostino RBS, Vasan RS. Serum gamma-glutamyl transferase and risk of heart failure in the community. Arterioscler Thromb Vasc Biol. 2010;30(9):1855–60.
113. Silvestre OM, Gonçalves A, Nadruz W, Claggett B, Couper D, Eckfeldt JH, et al. Ferritin levels and risk of heart failure—the atherosclerosis risk in communities study. Eur J Heart Fail. 2017;19(3):340–7.
114. Castiglione V, Aimo A, Vergaro G, Saccaro L, Passino C, Emdin M. Biomarkers for the diagnosis and management of heart failure. Heart Fail Rev. 2022;27(2):625–43.
115. Velagaleti RS, Gona P, Larson MG, Wang TJ, Levy D, Benjamin EJ, et al. Multimarker approach for the prediction of heart failure incidence in the community. Circulation. 2010;122(17):1700–6.
116. Brouwers FP, van Gilst WH, Damman K, van den Berg MP, Gansevoort RT, Bakker SJL, et al. Clinical risk stratification optimizes value of biomarkers to predict new-onset heart failure in a community-based cohort. Circ Heart Fail. 2014;7(5):723–31.
117. Wik L, Nordberg N, Broberg J, Björkesten J, Assarsson E, Henriksson S, et al. Proximity extension assay in combination with next-generation sequencing for high-throughput proteome-wide analysis. Mol Cell Proteomics. 2021;20:100168.
118. Stenemo M, Nowak C, Byberg L, Sundström J, Giedraitis V, Lind L, et al. Circulating proteins as predictors of incident heart failure in the elderly. Eur J Heart Fail. 2018;20(1):55–62.
119. Egerstedt A, Berntsson J, Smith ML, Gidlöf O, Nilsson R, Benson M, et al. Profiling of the plasma proteome across different stages of human heart failure. Nat Commun. 2019;10(1):5830.
120. Kuznetsova T, Mischak H, Mullen W, Staessen JA. Urinary proteome analysis in hypertensive patients with left ventricular diastolic dysfunction. Eur Heart J. 2012;33(18):2342–50.
121. Ferreira JP, Verdonschot J, Collier T, Wang P, Pizard A, Bär C, et al. Proteomic bioprofiles and mechanistic pathways of progression to heart failure. Circ Heart Fail. 2019;12(5):e005897.
122. Bayes-Genis A, Liu PP, Lanfear DE, de Boer RA, González A, Thum T, et al. Omics phenotyping in heart failure: the next frontier. Eur Heart J. 2020;41(36):3477–84.
123. Arnett DK, Claas SA. Omics of blood pressure and hypertension. Circ Res. 2018;122(10):1409–19.

Natriuretic Peptides in the Progression to Heart Failure in Hypertensive Patients

18

Massimo Volpe and Speranza Rubattu

18.1 Introduction

The family of natriuretic peptides (NPs) includes a group of hormones produced by the heart (atrial and brain natriuretic peptides (ANPs and BNPs, respectively)) and by the endothelium (C-type natriuretic peptides (CNPs)). NPs exhibit important cardiovascular properties and play a key role in the regulation of volume and electrolyte homeostasis, peripheral vascular resistance, and blood pressure levels [1].

NPs are also closely involved in the process of cardiovascular remodeling through their antiproliferative, anti-hypertrophic, and anti-fibrotic properties [2]. All these characteristics make them important players in many pathological conditions of the cardiovascular system and critical tools in the diagnosis and prognosis of cardiovascular diseases such as hypertension and heart failure (HF). They also represent attractive targets for therapeutic purposes [2, 3].

M. Volpe (✉)
Department of Clinical and Molecular Medicine, School of Medicine and Psychology, Sapienza University of Roma, Rome, Italy

IRCCS San Raffaele, Rome, Italy
e-mail: massimo.volpe@uniroma1.it

S. Rubattu (✉)
Department of Clinical and Molecular Medicine, School of Medicine and Psychology, Sapienza University of Roma, Rome, Italy

IRCCS Neuromed, Pozzilli, Isernia, Italy
e-mail: speranzadonatella.rubattu@uniroma1.it

M. Dorobantu et al. (eds.), *Hypertension and Heart Failure*, Updates in Hypertension and Cardiovascular Protection,
https://doi.org/10.1007/978-3-031-39315-0_18

18.2 The Biology of Natriuretic Peptides

The atrial natriuretic peptide (ANP) was the first component of the NP family to be discovered 40 years ago through the pioneering work of de Bold and colleagues in rat atrial cardiomyocytes [4, 5]. In their work, ANP was synthesized and secreted in response to myocardial stretch. The human ANP gene (*NPPA*) encodes a 151 amino acid peptide (pre-proANP), subsequently cleaved to release the pre-signal peptide. ProANP is stored in specific granules within atrial cardiomyocytes and is further processed by the corin enzyme to obtain the 1–28 amino acid carboxyl-terminal end (αANP or $ANP_{99–126}$) along with the 1–98 amino acid-terminal peptide (NT-proANP or $ANP_{1–98}$) [6]. The cleavage by corin takes place within the cellular plasma membrane during the process of secretion [7]. Corin, in turn, results from the cleavage of a precursor. Both carboxyl-terminal ($ANP_{99–126}$) and amino-terminal ($ANP_{1–98}$) forms circulate in the blood in a 1:1 molar ratio. Urodilatin is a peptide that is transcribed by *NPPA* but is differentially processed in the kidneys between amino acids 95 and 96 (instead of amino acids 98 and 99) and can be detected only in urine [8].

The brain natriuretic peptide (BNP) is encoded by *NPPB* and is synthetized by ventricular cardiomyocytes in response to hemodynamic stress. The genes encoding ANP and BNP map closely on the distal arm of chromosome 1 (1p36.2). The cleavage of the BNP precursor by furin leads to a 32 amino acid carboxyl-terminal peptide along with the amino-terminal $NT\text{-}proBNP_{1–76}$ [1, 2]. Both peptides circulate in the blood. Finally, CNP, encoded by *NPPC*, is of endothelial origin and is cleaved into a CNP-53 biological active peptide that is further processed to yield a 22 amino acid peptide with identical biological activity in vitro as CNP-53 [9]. CNP does not circulate in the blood.

Three main receptors mediate the NP functions: the particulate guanylyl cyclase type A (GC-A) and type B (GC-B) receptors and the natriuretic peptide clearance receptor (NPR-C). The latter is mainly involved in NP clearance by lysosomal degradation.

$ANP_{99–126}$ activates the GC-A receptor, also known as the type A natriuretic peptide receptor (NPR-A), with an affinity higher than that of the BNP carboxyl-terminal peptide [10]. The receptor's activation leads to the production of the second messenger cyclic guanosine monophosphate (cGMP) and to a consequent increase in protein kinase G, gated ion channels, and phosphodiesterase. Protein kinase G is the main effector of the biological functions of both αANP and $BNP_{1–32}$. On the other hand, NPR-B mediates the effects of the CNP. NPs exert diuretic, natriuretic, vasorelaxant, and anti-hypertrophic and anti-fibrotic effects on the cardiovascular system (Fig. 18.1).

Different molecules are responsible for αANP degradation, in particular NPR-C, the neprilysin (NEP) enzyme, and the insulin-degrading enzyme (IDE) [2]. The half-life of αANP is limited to a few minutes. The amino-terminal ($ANP_{1–98}$) has a longer half life (1 2 h) and is mainly excreted with urine.

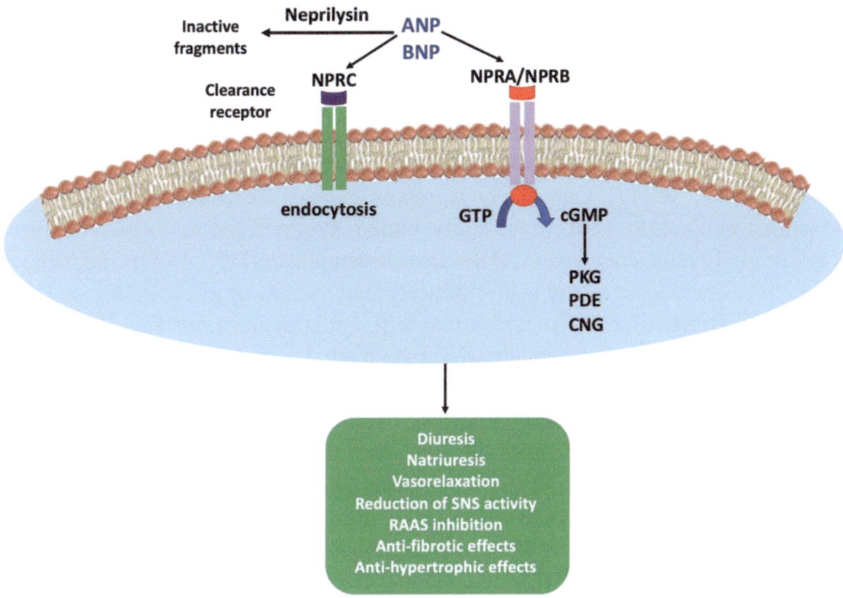

Fig. 18.1 A schematic representation of the natriuretic peptide family and their cardiovascular functions

NEP and NPR-C also mediate the clearance of the BNP carboxyl-terminal peptide, whereas the amino-terminal NT-proBNP$_{1-76}$ is cleared through the kidneys and has a longer half-life (22 min and 70 min, respectively).

Over the last two decades, the assay of the circulating NT-proBNP and BNP levels has become the most common and represents a widely used diagnostic and prognostic tool in clinical practice, mainly for guiding heart failure (HF) diagnosis (or excluding it) and for clinical management. The European Society of Cardiology considers a BNP level >35 pg/mL and/or an NT-proBNP >125 pg/mL suggestive of chronic HF. Higher values are required for a diagnosis in the acute setting: >100 pg/mL and >300 pg/mL for BNP and NT-proBNP, respectively [11]. In the context of HF associated with obesity, where the NP level is significantly reduced for both reduced synthesis and increased clearance [12], a lower cutoff can be proposed. In fact, a reduction of 50% has been suggested by the European Society of Cardiology [13]. Specifically, a cutoff level of 54 pg/ml for BNPs may be indicative of HF patients with a body mass index (BMI) >35 kg/m^2.

Despite having a longer half-life, NT-proANP has failed to emerge as a diagnostic marker due to the increased number of cleaved fragments that limit its detection [14]. The recent introduction of the mid-regional NT-proANP (MR-NT-proANP) measurement may offer more specific and appropriate clinical applications [15].

18.3 NPs in Hypertension and Hypertension-Mediated Organ Damage (HMOD)

Experimental and epidemiological studies have indicated that NPs, mostly ANPs, play a contributory role in the development of hypertension. In fact, the lack of the ANP gene (*Nppa*) and the NPR-A gene (*Npr1*) causes a salt-sensitive form of hypertension in mice [16, 17]. Conversely, overexpression of *Nppa* led to hypotension in the animal model [18]. Most importantly, human hypertension has been recognized as a state of NP deficiency where ANPs are relevant players [19]. As a further support to the clinical evidence, genetically induced lower levels of either ANPs or BNPs have been associated with hypertension in different human cohorts [20–22].

Within established hypertension, NPs play a contributory role in the pathogenesis of both cardiac and vascular damage. The involvement of NPs in HMOD can be in part explained by their hemodynamic and blood pressure-lowering effects, though a relevant contribution of other pleiotropic protective functions carried out by NPs within the heart and vessels cannot be excluded. Therefore, based on their properties, lower NP levels favor the development and progression of cardiovascular damage. On the other hand, therapeutic interventions, which are able to directly target the NP system and raise ANP levels, offer protection against HMOD in different experimental models of hypertension [23, 24]. Since no antihypertensive therapy based on NPs has been so far introduced into clinical practice, we are currently unable to translate to humans the experimental findings regarding HMOD protection based on the effect of increased NP levels [25].

Importantly, increased NP levels predict worse outcomes and mortality in hypertensive patients. This is consistent with the notion that the prognostic impact of NPs on patients with CVD reflects the underlying cardiovascular dysfunction and damage. Therefore, NPs can be viewed as markers of cardiac, vascular, and renal functions in hypertension. In this regard, the most striking evidence relates to the relationship with cardiac damage [2].

18.4 NPs and Left Ventricular Hypertrophy in Hypertension

The systemic hemodynamic effects of NPs contribute to maintaining cardiovascular balance in physiological conditions. In addition, these peptides exhibit local anti-hypertrophic and anti-fibrotic properties documented by excess cardiac mass and fibrosis in the absence of *Nppa*, *Npr1*, or *Nppb* (BNP gene) [1]. The secretion of both ANPs and BNPs depends on the left ventricular wall stress, which results from a complex interplay between left ventricular diameter and wall thickness. Both systolic and diastolic dysfunction affect NP levels. Moreover, extracardiac factors, such as increased ventricular afterload in hypertension and the presence of renal failure, responsible for reduced NP clearance, influence the circulating NP levels.

As stated above, ANPs are produced by the atria in adult life. The ventricles express ANPs in the developing embryo and fetus. ANP secretion by ventricular cardiomyocytes is reinduced in adult life through the activation of a fetal program only under pathological conditions, thus affecting the heart. Left ventricular hypertrophy (LVH) is a typical pathological condition associated with increased ANP and BNP secretion from the ventricles. Notably, levels of NPs did not change in response to the adaptive physiological cardiac hypertrophy induced by a 28-day swimming protocol in male Sprague Dawley rats [26]. The different responses of NPs to physiological and pathological cardiac hypertrophy further underscore the key role of these cardiac hormones as a defensive mechanism in pathology. Other common disease conditions associated with increased myocardial production of and release of NPs include heart failure (both HF with reduced ejection fraction (HFrEF) and HF with preserved EF (HFpEF)) and atrial fibrillation.

Many studies have indeed shown a correlation of BNP and/or NT-proBNP levels with indicators of left ventricular and right ventricular functions as assessed by echocardiography. Moreover, increased plasma BNP and NT-proBNP levels are associated with left ventricular systolic and diastolic dysfunction, left ventricular dilatation such as in cardiomyopathies, left ventricular hypertrophy, pulmonary hypertension, and significant valvular disease in asymptomatic or mildly symptomatic patients [27, 28].

The value of NT-proBNP levels to identify the presence of LVH was supported by several studies performed in hypertensive patients [29–32].

In particular, the study by Conen et al. established that echocardiographic LVH can be excluded based on a single blood sample for the determination of BNPs [30]. The optimal cutoff points for the diagnosis of LVH was 35 pg/ml for BNPs (sensitivity 73%, specificity 72%). Nishikimi et al. demonstrated that measurement of plasma ANP and BNP levels may be useful for the detection of concentric LVH in patients with essential hypertension [32].

In a cohort of 139 patients with moderate hypertension, the NP levels increased with an increasing number of markers of cardiac remodeling (LVH, left ventricular dysfunction (LVD), or left ventricular dilatation). In this cohort, NT-proBNP and BNP levels were revealed as useful markers to discriminate between patients with either the presence or absence of cardiac remodeling, and they proved to be superior to the NT-proANP level. Specifically, left ventricular hypertrophy, left ventricular dysfunction, or left ventricular dilatation (all indicators of cardiac remodeling) were independently correlated with logarithmic (ln) NT-proBNP and ln BNP, whereas only left ventricular diameter was independently correlated with ln NT-proANP [33].

Another study confirmed that BNPs reflect the cardiac remodeling process in hypertension [34]. Out of few tested biomarkers, BNPs were the only biochemical parameter that independently predicted interventricular septal diastolic diameter, left ventricular mass index, and the ratio of the velocity–time integrals of the E and A waves of the mitral inflow, thus reflecting the magnitude of LVH and diastolic function.

However, some uncertainties exist on this matter. Vasan et al. [35] and Nakamura et al. [36] raised some concerns about the appropriateness of NP measurement to detect LVH. In fact, they did not support the use of NP levels as a more powerful marker than echocardiography to detect LVH since the power of the cardiac hormones may not be sufficient for large-scale screenings of hypertensive patients.

18.5 NPs and Cardiac Dysfunction in Hypertension

Myocardial hypertrophy is the main mechanism of compensation for hemodynamic overload in hypertension. This process is accompanied by fibrosis, which may lead to diastolic and systolic dysfunction. Because of the association between LVH and higher NP levels in hypertension, it is not surprising to observe that LV diastolic dysfunction associates, in turn, with elevated NT-proBNP levels in patients with HFpEF, a frequent consequence of hypertension. Herein, LV diastolic dysfunction often reflects a ventricular dysfunction secondary to cardiac chamber stiffness and atrial overload associated with LVH. It has been reported that the NT-proBNP value progressively rises as a function of gradual deterioration of LV diastolic dysfunction among HFpEF patients [37].

The study by Talwar et al. assessed the effect of both LVH and left ventricular systolic dysfunction (LVSD) on plasma concentration of NT-proBNP [26]. The authors found that an increased NT-proBNP concentration in a hypertensive population reflected LVSD, but it did not provide further clinical information about LVH. Thus, measurement of NT-proBNP in hypertensive patients may provide information on the presence of LVH or LVSD.

18.6 Prediction of Cardiac Damage in Hypertension

Because of their functional effects on cardiovascular hemodynamics and structure, a predictive role of NPs in CVDs in the general population has been reported. NT-proBNP levels were able to predict future coronary heart disease (CHD), HF, and stroke in subjects free from CVDs at the time of enrollment in one of the largest cohorts of the general population that included more than 95,000 subjects [38]. Brouwers et al. demonstrated that NP levels predicted new-onset HF in a community-based cohort, particularly among individuals at a higher risk, including subjects with high BP levels [39].

Several efforts were made to assess the ability of NPs to specifically detect preclinical cardiac abnormalities. The results were not quite consistent, likely because of the complex interaction between left ventricular mass, diastolic and systolic dysfunction, high BP, and aortic stiffness, all underlying the cardiac damage. The prevalence of asymptomatic left ventricular dysfunction (ALVD) in subjects at risk of HF is about 5%. Screening for ALVD in subjects at risk of HF can affect clinical management. A study examined the role of NT-pro BNP in the diagnosis of ALVD in subjects with hypertension and diabetes from primary care and revealed that NT-proBNP levels can be used by general practitioners to rule out ALVD in hypertensive or diabetic patients [40]. These results highlight the ability of BNPs to identify early the cardiac damage in asymptomatic subjects.

Consistent with the above-reported evidence, in a study by Mueller et al., BNP and NT-proBNP were able to indicate the presence of cardiac structural disease in asymptomatic hypertensive patients [41]. However, the authors recognized the limited value of the BNP or NT-proBNP measurement as a screening strategy because of the low specificity at the cutoff values chosen in their setting.

Another investigation reported a gradual increase of NT-proBNP levels in the presence of HMOD. Consequently, NT-proBNP was proposed as a suitable surrogate to detect subclinical dysfunctions of the aorta, left ventricle, and kidneys before the onset of HF [42]. In the same investigation, a decrease in NT-proBNP levels under antihypertensive treatment was associated with a better cardiovascular outcome, mainly under the effect of calcium channel blockers or angiotensin receptor blockers, which reduce the vascular load and LVH more efficiently than do beta-blockers. This observation suggested the value of NT-proBNP as a potential biomarker to monitor the efficacy of antihypertensive treatment toward target organ damage (TOD), apart from being a good integrator of HMOD.

18.7 The Role of BNP-Based Strategies in Preventing HF in Hypertension

Due to the high impact of HF on national health systems for its heavy socioeconomic burden, a timely prediction of HF in subjects at the highest risk, such as patients affected by hypertension, has fundamental relevance to starting an effective and timely management. In particular, the transition from stage C to stage D of HF represents an irreversible process toward the end-stage disease. Patients should be identified at an earlier stage (stage A or B HF) [43]. Regarding this matter, the use of NPs may prove to be extremely helpful. In fact, an increase in NP levels has been shown to have prognostic implications for cardiovascular events and HF in both asymptomatic individuals from the general population and in a wide variety of CVD patients [2]. Therefore, the use of BNPs has been successfully introduced in programs aimed to prevent HF. The first trial, the St. Vincent's Screening to Prevent Heart Failure (STOP-HF), included asymptomatic individuals with one or more of the following disease conditions: hypertension, hyperlipidemia, obesity, vascular disease (coronary, cerebrovascular, peripheral arteriopathy), diabetes mellitus, arrhythmia, and valvular heart disease. Overall, subjects belonged to either stage A or stage B HF. The trial compared a BNP-based screening in combination with collaborative care between primary care physicians and cardiovascular specialists. The control group received the primary care physicians and specialist care when required. As a likely combination of several factors, i.e., facilitating targeted therapy changes, promoting adherence to therapy and lifestyle advice, along with an increased use of diagnostic tests, the intervention group showed a greater reduction of HF development [44].

A second trial, Nt-proBNP Guided Primary Prevention of CV Events in Diabetic Patients (PONTIAC), used the NP-based approach as part of a strategy to identify subjects at the highest risk of CV events, targeting treatment to these groups to prevent HF and other CV disorders. The study included 300 diabetic patients free from CVDs but with increased levels of NT-proBNP, who were randomized to either standard treatment or an intensified strategy involving a cardiac outpatient clinic. As a result, the intensified treatment led to a significant reduction of CVD development, HF, and hospitalizations [45]. Interestingly, the incorporation of NP levels into clinical information led to a superior risk prediction of HF, compared to a model only incorporating risk factors, among individuals with dysglycemia [46].

It should be noted that in both STOP-HF and PONTIAC studies, the reduction of blood pressure was similar in the two study groups (intervention and control groups). Thus, mechanisms different from blood pressure reduction caused the observed decreased occurrence of the end points in both studies.

In the context of hypertension, two studies demonstrated the role of NPs in the prediction of LVH and HF [33, 47]. In the latter study, plasma NT-proBNP levels provided incremental prognostic information beyond traditional risk factors (age, sex, ethnicity, systolic blood pressure, diabetes mellitus, smoking, estimated glomerular filtration rate, medications, LV mass index, interim myocardial infarction) and the magnetic resonance imaging-determined LV mass index for incident symptomatic HF in an asymptomatic multiethnic population.

Due the known equimolar amount between NT-proBNP and BNP in the circulation, other studies investigated the prognostic role of the NT-proBNP/BNP molar ratio in stable outpatients with cardiovascular risk factors, including hypertension. The results obtained in a Japanese population fulfilled the expectancy by demonstrating that stable outpatients with a high molar ratio of NT-proBNP/BNP developed HF-related events more frequently than did those with a low molar ratio [48]. The study indicated that the NT-proBNP/BNP molar ratio provides useful prognostic implications for HF management. Contradictory results were obtained about the simultaneous routine measurement of NT-proBNP and BNP in a stage A/B HF European Caucasian population as part of the STOP-HF follow-up study [49]. This study found that the lower baseline NT-proBNP/BNP molar ratio was associated with an elevated risk of HF or LVD over a 5-year follow-up period, independently from the NT-proBNP level and other covariates. However, the NT-proBNP/BNP ratio did not improve the C-statistic beyond the NT-proBNP level alone. Because of the significant variability of the ratio, the authors of the study did not support routine simultaneous measurement of BNP and NT-proBNP levels in programs aimed to prevent HF.

More recently, a genetic sub-study from the STOP-HF trial has investigated the role of the rs198389 single nucleotide polymorphism (SNP), which is associated with higher BNP levels [50], in the clinical phenotype and risk associated with chronic exposure to higher concentrations of this protective cardiac hormone in subjects at risk of HF. The study provided significant evidence that a lifelong, genetically induced elevation of BNPs reduces the risk of hypertension and of major adverse cardiovascular events. More importantly, the results of this study demonstrated that the favorable condition related to the genetically induced elevation of BNPs reduces the progression of left ventricular systolic dysfunction and the development of HF [51]. Therefore, a genetic setup responsible for increased BNP levels appears to be protective toward hypertension-related HF development, thus reinforcing the main results of the original STOP-HF trial.

Both the STOP-HF and PONTIAC randomized clinical trials are included in the 2014 Canadian Cardiovascular Society Heart Failure Management Guidelines as a recommendation for the use of NPs in at-risk individuals [52]. These guidelines recommend that NP levels should be used to implement strategies to prevent HF in individuals with risk factors for the development of HF. In the same guidelines, an increased level of BNP >100 pg/ml and of NT-proBNP >300 pg/ml is recommended to avoid overscreening. However, this recommendation does not appear to be based on any analysis and is not supported by the cost-effectiveness analysis carried out by the STOP-HF team. More recent guidelines from the United States [12] have recommended, for patients at risk of developing HF, an NP biomarker-based screening followed by team-based care, including a CV specialist optimizing guideline-directed medical therapy. This strategy can be useful for preventing the development of LV dysfunction (systolic or diastolic) or new-onset HF. However, there is a need for further studies to determine the cost-effectiveness and the risk of such screening as well as its impact on both quality of life and mortality rate.

18.8 Conclusions

As a result of an extensive literature investigating the involvement of NPs in hypertension and related cardiac damage, a key role of NP levels has emerged as a marker of progression to HF. NP levels start increasing as soon as the cardiac damage is established, and these levels represent a key feature of all subsequent steps of the disease. Consequently, early detection of an increase of NP levels in hypertensive patients can be used as a valuable tool to guide a prompt efficacious therapeutic strategy. This approach should help prevent the deleterious consequences of hypertensive cardiac damage and the development of HF (Fig. 18.2). It is also expected that a therapeutic antihypertensive strategy based on promoting increased NP levels could contribute to avoiding the establishment of a more severe cardiac disease and congestive HF in hypertensive patients.

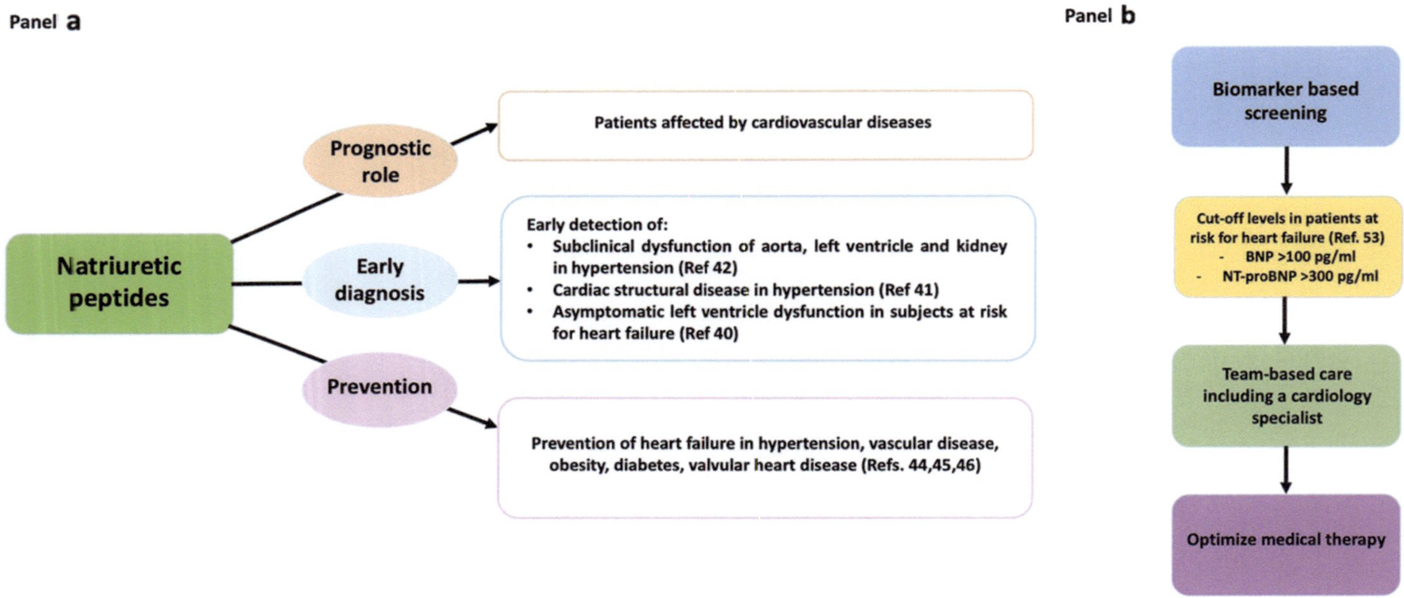

Fig. 18.2 (**a**) Natriuretic peptides (NPs) carry out prognostic, diagnostic, and preventive functions in cardiovascular diseases. Due to their ability to detect early signs of cardiac abnormalities, they can allow timely interventions and prevent the development of HF. (**b**) A biomarker-based screening is proposed as a valuable strategy to prevent HF development based on the results of the STOP-HF and PONTIAC studies. Specific cutoff values of NPs, such as those proposed by the 2014 Canadian guidelines, can be used to screen individuals more susceptible to developing HF with the aim to optimize medical interventions and monitoring

References

1. Goetze JP, Bruneau BG, Ramos HR, Ogawa T, de Bold MK, de Bold AJ. Cardiac natriuretic peptides. Nat Rev Cardiol. 2020;17:698–717.
2. Volpe M, Rubattu S, Burnett J Jr. Natriuretic peptides in cardiovascular diseases: current use and perspectives. Eur Heart J. 2014;35:419–25.
3. Rubattu S, Volpe M. Natriuretic peptides in the cardiovascular system: multifaceted roles in physiology, pathology and therapeutics. Int J Mol Sci. 2019;20:E3991.
4. de Bold AJ, Borenstein HB, Veress AT, Sonnenberg H. A rapid and potent natriuretic response to intravenous injection of atrial myocardial extract in rats. Life Sci. 1981;28:89–94.
5. de Bold AJ. Atrial natriuretic factor: a hormone produced by the heart. Science. 1985;230:767–70.
6. Chen, et al. PCSK6-mediated corin activation is essential for normal blood pressure. Nat Med. 2015;21:1048–53.
7. Yan W, Wu F, Morser J, Wu Q. Corin, a transmembrane cardiac serine protease, acts as a pro-atrial natriuretic peptide-converting enzyme. Proc Natl Acad Sci U S A. 2000;97:8525–9.
8. Forssmann W, Meyer M, Forssmann K. The renal urodilatin system: clinical implications. Cardiovasc Res. 2001;51:450–62.
9. Moyes AJ, Hobbs AJ. C-type natriuretic peptide: a multifaceted paracrine regulator in the heart and vasculature. Int J Mol Sci. 2019;20:2281.
10. Pandey KN. Molecular signaling mechanisms and function of natriuretic peptide receptor-A in the pathophysiology of cardiovascular homeostasis. Front Physiol. 2021;12:693099.
11. Ponikowski P, et al. 2016 ESC Guidelines for the diagnosis and treatment of acute and chronic heart failure: The Task Force for the diagnosis and treatment of acute and chronic heart failure of the European Society of Cardiology (ESC). Developed with the special contribution of the Heart Failure Association (HFA) of the ESC. Eur J Heart Fail. 2016;18:891–975.
12. Yancy CW, Jessup M, Bozkurt B, Butler J, Casey DE Jr, Colvin MM, Drazner MH, Filippatos GS, Fonarow GC, Givertz MM, Hollenberg SM, Lindenfeld J, Masoudi FA, McBride PE, Peterson PN, Stevenson LW, Westlake C. 2017 ACC/AHA/HFSA focused update of the 2013 ACCF/AHA guideline for the management of heart failure: a report of the American College of Cardiology/American Heart Association Task Force on Clinical Practice Guidelines and the Heart Failure Society of America. J Card Fail. 2017;23:628–51.
13. Zhai AB, Haddad H. The impact of obesity on heart failure. Curr Opin Cardiol. 2017;32:196–202.
14. Vesely DL. Atrial natriuretic hormones originating from the N-terminus of the atrial natriuretic factor prohormone. Clin Exp Pharmacol Physiol. 1995;22:108–14.
15. Chen Y, Wen Z, Peng L, Liu X, Luo Y, Wu B, Li S. Diagnostic value of MR-proANP for heart failure in patients with acute dyspnea: a meta-analysis. Acta Cardiol. 2020;75:68–74.
16. John SW, Krege JH, Oliver PM, Hagaman JR, Hodgin JB, Pang SC, Flynn TG, Smithies O. Genetic decreases in atrial natriuretic peptide and salt-sensitive hypertension. Science. 1995;267:679–81.
17. Lopez MJ, Wong SK, Kishimoto I, Dubois S, Mach V, Friesen J, Garbers DL, Beuve A. Salt-resistant hypertension in mice lacking the guanylyl cyclase-A receptor for atrial natriuretic peptide. Nature. 1995;378:65–8.
18. Melo LG, Steinhelper ME, Pang SC, Tse Y, Ackermann U. ANP in regulation of arterial pressure and fluid-electrolyte balance: lessons from genetic mouse models. Physiol Genomics. 2000;3:45–58.
19. Macheret F, Heublein D, Costello-Boerrigter LC, Boerrigter G, McKie P, Bellavia D, Mangiafico S, Ikeda Y, Bailey K, Scott CG, Sandberg S, Chen HH, Malatino L, Redfield MM, Rodeheffer R, Burnett J Jr, Cataliotti A. Human hypertension is characterized by a lack of activation of the antihypertensive cardiac hormones ANP and BNP. J Am Coll Cardiol. 2012;60:1558–65.
20. Salo PP, Havulinna AS, Tukiainen T, Raitakari O, Lehtimäki T, Kähönen M, Kettunen J, Männikkö M, Eriksson JG, Jula A, Blankenberg S, Zeller T, Salomaa V, Kristiansson K,

Perola M. Genome-wide association study implicates atrial natriuretic peptide rather than B-type natriuretic peptide in the regulation of blood pressure in the general population. Circ Cardiovasc Genet. 2017;10:e001713.
21. Cannone V, Scott CG, Decker PA, Larson NB, Palmas W, Taylor KD, Wang TJ, Gupta DK, Bielinski SJ, Burnett JC Jr. A favorable cardiometabolic profile is associated with the G allele of the genetic variant rs5068 in African Americans: the multi-ethnic study of atherosclerosis (MESA). PLoS One. 2017;12:e0189858.
22. Rubattu S, Evangelista A, Barbato D, Barba G, Stanzione R, Iacone R, Volpe M, Strazzullo P. Atrial natriuretic peptide (ANP) gene promoter variant and increased susceptibility to early development of hypertension in humans. J Hum Hypertens. 2007;21:822–4.
23. Rubattu S, Cotugno M, Forte M, Stanzione R, Bianchi F, Madonna M, Marchitti S, Volpe M. Effects of dual Angiotensin type 1 receptor/neprilysin inhibition versus Angiotensin type 1 receptor inhibition on target organ injury in the stroke-prone spontaneously hypertensive rat. J Hypertens. 2018;36:1902–14.
24. Altara R, da Silva GJJ, Frisk M, Spelta F, Zouein FA, Louch WE, Booz GW, Cataliotti A. Cardioprotective effects of the novel compound vastiras in a preclinical model of end-organ damage. Hypertension. 2020;75:1195–204.
25. Rubattu S, Gallo G. Natriuretic peptides for hypertension treatment. High Blood Pressure Cardiovasc Prevent. 2022;29:15–21.
26. Dunn ME, et al. Natriuretic peptides as differential biomarkers allowing for the distinction between physiologic and pathologic left ventricular hypertrophy. Toxicol Pathol. 2017;45:344–52.
27. Talwar S, Siebenhofer A, Williams B, Ng L. Influence of hypertension, left ventricular hypertrophy, and left ventricular systolic dysfunction on plasma N terminal proBNP. Heart. 2000;83:278–82.
28. Murakami Y, Shimada T, Inoue S, Shimizu H, Ohta Y, Katoh H, et al. New insights into the mechanism of the elevation of plasma brain natriuretic polypeptide levels in patients with left ventricular hypertrophy. Can J Cardiol. 2002;18:1294–300.
29. Mouly-Bertin C, et al. N-terminal pro-brain natriuretic peptide–a promising biomarker for the diagnosis of left ventricular hypertrophy in hypertensive women. Arch Cardiovasc Dis. 2008;101:307–15.
30. Morillas P, et al. Usefulness of NT-proBNP level for diagnosing left ventricular hypertrophy in hypertensive patients. A cardiac magnetic resonance study. Rev Esp Cardiol. 2008;61:972–5.
31. Conen D, Zeller A, Pfisterer M, Martina B. Usefulness of B-type natriuretic peptide and C-reactive protein in predicting the presence or absence of left ventricular hypertrophy in patients with systemic hypertension. Am J Cardiol. 2006;97:249–52.
32. Nishikimi T, et al. Relationship between left ventricular geometry and natriuretic peptide levels in essential hypertension. Hypertension. 1996;28:22–30.
33. Magnusson M, Jovinge S, Rydberg E, Dahlöf B, Hall C, Nielsen OW, Grubb A, Willenheimer R. Natriuretic peptides as indicators of cardiac remodeling in hypertensive patients. Blood Press. 2009;18:196–203.
34. Uusimaa P, et al. Plasma B-type natriuretic peptide reflects left ventricular hypertrophy and diastolic function in hypertension. Int J Cardiol. 2004;97:251–6.
35. Vasan RS, Benjamin EJ, Larson MG, Leip EP, Wang TJ, Wilson PW, Levy D. Plasma natriuretic peptides for community screening for left ventricular hypertrophy and systolic dysfunction: the Framingham heart study. JAMA. 2002;288:1252–9.
36. Nakamura T, Nakajima T, Suzuki N, Arai S, Suwa N. Left ventricular stiffness and chamber geometry in the pressure-overloaded hypertrophied heart. Tohoku J Exp Med. 1976;119:245–56.
37. Islam MN, Chowdhury MS, Paul GK, Debnath RC, Shakil SS. Association of diastolic dysfunction with N-terminal pro-B-type natriuretic peptide level in heart failure patients with preserved ejection fraction. Mymensingh Med J. 2019;28:333–46.
38. Willeit P, Kaptoge S, Welsh P, Butterworth AS, Chowdhury R, Spackman M, Math SA, Pennells L, Gao P, Burgess S, Freitag DF, Sweeting M, Wood AM, Cook NR, Judd S, Trompet S, Nambi V, Hecht Olsen H, Everett BM, Kee F, Arnlöv J, Salomaa V, Levy D, Kauhanen J, Laukkanen JA, Kavousi M, Ninomiya T, Casas JP, Daniels LB, Lind L, Kistorp CN, Rosenberg J, Mueller T, Rubattu S, Panagiotakos DB, Franco OH, de Lemos JA, Luchner A,

Kizer JR, Kiechl S, Salonen JT, Wannamethee G, de Boer RA, Nordestgaard BG, Andersson J, Jørgensen T, Melander O, Ballantyne CM, De Filippi C, Ridker PM, Cushman M, Rosamond WD, Thompson SG, Gudnason V, Sattar N, Danesh J, Di Angelantonio E. Natriuretic peptides and integrated risk assessment for cardiovascular disease: individual-participant data meta-analysis of 40 prospective studies. Lancet Diabetes Endocrinol. 2016;4:840–9.

39. Brouwers FP, van Gilst WH, Damman K, van den Berg MP, Gansevoort RT, Bakker SJ, Hillege HL, van Veldhuisen DJ, van der Harst P, de Boer RA. Clinical risk stratification optimizes value of biomarkers to predict new-onset heart failure in a community-based cohort. Circ Heart Fail. 2014;7:723–31.
40. Betti I, Castelli G, Barchielli A, Beligni C, Boscherini V, De Luca L, Messeri G, Gheorghiade M, Maisel A, Zuppiroli A. The role of N-terminal PRO-brain natriuretic peptide and echocardiography for screening asymptomatic left ventricular dysfunction in a population at high risk for heart failure. The PROBE-HF study. J Card Fail. 2009;15:377–84.
41. Mueller T, et al. Capability of B-type natriuretic peptide (BNP) and amino-terminal proBNP as indicators of cardiac structural disease in asymptomatic patients with systemic arterial hypertension. Clin Chem. 2005;51:2245–51.
42. Courand PY, Harbaoui B, Bècle C, Mouly-Bertin C, Lantelme P. Plasma NT-proBNP mirrors the deleterious cardiovascular and renal continuum in hypertension. Eur J Prev Cardiol. 2017;24:452–9.
43. American College of Cardiology; American Heart Association Task Force on Practice Guidelines; American College of Chest Physicians; International Society for Heart and Lung Transplantation; Heart Rhythm Society. ACC/AHA 2005 guideline update for the diagnosis and management of chronic heart failure in the adult—summary article: a report of the American College of Cardiology/American Heart Association Task Force on Practice Guidelines (Writing Commit-tee to Update the 2001 guidelines for the evaluation and management of heart failure). J Am Coll Cardiol. 2005, 46:1116–43.
44. Ledwidge M, et al. Natriuretic peptide-based screening and collaborative care for heart failure: the STOP-HF randomized trial. JAMA. 2013;310:66–74.
45. Huelsmann M, et al. PONTIAC (NT-proBNP selected prevention of cardiac events in a population of diabetic patients without a history of cardiac disease): a prospective randomized controlled trial. J Am Coll Cardiol. 2013;62:1365–72.
46. Segar MW, et al. Incorporation of natriuretic peptides with clinical risk-scores to predict heart failure among individuals with dysglycemia. Eur J Heart Fail. 2022;24:169–80.
47. Choi EY, Bahrami H, Wu CO, Greenland P, Cushman M, Daniels LB, et al. N-terminal pro-B-type natriuretic peptide, left ventricular mass, and incident heart failure: multi-ethnic study of atherosclerosis. Circ Heart Fail. 2012;5:727–34.
48. Suzuki S, Sugiyama S. The molar ratio of N-terminal pro-B-type natriuretic peptide/B-type natriuretic peptide for heart failure-related events in stable outpatients with cardiovascular risk factors. Intern Med. 2018;57:2621–30.
49. Sweeney C, Pharithi RB, Kerr B, Ryan C, Ryan F, Collins L, Halley C, Barrett M, Watson CJ, McDonald K, Ledwidge M. NT-proBNP/BNP ratio for prognostication in European Caucasian patients enrolled in a heart failure prevention programme. ESC Heart Fail. 2021;8:5081–91.
50. Seidelmann SB, Vardeny O, Claggett B, Yu B, Shah AM, Ballantyne CM, Selvin E, MacRae CA, Boerwinkle E, Solomon SD. An *NPPB* promoter polymorphism associated with elevated N-terminal pro-B-type natriuretic peptide and lower blood pressure, hypertension, and mortality. J Am Heart Assoc. 2017;6:e005257.
51. Cannone V, Ledwidge M, Watson C, McKie PM, Burnett JC Jr, McDonald K. STOP-HF trial: higher endogenous BNP and cardiovascular protection in subjects at risk for heart failure. JACC Basic Transl Sci. 2021;6:497–504.
52. Moe GW, Ezekowitz JA, O'Meara E, Lepage S, Howlett JG, Fremes S, Al-Hesayen A, Heckman GA, Abrams H, Ducharme A, Estrella-Holder E, Grzeslo A, Harkness K, Koshman SL, McDonald M, McKelvie R, Rajda M, Rao V, Swiggum E, Virani S, Zieroth S, Arnold JM, Ashton T, D'Astous M, Chan M, De S, Dorian P, Giannetti N, Haddad H, Isaac DL, Kouz S, Leblanc MH, Liu P, Ross HJ, Sussex B, White M. The 2014 Canadian Cardiovascular Society heart failure management guidelines focus update: anemia, biomarkers, and recent therapeutic trial implications. Can J Cardiol. 2015;31:3–16.

Part IV

Arrhythmic Complications of Hypertensive Heart Disease

Atrial Fibrillation, Hypertension, and Heart Failure

19

N. Papaioannou, M. S. Kallistratos, T. Zamfir, and A. J. Manolis

19.1 Introduction

Atrial fibrillation (AF) is the most common and clinically significant cardiac arrhythmia and, independently, its presentation, either symptomatic or silent, is a major risk factor for stroke. It occurs in 1–2% of the general population and accounts for one-third of all patient discharges, with arrhythmia as the principal diagnosis. Several risk factors, such as age, hypertension, diabetes mellitus, obesity, obstructive sleep apnea, left ventricular (LV) hypertrophy, coronary artery disease, heart failure (HF), and others, contribute to the development of AF [1, 2].

Arterial hypertension (HTN) is the most common cardiovascular disorder, affecting up to 50% of the adult population in developed countries, and is the main risk factor for cardiovascular mortality and morbidity. It affects approximately 1 billion adults in the general population, and this number is estimated to increase to more than 1.5 billion by 2025, with more than 7.5 billion deaths correlated with HTN yearly [3].

According to the recently published European Society of Cardiology/European Society of Hypertension (ESC/ESH) hypertension guidelines, hypertension is defined as systolic blood pressure (SBP) of 140 mmHg or more, diastolic blood pressure (DBP) of 90 mmHg or more, or reported use of antihypertensive medication [4]. In some cases, such as in the elderly or frail patients, these thresholds are more flexible.

Untreated high blood pressure may progress to many organ-specific changes, and these changes have been referred to in the new ESC/ESH hypertension guidelines as "hypertension-mediated organ damage (HMOD)."

N. Papaioannou
Cardiology Department, Asklepeion General Hospital, Athens, Greece

M. S. Kallistratos (✉) · T. Zamfir · A. J. Manolis
2nd Department of Cardiology, Metropolitan Hospital, Athens, Greece
e-mail: ajmanol@otenet.gr

M. Dorobantu et al. (eds.), *Hypertension and Heart Failure*, Updates in Hypertension and Cardiovascular Protection,
https://doi.org/10.1007/978-3-031-39315-0_19

19.2 Atrial Fibrillation and Arterial Hypertension

HTN frequently coexists with AF, especially in older individuals, being, at the same time, the most potent risk factor for cerebrovascular disease [5]. Long-standing hypertension, especially if suboptimally controlled, leads to left ventricular hypertrophy, enlargement of the left atrium, and heterogenicity of atrial conduction and fibrosis, all of which may contribute to the development of atrial fibrillation [6]. From the pathophysiological point of view, while hypertension-induced left ventricular hypertrophy and left atrial enlargement remain the two main routes for the development of AF in hypertensive patients, activation of the sympathetic nervous system and the renin–angiotensin–aldosterone system (RAAS) may also play a role in the occurrence of AF. The progressive remodeling of the atrial tissue due to increased pressures or fibrosis may degenerate the atrial cells and, subsequently, alter their electrical conductivity properties [7].

The term "hypertensive atrial cardiomyopathy" has been described [8] based on the abnormal expression of ion channels or junctional complexes in the atrial conductivity system, provoking a vulnerable atrial tissue, which results in the overfunctioning of the RAAS and slow but steady adaption to diastolic dysfunction of the heart muscle. Higher levels of angiotensin II have been observed in AF patients and have been associated with a proarrhythmic effect on the atrial tissue through inflammation mediators, such as interleukin-6, growth factors, and increased intracellular calcium [9]. A cyclic pattern arises, as described in Fig. 19.1.

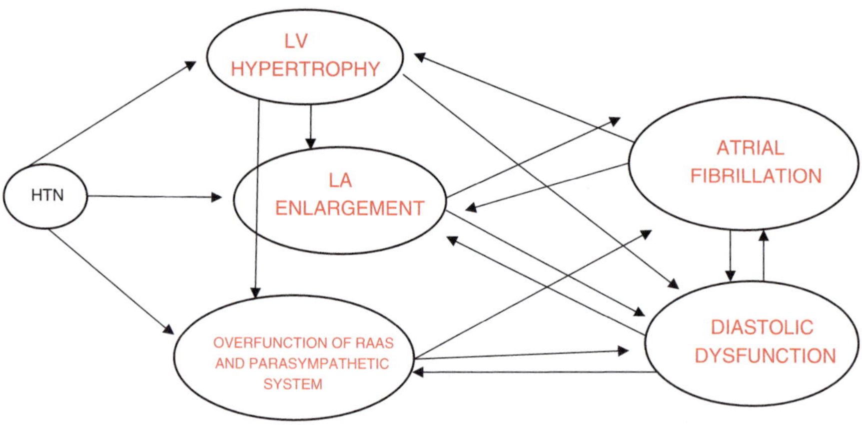

Fig. 19.1 Hypertensive atrial cardiomyopathy. The abnormal expression of ion channels or junctional complexes in the atrial conductivity system leads to a vulnerable atrial tissue, which results in overfunctioning of the renin–angiotensin–aldosterone system and slow but steady adaption to diastolic dysfunction of the heart muscle. Higher levels of angiotensin II have been observed in AF patients and have been associated with a proarrhythmic effect on the atrial tissue through inflammation mediators, such as interleukin-6, growth factors, and increased intracellular calcium. A cyclic pattern arises

The incidence of initial and recurrent stroke increases in proportion to an increase in blood pressure, and hypertension significantly contributes to the occurrence of brain infarction and hemorrhage at all ages [10]. The medical history of HTN in patients suffering from AF showed a threefold increase in the annual incidence of stroke while, in many studies and clinical trials on atrial fibrillation, 50–90% of patients revealed that they were suffering from hypertension [11].

19.3 Management of Atrial Fibrillation and Arterial Hypertension

The coexistence of AF and HTN has proved to be highly frequent, and their management appears to be even more demanding in clinical practice.

Although dealing with AF and HTN can be multifaceted, the use of the CHADS2 (Congestive heart failure, Hypertension, Age ≥ 75 years, Diabetes mellitus, prior Stroke or TIA or thromboembolism (doubled)) score and other scores has proved to be extremely helpful in the determination of stroke risk and antithrombotic therapy guidance, with respect to the hemorrhagic risk, which can be estimated with the HAS-BLED (Hypertension, Abnormal renal/liver function, Stroke, Bleeding history or predisposition, Labile INR, Elderly, Drugs/alcohol concomitantly) score.

The management of AF is aimed at reducing symptoms and preventing the consequences associated with AF and is based on controlling the risk factors, like HTN, ventricular rate, and antithrombotic therapy [2]. Antithrombotic therapy is the milestone for the primary and secondary prevention of stroke in AF patients [12], achieving almost a 50% reduction in the incidence of ischemic stroke [13]. The past decade, main antithrombotic therapy moves around novel oral anticoagulant drugs (NOACs) such as direct thrombin inhibitors (dabigatran) and direct factor Xa inhibitors (rivaroxaban, apixaban, edoxaban), which managed to overcome many drawbacks of warfarin..

The superiority of NOACs and their preference in clinical use against any other antithrombotic therapy is well-established and undoubted, with three overpopulated trials (RE-LY (Randomized Evaluation of Long-term anticoagulant therapY) for dabigatran), (ROCKET (Rivaroxaban Once Daily Oral Direct Factor Xa Inhibition Compared with Vitamin K Antagonism for Prevention of Stroke and Embolism Trial) for rivaroxaban), and (ARISTOTLE (The Apixaban for Reduction in Stroke and Other Thromboembolic Events in Atrial Fibrillation) for apixaban) showing highly promising results, however, with respect to the hemorrhagic risk that they pose. It is a fact that patients with intracranial, intraocular, spinal, retroperitoneal, or traumatic intra-articular bleeding were excluded from the trials, thus creating a cloudy field around the management of a patients with a history of bleeding. Moreover, episodes of intracranial hemorrhage increased in patients receiving NOACs, a fact that reminds us of the importance of balancing the "bleeding-thrombus" dilemma.

Antihypertensive therapy is also associated with significant reductions in the number of stroke episodes. Several meta-analyses of a large number of randomized trials have shown that a reduction in blood pressure is linearly related to the incidence of stroke, irrespective of the drugs employed [14, 15], and a 40% lower risk of a cerebrovascular event has been reported for 5- and 10-mmHg reductions in DBP and SBP, respectively [16]. It has also been shown that antihypertensive treatment may offer more effective protection against hemorrhagic stroke both at initially high and at initially normal blood pressure levels [17], with an overall 50–70% reduction in the risk of brain hemorrhage for reductions of SBP and DBP of about 12/4 mmHg [17, 18].

Some recent publications have commented on the role of antihypertensive medication in the recurrence of AF, or even its prevention, as some antihypertensive drugs like angiotensin-converting enzyme inhibitors (ACEIs), angiotensin receptor blockers (ARBs), beta-blockers (β-blockers), and calcium channel blockers (CCBs) play a major role either directly or indirectly in the atrial mechanical and electrical system.

Recent studies have shown that the direct effect of ACEIs and ARBs in patients dealing with AF, achieving a reduction in the risk of new-onset AF to 28, however, was limited in patients with left ventricular dysfunction of hypertrophy [19, 20]. Interestingly, they proved to be even more beneficial in patients with the highest rate of AF. As for secondary prevention, two randomized trials designed in hypertensive patients with preserved ejection fraction receiving ACEIs or ARBs after successful electrical cardioversion showed significant reduction in recurrent AF episodes [21].

Beta-blockers have a documented role in the AF rate control and in maintaining sinus rhythm, especially in patients with left ventricular dysfunction [22]. A recent study of five million participants in the UK has shown the superiority of β-blockers and ACEIs/ARBs over CCBSs in reducing the risk of AF [22, 23].

Dihydropyridine calcium channel blockers, like diltiazem and verapamil, are widely used to slow down ventricular responses in AF. Verapamil, especially, also plays an effective role in maintaining sinus rhythm after cardioversion. However, as mentioned previously, in the United States, in a retrospective study, 5500 patients treated for hypertension with an ACEI were compared to an equal number of matched patients treated with a CCB, and, in 4 years of follow-up, the incidence of new AF was significantly lower in the ACEI-treated patients with an hazard ratio (HR) of 0.85 (0.74–0.97) [24].

Diuretics, although they belong to the first-line drugs in the therapy of hypertension, have not yet any proved their efficacy in the management or prevention of AF.

19.4 Arterial Hypertension and Heart Failure

Is only the electrical conduction system of the heart muscle affected at the time of uncontrolled and elevated blood pressure? The mechanical and structural changes of the heart provoked by endothelial dysfunction, subclinical organ damage like microalbuminuria, increase of intima media thickness, chronic kidney disease, and hypertensive retinopathy may lead to remodeling and altered function of the left ventricle and blood vessels, ultimately resulting in heart failure with preserved or reduced ejection fraction [25].

According to the ESC heart failure (HF) guidelines, HF is a clinical syndrome characterized by typical symptoms (e.g., breathlessness, ankle swelling, and fatigue) that may be accompanied by signs (e.g., elevated jugular venous pressure, pulmonary crackles, and peripheral edema) caused by a structural and/or functional cardiac abnormality, resulting in a reduced cardiac output and/or elevated intracardiac pressure at rest or during stress [26].

Different trials have shown an association between long-standing hypertension and heart failure. A meta-analysis of 23 trials including 193,424 patients with hypertension or at a "high" cardiovascular risk with a predominant presence of hypertensive patients showed that 28.9% developed heart failure, which accounts for 9.1 events per 1000 patients. Moreover, heart failure development was increased in older subjects (>65 years). This analysis showed that heart failure development remains a major problem in hypertension [27].

Data from the same meta-analysis comment on the risk of developing CV disease, as the individuals with BP ≥ 130/85 mmHg presented a threefold increased risk compared with the participants with BP < 120/80 mmHg [28]. Moreover, data from the Framingham Study showed that 91% of the participants with HF had a history of HTN, and, when compared with normotensive patients, they presented a twofold increased risk of left ventricular dysfunction and, subsequently, heart failure for males and a threefold increased risk for females [29].

The risk of HF is increased by 50% in patients with blood pressure > 160/100 mmHg, compared with those with blood pressure < 140/90 mmHg [30]. In addition, diabetes and dyslipidemia frequently coexist with HTN [31], which further increase the risk of CV disease.

The term "hypertensive heart disease (HHD)" was introduced in the past decade to the medical world, to comment on cardiac remodeling due to increased pressure overload, which results in diastolic dysfunction and concentric LV hypertrophy [32]. HHD is divided in four different stages, which include isolated diastolic dysfunction of the left ventricle with the absence or presence of LV hypertrophy (stages I and II, respectively), clinical manifestations of HF with preserved ejection fraction (grade III), and eccentric hypertrophy of the left ventricle accompanied by a reduced ejection fraction (grade IV) [32, 33]. During grade I, isolated LV diastolic dysfunction is characterized by pathological alterations in diastolic filling or relaxation of the left ventricle [34]. Pressure and volume overload in hypertension lead to structural and functional changes known as remodeling, which is considered not only as an adaptation to chronic changes but also as an omen for circulatory disorders [35]. Cardiac remodeling is different; it either occurs due to pressure overload or volume overload. In the first case, the short-axis diameter of myocytes increases in response to concentric LV hypertrophy, whereas in the case of volume overload, cardiac adaptation is characterized by increasing myocyte length, accompanied by dilation of the left ventricle with thin walls, resulting in eccentric LV hypertrophy [36, 37]. Once LV hypertrophy is established (grade II), the possibility of HFpEF and HFrEF development is significantly increased. Chronic hypertension usually leads to HF with preserved ejection fraction (grade III); however, ischemic events like myocardial infarction or stroke may precipitate the progression to HF with reduced ejection fraction (grade IV) [38]. These chronic adaptive mechanisms play a role in the transition from HTN to HF. Overactivation of the renin–angiotensin–aldosterone system, endothelial dysfunction, sympathetic overdrive, and others occur during HTN [39], and, when combined with development of LV hypertrophy, myocardial fibrosis and vascular dysfunction create the conditions and thoughts that explain the progression from hypertension to heart failure [40].

The paradoxical fact though is that despite hypertension triggering acute incidents of HF, patients with "established" hypertension appeared with improved outcomes. Many trials have proven that that higher systolic blood pressure acts as a protective mechanism in the survival rate in patients with acute or chronic HF [41, 42], whereas data from the COPERNICUS (the Carvedilol Prospective Randomized Cumulative Survival) study in 2.289 patients with congestive heart failure mentioned that major clinical events were more frequent in patients who received lower pretreatment for systolic hypertension [43].

19.5 Management of Arterial Hypertension and Heart Failure

Dealing with HF and HTN can prove to be highly demanding from the view of the clinician, as the clinical manifestation of HF, regardless of the ejection fraction percentage, is usually similar. The therapeutic approach, however, is totally different in each case and should always be based on the blood pressure thresholds of the patient. Usually, patients with HFpEF present with uncontrolled hypertension requiring antihypertensive treatment. On the other hand, the majority of patients with HFrEF present with low blood pressure levels, making it even more difficult for the physician to consolidate a stable antihypertensive treatment. Data drawn from the OPTIMIZE-HF (Organized Program to Initiate Lifesaving Treatment in Hospitalized Patients with Heart Failure) concluded that patients with higher blood pressure measurements appeared with lower in-hospital mortality rates [44], whereas in contrast, low systolic BP was an independent predictor of morbidity and mortality in patients with a reduced and preserved ejection fraction [45].

According to the ESC/ESH hypertension guidelines, the following are proposed:

1. Patients with HFrEF should receive antihypertensive drug treatment when BP is >140/90 mmHg (if not yet initiated).
2. The ideal goals of the lowest BP levels are not clarified.
3. Low blood pressure can create poor outcomes for patients with HF, which suggests that it may be wise to avoid actively lowering BP to <120/70 mmHg [16].

Although these recommendations seem to be applicable in patients with HFpEF, patients with HFrEF rarely have BP levels >140 mmHg; the fact is that only when their BP levels are above 95–100 mmHg can physicians start drug treatment with the RAAS and/or beta-blockers (in low dose).

The following antihypertensive drugs can be used in patients with HF and hypertension: diuretics, beta-blockers, angiotensin-converting enzyme inhibitors (ACEIs), angiotensin receptor blockers (ARBs), mineralocorticoid receptor antagonists (MRAs), angiotensin receptor blockers, and even hydralazine/nitrates. Moreover, neprilysin inhibitors (angiotensin receptor neprilysin inhibitors (ARNIs)) and sodium-glucose cotransporter-2 (SGLT2) inhibitors are considered the first-line drugs for the treatment of patients with heart failure (and also exerts a potent and mild antihypertensive effect for ARNIs and SGLT2 inhibitors, respectively).

ACEIs and ARBs, administered to patients with HF, have been shown to reduce mortality and morbidity, increase cardiac output, and decrease congestive symptoms In case ACEIs are contraindicated or not tolerated, ARBs should be considered. ACEIs and ARBs should be up-titrated to the maximum tolerated dose in order to achieve adequate inhibition of the renin–angiotensin–aldosterone system (RAAS). ACEIs are also recommended in patients with asymptomatic LV systolic dysfunction to reduce the risk of HF development, HF hospitalization, and death. ACEIs and ARBs should be stopped only if there is a significant increase in plasma creatinine [26].

The role of beta-blockers in HF is based on achieving an improvement in systolic function and a reversal of cardiac remodeling. Trials have shown that adding a beta-blocker, on top of standard treatment, may result in approximately 25–35% mortality rate reduction and can be started with ACEIs together, as soon as the diagnosis of HFrEF is made [26]. Beta-blockers should be initiated in clinically stable patients at a low dose and gradually up-titrated to the maximum tolerated dose.

MRAs, spironolactone and eplerenone, are recommended for patients with symptomatic HF with an ejection fraction of 35% or less. Although the role of MRAs in patients with an ejection fraction greater than 35% has not yet been determined, they may benefit patients with comorbid hypertension or renal insufficiency and are also recommended as third- or fourth-line treatment for hypertension. The risk of hyperkalemia as a side effect of MRAs should always be taken into consideration [26].

ARNIs (angiotensin receptor neprilysin inhibitors) are approved for the treatment of patients with HF with reduced ejection fraction (HFrEF) with New York heart Association (NYHA) class II, III, or IV. ARNIs can be used instead of an ACEIs or ARBs, as they too have antihypertensive properties. According to the 2016 ESC Heart Failure Guidelines [26], ARNIs are now recommended in patients with chronic symptomatic HFrEF to reduce morbidity and mortality (class I recommendation).

In patients with hypertension, using diuretics as monotherapy or in combination with other classes of drugs is highly common. In HF, diuretics are used as symptomatic relief therapy from congestive symptoms.

SGLT2 inhibitors introduced as oral antihyperglycemics, however, have also proven to be beneficial to patients with HF who share an antihypertensive effect. In a recent meta-analysis of the two major outcome trials (DAPA-HF ((Dapagliflozin and Prevention of Adverse Outcomes in Heart Failure)) and EMPEROR-Reduced (Empagliflozin Outcome Trial in Patients with Chronic Heart Failure and a Reduced

Ejection Fraction)), empagliflozin or dapagliflozin, on top of standard treatment for HF, reduced all-cause mortality, cardiovascular death, and hospitalizations for HF. Interestingly, many studies and meta-analyses have shown that the administration of SGLT2 inhibitors to patients with type 2 diabetes mellitus (T2DM) with or without HTN has a positive effect on BP independently of the duration of treatment.

References

1. Kannel WB, Wolf PA, Benjamin EJ, Levy D. Prevalence, incidence, prognosis, and predisposing conditions for atrial fibrillation: population-based estimates. Am J Cardiol. 1998;82:2N–9N.
2. Manolis AJ, Kallistratos MS, Poulimenos LE. Recent clinical trials in atrial fibrillation in hypertensive patients. Curr Hypertens Rep. 2012;14:350–9. https://doi.org/10.1007/s11906-012-0268-5.
3. Keamy PM, Whelton M, Reynolds K, Muntner P, Whelton PK, He J. Global burden of hypertension: analysis of worldwide data. Lancet. 2005;365:217–23.
4. Williams B, Mancia G, Spiering W, Agabiti Rosei E, Azizi M, Burnier M, et al. 2018 ESC/ESH guidelines for the management of arterial hypertension. Eur Heart J. 2018;39(33):3021–104. https://doi.org/10.1093/eurheartj/ehy339. Erratum in: Eur Heart J 2019 Feb 1;40(5):475.
5. Manolis A, Doumas M, Poulimenos L, Kallistratos M, Mancia G. The unappreciated importance of blood pressure in recent and older atrial fibrillation trials. J Hypertens. 2013;31(11):2109–17. https://doi.org/10.1097/HJH.0b013e3283638194. discussion 2117.
6. Manolis AJ, Agabiti-Rosei E, Coca A, Cifkova R, Erdine S, Kjeldsen S, et al. Hypertension and atrial fibrillation: diagnostic approach, prevention, and treatment. Position paper of the working group 'hypertension, arrhythmias and thrombosis' of the European Society of Hypertension. J Hypertens. 2012;30:239–52.
7. Kalogeropoulos AS, Tsiodras S, Rigopoulos AG, Sakadakis EA, Triantafyllis A, Kremastinos DT, Rizos I. Novel association patterns of cardiac remodeling markers in patients with essential hypertension and atrial fibrillation. BMC Cardiovasc Disord. 2011;11:77. https://doi.org/10.1186/1471-2261-11-77. PMID: 22204652; PMCID: PMC3276440.
8. Goette A, Kalman JM, Aguinaga L, et al. EHRA/HRS/APHRS/SOLAECE expert consensus on atrial cardiomyopathies: definition, characterization, and clinical implication. Europace. 2016;18(10):1455–90. https://doi.org/10.1093/europace/euw161.
9. Kumagai K, Nakashima H, Urata H, Gondo N, Arakawa K, Saku K. Effects of angiotensin II type 1 receptor antagonist on electrical and structural remodeling in atrial fibrillation. J Am Coll Cardiol. 2003;41(12):2197–204. https://doi.org/10.1016/s0735-1097(03)00464-9.
10. Mancia G, De Backer G, Dominiczak A, Cifkova R, Fagard R, Germano G, et al. 2007 guidelines for the Management of Arterial Hypertension: the task force for the Management of Arterial Hypertension of the European Society of Hypertension (ESH) and of the European Society of Cardiology (ESC). J Hypertens. 2007;25:1105–87.
11. Lip GY, Tse HF. Management of atrial fibrillation. Lancet. 2012;370:604–18.
12. Hart RG, Benavente O, McBride R, Pearce LA. Antithrombotic therapy to prevent stroke in patients with atrial fibrillation: a meta-analysis. Ann Intern Med. 1999;131:492–501.
13. Lansberg MG, O'Donnell MJ, Khatri P, et al. Antithrombotic and thrombolytic therapy for ischemic stroke: antithrombotic therapy and prevention of thrombosis, 9th ed: American College of Chest Physicians Evidence-Based Clinical Practice Guidelines. Chest. 2012;141(2 Suppl):e601S–36S. https://doi.org/10.1378/chest.11-2302.
14. Turnbull F, Blood Pressure Lowering Treatment Trialists' Collaboration. Effects of different blood-pressure-lowering regimens on major cardiovascular events: results of prospectively-designed overviews of randomised trials. Lancet. 2003;362(9395):1527–35. https://doi.org/10.1016/s0140-6736(03)14739-3.

15. Reboldi G, Gentile G, Angeli F, Ambrosio G, Mancia G, Verdecchia P. Effects of intensive blood pressure reduction on myocardial infarction and stroke in diabetes: a meta-analysis in 73,913 patients. J Hypertens. 2011;29(7):1253–69. https://doi.org/10.1097/HJH.0b013e3283469976.19. Collins R, MacMahon S. Blood pressure, antihypertensive drug treatment.
16. Collins R, MacMahon S. Blood pressure, antihypertensive drug treatment and the risks of stroke and of coronary heart disease. Br Med Bull. 1994;50(2):272–98. https://doi.org/10.1093/oxfordjournals.bmb.a072892.
17. Arima H, Chalmers J, Woodward M, Anderson C, Rodgers A, Davis S, Macmahon S, Neal B, PROGRESS Collaborative Group. Lower target blood pressures are safe and effective for the prevention of recurrent stroke: the PROGRESS trial. J Hypertens. 2006;24(6):1201–8. https://doi.org/10.1097/01.hjh.0000226212.34055.86. PMID: 16685221.
18. PROGRESS Collaborative Group. Randomised trial of a perindopril-based blood-pressure-lowering regimen among 6,105 individuals with previous stroke or transient ischaemic attack. Lancet. 2001;358(9287):1033–41. https://doi.org/10.1016/S0140-6736(01)06178-5. Erratum in: Lancet 2001 Nov 3;358(9292):1556. Erratum in: Lancet 2002 Jun 15;359(9323):2120.
19. Healey JS, Baranchuk A, Crystal E, Morillo CA, Garfinkle M, Yusuf S, Connolly SJ. Prevention of atrial fibrillation with angiotensin-converting enzyme inhibitors and angiotensin receptor blockers: a meta-analysis. J Am Coll Cardiol. 2005;45(11):1832–9. https://doi.org/10.1016/j.jacc.2004.11.070.
20. Kalus JS, Coleman CI, White CM. The impact of suppressing the renin-angiotensin system on atrial fibrillation. J Clin Pharmacol. 2006;46(1):21–8. https://doi.org/10.1177/0091270005283284.
21. Madrid AH, Bueno MG, Rebollo JM, Marín I, Peña G, Bernal E, Rodriguez A, Cano L, Cano JM, Cabeza P, Moro C. Use of irbesartan to maintain sinus rhythm in patients with long-lasting persistent atrial fibrillation: a prospective and randomized study. Circulation. 2002;106(3):331–6. https://doi.org/10.1161/01.cir.0000022665.18619.83.
22. Nasr IA, Bouzamondo A, Hulot JS, Dubourg O, Le Heuzey JY, Lechat P. Prevention of atrial fibrillation onset by beta-blocker treatment in heart failure: a meta-analysis. Eur Heart J. 2007;28(4):457–62. https://doi.org/10.1093/eurheartj/ehl484. Epub 2007 Feb 8.
23. Schaer BA, Schneider C, Jick SS, Conen D, Osswald S, Meier CR. Risk for incident atrial fibrillation in patients who receive antihypertensive drugs: a nested case-control study. Ann Intern Med. 2010;152(2):78–84. https://doi.org/10.7326/0003-4819-152-2-201001190-00005.
24. L'Allier PL, Ducharme A, Keller PF, Yu H, Guertin MC, Tardif JC. Angiotensin-converting enzyme inhibition in hypertensive patients is associated with a reduction in the occurrence of atrial fibrillation. J Am Coll Cardiol. 2004;44(1):159–64. https://doi.org/10.1016/j.jacc.2004.03.056.
25. Oh GC, Cho HJ. Blood pressure and heart failure. Clin Hypertens. 2020;26:1. Published 2020 Jan 2. https://doi.org/10.1186/s40885-019-0132-x.
26. Ponikowski P, Voors AA, Anker SD, Bueno H, Cleland JGF, Coats AJS, et al. 2016 ESC guidelines for the diagnosis and treatment of acute and chronic heart failure: the task force for the diagnosis and treatment of acute and chronic heart failure of the European Society of Cardiology (ESC)developed with the special contribution of the heart failure association (HFA) of the ESC. Eur Heart J. 2016;37(27):2129–200. https://doi.org/10.1093/eurheartj/ehw128.
27. Tocci G, Sciarretta S, Volpe M. Development of heart failure in recent hypertension trials. J Hypertens. 2008;26(7):1477–86. https://doi.org/10.1097/HJH.0b013e3282fe1d3d.
28. Vasan RS, Larson MG, Leip EP, Evans JC, O'Donnell CJ, Kannel WB, Levy D. Impact of high-normal blood pressure on the risk of cardiovascular disease. N Engl J Med. 2001;345(18):1291–7. https://doi.org/10.1056/NEJMoa003417.
29. Levy D, Larson MG, Vasan RS, Kannel WB, Ho KK. The progression from hypertension to congestive heart failure. JAMA. 1996;275(20):1557–62.
30. Lloyd-Jones DM, Larson MG, Leip EP, Beiser A, D'Agostino RB, Kannel WB, Murabito JM, Vasan RS, Benjamin EJ, Levy D, Framingham Heart Study. Lifetime risk for developing congestive heart failure: the Framingham heart study. Circulation. 2002;106(24):3068–72. https://doi.org/10.1161/01.cir.0000039105.49749.6f.

31. de Boer IH, Bangalore S, Benetos A, Davis AM, Michos ED, Muntner P, Rossing P, Zoungas S, Bakris G. Diabetes and hypertension: a position statement by the American Diabetes Association. Diabetes Care. 2017;40(9):1273–84. https://doi.org/10.2337/dci17-0026.
32. Messerli FH, Rimoldi SF, Bangalore S. The transition from hypertension to heart failure: contemporary update. JACC Heart Fail. 2017;5(8):543–51. https://doi.org/10.1016/j.jchf.2017.04.012. Epub 2017 Jul 12. Erratum in: JACC Heart Fail. 2017 Dec;5(12):948.
33. Slivnick J, Lampert BC. Hypertension and heart failure. Heart Fail Clin. 2019;15(4):531–41. https://doi.org/10.1016/j.hfc.2019.06.007. Epub 2019 Jul 31.
34. Sorrentino MJ. The evolution from hypertension to heart failure. Heart Fail Clin. 2019;15(4):447–53. https://doi.org/10.1016/j.hfc.2019.06.005. Epub 2019 Aug 7.
35. Gibbons GH, Dzau VJ. The emerging concept of vascular remodeling. N Engl J Med. 1994;330(20):1431–8. https://doi.org/10.1056/NEJM199405193302008.
36. Nakamura M, Sadoshima J. Mechanisms of physiological and pathological cardiac hypertrophy. Nat Rev Cardiol. 2018;15(7):387–407. https://doi.org/10.1038/s41569-018-0007-y.
37. Gaasch WH, Zile MR. Left ventricular structural remodeling in health and disease: with special emphasis on volume, mass, and geometry. J Am Coll Cardiol. 2011;58(17):1733–40. https://doi.org/10.1016/j.jacc.2011.07.022.
38. de Simone G, Gottdiener JS, Chinali M, Maurer MS. Left ventricular mass predicts heart failure not related to previous myocardial infarction: the cardiovascular health study. Eur Heart J. 2008;29(6):741–7. https://doi.org/10.1093/eurheartj/ehm605. Epub 2008 Jan 19.
39. Sciarretta S, Paneni F, Palano F, Chin D, Tocci G, Rubattu S, Volpe M. Role of the renin-angiotensin-aldosterone system and inflammatory processes in the development and progression of diastolic dysfunction. Clin Sci (Lond). 2009;116(6):467–77. https://doi.org/10.1042/CS20080390.
40. Wright JW, Mizutani S, Harding JW. Pathways involved in the transition from hypertension to hypertrophy to heart failure. Treatment strategies. Heart Fail Rev. 2008;13(3):367–75. https://doi.org/10.1007/s10741-007-9060-z. Epub 2007 Nov 7.
41. Bhatia RS, Tu JV, Lee DS, et al. Outcome of heart failure with preserved ejection fraction in a population-based study. N Engl J Med. 2006;355:260–9.
42. Gheorghiade M, Abraham WT, Albert NM, et al. Systolic blood pressure at admission, clinical characteristics, and outcomes in patients hospitalized with acute heart failure. JAMA. 2006;296(18):2217–26. https://doi.org/10.1001/jama.296.18.2217.
43. Rouleau JL, Roecker EB, Tendera M, Mohacsi P, Krum H, Katus HA, Fowler MB, Coats AJ, Castaigne A, Scherhag A, Holcslaw TL, Packer M. Carvedilol prospective randomized cumulative survival study group. Influence of pretreatment systolic blood pressure on the effect of carvedilol in patients with severe chronic heart failure: the carvedilol prospective randomized cumulative survival (COPERNICUS) study. J Am Coll Cardiol. 2004;43(8):1423–9. https://doi.org/10.1016/j.jacc.2003.11.037.
44. Tsimploulis A, Lam PH, Arundel C, Singh SN, Morgan CJ, Faselis C, Deedwania P, Butler J, Aronow WS, Yancy CW, Fonarow GC, Ahmed A. Systolic blood pressure and outcomes in patients with heart failure with preserved ejection fraction. JAMA Cardiol. 2018;3(4):288–97. https://doi.org/10.1001/jamacardio.2017.5365. Erratum in: JAMA Cardiol 201;3(4):358. PMID: 29450487; PMCID: PMC5875342.
45. Vidán MT, Bueno H, Wang Y, Schreiner G, Ross JS, Chen J, Krumholz HM. The relationship between systolic blood pressure on admission and mortality in older patients with heart failure. Eur J Heart Fail. 2010;12(2):148–55. https://doi.org/10.1093/eurjhf/hfp195. PMID: 20083624; PMCID: PMC2807767.
46. Zhang Y, Wang C, Zhang J, Zhang H, Yin Z, Chen Y, Xie Q. Low systolic blood pressure for predicting all-cause mortality in patients hospitalised with heart failure: a systematic review and meta-analysis. Eur J Prev Cardiol. 2019;26(4):439–43. https://doi.org/10.1177/2047487318784092. Epub 2018 Jun 25.

Ventricular Arrhythmias, Hypertension, and Heart Failure

20

Alexandru Deaconu and Radu Vătășescu

Abbreviations

ACEI	Angiotensin-converting enzyme inhibitor
AF	Atrial fibrillation
APD	Action potential duration
ARB	Angiotensin receptor blocker
ARNI	Angiotensin receptor blocker/neprilysin inhibitor
BP	Blood pressure
CA	Catheter ablation
CAD	Coronary artery disease
CRS	Cardiorenal metabolic syndrome
CSD	Cardiac sympathetic denervation
Cx43	Gap junction connexin 43
ECG	Electrocardiography
ECM	Extracellular matrix
HF	Heart failure
HFpEF	Heart failure with preserved left ventricular ejection fraction
HFrEF	Heart failure with reduced left ventricular ejection fraction
HHD	Hypertensive heart disease
HTN	Hypertension
ICD	Implantable cardioverter defibrillator
IGF-1	Insulin-like growth factor-1
LIFE	Losartan Intervention for Endpoint Reduction in Hypertension
LV	Left ventricular

A. Deaconu · R. Vătășescu (✉)
University of Medicine and Pharmacy "Carol Davila" Bucharest, Bucharest, Romania

Emergency Clinical Hospital of Bucharest, Bucharest, Romania

M. Dorobantu et al. (eds.), *Hypertension and Heart Failure*, Updates in Hypertension and Cardiovascular Protection,
https://doi.org/10.1007/978-3-031-39315-0_20

LVEF	Left ventricular ejection fraction
LVH	Left ventricular hypertrophy
NSVT	Non-sustained ventricular tachycardia
NPY	Neuropeptide Y
OMT	Optimal medical therapy
PKC	Protein kinase C
RSDN	Renal sympathetic denervation
SCD	Sudden cardiac death
T2DM	Type 2 diabetes mellitus
VA	Ventricular arrhythmia
VF	Ventricular fibrillation
VPB	Ventricular premature beat
VT	Ventricular tachycardia

20.1 Introduction

Arterial hypertension (HTN) in combination with left ventricular hypertrophy (LVH) is an independent risk factor for ventricular arrhythmia (VA) even in patients without an established coronary artery disease (CAD) [1]. Each 20/10-mmHg rise in systolic/diastolic blood pressure (BP) was associated with an ~20% increased risk of sudden cardiac death (SCD) in a large cohort of individuals free from cardiovascular disease [2]. Moreover, the lifetime risk of SCD at the age of 30 was greater by 30% in patients with HTN [2]. Increased BP is not arrhythmogenic on its own, but it induces ventricular overload, which, in turn, leads to LVH and/or heart failure (HF) [3].

20.2 Mechanisms of Ventricular Arrhythmias in Hypertension

20.2.1 Left Ventricular Hypertrophy

Multiple arrhythmogenic factors may be responsible for the occurrence of VAs in hypertensive patients. LVH plays a central role in the development of VAs. A meta-analysis by Chatterjee et al. showed that the incidence of VAs was 5.5% compared with 1.2% in patients without LVH ($p < 0.001$). The occurrence of ventricular tachycardia (VT) or ventricular fibrillation (VF) was 2.8-fold greater in the presence of LVH [4].

Furthermore, the more hypertrophied the left ventricle, the higher the risk of ventricular ectopy [3, 5]. The relation between LV mass or wall thickness and VA is graded and continuous. The magnitude of LVH seems to be directly related to increased ventricular ectopy. A study by Ghali et al. showed that every 1-mm increase in the thickness of the interventricular septum or posterior wall was associated with a two to three times increase in the occurrence and complexity of VAs [6].

Asymmetric septal and eccentric hypertrophy is more often associated with VA than concentric LVH [7]. Analysis from the Framingham Heart Study showed that women develop more frequently concentric LVH, whereas men have eccentric hypertrophy [8]. The risk of SCD increases progressively with LV mass [3]. Messerli et al. reported that hypertensive patients with electrocardiographic (ECG) evidence of LVH had significantly more ventricular premature beats (VPBs) than did subjects without LVH or with normal BP [9]. However, the Framingham Heart Study showed that echocardiographic LVH was an independent risk factor for VA and that it is more prevalent and sensitive than ECG LVH [10].

20.2.2 Pathophysiological Cellular and Endocrine Mechanisms Leading to Fibrosis and Ventricular Arrhythmias

LVH is proarrhythmogenic in relation to the resulting fibrosis extension. There are two types of myocardial fibrosis: replacement and interstitial fibrosis [11]. The former is localized and occurs due to myocyte necrosis after myocardial infarction, whereas the latter is diffuse, and its subtypes include reactive and infiltrative interstitial fibrosis [12]. Hypertension is associated with reactive fibrosis. Myocardial fibrosis is the result of augmented myofibroblast activity and excessive extracellular matrix deposition [13]. Fibrosis alters electrophysiological homogeneity and is associated with increased sympathetic nervous system and renin–angiotensin–aldosterone system activation, the presence of early afterdepolarizations/delayed afterdepolarizations, prolongation and dispersion of repolarization, myocardial ischemia, and impaired cell-to-cell communication at the gap junctions [3, 5].

Patients with hypertensive heart disease (HHD) and arrhythmias exhibit higher values of myocardial collagen compared to those without arrhythmias, despite similar ejection fractions and lack of coronary artery disease [14]. Moreover, myofibroblasts may alter the electric activity of cardiomyocytes through either direct intercellular interactions or secretion of paracrine factors.

Hypertension is one of the pathophysiological states along with vascular injury, cardiorenal metabolic syndrome (CRS), type 2 diabetes mellitus (T2DM), and cancer associated with local tissue autocrine/paracrine angiotensin II (Ang II)-generating system (tissue renin–angiotensin system (tRAS)) activation [15]. In contrast to the constitutive, endocrine–hormonal circulatory RAS (cRAS), tRAS is more of a subacute maintenance/remodeling system, which leads to tissue remodeling (proliferation, hypertrophy, and differentiation) and extracellular matrix (ECM) remodeling repair and/or fibrosis [15]. tRAS is recognized as a relevant element of tissue injury and remodeling and acts in conjunction with and independently of cRAS [15].

On the other hand, sympathetic activation exerts calcium-loading effects on the sarcoplasmic reticulum, which can lead to delayed afterdepolarizations and thus initiate ventricular arrhythmias [16]. Action potential duration (APD) is shortened in areas of dense sympathetic innervation, and, due to heterogeneity of sympathetic innervation, APD dispersion increases [17]. Furthermore, sympathetic activation is associated with a neurally induced pro-inflammatory state, which promotes negative remodeling of the myocardium [18]. Moreover, in addition to cardiac sympathetic activity, renal sympathetic activity is upregulated in patients with heart failure [19].

The hypertrophic response resulting from hypertension is a continuum in which progressive changes in gene expression and the structure of cells and extracellular matrix may mediate the transition from a phase of compensated structural and functional adaptation to an increasingly maladaptive state building up to heart failure [20]. Remodeling of the tissue structure in hypertensive heart disease leads to modifications in the patterns of gap junctions that are likely to alter the conduction properties of the myocardium and contribute to arrhythmogenesis, independent of changes in the active membrane properties of individual cells [20]. Partial uncoupling in viable, but a structurally altered myocardium changes the conduction properties of the tissue and increases the possibility of developing arrhythmias dependent on alterations in conduction [20].

Hypertrophic growth of cardiomyocytes also induces mechanical stretch of cardiomyocytes, which is transduced across the cell membrane, but the mechanisms are unclear. They probably involve stretch-sensitive ion channels, an Na^+/H^+ exchanger, integrins, and integrin-interacting molecules in a complex network that links the extracellular matrix (ECM), the cytoskeleton, the sarcomere, Ca^{2+}-handling proteins, and the nucleus [14].

Changes in the expression and distribution of ion channels also lead to QTc prolongation [1]. Patients with LVH and complex VAs presented the most prolonged QTc intervals [21]. QTc duration predicted mortality risk in patients with ECG criteria for LVH, especially in those with QTc >500 ms [22]. Dispersion of the QTc interval is an index of inhomogeneity of repolarization and has been associated with VAs [23].

Moreover, patients with HTN are usually treated with diuretics that can lead to electrolyte imbalances, such as hypokalemia and hypomagnesemia, which can constitute an important triggering factor. On the other hand, effective control of BP may prevent the development of arrhythmias (Fig. 20.1) [3, 5]. Circadian variations and sudden increases in BP can trigger arrhythmia as a result of associated changes in pre- and post-charge [24]. Severe LVH may also be accompanied by abnormal coronary regulation and flow reserve, which increases the risk of microvascular ischemic events. Furthermore, HTN is a major risk factor for atherosclerosis and may affect the epicardial coronary arteries, leading to myocardial ischemia, which may induce VAs [5].

Hypertensive patients may have a wide range of VAs (mechanisms: early afterdepolarizations/delayed afterdepolarizations (EADs/DADs), abnormal automatism, reentry, cardiac magnetic resonance imaging (cMRI) for risk assessment), but VPBs are more common in the presence of LVH [5, 25]. Sustained VA can be triggered by simple or complex ventricular extrasystoles and is usually perpetuated by a reentry circuit [25]. Sustained VT and VF occur frequently in hypertensives with ischemic heart disease [5].

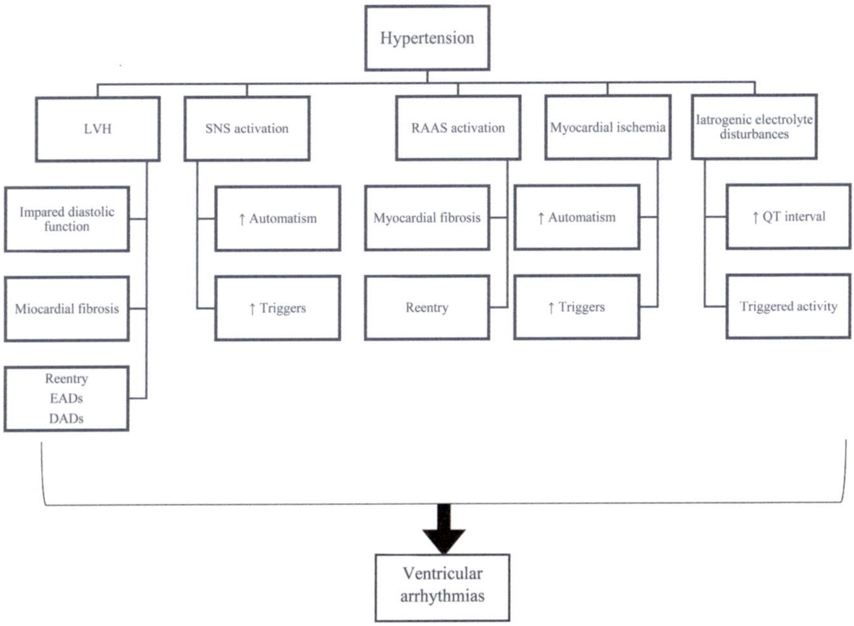

Fig. 20.1 The pathogenesis of ventricular arrhythmias in patients with hypertension. *LVH* left ventricular hypertrophy, *SNS* sympathetic nervous system, *RAAS* renin–angiotensin–aldosterone system, *EADs* early afterdepolarizations, *DADs* delayed afterdepolarizations

20.2.3 Atrial Fibrillation (AF): The Pathway to Ventricular Arrhythmias

Atrial fibrillation is an independent and underrecognized risk factor for ventricular arrhythmias and SCD [26]. Hypertensive patients with LVH on ECG who develop new-onset AF have an increased risk of SCD, especially in the setting of aggressive blood pressure lowering [27]. The increased risk of SCD associated with new-onset AF is independent of traditional risk factors [27]. Available data suggest that prevention of the development of new AF may be a therapeutic approach to reducing ventricular arrhythmias and SCD in hypertensive patients [27]. There are multiple underlying mechanisms through which AF can trigger ventricular arrhythmia: AF produces irregular ventricular activation, leading to marked short–long–short sequences of RR intervals, which can enable VAs [28]. Moreover, AF may directly contribute to left ventricular systolic dysfunction due to rapid rates, irregular ventricular rhythm, or loss of atrial systole [29]. Development of new AF may also suggest additional or worse underlying structural heart disease, which increases the risk of SCD [27].

20.2.4 Genetic Factors Associated with LVH-Induced Arrhythmogenesis

LV mass is, in part, under genetic control, with heritability estimates generally between 0.17 and 0.59 [30, 31]. Studies have documented a statistically significant association between LV mass and the angiotensin-converting enzyme (ACE) gene [32], the insulin-like growth factor (IGF-1) gene [33], and the neuropeptide Y (NPY) gene [34]. Human population studies and genome-wide association studies have identified genes that are involved in LVH and arrhythmogenesis, such as *KCNB1*, which encodes the voltage-gated potassium channel [35]. Animal studies suggest that abnormal distribution and/or downregulation of gap junction connexin 43 (Cx43) accompanied by altered protein kinase C (PKC) signaling in spontaneously hypertensive rats were associated with increased predisposition to ventricular fibrillation compared to normotensive rats. The maladaptive consequences of hypertension include abnormal myocardial distribution and downregulation of Cx43, which can contribute to the occurrence of malignant arrhythmias [36].

20.3 Hypertensive Heart Failure and Ventricular Arrythmias

HTN is a leading risk factor for the development of HF, with almost two-thirds of HF patients being hypertensive [37]. A pressure overload will lead to concentric hypertrophy, whereas a volume overload will more frequently develop into eccentric hypertrophy [38]. In HTN, LVH can be both concentric and eccentric, as it generates both pressure and volume overloads. The former is associated with HF with preserved ejection fraction (HFpEF), whereas the latter leads to HF with reduced ejection fraction (HFrEF) [38]. Patients with HFpEF present a lower incidence of ventricular arrhythmias and SCD compared to those with a reduced left ventricular ejection fraction (LVEF) [39, 40]. However, diastolic dysfunction is an independent risk factor for SCD. One prospective study concluded that diastolic dysfunction is associated with a significantly increased risk of arrhythmic death, independent from the degree of LVEF reduction, and is observed both in patients with an LVEF ≤35% and in those with an LVEF >35% [41]. Furthermore, SCD risk was significantly different between diastolic dysfunction grades and was the highest in patients with grade III dysfunction, suggesting that diastolic function grading might contribute to an improved risk stratification for arrhythmic death in patients with an LVEF >35% [41].

20.4 Workup and Diagnosis

Patients with frequent VPBs, couplets, or non sustained ventricular arrhythmias should be evaluated by physical examination, blood chemistry, a 12-lead ECG, and a 24-h Holter recording [3]. Symptoms include palpitations, light-headedness, nausea, vertigo, and cough. Sustained VAs may cause syncope or SCD [3]. A 12-lead

ECG and a 24-h Holter recording are often used to localize the site of origin and for VPB quantification [3]. Transthoracic echocardiography assesses the underlying heart disease and left ventricular systolic function, whereas cardiac magnetic resonance imaging provides a detailed analysis of the myocardial substrate and may assess the extent of myocardial fibrosis [3]. When CAD is suspected, exercise test, echo stress, perfusion scintigraphy, and coronary angiography may be indicated [3]. It is mandatory to search for transient correctable factors, such as electrolyte imbalances and excess of alcohol, caffeine, non-pharmacological stimulants or use of recreational drugs, and chronic exposure to high-stress conditions [3].

20.5 Management of Ventricular Arrhythmias in Hypertension

The management strategy for VAs in hypertensive patients is dependent on the severity of the issue. The most common ventricular arrhythmia encountered are VPBs [3], but non-sustained ventricular tachycardia (NSVT), monomorphic VT, and polymorphic VT may also be found. Nevertheless, sustained monomorphic VT is rare in the absence of structural heart disease [3].

20.5.1 Pharmacological Treatment for Ventricular Arrhythmias

Since LVH represents the most significant pathophysiological mechanism for the occurrence of VAs in hypertensive patients, it is consequently the main treatment target [5]. Prevention or regression of LVH with BP control improves prognosis in hypertensive patients [42]. Upstream therapy (lifestyle changes, anti-HTN, and/or HF drugs) that effectively induces reverse remodeling decreases the risk of VAs. Patients should be counseled to make appropriate lifestyle changes and eliminate common premature ventricular complex triggers, such as caffeine and tobacco, or to manage possible stress and anxiety. Furthermore, physical activity and aerobic exercise seem to contribute to the regression of premature ventricular complexes [43].

A combination of antihypertensive drug classes should be considered in order to achieve an adequate BP control [3]. In contrast, thiazide diuretics are associated with a dose-dependent increased risk of VA and SCD, probably due to undetected hypokalemia causing greater QTc dispersion [44, 45].

In patients with HTN and asymptomatic NSVT, there is no indication for prophylactic use of antiarrhythmic agents [3]. In the presence of clinically relevant VAs, beta-blockers and amiodarone are the main recommended drugs [25]. Use of class IC agents (flecainide, propafenone) is not recommended, especially in the presence of severe LVH or systolic dysfunction.

Angiotensin-converting enzyme inhibitors (ACEIs) or angiotensin receptor blockers (ARBs) seem to lower SCD risk independent of BP reduction [3]. The Losartan Intervention for Endpoint Reduction in Hypertension (LIFE) study demonstrated that the absence of ECG LVH during antihypertensive treatment led to a lower risk of SCD independently of BP reduction and other known predictors of SCD and suggested that losartan seemed to provide similar protection against SCD compared with atenolol [46].

Ranolazine is a novel antianginal agent, which additionally acts as a multichannel ion blocker. At low concentrations, it is a relatively selective blocker of the slowly inactivating component of the Na^+ current, referred to as "late" INa, which may be an antiarrhythmic target. At higher concentrations, it inhibits IKr and, to a lesser extent, ICaL [47, 48]. In theory, ranolazine might be effective against Ca^{2+} overload-related arrhythmias by lowering intracellular Na^+ and subsequently reducing excessive intracellular Ca^{2+} load. However, ranolazine has not yet been tested in these settings [49].

20.5.2 Prevention of Sudden Cardiac Death in Hypertensive Patients with Ventricular Arrhythmias

Hypertensive patients with sudden cardiac arrest (SCA) and/or sustained ventricular tachycardia (VT) without reversible causes should undergo secondary prophylaxis with an implantable cardioverter defibrillator (ICD) if they are expected to survive for >1 year with good functional status [37]. Hypertensive patients with LVH and/or non-sustained VAs should undergo risk stratification for SCD primary prophylaxis with ICD [5]. There is an unmet need for refining risk stratification for this group. The ECG criteria for LVH in a hypertensive patient may prove to be useful in the evaluation of SCD risk. The absence of in-treatment ECG LVH is linked to a reduced risk of SCD independently of treatment modality, blood pressure reduction, prevalent coronary heart disease, and other cardiovascular risk factors in hypertensive patients with LVH [46]. Cardiovascular magnetic resonance (CMR) with late gadolinium enhancement (LGE) may be helpful in improving the risk stratification strategies in hypertensive patients with LVH. In patients with hypertrophic cardiomyopathy (HCM), the amount of LGE seems to outperform the HCM risk SCD score and the American College of Cardiology Foundation/American Heart Association (ACCF/AHA) algorithm in the identification of those at an increased risk of SCD [50]. LGE exposes myocardial fibrosis, which is the pathophysiological substrate for malignant ventricular arrhythmias in HCM patients, similarly to hypertensive patients. Genetic testing may also be useful in young patients (<40 years) without structural heart disease who present with unexplained cardiac arrest or recurrent exertional syncope, in order to possibly identify an inherited arrhythmia syndrome as a likely cause [51].

Although, as discussed before, the degree of LVH is associated with a higher risk of VAs, traditional markers of risk such as a depressed LV ejection fraction, the presence of CAD, and a high VPB burden may be used [5]. Although not widely recommended in this population, electrophysiology study with programmed ventricular stimulation might provide additional information in selected patients. In patients with sustained VA, HTN does not influence the indications for ICD or ablation [3]. Furthermore, an ICD should be used in patients with symptomatic HF and a left ventricular ejection fraction (LVEF) $\leq 35\%$ of ischemic or nonischemic etiology despite ≥3 months of optimal medical therapy (OMT), provided they are expected to survive for >1 year with a good functional status [37].

20.5.3 Interventional Treatment for Ventricular Arrhythmias

VPB ablation should be considered in highly selected patients, with VPBs >20–25% daily beats or > 10,000 VPBs/24 h and left ventricular dysfunction, to potentially reverse tachycardia-induced cardiomyopathy [3]. The presence of significant LVH and/or multiple morphologies of VPBs are likely to reduce the procedural success. In the absence of a predominant VPB, medical management should be considered [52].

Furthermore, data suggest a possible beneficial role of renal sympathetic denervation (RSDN) in reducing the arrhythmic burden and, more specifically, VT episodes in hypertensives with ischemic heart disease or with resistant HTN [53]. The impact of renal denervation on the cardiac conduction system and cardiac arrhythmias seems to be independent of BP reduction [54]. Pathophysiological mechanisms such as modulation of sympathetic-vagal balance, reduction of sympathetic activity and angiotensin II levels, suppression of early afterdepolarizations and reentrant mechanisms, and increase of the VF threshold or shortening of the QT interval have been suggested [54]. Along with cardiac sympathetic, renal sympathetic activity is upregulated in patients with heart failure. Preclinical and clinical studies have shown that sympathetic modulation by RSDN might represent a viable therapeutic option in patients with recurrent VA and prior failed catheter ablation (CA) or in whom CA for VA is not an option [19].

However, robust data derived from adequately powered clinical trials are lacking [53]. In a meta-analysis, renal sympathetic denervation appears to be a safe and effective treatment strategy in patients with both ischemic and nonischemic cardiomyopathy with prior failed antiarrhythmic drugs and catheter ablation for recurrent ventricular arrhythmia and electrical storm [19].

Another possible therapeutic option is cardiac sympathetic denervation (CSD), which can be achieved with surgical removal of the stellate and T1–T4 ganglia via video-assisted thoracoscopic surgery. In a case series of 22 patients with long QT, catecholaminergic polymorphic VT, and idiopathic VT, 16 had a marked reduction in VT burden, with 12 having complete cessation at a median follow-up of 28 months following left CSD [55].

20.6 Conclusions

The occurrence of ventricular arrhythmias ranging from isolated VPBs to nonsustained ventricular tachycardia and even SCD may have important implications in the morbidity and mortality of hypertensive patients. Clinicians should acknowledge this scenario and encourage lifestyle modifications and treatments targeting LVH prevention or regression as well as actively search for VAs using a standardized workup.

References

1. Nadarajah R, Patel PA, Tayebjee MH. Is hypertensive left ventricular hypertrophy a cause of sustained ventricular arrhythmias in humans? J Hum Hypertens. 2021;35:492–8.
2. Rapsomaniki E, Timmis A, George J, et al. Blood pressure and incidence of twelve cardiovascular diseases: lifetime risks, healthy life-years lost, and age-specific associations in 1·25 million people. Lancet. 2014;383:1899–911.
3. Lip GYH, Coca A, Kahan T, et al. Hypertension and cardiac arrhythmias: a consensus document from the European heart rhythm association (EHRA) and ESC Council on hypertension, endorsed by the Heart Rhythm Society (HRS), Asia-Pacific Heart Rhythm Society (APHRS) and Sociedad Latinoamericana de Estimulación Cardíaca y Electrofisiología (SOLEACE). EP Europace. 2017;19:891–911.
4. Chatterjee S, Bavishi C, Sardar P, Agarwal V, Krishnamoorthy P, Grodzicki T, Messerli FH. Meta-analysis of left ventricular hypertrophy and sustained arrhythmias. Am J Cardiol. 2014;114:1049–52.
5. Varvarousis D, Kallistratos M, Poulimenos L, Triantafyllis A, Tsinivizov P, Giannakopoulos A, Kyfnidis K, Manolis A. Cardiac arrhythmias in arterial hypertension. J Clin Hypertens. 2020;22:1371–8.
6. Ghali JK, Kadakia S, Cooper RS, Liao Y. Impact of left ventricular hypertrophy on ventricular arrhythmias in the absence of coronary artery disease. J Am Coll Cardiol. 1991;17:1277–82.
7. Nunez BD, Lavie CJ, Messerli FH, Schmieder RE, Garavaglia GE, Nunez M. Comparison of diastolic left ventricular filling and cardiac dysrhythmias in hypertensive patients with and without isolated septal hypertrophy. Am J Cardiol. 1994;74:585–9.
8. Kuch B, Muscholl M, Luchner A, Döring A, Riegger GAJ, Schunkert H, Hense HW. Gender specific differences in left ventricular adaptation to obesity and hypertension. J Hum Hypertens. 1998;12:685–91.
9. Messerli FH, Ventura HO, Elizardi DJ, Dunn FG, Frohlich ED. Hypertension and sudden death: increased ventricular ectopic activity in left ventricular hypertrophy. Am J Med. 1984;77:18–22.
10. Levy D, Anderson KM, Savage DD, Balkus SA, Kannel WB, Castelli WP. Risk of ventricular arrhythmias in left ventricular hypertrophy: the Framingham heart study. Am J Cardiol. 1987;60:560–5.
11. Ambale-Venkatesh B, Lima JAC. Cardiac MRI: a central prognostic tool in myocardial fibrosis. Nat Rev Cardiol. 2015;12:18–29.
12. Burt JR, Zimmerman SL, Kamel IR, Halushka M, Bluemke DA. Myocardial T1 mapping: techniques and potential applications. Radiographics. 2014;34:377–95.
13. Liu T, Song D, Dong J, Zhu P, Liu J, Liu W, Ma X, Zhao L, Ling S. Current understanding of the pathophysiology of myocardial fibrosis and its quantitative assessment in heart failure. Front Physiol. 2017;8:238.
14. González A, Ravassa S, López B, Moreno MU, Beaumont J, San José G, Querejeta R, Bayés-Genís A, Díez J. Myocardial remodeling in hypertension. Hypertension. 2018;72:549–58.
15. Hayden MR, Sowers KM, Pulakat L, Joginpally T, Krueger B, Whaley-Connell A, Sowers JR. Possible mechanisms of local tissue renin-angiotensin system activation in the Cardiorenal metabolic syndrome and type 2 diabetes mellitus. Cardiorenal Med. 2011;1:193–210.
16. Myles RC, Wang L, Kang C, Bers DM, Ripplinger CM. Local β-adrenergic stimulation overcomes source-sink mismatch to generate focal arrhythmia. Circ Res. 2012;110:1454–64
17. Huang WA, Boyle NG, Vaseghi M. Cardiac innervation and the autonomic nervous system in sudden cardiac death. Card Electrophysiol Clin. 2017;9:665–79.
18. Jänig W. Sympathetic nervous system and inflammation: a conceptual view. Auton Neurosci. 2014;182:4–14.
19. Garg J, Shah S, Shah K, Turagam MK, Natale A, Lakkireddy D. Renal sympathetic denervation for the treatment of recurrent ventricular arrhythmias—ELECTRAM investigators. Pacing Clin Electrophysiol. 2021;44:865–74.

20. Saffitz JE, Schuessler RB, Yamada KA. Mechanisms of remodeling of gap junction distributions and the development of anatomic substrates of arrhythmias. Cardiovasc Res. 1999;42:309–17.
21. Kulan K, Ural D, Komsuoğlu B, Ağaçdiken A, Göldeli Ö, Komsuoğlu SŞ. Significance of QTc prolongation on ventricular arrhythmias in patients with left ventricular hypertrophy secondary to essential hypertension. Int J Cardiol. 1998;64:179–84.
22. Haugaa KH, Bos JM, Borkenhagen EJ, Tarrell RF, Morlan BW, Caraballo PJ, Ackerman MJ. Impact of left ventricular hypertrophy on QT prolongation and associated mortality. Heart Rhythm. 2014;11:1957–65.
23. Perkiömäki JS, Ikäheimo MJ, Pikkujämsä SM, Rantala A, Lilja M, Kesäniemi YA, Huikuri HV. Dispersion of the QT interval and autonomic modulation of heart rate in hypertensive men with and without left ventricular hypertrophy. Hypertension. 1996;28:16–21.
24. James MA, Jones JV. Systolic Wall stress and ventricular arrhythmia: the role of acute change in blood pressure in the isolated working rat heart. Clin Sci. 1990;79:499–504.
25. Baguet J-P, Erdine S, Mallion J-M. European Society of Hypertension Scientific Newsletter: update on hypertension management: hypertension and dysrhythmias. J Hypertens. 2006;24:409–11.
26. Shenasa M, Shenasa H. Hypertension, left ventricular hypertrophy, and sudden cardiac death. Int J Cardiol. 2017;237:60–3.
27. Okin PM, Bang CN, Wachtell K, Hille DA, Kjeldsen SE, Dahlöf B, Devereux RB. Relationship of sudden cardiac death to new-onset atrial fibrillation in hypertensive patients with left ventricular hypertrophy. Circ Arrhythm Electrophysiol. 2013;6:243–51.
28. Grönefeld GC, Mauss O, Li Y-G, Klingenheben T, Hohnloser SH. Association between atrial fibrillation and appropriate implantable cardioverter defibrillator therapy: results from a prospective study. J Cardiovasc Electrophysiol. 2000;11:1208–14.
29. Balasubramaniam R, Kistler PM. Atrial fibrillation in heart failure: the chicken or the egg? Heart. 2009;95:535–9.
30. Arnett DK, Hong Y, Bella JN, Oberman A, Kitzman DW, Hopkins PN, Rao DC, Devereux RB. Sibling correlation of left ventricular mass and geometry in hypertensive African Americans and whites: the HyperGEN study*. Am J Hypertens. 2001;14:1226–30.
31. Post WS, Larson MG, Myers RH, Galderisi M, Levy D. Heritability of left ventricular mass. Hypertension. 1997;30:1025–8.
32. Doolan G, Nguyen L, Chung J, Ingles J, Semsarian C. Progression of left ventricular hypertrophy and the angiotensin-converting enzyme gene polymorphism in hypertrophic cardiomyopathy. Int J Cardiol. 2004;96:157–63.
33. Nagy Z, Busjahn A, Bähring S, Faulhaber H-D, Gohlke H-R, Knoblauch H, Rosenthal M, Müller-Myhsok B, Schuster H, Luft FC. Quantitative trait loci for blood pressure exist near the IGF-1, the Liddle syndrome, the angiotensin II-receptor gene and the renin loci in man. J Am Soc Nephrol. 1999;10:1709–16.
34. Kuch-Wocial A, Ślubowska K, Kostrubiec M, Pasierski T, Januszewicz W, Świtalska H, Wocial B, Pruszczyk P. Plasma neuropeptide Y immunoreactivity influences left ventricular mass in pheochromocytoma. Clin Chim Acta. 2004;345:43–7.
35. Arnett DK, Li N, Tang W, Rao DC, Devereux RB, Claas SA, Kraemer R, Broeckel U. Genome-wide association study identifies single-nucleotide polymorphism in KCNB1 associated with left ventricular mass in humans: the HyperGEN study. BMC Med Genet. 2009;10:43.
36. Benova TE, Bacova BS, Viczenczova C, Barancik M, Tribulova N. Myocardial Connexin-43 is implicated in the prevention of malignant arrhythmia in rats suffering from essential hypertension. In: Update on essential hypertension. London: Intech; 2016. https://doi.org/10.5772/63456.
37. McDonagh TA, Metra M, Adamo M, et al. 2021 ESC guidelines for the diagnosis and treatment of acute and chronic heart failure: developed by the task force for the diagnosis and treatment of acute and chronic heart failure of the European Society of Cardiology (ESC) with the special contribution of the heart failure association (HFA) of the ESC. Eur Heart J. 2021;42:3599–726.
38. Oh GC, Cho H-J. Blood pressure and heart failure. Clin Hypertens. 2020;26:1.

39. McMurray JJV, Carson PE, Komajda M, McKelvie R, Zile MR, Ptaszynska A, Staiger C, Donovan JM, Massie BM. Heart failure with preserved ejection fraction: clinical characteristics of 4133 patients enrolled in the I-PRESERVE trial. Eur J Heart Fail. 2008;10:149–56.
40. Adabag S, Smith LG, Anand IS, Berger AK, Luepker RV. Sudden cardiac death in heart failure patients with preserved ejection fraction. J Card Fail. 2012;18:749–54.
41. Pezawas T, Burger AL, Binder T, Diedrich A. Importance of diastolic function for the prediction of arrhythmic death. Circ Arrhythm Electrophysiol. 2020;13:e007757.
42. Oktay AA, Lavie CJ, Milani RV, Ventura HO, Gilliland YE, Shah S, Cash ME. Current perspectives on left ventricular geometry in systemic hypertension. Prog Cardiovasc Dis. 2016;59:235–46.
43. Marcus GM. Evaluation and Management of Premature Ventricular Complexes. Circulation. 2020;141:1404–18.
44. Yiu K-H, Tse H-F. Hypertension and cardiac arrhythmias: a review of the epidemiology, pathophysiology and clinical implications. J Hum Hypertens. 2008;22:380–8.
45. Siscovick DS, Raghunathan TE, Psaty BM, Koepsell TD, Wicklund KG, Lin X, Cobb L, Rautaharju PM, Copass MK, Wagner EH. Diuretic therapy for hypertension and the risk of primary cardiac arrest. N Engl J Med. 1994;330:1852–7.
46. Wachtell K, Okin PM, Olsen MH, Dahlöf B, Devereux RB, Ibsen H, Kjeldsen SE, Lindholm LH, Nieminen MS, Thygesen K. Regression of electrocardiographic left ventricular hypertrophy during antihypertensive therapy and reduction in sudden cardiac death. Circulation. 2007;116:700–5.
47. Antzelevitch C, Belardinelli L, Zygmunt AC, Burashnikov A, Di Diego JM, Fish JM, Cordeiro JM, Thomas G. Electrophysiological effects of Ranolazine, a novel antianginal agent with antiarrhythmic properties. Circulation. 2004;110:904–10.
48. Schram G, Zhang L, Derakhchan K, Ehrlich JR, Belardinelli L, Nattel S. Ranolazine: ion-channel-blocking actions and in vivo electrophysiological effects. Br J Pharmacol. 2004;142:1300–8.
49. Antoons G, Oros A, Bito V, Sipido KR, Vos MA. Cellular basis for triggered ventricular arrhythmias that occur in the setting of compensated hypertrophy and heart failure: considerations for diagnosis and treatment. J Electrocardiol. 2007;40:S8–S14.
50. Freitas P, Ferreira AM, Arteaga-Fernández E, et al. The amount of late gadolinium enhancement outperforms current guideline-recommended criteria in the identification of patients with hypertrophic cardiomyopathy at risk of sudden cardiac death. J Cardiovasc Magn Reson. 2019;21:50.
51. Al-Khatib SM, Stevenson WG, Ackerman MJ, et al. 2017 AHA/ACC/HRS guideline for management of patients with ventricular arrhythmias and the prevention of sudden cardiac death: a report of the American College of Cardiology/American Heart Association task force on clinical practice guidelines and the Heart Rhythm Society. Heart Rhythm. 2018;15:e73–e189.
52. Zipes DP, Camm AJ, Borggrefe M, et al. ACC/AHA/ESC 2006 guidelines for management of patients with ventricular arrhythmias and the prevention of sudden cardiac death. Circulation. 2006;114:e385–484.
53. Bazoukis G, Korantzopoulos P, Tsioufis C. The impact of renal sympathetic denervation on cardiac electrophysiology and arrhythmias: a systematic review of the literature. Int J Cardiol. 2016;220:87–101.
54. Kosiuk J, Hilbert S, Pokushalov E, Hindricks G, Steinberg JS, Bollmann A. Renal denervation for treatment of cardiac arrhythmias: state of the art and future directions. J Cardiovasc Electrophysiol. 2015;26:233–8.
55. Hofferberth SC, Cecchin F, Loberman D, Fynn-Thompson F. Left thoracoscopic sympathectomy for cardiac denervation in patients with life-threatening ventricular arrhythmias. J Thorac Cardiovasc Surg. 2014;147:404–11.

Part V

Treatment of Heart Failure in Hypertensive Patients

Treating Hypertension Complicated with Heart Failure: Going Beyond the Numbers

21

Elisabeta Badila and Maria Dorobantu

21.1 Same BP Values, Different Risks

High blood pressure (BP) is the most important risk factor for death worldwide, causing an estimated 10.8 million of deaths every year, an increased number when compared with previous data [1]. Prospective cohort studies have reported a continuous linear association between BP and cardiovascular (CV) events from a value of 115/75 mmHg, with no apparent threshold [2]. Not only are high BP values associated with an increased risk to develop coronary heart disease and heart failure (HF) but also with a risk of developing atrial fibrillation, valve disease, aortic syndromes, cognitive impairment, and dementia [2, 3]. Hypertension is responsible for 54 percent of all strokes and 47 percent of all ischemic heart disease events globally [4]. The arbitrary use of cutoff BP values to define hypertension arose rather from the need to simplify the diagnosis and treatment of these patients. Hypertension is recognized as a risk factor for CV disease in all ethnic groups, irrespective of socioeconomic status. The increase of CV risk has been attributed mainly to high systolic BP values in patients after the age of 50 years, while high diastolic BP values are more commonly a predictor of CV risk in younger patients. As diastolic BP declines with age as a consequence of increased arterial stiffness, elevated pulse pressure is also associated with an enhanced risk of adverse CV events.

Hypertension rarely appears in isolation. It is often associated with other cardiovascular risk factors, such as cigarette smoking, glucose intolerance, and lipid abnormalities, leading to an increased risk of future adverse events. Among all CV

E. Badila (✉)
Carol Davila University of Medicine and Pharmacy, Bucharest, Romania

Colentina Clinical Hospital Bucharest, Bucharest, Romania

M. Dorobantu
Carol Davila University of Medicine and Pharmacy, Bucharest, Romania

M. Dorobantu et al. (eds.), *Hypertension and Heart Failure*, Updates in Hypertension and Cardiovascular Protection,
https://doi.org/10.1007/978-3-031-39315-0_21

risk factors, hypertension has the strongest causative effect, which implies that lowering BP values with antihypertensive therapy would contribute significantly to the reduction in overall CV risk. It is important to emphasize that for the same BP values, the risk is different depending on the presence or absence of other risk factors, the subclinical organ damage (hypertension-mediated organ damage—HMOD), or the associated clinical condition. As a consequence, quantification of total CV risk is an important aspect of the risk stratification process for hypertensive patients.

Several risk prediction models have been developed and validated in large population studies, such as the Framingham Risk Score or the Pooled Cohort Equations (PCE) for the Afro-American population, the QRISK2 (a cardiovascular disease risk algorithm) in the United Kingdom, and the SCORE2 (Systematic COronary Risk Evaluation) system for the European population aged between 40 and 69 years [5, 6]. It is important to note that these risk scores should be used in patients without previous documented CV disease. All of them take into account several risk factors that contribute to the total CV burden, including age, sex, hypertension, smoking, dyslipidemia, and diabetes mellitus, estimating the risk of developing fatal and non-fatal CV events in the next 10 years. The goal is to identify patients at "high risk" or "very high risk" of major CV events, who would then benefit most from a preventive strategy. Hypertensive patients with established atheromatous disease, diabetes mellitus, and chronic kidney disease (CKD) stages 3-5 and those with grade 3 hypertension are automatically categorized as "very high risk" or "high risk" so that there is no point in using formal risk scores in such cases [7]. The SCORE2 system has been adjusted to risk regions so that all European countries are grouped into four different risk categories according to their overall CV disease mortality rates. This means that, in the case of patients with similar risk factors, including similar BP values, the overall CV risk may range from low to high or even very high, depending on the region of interest. Countries such as Norway, France, or the United Kingdom are included in the list of low-risk regions, with less than 100 WHO (World Health Organization)-reported CV deaths per 100.000 people. A possible limitation of CV risk prediction models is that they do not include data on medication use, as many patients may already be under antihypertensive or lipid-lowering treatment at the time of their risk assessment. Additionally, physical activity, socioeconomic status, family history, or nutrition habits are also important factors that contribute to the development of CV diseases and are not included in the calculation of overall risk scores. This means that every clinician should interpret risk scores with caution, taking also into account the risk factors not included in the evaluation. Another major inconvenience is that markers of HMOD are not included in the overall estimation of CV risk models, except for CKD and established peripheral artery disease (PAD), which automatically categorize the patient as "very high risk" (PAD and severe CKD) or "high risk" (moderate CKD). This is particularly important, as many patients with HMOD, such as retinopathy, arterial stiffening, or left ventricular hypertrophy (LVH), may be completely asymptomatic but still have a negative prognosis, although they may not express other classical CV risk factors, and thus, their overall 10-year risk of developing a CV event may be underestimated by the current risk scores. This means that it is necessary to evaluate HMOD in every hypertensive patient in order to correctly estimate the total CV risk.

Current European guidelines recommend prompt BP-lowering treatment initiation in patients with grade 2 and 3 hypertension at any level of CV risk (class of recommendation I, level of evidence A). In patients with grade 1 hypertension and low-moderate CV risk and without evidence of HMOD, antihypertensive treatment is recommended if the patient remains hypertensive with lifestyle interventions; in case of grade 1 hypertension and with high estimated CV risk or evidence of HMOD, prompt initiation of treatment is recommended (class of recommendation I, level of evidence A). A major change in the latest guidelines is attributed to patients with high-normal BP, for whom lifestyle interventions are recommended, as well as drug treatment if their CV risk is very high (class of recommendation IIB, level of evidence A) [7].

Cardiovascular risk varies with age. In younger people, the absolute risk of CV disease is low, even in those with markedly modified risk profiles, such as high BP values so that 10-year risk prediction models perform poorly in these patients. Although the relative risk (RR) may be high in young hypertensive patients, treatment decision should be based on the absolute rather than the relative risk of an individual. It may be reasonable to consider estimating lifetime risk prediction models in younger people, as they take into account the consequences of long-term exposure to CV risk factors. This implies that some patients may need pharmacologic treatment from a young age, which indeed leads to greater costs but also to an improved life expectancy. Some authors advocate the use of lifetime risk profiles instead of 10-year risk profiles, but this strategy has not been adopted yet on a large scale. The LIFEtime-perspective CardioVascular Disease (LIFE-CVD) score, proposed by the European Society of Cardiology, may help to identify high-risk patients, who may then be counseled on lifestyle changes and even receive anti-hypertensive or statin treatment. Another possibility is to predict lifetime treatment benefit, measured in cardiovascular-free life years gained from pharmacological treatment [8, 9].

By contrast, the absolute risk of elderly people is high, even in the absence of traditional CV risk factors. Traditional 10-year CV disease risk score models are difficult to apply in elderly people because they do not take into account the competing risk of non-CV disease mortality, which becomes significant as the individuals advance with age. As a result, they may determine an overestimation of the overall risk, leading to unnecessary drug prescriptions, supplementary costs, and reduced quality of life. For this reason, the European Society of Cardiology developed a new algorithm for people aged above 70 years, the Systematic COronary Risk Evaluation 2—Older Persons (SCORE2-OP), aimed to evaluate the 5-year and 10-year risk of CV events in elderly individuals in the same four geographical regions, classified according to their CV disease mortality rates [10].

Finally, recent efforts have been made in the attempt to estimate CV disease risk score based on polygenic variants. Such algorithms are promising since genome-wide association studies performed in the last decade have confirmed the association between cardio-metabolic disorders and genetics. Polygenic risk scores comprise multiple single-nucleotide variants associated with CV diseases [11]. In the near future, 10-year or lifetime CV disease risk may be based both on traditional risk factors and on genome studies, which will further refine risks and lead to tailored preventive strategies.

21.2 Beneficial Effects of Blood Pressure Lowering Therapy Per Se in Hypertension

At this moment, it is very important to regard hypertension as a risk factor with a continuous relationship with heart failure (HF). In the Framingham Heart Study cohort, hypertension was responsible for about one-quarter of HF cases, whereas in the elderly population, as many as 68% of HF cases were attributed to hypertension. Compared with normotensive individuals, male and female hypertensive patients had a two- and three-fold increase in the risk of developing HF [12].

From a pathophysiological perspective, persistently elevated BP values cause structural and functional abnormalities of the heart, with increased left ventricular (LV) afterload, as well as neuroendocrine and vascular changes. Pressure- and volume-mediated left ventricular remodeling, with the development of ventricular hypertrophy, should be regarded as an initial compensatory mechanism in order to preserve adequate cardiac output. Left ventricular hypertrophy (LVH) is the cardinal manifestation of hypertensive HF. However, increased left ventricular stiffness is responsible for the elevations in left ventricular end-diastolic pressures, with the development of diastolic dysfunction. Hypertensive HF occurs in patients with persistent arterial hypertension and in the absence of any other causes of LVH, such as hypertrophic cardiomyopathy, Fabry disease, or cardiac amyloidosis. Chronic hypertension involves both concentric and eccentric LVH, associated with HF with either a preserved or reduced ejection fraction (EF). Large prospective studies and meta-analyses have demonstrated the negative prognostic impact of LVH in hypertensive HF. Studies have shown that LVH was associated with a 2.3-fold increase in the risk of CV disease mortality, as well as a 2.1-fold increase in the risk of all-cause mortality [13, 14]. For every increase of 39 mg/m^2 in LV mass, there is an associated independent increase of 40% in the risk of major adverse CV events in hypertensive HF patients [15].

Initiation of antihypertensive treatment based on the estimated absolute cardiovascular disease risk, rather than the cutoff value of 140/90 mmHg, has gained importance in the past two decades. In light of new evidence that a risk-based approach to treatment could lead to better outcomes, guidelines now incorporate predicted risk to guide treatment decisions in patients with mild hypertension. Once the 10-year risk of developing CV disease is accounted for, there seems to be little use in using BP thresholds to guide treatment. As predicted risk models incorporate BP and multiple other risk factors, it seems unsurprising that a risk-based approach is a better predictor of outcomes than BP alone. However, the safety and effectiveness of antihypertensive treatment at lower BPs remain controversial. The extent to which BPs should be lowered is also under debate. Benefits of antihypertensive treatment in relation to outcomes must also be weighed against the adverse effects induced by lower BP values (for example, aggravating chronic kidney disease). Although current hypertension management guidelines offer recommendations for special categories of the population and for certain therapeutic classes (compelling indications), the magnitude of BP lowering achieved is more important than the choice of a particular medication class.

Some authors suggest that the relationship between BP and mortality seems to follow a J-curve, with both lower and higher BP values at baseline associated with increased mortality. One study investigated the association between on-treatment BP and clinical outcomes in patients with HF. The nadir of systolic blood pressure (SBP) and diastolic blood pressure (DBP), associated with the lowest event rate, was 132 mmHg and 74 mmHg, respectively. This suggested that BP values less than 130/70 mmHg at discharge or during follow-up were associated with worse survival in HF patients so that the lowest possible BP value may not necessarily be the optimal target in hypertensive patients with HF [16]. Other authors have noted that the J-shaped curve is present only in patients aged over 60 years, in whom the decline in BP might be explained by marked vascular damage, severe left ventricular dysfunction, or other non-CV comorbidities [17]. However, we should take into account the important differences between the two study groups: the first study involved patients from the Korean Acute Heart Failure registry, whereas the other involved an English cohort. It is admitted that ethnicity may influence the total CV risk so that the differences in outcome could partly be explained by the different characteristics of the studied populations.

Many trials and meta-analyses of randomized controlled trials (RCT) have shown that a reduction in blood pressure values is associated with a significant reduction in heart failure cases. Thus, a decrease in systolic blood pressure by 10 mmHg and diastolic blood pressure by 5 mmHg can reduce the risk of developing HF by up to 40% [18, 19]. Relative risk reduction is present irrespective of baseline BP, the level of cardiovascular risk, comorbidities, age, sex, and ethnicity [7].

Before the adoption of lower BP targets <130/80 mmHg by hypertension guidelines, several studies proved that the absolute benefits of intensive BP lowering in high-risk individuals were significant. The decision to shift toward more aggressive BP targets in hypertensive patients came after the results of the Systolic Blood Pressure Intervention Trial (SPRINT), published in 2015. This was a RCT, comprising 9361 non-diabetic, hypertensive patients, conducted at 102 clinical sites in the United States, which aimed to compare the outcomes of intensive BP-lowering therapy below the target of 120 mmHg with treatment to a target below 140 mmHg. Patients were aged above 50 years, had a SBP in the range 130-180 mmHg, and expressed a high-risk CV profile. All major classes of antihypertensive agents were prescribed during the study. The primary endpoint was a composite of myocardial infarction, other acute coronary syndromes, stroke, acute decompensated heart failure, and death from CV causes. The study was stopped prematurely after a median follow-up of 3.26 years because the incidence of primary composite outcomes was significantly lower in the intensive treatment group than in the standard-treatment group (1.65% per year vs. 2.19% per year; hazard ratio with intensive treatment, 0.75; 95% confidence interval [CI], 0.64 to 0.89; $P < 0.001$). A total of 155 deaths occurred in the intensive-treatment group and 210 in the standard-treatment group (hazard ratio, 0.73; 95% CI, 0.60 to 0.90; $P = 0.003$). Regarding renal outcomes, different definitions were used for patients with pre-existing chronic kidney disease (decrease in eGFR [estimated Glomerular Filtration Rate] of at least 50% or the development of end-stage renal disease) and for those without (decrease in eGFR of at least 30%, to a value less than 60 mL/min/1.73m^2). Whereas no significant differences in outcomes were observed between those with established CKD, the incidence of outcomes was higher in the intensive-treatment group in patients with no CKD at baseline (1.21% per year vs. 0.35% per year; hazard ratio, 3.49; 95% CI, 2.44 to 5.10; P < 0.001). Importantly, serious adverse events of hypotension, syncope, electrolyte abnormalities, and acute kidney injury occurred more often in the intensive-treatment group than in the standard-treatment group, although the differences were not statistically significant [20].

Since the results of the SPRINT study, the 2017 American College of Cardiology/American Heart Association guidelines on hypertension recommend a BP target of less than 130/80 mmHg in hypertensive adults with known CV disease or a 10-year atherosclerotic CV disease (ASCVD) event risk of 10% or higher, assessed using the atherosclerotic CV disease (ASCVD) risk calculator [21].

One meta-analysis comprising 19 trials, excluding SPRINT, evaluated the results of more intensive BP-lowering therapy, demonstrating that intensive antihypertensive regimens reduced the risk of major CV events by 14% (95% CI, 4 to 22) compared with less intensive regimens. The risk of myocardial infarction and stroke decreased by 13% (95% CI, 0 to 24) and 22% (95% CI, 10 to 32), respectively. By contrast, the occurrence of heart failure was associated with a non-significant reduction in outcome in patients receiving intensive antihypertensive regimens (RR 0.85, 95% CI, 0.66 to 1.11) [22]. However, after the addition of SPRINT data, the cumulative risk of developing HF was reduced by 25% in the intensive-treatment arm [23].

Contrary to the outcomes observed during the intervention period of the SPRINT trial, postinterventional outcomes differed significantly with respect to acute decompensated heart failure events. A greater number of heart failure events were noted in the intensive-treatment group than in the standard-treatment group (1.42% per year vs. 0.59% per year; hazard ratio, 2.34; 95% CI, 1.25 to 4.64; $P = 0.007$). However, the benefits of intensive treatment remained significant in terms of the components of the primary outcome, excluding heart failure. The authors suggested that a lower-intensity antihypertensive treatment in the intensive-treatment group and an increased use of other agents in the standard-treatment group could explain the higher incidence of heart failure events in the first group [24].

The results of the SPRINT trial are in discordance with the previously published results of a similar trial. The Action to Control Cardiovascular Risk in Diabetes (ACCORD) blood pressure trial (ACCORD BP) examined identical SBP targets in 4733 patients with type 2 diabetes. The primary composite outcome of myocardial infarction, stroke, or CV mortality occurred at a lower rate in the intensive-therapy group, but with no significant difference. The only significant differences were reported in the rate of strokes: 0.32% per year in the intensive-therapy group and 0.53% per year in the standard therapy group (hazard ratio, 0.59; 95% CI, 0.39 to 0.89; P = 0.01). Thus, more intensive antihypertensive treatment did not demonstrate a significant reduction in the primary composite endpoint [25]. Additionally, other studies showed an increased CV risk if BP was lowered to less than 120/80 mmHg in patients with diabetes [26, 27]. Data from the Saxagliptin Assessment of Vascular Outcomes Recorded in patients with diabetes mellitus—Thrombolysis in Myocardial Infarction 53 (SAVOR-TIMI 53) trial indicated an adjusted risk of the composite endpoint of CV death, myocardial infarction, and ischemic stroke in a U-shaped relationship with baseline SBP and DPB. Nadir SBP and DBP were 130-140 mmHg and 80-90 mmHg, respectively, with values of DBP less than 60 mmHg associated with an increased risk of myocardial infarction (adjusted hazard ratio, 2.30; 95% CI, 1.50 to 3.53) in patients with type 2 diabetes and elevated CV risk [28]. Currently, the 2018 European guidelines recommend reducing BP levels close to 130 mmHg SBP (or lower if tolerated) and 70-79 mmHg DBP, while the 2017 American guidelines suggest achieving BP levels less than 130/80 mmHg in hypertensive patients with diabetes [7, 21].

A meta-analysis comprising 68 studies and 245,885 patients evaluated the extent of benefits of BP-lowering therapy in hypertensive patients. Investigators confirmed that HF and stroke are the events most effectively prevented by antihypertensive therapy. Reductions of 10 mmHg in SBP and 5 mmHg in DBP were found to reduce the risk of HF and stroke by 43% (RR 0.57, 95% CI, 0.46-0.72) and 36% (RR 0.64, 95% CI, 0.58-0.71), respectively. In terms of absolute risk, a lowering of 10/5 mmHg in SBP/DPB could prevent 17 strokes and 28 cardiovascular events, including HF, for every 1000 patients treated for 5 years (the number needed to treat [NNT] 58 and 36, respectively). The results confirmed that progressively greater BP reductions resulted in progressively greater risk reductions, with less adverse CV events, especially HF and stroke [19].

One of the largest meta-analyses, including 123 studies and 613,815 participants, showed that relative risk reductions for major cardiovascular disease events, stroke, heart failure, and all-cause mortality are proportional to the magnitude of the blood pressure reductions achieved [18]. Every 10-mmHg reduction in systolic BP significantly reduced the risk of heart failure by 28% (relative risk [RR] 0.72, 95% CI, 0.67-0.78), irrespective of baseline BP levels. This is of particular importance, as BP-lowering therapy provides great benefits even in patients with lower baseline systolic BP (<130 mmHg), which suggests that BP therapy should not be guided by rigid targets (i.e., <140/90 mmHg), but rather by risk-based targets. This means that one should take into consideration the interaction of multiple risk factors and, in high-risk patients with lower baseline BP values, consider initiating antihypertensive treatment. An interesting aspect of this meta-analysis is that a sub-analysis based on the presence of heart failure suggested that the reduction in BP might increase the risk of renal failure, but the results were based on just 84 renal failure events in two studies [18].

Overall, although evidence exists toward a lower target of BP treatment, controversy exists among physicians. Many concerns regard the potential deleterious effects of lower BP, such as alterations in renal hemodynamics leading to acute kidney injury or orthostatic hypotension. Taking into consideration recent evidence from studies, more intense BP-lowering therapy significantly reduces the risk of stroke, coronary events, major CV events, and CV mortality so that intensive antihypertensive algorithms should be more and more adopted in the routine clinical practice. Even a recent analysis of the PARADIGM-HF (Prospective Comparison of ARNI [Angiotensin Receptor-Neprilysin Inhibitor] with ACEI [Angiotensin-Converting-Enzyme Inhibitor] to Determine Impact on Global Mortality and

Morbidity in Heart Failure) trial revealed that the study drugs had benefit even in those patients with lower initial SB, encouraging physicians to prescribe neurohormonal-blocking agents even in those with lower BP [29].

21.3 Beneficial Effects of Different Antihypertensive Drug Classes in Patients with Heart Failure

The increased prevalence of hypertension among the general adult population makes it common in the evolution of the cardiovascular continuum from risk factors to the end stage of the disease—the heart failure. Thus, antihypertensive medication is frequently recommended for patients with heart failure both for the decrease in BP in those whose pressure limits exceed the cutoff arbitrarily established in the definition and for the benefits proven beyond the numbers.

There are ten classes of antihypertensive drugs available to lower the blood pressure. Each of them acts at different sites on one or more pathophysiological processes involved in the development and maintenance of hypertension.

In the latest European guidelines, five major drug classes were recommended for the treatment of hypertension: angiotensin-converting enzyme inhibitors (ACEIs), angiotensin II receptor blockers (ARBs), beta-blockers (BB), calcium channel blockers (CCBs), and diuretics (thiazides and thiazide-like diuretics such as chlortalidone and indapamide). These recommendations were based on the one hand on the ability to lower blood pressure and to reduce CV events, and on the other hand on evidence of broad equivalence on overall cardiovascular mortality and morbidity, noting that the benefit of their use derives mainly from the effect of BP lowering [7]. There are some differences between certain drug classes in terms of preventing heart failure, e.g., lower prevention for CCBs.

Most classes of antihypertensives have clear indications in patients with heart failure with reduced ejection fraction (HFrEF): ACEIs, ARBs, BB, mineralocorticoid receptor antagonists (MRAs), and diuretics. The newer class of drugs, the angiotensin receptor-neprilysin inhibitor (ARNi), with sacubitril-valsartan as the only drug on the market at present, is very useful in HFrEF and also lowers blood pressure. Another new pharmacological class, the sodium-glucose cotransporter 2 inhibitors (SGLT2i), is class I level A recommendation in the current guidelines for HFrEF management with proven benefits also in hypertensions.

Patients with heart failure with preserved ejection fraction (HFpEF) are often older, especially female, and have higher prevalence of hypertension, obesity, atrial fibrillation, and anemia, when compared with HFrEF [30]. Therapeutic targets in HFpEF are directed toward associated conditions, like control of hypertension and heart rate, control of congestion, etc. In this context, obtaining BP control through guidelines recommendations seems to be the most effective in the absence of specific therapies. Recently, a SGLT2 inhibitor (empagliflozin) has been shown to reduce the combined risk of cardiovascular death or hospitalization for heart failure in patients with HFpEF, regardless of the presence or absence of diabetes [31].

Neurohormonal activation in HF leads to hydrosaline retention, excessive tachycardia, systemic vasoconstriction, exacerbation of myocardial ischemia, alteration of the myocardial structure with fibrosis, and left ventricular hypertrophy. The long-term consequences are found in the progression of ventricular dysfunction, remodeling, and heart pump dysfunction, all leading to decreased survival. Thus, drug control of these systems is the main target of therapy in heart failure both to improve pump function with regression of symptoms and to slow the progression of the disease, regardless of the etiology of heart failure.

Antihypertensive medication plays an important role, both in the prevention of heart failure, by blood pressure lowering, and in the treatment of the disease, favorably influencing the remodeling of the left ventricle, disease progression, survival, improvement of symptoms and increased exercise tolerance, increased survival, and costs.

21.3.1 Renin-Angiotensin-Aldosterone System Inhibition

The benefits of inhibition of the activated renin-angiotensin-aldosterone system (RAAS) in HF derive from the pathological implications of the components of this system. Initially, the system was thought to be relatively simple as an intravascular mechanism, with a role in regulating blood pressure, renal function, and mineralocorticoid secretion. Subsequently, it has been shown that there are tissue renin-angiotensin systems in the heart, kidney, and vessels, which are not mediated by the conversion enzyme for the generation of angiotensin II, but through other enzymatic complexes—chymases, trypsin, and peptidases. Aldosterone further contributes to myocardial fibrosis and remodeling. On the other hand, there are complex interactions between RAAS and the sympathetic nervous system, bradykinin, prostaglandins, and nitric oxide (NO), the inhibition of RAAS also have beneficial effects through the responses of other systems.

(a) **Angiotensin-converting enzyme inhibitors** prevent the generation of angiotensin II and aldosterone, the main contributors to maladaptative vascular processes, with increases in arteriolar wall thickness and systemic vascular resistance. Additionally, angiotensin II affects matrix protein composition in the heart and blood vessels by promoting the synthesis and deposition of collagen and other structural proteins [30]. A part of the antihypertensive effect of ACEIs is due to the withdrawal of angiotensin's II effect on vascular smooth muscle. On the other hand, ACEIs prevent the degradation of bradykinin, increasing the level of nitric oxide (NO) and prostaglandins, with beneficial effects on endothelium and vascular function. By reducing aldosterone, ACEIs have an indirect natriuretic and potassium-sparing effect. Because angiotensin II also stimulates the release of norepinephrine and increases sympathetic tone, ACEIs also have an indirect antiadrenergic effect. Due to their significant vasodilating effect, ACEIs are one of the most widely used classes of antihypertensives in the world, with first-line recommendations in international guidelines.

Beyond the decrease in BP values, ACEIs have been shown to be very effective in relieving heart failure symptoms, increasing exercise tolerance, prolonging survival, and reducing the costs of managing patients with HF at various stages (symptomatic or asymptomatic; severe, mild-moderate, or asymptomatic LV dysfunction; after myocardial infarction). Several meta-analyses of trials with ACEIs reveal a significant reduction in mortality and mortality plus HF-related hospitalizations. Given the favorable results obtained in studies with more angiotensin-converting enzyme inhibitors, the beneficial effect is considered a class effect.

(b) **Angiotensin II receptor blockers.** ACEIs and ARBs represent two of the most often prescribed types of antihypertensive medications. Additionally, in terms of cardiovascular events and mortality results, they are comparable to each other and other major drug classes [7].

By binding to the angiotensin II subtype 1 receptor (AT1), ARBs have antagonizing effects on angiotensin II. All drugs from ARBs class have binding selectivity for the AT1 receptor, which are distributed in high concentration in a variety of tissues, especially in smooth muscle cells, kidneys, heart, and aorta. Although clinically utilized ARBs bind to the AT1 receptor in a competitive manner, their dissociation is gradual, they thus have a longer-lasting BP-lowering effect. Angiotensin II activates the AT1 receptor, an action that leads to cellular growth, proliferation, and contraction in vascular smooth muscle cells (VSMC) of small arteries. In addition, the aforementioned effects also take place in VSMC of large arteries, as well as in cardiac myocytes and fibroblasts. The reason behind ARBs' development was the need to address certain limitations posed by the mechanism of action of ACEIs. Indeed, ACEIs are unable to inhibit the generation of angiotensin II by alternative enzymatic pathways. These pathways include chymase and other tissue-based proteases, which may upregulate on the long run and diminish their blood pressure-lowering impact. Due to the powerful vasoconstrictor property of angiotensin II, blocking its activity at AT1 receptors results in dilatation of small resistance arteries, a decrease in total peripheral resistance, and a fall in BP [32]. Thus, chronic treatment with ARBs is linked to a decrease in both left ventricular hypertrophy and stiffness of large arteries, an enhancement of endothelial function, as well as remodeling of large and small arteries. Principally, the reduction in BP values is responsible for target organ protection, but there are a variety of studies that proved the beneficial effects of ARBs directly on the kidneys, heart, and arteries [33, 34].

Over the course of the last several years, the role that ARBs play in the treatment of HFrEF has shifted, and currently, they are being recommended to patients who cannot tolerate ACEI due to their severe adverse effects [35]. The CHARM (Candesartan in Heart failure: Assessment of Reduction in Mortality and morbidity)-Alternative study proved that candesartan decreases cardiovascular mortality and HF hospitalization rates in patients previously intolerant to ACEI [36]. Val-HeFT (Valsartan Heart Failure Trial) that studied the benefits of adding valsartan to usual therapy, which included ACE1, also showed a reduction in HF hospitalization [37]. Unfortunately, the all-cause mortality rate has not been reduced by any drug from ARBs class in any trial.

(c) **Mineralocorticoid receptor antagonists.** The interchange of sodium and water with potassium is obtained by aldosterone binding to the mineralocorticoid receptor (MR). This exchange takes place in several sites, such as distal tubule and the collecting duct of the nephron and results in elevated intravascular fluid retention, as well as volume overload. Moreover, aldosterone may also induce inflammation, collagen formation, vascular stiffening, fibrosis stimulation, and myocardial necrosis. By stimulating myocardial hypertrophy and remodeling, promoting fibrosis and apoptosis, causing endothelial dysfunction, and downsizing the perfusion of the myocardium, high levels of aldosterone may lead to a rise in the incidence rate of cardiovascular events. Electrical remodeling is another result of aldosterone-induced myocardial fibrosis, which plays an arrhythmogenic role in heart failure. Other detrimental consequences of aldosterone/MR activation include decreased baroreceptor sensitivity, sympathetic nervous system activation, and myocyte apoptosis. Thus, MRAs (mineralocorticoid receptor antagonists) provide several positive outcomes in patients with HF [43].

RALES (Randomized Aldactone Evaluation Study), EMPHASIS-HF (Eplerenone in Mild Patients Hospitalization And SurvIval Study in Heart Failure), and TOPCAT (Treatment of Preserved Cardiac Function Heart Failure with an Aldosterone Antagonist) are three landmark trials that aimed to evaluate the beneficial effects of MRAs in patients with HF. When comparing the effects of spironolactone with placebo, the RALES's study result regarding patients with HFrEF and severe symptoms was a 30% reduction in the relative rate of all-cause death, while EMPHASIS-HF that enrolled patients with HFrEF and mild symptoms showed a reduction of 24% of the same rate. However, eplerenone also substantially decreased (by 37%) the major composite outcome of risk of mortality from CV causes and first hospitalization for HF after a median follow-up of 21 months [44, 45]. The TOPCAT study did not support the same effects of spironolactone on HFpEF patients, showing no difference between the two arms of the study in the rates of CV death, chronic HF hospitalization, and resuscitated cardiac arrest over a period of 6 years [46]. Apart from ACEI and a BB, MRAs (spironolactone or eplerenone) are suggested in all patients diagnosed with HFrER in order to decrease mortality and the risk of hospitalization due to HF [35].

21.3.2 β-Blockers

When compared with placebo, RCTs and meta-analyses illustrate that beta-blockers notably decrease the risk of heart failure, stroke, and major CV events in hypertensive patients [19]. When analyzed in comparison to other BP-lowering drugs, BB effectiveness is generally equivalent in the prevention of major CV events, except for less effective stroke prevention. RCTs based on HMOD have also shown that BB are slightly less effective than RAS blockers and CCBs in preventing or diminishing LVH (left ventricular hypertrophy), carotid IMT (intima-media thickness), aortic stiffness, and small artery remodeling [7].

Chronic administration of BB leads to direct antagonism of catecholamine cardiotoxicity, upregulation, and re-establishment of β-receptors, resulting in increased inotropic reserve, inhibition of cardiotoxic neurohormonal systems (the renin-angiotensin-aldosterone and endothelin systems), and improved coronary flow and diastolic phase lengthening. As a result, the left ventricle's structure and function are improved (reverse remodeling). Moreover, in patients with HFrEF, BB also reduce both heart rate and BP, have anti-ischemic effects, and provide a decrease in atrial and ventricular arrhythmia. In addition, β-blockers have been shown to improve the contractility of viable but non-contractile myocardium in patients with both ischemic and non-ischemic (hibernating and stunning myocardium) HFrEF [38].

Besides therapy with an ACEI and diuretic, BB have been demonstrated to improve mortality and morbidity rates in patients who have HFrEF [35]. As far as Metoprolol is concerned, according to MERIT-HF (Metoprolol CR/XL Randomized Intervention Trial in Congestive Heart Failure) that enrolled 3991 patients with HFrEF [ejection fraction (EF) < 40%], there was a statistically significant reduction of 34% in the all-cause mortality rate, and also a smaller number of sudden deaths and deaths from aggravating HF when compared to the placebo group [39]. The COPERNICUS study (Carvedilol Prospective Randomized Cumulative Survival Study) evaluated the benefits of carvedilol in the treatment of HFrEF patients with EF < 25%. The follow-up of 10.4 months showed a decrease in mortality by 34% and also registered a reduction of 27% in the number of days of hospitalization for any cause and one of 40% for HF [40]. Bisoprolol's effects on HFrEF patients (EF < 35%) were studied in CIBIS II (The Cardiac Insufficiency Bisoprolol Study II), which included 2647 patients. When comparing the two arms of the study (bisoprolol vs. placebo), a significantly decreased rate of all-cause mortality in the bisoprolol group (11.8% vs. 17.3%) was reported. Moreover, the rate of sudden deaths was also reduced in this group (3.6% vs. 6.3%) [41]. Nebivolol was also studied for its effect on HFrEF patients in SENIORS (Study of Effects of Nebivolol Intervention on Outcomes and Rehospitalization in Seniors with Heart Failure) trial, which included 2128 patients with HF (EF < 35%). Although, by comparison, all-cause mortality and HF hospitalization rates were lower in the nebivolol group (31.1% vs. 35.3%), the difference was not statistically significant [42]. Thus, nebivolol was included as a treatment option for HFrEF only in the European guidelines with the mention that it has not been proven to decrease the rates of all-cause mortality in HF patients.

21.3.3 Diuretics

When it comes to antihypertensive treatment, diuretics have remained fundamental since their emergence in the 1960s. In RCTs and meta-analyses, their efficacy in preventing all forms of CV morbidities and mortality has been proven [19]. When dealing with heart failure prevention, diuretics also appear to be more effective than other drug classes [7].

The Na-Cl (sodium-chloride) cotransporter in the nephron's distal convoluted tubule represents the site where thiazide acts. It also counts up for about 5% of the total sodium reabsorption. Although their chemical structure is different, thiazide-like diuretics have the same mechanism of action. Thiazide diuretics have been shown to have an immediate beneficial effect on blood pressure by producing natriuresis, which in turn decreases venous return, extracellular volume, and lastly, cardiac output. However, with long-term thiazide administration, cardiac output returns to pretreatment levels and peripheral resistance decreases, thus maintaining the antihypertensive effect. The mechanism behind the decrease in total peripheral resistance is not elucidated [47].

According to a meta-analysis, when compared with placebo, loop and thiazide diuretics decrease the risk of death and aggravating HF in patients with HFrEF. Moreover, diuretics seem to enhance exercise tolerance in comparison to an active control [48]. Although they have synergic effect (sequential nephron blockage) and their combination represents a treatment option for diuretic resistance, loop diuretics generate a more powerful and shorter diuresis than thiazides [35].

21.3.4 Calcium Channel Blockers

CCBs have a stronger impact on stroke reduction than anticipated based on the decrease obtained in BP's values. However, they may be less effective at preventing HFrEF [19]. In terms of diminishing the development of carotid atherosclerosis and decreasing LVH and proteinuria, when compared to other classes of antihypertensive drugs, CCBs have been shown to be more useful than BB [7].

The two major pharmacological classes of CCB are dihydropyridines (DHP) and non-dihydropyridines (non-DHP). Both of them bind to the voltage-gated L-type calcium channel found in cardiac and smooth muscle. DHP-CCBs target vascular smooth muscle more specifically and have fewer effects on cardiac muscle than the non-DHP-CCBs. The effects of CCBs are multifaceted, such as vasodilatation, increased nitric oxide production, inhibition of blood vessel aggregation, antioxidant and anti-inflammatory effects, suppression of the migration and proliferation of vascular smooth cells, as well as plaque stabilization. They also decrease cardiac contraction and heart rate [49].

With the prevalence of 60-80%, hypertension is the leading etiology of HFpEF. While there have been studies that accounted for CCBs effect on reducing all-cause mortality in this category of patients, other clinical trials did not find any statistically significant correlation. However, when compared to the placebo group of patients, in a recent analysis from TOPCAT, there was an important reduction in the risk of mortality in patients with HFpEF, as well as a decrease in the risk of all-cause mortality for the group of patients taking CCBs. Due to the contradictory information from the available trials, the relationship between CCB treatment and the prognosis of HFpEF patients requires more multicenter, large-scale studies [50]. Bearing in mind that CCBs offer no benefit in terms of functionality or mortality and they may also aggravate the outcome, they are recommended to be avoided in HFrEF patients [35].

21.4 Conclusions

In conclusion, the last few years have marked the significant progress in our understanding of the predicted cardiovascular risk of an individual and how blood pressure should be interpreted in association with other risk factors and not solely be judged based on rigid values. This has translated into an integrative approach to treating hypertension even in patients with lower blood pressure values, but with a high cardiovascular risk profile. As a result, a risk-based strategy represents a better predictor of outcomes than blood pressure alone. Moreover, studies investigating more intensive versus standard blood pressure-lowering therapy have demonstrated a clear benefit on cardiovascular outcomes when lower blood pressure values were reached. Although controversy exists among physicians, especially concerning the risk of worsening renal function, orthostatic hypotension, syncope, and aggravating heart failure, more intensive antihypertensive treatment should be adopted with caution in all patients based on the individual absolute cardiovascular risk profile. The magnitude of BP lowering achieved is more important than the choice of a particular medication class. However, certain drug classes have proved particular benefits in patients with heart failure with reduced ejection fraction in addition to the blood pressure-lowering effect itself. Although some agents have recently been shown to improve outcomes in heart failure with preserved ejection fraction, treatment of associated conditions, including hypertension, remains the cornerstone strategy.

References

1. Vos T, Lim SS, Abbafati C, Abbas KM, Abbasi M, Abbasifard M, et al. Global burden of 369 diseases and injuries in 204 countries and territories, 1990–2019: a systematic analysis for the global burden of disease study 2019. Lancet. 2020;396:1204–22.
2. Lewington S, Clarke R, Qizilbash N, Peto R, Collins C. Age-specific relevance of usual blood pressure to vascular mortality: a meta-analysis of individual data for one million adults in 61 prospective studies. Lancet. 2002;360:1903–13.
3. Angeli F, Reboldi G, Verdecchia P. Hypertension, inflammation and atrial fibrillation. J Hypertens. 2014;32(3):480–3.
4. Lawes CMM, Vander HS, Rodgers A. Global burden of blood-pressure-related disease, 2001. Lancet. 2008;371(9623):1513–8.
5. D'Agostino RB, Vasan RS, Pencina MJ, Wolf PA, Cobain M, Massaro JM, et al. General cardiovascular risk profile for use in primary care: the Framingham heart study. Circulation. 2008;117(6):743–53.
6. SCORE2 working group and ESC Cardiovascular risk collaboration. SCORE2 risk prediction algorithms: new models to estimate 10-year risk of cardiovascular disease in Europe. Eur Heart J. 2021;42(25):2439–54.
7. Williams B, Mancia G, Spiering W, Rosei EA, Azizi M, Burnier M, et al. 2018 ESC/ESH guidelines for the management of arterial hypertension. Eur Heart J. 2018;39(33):3021–104. https://doi.org/10.1093/eurheartj/ehy339.
8. Visseren FLJ, MacH F, Smulders YM, Carballo D, Koskinas KC, Bäck M, et al. 2021 ESC guidelines on cardiovascular disease prevention in clinical practice. Eur Heart J. 2021;42(34):3227–337.
9. Rossello X, Dorresteijn JA, Janssen A, Lambrinou E, Scherrenberg M, Bonnefoy-Cudraz E, et al. Risk prediction tools in cardiovascular disease prevention: A report from the ESC

prevention of CVD Programme led by the European Association of Preventive Cardiology (EAPC) in collaboration with the acute cardiovascular care association (ACCA) and the Association of Cardiovascular Nursing and Allied Professions (ACNAP) clinical practice. Eur J Cardiovasc Nursing. 2019;18(7):534–44.
10. SCORE2-OP working group and ESC cardiovascular risk collaboration. SCORE2-OP risk prediction algorithms: estimating incident cardiovascular event risk in older persons in four geographical risk regions. Eur Heart J. 2021;42:2455–67.
11. O'Sullivan JW, Raghavan S, Marquez-Luna C, Luzum JA, Damrauer SM, Ashley EA, et al. Polygenic risk scores for cardiovascular disease: a scientific statement from the American Heart Association. Circulation. 2022;146(8):e93. https://doi.org/10.1161/CIR.0000000000001077.
12. Mahmood SS, Levy D, Vasan RS, Wang TJ. The Framingham heart study and the epidemiology of cardiovascular diseases: a historical perspective. Lancet. 2014;383(9921):999–1008. https://doi.org/10.1016/S0140-6736(13)61752-3.
13. Drazner MH. The progression of hypertensive heart disease. Circulation. 2011;123(3):327–34.
14. Hilton DJ. Echocardiographic left ventricular hypertrophy (LVH) and prognosis—a meta-analysis. J Med Clin Res Rev. 2018;2(1):1–5.
15. Verdecchia P, Carini G, Circo A, Dovellini E, Giovannini E, Lombardo M, et al. Left ventricular mass and cardiovascular morbidity in essential hypertension: the MAVI study. J Am Coll Cardiol. 2001;38(7):1829–35.
16. Lee SE, Lee HY, Cho HJ, et al. Reverse J-curve relationship between on-treatment blood pressure and mortality in patients with heart failure. JACC Heart Fail. 2017;5(11):810–9. https://doi.org/10.1016/j.jchf.2017.08.015.
17. Herrett E, Strongman H, Gadd S, Tomlinson L, Nitsch D, Bhaskaran K, et al. The importance of blood pressure thresholds versus predicted cardiovascular risk on subsequent rates of cardiovascular disease: a cohort study in English primary care. Lancet Healthy Longev. 2022;3(1):e22–30.
18. Ettehad D, Emdin CA, Kiran A, Anderson SG, Callender T, Emberson J, et al. Blood pressure lowering for prevention of cardiovascular disease and death: a systematic review and meta-analysis. Lancet. 2016;387(10022):957–67.
19. Thomopoulos C, Parati G, Zanchetti A. Effects of blood pressure lowering on outcome incidence in hypertension.1. Overview, meta-analyses, and meta-regression analyses of randomized trials. J Hypertens. 2014;32(12):2285–95. https://doi.org/10.1097/HJH.0000000000000378.
20. Wright JT, Williamson JD, Whelton PK, Snyder JK, Sink KM, Rocco MV, et al. A randomized trial of intensive versus standard blood-pressure control. N Engl J Med. 2015;373(22):2103–19.
21. Whelton PK, Carey RM, Aronow WS, Casey DE, Collins KJ, Dennison Himmelfarb C, et al. 2017 ACC/AHA/AAPA/ABC/ACPM/AGS/APhA/ASH/ASPC/NMA/PCNA guideline for the prevention, detection, evaluation, and management of high blood pressure in adults: a report of the American College of Cardiology/American Heart Association task force on clinical practice guidelines. J Am Coll Cardiol. 2018;71(19):e127–248.
22. Xie X, Atkins E, Lv J, Bennett A, Neal B, Ninomiya T, et al. Effects of intensive blood pressure lowering on cardiovascular and renal outcomes: updated systematic review and meta-analysis. Lancet. 2016;387(10017):435–43.
23. Verdecchia P, Angeli F, Gentile G, Reboldi G. More versus less intensive blood pressure-lowering strategy. Hypertension. 2016;68(3):642–53.
24. Lewis CE, Fine LJ, Beddhu S, Cheung AK, Cushman WC, Cutler JA, et al. Final report of a trial of intensive versus standard blood-pressure control. N Engl J Med. 2021;384(20):1921–51.
25. Cushman C, Evans GW, Byington RP, Goff DC, Grimm RH, Cutler JA, et al. Effects of intensive blood-pressure control in type 2 diabetes mellitus. N Engl J Med. 2010;362:1575–85.
26. Wan EYF, Yu EYT, Chin WY, Fung CSC, Fong DYT, Choi EPH, et al. Effect of achieved systolic blood pressure on cardiovascular outcomes in patients with type 2 diabetes: a population-based retrospective cohort study. Diabetes Care. 2018;41(6):1134–41.
27. Gomadam P, Shah A, Qureshi W, Yeboah PN, Freedman BI, Bowden D, et al. Blood pressure indices and cardiovascular disease mortality in persons with or without diabetes mellitus. J Hypertens. 2018;36(1):85–92.

28. Bergmark BA, Scirica BM, Steg PG, Fanola CL, Gurmu Y, Mosenzon O, et al. Blood pressure and cardiovascular outcomes in patients with diabetes and high cardiovascular risk. Eur Heart J. 2018;39(24):2255–62. https://doi.org/10.1093/eurheartj/ehx809.
29. McMurray JJ, Packer M, Desai AS, et al. Angiotensin-neprilysin inhibition versus enalapril in heart failure. N Engl J Med. 2014;371(11):993–1004. https://doi.org/10.1056/NEJMoa1409077.
30. Bhatt D. Opie's cardiovascular drugs: a companion to Braunwald's heart disease. 9th ed. Amsterdam: Elsevier; 2021.
31. Anker SD, Butler J, Filippatos G, et al. Empagliflozin in heart failure with a preserved ejection fraction. N Engl J Med. 2021;385(16):1451–61. https://doi.org/10.1056/NEJMoa2107038.
32. Laurent S. Antihypertensive drugs. Pharmacol Res. 2017;124:116–25. https://doi.org/10.1016/j.phrs.2017.07.026.
33. Laurent S, Boutouyrie P. Vascular mechanism collaboration. Dose-dependent arterial destiffening and inward remodeling after olmesartan in hypertensives with metabolic syndrome. Hypertension. 2014;64(4):709–16. https://doi.org/10.1161/HYPERTENSIONAHA.114.03282.
34. Nakamura T, Fujii S, Hoshino J, Saito Y, Mizuno H, Saito Y, et al. Selective angiotensin receptor antagonism with valsartan decreases arterial stiffness independently of blood pressure lowering in hypertensive patients. Hypertens Res. 2005;28(12):937–43.
35. McDonagh TA, Metra M, Adamo M, et al. 2021 ESC guidelines for the diagnosis and treatment of acute and chronic heart failure. Eur Heart J. 2021;42(36):3599–726. https://doi.org/10.1093/eurheartj/ehab368.
36. Granger CB, McMurray JJV, Yusuf S. Effects of candesartan in patients with chronic heart failure and reduced left-ventricular systolic function intolerant to angiotensin-converting-enzyme inhibitors: the CHARM-alternative trial. Lancet. 2003;362:772–6.
37. Cohn JN, Tognoni G. Valsartan heart failure trial investigators. A randomized trial of the angiotensin-receptor blocker valsartan in chronic heart failure. N Engl J Med. 2001;345(23):1667–75. https://doi.org/10.1056/NEJMoa010713.
38. Masarone D, Martucci ML, Errigo V, Pacileo G. The use of β-blockers in heart failure with reduced ejection fraction. J Cardiovasc Dev Dis. 2021;8(9):101. https://doi.org/10.3390/jcdd8090101.
39. MERIT-HF Study Group. Effect of metoprolol CR/XL in chronic heart failure: metoprolol CR/XL randomised intervention trial in congestive heart failure (MERIT-HF). Lancet. 1999;353:2001–7.
40. Eichhorn EJ, Bristow MR. The carvedilol prospective randomized cumulative survival (COPERNICUS) trial. Curr Control Trials Cardiovasc Med. 2001;2:20–3. https://doi.org/10.1186/cvm-2-1-020.
41. CIBIS-II Investigators and Committees. The cardiac insufficiency Bisoprolol study II (CIBIS-II): a randomised trial. Lancet. 1999;353(9146):9–13.
42. Shibata MC, Flather MD, Böhm M, et al. Study of the effects of Nebivolol intervention on outcomes and Rehospitalisation in Seniors with heart failure (SENIORS). Rationale and design. Int J Cardiol. 2002;86(1):77–85. https://doi.org/10.1016/s0167-5273(02)00321-2.
43. Vizzardi E, Regazzoni V, Caretta G, et al. Mineralocorticoid receptor antagonist in heart failure: Past, present and future perspectives. Int J Cardiol Heart Vessel. 2014;3:6–14. Published 2014 Mar 19. https://doi.org/10.1016/j.ijchv.2014.03.005.
44. Pitt B, Zannad F, Remme WJ, et al. The effect of spironolactone on morbidity and mortality in patients with severe heart failure. Randomized Aldactone evaluation study investigators. N Engl J Med. 1999;341(10):709–17. https://doi.org/10.1056/NEJM199909023411001.
45. Zannad F, McMurray JJ, Krum H, et al. Eplerenone in patients with systolic heart failure and mild symptoms. N Engl J Med. 2011;364(1):11–21. https://doi.org/10.1056/NEJMoa1009492.
46. Pitt B, Pfeffer MA, Assmann SF, et al. Spironolactone for heart failure with preserved ejection fraction. N Engl J Med. 2014;370(15):1383–92. https://doi.org/10.1056/NEJMoa1313731.
47. Sinha AD, Agarwal R. Clinical pharmacology of antihypertensive therapy for the treatment of hypertension in CKD. Clin J Am Soc Nephrol. 2019;14(5):757–64. https://doi.org/10.2215/CJN.04330418.

48. Faris R, Flather M, Purcell H, Henein M, Poole-Wilson P, Coats A. Current evidence supporting the role of diuretics in heart failure: a meta analysis of randomised controlled trials. Int J Cardiol. 2002;82(2):149–58. https://doi.org/10.1016/s0167-5273(01)00600-3.
49. Sueta D, Tabata N, Hokimoto S. Clinical roles of calcium channel blockers in ischemic heart diseases. Hypertens Res. 2017;40(5):423–8. https://doi.org/10.1038/hr.2016.183.
50. Wang X, Ju J, Chen Z, et al. Associations between calcium channel blocker therapy and mortality in heart failure with preserved ejection fraction. Eur J Prev Cardiol. 2022;29(9):1343–51. https://doi.org/10.1093/eurjpc/zwac004.

Reversibility of Cardiac Remodeling in Hypertensive Patients with Heart Failure

22

Anna Paini, Claudia Agabiti Rosei, Carolina De Ciuceis, Massimo Salvetti, and Maria Lorenza Muiesan

22.1 Introduction

Hypertensive patients show a fivefold to sixfold higher risk of developing HF than healthy subjects, and the lifetime heart failure (HF) risk grows directly proportional to blood pressure [1]. According to the most recent European Society of Cardiology guidelines, HF has been categorized according to ejection fraction values (reduced, mildly reduced, or preserved) [1]. In addition, the American College of Cardiology Foundation/American Heart Association classification has identified 4 stages, from A to D [2], based on clinical symptoms and cardiac structural alterations. Most hypertensive patients fall in stage A, since they have asymptomatic cardiac structural and/or function alterations, while the development of LV concentric hypertrophy represents a criterion for stage B HF.

Longstanding pressure overload on the LV is the most important and well-known mechanism of damage, which brings to the development of hypertensive heart disease, characterized by the impairment of LV relaxation and filling (so-called diastolic dysfunction), the development of left ventricular hypertrophy (LVH), and of HF with preserved ejection fraction (HFpEF). It should be noticed that hypertensive heart disease is also characterized by asymptomatic systolic dysfunction, even before the onset of clinical HF.

A. Paini · C. A. Rosei · C. De Ciuceis · M. Salvetti
Department of Clinical and Experimental Sciences-University of Brescia, Brescia, Italy

Department of Medicine ASST Spedali Civili 25121, Brescia, Italy

M. L. Muiesan (✉)
Department of Clinical and Experimental Sciences-University of Brescia, Internal Medicine ASST Spedali Civili, Brescia, Italy
e-mail: marialorenza.muiesan@unibs.it

M. Dorobantu et al. (eds.), *Hypertension and Heart Failure*, Updates in Hypertension and Cardiovascular Protection,
https://doi.org/10.1007/978-3-031-39315-0_22

Structural alteration of small coronary arteries, with increased collagen deposition and fibrosis (i.e., microvascular disease), is a key feature of cardiac chamber damage, detectable in the early stages of hypertensive heart disease. Chronic hyperactivation of neurohormonal pathways (i.e., renin-angiotensin-aldosterone system) has a pivotal role in arterial hypertension (AH), as does the persistent increase of proinflammatory cytokines (tumor necrosis factor-alpha, interleukin-1, and interleukin-6), growth factors (i.e., transforming growth factor-beta), and reactive oxygen species. This complex humoral activation leads to myocardial fibrosis, coronary microvascular endothelial inflammation and dysfunction, and reduction in coronary flow reserve. The decrease in oxygen supply induces ischemia and further fibrosis and remodeling. The increased pressure load may also favor the development of coronary artery disease and LV hypertrophy associated with enlargement of LV, with clear impairment of ejection fraction, leading to HF with reduced ejection fraction (HFrEF) (Fig. 22.1).

The longstanding neurohormonal and inflammatory activation represents a cause of the parallel remodeling and stiffening of large arteries. The increase of large arteries stiffness causally contributes to LV diastolic function, both ventricular-arterial uncoupling and microvascular damage [3] (Fig. 22.1). Approximately half of the patients with overt HF may display normal ejection fraction and marked impairment in diastolic function [1, 2]. Patients with HFpEF have a prevalence of hypertension up to 90% both in large randomized controlled trials and in epidemiological studies, and are usually older women, frequently obese, with diabetes mellitus, hyperlipidemia, and atrial fibrillation. They may have hypertensive pulmonary edema as an expression of HFpEF.

Treatment of hypertension reduces the risk of HF, especially in elderly and very elderly patients, and prevention of HF represents the largest benefit associated with BP-lowering drugs [4, 5]. However, at the present time, with increased aging of the

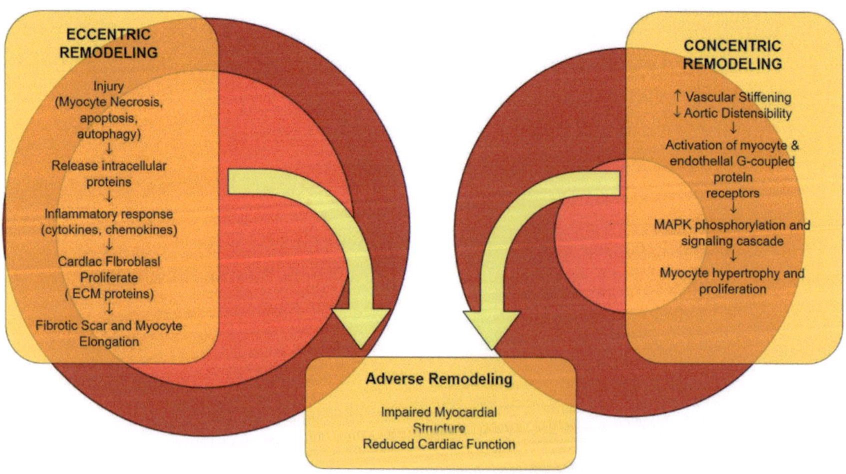

Fig. 22.1 Ventricular remodeling and cardiac dysfunction

general population, a persistent increase in morbidity and mortality associated with HF has been observed [1, 2]. This has been reported using all five major classes of antihypertensive drugs, with diuretics being more effective, at least in preventing symptoms leading to HF hospitalization, and calcium channel blockers apparently less effective in comparative trials, although mainly in those trials in which they replaced diuretics. A network meta-analysis by Sciarretta et al. [6] investigated the impact of different antihypertensive therapies in the prevention of HF, concluding that diuretics seem the most effective class of drugs in preventing the transition from AH to overt HF.

The degree of blood pressure reduction has been associated with the decrease of LVH and of aortic stiffness. In the SPRINT trial, hypertensive patients randomized to an intensive treatment strategy had a significantly lower incidence of HF, although this may be related at least in part to the higher prescription of diuretics [7].

Therefore, in hypertensive patients, the optimal intervention for improving prognosis is an aggressive approach to early-stage, i.e., Stage A and B HF, reducing blood pressure values and reducing the progression of cardiovascular structural alterations. In this regard, several clinical comparative studies and meta-analysis have demonstrated the superiority of some classes of drugs (RAAS blockers and CCB over beta-blockers) on LVH regression [8]. It should be underlined that, even in the presence of clinical HF, a reverse remodeling phenomenon may be obtained by the most appropriate use of cardiovascular drugs.

The changes in LV mass, volume, shape, and composition of the ventricle in response to the mechanical (stress and strain) and systemic neurohormonal activation have been defined as "LV remodeling." Remodeling can be observed in physiological conditions, such as normal growth of an organism or in response to exercise. Cardiac remodeling occurs also in response to sustained hypertension, diabetes, obesity, and in heart failure [4] and may be observed in any of the 4 cardiac chambers.

In 1995 Kass et al. proposed the term "reverse remodeling" when describing the change in LV end-diastolic pressure-volume curve of the failing heart after hemodynamic unloading [9]. Reverse remodeling is characterized by a decrease in LV size and geometry that persists when the treatment is stopped, suggesting that intrinsic biological changes, and not simply the reduction in LV end-diastolic pressure, have occurred.

Most studies evaluating the reversibility of cardiac remodeling have used 2D echocardiography, based on broader availability, with the evaluation of an increasing number of different diastolic and systolic function parameters. Other noninvasive modalities for the assessment of cardiac anatomic and functional changes, used in most recent studies, are cardiac magnetic resonance (CMR) and radionuclide imaging.

Patients with HF should receive a baseline therapy with an ACE inhibitor or an ARB, plus a b-blocker and a loop diuretic. This triple therapeutic regimen is aimed to improve diastolic and microvascular dysfunction in HFpEF and to preserve systolic function in HFrEF. In the presence of elevated BP a further reduction in afterload could be obtained switching to valsartan/sacubitril and choosing a vasodilating b-blocker [10].

22.2 Heart Failure with Reduced Ejection Fraction

The Cooperative North Scandinavian Enalapril Survival Study (CONSENSUS) [11], the Veteran's Administration Cooperative Vasodilator—Heart Failure Trial (V-HeFT I) [12, 13], the Studies of Left Ventricular Dysfunction (SOLVD) [14], and the Assessment of Treatment with Lisinopril and Survival (ATLAS) study [15] have demonstrated a reduction in mortality associated with ACE inhibitors treatment within a wide range of heart failure patients, from asymptomatic left ventricular dysfunction to severe heart failure. In all these studies the prevalence of hypertensive patients ranged from 40% (SOLVD) to 48% (V-HeFT I) or was not reported (CONSENSUS and ATLAS).

Few studies have evaluated changes in ejection fraction. In a SOLVD substudy (Studies of Left Ventricular Dysfunction), a significant difference in serial measures of LV volumes, detected by radionuclide ventricular imaging, was observed for patients randomized to the angiotensin-converting enzyme inhibitor enalapril versus those randomized to placebo; more interestingly those changes paralleled the enalapril effect on all-cause mortality and on heart failure hospitalization [16–18].

In another study comparing an ACE inhibitor and an angiotensin receptor blocker (ARB), it was confirmed that both captopril 150 mg daily and losartan 50 mg daily prevent LV dilation, i.e., adverse ventricular remodeling, while LV end-systolic volume index reduction (i.e., reverse remodeling) was obtained by captopril, but not losartan therapy [19].

Other studies have consistently shown that, compared to placebo, beta-blockers reduce both mortality and morbidity in symptomatic stable HFrEF patients who are on optimum medical therapies [1]. Patients had a LV ejection fraction range between 20 and 28%, and symptoms corresponding to NYHA class III–IV in 52–100%. In the CIBIS study, bisoprolol induced a 10% increase in EF, but not changes in the cardio-thoracic index [20].

In a group of patients with milder symptoms of heart failure, a similar improvement in ejection fraction (+10%) was observed during treatment with the vasodilator beta-blocker carvedilol, as compared with placebo [21]. Moreover, a decrease in both LV end-diastolic and end-systolic volume indexes, associated with an increase in LV ejection fraction, measured by CMR was found after 6 months of treatment with metoprolol as compared to placebo. LV mass index decreased in both groups of patients randomized to both placebo and metoprolol [22]. These results confirmed a previous study in a small number of patients, showing a slight improvement in ejection fraction after 3 months of treatment with metoprolol, while a decrease in LV mass was reached only after a longer period (18 months) [23] (Table 22.1).

A beneficial effect of HR control obtained by treatment with ivabradine was observed in the BEAUTIFUL Study, in patients with coronary artery disease and left ventricular systolic dysfunction, with a 71% prevalence of a history of hypertension. Treatment with ivabradine was associated with a decrease in LV end-systolic volume versus an increase with placebo ($P = 0.018$). An increase in LV ejection

Table 22.1 Changes in ejection fraction and function parameters in patients with reduced ejection fraction during treatment with different drugs

	Hypertensive pz %	Baseline	12 months	Relative change (n,%)	Trial
LVEF (%)					
RAAS blockade	42,8	25 ± 7	29 ± 8	4, + 13.8%	SOLVD
Beta-blockade	–	28.6 ± 0.9	34.1 ± 1.5	5.5, + 14.4%	ANZ
MRA[a]	65%	26.2 ± 0.6	28 ± 0.4	1.8, + 6.9%	NCT00082589
Arni	88%	28.2 (24.5 to 32.7)	37.8 (32.3 to 45.2)	9.4 (8.8 to 9.9)	PROVE-HF
SGLT2	37%	35 ± 9	37 ± 11	2.4 + 7.5%	EMPIRE HF[b]
	70.5%	31.7 ± 9.9	33.5 ±10.3	1.8 +5.7%	SUGAR-DM-HF[c]
LVESV index (mL/m^2)					
RAAS blockade[b]		106 ± 42	93 ± 37	13, −12.3%	SOLVD
Beta-blockade		72.9 ± 4.1	65 ± 4.5	7.9, −10.8%	ANZ
MRA [a]		124.3 ± 3.9	119.2 ± 1.5	5.1, −4.1%	NCT00082589
ARNI		61.68 (51.95 to 75.00)	45.5 (34.8 to 57.6)	−15.3 (−16.03 to −14.6)	PROVE-HF
SGLT2		54 ± 30	49 ± 23	−4.1, −13.1%	EMPIRE HF[b]
		80.8 ±3 7.2	72.9 ± 37	−7.9, −11.8%	SUGAR-DM-HF[c]
LVEDV index (mL/m^2)					
RAAS blockade[b]		140 ± 44	127 ± 37	13, −9.3%	SOLVD
Beta-blockade		100.2 ± 4.6	95.6 ± 4.9	4.6, −4.6%	ANZ
MRA [a]		167.0 ± 4.4	163,3 ± 1.8	3.7, −2.2%	NCT00082589
Arni		86.9 (76.2 to 100.4)	74.2 (63.5 to 86.3)	−12.3 (−12.9 to −11.6)	PROVE-HF
SGLT2		81 ± 36	75 ± 26	−4.2, −15.2%	EMPIRE HF[b]
		114.7 ±37	105.7 ± 37.6	−9, −12.4%	SUGAR-DM-HF[c]

LVEF left ventricular ejection fraction, *LVESV* left ventricular end-systolic volume, *LVEDV* left ventricular end-diastolic volume, *RAAS* renin angiotensin-aldosterone system, *MRA* mineralocorticoid receptor antagonist, *Arni* angiotensin receptor/neprilysin inhibitor, *SGLT2* sodium-glucose cotransporter-2 inhibitors

SOLVD Studies of Left Ventricular Dysfunction [18], ANZ Australia/New Zealand Collaborative Group [56], NCT00082589 [26], PROVE-HF Prospective Study of Biomarkers, Symptom Improvement, and Ventricular Remodeling During Sacubitril/Valsartan Therapy for Heart Failure [29], EMPIRE HF Empagliflozin in Heart Failure Patients with Reduced Ejection Fraction [57], SUGARD-DM-HF Effect of Empagliflozin on Left Ventricular Volumes in Patients with Type 2 Diabetes, or Prediabetes, and Heart Failure With Reduced Ejection Fraction [34]

[a]Data at 36-week therapy with MRA

[b]12% with type 2 diabetes

[c]Data at 36-week therapy with SGLT2 by cardiovascular magnetic resonance, 78.1% patients with type 2 diabetes

fraction was observed with ivabradine versus no change with placebo ($P = 0.009$). The decrease in LV end-systolic volume index was related to the degree of heart rate reduction with ivabradine, while blood pressure did not change during the study. Authors reported no differences in other echocardiographic parameters such as LV shape, wall motion scores, left atrial area, or mitral regurgitation severity [24]. The benefit could be explained by the improvement of oxygen balance, favoring the recruitment of contractile reserve in segments with ischemic and post-ischemic reversible ventricular dysfunction, and by the prolongation of diastolic filling in order to maintain cardiac output [25].

22.3 Mineral Corticoid Receptor Agonist (MRA)

In a study including a larger percentage of hypertensive patients (65% in the active treatment arm and 56% in the placebo group), no benefit of the aldosterone receptor blocker eplerenone on LV remodeling was reported in a group of patients with class II and III heart failure and decreased LV ejection fraction. This finding is in contrast with the favorable impact that this class of agent has exerted in patients with heart failure and decreased LV ejection fraction and could be explained by the optimal background treatment (95% ACE-I and BB therapy) [26] (Table 22.1).

22.4 Angiotensin Receptor-Neprilysin Inhibitor (Sacubitril/Valsartan, ARNI)

The use of LCZ696 angiotensin receptor-neprilysin inhibitor (sacubitril/valsartan, ARNI) significantly improves the prognosis of patients with HF with HFrEF [27, 28]. The prognostic benefits of ARNI treatment are associated with a reduction in N-terminal pro B-type natriuretic peptide (NT-proBNP) level and with cardiac reverse remodeling, which includes an improvement in the LV ejection fraction and a reduction in LV volume. In the prospective, 12-month, single-group, open-label study of patients with HFrEF (PROVE-HF study), 794 patients were enrolled, 654 completed the study and 88% of them had hypertension at medical history. At echocardiography changes in cardiac remodeling indices demonstrated a significant increase in LV ejection fraction and corresponding reduction in LV volumes as early as 6 months; such changes continued to 12 months. In addition, a reduction in left atrial volume index and LV mass index at 6 months, with a further decrease by 1 year, was observed. Treatment with sacubitril-valsartan was also associated with significant change in the E/e′ ratio both at 6 and 12 months, compared with baseline (both $P < 0.001$). The improvement in cardiac remodeling measures was weak, albeit significantly related to the reduction in NT-proBNP concentration, suggesting that the observed reverse cardiac remodeling could represent an explanation for the effects of sacubitril-valsartan in patients with HFrEF [29].

Another study (EVALUATE-HF, Effects of Sacubitril/Valsartan vs. Enalapril on Aortic Stiffness in Patients with Mild to Moderate HF with Reduced Ejection Fraction) aiming to assess central aortic stiffness in 464 patients with HFrEF on guidelines optimal therapy compared the effect of sacubitril/valsartan or enalapril therapy on changes in aortic stiffness. History of hypertension was considered among eligibility criteria for the study, and mean BP at baseline was 130/77 and 131/75 mmHg in the sacubitril/valsartan and enalapril groups, respectively. After 3 months of treatment with sacubitril/valsartan, no reduction in central aortic stiffness was observed, but a significant reduction in both atrial and ventricular volumes, improvement in diastolic function, associated with a reduction in NTproBNP as compared to those treated with enalapril was reported. No difference between enalapril and sacubitril/valsartan in terms of LV ejection fraction change was reported [30].

A meta-analysis conducted by Wang et al. included 20 studies (18 in patients with HFrEF) enrolling 10,175 patients to compare the effects of ARNI versus angiotensin-converting enzyme inhibitors or angiotensin receptor blockers on cardiac reverse remodeling parameters. The results show that ARNI was superior to angiotensin-converting enzyme inhibitors/angiotensin receptor blockers in terms of improvement of changes in LV ejection fraction, diameter, volumes, LV mass index, and left atrium volume in patients with HFrEF. The changes were observed after 3 months of treatment and became more significant with longer follow-up to 12 months [31] (Tables 22.1 and 22.2).

Table 22.2 Changes in ejection fraction and function parameters in patients with reduced ejection fraction during treatment with different drugs

	Hypertensive patients %	Baseline	3 months	6 months	12 months
LVEF (%)					
PROVE-HF	88%	28.2	–	34.1	37.8
EVALUATE-HF	99.5	34	36	–	–
LVEDVI (mL/m^2)					
PROVE-HF	88%	86.93	70.3	79.50	74.15
EVALUATE-HF	99.5%	75.1	–	–	–
LVESVI (mL/m^2)					
PROVE-HF	88%	61.68	–	52.25	45.46
EVALUATE-HF	99.5%	50.8	46.3		
LAVI (mL/m^2)					
PROVE-HF	88%	37.76	–	32.8	
EVALUATE-HF	99.5	30.4	28.2	–	–
E/E′ ratio					
PROVE-HF	88%	11.70	–	10.50	10.20
EVALUATE-HF	99.5	13.8	12.3	–	–

EVALUATE-HF Study of Effect of Sacubitril/Valsartan vs. Enalapril on Aortic Stiffness in patients With Mild to Moderate HF With Reduced Ejection Fraction [30], LAVI left atrial volume index, LVEDVI left ventricular end-diastolic volume index, LVEF left ventricular ejection fraction, LVESVI left ventricular end-systolic volume index, PROVE-HF Prospective Study of Biomarkers, Symptom Improvement, and Ventricular Remodeling During Sacubitril/Valsartan Therapy for Heart Failure [29]

22.5 Sodium-Glucose Cotransporter-2 Inhibitors (SGLT2i)

More recently, the impact of SGLT2i in patients with HFrEF was reported, with significant reductions in HF hospitalizations and CV mortality, in both patients with and without T2D [32, 33].

The results from the DAPA-HF [32] and EMPEROR [33] trials represent a revolutionary approach to HF treatment, reinforcing the rationale for the use of SGLT2 inhibitors in patients with HFrEF, with a significant impact on future clinical practice. In these trials, the benefit of the SGLT2i on the primary composite outcome was driven mainly by reducing hospitalizations for HF, on the background of optimal HF therapy.

Two recent trials SUGAR-DM-HF (Studies of Empagliflozin and Its Cardiovascular, Renal and Metabolic Effects) [34] and EMPATROPISM (Are the "Cardiac Benefits" of Empagliflozin Independent of Its Hypoglycemic Activity? [ATRU-4]) [35] have investigated the role of these drugs on cardiac reverse remodeling, as a possible mechanism explaining the clinical benefit, independently of the glycemic control. In the SUGAR-DM-HF, patients with both HFrEF and type 2 diabetes or prediabetes (and 70% prevalence of hypertension) had favorable reverse remodeling, assessed by CMR, with reductions in LV end-systolic and end-diastolic volumes over 36 weeks. No significant changes in the other CMR variables (LV ejection fraction, LV mass index, LV global function index, indexed left atrial volume, myocardial blood flow, or extracellular volume fraction) after 36 weeks of treatment with empagliflozin, in comparison with placebo, were observed [34].

In the REFORM (Research into the effect of SGLT2 inhibition on left ventricular remodeling in patients with heart failure and diabetes mellitus) trial, no improvement in LV remodeling with dapagliflozin on diabetic heart failure patients was demonstrated. In this trial 50% of patients were in NYHA functional class I with a mean baseline ejection fraction of 46%, including patients with both HFrEF and HFpEF. The overall prevalence of hypertension was 72% and of MI 50%, respectively [36].

On the contrary, in the EMPATROPISM trial empagliflozin significantly reversed and improved LV remodeling as shown by reduced LV volumes and LV hypertrophy, an improvement of the spherical shape of the LV, and by an increased LV ejection fraction in patients with HFrEF (hypertension prevalence of 70% and without diabetes). All these changes were assessed by MRI after 6 months of treatment with either empagliflozin or placebo (on the top of optimal medical treatment). In contrast to the REFORM study, patients enrolled in this study had a higher severity ejection fraction impairment and were all with HFrEF (mean ejection fraction 36% and NYHA functional class II or III) [35].

Few meta-analyses have further analyzed the results of these studies, comparing patients with stage A-B and C, patients with and without diabetes, or restricting the analysis to studies using CMR [37–39]. The results of the more recent systematic review [39] included 13 trials. Nine trials were placebo-controlled, 4 evaluated the effect of empagliflozin, 6 of dapagliflozin, 2 of canagliflozin, and one of luseogliflozin. In 4 studies, assessment of cardiac remodeling was based on cardiac magnetic resonance and in 7 on echocardiography and in 2 on CMR main study and prespecified echocardiographic substudy. Four trials included patients with HFrEF, 1 with HFpEF and 2 with both HFpEf and HFrEF, but only one trial shows results separately according to ejection fraction. All remaining studies were not limited to patients with HF. In HFrEF patients SGLT2i treatment significantly improved the LV ejection fraction, reduced left ventricular mass index and left atrial volume index. When the analysis was restricted to HFrEF patients, a significant improvement of ejection fraction, systolic and diastolic volumes, of LV mass index and of left atrial volume was confirmed (Table 22.1).

22.6 Heart Failure with Preserved Ejection Fraction

HFpEF is characterized by structural and functional alterations of left ventricular and atrial structure, diastolic function, and systolic function despite preserved LV ejection fraction [40–43]. The benefit of drug classes that are efficacious in HF with reduced ejection fraction has been lower in patients with higher LV ejection fraction, decreasing the risk of HF hospitalization but not reducing cardiovascular or all-cause death in HFpEF. Three large trials, including the DIG (Digitalis Investigation Group)-REF and DIG-PEF [44], the CHARM (Candesartan in Heart Failure: Assessment of Reduction in Mortality and Morbidity)-REF and CHARM-Preserved [45–47], and the I-PRESERVE (Irbesartan in Heart Failure with Preserved Systolic Function) trial [48, 49], have reported all-cause mortality and heart failure hospitalization rates per 1000 patient-years of follow-up, allowing comparison between these studies. The demographic characteristics of patients enrolled in the 3 studies have shown a high prevalence of obesity in all patients with HFpEF, and small differences between the DIG and the CHARM trials inclusion criteria; patients enrolled in the I-PRESERVE were older and had a lower prevalence of coronary heart disease.

These pieces of evidence have highlighted the importance of noncardiac comorbidities in the complex pathophysiological mechanisms, both cardiac and noncardiac, underlying and influencing the clinical HFpEF.

More recently treatment with other classes of drugs, i.e., an angiotensin-neprilysin inhibitor and ***SGLT2i***, has reduced the risk of composite cardiovascular events, driven by a reduction in HF hospitalizations, in patients with and without diabetes mellitus.

Comprehensive management of HFpEF includes exercise as well as treatment of risk factors and comorbidities possibly favoring the regression or at least limiting the progression of structural and functional cardiovascular abnormalities.

22.7 Angiotensin II Receptor Blockers

In the CHARM echocardiographic substudy (CHARMES), the assessment of cardiac structure and function was performed in 312 patients (10% of patients enrolled in the CHARM preserved study), 14 months after randomization to candesartan or placebo. No significant differences in LV diastolic and systolic function or size between the placebo and candesartan groups were observed; a similar distribution in diastolic dysfunction abnormalities was reported in both treatment groups (p 0.85), while patients receiving candesartan had lower LV mass and LV mass index ($p < 0.05$) and lower prevalence of LVH (45% vs. 59%, p 0.04) [50].

In the I-PRESERVE study, an increased LV mass, mass/volume ratio, and left atrial size were independently associated with an increased risk of both primary and heart failure events; no data on changes induced by treatment have been reported [40].

22.8 Mineral Corticoid Receptor Agonist (MRA)

In the Treatment of Preserved Cardiac Function Heart Failure with an Aldosterone Antagonist (TOPCAT) trial, cardiac structure and function were assessed by quantitative echocardiography at baseline and at 12- to 18-month follow-up in 239 out of the 3206 patients with heart failure and preserved ejection fraction (LV ejection fraction ≥45%) enrolled in the main study. In patients randomized to spironolactone compared with placebo no significant differences in changes of LV structure, LV systolic function, LV diastolic function, left atrial size, or right ventricular function and pulmonary pressure were observed. The decrease in left atrial volume at follow-up during treatment with spironolactone was associated with a lower risk of subsequent occurrence of the primary outcome (cardiovascular death, heart failure hospitalization, or aborted cardiac arrest) [41].

All these studies suggest that cardiac remodeling, diastolic dysfunction, or both might explain the difference in outcome among patients with HFpEF although whether changes in cardiac remodeling may be associated with improvement in outcome remains an unanswered question.

22.9 Sodium-Glucose Cotransporter-2 Inhibitors (SGLT2i)

Zhang et al. have performed a systematic review and meta-analysis on the effect of sodium-glucose cotransporter 2 inhibitors on reverse cardiac remodeling, including patients with type 2 diabetes mellitus (T2DM) and/or heart failure [39]. The data have been analyzed according to baseline ejection fraction (reduced or preserved). When the analysis was restricted to HFpEF patients [51] SGLT2i treatment with luseogliflozin did not reduce LV mass index, left atrial volume index, or E/e′ ratio. No significant changes in echocardiographic parameters related to the left ventricular systolic and diastolic function were observed between patients treated with canagliflozin or glimepiride [52]. No data are currently available on the effect of empagliflozin in the Emperor Preserved study.

Some potential mechanisms have been proposed to explain the benefit of SGLT2 inhibitors in HF patients, including (a) the reduction of cardiac preload due to the diuretic and natriuretic effect, with a decrease in LV stretch and wall stress and in LV volume; (b) the reduction in blood pressure and arterial stiffness; (c) the inhibition of activation of the nucleotide-binding domain-like receptor protein 3 (NLRP3) inflammasome and decreasing the calmodulin kinase II activity; lastly SGLT2i might have downstream epigenetic-oriented effects in cardiac cells to ameliorate cardiac remodeling. Other proposed effects include reduction of cardiac oxidative stress, improvement of myocardial energetics, and inhibition of the mammalian target of rapamycin pathway.

The different effect of the same pharmacological therapy in patients with HFrEF and HFpEF may have been due to the different signal transduction cascades of myocardial remodeling found in these two conditions. In addition, only two studies included HFpEF patients, and further trials are needed to confirm or not the results of the meta-analysis results.

22.10 Angiotensin Receptor-Neprilysin Inhibitor (Sacubitril/Valsartan, ARNI)

In the phase II PARAMOUNT (Prospective Comparison of ARNI with ARB on Management of Heart Failure with Preserved Ejection Fraction) trial, some results suggested the occurrence of reverse cardiac remodeling during treatment with sacubitril/valsartan therapy. Left atrial volume index was more significantly reduced in patients with HFpEF treated with sacubitril/valsartan versus valsartan monotherapy [53]. In a meta-analysis performed by Wang et al. [31], ARNI outperformed ACEIs/ARBs in patients with HFpEF [54] or LVH [55] in terms of LV mass index (− 3.25 g/m^2, 95% CI − 3.78 to −2.72) and LAV (− 7.20 mL, 95% CI − 14.11 to −0.29) reductions, without significant changes of other indices of reverse remodeling.

Further large sample trials are required to determine the effects of ARNI on cardiac reverse remodeling in HF with preserved ejection fraction patients.

References

1. McDonagh TA, Metra M, Adamo M, Gardner RS, Baumbach A, Böhm M, et al. ESC guidelines for the diagnosis and treatment of acute and chronic heart failure. Eur Heart J. 2021;2021:3599–726.
2. Yancy CW, Jessup M, Bozkurt B, Butler J, Casey DE, Colvin MM, et al. ACC/AHA/HFSA focused update of the 2013 ACCF/AHA guideline for the management of heart failure: a report of the American College of Cardiology/American Heart Association task force on clinical practice guidelines and the heart failure society of amer. Circulation. 2017;2017:e137–61.
3. Chirinos JA, Segers P, Hughes T, Townsend R. Large-artery stiffness in health and disease: JACC state-of-the-art review. J Am Coll Cardiol. 2019;74:1237–63.
4. Mann DL, Barger PM, Burkhoff D. Myocardial recovery and the failing heart. J Am Coll Cardiol. 2012;60:2465–72.
5. Williams B, Mancia G, Spiering W, Agabiti Rosei E, Azizi M, Burnier M, et al. 2018 ESC/ESH guidelines for the management of arterial hypertension. J Hypertens. 2018;1:1.
6. Sciarretta S, Palano F, Tocci G, Baldini R, Volpe M. Antihypertensive treatment and development of heart failure in hypertension: a Bayesian network meta-analysis of studies in patients with hypertension and high cardiovascular risk. Arch Intern Med. 2011;171:384–94.
7. Sobieraj P, Nilsson PM, Kahan T. Heart failure events in a clinical trial on arterial hypertension: new insights into the SPRINT trial. Hypertension. 2021;78:1241–7.

8. Salvetti M, Paini A, Bertacchini F, Stassaldi D, Aggiusti C, Agabiti Rosei C, et al. Changes in left ventricular geometry during antihypertensive treatment. Pharmacol Res. 2018;134:193–9.
9. Kass DA, Baughman KL, Pak PH, Cho PW, Levin HR, Gardner TJ, et al. Reverse remodeling from cardiomyoplasty in human heart failure: external constraint versus active assist. Circulation. 1995;91:2314–8.
10. Messerli FH, Rimoldi SF, Bangalore S. The transition from hypertension to heart failure: contemporary update. JACC Heart Fail. 2017;5:543–51.
11. The Consensus Trial Study Group. Effects of enalapril on mortality in severe congestive heart failure. N Engl J Med. 1987;316:1429–35.
12. Wong M, Johnson G, Shabetai R, Hughes V, Bhat G, Lopez B, et al. Echocardiographic variables as prognostic indicators and therapeutic monitors in chronic congestive heart failure. Veterans affairs cooperative studies V-HeFT I and II. V-HeFT VA cooperative studies group. Circulation. 1993;87(6 Suppl):VI65–70.
13. Wong M, Staszewsky L, Latini R, Barlera S, Glazer R, Aknay N, et al. Severity of left ventricular remodeling defines outcomes and response to therapy in heart failure: valsartan heart failure trial (Val-HeFT) echocardiographic data. J Am Coll Cardiol. 2004;43:2022–7; [cited 2022 Apr 11]. https://pubmed.ncbi.nlm.nih.gov/15172407/.
14. SOLVD Investigators*. Effect of enalapril on survival in patients with reduced left ventricular ejection fractions and congestive heart failure. N Engl J Med. 1991;325:67; [cited 2022 Apr 11]. https://pubmed.ncbi.nlm.nih.gov/2057034/.
15. Rydén L, Armstrong PW, Cleland JGF, Horowitz JD, Massie BM, Packer M, et al. Efficacy and safety of high-dose lisinopril in chronic heart failure patients at high cardiovascular risk, including those with diabetes mellitus. Results from the ATLAS trial. Eur Heart J. 2000;21:1967–78; [cited 2022 Apr 11]. https://pubmed.ncbi.nlm.nih.gov/11071803/.
16. Konstam MA, Rousseau MF, Kronenberg MW, Udelson JE, Melin J, Stewart D, et al. Effects of the angiotensin converting enzyme inhibitor enalapril on the long-term progression of left ventricular dysfunction in patients with heart failure. Circulation. 1992;86:431–8.
17. Konstam MA, Kronenberg MW, Rousseau MF, Udelson JE, Melin J, Stewart D, et al. Effects of the angiotensin converting enzyme inhibitor enalapril on the long-term progression of left ventricular dilatation in patients with asymptomatic systolic dysfunction. Circulation. 1993;88:2277–83.
18. Greenberg B, Quinones MA, Koilpillai C, Limacher M, Shindler D, Benedict C, et al. Effects of long-term enalapril therapy on cardiac structure and function in patients with left ventricular dysfunction results of the SOLVD echocardiography substudy. Circulation. 1995;91:2573–81.
19. Konstam MA, Patten RD, Thomas I, Ramahi T, La Bresh K, Goldman S, et al. Effects of losartan and captopril on left ventricular volumes in elderly patients with heart failure: results of the elite ventricular function substudy. Am Heart J. 2000;139:1081–7.
20. Léchât P, Escolano S, Golmard JL, Lardoux H, Witchitz S, Henneman JA, et al. Prognostic value of bisoprolol-induced hemodynamic effects in heart failure during the cardiac insufficiency bisoprolol study (CIBIS). Circulation. 1997;96:2197–205.
21. Colucci WS, Packer M, Bristow MR, Gilbert EM, Cohn JN, Fowler MB, et al. Carvedilol inhibits clinical progression in patients with mild symptoms of heart failure. US carvedilol heart failure study group. Circulation. 1996;94:2800–6.
22. Groenning BA, Nilsson JC, Sondergaard L, Fritz-Hansen T, Larsson HBW, Hildebrandt PR. Antiremodeling effects on the left ventricle during beta-blockade with metoprolol in the treatment of chronic heart failure. J Am Coll Cardiol. 2000;36:2072–80.
23. Hall SA, Cigarroa CG, Marcoux L, Risser RC, Grayburn PA, Eichhorn EJ. Time course of improvement in left ventricular function, mass and geometry in patients with congestive heart failure treated with beta-adrenergic blockade. J Am Coll Cardiol. 1995;25:1154–61.
24. Ceconi C, Freedman SB, Tardif JC, Hildebrandt P, McDonagh T, Gueret P, et al. Effect of heart rate reduction by ivabradine on left ventricular remodeling in the echocardiographic substudy of BEAUTIFUL. Int J Cardiol. 2011;146:408–14.
25. Messerli FH, Rimoldi SF, Bangalore S, Bavishi C, Laurent S. When an increase in central systolic pressure overrides the benefits of heart rate lowering. J Am Coll Cardiol. 2016;68:754–62.

26. Udelson JE, Feldman AM, Greenberg B, Pitt B, Mukherjee R, Solomon HA, et al. Randomized, double-blind, multicenter, placebo-controlled study evaluating the effect of aldosterone antagonism with eplerenone on ventricular remodeling in patients with mild-to-moderate heart failure and left ventricular systolic dysfunction. Circ Heart Fail. 2010;3:347–53.
27. McMurray JJV, Packer M, Desai AS, Gong J, Lefkowitz MP, Rizkala AR, et al. Angiotensin–neprilysin inhibition versus enalapril in heart failure. N Engl J Med. 2014;371:993–1004.
28. Solomon SD, Claggett B, McMurray JJV, Hernandez AF, Fonarow GC. Combined neprilysin and renin–angiotensin system inhibition in heart failure with reduced ejection fraction: a meta-analysis. Eur J Heart Fail. 2016;18:1238–43.
29. Januzzi JL, Prescott MF, Butler J, Felker GM, Maisel AS, McCague K, et al. Association of change in n-terminal pro-b-type natriuretic peptide following initiation of sacubitril-valsartan treatment with cardiac structure and function in patients with heart failure with reduced ejection fraction. JAMA. 2019;322:1085–95.
30. Desai AS, Solomon SD, Shah AM, Claggett BL, Fang JC, Izzo J, et al. Effect of sacubitril-valsartan vs enalapril on aortic stiffness in patients with heart failure and reduced ejection fraction: a randomized clinical trial. JAMA. 2019;322:1077–84.
31. Wang Y, Zhou R, Lu C, Chen Q, Xu T, Li D. Effects of the angiotensin-receptor neprilysin inhibitor on cardiac reverse remodeling: meta-analysis. J Am Heart Assoc. 2019;8:e012272.
32. McMurray JJV, Solomon SD, Inzucchi SE, Kober L, Kosiborod MN, Martinez FA, et al. Dapagliflozin in patients with heart failure and reduced ejection fraction. N Engl J Med. 2019;381:1995–2008.
33. Packer M, Anker SD, Butler J, Filippatos G, Pocock SJ, Carson P, et al. Cardiovascular and renal outcomes with empagliflozin in heart failure. N Engl J Med. 2020;383:1413–24.
34. Lee MMY, Brooksbank KJM, Wetherall K, Mangion K, Roditi G, Campbell RT, et al. Effect of empagliflozin on left ventricular volumes in patients with type 2 diabetes, or prediabetes, and heart failure with reduced ejection fraction (SUGAR-DM-HF). Circulation. 2021;143:516–25.
35. Santos-Gallego CG, Vargas-Delgado AP, Requena-Ibanez JA, Garcia-Ropero A, Mancini D, Pinney S, et al. Randomized trial of empagliflozin in nondiabetic patients with heart failure and reduced ejection fraction. J Am Coll Cardiol. 2021;77:243–55.
36. Singh JSS, Mordi IR, Vickneson K, Fathi A, Donnan PT, Mohan M, et al. Dapagliflozin versus placebo on left ventricular remodeling in patients with diabetes and heart failure: the reform trial. Diabetes Care. 2020;43:1356–9.
37. Dhingra NK, Mistry N, Puar P, Verma R, Anker S, Mazer CD, et al. SGLT2 inhibitors and cardiac remodelling: a systematic review and meta-analysis of randomized cardiac magnetic resonance imaging trials. ESC Heart Fail. 2021;8:4693–700.
38. Yu YW, Zhao XM, Wang YH, Zhou Q, Huang Y, Zhai M, et al. Effect of sodium–glucose cotransporter 2 inhibitors on cardiac structure and function in type 2 diabetes mellitus patients with or without chronic heart failure: a meta-analysis. Cardiovasc Diabetol. 2021;20:25.
39. Zhang N, Wang Y, Tse G, Korantzopoulos P, Letsas KP, Zhang Q, et al. Effect of sodium-glucose cotransporter-2 inhibitors on cardiac remodelling: a systematic review and meta-analysis. Eur J Prev Cardiol. 2021;28:1961–73.
40. Zile MR, Gottdiener JS, Hetzel SJ, McMurray JJ, Komajda M, McKelvie R, et al. Prevalence and significance of alterations in cardiac structure and function in patients with heart failure and a preserved ejection fraction. Circulation. 2011;124:2491–501.
41. Shah AM, Claggett B, Sweitzer NK, Shah SJ, Anand IS, O'Meara E, et al. Cardiac structure and function and prognosis in heart failure with preserved ejection fraction. Circ Heart Fail. 2014;7:740–51.
42. Shah AM, Claggett B, Sweitzer NK, Shah SJ, Anand IS, Liu L, et al. Prognostic importance of impaired systolic function in heart failure with preserved ejection fraction and the impact of spironolactone. Circulation. 2015;132:402–14.
43. Shah AM, Cikes M, Prasad N, Li G, Getchevski S, Claggett B, et al. Echocardiographic features of patients with heart failure and preserved left ventricular ejection fraction. J Am Coll Cardiol. 2019;74:2858–73.

44. THE DIGITALIS INVESTIGATION GROUP. The effect of digoxin on mortality and morbidity in patients with heart failure. N Engl J Med. 1997;336:525–33.
45. Yusuf S, Pfeffer MA, Swedberg K, Granger CB, Held P, McMurray JJV, et al. Effects of candesartan in patients with chronic heart failure and preserved left-ventricular ejection fraction: the CHARM-preserved trial. Lancet. 2003;362:777–81.
46. Granger CB, McMurray JJV, Yusuf S, Held P, Michelson EL, Olofsson B, et al. Effects of candesartan in patients with chronic heart failure and reduced left-ventricular systolic function intolerant to angiotensin-converting-enzyme inhibitors: the CHARM-alternative trial. Lancet. 2003;362:772–6.
47. McMurray JJV, Östergren J, Swedberg K, Granger CB, Held P, Michelson EL, et al. Effects of candesartan in patients with chronic heart failure and reduced left-ventricular systolic function taking angiotensin-converting-enzyme inhibitors: the CHARM-added trial. Lancet. 2003;362:767–71.
48. McMurray JJV, Carson PE, Komajda M, McKelvie R, Zile MR, Ptaszynska A, et al. Heart failure with preserved ejection fraction: clinical characteristics of 4133 patients enrolled in the I-PRESERVE trial. Eur J Heart Fail. 2008;10:149–56.
49. Massie BM, Carson PE, McMurray JJ, Komajda M, McKelvie R, Zile MR, et al. Irbesartan in patients with heart failure and preserved ejection fraction. N Engl J Med. 2008;359:2456–67.
50. Persson H, Lonn E, Edner M, Baruch L, Lang CC, Morton JJ, et al. Diastolic dysfunction in heart failure with preserved systolic function: need for objective evidence. Results from the CHARM echocardiographic substudy-CHARMES. J Am Coll Cardiol. 2007;49:687–94.
51. Ejiri K, Miyoshi T, Kihara H, Hata Y, Nagano T, Takaishi A, et al. Effect of luseogliflozin on heart failure with preserved ejection fraction in patients with diabetes mellitus. J Am Heart Assoc. 2020;9:9.
52. Tanaka A, Hisauchi I, Taguchi I, Sezai A, Toyoda S, Tomiyama H, et al. Effects of canagliflozin in patients with type 2 diabetes and chronic heart failure: a randomized trial (CANDLE). ESC Heart Fail. 2020;7:1585–94.
53. Solomon SD, Zile M, Pieske B, Voors A, Shah A, Kraigher-Krainer E, et al. The angiotensin receptor neprilysin inhibitor LCZ696 in heart failure with preserved ejection fraction: a phase 2 double-blind randomised controlled trial. Lancet. 2012;380:1387–95.
54. Solomon SD, Zile M, Pieske B, Voors A, Shah A, Kraigher-Krainer E, et al. The angiotensin receptor neprilysin inhibitor LCZ696 in heart failure with preserved ejection fraction: a phase 2 double-blind randomised controlled trial. Lancet (London, England). 2012;380:1387–95; [cited 2022 Apr 11]. https://pubmed.ncbi.nlm.nih.gov/22932717/.
55. Schmieder RE, Wagner F, Mayr M, Delles C, Ott C, Keicher C, et al. The effect of sacubitril/valsartan compared to olmesartan on cardiovascular remodelling in subjects with essential hypertension: the results of a randomized, double-blind, active-controlled study. Eur Heart J. 2017;38:3308–17; [cited 2022 Apr 11]. https://pubmed.ncbi.nlm.nih.gov/29029087/.
56. Randomised, placebo-controlled trial of carvedilol in patients with congestive heart failure due to ischaemic heart disease. Lancet. 1997;349:375–80.
57. Jensen J, Omar M, Kistorp C, Poulsen MK, Tuxen C, Gustafsson I, et al. Empagliflozin in heart failure patients with reduced ejection fraction: a randomized clinical trial (Empire HF). Trials. 2019;20:374.

New Antidiabetic Agents for the Treatment of Heart Failure in Hypertensive Patients

23

Peter M. Nilsson, Hannes Holm, and Martin Magnusson

23.1 Introduction

Heart failure (HF) constitutes a major clinical future threat for patients with the combination of type 2 diabetes (T2D) and hypertension. This will lead to increased morbidity and mortality, but also to increased costs for hospital care besides the loss of Quality-Adjusted Life Years (QUALY) for the patients affected.

HF is globally gaining in prevalence with an increasing burden on societies, both in terms of quality of life and health care expenditures [1]. The estimated prevalence of HF in developed countries ranges from 1% to 2% and despite that treatment of HF has improved over recent decades, the mortality is still high with a 5-year survival rate of 45%, depending on the severity [2]. Further, HF is one of

P. M. Nilsson (✉)
Lund University, Malmö, Sweden

Skane University Hospital, Malmö, Sweden

Department of Clinical Sciences, Malmö, Sweden
e-mail: peter.nilsson@med.lu.se

H. Holm
Lund University, Malmö, Sweden

Skane University Hospital, Malmö, Sweden
e-mail: hannes.holm_isholth@med.lu.se

M. Magnusson
Lund University, Malmö, Sweden

Skane University Hospital, Malmö, Sweden

Wallenberg Center for Molecular Medicine, Malmö, Sweden

Hypertension in Africa Research Team (HART), North West University, Potchefstroom, South Africa
e-mail: martin.magnusson@med.lu.se

M. Dorobantu et al. (eds.), *Hypertension and Heart Failure*, Updates in Hypertension and Cardiovascular Protection,
https://doi.org/10.1007/978-3-031-39315-0_23

the most frequent reasons for hospital admission among elderly [3]. Structural and functional cardiac changes are often present before clinical signs of HF appear and since occurrence of these changes is associated with increased mortality and comorbidity, it is important to identify them at an early stage [4]. Additionally, it has been suggested that treatments targeting the asymptomatic phase of HF may slow the progression to the symptomatic phase and reduce subsequent morbidity and mortality [5].

The emergence of newer glucose-lowering drugs for the treatment of T2D has provided rich opportunities for not only control of cardiovascular risk but also reduction of morbidity and mortality based on results from large clinical trials. Both the sodium glucose transporter 2 inhibitors (SGLT-2i) and the glucagon-like peptide 1 (GLP-1) receptor agonists/analogues (RA) have shown impressive effects on prevention of both macrovascular and microvascular endpoints [6–9], including slowing the progression of diabetic nephropathy which has been seen for SGLT-2 inhibitors in large clinical trials [10]. The prevention and treatment of HF has added to the benefits and was well documented, especially for the SGLT-2i, but also to some degree for GLP-1 RA, as recently reviewed [11]. In fact, these benefits have recently been expanded also to nondiabetic patients included in large endpoint studies [12–15].

23.2 Preventive Mechanisms: Still Not Fully Understood

The mechanisms underlying these clinical cardio-renal benefits in HF are still not well understood, as they go beyond the effects of traditional control of hyperglycemia and improved cardiac metabolism. Among the pleiotropic effects of these newer drugs, blood pressure (BP) lowering properties have been observed [13], both based on observations on conventional office BP measurements and by use of 24-h ambulatory BP monitoring (ABPM). In addition, beneficial effects on central hemodynamics and aortic stiffness have been reported [14]. These findings will here be briefly reviewed and discussed as they may be of importance for the prevention of HF, for example in patients with T2D and elevated BP combined. Hypertension is a common and important cardiovascular and renal risk factor in most T2D patients [15], often complicated by special features linked to diabetes such as masked hypertension [16], orthostatic reactions upon standing [17], or even resistant hypertension [18].

23.3 SGLT-2 Inhibitors

This is a class of drugs with several effects that may contribute to BP-lowering and thereby HF prevention, as well as improved central hemodynamics, including the lowering of aortic pulse wave velocity (aoPWV), a marker of arterial stiffness. Among these contributing mechanisms are weight reduction [19], increased natriuresis combined with glucosuria [20], and vascular relaxation [21]. The glucosuric effect will reduce hyperglycemia.

Already, the first large-scale intervention study (EMPA-REG) in 2015 could document lowering of office BP (3–4 mmHg depending on dosage of empagliflozin) by use of the SGLT-2i empagliflozin versus placebo in patients with T2D at high risk for cardiovascular events [22]. No reflex increase of heart rate was noted. This finding was later followed by a study showing similar BP-lowering effects based on 24-h ABPM, with or without background antihypertensive drug medication such as hydrochlorothiazide [23]. Also, other SGLT-2 inhibitors (canagliflozin, dapagliflozin, etc.) have documented similar effects that go in parallel with modest weight reduction. In one study, treatment with dapagliflozin significantly reduced ambulatory, brachial, and central BP levels and aoPWV in DM2 patients versus placebo [24].

More recently, a meta-analysis could show that treatment with SGLT-2i leads to an average reduction of systolic/diastolic BP of 3.62/1.70 mmHg in 24-h ambulatory BP. This BP-lowering effect remains unmodified regardless of the dose of SGLT-2i used and is comparable with the BP-lowering efficacy of low-dose hydrochlorothiazide [25], i.e., a mild diuretic effect.

In cases of resistant hypertension, SGLT-2i have been added for improved BP control [26], a new way to achieve this goal when previously the addition of an aldosterone antagonist was recommended (spironolactone, eplerenone, finerenone).

The utilization of SGLT-2 inhibitors has been documented to have a positive impact on arterial stiffness and aoPWV in individuals diagnosed with both type 1 and type 2 diabetes [27, 28]. This effect holds true even in cases where surrogate markers, rather than a direct measurement of aoPWV, were occasionally employed. The reduction of arterial stiffness could be secondary to BP-lowering, but it could be speculated that also other, more direct effects on the arterial wall could be present, for example improvement of endothelial function.

It is noteworthy that the rapidly occurring preventive effect on HF of SGLT2i in clinical trials (after a few weeks) could most likely be at least partly explained by a reduction of arterial stiffness and normalization of the reflex wave that impacts on cardiac function, a mechanism called ventriculo-arterial coupling (VAC) [29]. This provides another argument in favor of using SGLT-2i for HF prevention in many cardiovascular risk patients, irrespective of diabetes status, as increased arterial (aortic) stiffness is a very common phenomenon in many patients at risk.

The SGLT-2i drugs are well tolerated in general, but some adverse effects may affect a few patients like urinary tract infection and dermatitis caused by candida infection. Other rarer side effects are euglycemic ketoacidosis and worsening of peripheral artery disease (in one study) [30].

Although the beneficial cardiovascular effects of SGLT-2i are conclusive, the underlying mechanisms are still debated. Data from randomized studies shows that the positive effects of SGLT-2i on cardiovascular outcome in HF cannot solely be attributed to the glucose-lowering effect [31]. In the cardiomyocyte membrane, SGLT-2i inhibit cardiac sodium transporters with a subsequent decrease of intracellular sodium concentration. Thus, the multiple downstream signaling pathways known to be active in HF due to sodium excess are suppressed [32]. Further, by stimulating the conversion of energy sources to ketone bodies, Empagliflozin has shown to improve the myocardial energy metabolism and systolic function by reversing cardiac remodeling in HF patients [33].

23.4 GLP-1 Receptor Agonists/Analogues

These drugs increase the activity of glucagon-like peptide 1 (GLP-1), a secreted incretin from the gastrointestinal mucosa, and thereby facilitating insulin secretion from pancreatic beta-cells. One of the main effects is weight reduction [34] that can sometimes be substantial with more recent incretin-active drugs such as tirzepatide that combines GLP-1 receptor agonism (RA) and gastric inhibitory peptide (GIP) RA, another incretin secreted from the gastrointestinal mucosa [35]. There is even a triple drug combination with the two incretin-active drugs plus glucagon that may further reduce body weight [36]. This weight reduction is mainly caused by a reduction of gastric emptying and increased satiety due to central nervous effects [37].

Besides weight loss that promotes BP reduction [38], these injectable drugs also improve vascular relaxation and are supposed to prevent atherosclerosis, an effect also achieved with newer oral preparations [39]. The slowing of atherosclerosis could be linked to beneficial effects on chronic inflammation in the vascular wall [40]. Secondary to vasodilatation an increase of heart rate has been noticed, but this is still debated whether it can increase cardiovascular risk or not. Taken together, the benefits achieved for reduction of cardiovascular disease manifestations in clinical studies overshadow the untoward effect of a slight increase in heart rate.

Among side effects of GLP-1 RA drugs are noticed most importantly nausea and vomiting that may affect up to one third of all patients, at least during the initiation of drug therapy. In most cases such initial side effects will disappear after some time. In a few cases also pancreatitis has been encountered, but the risk seems to be very low.

23.5 Effects on Cardiac Metabolism

It is a well-known fact that intensive glycemic control has no effect on major macrovascular events or cardiovascular mortality, but only in microvascular events [41]. Hence it is postulated that the GLP-1 agonists may execute their CVD beneficial effects by effecting other risk factors.

Experimental studies have shown the beneficial effect of liraglutide treatment with the inhibition of adhesion molecules and promoted nitric oxide production synthase activity [42]. Indeed, epidemiological data showing independent associations between physiological high levels of plasma GLP-1 and decreased intima media thickness of the carotid arteries supports the role of protective atherosclerotic properties of GLP-1 [43]. Interestingly, treatment with GLP-1 agonists has also been associated with an improvement of dyslipidemia. The mechanisms behind these findings are not known, but it is speculated that enhanced glycemic control improves hepatic lipid metabolism through a decrease in insulin resistance [44].

23.6 Combination Therapy for Control of Risk

As many subjects with the metabolic syndrome have hyperglycemia linked to insulin resistance and abdominal obesity, they are prone to develop different states of pre-diabetes or even overt T2D, when these new classes of glucose-lowering drugs may prove to be useful, often in combination with first-line drug metformin according to current guidelines. Other useful combinations could be with blockers of the renin-angiotensin system [45], or even a combination of SGLT-2i with GLP-1 RA drugs [46]. A recent randomized study reported that a 12-month treatment with GLP-1 RA, SGLT-2i, and their combination showed a greater improvement of vascular markers and effective cardiac work than insulin treatment in T2D. The combined therapy as a second-line strategy was superior to either insulin or GLP-1 RA and SGLT-2i prescribed separately [47].

This way of thinking paves the way for a fixed drug combination (FDC) in patients with T2D, when also a statin and antihypertensive drugs can be added. Such fixed combinations based on a statin, antihypertensive drugs, and aspirin (PolyPill, PolyCaps) have recently been shown to be efficient based on a meta-analysis [48]. The addition of one or more glucose-lowering drugs [49] to such a FDC could become a more cost-effective way in the future to help many T2D patients due to the simplification of treatment and improved compliance.

23.7 SGLT-2i and GLIP-1 RA for Prevention and Treatment of Heart Failure

The EMPA-REG OUTCOME TRIAL, DECLARE-TIMI, CANVAS, and VERTIS-CV studies have all shown that SGLT2 inhibitors empagliflozin, dapagliflozin, canagliflozin, and ertugliflozin reduce HF hospitalization, and the DAPA-HF study and the EMPEROR-REDUCED TRIAL have further shown that dapagliflozin and empagliflozin could be used to treat patients with heart failure, with or without diabetes [50]. In addition the DELIVER and EMPEROR-preserved trial have shown equal beneficial effects in patients with heart failure with mildly reduced and preserved ejection fraction [13, 15]. As for the effects of GLP-1 and risk of HF hospitalization, meta-analyses of GLP-1 RAs studies showed a neutral effect on the risk of hospitalization for heart failure [51]. On the other hand, liraglutide showed a reduction of risk of heart failure hospitalization and a significant decrease of cardiovascular and all-cause mortality in diabetic patients with a high risk of cardiovascular disease [52].

Based on systematic reviews of these newer glucose-lowering drugs, the evidence base for their use has been substantially strengthened [53–57]. In one recent analysis 336,334 people with T2D patients free of cardiovascular disease were included, of whom 18,531 (5.5%) experienced a major adverse cardiovascular event (MACE) [11]. In a cohort of 411,206 with T2D and without HF, 17,451 (4.2%) experienced a HF event. Compared with other combination regimens, the adjusted pooled OR and 95% CI for MACE associated with SGLT2i regimens was 0.82 (0.73, 0.92), with GLP-1RA regimens 0.93 (0.81, 1.06), and with the SGLT2i/GLP-1RA combination 0.70 (0.50, 0.98). Corresponding data for HF were SGLT2i 0.49 (0.42, 0.58), GLP-1RA 0.82 (0.71, 0.95), and SGLT2i/GLP-1RA combination 0.43 (0.28, 0.64). The authors concluded that SGLT2i and SGLT2i/GLP-1RA combination regimens may be beneficial in primary prevention of MACE and HF and GLP-1RA for HF [11].

In addition, the cost-effectiveness of these newer antidiabetes drugs has been studied [53]. Recently practice recommendations for use of SGLTi drugs also in nondiabetic subjects at risk of HF were published [54].

23.8 Conclusions

The newer glucose-lowering drugs for T2D (i.e., SGLT-2i and GLP-1 RA) have contributed to our understanding of how to improve prevention of macro- and microvascular complications in these patients, including HF. Among the beneficial effects of these drugs, also BP-lowering and improvement of central hemodynamics can be of great importance to explain the reduction of cardiovascular events. For the SGLT-2i there is increased natriuresis, beneficial renal effects, and weight loss, and for the GLP-1 RA drugs there is most of all weight loss but also improvement of endothelial function. However, the exact mechanisms behind the hemodynamic changes of the two classes of drugs are not fully known. As hypertension poses a substantial risk for these patients, the use of antihypertensive drugs in combination with these newer glucose-lowering drugs may prove to be important for BP control, normalization of central hemodynamics, and lowering of the BP-dependent risk. Additionally, both SGLT-2i and GLP-1 RA have been recognized with direct beneficial effects on the myocardium in HF patients.

In the future it is expected that more fixed drug combinations (FDC) will be used, and this may involve also these two promising classes of drugs to treat T2D patients and lower their cardiovascular and renal risk, including the prevention and treatment of HF.

Acknowledgments This review was supported by the Research Council of Sweden and the Heart and Lung Foundation of Sweden with grants to PMN.

MM was supported by grants from the Swedish Heart and Lung Foundation, the Medical Faculty of Lund University, Skåne University Hospital, the Crafoord Foundation, Region Skane, and the Wallenberg Centre for Molecular Medicine at Lund University.

References

1. Groenewegen A, et al. Epidemiology of heart failure. Eur J Heart Fail. 2020;22(8):1342–56.
2. Taylor CJ, et al. Trends in survival after a diagnosis of heart failure in the United Kingdom 2000-2017: population based cohort study. BMJ. 2019;364:l223.
3. Neumann T, et al. Heart failure: the commonest reason for hospital admission in Germany: medical and economic perspectives. Dtsch Arztebl Int. 2009;106(16):269–75.
4. Echouffo-Tcheugui JB, et al. Assessing the risk of progression from asymptomatic left ventricular dysfunction to overt heart failure: a systematic overview and meta-analysis. JACC Heart Fail. 2016;4(4):237–48.
5. Reed BN, Sueta CA. Stage B: what is the evidence for treatment of asymptomatic left ventricular dysfunction? Curr Cardiol Rev. 2015;11(1):18–22.
6. Marso SP, et al. Liraglutide and cardiovascular outcomes in type 2 diabetes. N Engl J Med. 2016;375(4):311–22.
7. Marx N, et al. Guideline recommendations and the positioning of newer drugs in type 2 diabetes care. Lancet Diabetes Endocrinol. 2021;9(1):46–52.
8. Zinman B, Lachin JM, Inzucchi SE. Empagliflozin, cardiovascular outcomes, and mortality in type 2 diabetes. N Engl J Med. 2016;374(11):1094.
9. Marso SP, et al. Semaglutide and cardiovascular outcomes in patients with type 2 diabetes. N Engl J Med. 2016;375(19):1834–44.
10. Wheeler DC, et al. Effects of dapagliflozin on major adverse kidney and cardiovascular events in patients with diabetic and non-diabetic chronic kidney disease: a prespecified analysis from the DAPA-CKD trial. Lancet Diabetes Endocrinol. 2021;9(1):22–31.
11. Wright AK, et al. Primary prevention of cardiovascular and heart failure events with SGLT2 inhibitors, GLP-1 receptor agonists, and their combination in type 2 diabetes. Diabetes Care. 2022;45(4):909–18.
12. Anker SD, et al. Evaluation of the effects of sodium-glucose co-transporter 2 inhibition with empagliflozin on morbidity and mortality in patients with chronic heart failure and a preserved ejection fraction: rationale for and design of the EMPEROR-preserved trial. Eur J Heart Fail. 2019;21(10):1279–87.
13. Liakos CI, et al. Blood pressure-lowering effect of newer antihyperglycemic agents (SGLT-2 inhibitors, GLP-1 receptor agonists, and DPP-4 inhibitors). Am J Cardiovasc Drugs. 2021;21(2):123–37.
14. Batzias K, et al. Effects of newer antidiabetic drugs on endothelial function and arterial stiffness: a systematic review and meta-analysis. J Diabetes Res. 2018;2018:1232583.
15. Jia G, Sowers JR. Hypertension in diabetes: an update of basic mechanisms and clinical disease. Hypertension. 2021;78(5):1197–205.
16. Wijkman M, et al. Masked nocturnal hypertension—a novel marker of risk in type 2 diabetes. Diabetologia. 2009;52(7):1258–64.
17. Freeman R, et al. Orthostatic hypotension: JACC state-of-the-art review. J Am Coll Cardiol. 2018;72(11):1294–309.
18. Oliveras A, de la Sierra A. Resistant hypertension: patient characteristics, risk factors, co-morbidities and outcomes. J Hum Hypertens. 2014;28(4):213–7.
19. Cho YK, Kim YJ, Jung CH. Effect of sodium-glucose cotransporter 2 inhibitors on weight reduction in overweight and obese populations without diabetes: a systematic review and a meta-analysis. J Obes Metab Syndr. 2021;30(4):336–44.
20. Thomas MC, Cherney DZI. The actions of SGLT2 inhibitors on metabolism, renal function and blood pressure. Diabetologia. 2018;61(10):2098–107.
21. Tikkanen I, et al. Empagliflozin reduces blood pressure in patients with type 2 diabetes and hypertension. Diabetes Care. 2015;38(3):420–8.
22. Mancia G, et al. Impact of empagliflozin on blood pressure in patients with type 2 diabetes mellitus and hypertension by background antihypertensive medication. Hypertension. 2016;68(6):1355–64.

23. Papadopoulou E, et al. Dapagliflozin decreases ambulatory central blood pressure and pulse wave velocity in patients with type 2 diabetes: a randomized, double-blind, placebo-controlled clinical trial. J Hypertens. 2021;39(4):749–58.
24. Georgianos PI, Agarwal R. Ambulatory blood pressure reduction with SGLT-2 inhibitors: dose-response meta-analysis and comparative evaluation with low-dose hydrochlorothiazide. Diabetes Care. 2019;42(4):693–700.
25. Alqudsi M, Velez JCQ, Navarrete J. Medical management of resistant hypertension: the role of sodium-glucose cotransporter 2 inhibitors (SGLT2i). Curr Opin Cardiol. 2021;36(4):420–8.
26. Bosch A, et al. How does empagliflozin improve arterial stiffness in patients with type 2 diabetes mellitus? Sub analysis of a clinical trial. Cardiovasc Diabetol. 2019;18(1):44.
27. Katakami N, et al. Effect of tofogliflozin on arterial stiffness in patients with type 2 diabetes: prespecified sub-analysis of the prospective, randomized, open-label, parallel-group comparative UTOPIA trial. Cardiovasc Diabetol. 2021;20(1):4.
28. Lamacchia O, Sorrentino MR. Diabetes mellitus, arterial stiffness and cardiovascular disease: clinical implications and the influence of SGLT2i. Curr Vasc Pharmacol. 2021;19(2):233–40.
29. Saeed S, Holm H, Nilsson PM. Ventricular-arterial coupling: definition, pathophysiology and therapeutic targets in cardiovascular disease. Expert Rev Cardiovasc Ther. 2021;19(8):753–61.
30. Neal B, Perkovic V, Matthews DR. Canagliflozin and cardiovascular and renal events in type 2 diabetes. N Engl J Med. 2017;377(21):2099.
31. Herrington WG, et al. Cardiac, renal, and metabolic effects of sodium-glucose co-transporter 2 inhibitors: a position paper from the European Society of Cardiology ad-hoc task force on sodium-glucose co-transporter 2 inhibitors. Eur J Heart Fail. 2021;23(8):1260–75.
32. Xiang B, Zhao X, Zhou X. Cardiovascular benefits of sodium-glucose cotransporter 2 inhibitors in diabetic and nondiabetic patients. Cardiovasc Diabetol. 2021;20(1):78.
33. Gaborit B, et al. Effect of empagliflozin on ectopic fat stores and myocardial energetics in type 2 diabetes: the EMPACEF study. Cardiovasc Diabetol. 2021;20(1):57.
34. Frias JP, et al. Tirzepatide versus semaglutide once weekly in patients with type 2 diabetes. N Engl J Med. 2021;385(6):503–15.
35. Bossart M, et al. Effects on weight loss and glycemic control with SAR441255, a potent unimolecular peptide GLP-1/GIP/GCG receptor triagonist. Cell Metab. 2022;34(1):59–74.e10.
36. van Bloemendaal L, et al. Effects of glucagon-like peptide 1 on appetite and body weight: focus on the CNS. J Endocrinol. 2014;221(1):T1–16.
37. Goud A, et al. GLP-1 agonists and blood pressure: a review of the evidence. Curr Hypertens Rep. 2016;18(2):16.
38. Saraiva JFK, Franco D. Oral GLP-1 analogue: perspectives and impact on atherosclerosis in type 2 diabetic patients. Cardiovasc Diabetol. 2021;20(1):235.
39. Gallego-Colon E, Wojakowski W, Francuz T. Incretin drugs as modulators of atherosclerosis. Atherosclerosis. 2018;278:29–38.
40. Cosentino F, et al. 2019 ESC guidelines on diabetes, pre-diabetes, and cardiovascular diseases developed in collaboration with the EASD. Eur Heart J. 2020;41(2):255–323.
41. ADVANCE Collaborative Group, et al. Intensive blood glucose control and vascular outcomes in patients with type 2 diabetes. N Engl J Med. 2008;358(24):2560–72.
42. Del Olmo-Garcia MI, Merino-Torres JF. GLP-1 receptor agonists and cardiovascular disease in patients with type 2 diabetes. J Diabetes Res. 2018;2018:4020492.
43. Jujic A, et al. Glucose-dependent insulinotropic peptide in the high-normal range is associated with increased carotid intima-media thickness. Diabetes Care. 2021;44(1):224–30.
44. Sposito AC, et al. GLP-1RAs in type 2 diabetes: mechanisms that underlie cardiovascular effects and overview of cardiovascular outcome data. Cardiovasc Diabetol. 2018;17(1):157.
45. Tian B, et al. Efficacy and safety of combination therapy with sodium-glucose cotransporter 2 inhibitors and renin-angiotensin system blockers in patients with type 2 diabetes: a systematic review and meta-analysis. Nephrol Dial Transplant. 2022;37(4):720–9.
46. Doumas M, et al. Combination of SGLT-2 inhibitors and GLP-1 receptor agonists: potential benefits in surrogate and hard endpoints. Curr Pharm Des. 2018;24(17):1879–86.

47. Ikonomidis I, et al. Effects of glucagon-like peptide-1 receptor agonists, sodium-glucose cotransporter-2 inhibitors, and their combination on endothelial glycocalyx, arterial function, and myocardial work index in patients with type 2 diabetes mellitus after 12-month treatment. J Am Heart Assoc. 2020;9(9):e015716.
48. Joseph P, et al. Fixed-dose combination therapies with and without aspirin for primary prevention of cardiovascular disease: an individual participant data meta-analysis. Lancet. 2021;398(10306):1133–46.
49. Bohm AK, et al. Regimen simplification and medication adherence: fixed-dose versus loose-dose combination therapy for type 2 diabetes. PLoS One. 2021;16(5):e0250993.
50. Razuk V, et al. SGLT-2 inhibitors and cardiovascular outcomes in patients with and without a history of heart failure: a systematic review and meta-analysis. Eur Heart J Cardiovasc Pharm. 2022;8:557.
51. Scheen AJ. GLP-1 receptor agonists and heart failure in diabetes. Diabetes Metab. 2017;43 Suppl 1:2S13–9.
52. Bahtiyar G, Pujals-Kury J, Sacerdote A. Cardiovascular effects of different GLP-1 receptor agonists in patients with type 2 diabetes. Curr Diab Rep. 2018;18(10):92.
53. Cavallari I, et al. SGLT-2 inhibitors on top of current pharmacological treatments for heart failure: a comparative review on outcomes and cost effectiveness. Am J Cardiovasc Drugs. 2022;22(3):263–70.
54. Giaccari A, Pontremoli R, Perrone Filardi P. SGLT-2 inhibitors for treatment of heart failure in patients with and without type 2 diabetes: a practical approach for routine clinical practice. Int J Cardiol. 2022;351:66–70.
55. Herrera Comoglio R, Vidal Guitart X. Cardiovascular outcomes, heart failure and mortality in type 2 diabetic patients treated with glucagon-like peptide 1 receptor agonists (GLP-1 RAs): a systematic review and meta-analysis of observational cohort studies. Int J Clin Pract. 2020;74(9):e13553.
56. Palmer SC, et al. Sodium-glucose cotransporter protein-2 (SGLT-2) inhibitors and glucagon-like peptide-1 (GLP-1) receptor agonists for type 2 diabetes: systematic review and network meta-analysis of randomised controlled trials. BMJ. 2021;372:m4573.
57. Santos-Ferreira D, Goncalves-Teixeira P, Fontes-Carvalho R. SGLT-2 inhibitors in heart failure and type-2 diabetes: hitting two birds with one stone? Cardiology. 2020;145(5):311–20.

24 Nonpharmacological Treatment in Hypertensive Patients with Heart Failure

Tomasz Zieliński and Andrzej Januszewicz

24.1 Introduction

Nonpharmacological treatment methods are actually widely accepted as preventive or therapeutic measures in prehypertension or hypertension. They can be used as a substitute for pharmacological therapy, especially in patients with prehypertension or borderline hypertension, or as an adjunct therapy to drug or interventional therapy in almost all hypertensive patients. There are many similarities in nonpharmacological treatment methods of hypertension and heart failure so many forms of therapy can be safely recommended for patients with coexisting hypertension and heart failure. Arterial hypertension is a recognized risk factor for the development of heart failure in all recognized phenotypes of heart failure: with preserved ejection fraction (HFpEF), with reduced ejection fraction (HFrEF), or in mildly reduced ejection fraction (HFmrEF). In the most advanced stages of heart failure (stage D—according to the AHA classification) especially in patients with heart failure with reduced ejection fraction, patients rarely are hypertensive, but in less advanced stages of heart failure, the coexistence of heart failure and hypertension can be common and aggravate and accelerate the progression of heart failure. So, the treatment of both diseases also with nonpharmacological methods is particularly important.

T. Zieliński (✉)
Department of Heart Failure and Transplantology, National Institute of Cardiology, Warszawa, Poland
e-mail: tzielin@ikard.pl

A. Januszewicz
Department of Hypertension, National Institute of Cardiology, Warszawa, Poland

M. Dorobantu et al. (eds.), *Hypertension and Heart Failure*, Updates in Hypertension and Cardiovascular Protection,
https://doi.org/10.1007/978-3-031-39315-0_24

24.2 Dietary Interventions

Many forms of dietary interventions were claimed to be "healthy," but only few were adequately scientifically tested and are approved as effective in reducing hypertension or heart failure incidence.

One of the most tested and approved diets consisting of consumption of whole grains and an increased amount of vegetables and fruits is currently recommended and is called the Dietary Approaches to Stop Hypertension Diet (DASH). These diet rules also suggest a decrease in the consumption of saturated fats, red meat, dairy products, and sweets, including sweetened beverages. The results of the DASH trial [1] demonstrated that patients on a diet rich in fruits, vegetables, and low-fat dairy products and with reduced saturated and total fat obtained a reduction in systolic blood pressure of 5.5 mmHg and diastolic blood pressure of 3.0 mmHg more than those on a control diet. This effect was less pronounced with a reduction in systolic blood pressure by 2.8 mmHg and diastolic blood pressure by 1.1 mmHg in the group with dietary intervention limited to an increase in the consumption of vegetables and fruits, but with no reduction of saturated and total fat consumption.

The results of the DASH trial clearly demonstrated that complex dietary interventions aimed at multiple components may be more efficient in reducing arterial pressure.

Interestingly, there is also evidence that the benefits of the DASH diet are not limited to reducing arterial blood pressure, but also have a beneficial effect on the incidence of heart failure in patients under 75 years of age, as shown in the multi-ethnic study of Atherosclerosis (MESA study) [2]. Also in the REasons for Geographic and Racial Differences in Stroke cohort (REGARDS study), adherence to the DASH diet was inversely associated with incident HF, specifically among individuals <75 years of age [3].

A large cohort study in the Swedish middle-aged and elderly population that included more than 70,000 participants observed for 20 years clearly demonstrated that an adherence to the DASH diet resulted in a lower risk of heart failure. Important from a practical point of view and possible advice to patients is that replacement of one meal per day of red or processed meat with one DASH-style meal (fruits, vegetables, low-fat dairy, legumes, nuts, or whole grains) was associated with an approximately 10% lower risk of developing heart failure [4].

Another diet that has some scientific studies on the efficacy on decreasing the incidence of cardiovascular events, hypertension, and heart failure is a Mediterranean diet. The term Mediterranean diet was proposed due to the lower incidence of coronary events in populations living in the Mediterranean region in Europe and refers to the dietary habits and lifestyle of this region.

The proposed diet recommends olive oil (> = 4 tablespoons of olive for cooking or seasoning a day), at least 3 servings of vegetables/week, and 3 servings of fish or seafood a week, white meat instead of processed meat or red meat, with the accepted use of red wine. A sauce with garlic, tomatoes, and olive oil served with vegetables, rice, seafood, and pasta is acceptable. Consumption of soft drinks in excess of 1 drink a day, bakery products more than 2 servings/week, and fats added more than 1 serving/day was discouraged [5].

The Mediterranean diet has been shown to moderately reduce systolic arterial blood pressure by approximately 2.5 to 3.1 mmHg and diastolic blood pressure by 1.3 to 1.9 mmHg [6]. There are also some possible effects in the adherence to the Mediterranean diet on a lower incidence of heart failure and mortality [7]; however its effects on all causes of mortality in patients with heart failure were not demonstrated in all analyzed populations [8].

Some other diets that show some importance in reducing the incidence of coronary heart disease, such as the low-fat diet, were not fully evaluated in patients with heart failure.

Despite some differences in composition and perhaps efficacy of proposed diet regimens, it is generally approved in the supplementary material to the ESC guidelines for cardiovascular disease by 2021 that adopting diet modifications gives health benefits by reducing the energy density and salt and refined sugar content in foods and drinks, as well as replacing trans and saturated fats with unsaturated fat [9].

24.3 Sodium Diet Restrictions

There is large experimental and clinical evidence that reduced sodium intake reduces blood pressure and can be desirable in patients with heart failure by reducing fluid retention.

In most urbanized population, the daily consumption of sodium exceeds 4 g per day, so it is well above the daily needs.

Hypertension in some groups is "salt sensitive." Salt sensitivity is found in older adults, in patients with kidney disease, diabetes mellitus, metabolic syndrome, or with a higher degree of hypertension or black patients; due to those findings a reduction in sodium consumption is often suggested.

The mean reduction in systolic blood pressure with a diet containing less than 2.4 g of sodium is 2 mm, and the mean diastolic pressure is 1 mmHg; with a higher sodium restriction at 1.5 g/day, the mean systolic blood pressure is reduced by 7 mmHg and the mean diastolic blood pressure is reduced by 3 mmHg [6].

Due to the ubiquitous presence of sodium in food products, especially processed food, an exact calculation of the sodium content in the diet can be exceedingly difficult for a patient, and the true amount of dietary sodium consumption is falsely lowered. The easiest way to describe sodium content is the monitoring of salt consumption in the diet. In 2.5 g of salt, there is 1 g of sodium.

Current European [10] and US Veterans Administration [11] guidelines for the treatment of hypertension recommend reducing the intake of dietary sodium from 2000 mg/day to 2300 mg/day respectively (it is equivalent to approximately 5 g of salt (about one teaspoon)). There is still no unambiguous evidence of the beneficial effects of low sodium diet on the clinical course of patients with heart failure. The recently completed SODIUM-HF trial did not show a benefit in terms of death or hospital admissions for heart failure in patients on a low sodium diet (<1.5 g day) compared to patients with a daily sodium intake greater than 2 g/day, but noted a positive effect on NYHA class and quality of life [12]. An additional important point in considering strict dietary salt restrictions is to take into account the high burden to comply with the advice for the patients [13]. An additional positive effect has been observed on the reduction in blood pressure in hypertensive patients (SBP by 6,8 and DBP by 4,6) by potassium supplementation (75 to 125 mmol/day) [6]. This effect was highly pronounced in high sodium consumers [14].

Taking into account the current still limited knowledge on the effects of salt restriction on the clinical course and risk of death in patients with heart failure, current 2021 ESC guidelines recommend that diets with salt intake >5 g/day (sodium intake 2.5 g) should be avoided [15].

24.4 Overweight Reduction

The most widely used parameter to define overweight or obesity is the body mass index (BMI). Due to some limitations of this parameter, which cannot fully estimate a true risk factor, which is excess Fat Mass (FM), additional measures such as waist circumference (WC) or waist-hip ratio (WHR) were proposed. Obesity is correlated with the incidence of hypertension and has been shown to be a risk factor for developing heart failure, especially heart failure with preserved ejection fraction (HFpEF) [16].

Approximately, a reduction in overweight by 1 kg results in a decrease in systolic blood pressure of 1 mmHg [17].

The relationship between obesity and prognosis in patients with already diagnosed heart failure is more complex. Patents with overweight or class I obesity (BMI between 30 and 35) and HF often have a better prognosis than underweight or even normal weight patients with diagnosed heart failure. This is called the "obesity paradox." Not all aspects of the obesity paradox are fully explained, but it can be partially attributed to the limitations of the BMI classification which do not differentiate between obesity due to real fat accumulation and BMI increase due to increased lean mass (LM) which improves cardiovascular function [18].

Visceral fat, defined to some extent by a waist-hip ratio (WHR) (exceeding 0,9 in men and 0,85 in women), is particularly undesirable due to the hormonal activity of these adipocytes.

Obese patients, especially those with accumulated visceral fat with hypertension and heart failure, should be advised to systemically lose weight to achieve a better control of hypertension and decrease the workload of the failing heart.

The most important aspect in weight reduction programs is good control and reduction in daily caloric intake. The weight reduction program should define the proposed daily caloric intake adapted to the speed of the planned weight loss and take into account and encourage the physical activity of the patient. For long-term success, it is important that caloric restrictions are moderate and acceptable for patients. Co-operation with the dietitian to define the goals and speed of weight reduction and advice on diet components is helpful. Patients for self-control can also use some smartphone or computer-based programs to help an easy calculation of calories consumed in meals and step counters to know the caloric expenditure during physical activity or exercise.

The recently completed trial on the effects of the lifestyle modification program on hypertension HERB-DH1 trial demonstrated the effectiveness of smartphone applications in the group actively assisted in lifestyle modification using a smartphone application program. The effects on hypertension were significantly higher in the assisted group than in standard advice group—difference in mean systolic blood reduction between groups of 2.4 mmHg [19].

It is important in weight reduction programs in patients with heart failure to achieve a reduction of fat excess, especially abdominal fat, but not a reduction in lean mass leading to sarcopenia. Most effective programs are made up of calories intake reduction and control and adapted to the patient possibilities of programs of physical activity or exercise programs.

Patients with obesity and night snoring should be diagnosed for the presence of obstructive sleep apnea (OSA), which is linked to a higher incidence of hypertension and heart failure.

Patients with class III obesity (BMI >40) should be informed about the possibility of bariatric surgical interventions to reduce weight.

24.5 Physical Activity: Leisure Activity and Training

All patients with hypertension and stable heart failure should be encouraged to exercise, at least moderate physical aerobic activity. In the 2021 ESC guidelines for the treatment of heart failure, it is recommended that all patients who are able to exercise do it to improve exercise capacity, quality of life, and reduce HF hospitalizations (a IA level recommendation) [15]. Patients should perform moderate to vigorous aerobic activity for 40 min at least 4 times a week to lower blood pressure. Those who are unable to complete this program should be encouraged to try to achieve these goals and avoid sedentary lifestyle. Even patients with frailty and comorbidities that limit the possibility of exercising should have special programs of physical activity or supervised activity proposed, including adapted isometric exercise or resistance training. Any physical activity is always better than no physical activity. Patients with advanced heart failure can also profit safely from moderate aerobic exercise programs like "Nordic walking" activities [20]. Supervised exercise training in patients with the NYHA I to IV class and a decrease in the ejection fraction has been shown to be safe and efficient in terms of a decrease in adjusted covariate mortality and increased peak Vo2 max after 3 months of training [21]. Furthermore, exercise training has been effective in patients with heart failure and preserved ejection fraction [22].

The proposed intensity of training sessions for patients with heart failure is 70% of the heart rate reserve, and the perceived exertion (RPE) rating on the "Borg scale" below 14/20 or the RPE on the modified Borg scale up to 5/10.

Physical activity is necessary in all effective strategies for weight reduction programs.

Patients with hypertension and heart failure should be advised to avoid competitive strenuous sports (e.g., hiking, high mountain climbing, competitive running and cycling, fighting sports) and heavy isometric exercise (e.g., weightlifting).

The use of modern smartphones and smartwatches can help patients self-monitor levels of physical activity by measuring exercise, number of steps, covered distance, and caloric expenditure, as the well as the safety of exercise by monitoring pulse rate, heart rhythm, and pulse oximetry.

Other forms of physical exercise, such as Tai chi or Quigong, are preferred in countries which have a long tradition and cultural background to apply those techniques and have been shown to be effective in nonpharmacological treatment of prehypertension, hypertension, and heart failure [22, 23].

24.6 Substitution of Alcohol Consumption for Nonalcoholic Beverages Without Sugar

Mostly from epidemiological cross-sectional studies, there is accumulated evidence that increased alcohol consumption is correlated with higher values of blood pressure values. As described in the meta-analysis of trials with more than 361,000 participants, alcohol consumption in men was associated with an increase in the risk of hypertension; in women no risk was observed when alcohol consumption was 1 to 2 drinks a day (1 drink = 12 g of pure alcohol) [24]. Other studies have shown that reduction of alcohol intake in heavy drinkers can result in a mean reduction in systolic blood pressure up to 5 mmHg and diastolic blood pressure of approximately 3 mmHg. These effects are even more pronounced in those with initially higher blood pressure values. The correlation between alcohol consumption and the risk of heart failure is not so clearly proven. However, many authors recognize a special form of alcohol-induced heart damage called alcohol-dilated cardiomyopathy leading to heart failure. To develop fully symptomatic disease characterized by progressive heart failure, cardiac chamber dilation, low left ventricular ejection fraction, and rhythm disturbances, an interplay between individual susceptibility to the toxic effects of alcohol, a lifetime total load of alcohol consumption, and concomitant deficiencies in vitamins and ions are involved [25]. A recent study using advanced statistical risk analysis techniques and biobank data from 370,000 patients demonstrated that the so-called protective effect of low-dose alcohol can be attributed to the effect of confounders due to other factors from healthy lifestyle factors that were found more frequently in those participants [26]. The results of this genetic epidemiologic study suggested that alcohol consumption of all amounts was associated with increased cardiovascular risk, but important risk differences were observed between levels of intake, greater risk with increased levels of alcohol consumption.

When dietetic and weight reduction programs are instituted, it is worth noting that 12 g of alcohol (the amount of pure alcohol in one drink) contains 85 calories.

Patients with hypertension and heart failure should clearly be advised to avoid alcohol consumption and substitute for it with nonalcoholic beverages and not sweetened beverages. Total alcohol abstinence should be advised for patients with suspected alcoholic dilated cardiomyopathy.

24.7 Relaxation Techniques

In fact, there is an extensive list of relaxation techniques. Some of them presented briefly below were positively tested as therapy for prehypertension or hypertension and summarized in the meta-analysis by Fu [23], but only a few as therapy for heart failure. These techniques include transcendental meditation for stress reduction under the guidance of a professional instructor, sessions of 20 min twice a day. Another form is breathing control, usually with the use of a device allowing patients to train slow and regular breathing to achieve less than 10 breath/min. The intensity is 15 min of breath control practice per day. An increasingly popular technique is yoga. After professional instruction from the Yoga trainer, patients should practice at home at least 30 min a day for 3 days a week. Heart failure patients can safely practice less vigorous styles of yoga when properly instructed. The metabolic equivalent of yoga is between 1.9 and 6.7 Mets. Improvement in quality of life and functional tests has been observed [27].

24.8 Lifestyle Modification and Medical Advice on Healthy Lifestyle

All the nonpharmacological methods presented above for hypertension and heart failure can be summarized as a lifestyle modification for a healthier life. These methods can be easily adapted by almost all willing patients with prehypertension, hypertension, and heart failure. Rarely can they be used as a standalone measure, but more often as a companion to pharmacologic or interventional therapy. Therapies were evaluated in patients with hypertension and heart failure, but less often in patients with a diagnosed coexistence of both. Interestingly, similar nonpharmacological methods proved its efficiency. It is also reflected in the multiple similarities in the drugs used in therapy of hypertension and heart failure reflecting the multiple common pathways of these diseases.

The problem of nonpharmacological therapy is often addressed only marginally during the medical consultation and unfortunately perceived so by most of the patients.

Many of the methods presented are complementary, and the best effects can be obtained when used together, such as weight reduction programs with exercise activity and other components of a healthy lifestyle.

From a practical point of view, it is important to determine the main problem of the patient and the best method is best accepted. At the start of a long process of lifestyle modification, it seems to be preferable to address one problem at a time and introduce other elements gradually during follow-up visits.

To achieve the best acceptance from patients, they should have a profound motivation for a change in their actual lifestyle, perceive proposed programs as attractive, attractive, easy to follow together with other patients, friends, or family members, with elements of competition or play and with clearly defined timelines to achieve individual goals. Those goals should be formulated clearly in sequential steps and remain realistic- possible to reach for the majority of patients.

References

1. Appel LJ, Moore TJ, Obarzanek E, Vollmer WM, Svetkey LP, Sacks FM, et al. A clinical trial of the effects of dietary patterns on blood pressure. N Engl J Med. 1997;336(16):1117–24.
2. Campos CL, Wood A, Burke GL, Bahrami H, Bertoni AG. Dietary approaches to stop hypertension diet concordance and incident heart failure: the multi-ethnic study of atherosclerosis. Am J Prev Med. 2019;56(6):819–26.
3. Goyal P, Balkan L, Ringel JB, Hummel SL, Sterling MR, Kim S, et al. The dietary approaches to stop hypertension (DASH) diet pattern and incident heart failure. J Card Fail. 2021;27(5):512–21.
4. Ibsen DB, Levitan EB, Akesson A, Gigante B, Wolk A. The dietary approach to stop hypertension (DASH) diet is associated with a lower risk of heart failure: a cohort study. Eur J Prev Cardiol. 2022;29:1114.
5. Sanches Machado d'Almeida K, Ronchi Spillere S, Zuchinali P, Correa SG. Mediterranean diet and other dietary patterns in primary prevention of heart failure and changes in cardiac function markers: a systematic review. Nutrients. 2018;10(1):58.
6. Verma N, Rastogi S, Chia YC, Siddique S, Turana Y, Cheng HM, et al. Non-pharmacological management of hypertension. J Clin Hypertens. 2021;23(7):1275–83.
7. Patel YR, Robbins JM, Gaziano JM, Djousse L. Mediterranean, DASH, and alternate healthy eating index dietary patterns and risk of death in the Physicians' health study. Nutrients. 2021;13(6):1893.
8. Chang C-Y, Lee C-L, Liu W-J, Wang J-S. Association of adherence to the mediterranean diet with all-cause mortality in subjects with heart failure. Nutrients. 2022;14(4):842.
9. Visseren FLJ, Mach F, Smulders YM, Carballo D, Koskinas KC, Back M, et al. 2021 ESC guidelines on cardiovascular disease prevention in clinical practice. Eur Heart J. 2021;42(34):3227–337.
10. Williams B, Mancia G, Spiering W, Agabiti Rosei E, Azizi M, Burnier M, et al. 2018 ESC/ESH guidelines for the management of arterial hypertension: the task force for the management of arterial hypertension of the European Society of Cardiology and the European Society of Hypertension: the task force for the management of arterial hypertension of the European Society of Cardiology and the European Society of Hypertension. J Hypertens. 2018;36(10):1953–2041.
11. Tschanz CMP, Cushman WC, Harrell CTE, Berlowitz DR, Sall JL. Synopsis of the 2020 U.S. Department of Veterans Affairs/U.S. Department of Defense clinical practice guideline: the diagnosis and management of hypertension in the primary care setting. Ann Intern Med. 2020;173(11):904–13.
12. Ezekowitz JA, Colin-Ramirez E, Ross H, Escobedo J, Macdonald P, Troughton R, et al. Reduction of dietary sodium to less than 100 mmol in heart failure (SODIUM-HF): an international, open-label, randomised, controlled trial. Lancet. 2022;399(10333):1391–400.
13. Damy T, Benedyga V, Pezel T, Berthelot E, Gauthier J, Habib G, et al. Prescription, compliance, and burden associated with salt-restricted diets in heart failure patients: results from the French national OFICSel observatory. Nutrients. 2022;14(2):308.
14. Iqbal S, Klammer N, Ekmekcioglu C. The effect of electrolytes on blood pressure: a brief summary of meta-analyses. Nutrients. 2019;11(6):1362.
15. McDonagh TA, Metra M, Adamo M, Gardner RS, Baumbach A, Bohm M, et al. 2021 ESC guidelines for the diagnosis and treatment of acute and chronic heart failure. Eur Heart J. 2021;42(36):3599–726.

16. Pandey A, LaMonte M, Klein L, Ayers C, Psaty BM, Eaton CB, et al. Relationship between physical activity, body mass index, and risk of heart failure. J Am Coll Cardiol. 2017;69(9):1129–42.
17. Neter JE, Stam BE, Kok FJ, Grobbee DE, Geleijnse JM. Influence of weight reduction on blood pressure: a meta-analysis of randomized controlled trials. Hypertension. 2003;42(5):878–84.
18. Carbone S, Canada JM, Billingsley HE, Siddiqui MS, Elagizi A, Lavie CJ. Obesity paradox in cardiovascular disease: where do we stand? Vasc Health Risk Manag. 2019;15:89–100.
19. Kario K, Nomura A, Harada N, Okura A, Nakagawa K, Tanigawa T, et al. Efficacy of a digital therapeutics system in the management of essential hypertension: the HERB-DH1 pivotal trial. Eur Heart J. 2021;42(40):4111–22.
20. Piotrowicz E, Zielinski T, Bodalski R, Rywik T, Dobraszkiewicz-Wasilewska B, Sobieszczanska-Malek M, et al. Home-based telemonitored nordic walking training is well accepted, safe, effective and has high adherence among heart failure patients, including those with cardiovascular implantable electronic devices: a randomised controlled study. Eur J Prev Cardiol. 2015;22(11):1368–77.
21. O'Connor CM, Whellan DJ, Lee KL, Keteyian SJ, Cooper LS, Ellis SJ, et al. Efficacy and safety of exercise training in patients with chronic heart failure: HF-ACTION randomized controlled trial. JAMA. 2009;301(14):1439–50.
22. Aggarwal M, Bozkurt B, Panjrath G, Aggarwal B, Ostfeld RJ, Barnard ND, et al. Lifestyle modifications for preventing and treating heart failure. J Am Coll Cardiol. 2018;72(19):2391–405.
23. Fu J, Liu Y, Zhang L, Zhou L, Li D, Quan H, et al. Nonpharmacologic interventions for reducing blood pressure in adults with prehypertension to established hypertension. J Am Heart Assoc. 2020;9(19):e016804.
24. Roerecke M, Tobe SW, Kaczorowski J, Bacon SL, Vafaei A, Hasan OSM, et al. Sex-specific associations between alcohol consumption and incidence of hypertension: a systematic review and meta-analysis of cohort studies. J Am Heart Assoc. 2018;7(13):e008202.
25. Fernandez-Sola J. The effects of ethanol on the heart: alcoholic cardiomyopathy. Nutrients. 2020;12(2):572.
26. Biddinger KJ, Emdin CA, Haas ME, Wang M, Hindy G, Ellinor PT, et al. Association of habitual alcohol intake with risk of cardiovascular disease. JAMA Netw Open. 2022;5(3):e223849.
27. Pullen PR, Seffens WS, Thompson WR. Yoga for heart failure: a review and future research. Int J Yoga. 2018;11(2):91–8.

The Role of Drug Therapy in Lowering Mortality and Morbidity: From High-Risk Hypertension to Established Heart Failure

25

Aurora Rist, Kaja Sevre, Nisha Mistry, Sverre E. Kjeldsen, and Arne S. Westheim

Abbreviations

ACEi	Angiotensin-converting enzyme inhibitor
ARB	Angiotensin receptor blocker
ARNi	Angiotensin receptor-neprilysin inhibitor
CCB	Calcium channel blocker
MRA	Mineralocorticoid receptor antagonist
SGLT2i	Sodium-glucose co-transporter 2 inhibitor

Acronyms

CHARM	Candesartan in heart failure assessment of reduction in mortality and morbidity
CONSENSUS	Cooperative north Scandinavian enalapril survival study
EMPHESUS	Eplerenone in patients with heart failure due to systolic dysfunction complicating acute myocardial infarction
EPHESUS	Eplerenone post-AMI heart failure efficacy and survival study

A. Rist · K. Sevre
University of Oslo, Faculty of Medicine, Institute for Clinical Medicine, Oslo, Norway
e-mail: a.c.p.rist@studmed.uio.no; k.h.sevre@studmed.uio.no

N. Mistry · A. S. Westheim
Department of Cardiology, University of Oslo Hospital, Ullevaal, Oslo, Norway

S. E. Kjeldsen (✉)
University of Oslo, Faculty of Medicine, Institute for Clinical Medicine, Oslo, Norway

Department of Cardiology, University of Oslo Hospital, Ullevaal, Oslo, Norway
e-mail: s.e.kjeldsen@medisin.uio.no

M. Dorobantu et al. (eds.), *Hypertension and Heart Failure*, Updates in Hypertension and Cardiovascular Protection,
https://doi.org/10.1007/978-3-031-39315-0_25

I-PRESERVE	Irbesartan in heart failure with preserved ejection fraction
RALES	Randomized aldactone evaluation study
SAVE	Survival and ventricular enlargement
SENIORS	Study of the effect of nebivolol on mortality and cardiovascular hospital admission in elderly patients with heart failure
SOLVD	Studies of left ventricular dysfunction
TOPCAT	Treatment of preserved cardiac function heart failure with an aldosterone antagonist

25.1 Introduction

Hypertension is the dominating cause of heart failure [1]. Hypertension is an independent risk factor for developing heart failure, as well as being associated with other direct causes such as coronary heart disease and myocardial infarction, atrial fibrillation, aortic stenosis, and type-2 diabetes. Since a high fraction of the general population has *masked hypertension*, the extent of heart failure developing as a consequence of hypertension may be even more extensive [2].

The mechanism by which hypertension triggers heart failure is either through concentric hypertrophy leading to mural ventricular thickening and reduced cavity or reduced mural elasticity and dilated left atrium, or through eccentric hypertrophy producing a dilated, thin-walled ventricle.

Interventions aimed at improving the prognosis of heart failure overlap to a great extent with measures for treatment of hypertension and lowering of blood pressure. Treating high blood pressure will therefore be of advantage to heart failure prognosis, and vice versa. In clinical practice the recommended treatment for heart failure differs widely dependent on *ejection fraction*—despite the similar burden of disease and poor prognosis among different levels of ejection fraction [1]. The difference in treatment approach has been justified by the lack of evidence supporting a standardized treatment regime, regardless of ejection fraction. This chapter aims to review the treatment options for heart failure, their scientific foundation, and associated consequences for subgroups of the heart failure population.

25.1.1 Concentric vs. Eccentric Hypertrophy

Traditionally concentric hypertrophy has been associated with diastolic dysfunction. A heart with diastolic dysfunction is characterized as stiff and inflexible, with myocyte hypertrophy and mural fibrosis. These structural anomalies lead to decreased filling and elevated end-diastolic pressures. Nevertheless, the pathophysiological mechanisms and consequences of concentric hypertrophy and diastolic dysfunction are more complex. First, the reduced end-diastolic volume may lead to a reduced stroke volume. Second, the high pressures and fibrosis initiate ischemic vascular remodeling in the inner myocardial layers leading to decreased contraction in the longitudinal axis. All the previous considered, a heart with diastolic dysfunction will have a preserved or high-normal ejection fraction, and both reduced diastolic and systolic function [3].

Eccentric hypertrophy is usually associated with systolic dysfunction. This is a result of thinning of the ventricle and myocyte mass producing a dilated cavity with abnormal diastolic filling and increased end-diastolic pressure [4].

25.2 Phenotypes and Ejection Fraction

Left ventricular ejection fraction (EF) expresses the percentage of blood the left ventricle pumps out with each contraction. It is one of the most featured clinical measurements in heart failure and has roots back to the second generation treatment trials that included patients with EF ≤ 40%. The Scandinavian CONSENSUS study [5] included nonselected heart failure patients (EF was not measured) and showed for the first time the significant clinical benefit of angiotensin-converting enzyme (ACE) inhibitors in severe heart failure. Subsequent studies, such as SOLVD, SAVE, and more, included HFrEF patients, probably because reduced EF was perceived as visually compelling and intuitively correct for researchers.

Since the 1980s, it has been recognized that heart failure exists along the entire spectrum of EF values, that is, one has learned that heart failure is not synonymous with reduced EF. Consequently, EF has been used as a method to classify different subgroups of heart failure though its prognostic properties had limited relevance.

25.2.1 Characteristics of Heart Failure Phenotypes

The different subgroups of left-sided heart failure share many features in terms of symptoms, signs, and comorbidities. According to ESC Guidelines 2021 [6], typical symptoms are breathlessness, orthopnea, paroxysmal nocturnal dyspnea, reduced exercise tolerance, fatigue, tiredness, increased time to recover after exercise, and ankle swelling. Regarding signs, the most common are elevated jugular venous pressure, hepatojugular reflux, the presence of a third heart sound (gallop rhythm), and laterally displaced apical impulse.

For the diagnosis of HFmrEF (heart failure with mid-range EF) or HFpEF, the presence of other evidence of structural heart disease (e.g., increased left atrial size or LV hypertrophy) or abnormalities makes the diagnosis more likely [6] (Fig. 25.1).

There is considerable overlap between the heart failure phenotypes when it comes to comorbidities, especially hypertension, hyperlipidemia, coronary heart disease, and atrial fibrillation, which are also perceived by many as pure predisposing factors for the development of heart failure itself. This etiological overlap raises the question of whether heart failure should be perceived as a continuum rather than a blanket term for distinct phenotypes, which may primarily represent different phases of the same clinical syndrome. This must be seen in the light of ejection fraction as a parameter, which we will discuss in the next section.

	Criteria	Etiology
HFrEF	1. Symptoms ± Signs 2. LVEF ≤ 40%	Ischemic heart disease Arterial hypertension Idiopathic dilated cardiomyopathy
HFmrEF	1. Symptoms ± Signs 2. LVEF 40-50%	
HFpEF	1. Symptoms ± Signs 2. LVEF ≥50% 3. Objective evidence of cardiac structural and/or functional abnormalities consistent with the presence of LV diastolic dysfunction/raised LV filling pressures, including raised natriuretic peptides	Arterial hypertension Coronary artery disease Atrial fibrillation Valvular heart disease Depositional diseases

HF = heart failure; HFrEF = heart failure with reduced ejection fraction; HFmrEF = heart failure with mildly reduced ejection fraction; HFpEF = heart failure with preserved ejection fraction

Fig. 25.1 Heart failure phenotypes

25.3 The Problem with Ejection Fraction

The EF value is decisive for the subtype of heart failure, which in turn dictates the choice of treatment according to guidelines; however, temporary fall in EF in conjunction with unstable conditions such as acute myocardial infarction secondary to reduced stroke volume is a well-known concept. Hence EF is not a static parameter. Several studies in recent years have shown, however, that the EF value among heart failure patients also fluctuate under stable conditions and between the phenotype ranges. Clarke et al. [7] estimated transition probabilities at several points of follow-up in patients with a primary discharge diagnosis of heart failure and ≥ 2 EF measurements separated by at least 30 days. They found that the probability for HFpEF to HFrEF transition was 45% and 50% in 1 year and 2 years, respectively. Likewise, the probability for HFrEF to HFpEF transition was 18% and 20% [7]. The findings are consistent with Dunlay et al. [8], who found that almost 40% of heart failure patients had EF compatible with an altered phenotype on new measurement 1 year after the initial diagnosis. These clinically relevant transitions are especially concerning because the guidelines operate with different cut-offs, and it is uncertain to what extent the guidelines can accommodate such a dynamic parameter.

Systolic left ventricular function is considered important for the prognosis of many heart diseases, which largely justifies the extensive use of EF as a surrogate marker for systolic function. However, the association with cardiovascular events and mortality primarily concerns moderately to severely impaired ventricles [9]. Additionally, EF may be preserved in early-stage systolic dysfunction when there is predominately reduced long-axis shortening, questioning the prognostic value of

EF. As the need for a better marker has been recognized, other methods such as global longitudinal strain (GLS) have increasingly been used in clinical practice for more accurate definition of myocardial deformation. Among other finding, the systematic review by Kalam et al. revealed significant advantages of using GLS over EF when the prognostic values of the parameters were compared with respect to major adverse cardiac events and all-cause mortality [9]. They found that the hazard ratio (HR) per SD change in GLS was associated with a reduction in mortality 1.62 ($p = 0.009$) times greater than the HR per SD change in LVEF, and propose that GLS may replace EF in the future [9].

There are several ways to measure EF, for example echocardiography, cardiac magnetic resonance, and gated single-photon emission computed tomography, the former being the most common. However, the correlation among modalities is shown to be modest and, in the case of measuring EF with echocardiography, reproducibility is strongly dependent on experienced echocardiographers [10].

Furthermore, the normal ranges for EF vary with the method used. The lower normal limit ranges from 46 to 57% and 46 to 59% for men and women, respectively [11], which underlines why it is unfortunate to categorize heart failure phenotypes based on cut-offs from a continuous variable in the first place.

25.4 Current Goals and Treatment for Heart Failure

The corner stones of drug therapy for heart failure with *reduced* ejection fraction include renin-angiotensin-system (RAS) blockers/angiotensin receptor-neprilysin inhibitor (ARNi), beta-blockers, mineralocorticoid receptor antagonists (MRA), and the recently added sodium-glucose co-transporter-2 (SGLT-2) inhibitors. The current recommendations for the treatment of HFpEF is to treat symptoms and etiology (e.g., hypertension, atrial fibrillation, etc.) [12]—which ESC justifies by the alleged lack of evidence on the effect of heart failure medications on cardiovascular mortality and morbidity in the HFpEF population. There has been conducted several randomized trials of HFpEF patients [13–19], and in evaluation of these it is underlined that the principles for treating hypertensive heart failure should apply regardless of ejection fraction [20].

A few patients have more specific causes of heart failure like depositional diseases or other rare genetic cardiomyopathies—in this small group most heart failure medications are not beneficial in terms of improving symptoms or prognosis.

The aims of treatment for hypertensive heart failure are to lower blood pressure, increase quality of life, and improve life expectancy. The life expectancy accompanying the diagnosis is particularly poor but can be increased if handled according to evidence-based treatment principles. In this chapter, we want to emphasize that refraining from doing so clinicians are worsening the prognosis of the substantial group of patients with heart failure with preserved ejection fraction.

25.5 Clinical Trials, Statistical Power, and Significance

25.5.1 Evaluating the Evidence

The current guidelines for treating heart failure are based on the available evidence for the existing drugs. However, the foundation used to form treatment guidelines for HFpEF should be re-evaluated. The relevant trials have investigated the efficacy of different heart failure medications on a composite primary outcome of cardiovascular death or hospitalization.

25.5.2 External Validity: The Connecting Link Between Trial and Target Population

External validity is ultimately the degree of overlap between the sample selected for trial and the target population intended to use the medication. In the trials conducted on the HFpEF population, the properties of the selected samples show in part wide discrepancies to the target populations (Fig. 25.2). The reasons for the discrepancy are complex—consisting of both ethical concerns in use of older and frailer, although more representative, patients in research and practical challenges such as the lack of sufficient sensitivity and specificity of the inclusion and exclusion criteria used to select proper patients for the sample. The result of which being low external validity, recently illustrated in an article published from Lund et al. [21], which concluded that in a real-world scenario, assessed through the Karolinska-Rennes study (KaRen) [22], only 34% of HFpEF patients were eligible to the PARAGON-HF trial.

25.5.3 Statistical Power: The Probability of Displaying the "True Effect"

In the evidence on efficacy of heart failure medications for HFpEF, it seems to be a recurring issue of overestimating the expected efficiency of the drugs. This is commonly due to an a priori overestimation of the morbidity and mortality of the sample, combined with underestimation of the extent of background medication and its effect on the measured outcome (Fig. 25.3). Many of the trials have therefore not reached the desired statistical power of 70–90%—possessing low probability of displaying an existing effect.

25.5.4 Uncovering a Significant Difference with Little Difference Left to Be Uncovered?

An important issue of the current evidence has been the extent of open label drug use. The HFpEF population is a group with high comorbidity and level of symptoms (Fig. 25.3), adding to the high risk of cardiovascular mortality. Accompanying the severe condition and ESC policy is, as might be expected, excessive use (and need)

Characteristics		KaRen (2014)	CHARM (2003)	SENIORS (2005)	PEP-CHF (2006)	I-PRESERVE (2008)	TOPCAT (2014)	PARAGON (2019)	EMPEROR PRESERVED (2021)
Age (years)		77 ± 9	67 ± 11	75	75	72 ± 7	68.7	72år ± 8	71 ± 9
Women (%)		57	40	37	55.5	60	51.5	52	45
LVEF (%)		62 ± 7	41-49 (35%) 50-59 (35%)	33.5**	64.5	59.5	56	57.5	54.3
NYHA (%)	I	12	-	3	75.5	-	3	3	0.1
	II	62	61	56		21.5	63	77	81.5
	III	23	38	39	24.5	76.5	32.5	19.5	18
	IV	3.0	1.5	2		3	0.4	0.4	0.3
NT-proBNP (pg/mL)		1409 [517-2635]	-	-	335-453	320-360	887-1017	910	970
Systolic BP* (mmHg)		138 ± 24	136	139	139	136	130	131	131± 15.5
Comorbidity (%)	AF*	39**	29	34	21	29	32-51	33	51
	DM*	30	28	26	20.5	27.5	13-34	43	49
	HT*	78	64	62	79	88.5	95	95	90
	MI*	15 (CAD* 29)	44	43.5	26	23	22-40	22.5	-

*LVEF: Left ventricular ejection fraction; BP: Blood pressure; AF: Atrial fibrillation; DM: Diabetes mellitus; HT: Hypertension; MI: Myocardial infarction; CAD: Coronary artery disease;

** SENIORS included patients with heart failure regardless of ejection fraction

*** Atrial arrhythmia

Fig. 25.2 Characteristics of the HFpEF samples compared with the characteristics of the KaRen cohort

	CHARM (2003)	SENIORS (2005)	PEP-CHF (2006)	I-PRESERVE (2008)	TOPCAT (2014)	PARAGON (2019)	EMPEROR PRESERVED (2021)
Predefined conditions for statistical power							
a-level	Two-sided, 0.05						
Expected ER (%) or TNE (n)	ER 11 (annualized)	ER 25 (annualized)	TNE 451	ER 18	ER 17.4	TNE 1847	TNE 841, 10% ER
Planned number of participants (n)	2900	2000**	1000	4100***	1102 or**** 1260	4600	4126 (option to increase to 6000)
Planned duration of follow-up (months)	24	Up to 40	12	24	18	26	20
RRR or HR* (statistical power %)	RR 18% (80) RR 16% (70)	RR 25% (90)	RR 40% (90)	RR 14.5% (90)	RR 20% (80 or 85)	RR 22% (95) RR 19% (80)	HR 0.8 (90)
Other a-priori considerations	-	ENCR 20%	-	-	-	-	-
Achieved conditions							
Primary events (n) or Actual event rate (%)	699 PE*	707 PE	207 PE	1505 PE	671 PE	1903 PE	926 PE
Number of included participants (n)	3023	2128	852	4128	3445	4796	5988
Actual duration of follow-up (months)	36.6	21	26.2	49.5	21.6	35	
Detected RR							26.6
Background medication (%) Diuretika	75	85.5	50	83	62.5-90	95.5	-
ACEi	19	82	-	25.5	47.6-70.6	86	80%
ARBs	-	6.5	-	-	13-33		
MRAs	11.5	-	10	15	-	25.5	37
b-blocker	56	-	54.5	58.5	79-87	80	-
Other challenges	-	-	DR at end of trial 35-37%	DR at end of trial 33-34%	-	-	-

***ER:** Event rate; **HR:** hazard ratio; **TNE:** target number of events; **RRR:** relative risk reduction **ENCR:** expected non-compliance rate; **PE:** primary events; **DR:** drop-out rate

** adjusted from 1700 due to miscalculated annualized ER and ENCR

***Adjusted from 3600 to 4100 due to miscalculated ER

**** corresponding to different statistical power

Fig. 25.3 Statistical power in the HFpEF trials

of medications—including many substances, which are directly affecting the common primary endpoint of the HFpEF clinical trials—cardiovascular mortality or hospitalization. This has also been an issue of the conducted trials—the proportion of, for instance, hypertension ranging between 62 and 95%, with a corresponding share using antihypertensive drugs.

Furthermore, the participants of several of the relevant trials were not only receiving drugs affecting the primary endpoint, but they were also, to a great extent, receiving medications targeting the same physiological systems, or even the exact same mechanism as the study drug. This leaves an open question to how much treatment effect there is left (Fig. 25.4) to be uncovered in such trials?

25.5.5 The Ability of Natriuretic Peptides to Select Patients with True HFpEF

As mentioned, one of the dominating challenges in conducting trials on the HFpEF population has been the difficulty of designing inclusion and exclusion criteria with the ability of selecting patients with preserved EF *and* true heart failure, ensuing external validity and statistical power. Natriuretic peptides (NP) are suggested as a solution to this issue, due to their objective nature and connection to the pathophysiological mechanisms of heart failure.

In the last decade, elevated natriuretic peptides have been established as standard stratum for inclusion of participants in medical trials on heart failure. However, most of the existing evidence on medical therapy for HFpEF consists of trials conducted before the last decade. PARAGON, TOPCAT, and EMPEROR-PRESERVED are the only HFpEF trials, where elevated NPs were used as an absolute or optional inclusion criterion (Fig. 25.5).

A recent reevaluation of the PARAGON trial illustrated the importance of using elevated NT-pro-BNP to include patients [23]. The background for this reanalysis was that PARAGON initially included patients either recently hospitalized for heart

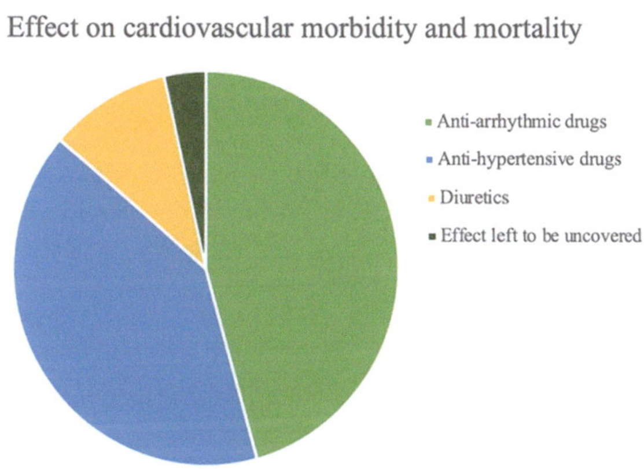

Fig. 25.4 Illustrating the effect of medical intervention on cardiovascular morbidity and mortality in heart failure

Trial	Elevated NPs as absolute inclusion criterion	Elevated NPs as an optional inclusion criterion	Elevated NPs not an inclusion criterion
CHARM (2003)			X
SENIORS (2005)			X
PEP-CHF (2006)			X
I-PRESERVE (2008)			X
TOPCAT (2014)		X	
PARAGON 2019	X*		
EMPEROR-PRESERVED (2021)	X		

NP: *Natriuretic peptide*
**Initially only optional criterion*

Fig. 25.5 Elevated NPs as inclusion criterion in HFpEF trials

failure or having elevated NPs. After initiating recruitment, the researchers received the results from TOPCAT leading to a study protocol amendment, removing the option of entering the study due to hospitalization. As a compromise, the patients already included in the trial due to hospitalization were allowed to proceed their participation on a lower NP threshold. The delayed change of course has facilitated a reanalysis investigating the influence of enrolment on elevated natriuretic peptides by comparing the group of participants included on the absolute NP criterion with the group included on hospitalization displaying a significant effect on the composite primary endpoint of cardiovascular hospitalization or death in the group included on elevated NPs. A similar reanalysis of COMMANDER-HF correspondingly illustrated how elevated NPs as an inclusion criterion heightened statistical power by increasing the event rate in the study selection [24].

25.6 The Renin-Angiotensin System Should Be Blocked in Case of Heart Failure

All patients with hypertension and heart failure should receive renin-angiotensin system (RAS) blockers, i.e., ACE inhibitors or angiotensin receptor blockers (ARBs). Both drug classes include many equivalent alternatives. First of choice should be preparations thoroughly documented in randomized trials. The dose needs to be adjusted to tolerable levels, also ensuring compatibility with other life-prolonging, antihypertensive medications. The lower tolerable blood pressure threshold of a patient is dictated by kidney function and orthostatic disorders. In patients with HFrEF, the evidence of ACE inhibitors being life-prolonging, reducing hospitalization, and improving function is substantial. The first ACE inhibitors became available in the early 1980s, 10–15 years before the development of ARBs. Therefore, ACE inhibitors are prioritized as the first choice, only being replaced by ARBs in case of adverse effects such as dry cough, or the more severe angioedema—both of which result from bradykinin degradation. Liver and kidney function as well as age are the important parameters in assessing dosage.

25.6.1 ACE Inhibitors (ACEi)

There are limited studies on the effects of ACE inhibitors in HFpEF, with the exception of a relatively small trial with 852 participants that investigated perindopril vs. placebo [14]. This trial had multiple challenges regarding study design and statistical power—a priori overestimating the event rate and underestimating the discontinuation rate. Due to this, the 1-year interim analysis displayed an unexpected low event rate and widespread discontinuation with the use of ACE inhibitors outside of protocol (Fig. 25.3). This made the Steering committee cease further inclusion of patients and the researchers did not reach the estimated number of participants. The 1-year analysis, however, did display a reduction in hospitalization, and a borderline significant reduction in primary endpoint. Thus, the study managed to illustrate a preventive effect on hospitalization for HFpEF, and, in light of the unfavorable preconditions, should be considered positive.

25.6.2 Angiotensin II Receptor Blockers (ARBs)

The evidence for the effects of ARBs is more comprehensive, including the evaluation of candesartan in the CHARM study and irbesartan in I-PRESERVE. These trials were both marked by weak statistical power because of the widespread use of other heart failure medications such as RAS blockers, beta-blockers, and aldosterone antagonists. Despite this, both trials had borderline significant results in favor of the study drug. Retrospective reanalysis published by the original researchers have showed a clear significant result in favor of ARB. CHARM was reanalyzed with a repeat-event analysis [25], and I-PRESERVE was reanalyzed with adjustment for common baseline variables [26]. In principle it is not desired to base on retrospective analysis in questions of effect. In this case, however, it can be defended given the weaknesses in conduction of the original trials. The data illustrates how candesartan and irbesartan reduce hospitalization and death in patients with HFpEF—most of whom also hypertensive (Fig. 25.2).

25.6.3 Angiotensin Receptor-Neprilysin Inhibitors (ARNIs)

The novel medication angiotensin receptor-neprilysin inhibitor (ARNi, sacubitril-valsartan) is through the combination of the neprilysin inhibitor sacubitril and ARB valsartan, the most potent existing antihypertensive drug. The first clinical trial for this medication investigated the effect on hypertensive patients [27]. However, there has since the initial trial never been conducted hypertension-related endpoint studies (phase 3), and in the lack of this, its indication does not include hypertension, only HFrEF.

The effect of sacubitril-valsartan on HFrEF was investigated in the PARADIGM-HF [28]. In this trial the study drug was compared to the ACE inhibitor enalapril, and the trial showed a significant reduction in hospitalization, as well as a significant life-prolonging effect.

Regarding HFpEF, the effect was investigated in the PARAGON trial. This trial was designed using the active comparator valsartan instead of placebo for ethical reasons. Despite this, the results on hospitalization and cardiovascular death were borderline significant. The US Food and Drug Administration (FDA) has in their interpretation of these results considered the limitations of using an active comparator, and therefore approved sacubitril-valsartan as treatment for HFpEF.

As illustrated in Fig. 25.2, 95% of participants had hypertension, and as expected, a corresponding share were treated according to current guidelines (Fig. 25.3), including extensive treatment with other RAS blockers. Sacubitril-valsartan might therefore be seen as an equivalent and perhaps better alternative to ACE inhibitor and ARB in all subgroups of hypertensive heart failure—particularly when blood pressure is elevated. ARNi possesses stronger reno-protective properties than conventional RAS blockers, which is of benefit to the high share of heart failure patients with impaired renal function [1]. A prerequisite to tolerate this treatment is hemodynamic stability and absence of hypotension.

25.7 Diuretics and Fluid Balance

Thiazides are included as standard treatment for hypertension, preferably given with ACE inhibitors or ARBs, which potentiate the blood pressure-lowering effect. Similarly, thiazides are beneficial in the treatment of heart failure since both hypertension and fluid retention are common features for this patient group. For patients with renal impairment and fluid retention loop diuretics such as bumetanide and furosemide are the alternatives, especially regarding over-hydration and as a part of avoiding hospitalization. In the long run, diuretics help to stabilize the fluid balance and the doses can in many cases be adjusted by the patient through careful weight control.

25.8 The Role of Beta-Blockers in Heart Failure Treatment

High sympathetic activity is prevalent in patients with heart failure, and increased levels of circulating noradrenaline has been shown to predict mortality and heart failure-related hospitalization in patients with HFrEF [29]. European Hypertension guidelines (2018) recommend beta-blockers at each step of antihypertensive treatment whenever specifically indicated, and the benefits are comparable to the life-prolonging effects seen with RAS inhibitors. It applies to patients with hypertension, with or without heart failure, and has been documented for both beta-1 selective bisoprolol, metoprolol, and nebivolol, as well as for the nonselective vasodilating carvedilol. As with RAS inhibitors, the existing literature on beta-blockers also mainly deals with HFrEF patients. An exception is the SENIORS trial on nebivolol [14], which included nonselected heart failure patients, and demonstrated a life-prolonging effect independent of EF in a prespecified subgroup analysis. However, the lack of documentation has not prevented the prescribing of beta-blockers in the HFpEF group, as illustrated by the 55–87% open label use of beta-blockers in clinical trials [20], which, in addition to the other drugs, make it difficult to show significant effects of new ones. In addition to counteracting unfavorable myocardial remodeling, beta-blockers reduce the heart frequency which is associated with increased cardiovascular risk from 80 bpm and higher [30]. Thus, beta-blockers are favorable to add in case of heart failure demanding rate control and can be combined with other antihypertensive drug categories.

25.9 Mineralocorticoid Receptor Antagonists (MRAs)

There is no existing evidence for the effect on mortality and morbidity in hypertension without heart failure using Mineralocorticoid Receptor Antagonists (MRA). In spite of this, the ESC/ESH 2018 guidelines for management of hypertension [30] have included low-dose spironolactone as the fourth medication used for patients with difficult to treat hypertension. The foundation for this recommendation is an extrapolation of the findings from HFrEF trials such as RALES, EMPHESUS, and EPHESUS to patients with difficult controllable hypertension and high risk of developing heart failure. The mentioned trials have displayed reduced hospitalization and a life-prolonging effect of spironolactone and eplerenone on top of RAS inhibition and beta-blocker in HFrEF. It is emphasized to observe potassium levels, which are increased by treatment with MRA, particularly with compromised kidney function. This issue can now be addressed, and life-prolonging MRA treatment can be facilitated through enabling therapy with newly developed oral potassium binders.

The effect of spironolactone in HFpEF was investigated in the multicenter trial TOPCAT [17] conducted in the USA and South America, as well as Russia and Georgia. In the aftermath, this trial is considered a tragic chapter in the history of heart failure research—the grounds and extent of which have been newly thoroughly examined by some of its top researchers [31]. The study protocol allowed for inclusion based on either hospitalization due to heart failure in the last 12 months or for having elevated NT-pro-BNP at any time. In retrospect it became apparent that there were systematical regional differences to which criterion was preferably used by the researchers in recruitment—participants in the Americas were to a far greater extent included on elevated NT-pro-BNP. Furthermore, in the Russian-Georgian part of the trial, the event rate for primary endpoint was comparable to the corresponding event rate of the healthy population—underlining an improbability of these patients having had any heart failure at all. Adding to this, a post-hoc analysis of blood samples showed that a lot of participants in this subgroup did not receive their randomized treatment with spironolactone.

The total tragedy of TOPCAT became evident when the trial as total did not manage to show statistically significant benefit of spironolactone, but a retrospective subgroup analysis showed a significant reduction of hospitalization in the American countries. Since the regional subgroup analysis was not predefined, it was not accepted by New England Journal of Medicine to be part of the original main publication (appeared in the online supplement, only), and the trial was considered "negative." Fortunately, the FDA recently decided to base themselves on the American subgroup and disregard the Russian-Georgian part of the trial—expanding the indication for spironolactone to HFpEF.

Finerenone is a novel selected, nonsteroidal mineralocorticoid receptor antagonist clinically tested on patients with chronic kidney disease and type 2 diabetes [32]. The conducted randomized trials have included participants with combined cardiorenal failure, with promising results also in this subgroup, including reduction of clinically important cardiovascular outcomes. However, further evidence on the effect of finerenone in heart failure must be retrieved in order for it to be included in the treatment principles for hypertensive heart failure.

25.10 Sodium-Glucose Co-Transporter-2 (SGLT2) Inhibitors

The initially intended to be blood glucose lowering Sodium-Glucose co-Transporter-2 Inhibitors (SGLT2i) have proved to be antihypertensive as well. Randomized, controlled testing has illustrated that empagliflozin, canagliflozin, dapagliflozin, ertugliflozin, and sotagliflozin (combined SGLT2 and SGLT1 inhibitor) are efficient in reducing the risk of heart failure in patients with diabetes mellitus, but also in patients with established cardiovascular disease or with high cardiovascular risk [33]. The target of these medications is the sodium-glucose co-transporter-2 in the proximal tubule in the kidney, where they reduce reuptake of uric acid, glucose, and sodium. Many mechanisms for the potent and beneficial effects on heart, kidneys, and vessels have been proposed [34]. However, few of these have been confirmed, and the mechanisms are therefore partly unknown. For HFrEF (with or without diabetes) and chronic kidney failure, a reduction in mortality and hospitalization has been shown in major clinical endpoint trials. Recently the researchers of EMPEROR-PRESERVED published their result—a significant reduction in cardiovascular mortality and hospitalization for HFpEF. The gathered clinical evidence has led to a recent approval by the American Food and Drug Administration (FDA) of SGLT2 inhibitors *regardless of ejection fraction*—a standpoint also adopted by the European Medicines Agency (EMA). The discovery of the drug is being compared to those of penicillin and statins [35]—there are in other words tremendous expectations for the potential impact from these medications. It is, among other, suggested that SGLT2 inhibitors can reduce the need of diuretics for patients with heart failure. The international heart failure society is still awaiting results from ongoing trials that will give further documentation on the effect of these drugs in HFpEF patients with or without diabetes.

On the fifth of May this year, a press release announced that the RCT with the acronym DELIVER (Dapagliflozin Evaluation to Improve the Lives of Patients with Preserved Ejection Fraction Heart Failure) was positive for the primary endpoint. The SGLT2i investigated in the study prevented both cardiovascular death and worsening of heart failure compared to placebo on top of quite extensive therapy with other heart failure medications in 6263 randomized patients with ejection fraction >40%. Of these patients 89% had a history of hypertension, and more than half of them had a history of coronary heart disease and the same for atrial fibrillation and less than half had diabetes. As many as 76% of these patients were on treatment with beta-blocker at baseline and large fractions also on the other heart failure medication discussed in this chapter [37]. The study was announced early because it is a stock sensitive issue and was published shortly later [38].

25.11 Entirety of the Regimen Treatment of Hypertensive Heart Failure

The underlying principle of heart failure treatment is the use of "tolerable" doses of existing, documented medications. This indicates the gradual buildup of treatment in response to achieved effect and occurrence of side effects—according to the order in which the medications have been discussed in this chapter. If symptomatic hypotension is developed, a more careful and slower approach to the gradual buildup must be made, and vice versa. It is furthermore important to limit extensive polypharmacy by massive comorbidity, frailty, and significantly reduced life expectancy The entirety of the drug treatment of hypertensive heart failure has been extensively reviewed and updated in the 2023 European Hypertension Guidelines issues by the European Society of Hypertension [39].

25.12 Hypertension Despite Established Treatment According to Medical Treatment Principles?

As a result of the antihypertensive effect exerted by all the life-prolonging medications in heart failure, these patients rarely possess remaining hypertension if treated according to the described principles. However, in some cases hypertension is a remaining issue after establishing the standardized regimen for heart failure. According to guidelines for management of hypertension [30] and a position paper describing how to deal with persistent hypertension in heart failure [36], further antihypertensive needs can be met through a calcium antagonist or an alpha-blocker. Nevertheless, it must be emphasized that, as a consequence of its negative inotropic effect, it is generally advised against calcium antagonists in the treatment of HFrEF.

25.13 Summary and Conclusions

Current treatment for hypertensive heart failure consists of combining tolerable doses of all existing life-prolonging medications.

In evaluation of the existing evidence for the different treatment principles, it becomes evident that conclusions for current treatment recommendations have been drawn on variable grounds and that the lack of significant results in HFpEF trials in greater part relies on technical weaknesses of the conducted trials, than on a true effect of the medications not being present. With few prospects of future attempts to recover these errors, we offer a compromise to limit the potential damage accompanying them and have accordingly described how the treatment principles for hypertensive heart failure consist of RAS inhibition or ARNi, beta-blocker and mineralocorticoid receptor antagonist, and, if necessary, application of diuretics for fluid control. Furthermore, SGLT2 inhibitors are proven to prevent hospitalization and death throughout the entire EF spectrum. In conclusion, these principles should apply to all patients with heart failure regardless of ejection fraction.

References

1. Conrad N, Judge A, Tran J, Mohseni H, Hedgecott D, Crespillo A, et al. Temporal trends and patterns in heart failure incidence: a population-based study of 4 million individuals. Lancet. 2018;391:572–80.
2. Franklin SS, O'Brien E, Thijs L, Asayama K, Staessen JA. Masked hypertension: a phenomenon of measurement. Hypertension. 2015;65(1):16–20.
3. Kraigher-Krainer E, Shah AM, Gupta DK, Santos A, Claggett B, Pieske B, et al. Impaired systolic function by strain imaging in heart failure with preserved ejection fraction. J Am Coll Cardiol. 2014;63(5):447–56.
4. Smiseth OA, Morris DA, Cardim N, Cikes M, Delgado V, Donal E, et al. Multimodality imaging in patients with heart failure and preserved ejection fraction: an expert consensus document of the European Association of Cardiovascular Imaging. Eur Heart J Cardiovasc Imaging. 2021;23(2):e34–61.
5. Group TCTS. Effects of enalapril on mortality in severe heart failure: results of the Cooperative North Scandinavian Enalapril Survival Study (CONSENSUS). N Engl J Med. 1987;316:1429–35.
6. Cardiology ESo. ESC guidelines for the diagnosis and treatment of acute and chronic heart failure. Eur Heart J. 2021;2021(42):3599–726.
7. Clarke CL, Grunwald GK, Allen LA, Barón AE, Peterson PN, Brand DW, et al. Natural history of left ventricular ejection fraction in patients with heart failure. Circ Cardiovasc Qual Outcomes. 2013;6:680–6.

8. Dunlay S, Roger V, Weston S, Jiang R, Redfield M. Longitudinal changes in ejection fraction in heart failure patients with preserved and reduced ejection fraction. Circ Heart Fail. 2012;5(6):720–6.
9. Kalam K, Otahal P, Marwich TH. Prognostic implications of global LV dysfunction: a systematic review and meta-analysis of global longitudinal strain and ejection fraction. Heart. 2014;100:1673–80.
10. Triposkiadis F, Butler J, Abboud F, Armstrong P, Adamopoulos S, Atherton J, et al. The continuous heart failure spectrum: moving beyond an ejection fraction classification. Eur Heart J. 2019;00:1–11.
11. Marwick T. Ejection fraction pros and cons. J Am Coll Cardiol. 2018;72(19):2360–79.
12. McDonagh TA, Metra M, Adamo M, Gardner RS, Baumbach A, Böhm M, et al. 2021 ESC guidelines for the diagnosis and treatment of acute and chronic heart failure: developed by the task force for the diagnosis and treatment of acute and chronic heart failure of the European Society of Cardiology (ESC) with the special contribution of the Heart Failure Association (HFA) of the ESC. Eur Heart J. 2021;42(36):3599–726.
13. Yusuf S, Pfeffer MA, Swedberg K, Granger CB, Held P, McMurray JJ, et al. Effects of candesartan in patients with chronic heart failure and preserved left-ventricular ejection fraction: the CHARM-preserved trial. Lancet. 2003;362(9386):777–81.
14. Flather M, Shibata M, Coats A, van Veldhuisen D, Parkhomenko A, Borbola J, et al. Randomized trial to determine the effect of nebivolol on mortality and cardiovascular hospital admission in elderly patients with heart failure (SENIORS). Eur Heart J. 2005;26(3):215–25.
15. Cleland JG, Tendera M, Adamus J, Freemantle N, Polonski L, Taylor J. The perindopril in elderly people with chronic heart failure (PEP-CHF) study. Eur Heart J. 2006;27(19):2338–45.
16. Massie BM, Carson PE, McMurray JJ, Komajda M, McKelvie R, Zile MR, et al. Irbesartan in patients with heart failure and preserved ejection fraction. N Engl J Med. 2008;359(23):2456–67.
17. Pitt B, Pfeffer M, Assmann S, Boineau R, Anand I, Claggett B, et al. Spironolactone for heart failure with preserved ejection fraction. N Engl J Med. 2014;370(15):1383–92.
18. Solomon SD, McMurray JJV, Anand IS, Ge J, Lam CSP, Maggioni AP, et al. Angiotensin–neprilysin inhibition in heart failure with preserved ejection fraction. N Engl J Med. 2019;381(17):1609–20.
19. Anker SD, Butler J, Filippatos G, Ferreira JP, Bocchi E, Böhm M, et al. Empagliflozin in heart failure with a preserved ejection fraction. N Engl J Med. 2021;385(16):1451–61.
20. Kjeldsen SE, von Lueder TG, Smiseth OA, Wachtell K, Mistry N, Westheim AS, et al. Medical therapies for heart failure with preserved ejection fraction. Hypertension. 2020;75:23–32.
21. Lund LH, Savarese G, Venkateshvaran A, Benson L, Lundberg A, Donal E, et al. Eligibility of patients with heart failure with preserved ejection fraction for sacubitril/valsartan according to the PARAGON-HF trial. ESC Heart Fail. 2022;9(1):164–77.
22. Donal E, Lund LH, Oger E, Hage C, Persson H, Reynaud A, et al. Baseline characteristics of patients with heart failure and preserved ejection fraction included in the Karolinska Rennes (KaRen) study. Arch Cardiovasc Dis. 2014;107(2):112–21.
23. Pabon MA, Cunningham JW, Claggett BL, Packer M, Zile M, Pfeffer MA, et al. Natriuretic peptide-based inclusion criteria in heart failure with preserved ejection fraction clinical trials: insights from PARAGON-HF. Eur J Heart Fail. 2022;26:26.
24. Cunningham JW, Ferreira JP, Deng H, Anker SD, Byra WM, Cleland JGF, et al. Natriuretic peptide-based inclusion criteria in a heart failure clinical trial. JACC Heart Fail. 2020;8(5):359–68.
25. Rogers J, Pocock S, McMurray J, Granger C, Michelson E, Östergren J, et al. Analysing recurrent hospitalisations in heart failure: a review of statistical methodology, with application to CHARM-preserved. Eur J Heart Fail. 2014;16(1):33–40.
26. Ferreira J, Dewan P, Jhund P, Lorenzo-Almorós A, Duarte K, Petrie M, et al. Covariate adjusted reanalysis of the I-preserve trial. Clin Res Cardiol. 2020;109:1358–65.
27. Ruilope LM, Dukat A, Böhm M, Lacourcière Y, Gong J, Lefkowitz MP. Blood-pressure reduction with LCZ696, a novel dual-acting inhibitor of the angiotensin II receptor and

neprilysin: a randomised, double-blind, placebo-controlled, active comparator study. Lancet. 2010;375(9722):1255–66.

28. McMurray JJV, Packer M, Desai AS, Gong J, Lefkowitz MP, Rizkala AR, et al. Angiotensin–neprilysin inhibition versus enalapril in heart failure. N Engl J Med. 2014;371(11):993–1004.
29. Investigators TS. Effect of enalapril on mortality and the development of heart failure in asymptomatic patients with reduced left ventricular ejection fractions. N Engl J Med. 1992;327(10):685–91.
30. Williams B, Mancia G, Spiering W, Agabiti Rosei E, Azizi M, Burnier M, et al. 2018 ESC/ESH guidelines for the management of arterial hypertension: the task force for the management of arterial hypertension of the European Society of Cardiology (ESC) and the European Society of Hypertension (ESH). Eur Heart J. 2018;39(33):3021–104.
31. Pfeffer MA, Claggett B. Behind the scenes of TOPCAT—bending to inform. NEJM Evidence. 2022;1(1):EVIDctcs2100007.
32. Agarwal R, Filippatos G, Pitt B, Anker SD, Rossing P, Joseph A, et al. Cardiovascular and kidney outcomes with finerenone in patients with type 2 diabetes and chronic kidney disease: the FIDELITY pooled analysis. Eur Heart J. 2022;43(6):474–84.
33. Zinman B, Wanner C, Lachin JM, Fitchett D, Bluhmki E, Hantel S, et al. Empagliflozin, cardiovascular outcomes, and mortality in type 2 diabetes. N Engl J Med. 2015;373(22):2117–28.
34. Pabel S, Hamdani N, Luedde M, Sossalla S. SGLT2 inhibitors and their mode of action in heart failure-has the mystery been unravelled? Curr Heart Fail Rep. 2021;18(5):315–28.
35. Braunwald E. SGLT2 inhibitors: the statins of the 21st century. Eur Heart J. 2021;43(11):1029–30.
36. Kasiakogias A, Rosei EA, Camafort M, Ehret G, Faconti L, Ferreira JP, et al. Hypertension and heart failure with preserved ejection fraction: position paper by the European Society of Hypertension. J Hypertens. 2021;39(8):1522–45.
37. Solomon SD, Vaduganathan M, Claggett BL, de Boer RA, DeMets D, Hernandez AF, et al. Baseline characteristics of patients with HF with mildly reduced and preserved ejection fraction: DELIVER trial. JACC Heart Fail. 2022;10:184–97.
38. Solomon SD, McMurray JJV, Claggett B, de Boer RA, DeMets D, Hernandez AF, et al. For the DELIVER trial committees and investigators. Dapagliflozin in heart failure with mildly reduced or preserved ejection fraction. N Engl J Med 2022;387:1089–1098. https://doi.org/10.1056/NEJMoa2206286.
39. Mancia G, Kreutz R, Brunström M, Burnier M, Grassi G, Januszewicz A, et al. 2023 ESH guidelines for the management of arterial hypertension. Journal of hypertension 2023;41:1874-2071.) also endorsed by the International Society of Hypertension (ISH) and the European renal association (ERA).

Unmet Needs in Drug Treatment of Heart Failure in Hypertension

26

Pierre Boutouyrie and Antoine Fayol

26.1 Introduction

Treatment for heart failure (HF) has evolved since the time of bleeding for dropsy, although bleeding addressed indirectly the question of volume overload and post-load. This cumbersome method, although effective in refractory pulmonary edema, is not a convenient therapy in practice, and "internal bleeding" through sodium excretion by diuretic reduces preload, congestion, and symptoms of HF. The progress came from a better understanding of the physiological basis of HF by splitting into the concept of HF with reduced ejection fraction (HFrEF; LVEF under 40%), in which the primary alteration is the contractile (systolic) function, as opposed to HF with preserved ejection fraction (HFpEF; LVEF value over 50%), in which the primary alteration is relaxation: the heart is unable to withstand increasing preload. Patients with an LVEF value between these two classes are considered as patients with Heart Failure with mildly reduced Ejection Faction (HFmrEF).

Thus, either the pump function is altered because of reduction in the contractile component (decreased in the number and/or the functionality cardiomyocytes), or because of a reduction in the filling capacity of the heart (either by an alteration of the relaxation of cardiomyocytes), and/or a fibrosis of the extracellular matrix. Both forms of HF lead to similar contraregulations, with activation of the sympathetic system, and the renin angiotensin aldosterone system, leading to increased volemia,

P. Boutouyrie (✉)
Pharmacology Department, Hôpital Européen Georges Pompidou, Paris, France

Pharmacology, Université de Paris Cité, Paris, France
e-mail: pierre.boutouyrie@aphp.fr

A. Fayol
Cardiology (CARTE) Department, Hôpital Européen Georges Pompidou, Paris, France
e-mail: antoine.fayol@aphp.fr

M. Dorobantu et al. (eds.), *Hypertension and Heart Failure*, Updates in Hypertension and Cardiovascular Protection,
https://doi.org/10.1007/978-3-031-39315-0_26

sodium retention, and inappropriate increase in postload. Reducing congestion will alleviate symptoms on both forms; however, it may lead to reductions in cardiac output when high filling pressures are needed.

The information carried by EF varies markedly with the threshold [1]. HF classifications based on LVEF are highly convenient from a pathophysiological point of view; this splitting is an oversimplification since HFpEF and HFrEF are almost always associated. Indeed, for instance, hypertension, aging, and chronic kidney diseases are the 3 main causes of HFpEF, but also of myocardial infarction, itself the main cause of HFrEF. So ischemic cardiopathy in older hypertensives leads to HFrEF and HFpEF, which means that defining HFrEF by EF <40% or pEF by EF > 50% is clearly an overstatement.

A better comprehension of the physiological bases of pharmacology in heart failure came by identifying contraregulations as true targets for treatment, rather than treating the original mechanisms. For instance, tachycardia and sodium retention, being the cause of congestion, result from the strong reflex activation of the sympathetic nervous system and renin angiotensin aldosterone systems, respectively. Since tachycardia reduces the filling time of the ventricles, and congestion aggravates both relaxation and may precipitate heart failure by overloading cardiomyocytes (flat part of the Starling relation), both components might constitute adequate targets for treating HF. Last, identification of the extreme vulnerability of the failing myocardium to arrhythmias explained why some medications might aggravate HF, such as antiarrhythmic drugs, vasodilating drugs activating the baroreflex (direct vasodilators or alpha-blockers) or potentially having proarrhythmic properties. The literature concerning pharmacological treatment of heart failure is massive, and different reviews and consensus guidelines are available [2, 3]. Since a vast majority of HF patients have a history of hypertension, dealing with issues related to hypertension in HF is equivalent to dealing with HF at large. Thus, the aim of the present chapter is to stress on gaps in knowledge rather than reviewing extensively the literature.

26.2 Remaining Problems in Pharmacological Treatment of Heart Failure

In the past ½ century, most of the progress in treatment of HF was restricted to HFrEF; HFpEF has long been considered as a hopeless condition. Modern treatment of heart failure has led to massive improvement of prognosis. In early reports, before angiotensin-converting enzyme inhibition (ACEi) treatments, stage 3–4 heart failure had 50% yearly risk of death at 6 months, as shown in placebo arms of early trials [4, 5]. Today, it is less than 5% [6]. Translated into life expectancy, modelization coming from network meta-analysis suggests that the gain is considerable, up to 8.5 year [7]. This does not mean that HF has disappeared, but simply that patients survive longer, and have more chance to be hospitalized at an older age [8].

Despite symptoms were obviously alleviated by sodium depletion, and vasodilatation, this did not translate to improved hard outcomes [9, 10]. The introduction of angiotensin receptor-neprilysin inhibitors (ARNi), was a first come-through [11–13]. Although not truly speaking an innovation -previous attempts at blocking both RAS system and neutral endopeptidase were limited by the frequency of Angioedema

[14]. ARNi is peculiar since it combines valsartan (angiotensin 2 receptor antagonist, ARB) with sacubitril (neutral endopeptidase antagonist) in a single original molecule. This molecule proved to be more effective than ACEi or ARBs, and as well tolerated [13]. ARNi would be interesting in hypertension; however, they are not approved in this indication except in Japan.

Mineralocorticoids are well-established treatments of HF [15–17], together with effective treatments of hypertension [18]. They are positioned as third-line treatments, mainly because of side effects (spironolactone), the risk of hyperkalemia [19], or because their positive effect was tested on top of other effective treatments. Finerenone is a non-steroidal, selective mineralocorticoid receptor antagonist, which reduced albuminuria in short-term trials involving patients with chronic kidney disease (CKD) and type II diabetes. In large clinical trials, use of Finerenone [20] improved cardiovascular outcomes including a marginal improvement in heart failure and slower CKD progression despite a modest effect on blood pressure, with a mean decrease of 3 mmHg. Finerenone may represent an advance in the prevention and management of heart failure, thus minimizing the considerable health care burdens in patients with CKD and type 2 diabetes.

The emergence of sodium glucose cotransport 2 inhibitors (SGLT2i) has totally changed this paradigm. sGLT2i were initially positioned for the treatment of type 2 diabetes. The approach was kind of "low tech," because these drugs, by inhibiting glucose cotransport 2 in the kidney, induce a 100 to 150 g/day loss of glucose in urine, and thus decrease glycemia and HbA1C, at the cost of minor side effects (genitourinary infections). One surprise came from the initial phase of development, with less cases of heart failure in treated diabetic patients. SGLT2i induce moderate natriuresis, and small decrease in blood pressure (−5 mmHg), insufficient to explain the reduction in heart failure. Hence, drug companies initiated a phase 3B development in heart failure, first with HFrEF, then with HFpEF.

Results came as a shock since, on top of maximal treatment (control of congestion, ACEi, ARB or ARNi, beta-blockers and aldosterone antagonists), SGLT2i conferred an additional reduction of hard outcome (hospitalization or CV death by 30 to 40%), for HFrEF [6, 21] and HEpEF [21, 22]. Now, SGLT2 antagonists have a prominent place in the armamentarium [23] in HF. Both European and Americans guidelines of heart failure management now recommend an early introduction of sGLT2i. The mechanisms by which SGLT2 improves cardiovascular outcome remain incompletely elucidated. Since SGLT2 is not expressed at the cardiomyocyte and mitochondrial membranes, the mechanism of cardiac protection must be indirect. Besides the positive effect on sodium balance, SGLT2i appear to have pleiotropic effects, including reduced inflammation and oxidative stress, fibrosis, intraglomerular pressure, and sympathetic nervous system activation. It may also improve myocardial efficiency by modulation of the energy balance at the mitochondrial level. Overall, SGLT2i directly affect cardiomyocytes by reducing [Na+]c and [Ca2+]c, by inhibition of Sodium–hydrogen antiporter 1(NHE-1), and by increasing mitochondrial function and cell viability. The exact membrane receptors/transporters mediating these effects in cardiomyocytes have not yet been identified, with the exemption of NHE-1 and L-type calcium channel complex (LTCC) [24, 25]. An overlooked possibility is that inhibiting glucose reuptake shifts cardiac metabolism to short-arm phospholipids [26]. SGLT2i can also improve cardiac efficiency by vascular mechanisms, both on coronary arteries [27], and also by improving heart/vessel coupling [28].

26.3 Escalade of Treatments

Evidence came from clinical trials, which means that drug implementation was additive rather than subtractive. For instance, early trials of ACEi were done compared to usual treatment (digitalis and/or diuretics); however background treatment was usually maintained. Once demonstrated, RAS blockers were considered as background treatments, and new drugs tested against background. This method does not help building a strategy for implementation of treatments, for instance to decide whether to use loop diuretics systematically if congestion is clinically absent. Two exceptions are ARNi, which was tested head to head against ACEi, and SGLT2i, which were tested against no treatment, at least in diabetic patients. Recent network meta-analyses provide some insight in the best combination of drugs associated with optimal survival [7]. Figure 26.1 shows that the combination of ARNi, BB, MRA, and

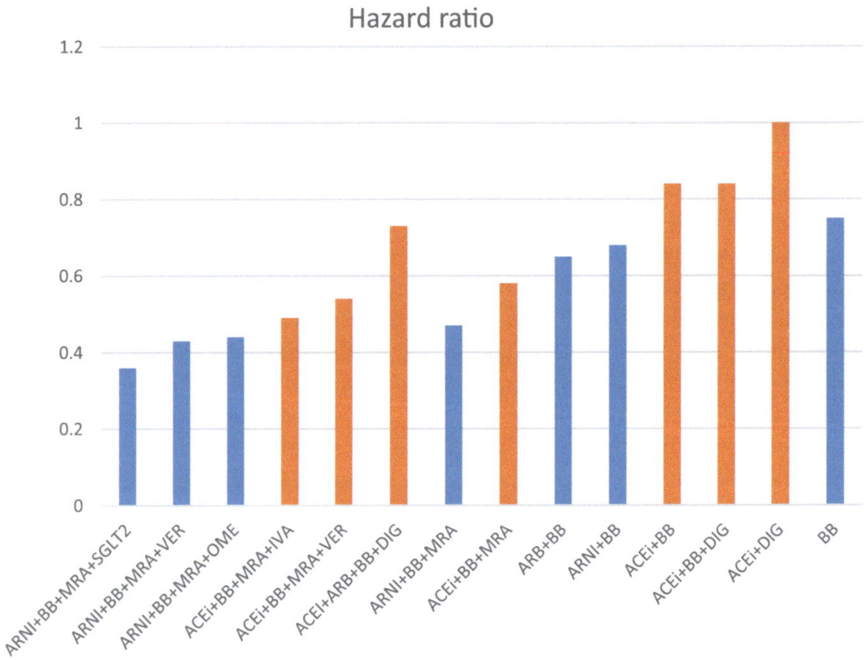

Fig. 26.1 From [7]: Hazard ratios in clinical trials for treatment escalade in heart failure (derived from ref. [41]). *MRA* mineralocorticoids receptor antagonists, *SGLT2i* sodium glucose transporter 2 inhibitors, *HR* heart rate, *BP* blood pressure, *ARNi* angiotensin 2 receptor/neprilysin inhibitors, *BB* betablockers, *IVA* ivabradine, *H-ISMN* hydralazine/isosorbide dinitrate, *DIG* digoxin. For convenience, confidence intervals are not represented. All 95% confidence intervals are significant. Orange bars represent ACEi-based strategies, blue bars represent ARNi-based strategies

SGLT2i provides the best protection, with a HR of 0.39, compared to older treatment or no treatment. It remains unclear whether this is representative of true life.

Proposed drug-ranking algorithms are based on good sense and pathophysiological considerations rather than on clear medical evidence. It is also evident that tolerance of drugs often dictates the drug escalade. Renal function often limits the use of aldosterone antagonists and SGLT2i; hypotension limits RAS blockers and ARNi; fatigue and slow heart rate might limit beta-blockers. The strategy is often dependent on good sense, which is not a warrant of efficacy [29].

26.4 All Patients Should Benefit from Pharmacologically Proven Efficient Drug Regimen

The optimal drug regimen is known, the additive benefit of each class of medically directed therapy in reducing all-cause mortality and improving survival, estimating that a 70-year-old patient would gain 5.0 years or more if treated with a combination of 4 therapies (ARNI, BB, AA, SGLT2) than with dual therapy. This finding builds on previous work estimating the mortality risk of delaying medically directed therapy in HFrEF, and it provides a strong motivation to improve treatment of these patients [7].

Unfortunately, not all patients benefit from this optimized treatment regimen. Contemporary studies show that few patients achieve target doses of Guided Medical Therapy. The US CHAMP-HF (Change the Management of Patients with Heart Failure) registry, including 3518 patients from 150 different practice sites, found that only 1% of patients were receiving target doses of the 3 major classes of acute heart failure medications [30]. Reasons for inadequate use of pharmacological treatment are complex: they combine (Fig. 26.2), and combine patients' selection in clinical studies (those who tolerate drugs), exclusion of patients with hypotension, known intolerance, comorbidities, socioeconomic considerations in low-middle income countries and in the USA. Tailoring treatments, using combo drugs and generic drugs to lower the price, and performing clinical studies in patients not fulfilling inclusion criteria would be needed. We also need to underline the fact that these combinations are all effective treatments in HFrEF and not in the other categories.

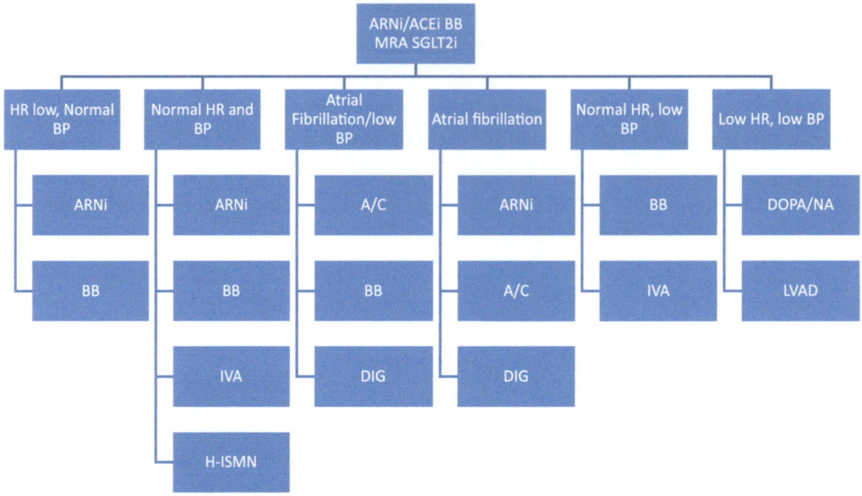

Fig. 26.2 From [41]: General algorithm for treatment escalade in heart failure. *MRA* mineralocorticoids receptor antagonists, *SGLT2i* sodium glucose transporter 2 inhibitors, *HR* heart rate, *BP* blood pressure, *ARNi* angiotensin 2 receptor/neprilysin inhibitors, *BB* betablockers, *IVA* ivabradine, *H-ISMN* hydralazine/isosorbide dinitrate, *DIG* digoxin, *DOPA* dopamine, *NA* noradrenalin, *LVAD* left ventricle assistance devices

26.5 Is There Any Place for Positive Inotropic Agents for Treatment of Chronic Heart Failure?

Digoxin and derivate were classical inotropic agents. Although they directly increase Ca++ bioavailability in cardiomyocytes through inhibition of Na/K ATPase, they are the only inotrope slowing the heart rate, probably through sensitization of baroreceptors. Although stopping digoxin leads to aggravation of symptoms, there is no evidence for efficacy, and digitalics are only used for heart rate control in Atrial Fibrillation (AF). Digoxin was used as neutral treatment in the most recent meta-analysis [7]. Trials with oral inotrope have led to overall failure [31]. One recent exception is omecamtiv. This compound has global inotrope positive action by improving the coupling of actin to myosin, and thus can improve the cardiomyocyte efficiency without increasing oxygen consumption. This tempting concept has been validated in hard outcome clinical trial [32] in HFrEF, with an 8% reduction in hospitalized heart failure or CV death. Whether this translates to clinical practice and the selection of patients likely to benefit from this treatment remains to be determined.

26.6 Atrial Fibrillation and Rhythmic Disorders

AF has a strong impact on HF. How to prevent AF in patients at high risk? Some drugs, including RAAS blockers, appear to be more effective than others [33, 34]. Intensive control of blood pressure using this class in priority is likely to decrease the occurrence of AF; however the level of proof is low. SGLT2 inhibitors may confer a specific AF/ALF (ALF: Atrial flutter) reduction benefit in the susceptible type 2 diabetes population, independent of age, body weight, HbA1c, and systolic blood pressure at baseline. This benefit in terms of AF/ALF reduction can be partially attributed to the pharmacological effects on reducing HbA1c, body weight, blood pressure, and the development of heart failure [35].

26.7 Drug Adherence in HF

Nonadherence to drug therapy is very frequent in chronic diseases, even if those diseases have severe symptoms or prognosis. In hypertension, up to 50% of patients presenting with resistant hypertension are partially or fully nonadherent when using objective methods such as drug detection in biological fluids [36, 37]. Similar figures are found for heart failure although less studies are available [38], and even fewer use reference methods [39].

A recent publication evaluated nonadherence in HF patients: nonadherence was biologically confirmed in one of five patients evaluated 4–6 weeks after hospitalization with HF. Nonadherence in chronic heart failure is probably one of the least adopted or detected areas in the management of HF patients [40]. In addition to crude adherence, the level of exposure to drug, dependent on morphology, renal dysfunction, etc., is seldom taken into account. It is likely that the common dosage of drugs used in clinical studies is not adequate in real life, such as older elderly (>85 years), frail subjects, and subjects with cardiorenal syndrome.

26.8 Place of Biomarkers

Biomarker tailoring of drug treatment is recommended [41], despite little evidence to support this personalization of treatments. Endocrine substances secreted by the heart under stress (ANP, BNP, and derivative) are extremely useful for the diagnostic of heart failure. They also are powerful predictors of outcome, both in HFpEF and HFrEF [42]; however, they did not prove so useful for monitoring response to treatment [43]. So is EF, which usually remains low, even in well-treated, asymptomatic patients. Improved EF under treatment is associated with better outcome [44]; however, whether treatments should be monitored through EF is unknown. Markers of sodium overload are also crude (peripheral edema), and quantification of central edema (by pulmonary ultrasound scans for instance) have proven to be useful in clinical studies [45], but is not currently recommended or used. Even monitoring of weight and salt restriction have proven to be useful [46], but is quite difficult to implement and seldom done. Novel biomarkers might be useful. For instance, arterial stiffness is increased in HF patients [47]; it is a strong predictor of incident heart failure [48]. We have shown that arterial stiffness could be used to improve treatment of hypertension[49]. It would be interesting to see whether titrating drugs using arterial stiffness could improve prognosis.

26.9 Role of Telemedicine in Treatment of HF

Patients or caregivers empowerment is a powerfull tool for adjusting therapies during real life according to symptoms or simple biomarkers. HF was (beside type 1 diabetes) the first application of telemedicine. Telehealth-mediated self-management was not consistently superior to usual care; none of the reviews reported adverse effects, suggesting that telehealth is a safe option for providing self-management support, particularly in conditions such as heart failure, where the evidence base is more developed. Larger-scale trials of telehealth-supported self-management based on explicit self-management theory are needed before the extent to which telehealth technologies can be harnessed to support self-management can be established [50].

26.10 Prevention of Heart Failure

Adequate treatment of cardiovascular risk factors for MI (lipids, hypertension, smoking, sedentarity), and more specifically for HFpEF (hypertension) is mandatory for prevention of HF [51]). Similarly, the improvement in prognosis in the HYVET (hypertension control in very elderly) trial was mainly due to a reduction in stroke (39%) and heart failure (−64%), and mainly imputable to indapamide (a thiazide-like diuretic) [52]. Thus, treating hypertension correctly with strict BP targets is the most effective means of decreasing heart failure among other CV diseases, and apparently with no lower threshold for treatment [53]. If combined with the prevention of ischemic heart disease (through smoking cessation, physical activity, reduction in blood lipids, control of obesity), this could lead to a massive reduction in cardiovascular disease as a whole, and particularly heart failure [54]. This cannot be achieved by doctors and caregivers alone, but must involve all strata of society to change mentalities, starting with education and legislation.

26.11 Women and Heart Failure

Women have particular forms of heart failure. Some causes of cardiomyopathies are specific to women (postpartum, Takotsubo [55], spontaneous coronary dissection) [56]. Cardiac remodeling following hypertension is more often concentric and hypertrophic; thus later heart failure is more often pEF than rEF. Acute coronary syndromes in women are treated later, less aggressively, and lead to larger MI. Despite this, women are still massively underrepresented in heart failure trials [56], and since HFpEF is the main form of heart failure in women, it is less effectively treated. Here again, massive changes in mentality, starting in every person, for considering the specificities of women in medicine are needed for promoting equality in care between sexes.

26.12 Heart Failure in Developing Countries

The burden of cardiovascular diseases becomes the major health issue in developing countries. The rise of CV risk factors, including hypertension, obesity, and type 2 diabetes, is very fast because of changes in lifestyle [53, 56]. Heart failure due to malignant hypertension is common in young patients, especially in black patients, [57]. Life style and economic status do not explain wholly the difference. Racism further amplifies the vicious circle by exacerbating hypertension [58], adherence to treatment [59], and long-term consequences [60]. What is needed is a change of paradigm worldwide, changing from a rampant racial discrimination to a true equipoise. Reduced access to drug therapy is amplified in, but not specific to developing countries, poorer and more socially deprived populations being more exposed [61]. Proactive primordial and primary prevention targeted to high-risk populations in developing and developed countries is the only way to durably reduce the burden of hypertension and HF [62].

26.13 Conclusion

Early identification and effective treatment of CV risk factors associated with HF, principally hypertension, blood lipids, and smoking, should lead to a massive decrease in future heart failure. Once HF is established, implementation of medical treatment in HFrEF and HFpEF patients is often difficult because of patient characteristics, including physiologic parameters, and comorbidities, including blood pressure, limiting the use of vital medications. Patient phenotyping can guide the personalization of pharmacotherapy to deliver matched treatment to patients while using all drug classes that have been shown to be effective in improving prognosis. However, the best combination of biomarkers (either clinical, biological, and imaging) and the best combination of drugs is unknown. Future clinical trials should reevaluate treatment algorithms for HFrEF and HFpEF patients to allow rapid implementation of all HF drugs into treatment. In the future, innovative, pragmatic clinical trials will be needed to determine the best strategy for drug use guided by the ideal combination of biomarkers. Advanced therapies, such as resynchronization, implantable defibrillators, transplantation, or circulation assistance should be considered as failures of prevention, rather than achievements. As compared with CV risk factors, complex treatment regimens are associated with poor adherence and may be improved by combining treatment in a polypill. A similar approach could be used for polychemotherapy in HFrEF and HFpEF so that patients could benefit from the theoretical maximum increase in life expectancy.

References

1. Ho JE, Zern EK, Wooster L, Bailey CS, Cunningham T, Eisman AS, Hardin KM, Zampierollo GA, Jarolim P, Pappagianopoulos PP, Malhotra R, Nayor M, Lewis GD. Differential clinical profiles, exercise responses, and outcomes associated with existing HFpEF definitions. Circulation. 2019;140:353–65.
2. Maddox TM, Januzzi JL, Allen LA, Breathett K, Butler J, Davis LL, Fonarow GC, Ibrahim NE, Lindenfeld J, Masoudi FA, Motiwala SR, Oliveros E, Patterson JH, Walsh MN, Wasserman A, Yancy CW, Youmans QR. 2021 update to the 2017 ACC expert consensus decision pathway for optimization of heart failure treatment: answers to 10 pivotal issues about heart failure with reduced ejection fraction. J Am Coll Cardiol. 2021;77:772–810.
3. Pfeffer MA, Shah AM, Borlaug BA. Heart failure with preserved ejection fraction in perspective. Circ Res. 2019;124:1598–617.
4. The Consensus Trial Study Group. Effects of Enalapril on mortality in severe congestive heart failure. N Engl J Med. 1987;316:1429–35.
5. Annane D, Bellissant E, Pussard E, Asmar R, Lacombe F, Lanata E, Madonna O, Safar M, Giudicelli J, Gajdos P. Placebo-controlled, randomized, double-blind study of intravenous enalaprilat efficacy and safety in acute cardiogenic pulmonary edema. Circulation. 1996;94:1316–24.
6. McMurray JJV, Solomon SD, Inzucchi SE, Køber L, Kosiborod MN, Martinez FA, Ponikowski P, Sabatine MS, Anand IS, Bělohlávek J, Böhm M, Chiang C-E, Chopra VK, de Boer RA, Desai AS, Diez M, Drozdz J, Dukát A, Ge J, Howlett JG, Katova T, Kitakaze M, Ljungman CEA, Merkely B, Nicolau JC, O'Meara E, Petrie MC, Vinh PN, Schou M, Tereshchenko S, Verma S, Held C, DeMets DL, Docherty KF, Jhund PS, Bengtsson O, Sjöstrand M, Langkilde A-M. Dapagliflozin in patients with heart failure and reduced ejection fraction. N Engl J Med. 2019;381:1995–2008.
7. Tromp J, Ouwerkerk W, van Veldhuisen DJ, Hillege HL, Richards AM, van der Meer P, Anand IS, Lam CSP, Voors AA. A systematic review and network meta-analysis of pharmacological treatment of heart failure with reduced ejection fraction. JACC Heart Fail. 2022;10:73–84.
8. Clark KAA, Reinhardt SW, Chouairi F, Miller PE, Kay B, Fuery M, Guha A, Ahmad T, Desai NR. Trends in heart failure hospitalizations in the US from 2008 to 2018. J Card Fail. 2022;28:171–80.
9. Yusuf S, Pfeffer MA, Swedberg K, Granger CB, Held P, McMurray JJV, Michelson EL, Olofsson B, Ostergren J, CHARM Investigators and Committees. Effects of candesartan in patients with chronic heart failure and preserved left-ventricular ejection fraction: the CHARM-preserved trial. Lancet Lond Engl. 2003;362:777–81.
10. Zheng SL, Chan FT, Nabeebaccus AA, Shah AM, McDonagh T, Okonko DO, Ayis S. Drug treatment effects on outcomes in heart failure with preserved ejection fraction: a systematic review and meta-analysis. Heart. 2018;104:407–15.
11. McMurray JJV, Packer M, Desai AS, Gong J, Lefkowitz MP, Rizkala AR, Rouleau JL, Shi VC, Solomon SD, Swedberg K, Zile MR, PARADIGM-HF Investigators and Committees. Angiotensin-neprilysin inhibition versus enalapril in heart failure. N Engl J Med. 2014;371:993–1004.
12. Velazquez EJ, Morrow DA, DeVore AD, Duffy CI, Ambrosy AP, McCague K, Rocha R, Braunwald E. Angiotensin–neprilysin inhibition in acute decompensated heart failure. N Engl J Med. 2019;380:539–48.
13. Solomon SD, McMurray JJV, Anand IS, Ge J, Lam CSP, Maggioni AP, Martinez F, Packer M, Pfeffer MA, Pieske B, Redfield MM, Rouleau JL, van Veldhuisen DJ, Zannad F, Zile MR, Desai AS, Claggett B, Jhund PS, Boytsov SA, Comin-Colet J, Cleland J, Düngen H-D, Goncalvesova E, Katova T, Kerr Saraiva JF, Lelonek M, Merkely B, Senni M, Shah SJ, Zhou J, Rizkala AR, Gong J, Shi VC, Lefkowitz MP. Angiotensin–neprilysin inhibition in heart failure with preserved ejection fraction. N Engl J Med. 2019;381:1609–20.
14. Kostis JB, Packer M, Black HR, Schmieder R, Henry D, Levy E. Omapatrilat and enalapril in patients with hypertension: the Omapatrilat cardiovascular treatment vs. Enalapril (OCTAVE) trial. Am J Hypertens. 2004;17:103–11.

15. Pitt B, Zannad F, Remme WJ, Cody R, Castaigne A, Perez A, Palensky J, Wittes J. The effect of spironolactone on morbidity and mortality in patients with severe heart failure. N Engl J Med. 1999;341:709–17.
16. Pitt B, Pfeffer MA, Assmann SF, Boineau R, Anand IS, Claggett B, Clausell N, Desai AS, Diaz R, Fleg JL, Gordeev I, Harty B, Heitner JF, Kenwood CT, Lewis EF, O'Meara E, Probstfield JL, Shaburishvili T, Shah SJ, Solomon SD, Sweitzer NK, Yang S, McKinlay SM. Spironolactone for heart failure with preserved ejection fraction. N Engl J Med. 2014;370:1383–92.
17. Zannad F, JJV MM, Krum H, van Veldhuisen DJ, Swedberg K, Shi H, Vincent J, Pocock SJ, Pitt B, EMPHASIS-HF Study Group. Eplerenone in patients with systolic heart failure and mild symptoms. N Engl J Med. 2011;364:11–21.
18. Williams B, Mancia G, Spiering W, Agabiti Rosei E, Azizi M, Burnier M, Clement DL, Coca A, de Simone G, Dominiczak A, Kahan T, Mahfoud F, Redon J, Ruilope L, Zanchetti A, Kerins M, Kjeldsen SE, Kreutz R, Laurent S, Lip GYH, McManus R, Narkiewicz K, Ruschitzka F, Schmieder RE, Shlyakhto E, Tsioufis C, Aboyans V, Desormais I, ESC Scientific Document Group, De Backer G, Heagerty AM, Agewall S, Bochud M, Borghi C, Boutouyrie P, Brguljan J, Bueno H, Caiani EG, Carlberg B, Chapman N, Cífková R, Cleland JGF, Collet J-P, Coman IM, de Leeuw PW, Delgado V, Dendale P, Diener H-C, Dorobantu M, Fagard R, Farsang C, Ferrini M, Graham IM, Grassi G, Haller H, Hobbs FDR, Jelakovic B, Jennings C, Katus HA, Kroon AA, Leclercq C, Lovic D, Lurbe E, Manolis AJ, McDonagh TA, Messerli F, Muiesan ML, Nixdorff U, Olsen MH, Parati G, Perk J, Piepoli MF, Polonia J, Ponikowski P, Richter DJ, Rimoldi SF, Roffi M, Sattar N, Seferovic PM, Simpson IA, Sousa-Uva M, Stanton AV, van de Borne P, Vardas P, Volpe M, Wassmann S, Windecker S, Zamorano JL, Windecker S, Aboyans V, Agewall S, Barbato E, Bueno H, Coca A, Collet J-P, Coman IM, Dean V, Delgado V, et al. 2018 ESC/ESH guidelines for the management of arterial hypertension. Eur Heart J. 2018;39:3021–104.
19. Juurlink DN, Mamdani MM, Lee DS, Kopp A, Austin PC, Laupacis A, Redelmeier DA. Rates of hyperkalemia after publication of the randomized Aldactone evaluation study. N Engl J Med. 2004;351:543–51.
20. Bakris GL, Agarwal R, Anker SD, Pitt B, Ruilope LM, Rossing P, Kolkhof P, Nowack C, Schloemer P, Joseph A, Filippatos G. Effect of Finerenone on chronic kidney disease outcomes in type 2 diabetes. N Engl J Med. 2020;383:2219–29.
21. Packer M, Anker SD, Butler J, Filippatos G, Pocock SJ, Carson P, Januzzi J, Verma S, Tsutsui H, Brueckmann M, Jamal W, Kimura K, Schnee J, Zeller C, Cotton D, Bocchi E, Böhm M, Choi D-J, Chopra V, Chuquiure E, Giannetti N, Janssens S, Zhang J, Gonzalez Juanatey JR, Kaul S, Brunner-La Rocca H-P, Merkely B, Nicholls SJ, Perrone S, Pina I, Ponikowski P, Sattar N, Senni M, Seronde M-F, Spinar J, Squire I, Taddei S, Wanner C, Zannad F. Cardiovascular and renal outcomes with Empagliflozin in heart failure. N Engl J Med. 2020;383:1413–24.
22. Solomon SD, JJV MM, Claggett B, de Boer RA, DeMets D, Hernandez AF, Inzucchi SE, Kosiborod MN, CSP L, Martinez F, Shah SJ, Desai AS, Jhund PS, Belohlavek J, Chiang C-E, CJW B, Comin-Colet J, Dobreanu D, Drozdz J, Fang JC, Alcocer-Gamba MA, Al Habeeb W, Han Y, Cabrera Honorio JW, Janssens SP, Katova T, Kitakaze M, Merkely B, O'Meara E, JFK S, Tereshchenko SN, Thierer J, Vaduganathan M, Vardeny O, Verma S, Pham VN, Wilderäng U, Zaozerska N, Bachus E, Lindholm D, Petersson M, Langkilde AM. Dapagliflozin in heart failure with mildly reduced or preserved ejection fraction. N Engl J Med. 2022;387(12):1089–98. https://doi.org/10.1056/NEJMoa2206286.
23. Chiang B, Chew DP, De Pasquale CG. Outcome trial data on sodium glucose cotransporter-2 inhibitors: putting clinical benefits and risks in perspective. Int J Cardiol. 2022;349:96–8.
24. Zelniker TA, Braunwald E. Mechanisms of cardiorenal effects of sodium-glucose cotransporter 2 inhibitors. J Am Coll Cardiol. 2020;75:422–34.
25. Uthman L, Baartscheer A, Schumacher CA, Fiolet JWT, Kuschma MC, Hollmann MW, Coronel R, Weber NC, Zuurbier CJ. Direct cardiac actions of sodium glucose cotransporter 2 inhibitors target pathogenic mechanisms underlying heart failure in diabetic patients. Front Physiol. 2018;9:1575.
26. Yamamoto T, Sano M. Deranged myocardial fatty acid metabolism in heart failure. Int J Mol Sci. 2022;23:996.

27. Adingupu DD, Göpel SO, Grönros J, Behrendt M, Sotak M, Miliotis T, Dahlqvist U, Gan L-M, Jönsson-Rylander A-C. SGLT2 inhibition with empagliflozin improves coronary microvascular function and cardiac contractility in prediabetic Ob/Ob−/− mice. Cardiovasc Diabetol. 2019;18:16.
28. Solini A, Giannini L, Seghieri M, Vitolo E, Taddei S, Ghiadoni L, Bruno RM. Dapagliflozin acutely improves endothelial dysfunction, reduces aortic stiffness and renal resistive index in type 2 diabetic patients: a pilot study. Cardiovasc Diabetol. 2017;16:138.
29. Cardiac Arrhythmia Suppression Trial (CAST) Investigators. Preliminary report: effect of encainide and flecainide on mortality in a randomized trial of arrhythmia suppression after myocardial infarction. N Engl J Med. 1989;321:406–12.
30. Greene SJ, Butler J, Albert NM, DeVore AD, Sharma PP, Duffy CI, Hill CL, McCague K, Mi X, Patterson JH, Spertus JA, Thomas L, Williams FB, Hernandez AF, Fonarow GC. Medical therapy for heart failure with reduced ejection fraction. J Am Coll Cardiol. 2018;72:351–66.
31. Ahmad T, Miller PE, McCullough M, Desai NR, Riello R, Psotka M, Böhm M, Allen LA, Teerlink JR, Rosano GMC, Lindenfeld J. Why has positive inotropy failed in chronic heart failure? Lessons from prior inotrope trials. Eur J Heart Fail. 2019;21:1064–78.
32. Teerlink JR, Diaz R, Felker GM, JJV MM, Metra M, Solomon SD, Adams KF, Anand I, Arias-Mendoza A, Biering-Sørensen T, Böhm M, Bonderman D, JGF C, Corbalan R, Crespo-Leiro MG, Dahlström U, Echeverria LE, Fang JC, Filippatos G, Fonseca C, Goncalvesova E, Goudev AR, Howlett JG, Lanfear DE, Li J, Lund M, Macdonald P, Mareev V, Momomura S-I, O'Meara E, Parkhomenko A, Ponikowski P, FJA R, Serpytis P, Sliwa K, Spinar J, Suter TM, Tomcsanyi J, Vandekerckhove H, Vinereanu D, Voors AA, Yilmaz MB, Zannad F, Sharpsten L, Legg JC, Varin C, Honarpour N, Abbasi SA, Malik FI, Kurtz CE, GALACTIC-HF Investigators. Cardiac myosin activation with omecamtiv mecarbil in systolic heart failure. N Engl J Med. 2021;384:105–16.
33. Chaugai S, Meng WY, Ali SA. Effects of RAAS blockers on atrial fibrillation prophylaxis: an updated systematic review and meta-analysis of randomized controlled trials. J Cardiovasc Pharmacol Ther. 2016;21:388–404.
34. Ma H, Jiang H, Feng J, Gan Y. Angiotensin receptor blocker and calcium channel blocker preventing atrial fibrillation recurrence in patients with hypertension and atrial fibrillation: a meta-analysis. Cardiovasc Ther. 2021;2021:1–8.
35. Li W, Chen X, Xu L, Li Y, Luo B. SGLT2 inhibitors and atrial fibrillation in type 2 diabetes: a systematic review with meta-analysis of 16 randomized controlled trials. Cardiovasc Diabetol. 2020;19:130.
36. Burnier M, Egan BM. Adherence in hypertension. Circ Res. 2019;124:1124–40.
37. Kably B, Billaud EM, Boutouyrie P, Azizi M. Is there any hope for monitoring adherence in an efficient and feasible way for resistant hypertension diagnosis and follow-up. Curr Hypertens Rep. 2020;22:22.
38. Andrews AM, Russell CL, Cheng A-L. Medication adherence interventions for older adults with heart failure: a systematic review. J Gerontol Nurs. 2017;43:37–45.
39. Krueger K, Griese-Mammen N, Schubert I, Kieble M, Botermann L, Laufs U, Kloft C, Schulz M. In search of a standard when analyzing medication adherence in patients with heart failure using claims data: a systematic review. Heart Fail Rev. 2018;23:63–71.
40. Simpson J, Jackson CE, Haig C, Jhund PS, Tomaszewski M, Gardner RS, Tsorlalis Y, Petrie MC, McMurray JJV, Squire IB, Gupta P. Adherence to prescribed medications in patients with heart failure: insights from liquid chromatography-tandem mass spectrometry-based urine analysis. Eur Heart J Cardiovasc Pharm. 2021;7:296–301.
41. Rosano GMC, Allen LA, Abdin A, Lindenfeld J, O'Meara E, Lam CSP, Lancellotti P, Savarese G, Gottlieb SS, Teerlink J, Wintrich J, Böhm M. Drug layering in heart failure: phenotype-guided initiation. JACC Heart Fail. 2021;9:775–83.
42. Kang S-H, Park JJ, Choi D-J, Yoon C-H, Oh I-Y, Kang S-M, Yoo B-S, Jeon E-S, Kim J-J, Cho M-C, Chae SC, Ryu K-H, Oh B-H. Prognostic value of NT-proBNP in heart failure with preserved versus reduced EF. Heart. 2015;101:1881–8.
43. Persson H, Erntell H, Eriksson B, Johansson G, Swedberg K, Dahlström U. Improved pharmacological therapy of chronic heart failure in primary care: a randomized study of NT-proBNP

guided management of heart failure–SIGNAL-HF (Swedish intervention study—guidelines and NT-proBNP AnaLysis in heart failure). Eur J Heart Fail. 2010;12:1300–8.
44. Ghimire A, Fine N, Ezekowitz JA, Howlett J, Youngson E, McAlister FA. Frequency, predictors, and prognosis of ejection fraction improvement in heart failure: an echocardiogram-based registry study. Eur Heart J. 2019;40:2110–7.
45. Mhanna M, Beran A, Nazir S, Sajdeya O, Srour O, Ayesh H, Eltahawy EA. Lung ultrasound–guided management to reduce hospitalization in chronic heart failure: a systematic review and meta-analysis. Heart Fail Rev. 2022;27:821–6.
46. Philipson H, Ekman I, Forslund HB, Swedberg K, Schaufelberger M. Salt and fluid restriction is effective in patients with chronic heart failure. Eur J Heart Fail. 2013;15:1304–10.
47. Marti CN, Gheorghiade M, Kalogeropoulos AP, Georgiopoulou VV, Quyyumi AA, Butler J. Endothelial dysfunction, arterial stiffness, and heart failure. J Am Coll Cardiol. 2012;60:1455–69.
48. Vasan RS, Pan S, Xanthakis V, Beiser A, Larson MG, Seshadri S, Mitchell GF. Arterial stiffness and long-term risk of health outcomes: the Framingham heart study. Hypertens. 1979;2022(79):1045–56.
49. Laurent S, Chatellier G, Azizi M, Calvet D, Choukroun G, Danchin N, Delsart P, Girerd X, Gosse P, Khettab H, London G, Mourad J-J, Pannier B, Pereira H, Stephan D, Valensi P, Cunha P, Narkiewicz K, Bruno R-M, Boutouyrie P. SPARTE study: normalization of arterial stiffness and cardiovascular events in patients with hypertension at medium to very high risk. Hypertens. 1979;2021(78):983–95.
50. Hanlon P, Daines L, Campbell C, McKinstry B, Weller D, Pinnock H. Telehealth interventions to support self-management of long-term conditions: a systematic metareview of diabetes, heart failure, asthma, chronic obstructive pulmonary disease, and cancer. J Med Internet Res. 2017;19:e172.
51. The SPRINT Research Group. A randomized trial of intensive versus standard blood-pressure control. N Engl J Med. 2015;373:2103–16.
52. Beckett NS, Peters R, Fletcher AE, Staessen JA, Liu L, Dumitrascu D, Stoyanovsky V, Antikainen RL, Nikitin Y, Anderson C, Belhani A, Forette F, Rajkumar C, Thijs L, Banya W, Bulpitt CJ. Treatment of hypertension in patients 80 years of age or older. N Engl J Med. 2008;358:1887–98.
53. Rahimi K, Bidel Z, Nazarzadeh M, Copland E, Canoy D, Wamil M, Majert J, McManus R, Adler A, Agodoa L, Algra A, Asselbergs FW, Beckett NS, Berge E, Black H, Boersma E, Brouwers FPJ, Brown M, Brugts JJ, Bulpitt CJ, Byington RP, Cushman WC, Cutler J, Devereaux RB, Dwyer JP, Estacio R, Fagard R, Fox K, Fukui T, Gupta AK, Holman RR, Imai Y, Ishii M, Julius S, Kanno Y, Kjeldsen SE, Kostis J, Kuramoto K, Lanke J, Lewis E, Lewis JB, Lievre M, Lindholm LH, Lueders S, MacMahon S, Mancia G, Matsuzaki M, Mehlum MH, Nissen S, Ogawa H, Ogihara T, Ohkubo T, Palmer CR, Patel A, Pfeffer MA, Pitt B, Poulter NR, Rakugi H, Reboldi G, Reid C, Remuzzi G, Ruggenenti P, Saruta T, Schrader J, Schrier R, Sever P, Sleight P, Staessen JA, Suzuki H, Thijs L, Ueshima K, Umemoto S, van Gilst WH, Verdecchia P, Wachtell K, Whelton P, Wing L, Woodward M, Yui Y, Yusuf S, Zanchetti A, Zhang Z-Y, Anderson C, Baigent C, Brenner BM, Collins R, de Zeeuw D, Lubsen J, Malacco E, Neal B, Perkovic V, Rodgers A, Rothwell P, Salimi-Khorshidi G, Sundström J, Turnbull F, Viberti G, Wang J, et al. Age-stratified and blood-pressure-stratified effects of blood-pressure-lowering pharmacotherapy for the prevention of cardiovascular disease and death: an individual participant-level data meta-analysis. Lancet. 2021;398:1053–64.
54. Lonn E, Bosch J, Teo KK, Pais P, Xavier D, Yusuf S. The polypill in the prevention of cardiovascular diseases: key concepts, current status, challenges, and future directions. Circulation. 2010;122:2078–88.
55. Templin C, Ghadri JR, Diekmann J, Napp LC, Bataiosu DR, Jaguszewski M, Cammann VL, Sarcon A, Geyer V, Neumann CA, Seifert B, Hellermann J, Schwyzer M, Eisenhardt K, Jenewein J, Franke J, Katus HA, Burgdorf C, Schunkert H, Moeller C, Thiele H, Bauersachs J, Tschöpe C, Schultheiss H-P, Laney CA, Rajan L, Michels G, Pfister R, Ukena C, Böhm M, Erbel R, Cuneo A, Kuck K-H, Jacobshagen C, Hasenfuss G, Karakas M, Koenig W, Rottbauer

W, Said SM, Braun-Dullaeus RC, Cuculi F, Banning A, Fischer TA, Vasankari T, Airaksinen KEJ, Fijalkowski M, Rynkiewicz A, Pawlak M, Opolski G, Dworakowski R, MacCarthy P, Kaiser C, Osswald S, Galiuto L, Crea F, Dichtl W, Franz WM, Empen K, Felix SB, Delmas C, Lairez O, Erne P, Bax JJ, Ford I, Ruschitzka F, Prasad A, Lüscher TF. Clinical features and outcomes of Takotsubo (stress) cardiomyopathy. N Engl J Med. 2015;373:929–38.
56. Vogel B, Acevedo M, Appelman Y, Bairey Merz CN, Chieffo A, Figtree GA, Guerrero M, Kunadian V, Lam CSP, Maas AHEM, Mihailidou AS, Olszanecka A, Poole JE, Saldarriaga C, Saw J, Zühlke L, Mehran R. The lancet women and cardiovascular disease commission: reducing the global burden by 2030. Lancet. 2021;397:2385–438.
57. Carnethon MR, Pu J, Howard G, Albert MA, CAM A, Bertoni AG, Mujahid MS, Palaniappan L, Taylor HA, Willis M, Yancy CW. Cardiovascular Health in African Americans: a scientific statement from the American Heart Association. Circulation. 2017;136:e393. https://doi.org/10.1161/CIR.0000000000000534.
58. Brondolo E, Love EE, Pencille M, Schoenthaler A, Ogedegbe G. Racism and hypertension: a review of the empirical evidence and implications for clinical practice. Am J Hypertens. 2011;24:518–29.
59. Daugherty SL, Helmkamp L, Vupputuri S, Hanratty R, Steiner JF, Blair IV, Dickinson LM, Maertens JA, Havranek EP. Effect of values affirmation on reducing racial differences in adherence to hypertension medication: the HYVALUE randomized clinical Trial. JAMA Netw Open. 2021;4:e2139533.
60. Cruickshank JK, Silva MJ, Molaodi OR, Enayat ZE, Cassidy A, Karamanos A, Read UM, Faconti L, Dall P, Stansfield B, Harding S. Ethnic differences in and childhood influences on early adult pulse wave velocity: the determinants of adolescent, now young adult, social well-being, and health longitudinal study. Hypertens. 1979;2016(67):1133–41.
61. Psotka MA, Fiuzat M, Solomon SD, Chauhan C, Felker GM, Butler J, Teerlink JR, Sinha SS, O'Connor CM, Konstam MA. Challenges and potential improvements to patient access to pharmaceuticals: examples from cardiology. Circulation. 2020;142:790–8.
62. Yusuf S, Joseph P, Rangarajan S, Islam S, Mente A, Hystad P, Brauer M, Kutty VR, Gupta R, Wielgosz A, AlHabib KF, Dans A, Lopez-Jaramillo P, Avezum A, Lanas F, Oguz A, Kruger IM, Diaz R, Yusoff K, Mony P, Chifamba J, Yeates K, Kelishadi R, Yusufali A, Khatib R, Rahman O, Zatonska K, Iqbal R, Wei L, Bo H, Rosengren A, Kaur M, Mohan V, Lear SA, Teo KK, Leong D, O'Donnell M, McKee M, Dagenais G. Modifiable risk factors, cardiovascular disease, and mortality in 155 722 individuals from 21 high-income, middle-income, and low-income countries (PURE): a prospective cohort study. Lancet Lond Engl. 2020;395:795–808.

Device-Based Treatment in Hypertension and Heart Failure

27

Seravalle Gino and Guido Grassi

27.1 Introduction

Arterial hypertension and heart failure are two pathophysiologic aspects of the cardiovascular continuum in which uncontrolled blood pressure (BP) represents the first actor in the majority of the cases.

A significant number of observations have clearly shown that the progressive increase in blood pressure levels, starting from high normal values to resistant hypertension, is accompanied by a progressive increase in the adrenergic tone [1–4] (Fig. 27.1). A further increase in the adrenergic tone is evident when the hypertensive state is associated with end organ damage and in particular when left ventricular dimensions increase inducing an impairment in cardiac function [5, 6].

In the last decades despite growing awareness of the importance of blood pressure goals and therapies to improve the clinical course of heart failure patients, morbidity and mortality and healthcare costs remain unacceptably high [7]. A significant number of hypertensive subjects fail to achieve adequate BP control despite adherence to maximal doses of several antihypertensive drugs and combination treatments. More intriguing is the recent interest for invasive procedures capable of reducing the pressor or to increase the depressor influences that physiologically modulate BP. A reduction of the pressure influences can be obtained by renal artery denervation that allows to reduce vasoconstriction in a district that accounts for a

S. Gino (✉)
Cardiology Department, IRCCS Istituto Auxologico Italiano, S. Luca Hospital, Milano, Italy

G. Grassi
University of Milano-Bicocca, Milano, Italy
e-mail: guido.grassi@unimib.it

M. Dorobantu et al. (eds.), *Hypertension and Heart Failure*, Updates in Hypertension and Cardiovascular Protection,
https://doi.org/10.1007/978-3-031-39315-0_27

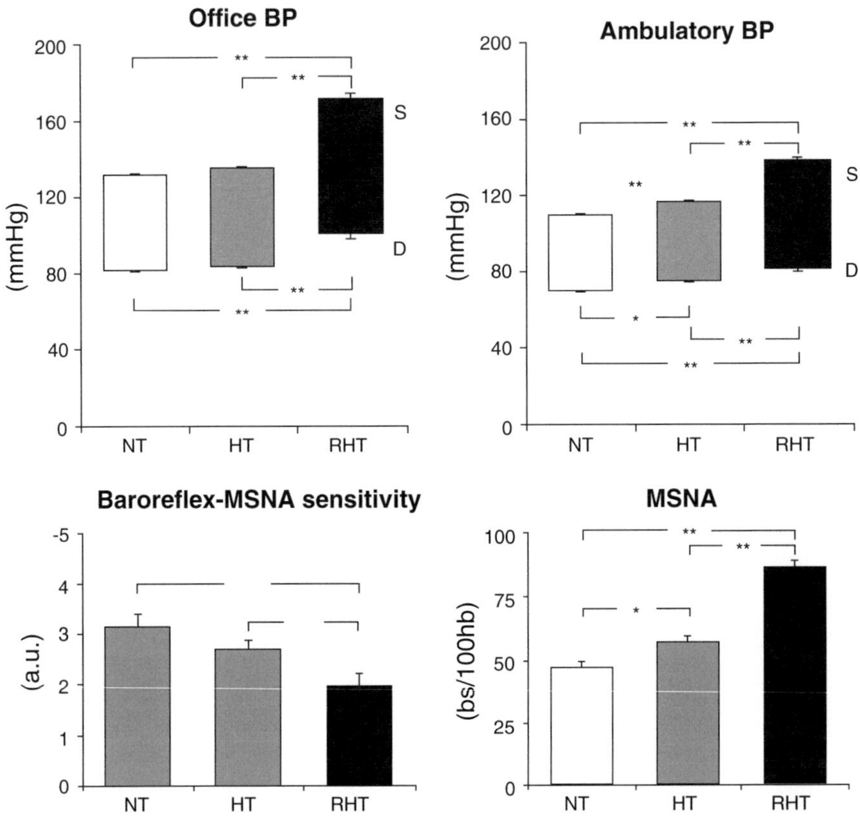

Fig. 27.1 Office and ambulatory blood pressure (BP), muscle sympathetic nerve activity (MSNA) and baroreflex-MSNA sensitivity in normotensive (NT), non-resistant essential hypertensive (HT) and resistant hypertensive (RHT) patients. Data expressed as mean ± SEM. Asterisks refer to the statistical significance between groups: *$p < 0.05$, **$p < 0.01$. (From Ref [19], with permission)

considerable proportion of systemic vascular resistance [8]. An enhancement of the depressor influences can be obtained by stimulation of carotid baroreceptors [9], thus inducing beneficial effects not only in patients with hypertension [10–12] but also in heart failure patients [13–15]. Heart failing has been the center of several nonpharmacological approaches from stimulation-based therapies to mechanical circulatory support devices. This chapter will briefly review the different device-based approaches to treat resistant hypertension and heart failure patients.

27.2 Devices-Based Treatment in Hypertension

The interest of devices-based treatment in the hypertensive state, in particular in resistant hypertension, comes from the evidence of a marked increase in the sympathetic activity associated with this condition. Activation of efferent sympathetic nerves to the kidneys stimulates renin release, enhances tubular reabsorption of sodium and water, and decreases renal blood flow [16]. It has been also observed that activation of afferent nerves increases central nervous system outflow in key organs [17]. On the other side, a renewed interest has been focused on the possibility to act on carotid sinus baroreceptors to obtain a reduction in sympathetic tone and a reduction in BP values [18, 19].

27.2.1 Renal Denervation

Catheter-based renal denervation (RDN) has been introduced in the first decade of the new millennium as a new interventional approach targeting sympathetic nerve activity in hypertensive subjects. The first non-randomized, proof-of-concept study (Symplicity HTN-1) [8] on resistant hypertensives allowed to observe a BP reduction of 27/17 mmHg at 6 months from the RDN procedure. The second randomized controlled study (Symplicity HTN-2) [20] was performed in subjects with office BP ≥ 160 mmHg after a 2-week observation period. The active group showed a BP reduction of 32/12 mmHg. A 36 months follow-up of both studies showed that BP reduction was maintained. This suggested that along this period a reinnervation did not occur [21, 22]. To overcome the deficiencies of these studies (lack of 24-h BP monitoring, unblinded design, white coat effect not excluded), a new prospective, single-blind, randomized, sham-controlled study (Symplicity HTN-3) was performed [23]. This study confirmed the safety of the procedure but did not reach the end point of BP reduction. BP was significantly reduced from baseline but active and sham groups did not show significant differences at 6 months follow-up [23]. A possible interpretation of this negative result could be due to the reduced number of patients with a complete 4 quadrant ablation [24]. There were also differences in the population enrolled with a great amount of African Americans in this third study. It is well known the difference in the response to BP treatment according to race suggesting different antihypertensive drug strategies in African American versus other racial groups [25]. Studies performed more recently with different catheters capable of reducing catheter manipulation and decreasing procedure time obtained similar BP effects. It is necessary to remember that most of these trials were uncontrolled (Fig. 27.2) [8, 20, 26–38]. The technical approaches to RDN are constantly in evolution with different success and complication rates. Catheters have been developed to introduce microinjection of neurotoxin (e.g., alcohol) to chemically ablate renal nerves, facilitating deeper nerve injury while avoiding damage to endothelium and intimal layers [39]. RDN can be also delivered via transurethral approach to ablate the richly innervated renal pelvis [40]. A wholly noninvasive system based on extracorporeal high-intensity focused ultrasound has been tested in the WAVE trials [41] showing in the WAVE III, with an improved protocol of energy delivered, an office BP reduction of 29.6/11.8 mmHg 3 months after the procedure. This result has not been confirmed at 24 weeks when compared to sham group in the WAVE IV [42].

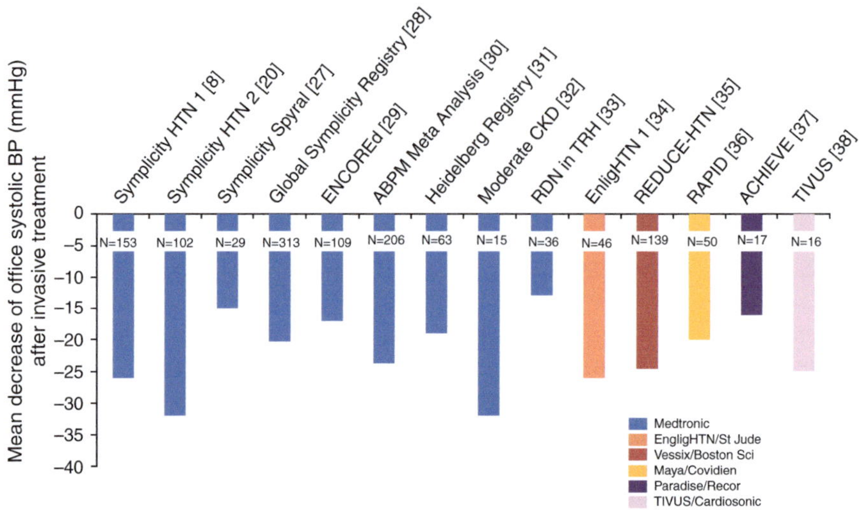

Fig. 27.2 Studies of the effects of renal denervation on blood pressure. (Modified from Ref [26], with permission)

27.2.2 Baroreflex Activation Therapy

Due to the pathogenetic role of adrenergic tone in the hypertensive state, it has been suggested that a sympathetic deactivation should be one of the major goals of hypertensive therapy [1–6]. In the past the role of the carotid baroreflex was demonstrated for short-term BP regulation, but continued refinements in implantable medical technology have led to a renewal of interest in carotid sinus stimulation for long-term BP control [18, 19].

The first generation device (Rheos System; CVRx Inc., Minneapolis, MN) requiring bilateral electrode stimulation was evaluated in a prospective non-randomized trial (DEBuT-HT). Eighteen resistant hypertensives of the 45 enrolled successfully completed the 58 ± 6 months follow-up [43]. Office systolic and diastolic BP showed a progressive and significant reduction (21 ± 4/12 ± 2 mmHg, 30 ± 6/20 ± 4 mmHg, 33 ± 48/22 ± 26 mmHg at 3, 12, and 24 months, respectively). This was accompanied by some other positive results: (1) a significant reduction was observed for 24-h BP monitoring, (2) 72% of the patients showed a reduction of more than 30 mmHg in systolic blood pressure at 4 years, (3) 67% of the patients measuring SBP < 140 mmHg at the end of the follow-up, (4) a reduction in left ventricular mass index (−24.1 ± 18.7 g/m^2) paralleled the BP reduction. A substudy of 12 patients in which adrenergic tone was directly measured with microneurographic approach allowed to observe a reduction in sympathetic tone and BP during activation of the device providing evidence that reduction of sympathetic outflow is the primary mechanism for BP reduction with baroreflex activation therapy (BAT) [44]. The Rheos Pivotal Trial [10] was a double-blind, randomized, parallel design study that enrolled 256 resistant hypertension patients.

The post hoc analysis utilizing change from pre-implant showed a decrease in SBP of 26 ± 30 mmHg and 35 ± 28 mmHg at 6 and 12 months, respectively. The procedural primary safety end point was not met, mainly because of surgical complications, but the criteria for BAT and device safety were met. A persistent BP reduction (−36/−16 mmHg) was also observed in a group of participants that continued in an open-label, non-randomized follow-up for an average of 28 months [45]. The new unilateral and less invasive system, Barostim neo, has been used starting from 2010. The reduction in BP was superimposable to that observed with the first generation device: −26.0 ± 4.4/−12.4 ± 2.5 mmHg in office BP after 6 months, −7/−5 mmHg at 24-h BP monitoring and 43% of patients reaching an office SBP < 140 mmHg [12, 46].

27.2.3 Endovascular Baroreflex Activation

A different approach for stimulating the carotid baroreceptor has been developed in the USA. This consists of a self-expanding nitinol system placed into the internal carotid artery by a special delivery catheter and capable of increasing the wall strain or vessel stretch during the systolic phase while in the diastolic phase of the cardiac cycle it returns to the baseline; the net result is an increased carotid sinus pulsatile stretch [47]. The European arm of the study showed a reduction in office BP by 38/23 mmHg, 27/14 mmHg, 22/12 mmHg, 24/11 mmHg, and 24/12 mmHg at 24 h, 1 week, 1 month, 3 months, and 6 months, respectively. Hypotensive crises may represent adverse events that are under observation in ongoing trials with endovascular baroreflex activation [9].

27.2.4 Central Iliac Arterio-Venous Anastomosis

In contrast with previous sympatho-modulatory devices, the central iliac arterio-venous anastomosis was developed to reduce effective arterial volume, systemic vascular resistance, and cardiac afterload, thus lowering BP. This consists of a 4-mm fixed caliber conduit between the proximal arterial and low resistance venous circulation using a nitinol stent-like device placed under fluoroscopic guidance [48]. The anastomosis induces an immediate and significant reduction of systemic vascular resistance and BP but there is also a sympatho-modulatory effect through increasing venous oxygenation and increasing right heart stretch through increased pre-load [49, 50]. The first investigation with this technique was the randomized controlled, open-label, ROX CONTROL HTN trial [51]. At 6 months, office and ambulatory BP were reduced in the active group by 27/20 and 14/14 mmHg, respectively. This was accompanied by a decrease in medications. The main complication reported was a low incidence of ipsilateral venous stenosis, which was successfully managed by venoplasty and/or stenting. This approach appears to be useful in cases where sympatho-modulation has failed but it needs data on long-term safety and efficacy.

27.2.5 Median Nerve and Vagal Nerve Stimulation

The median nerve stimulator is a 2 cm in diameter subcutaneously placed system for resistant hypertension. Actually data are limited and principally come from animal studies [52]. Unpublished data showed a 16.7 mmHg reduction in systolic BP in the active group when compared to the sham group (not activated implanted device) (http://valenciatechnologies.com/first-application/).

The vagus nerve and thoracic ganglia are the principal source of parasympathetic innervation of the heart with resultant negative inotropic and chronotropic effects. It has been shown that vagal nerve stimulation is able to reduce BP [53] without inducing bradycardia or apnea. These side effects can be observed following selective vagal nerve stimulation with beta blockers [54]. Reversible episodes of atrioventricular block and ventricular asystole can be observed at higher current and frequency [55]. Also for this approach more data are needed.

27.3 Device-Based Treatment in Heart Failure

It is well recognized that heart failure (HF) is not merely the result of cardiac structure abnormalities. In patients with HF, neurohormonal activation [5, 6] and increased levels of inflammatory mediators promote ventricular remodeling, vascular dysfunction, and development of HF, thus representing the key mechanisms for management. Hemodynamic monitoring strategies aimed to optimize volume/pressures and personalized medical intervention represent some of the few successful strategies for HF.

27.3.1 Neuromodulation in HF

Percutaneous transluminal ablation of renal nerves with its capacity to lower sympathetic activity has been employed not only in resistant hypertensive patients but also in HF patients. The few clinical studies performed so far did not produce unequivocal results, with some showing benefits (particularly in term of quality of life), some showing no substantial advantages, and other being prematurely stopped due to difficulties in patient recruitment [56–60]. Although a recent meta-analysis examining the results of four major published studies provides evidence on the favorable effects of the procedure on left ventricular function [61], further data are needed to achieve conclusive information on the use of renal nerve ablation as a potential therapeutic intervention for congestive heart failure [62].

The second neuromodulatory approach, as seen for resistant hypertension, is represented by the baroreflex activation therapy [9–15]. The first single-center, open-label, single-arm proof-of-concept study performed in 11 HF patients with left ventricular ejection fraction (LVEF) < 40% and NYHA class III on optimal medical therapy [63] has shown a consistent reduction in the primary study endpoint, i.e., muscle sympathetic nerve traffic (about 30% at 6 months), which was associated with a significant improvement in the severity of HF state, quality of life, and baroreflex sensitivity (Fig. 27.3). Blood pressure and renal function were unaffected by the procedure and a reduction in the cumulative number of hospitalizations were observed in the 6 months follow-up. These results were by and large confirmed in the assessment performed after 21 months [14]. Recent trials have shown that during the long-term follow-up period (on average more than 3 years) unilateral carotid baroreceptor stimulation is capable of improving in HF patients belonging to NYHA class III-IV LVEF, brain natriuretic peptides, exercise capacity, and quality of life [13, 64, 65]. Many of these effects are likely to be mediated by the sympatho-inhibitory properties of the procedure, which are consistent in magnitude, long-lasting, and dependent on the restoration of the modulatory effects of baroreceptors on sympathetic neural drive [14, 63, 66].

27.3.2 Splanchnic Nerve Modulation

It has been shown by continuous monitoring that intracardiac pressure elevation precedes any significant weight gain by several weeks [67]. Thus volume redistribution to the splanchnic compartment may be particularly useful in HF. To this aim splanchnic nerves, which control arterial and venous vascular tone [68], may be

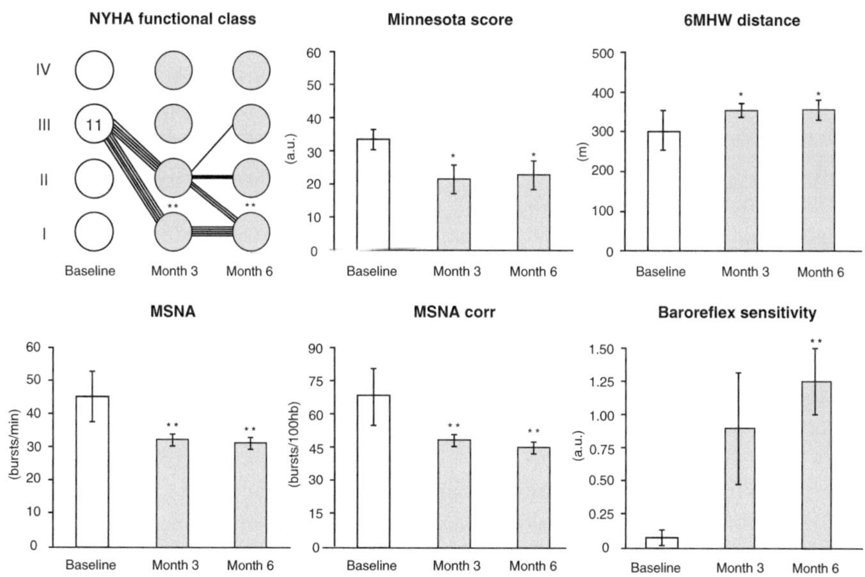

Fig. 27.3 Effects of baroreceptor activation therapy on NYHA functional class, Minnesota score, 6-min walking test, muscle sympathetic nerve activity (MSNA) expressed as burst incidence over time (bs/min) and corrected for heart rate (bs/100hb), and baroreflex sensitivity in congestive heart failure patients. Data are shown as mean ± SD. Asterisks refer to the statistical significance at 3 and 6 months after baroreceptor activation therapy as compared to the pre-intervention baseline values: $*p < 0.05$, $**p < 0.01$. (From Ref [15], with permission)

identified as a potential therapeutic target in HF. Some studies investigated the physiological effects of short-term pharmacological splanchnic nerve block. Bilateral nerve block with lidocaine reduced resting cardiopulmonary filling pressures and improved the cardiac output without complications [69, 70]. Changes in intracardiac pressures are accompanied by an improvement in the cardiac index and peak oxygen consumption. Right-side great splanchnic nerve resection has shown sustained beneficial effects and represents a potential therapeutic and safe approach in HF. Actually data are lacking about randomized controlled trials.

27.3.3 Atrial Shunt Devices

Atrial shunt devices (ASD) are designed to lower elevated left atrial pressure by creating a passage from the left atrium to other structures. Atrial shunts are currently being evaluated in clinical trials [71–79] and hold promise for improving symptoms of pulmonary congestion, exercise capacity, and quality of life. This approach needs anticoagulation therapy and due to the extra stress on the right heart makes these devices not suitable for patients with right ventricular dysfunction, valvular defects, or severe restrictive/obstructive lung disease. The use of interatrial shunt device (IASD) [71] allowed to obtain, both at 30 days and 1 year follow-up, a reduction in left ventricular filling pressure, left ventricular end diastolic volume, and work-indexed pulmonary capillary wedge pressure (PCWP), alongside sustained improvements in NYHA functional class and 6-min walking distance. After 1 year no major peri-procedural or adverse events were observed and a survival of 95% was evident [72]. The V-Wave Shunt [73] is a unidirectional left-to-right atrium device that allows flow when the pressure gradient exceeds 5 mmHg. The device remained patent for the duration of the 12-week follow-up and improved left ventricular filling pressure (14 vs. 22 mmHg), LVEF (42 vs. 25%), and survival (93 vs. 57%) when compared with controls [74]. Following enhancements in the device design to improve patency, several trials were launched to evaluate safety and efficacy in patients pulmonary arterial hypertension and HF. Preliminary results [75] have shown patency and absence of stenosis at 1-year follow-up and significant improvement in NYHA class and 6-min walking distance. The Occlutech atrial flow regulator (AFR) is a self-expandable nitinol mesh braided into 2 flat discs designed to allow interatrial bidirectional flow and implanted in the interatrial septum [76]. A non-randomized pilot study with AFR implanted in patients with severe pulmonary arterial hypertension reported an improvement in 6-min walking distance, cardiac index (2.89 ± 0.56 vs. 2.36 ± 0.52 L/min/m^2, $p = 0.04$) with a patency of the device maintained in all patients [77]. Results from the PRELIEVE (Pilot Study to Assess Safety and Efficacy of a Novel Atrial Flow Regulator in Heart Failure Patients) trial [78] confirmed device patency in 92% of the cases at 3 and 12 months, and a drop of 5 mmHg in PCWP. The study did not show significant changes in NYHA functional class, 6-min walking distance, and quality of life and was accompanied by two major adverse events (embolization and post-procedure bleeding and syncope). The last approach to reduce PWCP and to overcome the risk of embolization is a percutaneously implanted bare-nitinol device allowing a shunt from the left atrium to the coronary sinus [79]. Preliminary data on a small group of HF patients have shown improvement in NYHA class, a clinically significant improvement in 6-min walking distance, a 9 mmHg reduction in PCWP, stable echocardiographic parameters (LVEF, left atrial volume index, right ventricular function), and patency of all devices.

27.3.4 Left Ventricular Expanders

Left ventricular expanders (LVEx) are spring-like devices that apply outward forces to the left ventricle to enhance filling performance. These devices store elastic energy during cardiac contraction and transfer it to the cardiac wall in the diastolic phase [80]. The ImCardia and the CORolla transapical approach (TAA) devices are the only LVEx developed for implantation on the pericardial and endocardial wall, respectively.

In vivo animal studies with the ImCardia have shown device safety and efficacy in improving filling dynamics [81]. An open-label, parallel, non-controlled trial has been conducted in patients with preserved HF undergoing aortic valve replacement. The active group maintained systolic function and improved left ventricular mass and left atrial area. However, the invasive nature of the implantation induced to terminate the trial [82].

Differently from ImCardia, the CORolla TTA requires a minimally invasive implantation procedure. The cone-like left ventricular expander is placed around the apex and anchored through a fixation suture [83]. Preliminary results on one patient showed at a 12-months follow-up a reduction in LV mass index, left atrial volume index, and LVEDV index, and an improvement in NYHA, 6-min walking test, and Minnesota Living questionnaire. A lower efficacy was observed in the continuation of the follow-up with a drop in LVEF raising concerns about long-term progression to reduced HF [84]. The efficacy of these approaches needs more extensive clinical evaluation.

27.3.5 Cardiac Contractility Modulation

The goal of this approach is to augment native cardiac contractility by application of a high-voltage (about 7.5 V), long duration (about 20 ms), biphasic stimulation to the right ventricular septum [85]. This favors the influx of calcium ions into the cardiomyocytes increasing contractility without raising myocardial oxygen consumption [86]. The Optimizer Smart System (impulse Dynamics) has been evaluated in several trials, showing safety and an improved exercise tolerance, NYHA functional class, and quality of life [87, 88]. The better effects were observed for LVEF greater than 35%.

27.3.6 Cardiac Resynchronization Therapy (CRT)

These pulse-generating devices made it possible to radically change the therapeutic approach to HF [89]. The device inserted subcutaneously in the chest wall consists of 3 wire leads to deliver electrical impulses to the right atrium and to the right and left ventricles. They are contraindicated in patients with a QRS duration <130 ms and are recommended for HFrEF patients with NYHA class III or IV despite optimal medical treatment and evidence of left ventricular dilation. CRT improves atrioventricular delay and corrects ventricular dyssynchrony favoring an improvement in left ventricular performance and mitral regurgitation [90–94]. A meta-analysis of 14 randomized controlled trials with 4420 patients with LV dysfunction (LVEF 21–30%) prolonged QRS duration (mean range 155–209 ms) and NYHA class III-IV symptoms despite being on optimal medical therapy showed a reduction in hospitalization by 37% and all-cause mortality by 22% [95]. The recent revision of European Society of Cardiology guidelines for HF patients with LVEF of no more than 35%, sinus rhythm, left bundle branch block with a QRS duration of at least 150 ms, and NYHA class >II who are already on optimal medical therapy has given CRT a class I indication [96].

27.3.7 Mechanical Circulatory Support (MCS)

MCS devices provide hemodynamic support to underperforming left and/or right ventricles, aiming to improve the quality of life of patients with end-stage HF, mostly inotrope-dependent and in critical conditions. In the last decades there has been significant progress in all aspects of left ventricular assist device therapy, from patient selection to device design and postoperative management. The devices have transitioned from pulsatile to continuous flow design and are now smaller, with enhanced durability more suited for long-term support [97, 98]. The mean survival following left ventricular assist device implantation is nearly 5 years. Some complications including bleeding, infection, pump thrombosis, and stroke limit the potential of the therapy [97, 98].

27.4 Conclusion

There has been very substantial progress in the field of device-based therapy for both hypertension and heart failure. The future will focus on further investigation into modifications in device design, safety and patency, duration, remote charging of batteries, and telemetric assessment of device function.

References

1. Mancia G, Grassi G, Giannattasio C, Seravalle G. Sympathetic activation in the pathogenesis of hypertension and progression of organ damage. Hypertension. 1999;34:724–8.
2. Seravalle G, Lonati L, Buzzi S, et al. Sympathetic nerve traffic and baroreflex function in optimal, normal, and high-normal blood pressure states. J Hypertens. 2015;33:1411–7.
3. Grassi G, Pisano A, Bolignano D, et al. Sympathetic nerve traffic activation in essential hypertension and its correlates. Systematic reviews and meta-analyses. Hypertension. 2018;72:483–91.

4. Grassi G, Seravalle G, Brambilla G, et al. Marked sympathetic activation and baroreflex dysfunction in true resistant hypertension. Int J Cardiol. 2014;177:1020–5.
5. Seravalle G, Quarti-Trevano F, Dell'Oro R, et al. Sympathetic and baroreflex alterations in congestive heart failure with preserved, midrange and reduced ejection fraction. J Hypertens. 2019;37:443–8.
6. Grassi G, D'Arrigo G, Pisano A, et al. Sympathetic neural overdrive in congestive heart failure and its correlates: systematic reviews and meta-analysis. J Hypertens. 2019;37:1746–56.
7. Go AS, Mozaffarian D, Roger VL, et al. American Heart Association statistics committee and stroke statistic subcommittee. Heart disease and stroke statistics-2014 update: a report from the American Heart Association. Circulation. 2014;129:e28–292.
8. Krum H, Schlaich M, Withbourn R, et al. Catheter-based renal sympathetic denervation for resistant hypertension: a multicenter safety and proof-of-principle cohort study. Lancet. 2009;373:1275–81.
9. Seravalle G, Dell'Oro R, Grassi G. Baroreflex activation therapy systems: current status and future prospects. Expert Rev Med Devices. 2019;16:1025–33.
10. Bisognano JD, Bakris G, Nadim MK, et al. Baroreflex activation therapy lowers blood pressure in patients with resistant hypertension: results from double-blind, randomized, placebo-controlled Rheos pivotal trial. J Am Coll Cardiol. 2011;58:765–73.
11. Bakris GL, Nadim MK, Haller H, et al. Baroreflex activation therapy provides durable benefit in patients with resistant hypertension: results of long-term follow-up in the Rheos pivotal trial. J Am Soc Hypertens. 2012;6:152–8.
12. Hoppe UC, Brandt MC, Watcher R, et al. Minimally invasive system for baroreflex activation therapy chronically lowers blood pressure with pacemaker-like safety profile: results from the Barostim neo trial. J Am Soc Hypertens. 2012;6:270–6.
13. Zile MR, Abraham WT, Weaver FA, et al. Baroreflex activation therapy for the treatment of heart failure with a reduced ejection fraction: safety and efficacy in patients with and without cardiac resynchronization therapy. Eur J Heart Fail. 2015;17:1066–74.
14. Gronda E, Seravalle G, Quarti-Trevano F, et al. Long-term chronic baroreflex activation: persistent efficacy in patients with heart failure and reduced ejection fraction. J Hypertens. 2015;33:1704–8.
15. Grassi G, Brambilla G, Prata Pizzalla D, Seravalle G. Baroreflex activation therapy in congestive heart failure: novel findings and future insights. Curr Hypertens Rep. 2016;18:60.
16. DiBona GF, Kopp UC. Neural control of renal function. Physiol Rev. 1997;77:75–197.
17. Stella A, Zanchetti A. Functional role of renal afferents. Physiol Rev. 1991;71:659–82.
18. Lohmeier TE, Irwin ED, Rossing MA, et al. Prolonged activation of the baroreflex produces sustained hypotension. Hypertension. 2004;43:306–11.
19. Seravalle G, Grassi G. Carotid baroreceptor stimulation in resistant hypertension and heart failure. High Blood Press Cardiovasc Prev. 2015;22:233–9.
20. Esler MD, Krum H, Sobotka PA, Symplicity HTN-2 Investigators, et al. Renal sympathetic denervation in patients with treatment resistant hypertension: a randomized controlled trial. Lancet. 2010;376:1903–9.
21. Krum H, Schlaich MP, Sobotka PA, et al. Percutaneous renal denervation in patients with treatment-resistant hypertension: final 3-year report of the symplicity HTN-1 study. Lancet. 2014;383:622–9.
22. Esler MD, Bohm M, Sievert H, et al. Catheter-based renal denervation for treatment of patients with treatment-resistant hypertension: 36 months results from the symplicity HTN-2 randomized clinical trial. Eur Heart J. 2014;35:1752–9.
23. Bhatt DL, Kandzari DE, O'Neill WW, Symplicity HTN-3 Investigators, et al. A controlled trial of renal denervation for resistant hypertension. N Engl J Med. 2014;370:1393–401.
24. Bhatt DL, Bakris GL. Renal denervation for resistant hypertension. N Engl J Med. 2014;371:184.
25. Weber MA, Schiffrin EL, White WB, et al. Clinical practice guidelines for the management of hypertension in the community: a statement by the American Society of Hypertension and the International Society of Hypertension. J Clin Hypertens. 2014;16:14–26.

26. Ott C, Schmieder RE. Invasive treatment of resistant hypertensin: present and future. Curr Hypertens Rep. 2014;16:488.
27. Whitbourn R, Harding SA, Walton A. Symplicity multi-electrode radiofrequency renal denervation system feasibility study. EuroIntervention. 2015;11:104–9.
28. Mahfoud F, Mancia G, Schmieder R, et al. Renal denervation in high-risk patients with hypertension. J Am Coll Cardiol. 2020;75:2879–88.
29. Persu A, Jin Y, Azizi M, et al. European Network COordinating Research on Renal Denervation (ENCOReD). Blood pressure changes after renal denervation at 10 European expert centers. J Hum Hypertens. 2014;28:150–6.
30. Mahfoud F, Ukena C, Schmieder RE, et al. Ambulatory blood pressure changes after renal sympathetic denervation in patients with resistant hypertension. Circulation. 2013;128:132–40.
31. Vogel B, Kirchberger M, Zeier M, et al. Renal sympathetic denervation therapy in the real world: results from the Heidelberg registry. Clin Res Cardiol. 2014;103:117–24.
32. Hering D, Mahfoud F, Walton AS, et al. Renal denervation in moderate to severe CKD. J Am Soc Nephrol. 2021;23:1250–7.
33. Ott C, Mahfoud F, Schmid A, et al. Renal denervation in moderate treatment-resistant hypertension. J Am Coll Cardiol. 2013;62:1880–6.
34. Papademetriou V, Tsioufis CP, Sinhal A, et al. Catheter-based renal denervation for resistant hypertension: 12-months results of the EnligHTN 1 first-in-human study using a multielectrode ablation system. Hypertension. 2014;64:565–72.
35. Sievert H, Schofer J, Ormiston J, et al. Bipolar radiofrequency renal denervation with the Vessix catheter in patients with resistant hypertension: 2-year results from the REDUCE-HTN trial. J Hum Hypertens. 2017;31:366–8.
36. Verheye S, Ormiston J, Bergmann MW, et al. Twelve-month results of the rapid renal sympathetic denervation for resistant hypertension using the OneShot™ ablation system (RAPID) study. EuroIntervention. 2015;10:1221–9.
37. Daemen J, Mahfoud F, Kuck KH, et al. Safety and efficacy of endovascular ultrasound renal denervation in resistant hypertension: 12-month results from the ACHIEVE study. J Hypertens. 2019;37:1906–12.
38. Chernin G, Szwarcfiter I, Steinert D, et al. First-in-man experience with a novel catheter-based renal denervation system of ultrasonic ablation in patients with resistant hypertension. J Vasc Interv Radiol. 2018;29:1158–66.
39. Fischell TA, Ebner A, Gallo S, et al. Transcatheter alcohol-mediated perivascular renal denervation with the peregrine system: first-in-human experience. JACC Cardiovasc Interv. 2016;9:589–98.
40. Heuser RR, Mahtre AU, Buelna TJ, et al. A novel non-vascular system to treat resistant hypertension. EuroIntervention. 2013;9:135–9.
41. Ormiston JA, Anderson T, Brinton TJ, et al. TCT-412 non-invasive renal denervation using externally delivered focused ultrasound: early experience using Doppler based imaging tracking and targeting for treatment. J Am Coll Cardiol. 2014;64:11-S. https://doi.org/10.1016/j.jacc.2014.07.461.
42. Schmieder RE, Ott C, Toennes SW, et al. Phase II randomized sham-controlled study of renal denervation for individuals with uncontrolled hypertension—WAVE IV. J Hypertens. 2018;36:680–9.
43. Kroon A, Schmidli J, Sheffers I, et al. Chronically implanted system: 4-year data of Rheos DEBuT-HT study in patients with resistant hypertension. J Hypertens. 2012;28(suppl A):e441.
44. Heusser K, Tank J, Engeli S, et al. Carotid baroreceptor stimulation, sympathetic activity, baroreflex function, and blood pressure in hypertensive patients. Hypertension. 2010;55:619–26.
45. Bakris GL, Nadim MK, Haller H, et al. Baroreflex activation therapy provides durable benefits in patients with resistant hypertension: results of long-term follow up in the Rheos pivotal trial. J Am Soc Hypertens. 2012;6:152–8.
46. Wallbach M, Lehning LY, Schroer C, et al. Effects of baroreflex activation therapy on ambulatory blood pressure in patients with resistant hypertension. Hypertension. 2016;67:701–9.

47. Spiering W, Williams B, van der Heyden J, CALM-FIM-EUR Investigators, et al. Endovascular baroreflex amplification for resistant hypertension: a safety and proof-of-principle clinical study. Lancet. 2017;390:2655–61.
48. Foran JP, Jain AK, Casserly JP, et al. The ROX coupler: creation of a fixed iliac arteriovenous anastomosis for the treatment of uncontrolled systemic arterial hypertension, exploiting the physical properties of the arterial vasculature. Catheter Cardiovasc Interv. 2015;85:880–6.
49. Burchell AE, Lobo MD, Sulke N, et al. Arteriovenous anastomosis: is this the way to control hypertension? Hypertension. 2014;64:6–12.
50. Kapil V, Sobotka PA, Saxena M, et al. Central iliac arteriovenous anastomosis for hypertension: targeting mechanical aspects of the circulation. Curr Hypertens Rep. 2015;17:585.
51. Lobo MD, Sobotka PA, Stanton A, et al. Central arteriovenous anastomosis for the treatment of patients with uncontrolled hypertension (the ROX CONTROL HTN study): a randomized controlled trial. Lancet. 2015;385:1634–41.
52. Li P, Tjen-A-Looi SC, Cheng L, et al. Long-lasting reduction of blood pressure by electroacupuncture in patients with hypertension: randomized controlled trial. Med Acupunct. 2015;27:253–66.
53. Annoni EM, Xie X, Lee SW, et al. Intermittent electrical stimulation of the right cervical vagus nerve in salt-sensitive hypertensive rats: effects on blood pressure, arrhythmias, and ventricular electrophysiology. Physiol Rep. 2015;3:e12476.
54. Gierthmuehlen M, Plachta DT. Effects of selective vagal nerve stimulation on blood pressure, heart rate and respiratory rate in rats under metoprolol medication. Hypertens Res. 2016;39:79–87.
55. Mirkovic T, Knezevic I, Radan I, et al. Frequency dependent effect of selective biphasic left vagus nerve stimulation on heart rate and arterial pressure. Signa Vitae. 2012;7:63–8.
56. Davies JE, Manisty CH, Petraco E, et al. First-in-man safety evaluation of renal denervation for chronic systolic heart failure: primary outcome from REACH-pilot study. Int J Cardiol. 2013;162:189–92.
57. Patel HC, Rosen SD, Hayward C, et al. Renal denervation in heart failure with preserved ejection fraction (RDT-PEF): a randomized controlled trial. Eur J Heart Fail. 2016;18:703–12.
58. Dai Q, Lu J, Wang B, Ma G. Effect of percutaneous renal sympathetic nerve radiofrequency ablation in patients with severe heart failure. Int J Clin Exp Med. 2015;8:9779–85.
59. Chen W, Ling Z, Xu Y, et al. Preliminary effects of renal denervation with saline irrigated catheter on cardiac systolic function in patients with heart failure: a prospective, randomized, controlled, pilot study. Catheter Cardiovasc Interv. 2017;89:E153–61.
60. Gao JQ, Xie Y, Yang W, et al. Effects of percutaneous renal sympathetic denervation on cardiac function and exercise tolerance in patients with chronic heart failure. Rev Port Cardiol. 2017;36:45–51.
61. Fukuta H, Goto T, Wakami K, Ohte N. Effects of catheter-based renal denervation on heart failure with reduced ejection fraction: a systematic review and meta-analysis. Heart Fail Rev. 2017;22:657–64.
62. Lu Y, Zhang L, Zhou X, Tang B. Renal sympathetic denervation: a potential alternative strategy for patients with heart failure. Int J Cardiol. 2015;201:140–1.
63. Gronda E, Seravalle G, Brambilla G, et al. Chronic baroreflex activation effects on sympathetic nerve traffic, baroreflex function and cardiac haemodynamics in heart failure. A proof-of-concept study. Eur J Heart Fail. 2014;16:977–83.
64. Abraham WT, Zile MR, Weaver FA, et al. Baroreflex activation therapy for the treatment of heart failure with a reduced ejection fraction. J Am Coll Cardiol Heart Fail. 2015;3:487–96.
65. Hallbach M, Abraham WT, Butter C, et al. Baroreflex activation therapy for the treatment of heart failure with reduced ejection fraction in patients with and without coronary artery disease. Int J Cardiol. 2018;266:187–92.
66. Dell'Oro R, Gronda G, Seravalle G, et al. Restoration of normal sympathetic neural function in heart failure following baroreflex activation therapy: final 43-month study report. J Hypertens. 2017;35:2532–6.

67. Zile MR, Bennett TD, St John Sutton M, et al. Transition from chronic compensated to acute decompensated heart failure: pathophysiological insights obtained from continuous monitoring of intracardiac pressures. Circulation. 2008;118:1433–41.
68. Barnes RJ, Bower EA, Rink TJ. Haemodynamic responses to stimulation of the splanchnic and cardiac sympathetic nerves in the anaesthetized cat. J Physiol. 1986;378:417–36.
69. Fudim M, Ganesh A, Green C, et al. Splanchnic nerve block for decompensated chronic heart failure: splanchnic-HF. Eur Heart J. 2018;39:4255–6.
70. Fudim M, Jones WS, Boortz-Marx RL, et al. Splanchnic nerve block for acute heart failure. Circulation. 2018;138:951–3.
71. Sondegaard L, Reddy V, Kaye D, et al. Transcatheter treatment of heart failure with preserved or mildly reduced ejection fraction using a novel interatrial implant to lower left atrial pressure. Eur J Heart Fail. 2014;16:796–801.
72. Hasenfuß G, Hayward C, Burkhoff D, et al. A transcatheter intracardiac shunt device for heart failure with preserved ejection fraction (REDUCE LAP-HF): a multicenter, open-label, single arm, phase 1 trial. Lancet. 2016;387:1298–304.
73. Del Trigo M, Bergeron S, Bernier M, et al. Unidirectional left-to-right interatrial shunting for treatment of patients with heart failure with reduced ejection fraction: a safety and proof-of-principle cohort study. Lancet. 2016;387:1290–7.
74. Rodes-Cabau J, Bernier M, Amat-Santos IJ, et al. Interatrial shunting for heart failure: early and late results from the first-in-human experience with the V-Wave system. J Am Coll Cardiol Interv. 2018;11:2300–10.
75. Guimaraes L, Bergeron S, Bernier M, et al. Interatrial shunt with the second generation V-Wave system for patients with advanced chronic heart failure. EuroIntervention. 2002;15:1426–8.
76. Patel MB, Samuel BO, Girgis RE, et al. Implantable atrial flow regulator for severe, irreversible pulmonary arterial hypertension. EuroIntervention. 2015;11:706–9.
77. Rajeshkumar R, Pavithran S, Sivakumar K, et al. Atrial septostomy with a predefined diameter using a novel occlutech atrial flow regulator improves symptoms and cardiac index in patients with severe pulmonary arterial hypertension. Catheter Cardiovasc Interv. 2017;9:1145–53.
78. Paitazoglou C, Bergmann MW, Ozdemir R, et al. One year results of the first-in-man study investigating the atrial flow regulator for left atrial shunting in symptomatic heart failure patients: the PRELIEVE study. Eur J Heart Fail. 2021;23:800–10.
79. Simard T, Labinaz M, Zahr R, et al. Percutaneous atriotomy for levoatrial-to-coronary sinus shunting in symptomatic heart failure: first-in-human experience. J Am Coll Cardiol Interv. 2020;13:1236–47.
80. Feld Y, Dubi S, Reisner Y, et al. Future strategies for the treatment of diastolic heart failure. Acute Card Care. 2006;8:13–20.
81. Feld Y, Dubi S, Reisner Y, et al. Energy transfer from systole to diastole: a novel device-based approach for the treatment of diastolic heart failure. Acute Card Care. 2011;13:232–42.
82. ImCardia TM for DHF to Treat Diastolic Heart Failure (DHF) Patient a Pilot Study (ImCardia). https://clinicaltrials.gov/ct2/show/NCT01347125. Accessed 1 May 2021.
83. Feld Y, Reisner Y, Meyer-Brodnitz G, et al. The CORolla device for energy transfer from systole to diastole: a novel treatment for heart failure with preserved ejection fraction. Heart Fail Rev. 2021;28:307. https://doi.org/10.1007/s10741-021-10104-x.
84. CORolla TAA for Heart Failure With Preserved Ejection Fraction (HFpEF) and Diastolic Dysfunction (DD). https://clinicaltrials.gov/ct2/show/NCTO2499601. Accessed 1 May 2021.
85. Borggrefe M, Mann DL. Cardiac contractility modulation in 2018. Circulation. 2018;138:2738–40.
86. Imai M, Rastogi S, Gupta RC, et al. Therapy with cardiac contractility modulation electrical signals improves left ventricular function and remodeling in dogs with chronic heart failure. J Am Coll Cardiol. 2007;49:2120–8.
87. Kuschyk J, Roger S, Schneider R, et al. Efficacy and survival in patients with cardiac contractility modulation: long-term single center experience in 81 patients. Int J Cardiol. 2015;183:76–81.

88. Abraham WT, Kuck K-H, Goldsmith RL, et al. A randomized controlled trial to evaluate the safety and efficacy of cardiac contractility modulation. JACC Heart Fail. 2018;6:874–83.
89. Abraham WT. Cardiac resynchronization therapy. In: Semirgran M, Shin JT, editors. Heart failure. 2nd ed. CRC Press; 2012.
90. Auricchio A, Stelibrink C, Sack S, et al. Long-term clinical effect of hemodynamically optimized cardiac resynchronization therapy in patients with heart failure and ventricular conduction delay. J Am Coll Cardiol. 2002;39:2026–33.
91. Bradley DJ, Bradley EA, Baughman KL, et al. Cardiac resynchronization and death from progressive heart failure: a meta-analysis of randomized controlled trials. JAMA. 2003;289:730–40.
92. Cleland JG, Daubert JC, Erdmann E, et al. The effect of cardiac resynchronization on morbidity and mortality in heart failure. N Engl J Med. 2005;352:1539–49.
93. St John Sutton MG, Plappert T, Abraham WT, et al. Effects of cardiac resynchronization therapy on left ventricular size and function in chronic heart failure. Circulation. 2003;107:1985–90.
94. Moss AJ, Hall WJ, Cannon DS, et al. Cardiac resynchronization therapy for the prevention of heart failure events. N Engl J Med. 2009;361:1329–38.
95. McAlister FA, Ezekowitz J, Hoorton N, et al. Cardiac resynchronization therapy for patients with left ventricular systolic dysfunction: a systematic review. JAMA. 2007;297:2502–14.
96. McDonagh TA, Metra M, Adamo M, et al. 2021 ESC guidelines for the diagnosis and treatment of acute and chronic heart failure. Eur Heart J. 2021;42:3599–726.
97. Miller LW, Rogers JG. Evolution of left ventricular assist device therapy for advanced heart failure. A review. JAMA Cardiol. 2018;3:650–8.
98. Frigerio M. Left ventricular assist device. Indication, timing, and management. Heart Fail Clin. 2021;17:619–34.

Renal Denervation: For the Prevention of Heart Failure in Hypertensive Patients

28

Konstantinos Tsioufis, Fotis Tatakis, and Panagiotis Tsioufis

28.1 Introduction

Over the past decade, the potential utility of device-based hypertension therapies has evolved from enthusiasm to skepticism and then renewed promise again. After a series of observational studies and an unblinded randomized trial [1] suggesting clinically significant reductions in BP among patients with severe treatment-resistant hypertension (RH) [2–4], renal denervation (RDN) is considered an important new treatment option with global public health benefit, and early enthusiasm for clinical adoption outpaced the available data. After the lack of demonstrable effectiveness compared with a sham procedure and antihypertensive therapy in the SYMPLICITY HTN-3 trial (Renal Denervation in Patients With Uncontrolled Hypertension) [5], however, the prospect of RDN's role in clinical practice diminished and appeared to validate earlier criticisms of trial conduct and, more broadly, the prospects for device-based hypertension treatments.

Advances in trial design and conduct to account for confounding variables of procedural technique, medication variability, and selection of both patients and outcomes have substantially informed subsequent studies [6]. Additional exploratory preclinical and clinical studies continued, and signals of effectiveness with RDN persisted, motivating further investigations. To date, results from 5 sham-controlled, randomized RDN trials have demonstrated significant BP decrease in patients on or off concomitant antihypertensive treatment (Fig. 28.1) [7–11], supporting a biological proof of principle for this novel treatment.

K. Tsioufis (✉)
National and Kapodistrian University, Athens, Greece

Georgetown University and VA Medical Center, Washington, DC, USA

F. Tatakis · P. Tsioufis
National and Kapodistrian University, Athens, Greece

M. Dorobantu et al. (eds.), *Hypertension and Heart Failure*, Updates in Hypertension and Cardiovascular Protection,
https://doi.org/10.1007/978-3-031-39315-0_28

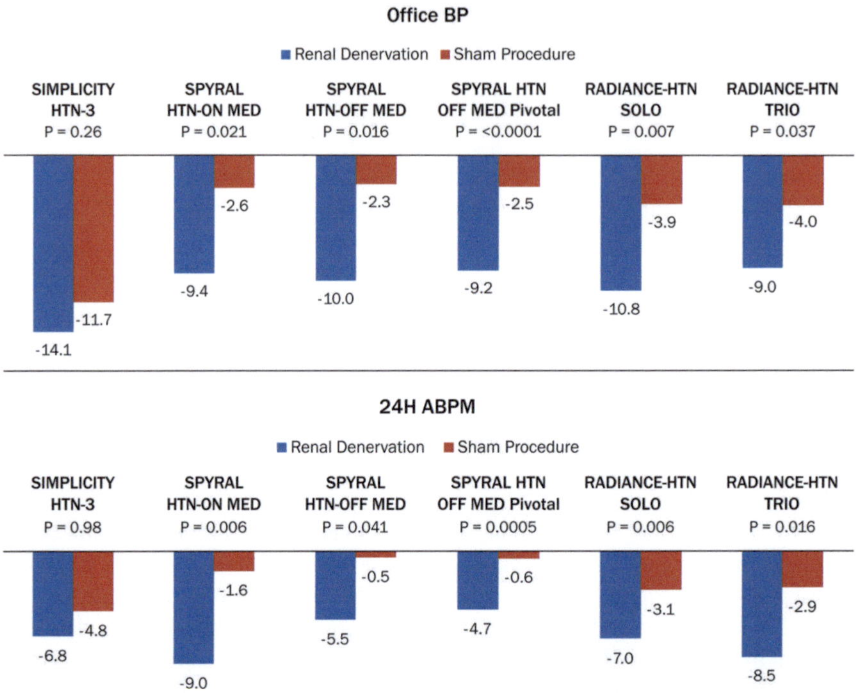

Fig. 28.1 Change in systolic blood pressure (mmHg) after renal denervation in 6 prospective, randomized, sham-controlled trials. ABPM indicates ambulatory blood pressure measurement; BP, blood pressure; RADIANCE HTN SOLO, Study of the ReCor Medical Paradise System in Clinical Hypertension in Absence of Hypertension Medications; RADIANCE-HTN TRIO, Study of the ReCor Medical Paradise System in Clinical Hypertension in the Presence of a Stabilized, Single Pill, Triple, Fixed Dose Antihypertensive Medication Regimen; SYMPLICITY HTN-3 trial, Renal Denervation in Patients with Uncontrolled Hypertension; SPYRAL HTN-OFF MED, Global Clinical Study of Renal Denervation with the Symplicity Spyral Multi-Electrode Renal Denervation System in Patients with Uncontrolled Hypertension in the Absence of Antihypertensive Medications; SPYRAL HTN-OFF PIVOTAL, Global Clinical Study of Renal Denervation with the Symplicity Spyral Multi-Electrode Renal Denervation System in Patients with Uncontrolled Hypertension in the Absence of Antihypertensive Medications Pivotal; and SPYRAL HTN ON MED, Global Clinical Study of Renal Denervation with the Symplicity Spyral Multi-Electrode Renal Denervation System in Patients with Uncontrolled Hypertension on Standard Medical Therapy. (Adapted from references [3, 5–9])

It is widely known that hypertension-related target organ damage (TOD) is strongly associated with adverse cardiovascular outcome, whereas blood pressure (BP) reduction is accompanied by organ damage regression and subsequent cardiovascular protection [12, 13]. Several prospective observational studies have suggested that renal denervation is associated with a reduction in early TOD [14–30]. All these studies lack either a control group or a sham procedure, but nevertheless, deserve attention as the TOD has been evaluated in a blind fashion. It remains to be elucidated whether this is causally related to a reduction in BP alone or because of the reduced sympathetic outflow to the target organs. On the basis of the above, we aimed to evaluate the effect of RDN on target organ damage by pooling currently available data, and to investigate its relation to BP-lowering following RDN.

Observational studies and randomized clinical trials assessing the effect of RDN on prespecified outcomes of TOD were included in the present meta-analysis. The prespecified outcomes of TOD for vessels were central augmentation index (AIx) and carotid-femoral pulse wave velocity (PWV), whereas the prespecified outcomes of TOD for heart structure and diastolic function were left ventricular mass index (LVMI), left atrial volume index (LAVI), E wave to A wave velocity of trans-mitral flow (E/A), and E wave velocity of trans-mitral flow to Em velocity of tissue Doppler imaging of mitral annulus (E/Em).

After executing the carefully designed search strategy and applying specific inclusion criteria, 17 studies ($n = 698$ patients) were incorporated in the present meta-analysis. Table 28.1 summarizes the characteristics of studies and recruited patients. Six studies had a control group and in 13 studies, the evaluation of endpoints was blind. The sample size ranged from 12 to 110 patients and most studies had a snapshot at 6 or 12 months of follow-up. The recruited population had RH in all the studies, with accompanying chronic kidney disease in one of them ($n = 45$ patients) [29]. The mean age of the patients ranged from 53.9 ± 11.3 to 67.1 ± 8.5 years and about 61% of the total population were men. The mean BMI in all studies that was reported ranged from overweight to Class I obesity.

28.1.1 Left Ventricular Mass Index

The echocardiographic estimation of LVM indexed to body surface area before and after RDN was reported in eight studies ($n = 250$ patients). There was modest heterogeneity among the reviewed trials ($I^2 = 3.5\%$, $Q = 7.25$, $P = 0.4$). The follow-up duration was 6 months in seven of them ($n = 219$) and 12 months in one ($n = 31$). RDN led to a regression of LVMI (baseline mean = 130.1 g/m^2) by 14.17 g/m^2 (10.9% reduction) at the end of follow-up (95% CI −18.33 to −10.01, $P < 0.001$; Fig. 28.2). Sensitivity analysis after excluding the only study with 12 months follow-up was confirmatory of the initial result [−13.842 g/m^2 (95% CI −18.662 to −9.022, $P < 0.001$)].

Echocardiographic LVM indexed to height was a reported outcome in three studies ($n = 130$ patients) with 6 months follow-up in all of them and modest heterogeneity ($I^2 = 14.7\%$, $Q = 2.34$, $P = 0.31$). The pooled effect size was also in favor of RDN, estimated at −7.92 g/m$^{2.7}$ (95% CI −9.09 to −6.75, $P < 0.001$; baseline mean = 58.19 g/m$^{2.7}$, 13.61% reduction; Fig. 28.2).

Cardiac magnetic resonance (CMR) was used to evaluate LVM, indexed to body surface area, in four studies ($n = 130$ patients). Three of these studies had a follow-up duration of 6 months ($n = 84$ patients) whereas one had a follow-up duration of 12 months ($n = 46$ patients). Modest heterogeneity was estimated among the reviewed trials ($I^2 = 0\%$, $Q = 2.34$, $P = 0.5$). A regression of LVMI by −4.75 g/m^2 was attributed to the effect of RDN (95% CI −7.83 to −1.67, $P = 0.003$; baseline mean = 78.61 g/m^2, 6.04% reduction). Sensitivity analysis after removing the trial

Table 28.1 Characteristics of included studies and patients

Year	First author	Control group	Blind endpoint analysis	Included population	RDN arm (n)	Age (years)	Sex (%males)	BMI (kg/m^2)	Follow-up duration (months)	Outcome measures
2012	Brandt	Yes	Yes	RH	110	63,6 ± 9,9	70	29,4 ± 4,5	6	Aix, PWV
2012	Brandt	Yes	Yes	RH	46	63,1 ± 10.2	67	28.6 ± 3.4	1, 6	LVMIecho, E/A, E/Em
2012	Mortensen	No	No	RH	21	64,4 ± 12,3	61,9	27,9 ± 4,1	6	Aix, PWV
2013	Hering	Yes	Yes	RH	40	60 ± 11	77,5	NR	3	Aix
2014	Doltra	Yes	No	RH	23	67,1 ± 8,5	60,9	28,1 ± 3,6	6	LVMIcmr
2014	Dores	No	No	RH	34 (22[a])	62,7 ± 7,6	50	30,9 ± 5,3	6	LVMIecho, LAVI, E/A, E/Em
2014	Mahfoud	Yes	Yes	RH	55	65 ± 10	71	29,2 ± 4,3	6	LVMIcmr
2014	Schirmer	No	Yes	RH	66	63,5 ± 1,2	54,5	29,4 ± 0,6	6	LVMIecho, E/A, E/Em
2015	Baroni	No	No	RH	12	69 ± 10	75	NR	1, 3, 6, 12	PWV
2015	McLellan	No	Yes	RH	14	64 ± 9	67	31 ± 3	6	LVMIcmr, echo, E/Em
2015	Ripp	No	Yes	RH	60	54,6 ± 9,5	55	32,9 ± 6,2	6	LVMIecho, LAVI, E/A, E/Em
2015	Schirmer	No	Yes	RH	66	63,5 ± 1,2	55	29,4 ± 0,6	6	LVMIecho, LAVI, E/A
2015	Tsioufis	Yes	Yes	RH	18	56 ± 10	67	33,6 ± 5,4	6	LVMIecho, E/A, E/Em
2015	Verloop	No	Yes	RH	54	58 ± 10	50	29,2 ± 5,2	12	LVMIcmr, Aix, PWV
2016	de Sousa Almeida	No	Yes	RH	31	65 ± 7	48,4	31,8 ± 5,5	12	LVMIecho, LAVI, E/A, E/Em
2016	Kiuchi	No	Yes	RH + CKD	45	53,9 ± 11,3	58	30,2 ± 4,3	6	LVMIecho
2016	Palionis	No	Yes[b]	RH	15	54 ± 7,2	53,3	34,5 ± 3,5	6	LVMIcmr, echo, LAVI, E/A, E/Em, Aix, PWV
2016	Patel	Yes	Yes	HFpEF	17 (14[a])	74,3 ± 6,1	60	30,6 ± 5,5	3, 12	LVMIcmr, echo, LAVI, E/Em

Aix augmentation index, *BMI* body mass index, *CKD* chronic kidney disease, *CMR* cardiac magnetic resonance, *HFpEF* heart failure with preserved ejection fraction, *LAVI* left atrium volume indexed, *LVMI* left ventricular mass index, *NR* not reported, *PWV* pulse wave velocity, *RDN* renal denervation, *RH* resistant hypertension

[a]Number of patients that completed follow-up

[b]Blind analysis of echocardiographic outcomes

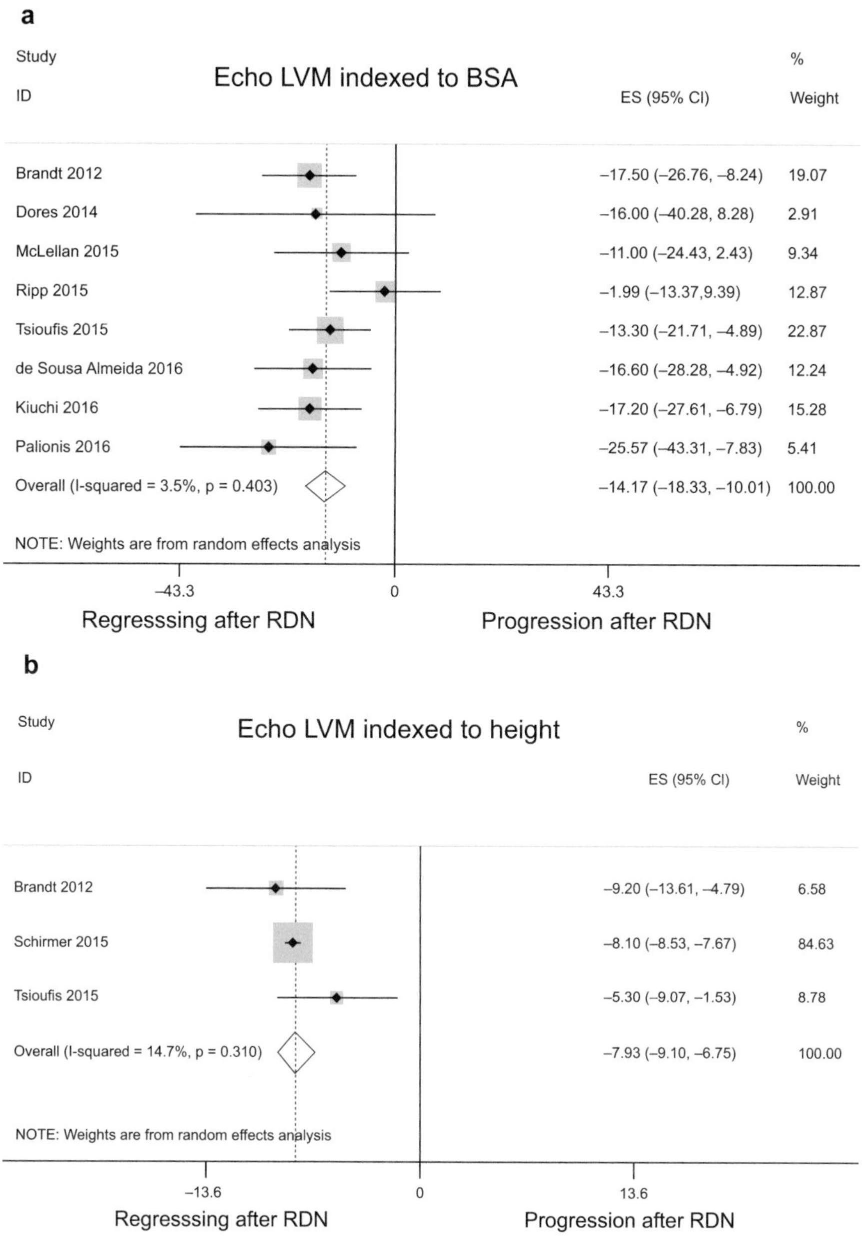

Fig. 28.2 Forest plots demonstrating the changes in various indices of LVM among prospective studies following RDN. (**a**) Echocardiographic LVM indexed to BSA. (**b**) Echocardiographic LVM indexed to height. (**c**) CMR LVM indexed to BSA. (**d**) CMR LVM indexed to height. *BSA* body surface area, *CI* confidence interval, *CMR* cardiac magnetic resonance, *Echo* echocardiographic, *ES* effect size, *LVM* left ventricular mass, *RDN* renal denervation

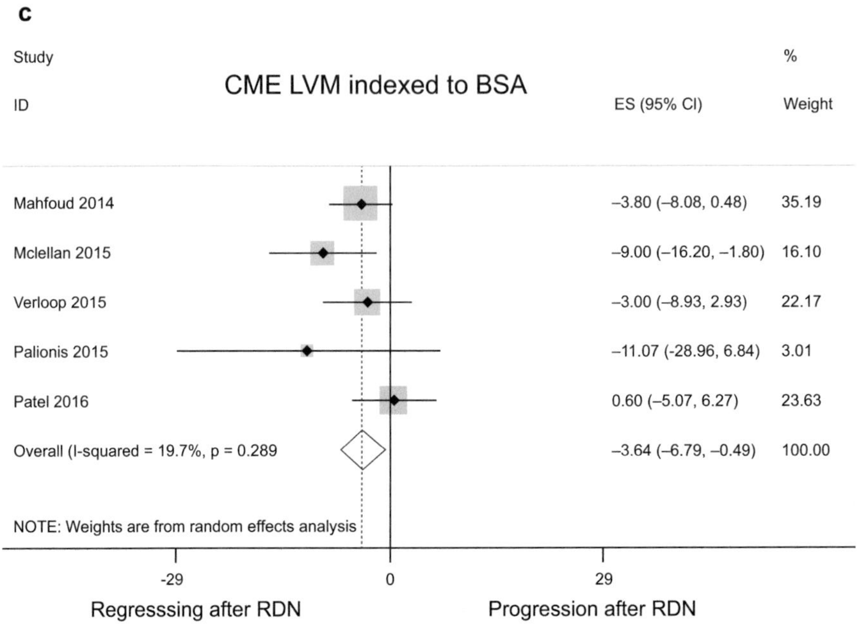

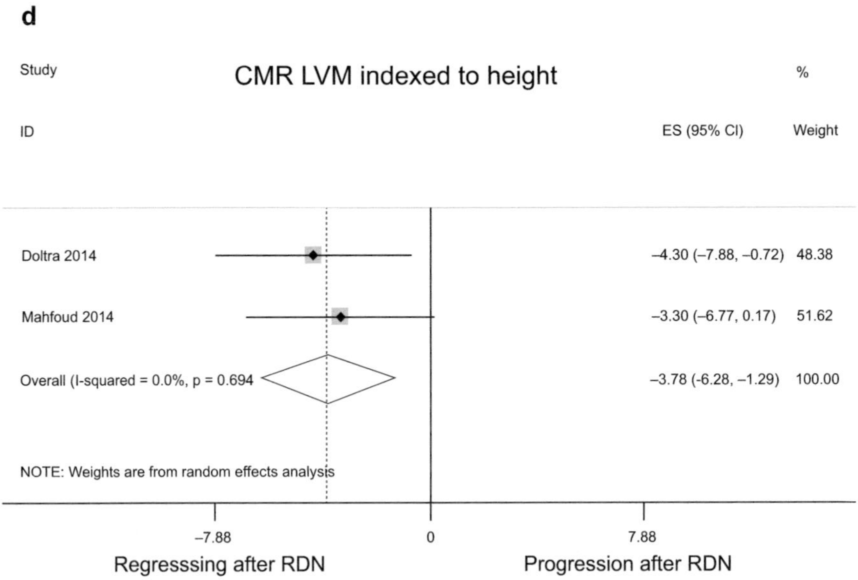

Fig. 28.2 (continued)

with follow-up duration of 12 months revealed a consistency in the estimated effect size [−5.4 (95% CI −9.01 to −1.79, $P = 0.003$)] (Fig. 28.2).

LVM indexed `to height using CMR was an outcome in two studies with a total population of 78 patients, follow-up duration of 6 months after RDN and no heterogeneity ($I^2 = 0\%$, $Q = 0.15$, $P = 0.69$). The pooled effect was 3.78 g/m$^{2.7}$ (95% CI −6.28 to −1.29, $P = 0.003$; baseline mean = 44.24 g/m$^{2.7}$, 8.54% reduction; Fig. 28.2).

28.1.2 Left Atrium Volume Index

Five studies ($n = 194$ patients) with substantial heterogeneity ($I^2 = 87.9\%$, $Q = 33.15$, $P < 0.001$) were reporting data on echocardiographic estimation of LAVI before and after RDN. Four studies had a follow-up duration of 6 months ($n = 163$) whereas one study followed up the patients for 12 months ($n = 31$). The pooled estimator found no effect of RDN on LAVI [−1.67 mL/m^2 (95% CI −4.6 to 1.27, $P = 0.27$)]. Sensitivity analysis for the follow-up duration was interesting as it revealed a regression of LAVI [−4.05 mL/m^2 (95% CI −4.44 to −3.66, $P < 0.001$; baseline mean = 40.62 mL/m^2, 4.11% reduction)] after excluding the only study with a follow-up duration of 12 months (representing the 24.93% of the weight in the initial analysis; Fig. 28.3).

28.1.3 Diastolic Dysfunction

E/A was reported at baseline and after RDN in seven studies ($n = 258$ patients). All but one of these studies had a follow-up duration of 6 months ($n = 227$), whereas de Sousa Almeida et al. [20] followed their cohort for 12 months ($n = 31$). The variation in effect size attributed to heterogeneity across studies was substantial ($I^2 = 79.9\%$, $Q = 29.78$, $P < 0.001$). Application of random effect models did not reveal a difference in E/A before and after RDN [0.043 (95% CI −0.031 to 0.117, $P = 0.252$)] (Fig. 28.4). Sensitivity analysis after excluding the study with 12 months follow-up confirmed the absence of difference. E/Em was reported as an outcome variable in eight studies ($n = 269$ patients). The follow-up duration was 6 months in seven of them ($n = 238$) and 12 months in one ($n = 31$). The estimated heterogeneity across reviewed trials was substantial ($I^2 = 71.5\%$, $Q = 24.54$, $P = 0.001$). The pooled effect was in favor of RDN revealing a reduction of 0.73 in E/Em (95% CI −1.38 to −0.08, $P = 0.03$; baseline mean = 10.76, 6.78% reduction) (Fig. 28.4). The estimated effect size was on the same direction after excluding the study with 12 months follow-up duration [−0.78 (95% CI −1.52 to −0.03, $P = 0.04$)].

28.1.4 Augmentation Index

Four studies were reporting results regarding AIx before and after RDN ($n = 186$ patients). The follow-up duration was 6 months in three of them ($n = 146$ patients) and 3 months in one ($n = 40$ patients). There was no heterogeneity among the

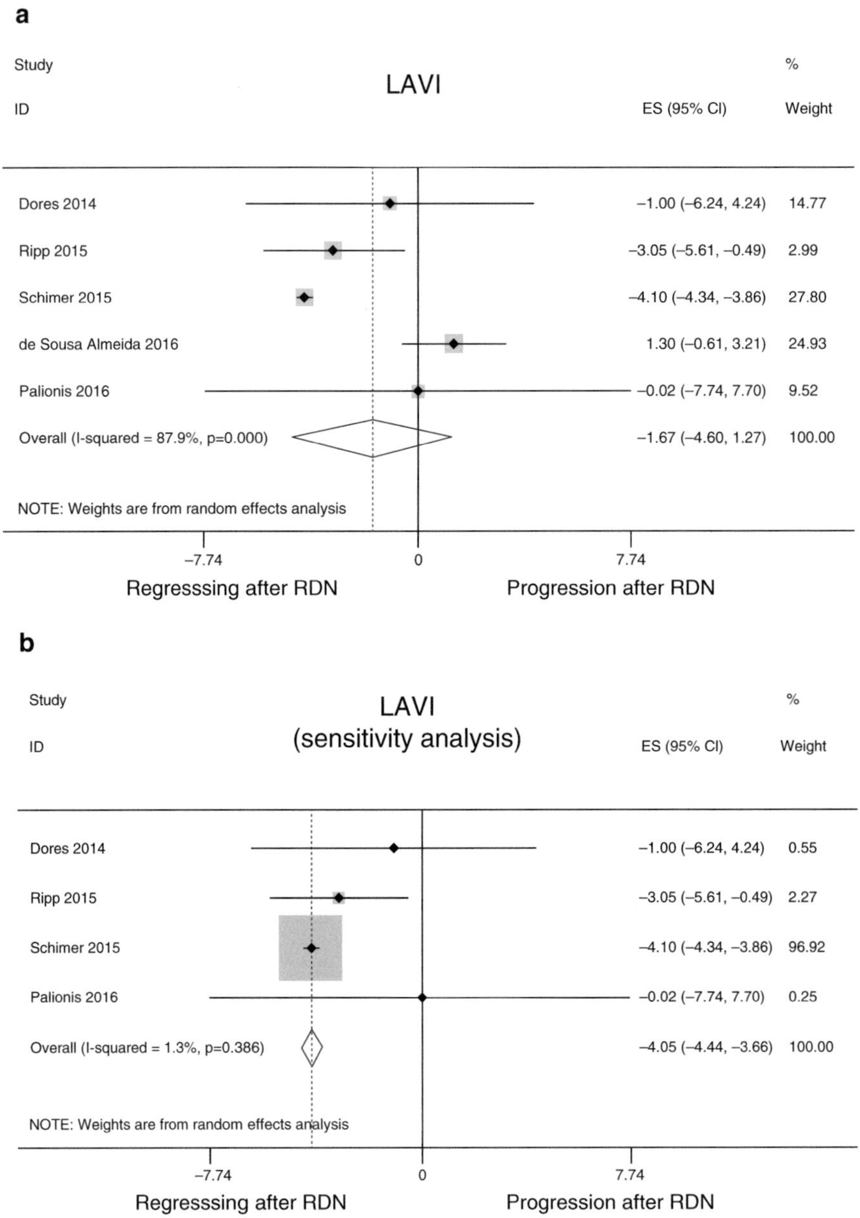

Fig. 28.3 (**a**) Forest plot demonstrating changes in LAVI among prospective studies following RDN. (**b**) Forest plot demonstrating changes in LAVI among prospective studies following RDN after excluding the only study with a follow-up duration of 12 months, a predefined sensitivity analysis for follow-up duration. *CI* confidence interval, *ES* effect size, *LAVI* left atrium volume index, *RDN* renal denervation

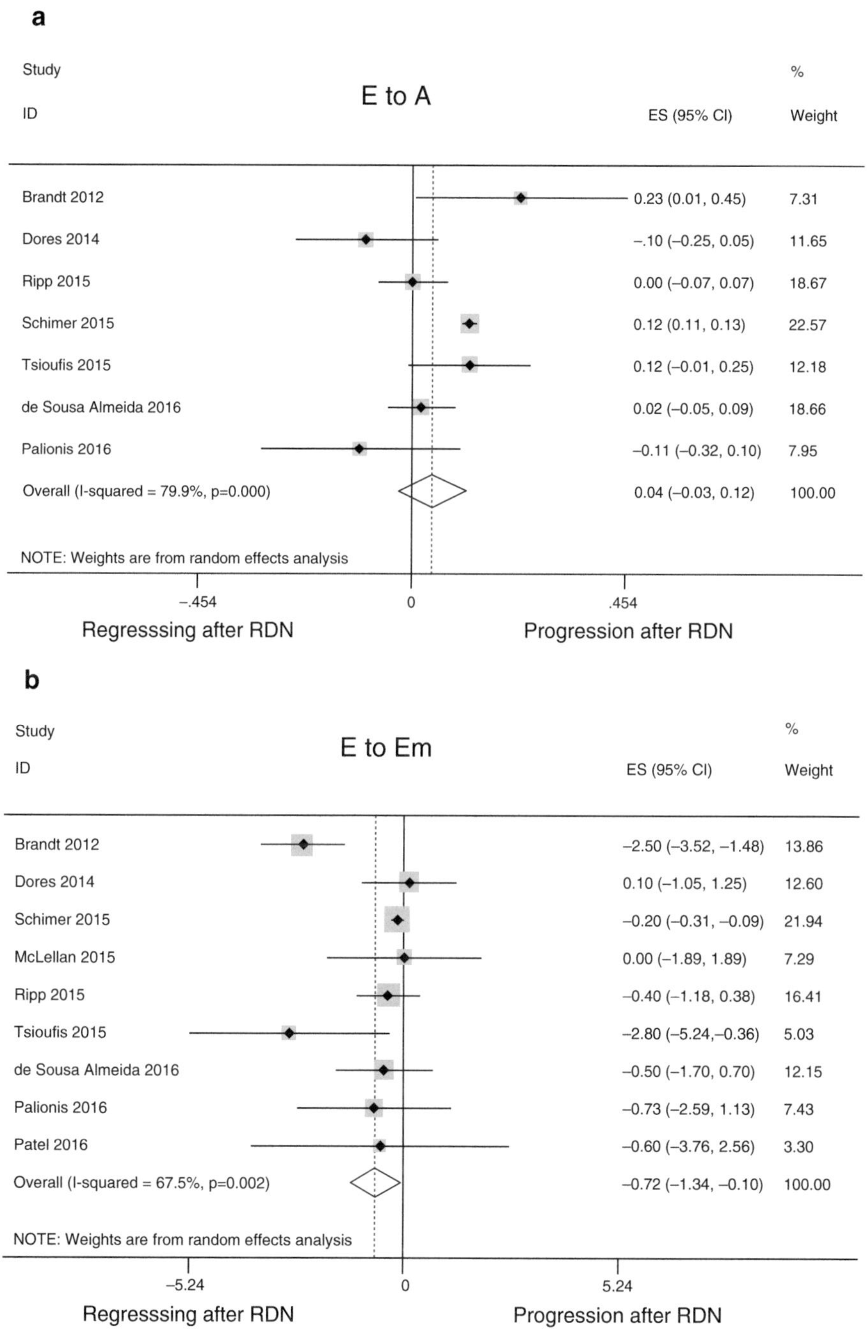

Fig. 28.4 Forest plots of indices of diastolic dysfunction. (**a**) Forest plot demonstrating the changes in E to A transmitral inflow waves velocity ratio among prospective studies following RDN. (**b**) Forest plot demonstrating the changes in E to Em velocities ratio among prospective studies following RDN. *CI* confidence interval, *ES* effect size, *RDN* renal denervation

reviewed trials ($I^2 = 0\%$, $Q = 1.52$, $P = 0.677$). The pooled effect size of RDN on AIx was −7.05 (95% CI −9.12 to −4.98, $P < 0.001$; baseline mean = 48.98, 14.39% reduction; Fig. 28.5). Sensitivity analysis after excluding the only study with 3 months follow-up was confirmatory of the initial result.

28.1.5 Pulse Wave Velocity

PWV measurements before and after RDN were reported in four studies ($n = 158$). The follow-up duration was 6 months in all of them, whereas the study of Baroni et al. was reporting measurements both in 6 and 12 months. The outcome of 6 months was used for the current analysis for homogeneity reasons. The variation in effect size attributable to heterogeneity was moderate ($I^2 = 53.2\%$, $Q = 6.42$, $P = 0.093$). RDN treatment reduced PWV by 1.54 m/s (95% CI −2.16 to −0.92, $P < 0.001$; baseline mean = 11.35 m/s, 13.57% reduction; Fig. 28.5).

28.1.6 Blood Pressure

The pooled effect of RDN on BP in the included studies was analyzed after excluding data from duplicated cohorts. In 12 studies reporting data on SBP ($n = 473$ patients), with substantial heterogeneity ($I^2 = 81.2\%$, $Q = 58.6$, $P < 0.001$), RDN led to a decrease of 24.13 mmHg (95% CI −27.62 to −20.64, $P < 0.001$; baseline mean = 173.9 mmHg, 13.88% reduction). DBP data were reported in 11 studies ($n = 363$ patients), also with substantial heterogeneity ($I^2 = 77.8\%$, $Q = 45$, $P < 0.001$), and a decrease of 9.8 mmHg was noted (95% CI −12.26 to −7.34, $P < 0.001$; baseline mean = 93.54 mmHg, 10.48% reduction; Fig. 28.6).

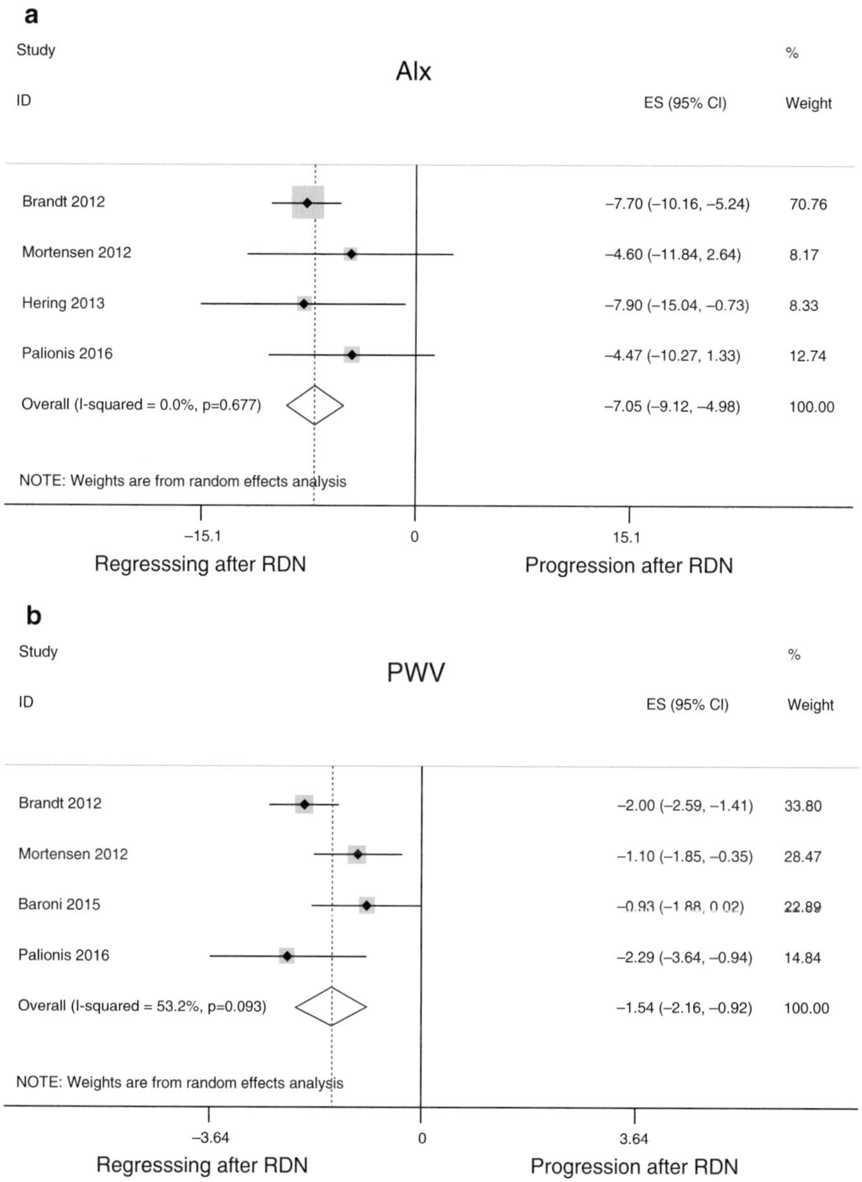

Fig. 28.5 Forest plots of indices of arterial stiffness. (**a**) Forest plot demonstrating the changes in Aix among prospective studies following RDN. (**b**) Forest plot demonstrating the changes in PWV among prospective studies following RDN. *AIx* augmentation index, *CI* confidence interval, *ES* effect size, *PWV* pulse wave velocity, *RDN* renal denervation

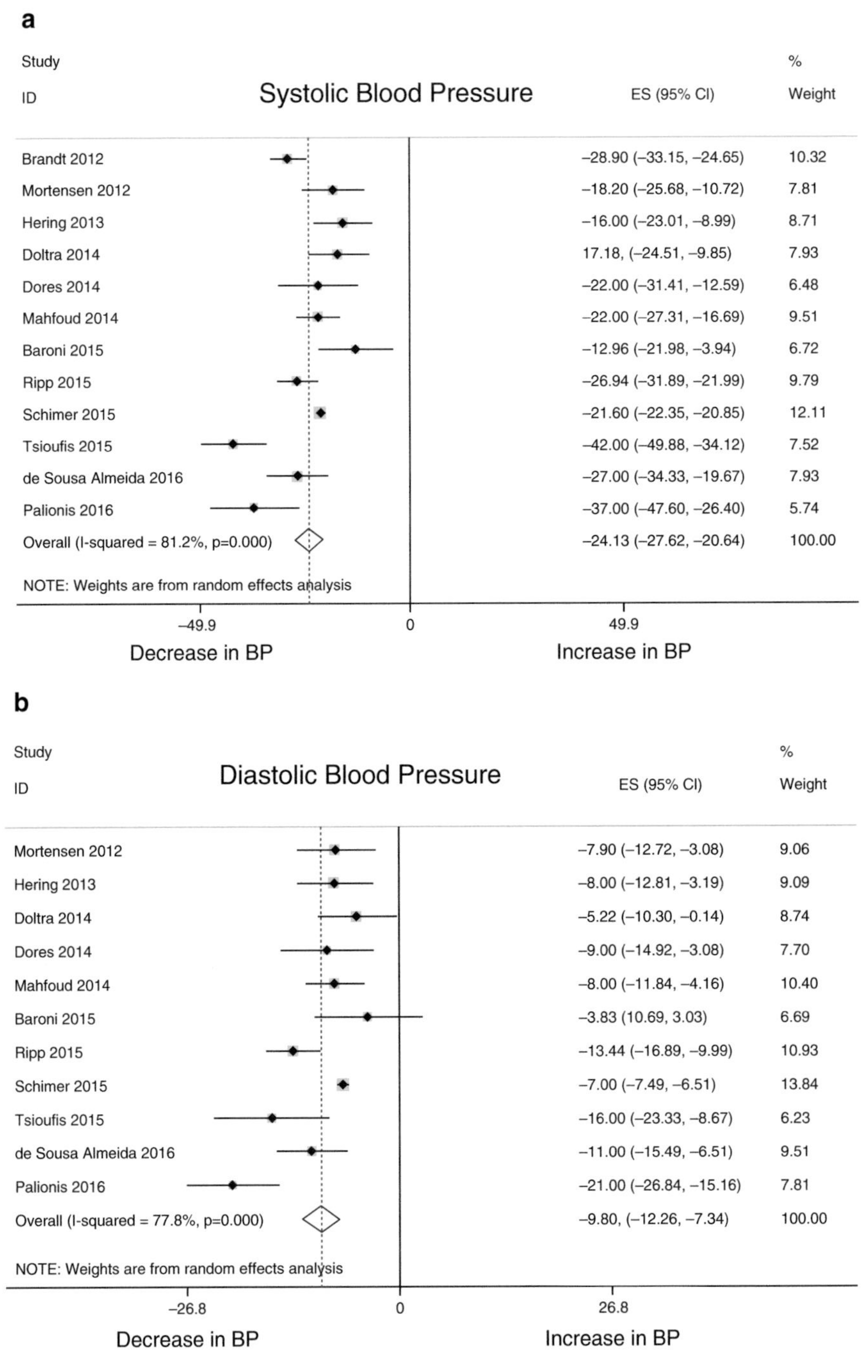

Fig. 28.6 (**a**) Forest plot demonstrating the changes in systolic blood pressure among prospective studies following RDN. (**b**) Forest plot demonstrating the changes in diastolic blood pressure among prospective studies following RDN. *BP* blood pressure, *CI* confidence interval, *ES* effect size, *RDN* renal denervation

28.2 Discussion

The aforementioned systematic review and meta-analysis of prospective observational studies demonstrated significantly beneficial effects of RDN on cardiovascular organ damage in patients with RH. First of all, this is the first meta-analysis of paired data that evaluated the effect of RDN on TOD taking into consideration the within-arm correlation of the outcome variance. Accordingly, a regression of cardiovascular adaptations after RDN was demonstrated. Secondly, there is no doubt that RDN has a solid anatomical and physiological background, which has to be more thoroughly explored focusing on the so-called "pleiotropic" benefits of this new therapeutic technique on TOD as well as in other disease conditions characterized by sympathetic overactivity [12, 31–36].

28.3 Cardiac Adaptations

RDN improves LV structure and function as indicated by a significant decrease in LVM (echo or CMR-assessed) and a decrease in E/Em. Meta-regression analysis demonstrated that RDN reduced echo or CMR-assessed LV mass irrespective of BP reduction. The effect of RDN on CMR-assessed LV mass was less pronounced mostly because of the more precise and reproducible assessment of cardiac mass using CMR and the blinded investigation [37]. Regarding blinding, we have to comment that blind endpoint analysis is only one aspect of blinding, whereas unblinded registration by the investigators may also influence the results. In addition to inducing LV mass regression, RDN significantly improved cardiac functional parameters expressed by a decrease in LV filling pressures as indicated by a reduced E/Em and left atrial size (in sensitivity analysis), along with the unaffected RDN transmitral inflow ratio E/A [38]. The latter, per se, represents the least objective index to assess diastolic function, as there is a significant overlay of E/A values across the spectrum of diastolic dysfunction. A global assessment of E/A in combination with tissue Doppler velocities and left atrial size is mandatory to reveal increased pressures [38]. From a pathophysiologic point of view, the fact that cardiac alterations are probably not only a result of BP reduction, as indicated by the results of the meta-regression analysis, points out a direct role of reduced sympathetic nervous system activity in patients after RDN. Catheter-based RDN reduces not only renal sympathetic efferent activity but also whole-body sympathetic activation by the ablation of afferent renal nerves, which stimulate sympathetic outflow in the hypothalamus [39]. It is established that sympathetic overactivity is a key component of the signaling pathways altered in hypertension-related cardiac remodeling even with experimental data indicating direct cardiomyocyte alpha-adrenergic receptors stimulation independent of hemodynamic load [40]. Indeed, a BP independent effect of sympathectomy on LV hypertrophy has been described in an animal model of abdominal aortic banding, where BP was unaffected by sympathectomy but LV hypertrophy decreased significantly [41].

28.4 Vascular Adaptations

In the current meta-analysis, central hemodynamics and arterial stiffness, which serve as better predictors for cardiovascular risk than peripheral brachial pressures, are improved in patients with RH after RDN [42, 43]. In line with cardiac damage regression, the reduction in PWV and AIx was independent of BP-lowering, pointing toward a RDN-related reduction of both total body (measured by muscle sympathetic nerve activity) and renal sympathetic activity (measured by norepinephrine spillover) [44]. AIx is a composite measure of the magnitude (e.g., vasoconstriction of peripheral blood vessels, which is under control of sympathetic system) and the timing of wave reflection, which itself is affected by an earlier return of the pulse wave as a result of an increased PWV because of arterial stiffness. RDN affects pulse wave reflection via a reduction in peripheral resistance, also resulting in a reduction of central BP [43, 44]. In addition to RDN-induced reduction of vasoconstrictive factors, improvement in endothelial function and intramural vascular remodeling processes may have additionally contributed to reduction in arterial stiffness and wave reflections [45]. In analogy with antihypertensive strategies that target the renin–angiotensin system and improve arterial stiffness, modification of sympathetic tone by RDN may induce favorable structural and functional alterations independent of BP lowering [46].

28.5 Limitations

The present meta-analysis is subject to several limitations. First, it is a meta-analysis of observational studies, hence vulnerable to systematic bias and the inability to control for confounding factors. Second, because of the design and the presented data of the included studies, we had to impute the correlation of measured outcomes between baseline and follow-up. However, sensitivity analysis using different levels of imputed correlation was confirmatory of the main analysis. Third, the majority of the studies had small sample size and several differences in methodology (type of catheters used, technical aspects of ablation, antihypertensive medication, adherence control). Fourth, the variability in quantification and presentation of measured outcomes (use of multiple imaging modalities, not indexed values or indexed to different variables) was substantial leading to multiple analyses with a small number of studies. Fifth, follow-up duration of the included studies was limited to 12 months, so no conclusions regarding the long-term effect of RDN on TOD regression can be drawn. Sixth, because of presented aggregate data, it was not possible to control for the effect of established protective medication on pooled effect. Last, the associations derived from meta-regressions are based on averages of patients' characteristics, containing variability of measurements or variability of measurement errors, and have a weaker interpretation for causal relationships.

28.6 Conclusion

In conclusion, RDN induces cardiovascular damage regression and may supposingly improve cardiovascular morbidity and mortality, while RDN has the potential to prevent the development of heart failure in hypertensive patients. The future of RDN depends on the technological advances that secure a complete or at least near-complete injury of renal sympathetic fibers and proof of long-term cardiovascular benefits in patients with RH and/or other diseases. Thus, well-designed studies that include as endpoints not only BP levels but also surrogates of target organ damage are necessary toward this direction.

References

1. Esler MD, Krum H, Sobotka PA, Schlaich MP, Schmieder RE, Böhm M, Symplicity HTN-2 Investigators. Renal sympathetic denervation in patients with treatment-RESISTANT HYPERTENSION (the symplicity HTN-2 trial): a randomised controlled trial. Lancet. 2010;376:1903–9. https://doi.org/10.1016/S0140-6736(10)62039-9.
2. Krum H, Schlaich M, Whitbourn R, Sobotka PA, Sadowski J, Bartus K, Kapelak B, Walton A, Sievert H, Thambar S, et al. Catheter-based renal sympathetic denervation for resistant hypertension: a multicentre safety and proof-of-principle cohort study. Lancet. 2009;373:1275–81. https://doi.org/10.1016/S0140-6736(09)60566-3.
3. Sievert H, Schofer J, Ormiston J, Hoppe UC, Meredith IT, Walters DL, Azizi M, Diaz-Cartelle J, Cohen-Mazor M. Renal denervation with a percutaneous bipolar radiofrequency balloon catheter in patients with resistant hypertension: 6-month results from the REDUCE-HTN clinical study. EuroIntervention. 2015;10:1213–20. https://doi.org/10.4244/EIJY14M12_01.
4. Worthley SG, Tsioufis CP, Worthley MI, Sinhal A, Chew DP, Meredith IT, Malaiapan Y, Papademetriou V. Safety and efficacy of a multi-electrode renal sympathetic denervation system in resistant hypertension: the EnligHTN I trial. Eur Heart J. 2013;34:2132–40. https://doi.org/10.1093/eurheartj/eht197.
5. Bhatt DL, Kandzari DE, O'Neill WW, D'Agostino R, Flack JM, Katzen BT, Leon MB, Liu M, Mauri L, Negoita M, SYMPLICITY HTN-3 Investigators, et al. A controlled trial of renal denervation for resistant hypertension. N Engl J Med. 2014;370:1393–401. https://doi.org/10.1056/NEJMoa1402670.
6. Kandzari DE, Bhatt DL, Brar S, Devireddy CM, Esler M, Fahy M, Flack JM, Katzen BT, Lea J, Lee DP, et al. Predictors of blood pressure response in the SYMPLICITY HTN-3 trial. Eur Heart J. 2015;36:219–27. https://doi.org/10.1093/eurheartj/ehu441.
7. Townsend RR, Mahfoud F, Kandzari DE, Kario K, Pocock S, Weber MA, Ewen S, Tsioufis K, Tousoulis D, ASP S, SPYRAL HTN- OFF MED Trial Investigators*, et al. Catheter-based renal denervation in pa- tients with uncontrolled hypertension in the absence of antihypertensive medications (SPYRAL HTN-OFF MED): a randomised, sham-controlled, proof-of-concept trial. Lancet. 2017;390:2160–70. https://doi.org/10.1016/S0140-6736(17)32281-X.
8. Kandzari DE, Böhm M, Mahfoud F, Townsend RR, Weber MA, Pocock S, Tsioufis K, Tousoulis D, Choi JW, East C, SPYRAL HTN-ON MED Trial Investigators, et al. Effect of renal denervation on blood pressure in the presence of antihypertensive drugs: 6-month efficacy and safety results from the SPYRAL HTN-ON MED proof-of-concept randomised trial. Lancet. 2018;391:2346–55. https://doi.org/10.1016/S0140-6736(18)30951-6.
9. Azizi M, Schmieder RE, Mahfoud F, Weber MA, Daemen J, Davies J, Basile J, Kirtane AJ, Wang Y, Lobo MD, RADIANCE-HTN Investigators, et al. Endovascular ultrasound renal denervation to treat hypertension (RADIANCE-HTN SOLO): a multicentre, international,

single-blind, randomised, sham-controlled trial. Lancet. 2018;391:2335–45. https://doi.org/10.1016/S0140-6736(18)31082-1.
10. Böhm M, Kario K, Kandzari DE, Mahfoud F, Weber MA, Schmieder RE, Tsioufis K, Pocock S, Konstantinidis D, Choi JW, SPYRAL HTN-OFF MED Pivotal Investigators, et al. Efficacy of catheter-based renal denervation in the absence of antihypertensive medications (SPYRAL HTN-OFF MED Pivotal): a multicentre, randomised, sham-controlled trial. Lancet. 2020;395:1444–51. https://doi.org/10.1016/S0140-6736(20)30554-7.
11. Azizi M, Sanghvi K, Saxena M, Gosse P, Reilly JP, Levy T, Rump LC, Persu A, Basile J, Bloch MJ, RADIANCE-HTN Investigators, et al. Ultrasound renal denervation for hypertension resistant to a triple medication pill (RADIANCE-HTN TRIO): a randomised, multicentre, single-blind, sham-controlled trial. Lancet. 2021;397:2476–86. https://doi.org/10.1016/S0140-6736(21)00788-1.
12. Mancia G, Fagard R, Narkiewicz K, Redon J, Zanchetti A, Bo¨hm M, et al. ESH/ESC guidelines for the management of arterial hypertension: the task force for the management of arterial hypertension of the European Society of Hypertension (ESH) and of the European Society of Cardiology (ESC). J Hypertens. 2013;2013(31):1925–38.
13. Tsioufis C, Vezali E, Tsiachris D, Dimitriadis K, Taxiarchou E, Chatzis D, et al. Left ventricular hypertrophy versus chronic kidney disease as predictors of cardiovascular events in hypertension: a Greek 6-year follow-up study. J Hypertens. 2009;27:744–52.
14. Brandt MC, Mahfoud F, Reda S, Schirmer SH, Erdmann E, Bo¨hm M, et al. Renal sympathetic denervation reduces left ventricular hypertrophy and improves cardiac function in patients with resistant hypertension. J Am Coll Cardiol. 2012;59:901–9.
15. Brandt MC, Reda S, Mahfoud F, Lenski M, Bo¨hm M, Hoppe UC. Effects of renal sympathetic denervation on arterial stiffness and central hemodynamics in patients with resistant hypertension. J Am Coll Cardiol. 2012;60:1956–65.
16. Mortensen K, Franzen K, Himmel F, Bode F, Schunkert H, Weil J, et al. Catheter-based renal sympathetic denervation improves central hemodynamics and arterial stiffness: a pilot study. J Clin Hypertens. 2012;14:861–70.
17. Hering D, Lambert EA, Marusic P, Ika-Sari C, Walton AS, Krum H, et al. Renal nerve ablation reduces augmentation index in patients with resistant hypertension. J Hypertens. 2013;31:1893–900.
18. Doltra A, Messroghli D, Stawowy P, Hassel JH, Gebker R, Leppa¨nen O, et al. Potential reduction of interstitial myocardial fibrosis with renal denervation. J Am Heart Assoc. 2014;3:e001353.
19. Dores H, de Sousa AM, de Arau'jo Gonc¸alves P, Branco P, Gaspar A, Sousa H, et al. Renal denervation in patients with resistant hypertension: six-month results. Rev Port Cardiol. 2014;33:197–204.
20. Mahfoud F, Urban D, Teller D, Linz D, Stawowy P, Hassel J, et al. Effect of renal denervation on left ventricular mass and function in patients with resistant hypertension: data from a multicentre cardiovascular magnetic resonance imaging trial. Eur Heart J. 2014;35:2224–31.
21. Schirmer S, Sayed M, Reil J, Ukena C, Linz D, Kindermann M, et al. Improvements in left ventricular hypertrophy and diastolic function following renal denervation: effects beyond blood pressure and heart rate reduction. J Am Coll Cardiol. 2014;63:1916–23.
22. Baroni M, Nava S, Giupponi L, Meani P, Panzeri F, Varrenti M, et al. Effects of renal sympathetic denervation on arterial stiffness and blood pressure control in resistant hypertensive patients: a single centre prospective study. High Blood Press Cardiovasc Prev. 2015;22:411–6.
23. McLellan A, Schlaich M, Taylor A, Prabhu S, Hering D, Hammond L, et al. Reverse cardiac remodeling after renal denervation: atrial electrophysiologic and structural changes associated with blood pressure lowering. Heart Rhythm. 2015;12:982–90.
24. Ripp T, Mordovin V, Pekarskiy S, Ryabova T, Zlobina M, Baev A, et al. Predictors of renal denervation efficacy in the treatment of resistant hypertension. Curr Hypertens Rep. 2015;17:90.

25. Schirmer S, Sayed M, Reil J, Lavall D, Ukena C, Linz D, et al. Atrial remodeling following catheter-based renal denervation occurs in a blood pressure- and heart rate-independent manner. JACC Cardiovasc Interv. 2015;8:972–80.
26. Tsioufis C, Papademetriou V, Dimitriadis K, Tsiachris D, Thomopoulos C, Kasiakogias A, et al. Effects of multielectrode renal denervation on cardiac and neurohumoral adaptations in resistant hypertension with cardiac hypertrophy: an EnligHTN I substudy. J Hypertens. 2015;33:346–53.
27. Verloop W, Vink E, Spiering W, Blankestijn P, Doevendans P, Bots M, et al. Effects of renal denervation on end organ damage in hypertensive patients. Eur J Prev Cardiol. 2015;22:558–67.
28. de Sousa AM, de Arau'jo Gonc¸alves P, Branco P, Mesquita J, Carvalho MS, Dores H, et al. Impact of renal sympathetic denervation on left ventricular structure and function at 1-year follow-up. PLoS One. 2016;11:e0149855.
29. Kiuchi MG, Mion D Jr, Graciano ML, de Queiroz Carreira MA, Kiuchi T, Chen S, et al. Proof of concept study: improvement of echocardiographic parameters after renal sympathetic denervation in CKD refractory hypertensive patients. Int J Cardiol. 2016;207:6–12.
30. Palionis D, Berukstis A, Misonis N, Ryliskyte L, Celutkiene J, Zakarkaite D, et al. Could careful patient selection for renal denervation warrant a positive effect on arterial stiffness and left ventricular mass reduction? Acta Cardiol. 2016;71:173–83.
31. Tsioufis C, Kordalis A, Flessas D, Anastasopoulos I, Tsiachris D, Papademetriou V, et al. Pathophysiology of resistant hypertension: the role of sympathetic nervous system. Int J Hypertens. 2011;2011:642416, 1.
32. Kordalis A, Tsiachris D, Pietri P, Tsioufis C, Stefanadis C. Regression of organ damage following renal denervation in resistant hypertension: a meta-analysis. J Hypertens. 2018;36(8):1614–21. https://doi.org/10.1097/HJH.0000000000001798; PMID: 29846327.
33. Follmann D, Elliott P, Suh I, Cutler J. Variance imputation for overviews of clinical trials with continuous response. J Clin Epidemiol. 1992;45:769–73.
34. Balk EM, Earley A, Patel K, Trikalinos TA, Dahabreh IJ. Empirical assessment of within-arm correlation imputation in trials of continuous outcomes. Rockville (MD): Agency for Healthcare Research and Quality (US); 2012. report no.: 12(13)-EHC141-EF. AHRQ Methods for Effective Healthcare.
35. Moss JG, Belli AM, Coca A, Lee M, Mancia G, Peregrin JH, et al. Executive summary of the joint position paper on renal denervation of the Cardiovascular and Interventional Radiological Society of Europe (CIRSE) and the European Society of Hypertension (ESH). Cardiovasc Intervent Radiol. 2016;39:1681–3.
36. Papademetriou V, Doumas M, Tsioufis C. Renal sympathetic denervation: hibernation or resurrection? Cardiology. 2016;135:87–97.
37. Rudolph A, Abdel-Aty H, Bohl S, Boye´ P, Zagrosek A, Dietz R, et al. Noninvasive detection of fibrosis applying contrast-enhanced cardiac magnetic resonance in different forms of left ventricular hypertrophy relation to remodeling. J Am Coll Cardiol. 2009;53:284–91.
38. Mitter SS, Shah SJ, Thomas JD. A test in context: E/a and E/e' to assess diastolic dysfunction and LV filling pressure. J Am Coll Cardiol. 2017;69:1451–64.
39. Krum H, Sobotka P, Mahfoud F, Bohm M, Esler M, Schlaich M. Device-based antihypertensive therapy: therapeutic modulation of the autonomic nervous system. Circulation. 2011;123:209–15.
40. Mancia G, Grassi G, Giannattasio C, Seravalle G. Sympathetic activation in the pathogenesis of hypertension and progression of organ damage. Hypertension. 1999;34(4 Pt 2):724–8.
41. Perlini S, Palladini G, Ferrero I, Tozzi R, Fallarini S, Facoetti A, et al. Sympathectomy or doxazosin, but not propranolol, blunt myocardial interstitial fibrosis in pressure overload hypertrophy. Hypertension. 2005;46:1213–8.
42. Vlachopoulos C, Aznaouridis K, Stefanadis C. Prediction of cardiovascular events and all-cause mortality with arterial stiffness: a systematic review and meta-analysis. J Am Coll Cardiol. 2010;55:1318–27.

43. Vlachopoulos C, Aznaouridis K, O'Rourke MF, Safar ME, Baou K, Stefanadis C. Prediction of cardiovascular events and all-cause mortality with central haemodynamics: a systematic review and meta-analysis. Eur Heart J. 2010;31:1865–71.
44. Boutouyrie P, Lacolley P, Girerd X, Beck L, Safar M, Laurent S. Sympathetic activation decreases medium-sized arterial compliance in humans. Am J Phys. 1994;267:H1368–76.
45. Hering D, Lambert EA, Marusic P, Walton AS, Krum H, Lambert GW, et al. Substantial reduction in single sympathetic nerve firing after renal denervation in patients with resistant hypertension. Hypertension. 2013;61:457–64.
46. Vlachopoulos C, Aznaouridis K, Stefanadis C. Clinical appraisal of arterial stiffness: the argonauts in front of the Golden fleece. Heart. 2006;92:1544–50.

Part VI

Special Populations

Hypertension in Young Adults

29

Ana-Maria Vintilă and Maria Dorobanțu

Abbreviations

ABI	Ankle-brachial index
ABPM	Ambulatory blood pressure monitoring
ACC	American college of cardiology
ACEi	Angiotensin-converting enzyme inhibitor
AHA	American heart association
ARBs	Angiotensin receptor blocker
BB	Beta-blocker
BMI	Body mass index
BP	Blood pressure
CAD	Coronary artery disease
CCB	Calcium channel blocker
CKD	Chronic kidney disease
CV	Cardiovascular
CVD	Cardiovascular disease
DBP	Diastolic blood pressure
ESC	European society of cardiology
HBPM	Home blood pressure monitoring
HTN	Arterial hypertension
IDH	Isolated diastolic hypertension

A.-M. Vintilă (✉)
Carol Davila University of Medicine and Pharmacy, Bucharest, Romania

Internal Medicine and Cardiology Department, Colțea Clinical Hospital, Bucharest, Romania

M. Dorobanțu
Carol Davila University of Medicine and Pharmacy, Bucharest, Romania

M. Dorobantu et al. (eds.), *Hypertension and Heart Failure*, Updates in Hypertension and Cardiovascular Protection,
https://doi.org/10.1007/978-3-031-39315-0_29

IMT	Intima-media thickness
ISH	Isolated systolic hypertension
LDL	Low-density lipoprotein
LV	Left ventricle
LVH	Left ventricular hypertrophy
MACE	Major adverse cardiovascular event
MBP	Mean blood pressure
NNT	Number needed to treat
PWV	Pulse wave velocity
RAAS	Renin-angiotensin-aldosterone system
SBP	Systolic blood pressure
SDH	Systolic and diastolic hypertension
SNS	Sympathetic nervous system
VLDL	Very-low-density lipoprotein

29.1 Prevalence of Hypertension in Young Adults

Arterial hypertension in young adults is not well defined throughout medical literature or hypertension guidelines. The age interval characterizing young adulthood varies widely according to publication: 18–30 years in the CARDIA study [1], 20–49 years in the Framingham Offspring group [2], <50 years in the ESC Hypertension guidelines [3]. However, ages 18 through 45 seem to be the most reasonable. When we discuss hypertension in young adults, irrespective of source, there seem to be some common elements like sympathetic overactivity, greater influence of metabolic changes in general and obesity in particular, a higher prevalence of secondary hypertension, lower disease awareness, and adherence to treatment.

Since the age interval for "young adults" is difficult to pinpoint, so too is the prevalence of hypertension among this age group [4]. What is certain, however, is that its prevalence had been on the rise during the past decades [1, 4–6].

In 1987, a study published by the Framingham Offspring group found a prevalence of hypertension among patients aged 20–49 of 12.9% and 14.2% for women and men, respectively [2]. Later, in 2007, based on data from the Coronary Artery Risk Development in Young Adults (CARDIA study), a prevalence of 20% among patients aged 18–30 was reported [7]. As recently as 2020, the National Center for Health Statistics (NCHS), part of the Center for Disease Control and Prevention (CDC) in the United States of America, published a report on the prevalence of arterial hypertension in young adults: 22.4% among patients aged 18–39, higher among the young men (31.2% vs. 13.0%) [8]. African-American are twice more likely to develop arterial hypertension by age 25 [1].

Gender differences are more likely to appear after puberty with more young and middle-aged men being hypertensive and either disappear or are reversed in old age [9–11].

The prevalence of hypertension among young adults is not distributed uniformly across the globe. A 2003 study evaluated the geographic distribution of hypertension in young adults (35–44 years) and found a prevalence of 14% in North America and 27% in Europe [12]. The prevalence of hypertension seems to be even higher in Eastern Europe according to data provided by the World Health Organization through National STEPS reports. Thus, the Ukraine has a hypertension prevalence of 10.8% and 15.7% in patients aged 18–29 and 30–44 respectively [13], in Belarus, patients already undergoing antihypertensive treatment are as many as 17.9% and 33.5% for age ranges 18–29 and 30–44 respectively [14] while, in Romania, according to the SEPHARIII survey hypertension prevalence is 15.4%, 21.6%, and 34.6% for 18–24 years, 25–34 years, and 35–44 years, respectively [15].

29.2 Cardiovascular Risk Associated with Hypertension in Young Adults

The prevalence of hypertension in young adults is closely tied to cardiovascular and cerebrovascular risk. Insulin resistance, obesity, and smoking aggregate in young adults with arterial hypertension [16], promoting atherosclerosis and leading to an increased risk of stroke and cardiovascular disease [7]. Moreover, the risk of cardiovascular events remains above average in spite of correct antihypertensive treatment [2].

A subanalysis of the Strong Heart Study performed on patients aged 14–39 showed hypertensive patients to be more likely to have higher weight, body mass index (BMI), and a propensity for central obesity. Hypertensive patients also had higher total cholesterol, LDL cholesterol, and insulin levels [11], a finding also supported by data from the CARDIA study. Higher fasting insulin levels are correlated with an increase in cardiovascular disease (CVD) risk probably by the promotion of plaque formation [17].

Left ventricular hypertrophy, LV mass index, and increased peripheral resistance have been associated with cardiovascular events [1], and hypertensive patients seem to have more of each when compared to non-hypertensive patients. The Strong Heart Study also reports a threefold increase in the prevalence of left ventricular hypertrophy in hypertensive patients when compared to normotensive patients [11].

A meta-analysis published in 2020 evaluated the cardiovascular risk associated with arterial hypertension in 4.5 million young adults aged 18–45. After a mean follow-up of 14 years, it was estimated that the risk of cardiovascular events increased with the degree of hypertension (patients were categorized into optimal blood pressure, normal blood pressure, high-normal blood pressure, grade 1 hypertension, and grade 2 hypertension). In patients with grade 1 and 2 hypertension, the risk of cardiovascular events increased by twofold and threefold compared to normotensive patients; a 4–5% increase in risk was attributed to every 10 mmHg increase in systolic blood pressure and every 5 mmHg increase in diastolic blood pressure. Cardiovascular risk was further stratified into risk of developing coronary artery disease, stroke, and all-cause mortality with coronary artery disease being 2.27 times more likely in grade 2 hypertensive patients [18].

A 2018 paper, published by Qi et al. [19] in light of the 2017 ACC/AHA Hypertension guideline, assessed cardiovascular risk associated with grade 1 hypertension stratified by age groups: patients aged 35–44 with stage 1 hypertension were at an increased risk of CVD, stroke, and CVD mortality. Interestingly, older patients (≥ 60 years) with stage 1 hypertension did not exhibit an increased risk of CVD [19]. It may be that the more substantial contribution of hypertension to CV risk among young patients is due to fewer confounding comorbidities than for older adults [18].

The systolic and diastolic components of blood pressure seem to have different outcomes on cardiovascular risk—while high systolic blood pressure increases the risk of stroke and coronary artery disease in a similar fashion, high diastolic blood pressure carries a heavier risk of stroke compared to CAD [18].

A paper from a Swedish group led by Sundstrom evaluated cardiovascular mortality in a national cohort of 1.2 million men (mean age 18.4 years) with a 24-year follow-up. Diastolic blood pressure (BP) seemed to better correlate with cardiovascular mortality than systolic BP and had an apparent risk threshold around 90 mmHg. Moreover, systolic BP had a U-shaped relationship to cardiovascular mortality with the lowest risk corresponding to 130 mmHg [20].

In another 2018 paper in reference to the 2017 Guidelines, a Korean group published a paper evaluating CV risk in a cohort of nearly 2.5 million patients aged 20–39. Similar to the study led by Qi et al., there was also an increase in cardiovascular disease risk associated with above normal systolic (≥120 mmHg) and diastolic (≥80 mmHg) blood pressure values. Patients with stage 1 hypertension had an additional 5.1 and 3.9 cardiovascular disease risk per 1000 participants for men and women respectively, 3.0 and 1.4 for coronary artery disease risk, and 3.4 and 2.8 for stroke risk [21].

A subanalysis of the CARDIA study showed that even in normotensive patients, an inappropriate rise in systolic blood pressure in response to exercise (≥190–210 mmHg) may lead to the development of hypertension at follow-up (15 years later) [1, 22].

29.2.1 Cardiovascular Risk of Isolated Systolic Hypertension

In young individuals, including children, adolescents, and young adults, isolated systolic hypertension (SBP ≥ 140–159 mmHg and DBP < 90 mmHg) is the most common form of essential hypertension [23].

The significance of isolated systolic hypertension (SIH) in young adults has been a subject of debate. On the one hand, ISH is frequently associated with obesity, insulin resistance, metabolic syndrome, and smoking and leads to aortic stiffening with a subsequent increase in heart rate, cardiac output, and stroke volume, while on the other hand, ISH is also commonly found in tall, nonsmoking, physically active young men and has been attributed to the elasticity of large vessels which overly amplify pulse pressure from central to peripheral arteries [24].

The prevalence of ISH among young adults (< 40 years) in the United States of America was reported as 0.7% between 1988 and 1994 and doubled to 1.6% between 1999 and 2004 [25].

In 2015, Yano et al. published a paper based on a cohort of young adults (18–49 years) from the Chicago Heart Association Detection Project in Industry (CHA) study assessing CVD risk during a 31-year follow-up period. The prevalence of ISH, IDH, and SDH was 25.3%, 3.7%, and 19.8%, respectively. Isolated systolic hypertension patients had higher BMI, smoking rate, and total cholesterol levels as well as a higher CVD risk and coronary artery disease mortality than patients with IDH or normal high-normal BP. Patients with SDH, however, had the highest CVD risk at follow-up [25].

Based on the publication by Yano et al. and the assumption that ISH is secondary to pulse pressure, in 2017, Saladini et al. assessed the prognostic significance of elevated pulse pressure in a cohort of young adults (aged 18–45 years) with stage 1 hypertension from HARVEST (Hypertension Ambulatory Recording Venetia Study) after 12 years of follow-up. Patients with high office mean arterial blood pressure were three times more likely to suffer major adverse cardiovascular events (hazard ratio: 3.06, 95% confidence interval: 1.32–7.09; $P = 0.009$) while those with high pulse pressure had a low risk of MACE (HR: 0.35, 95% CI: 0.17–0.73; $P = 0.005$). It should be noted that patients at high cardiovascular risk (including heavy smokers), patients with diabetes, chronic kidney disease, and those with pre-existing cardiovascular disease were excluded from the analysis [26].

A 2020 study by Lee et al. evaluated cardiovascular outcomes in 6.4 million Koreans aged 20–39 and found that stage 1 isolated systolic hypertension (ISH), isolated diastolic hypertension (IDH), and systolic and diastolic hypertension (SDH) were all associated with a higher risk of CVD and CV mortality. The CV risk for stage 1 ISH and IDH was slightly lower than stage 2 ISH or IDH and stage 1 SDH. In this cohort, the cardiovascular risk associated with nonoptimal blood pressure was reported as higher in women than men [27].

In 2021, Seryan et al. published a study evaluating cardiovascular outcomes in patients 25–45 years of age from the prospective population-based MONICA/KORA cohort after a 25-year follow-up. Results showed that patients with IDH, ISH, or SDH had higher absolute CVD mortality compared to non-hypertensive patients (1.3, 1.5, and 2.6 vs. 0.6 cases/1000 person-years). Moreover, CVD mortality was almost two times higher in men with ISH than women with ISH [24].

Young adults with arterial hypertension are more likely to be obese, have higher cholesterol and fasting insulin values, as well as LV hypertrophy and increased peripheral resistance, thus at an increased risk of developing cardiovascular disease. The importance of hypertension in assessing CVD risk is higher for young patients due to fewer confounding risk factors than in older populations. Moreover, isolated systolic hypertension increases the risk of CVD and CV mortality, though to a lower extent than systolic and diastolic hypertension. However, since it is the most common form of HTN among children, adolescents, and young adults and its incidence has been rising steadily through the past decades, the increase in risk should not be overlooked.

29.3 Pathogenic Relationship: Obesity and Hypertension in Young Adults

Obesity has a paramount role in the development of arterial hypertension in young adults. Arterial stiffness increase in childhood correlates with the development of obesity and hypertension in young adults. From a pathogenic standpoint, adipokines produced by visceral fat activate the sympathetic nervous system leading to increased glomerular filtration, sodium reabsorption and renin as well as angiotensin-II release, while at the same time promoting insulin secretion and resistance. Moreover, the accumulation of fat within various tissues has a lipotoxic and proinflammatory effect, which leads to endothelial dysfunction through nitric oxide release inhibition. As obesity leads to an increase in BP values, with time, target organ damage occurs and further exacerbates arterial hypertension.

29.3.1 Central Arterial Stiffness

Central arterial stiffness is considered one of the earliest identifiable manifestations of vascular insult [28] and is the most important pathophysiological determinant of isolated systolic hypertension [3]. It also correlates strongly with cardiovascular disease and has a high predictive value for adverse cardiovascular events [28]. In order to assess arterial stiffness, noninvasive measurements such as pulse wave velocity (PWV) and intima-media thickness (IMT) are used, though not routinely in clinical practice [29].

A meta-analysis published in 2015 analyzed the association between arterial stiffness measurements and obesity over 15 studies with a total number of 2237 participants (aged 5–24). Thirteen of the 15 studies reported greater arterial stiffness in obese participants compared to control groups, while the other two found an association between PWV and blood pressure, irrespective of BMI [28].

In 2005 McEniery et al. published a paper aiming to assess the association of arterial stiffening and essential hypertension (in particular ISH) in a cohort of 1028 healthy young adults (aged 17–27 years) from the ENIGMA study. The authors believe a difference in pathophysiological mechanism exists between SDH and ISH based on measures of arterial stiffness. Patients with ISH had higher aortic pulse wave velocity, cardiac output, and stroke volume compared to non-hypertensive patients while SDH patients had higher peripheral vascular resistance, decreased cardiac output, and normal isobaric aortic stiffness [30].

A 2022 paper reported data on a prospective trial of 3862 participants from the ALSPAC (Avon Longitudinal Study of Parents and Children) birth cohort, which evaluated arterial stiffness, measured by pulse wave velocity and intima-media thickness at 17.7 years and prevalence of obesity and hypertension at 24.5 years. A higher PWV at enrollment predicted hypertension and obesity at follow-up, and a PWV increase over the 7-year gap predicted an increment in systolic and diastolic BP. Moreover, IMT progression was associated with BMI, obesity, and increased diastolic BP. It was concluded that increased arterial stiffness in adolescence predates obesity and hypertension in adulthood [29].

Mild fluctuations in BP occur in response to caloric intake: fasting reduces BP while weight gain increases BP. In patients with obesity (visceral obesity in particular), however, there is a vicious cycle where weight gain leads to an increase in BP and chronically elevated BP further leads to target organ damage and exacerbates hypertension [31].

29.3.2 Overactivation of the Sympathetic Nervous System

It seems that in young adults, the pathogenic link between obesity and hypertension lies in the overactivation of the sympathetic nervous system (SNS) and renin-angiotensin-aldosterone system (RAAS). Visceral fat produces leptin and a variety of other adipokines which activate the SNS with subsequent renal sodium reabsorption and renin release. Renin promotes further SNS stimulation through angiotensin-II [31, 32]. Moreover, retroperitoneal fat compresses the kidney and mechanically stimulates the RAAS.

Overstimulation of the sympathetic nervous system leads in time to a downregulation of β-receptors which has deleterious effects on the cardiovascular system by promoting myocardial necrosis, ventricular fibrillation, atherosclerosis, and ventricular hypertrophy. Furthermore, SNS stimulation increases stroke volume, heart rate, and cardiac output while endothelial dysfunction leads to impaired vasodilation and increased peripheral resistance and together promote hypertension [32].

A small study performed on 163 young adults (aged 15–45 years) with grade 1 hypertension aimed to assess if a high sympathetic or a low parasympathetic tone (assessed through heart rate variability) influences blood pressure values and serum lipid levels. Results showed that patients with sympathetic predominance and reduced heart rate variability progressed to grade 2 HTN and developed hypercholesterolemia by the time of the 6-year follow-up [33].

29.3.3 Systemic Inflammation and Hypertension

Systemic inflammation is a product of obesity through multiple mechanisms. Firstly, visceral obesity in itself leads to a systemic inflammatory response—elevated levels of C-reactive protein, erythrocyte sedimentation rate, TNFα and IL-6 [34]. Secondly, tissue accumulation of lipids has lipotoxic effects which in turn exacerbate inflammation through oxidative stress and prevent nitric oxide release, thus leading to endothelial dysfunction. Finally, adipokines promote insulin secretion and insulin resistance [31, 32, 34].

The initial effects of obesity on the kidney, aside from tubular sodium reabsorption and renin release, include an increase in glomerular filtration rate (hyperfiltration). However, with passing time, hyperfiltration leads to kidney injury—podocyte injury with microalbuminuria, expansion of mesangial matrix, and varying degrees of fibrosis up to focal segmental glomerular sclerosis—and ultimately a decline in glomerular filtration rate through loss of functioning nephrons [31, 32, 34].

The pooled results of multiple reports seem to point at hyperuricemia as a predictor of (severe) hypertension in adolescents and young adults. An experimental study found that uric acid levels correlated with both systolic and diastolic BP and drugs used to treat hyperuricemia also improved BP control rates. Pathogenically, renal physiology alterations caused by excess uric acid include production of inflammatory cytokines ultimately leading to sodium-sensitive hypertension. It should be noted that obesity increases the risk of hyperuricemia by threefold [35].

29.4 Diagnosis of Hypertension in Young Adults

Essential hypertension, in the form of isolated systolic or diastolic hypertension, is much more common than secondary hypertension over the age of 10, especially when the patient is also overweight and/or has immediate family already diagnosed with HTN [4].

White coat hypertension is defined by elevated office blood pressure in untreated patients with normal home or ambulatory BP measurements. White coat hypertension is more prevalent among children and young adults with a positive family history of HTN and mild elevation of blood pressure; ambulatory BP measurements are recommended in order to establish the diagnosis and differentiate sustained hypertension from white coat and masked hypertension [36]. Patients with white coat hypertension during childhood have an increased risk of developing sustained arterial hypertension as young adults and are at an increased risk of developing CVD [37]. A 2006 paper showed that adolescents and young adults (11–21 years) with either white coat hypertension or sustained hypertension had higher BMI and lower baroreflex sensitivity when compared to normotensive controls [38].

The prevalence of secondary hypertension is, however, higher in young patients in comparison to older adults—as high as 10% under the age of 50—and should be suspected for grade 2 or 3 hypertension, signs of endocrinopathy, sleep apnea, or known kidney disease [3].

A retrospective, cross-sectional study performed on over 6300 Korean hypertensive men aged 19–29 identified secondary hypertension in 2.2% of the cohort. The most common causes of HTN identified were autosomal dominant polycystic kidney disease (0.74%), renal parenchymal disease (0.38%), hyperthyroidism (0.36%), and subclinical hypercorticism (0.19%) [39]. A retrospective analysis on the prevalence of secondary hypertension performed on young (18–40 years), consecutive, hypertensive patients reported a prevalence of secondary hypertension as high as 29.6% and included patients with primary hyperaldosteronism (7.4%), fibromuscular dysplasia (5.8%), and pheochromocytoma (3.9%). It should be noted that the latter analysis was carried out at a tertiary center for hypertension management, thus explaining a higher prevalence of secondary hypertension when compared to the general population [40].

Diagnosis of secondary hypertension may lead to identification of potentially curable causes with improved BP control and lower CVD risk after appropriate treatment [41]. According to the ESC hypertension guidelines, the most common causes of secondary hypertension in young adults are renal parenchymal disease, fibromuscular dysplasia, and monogenic disorders [3].

The initial workup for hypertensive young adults should include routine laboratory assays (complete blood count, creatinine, urea, serum electrolytes, and urine analysis including microalbuminuria) as well as an electrocardiogram and echocardiography in order to evaluate target organ damage. Further investigations for causes of secondary hypertension should be restricted to selected patients where the index of clinical suspicion is high [4]. The ESC guidelines recommend screening for secondary hypertension in all young adults (<40 years) with grade 2 or greater hypertension while the ACC/AHA guidelines recommend screening in patients <30 years with elevated BP [3, 41, 42]—Fig. 29.1.

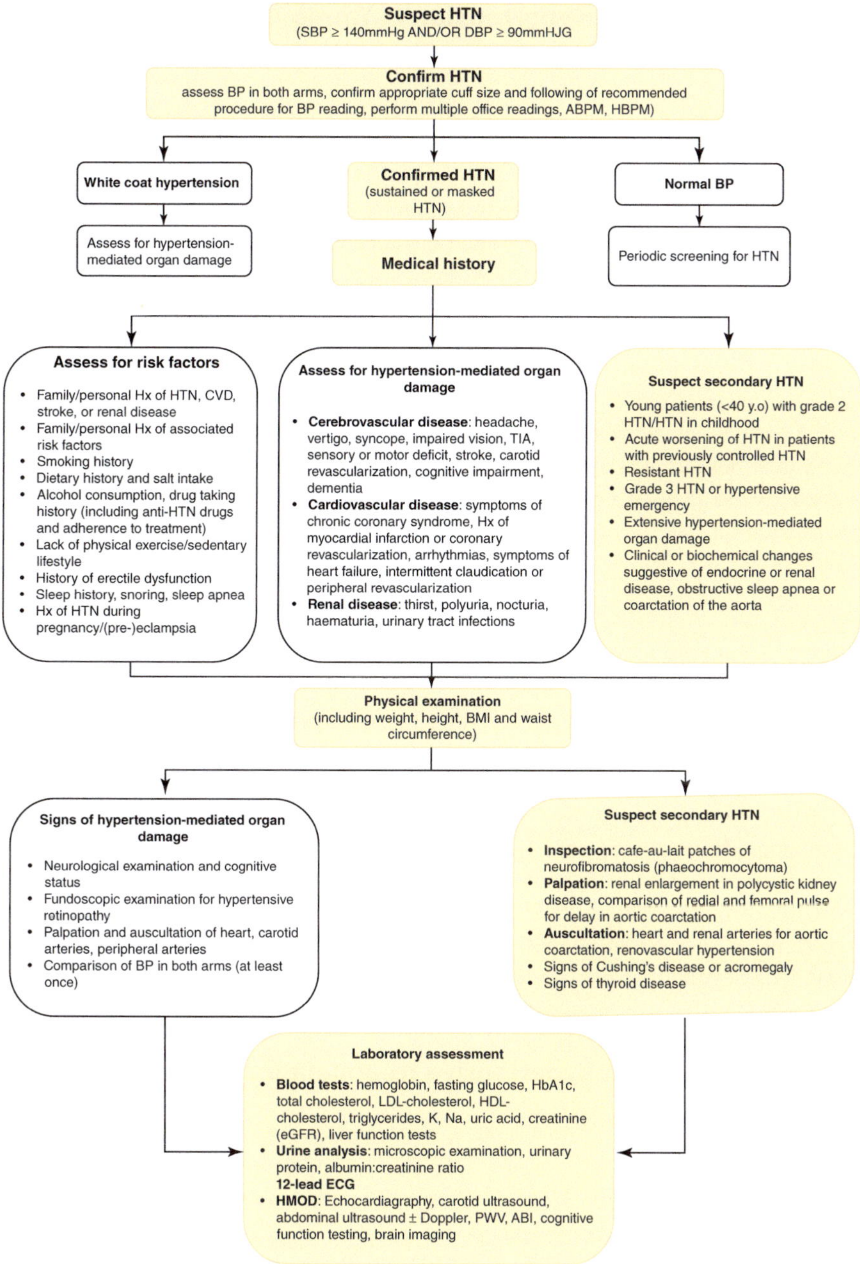

Fig. 29.1 Algorithm for the evaluation of patients with suspected hypertension—marked in yellow is the path for diagnosis of secondary hypertension (based on ESC Hypertension Guideline [3]). *HTN* arterial hypertension, *BP* blood pressure, *SBP* systolic blood pressure, *DBP* diastolic blood pressure, *ABPM* ambulatory blood pressure monitoring, *HBPM* home blood pressure monitoring, *CVD* cardiovascular disease, *Hx* history, *TIA* transient ischemic attack, *BMI* body mass index, *PWV* pulse wave velocity, *ABI* ankle-brachial index

29.5 Treatment of Hypertension in Young Adults

Hypertension awareness is the lowest among young adults, with a low likelihood of early diagnosis [1, 3, 7, 21, 41]. Consequently, hypertensive young adults have the least antihypertensive medication (49%) and lowest BP control rates (38%) when compared to middle-aged or older adults [43].

Treatment of hypertension in young adults is not much different from what guidelines recommend for middle-aged or older adults. However, risk assessment models for cardiovascular mortality were developed based on older populations and underestimate CV risk in young patients since the absolute 10-year risk is low but there is an increased likelihood of lifetime cardiovascular events [3, 41].

The ESC guidelines recommend treatment be considered for grade 1 SDH in order to prevent target organ damage and CVD, while it is unclear whether isolated systolic hypertension should be treated or not. The ACC/AHA guidelines recommend treatment in all stage 2 HTN patients as well as stage 1 HTN patients with associated risk factors (i.e., diabetes, CKD). The NICE guidelines recommend ISH treatment after risk stratification by QRISK3 which was validated for patients aged 25–84 [3, 41, 42, 44].

29.5.1 Lifestyle Interventions

Lifestyle interventions are considered first-line treatment for young adults with mild hypertension in the absence of target organ damage. A 2016 study by Williamson et al. showed that physical activity interventions decrease BP by 4-5 mmHg during the first 3–6 months but lack any significant effect on the long term—beyond 12 months [7]. However, when lifestyle interventions include weight loss, BP control may be more readily attainable as shown by a 1985 study comparing pharmacologic intervention (metoprolol 200 mg) to weight loss (average 7 kg) in a cohort of hypertensive overweight patients younger than 55 years—although follow-up did not extend beyond 21 weeks [45].

29.5.2 Choice of Antihypertensive Treatment in Young Adults

Regarding the choice of antihypertensive agents, neither the American nor European guidelines advocate for the use of a specific class of BP-lowering drugs in young adults. The British guidelines, however, recommend treatment initiation with ACEi/ARBs in patients <55 years (except for patients of Black African or African-Caribbean descent who should be initiated on calcium channel blockers) [44].

In 1999 Dickerson et al. performed a study aiming to identify the best antihypertensive treatment in young adults (aged 22–51 years) through treatment crossover rotation of four major classes (ACEi—lisinopril, BB—bisoprolol, CCB—nifedipine, diuretic—hydrochlorothiazide + triamterene). Patients achieved better BP control with either ACEi or BB. The authors further suggested that young hypertensive patients be initiated on either ACEi or BB and switched to CCB or diuretic if target BP is not achieved. However, the number of patients included in the study was very low and only 36 patients completed all 4 cycles of treatment [46].

Three years later, in 2002, Deary et al. published the results of a double-blind, placebo-controlled crossover study comparing five classes of antihypertensive drugs in a small cohort (34 patients) of young adults (aged 28–55 years). The five classes of antihypertensive drugs were ACEi—lisinopril, BB—bisoprolol, CCB—amlodipine, diuretic—bendrofluazide, and α-adrenergic blocker—doxazosin, and results were concordant with the previous paper. The authors hypothesize that age should be considered a surrogate for renin secretion and further claim that young patients may respond better to ACEi or BB and only in case of a suboptimal response should there be an upgrade to a two-drug regimen by addition of a CCB or diuretic [47].

Overactivity of the sympathetic nervous system in hypertensive young adults confers beta-blockers a therapeutic advantage. As previously mentioned, the pathogenic mechanisms behind arterial hypertension in young adults are related to peripheral resistance and increased renin secretion. Beta-blockers may act on either of the above-mentioned mechanism: BB with intrinsic sympathomimetic activity (acebutolol, pindolol, labetalol) increase nitric oxide release while BB without intrinsic sympathomimetic activity (atenolol, metoprolol, bisoprolol, etc.) reduce renin release and, consequently, heart rate, stroke volume, and cardiac output. BB with intrinsic sympathomimetic activity are less effective in lowering blood pressure than BB without it [32].

BB without intrinsic sympathomimetic activity act in two stages: initially, the reduction in heart rate and cardiac output leads to a reflex increase in peripheral resistance and BP is either unchanged or only slightly lowered; however, with time, peripheral resistance reverts to previous levels, and the antihypertensive effect becomes more pronounced. In comparison, BB with intrinsic sympathomimetic activity do not produce a reduction in heart rate or cardiac output in patients with low sympathetic tone and thus are less likely to be of use in treating young hypertensive adults [48].

Since an elevated resting heart rate seems to be an independent predictor for sudden cardiac death, beta-blocker treatment as well as exercise conditioning should be considered in order to reduce cardiovascular mortality [49].

Left ventricular remodeling is reversed by ACEi treatment, and, to a lesser extent, by beta-blockers [32]. However, bisoprolol seems to be just as effective as enalapril in a 1990 study on 56 hypertensive patients who had similar blood pressure and LV mass index reduction at follow-up [50].

29.6 Antihypertensive Treatment in Special Circumstances

29.6.1 Pregnancy

Cardiovascular risk is higher for women who develop pregnancy-related hypertension. Out of a cohort of 5960 normotensive women, 6% developed hypertensive disorders during pregnancy—60.1% of whom later developed (pre-)eclampsia. Patients with (pre-)eclampsia were twice more likely to develop hypertension in the year following delivery [51]. Furthermore, a population-based retrospective study performed in a cohort of 1.03 million pregnant women (aged 28.2 ± 5.5 years) without history of cardiovascular disease found that maternal placental syndrome developed in 7% of the population and these patients had a higher risk of premature CVD [52].

Blood pressure targets in pregnant women have yet to be established; however the ESC guidelines recommend treatment initiation for BP > 150/95 mmHg and lower (140/90 mmHg) in women with gestational hypertension, preexisting HTN with superimposed gestational hypertension, and HTN with signs or target organ damage during pregnancy [53]. An open, multicentric, international trial evaluated the effects of less-tight (target DBP 100 mmHg) versus tight control (target DBP 85 mmHg) of hypertension on pregnancy complications in a cohort of 987 women with nonproteinuric preexisting or gestational hypertension. Severe hypertension was more frequently encountered in the less-tight control group, but otherwise, pregnancy loss, extended high-level neonatal care, and maternal complications had similar rates between the two groups [54].

A study recently communicated at ACC 22 meeting investigated the benefits and safety of antihypertensive therapy for mild chronic hypertension (SBP140–159 or DBP 90–104) during pregnancy, to a target of less than 140/90 mmHg, in 2408 pregnant women, mean age 32 years. The study showed a statistically significant 18% relative risk reduction in primary outcome (a composite of pre-eclampsia with severe features, medically indicated preterm birth at <35 weeks' gestation, placental abruption, or fetal/neonatal death), with a NNT of 14.7 and no safety issues. It should be noted that 48% of the women enrolled were non-Hispanic black women, so generalization to other types of population should be cautiously made [55].

It bears mention that diuretics should be avoided during pregnancy as they decrease blood flow to the placenta [4] and ACEi and ARBs are contraindicated due to fetal toxicity. Young women who want to become pregnant or already are should be switched to another antihypertensive treatment such as calcium channel blockers (nifedipine), beta-blockers (labetalol), or methyldopa [4, 41, 53]. Beta-blockers are less effective than CCB and may be the cause of fetal bradycardia, growth retardation, and hypoglycemia—the type and dose should be considered carefully, and most studies recommend the use of labetalol and avoidance of atenolol. After delivery, methyldopa should be avoided since it increases the risk of developing postpartum depression. Most antihypertensive therapies are excreted minimally in breast milk, except for propranolol and nifedipine where breast milk concentration mirrors maternal serum concentration [53].

29.6.2 Athletes

The prevalence of hypertension among young athletes was reported at 3%—with a prevalence of secondary hypertension of 0.2%—in a cohort of competitive athletes aged 18–40. Hypertensive athletes were more likely to have a family history of hypertension, be overweight, and have larger left ventricular hypertrophy [56], as well as use supplements, energy drinks, anti-inflammatory drugs, or performance-enhancing drugs (erythropoietin, anabolic steroids)—potential causes of secondary hypertension [57, 58].

In hypertensive athletes, an acute increase in BP during physical exercise could lead to plaque rupture and ultimately to cardio- or cerebrovascular adverse events. Therefore, exercise testing should be performed routinely in order to evaluate for exercise-induced hypertension; though exclusion from competitive sports is not warranted, special attention should be taken regarding cardiovascular risk factors and ABPM measurements in athletes with exercise-induced hypertension [58].

Young hypertensive athletes (<65 years) should, for the first 3 months after the initial diagnosis of HTN, be subjected to non-pharmacological measures such as restriction of salt intake and alcohol consumption, weight reduction if applicable, a balanced diet and cessation of smoking, supplements or anti-inflammatory drugs. After 3 months, if BP elevation persists, pharmacological therapy should be initiated—except for grade 3 HTN where antihypertensive medication should be started immediately. The antihypertensive drugs of choice are ACEi, ARBs, and CCB—BB are generally avoided since they may lower aerobic-exercise capacity [58].

Diuretics are prohibited in all competitive sports since they may mask performance-enhancing drugs [57, 58]. Beta-blockers are prohibited by the World Anti-Doping Association for a variety of competitive sports where control of tremor is required (archery, billiards, golf, darts, shooting, skiing, and underwater sports) [59].

29.7 Treatment Adherence and Blood Pressure Control Rate in Young Adults

Treatment adherence and HTN control rates were evaluated in a 2008 paper by Bailey et al. in a cohort of 2587 hypertensive patients aged 15–98 years, by means of Kaplan-Meyer curves. Control rates seem to be more easily achieved in younger patients (15–39 years) compared to other age groups (61.5% vs. 33.0%). Moreover, younger patients seem to be less likely to abandon dose incrementation before achieving BP control (20.7% vs. 35.6%). Interestingly, HTN control rate seemed to decrease by 10% in men and 20% in women every 10 years [60]. It should be noted that the study enrolled only patients followed in Mayo Clinic, either for long-standing hypertension or referred for difficult-to-control hypertension and secondary forms of hypertension. Newer data shows lower BP control rates in young adults when compared to middle-aged or older adults [43].

A nonblinded randomized trial on hypertensive adults (18–84 years) followed patients for 6 months while observing the difference in adherence and BP control rates between the intervention (artificial intelligence coaching app providing tracking, reminders, and encouragement) and control group (BP tracking app). At 6-months follow-up both the intervention and control group had lower BP (−8.3 mmHg vs. −6.8 mmHg, $p = 0.16$). It seems that auto-monitoring of BP values and tracking via a digital app had a deeper effect on adherence and BP control than directive messages from the coaching app [61, 62].

A randomized controlled trial on 622 treated but poorly controlled hypertensive patients with a mean age of 66 years was performed in the UK in 2021. The digital intervention for HTN management included BP self-monitoring, medication titration based on self-monitored values, lifestyle advice, and referral to healthcare professionals when needed. At 12-month follow-up, there was a greater BP decrease in the digital intervention group (SBP: −3.4 mmHg, DBP: −0.5 mmHg). It was estimated that the produced effect could translate into a lower rate of coronary events and stroke by as much as 5–10% and 10–15%, respectively [63].

A prospective, open-label, randomized controlled trial performed in Japan on 390 previously untreated hypertensive patients aged 20–64 years evaluated the increase in HTN control rates in patients using the HERB system (a digital app designed to facilitate disease management). Compared to the control group, the patients who used the app had a greater decrease in blood pressure, heart rate, body weight, and salt intake [64].

It would seem that digital interventions through smartphone apps hold a promising perspective in improving treatment adherence, blood pressure control, and the risk of cardiovascular disease.

In conclusion, arterial hypertension is often overlooked in young adults, despite its growing prevalence during the last couple of decades. Unfortunately, its long-term effects are deleterious, leading to cardiovascular and cerebrovascular diseases, as well as increased mortality. Lifestyle interventions and pharmacologic treatment should be aimed at lowering the overactivity of the sympathetic nervous and renin-angiotensin-aldosterone systems, especially through ACEi and beta-blockers without sympathomimetic activity. In young individuals, essential hypertension in general and isolated systolic hypertension in particular are the most common forms of hypertension though secondary hypertension is more common when compared to older adults, particularly in the setting of severe hypertension, endocrinopathy, or chronic kidney disease. Screening for arterial hypertension should be performed in selected young adults in order to instate antihypertensive treatment earlier and lessen the disease burden. Hypertension awareness and blood pressure control rates remain lowest in young adults compared to other population groups.

References

1. De Venecia T, Lu M, Figueredo VM. Hypertension in young adults. Postgrad Med. 2016;128(2):201–7.
2. Garrison RJ, Kannel WB, Stokes J III, Castelli WP. Incidence and precursors of hypertension in young adults: the Framingham offspring study. Prev Med. 1987;16(2):235–51.
3. Williams B, Mancia G, Spiering W, Agabiti Rosei E, Azizi M, Burnier M, et al. 2018 ESC/ESH guidelines for the management of arterial hypertension: the task force for the management of arterial hypertension of the European Society of Cardiology (ESC) and the European Society of Hypertension (ESH). Eur Heart J. 2018;39(33):3021–104.
4. Battistoni A, Canichella F, Pignatelli G, Ferrucci A, Tocci G, Volpe M. Hypertension in young people: epidemiology, diagnostic assessment and therapeutic approach. High Blood Press Cardiovasc Prev. 2015;22(4):381–8.

5. Zhou B, Carrillo-Larco RM, Danaei G, Riley LM, Paciorek CJ, Stevens GA, et al. Worldwide trends in hypertension prevalence and progress in treatment and control from 1990 to 2019: a pooled analysis of 1201 population-representative studies with 104 million participants. Lancet. 2021;398(10304):957–80.
6. Mills KT, Bundy JD, Kelly TN, Reed JE, Kearney PM, Reynolds K, et al. Global disparities of hypertension prevalence and control: a systematic analysis of population-based studies from 90 countries. Circulation. 2016;134(6):441–50.
7. Williamson W, Foster C, Reid H, Kelly P, Lewandowski AJ, Boardman H, et al. Will exercise advice be sufficient for treatment of young adults with prehypertension and hypertension? A systematic review and meta-analysis. Hypertension. 2016;68(1):78–87.
8. Loria CM, Liu K, Lewis CE, Hulley SB, Sidney S, Schreiner PJ, et al. Early adult risk factor levels and subsequent coronary artery calcification: the CARDIA study. J Am Coll Cardiol. 2007;49(20):2013–20.
9. Ostchega Y, Fryar CD, Nwankwo T, Nguyen DT. Hypertension prevalence among adults aged 18 and over: United States, 2017–2018. 2020.
10. Whelton PK. Epidemiology of hypertension. Lancet (London, England). 1994;344(8915):101–6.
11. Drukteinis JS, Roman MJ, Fabsitz RR, Lee ET, Best LG, Russell M, et al. Cardiac and systemic hemodynamic characteristics of hypertension and prehypertension in adolescents and young adults: the strong heart study. Circulation. 2007;115(2):221–7.
12. Wolf-Maier K, Cooper RS, Banegas JR, Giampaoli S, Hense H-W, Joffres M, et al. Hypertension prevalence and blood pressure levels in 6 European countries, Canada, and the United States. JAMA. 2003;289(18):2363–9.
13. WHO. STEPS: prevalence of noncommunicable disease risk factors in Ukraine 2019. Copenhagen: WHO Regional Office for Europe; 2020; Licence: CC BY-NC-SA 3.0 IGO.
14. WHO. STEPS: prevalence of noncommunicable disease risk factors in Belarus 2016. Copenhagen: WHO Regional Office for Europe; 2017.
15. Dorobantu M, Tautu O-F, Dimulescu D, Sinescu C, Gusbeth-Tatomir P, Arsenescu-Georgescu C, et al. Perspectives on hypertension's prevalence, treatment and control in a high cardiovascular risk east European country: data from the SEPHAR III survey. J Hypertens. 2018;36(3):690–700.
16. Vasan RS. High blood pressure in young adulthood and risk of premature cardiovascular disease: calibrating treatment benefits to potential harm. JAMA. 2018;320(17):1760–3.
17. Manolio TA, Savage PJ, Burke GL, Liu K, Wagenknecht LE, Sidney S, et al. Association of fasting insulin with blood pressure and lipids in young adults. The CARDIA study. Arteriosclerosis. 1990;10(3):430–6.
18. Luo D, Cheng Y, Zhang H, Ba M, Chen P, Li H, et al. Association between high blood pressure and long term cardiovascular events in young adults: systematic review and meta-analysis. BMJ. 2020;370:m3222.
19. Qi Y, Han X, Zhao D, Wang W, Wang M, Sun J, et al. Long-term cardiovascular risk associated with stage 1 hypertension defined by the 2017 ACC/AHA hypertension guideline. J Am Coll Cardiol. 2018;72(11):1201–10.
20. Sundström J, Neovius M, Tynelius P, Rasmussen F. Association of blood pressure in late adolescence with subsequent mortality: cohort study of Swedish male conscripts. BMJ. 2011;342:342.
21. Son JS, Choi S, Kim K, Kim SM, Choi D, Lee G, et al. Association of blood pressure classification in Korean young adults according to the 2017 American College of Cardiology/American Heart Association guidelines with subsequent cardiovascular disease events. JAMA. 2018;320(17):1783–92.
22. Manolio TA, Burke GL, Savage PJ, Sidney S, Gardin JM, Oberman A. Exercise blood pressure response and 5-year risk of elevated blood pressure in a cohort of young adults: the CARDIA study. Am J Hypertens. 1994;7(3):234–41.

23. Unger T, Borghi C, Charchar F, Khan NA, Poulter NR, Prabhakaran D, et al. 2020 International Society of Hypertension global hypertension practice guidelines. Hypertension. 2020;75(6):1334–57.
24. Seryan A, Martin M, Hamimatunnisa J, Annette P, Margit H, Karl-Heinz L. Cardiovascular mortality risk in young adults with isolated systolic hypertension: findings from population-based MONICA/KORA cohort study. J Hum Hypertens. 2021;36:1–7.
25. Yano Y, Stamler J, Garside DB, Daviglus ML, Franklin SS, Carnethon MR, et al. Isolated systolic hypertension in young and middle-aged adults and 31-year risk for cardiovascular mortality: the Chicago heart association detection project in industry study. J Am Coll Cardiol. 2015;65(4):327–35.
26. Saladini F, Fania C, Mos L, Mazzer A, Casiglia E, Palatini P. Office pulse pressure is a predictor of favorable outcome in young-to middle-aged subjects with stage 1 hypertension. Hypertension. 2017;70(3):537–42.
27. Lee H, Yano Y, Cho SMJ, Park JH, Park S, Lloyd-Jones DM, et al. Cardiovascular risk of isolated systolic or diastolic hypertension in young adults. Circulation. 2020;141(22):1778–86.
28. Cote AT, Phillips AA, Harris KC, Sandor GG, Panagiotopoulos C, Devlin AM. Obesity and arterial stiffness in children: systematic review and meta-analysis. Arterioscler Thromb Vasc Biol. 2015;35(4):1038–44.
29. Agbaje AO, Barker AR, Tuomainen T-P. Effects of arterial stiffness and carotid intima-media thickness progression on the risk of overweight/obesity and elevated blood pressure/hypertension: a cross-lagged cohort study. Hypertension. 2022;79(1):159–69.
30. McEniery CM, Yasmin WS, Maki-Petaja K, McDonnell B, Sharman JE, et al. Increased stroke volume and aortic stiffness contribute to isolated systolic hypertension in young adults. Hypertension. 2005;46(1):221–6.
31. Hall ME, Cohen JB, Ard JD, Egan BM, Hall JE, Lavie CJ, et al. Weight-loss strategies for prevention and treatment of hypertension: a scientific statement from the American Heart Association. Hypertension. 2021;78(5):e38–50.
32. Cruickshank J. Are we misunderstanding beta-blockers. Int J Cardiol. 2007;120(1):10–27.
33. Palatini P, Longo D, Zaetta V, Perkovic D, Garbelotto R, Pessina AC. Evolution of blood pressure and cholesterol in stage 1 hypertension: role of autonomic nervous system activity. J Hypertens. 2006;24(7):1375–81.
34. Cohen JB. Hypertension in obesity and the impact of weight loss. Curr Cardiol Rep. 2017;19(10):1–8.
35. Feig DI. The role of uric acid in the pathogenesis of hypertension in the young. J Clin Hypertens (Greenwich). 2012;14(6):346–52.
36. Stergiou GS, Yiannes NJ, Rarra VC, Alamara CV. White-coat hypertension and masked hypertension in children. Blood Press Monit. 2005;10(6):297–300.
37. Westerståhl M, Forss M, Persson L, Bouma K, Gustavsson T, Wühl E, et al. Hypertension outcomes and cardiovascular status in young adults with childhood-diagnosed white coat hypertension. Arch Dis Child. 2018;103(1):113–4.
38. Honzíková N, Nováková Z, Závodná E, Paděrová J, Lokaj P, Fišer B, et al. Baroreflex sensitivity in children, adolescents, and young adults with essential and white-coat hypertension. Klin Paediatr. 2006;218(04):237–42.
39. Kim K, Kim J-Y, Choi E-Y, Kwon H-M, Rim S-J. Prevalence and risk factors for secondary hypertension among young Korean men. 2020.
40. Camelli S, Bobrie G, Na P-V, Azizi M, Plouin P, Amar L. LB01. 11: prevalence of secondary hypertension in young hypertensive adults. J Hypertens. 2015;33:e47.
41. Hinton TC, Adams ZH, Baker RP, Hope KA, Paton JF, Hart EC, et al. Investigation and treatment of high blood pressure in young people: too much medicine or appropriate risk reduction? Hypertension. 2020;75(1):16–22.

42. Whelton PK, Carey RM, Aronow WS, Casey DE, Collins KJ, Dennison Himmelfarb C, et al. 2017 ACC/AHA/AAPA/ABC/ACPM/AGS/APhA/ASH/ASPC/NMA/PCNA guideline for the prevention, detection, evaluation, and management of high blood pressure in adults: a report of the American College of Cardiology/American Heart Association task force on clinical practice guidelines. J Am Coll Cardiol. 2018;71(19):e127–248.
43. Johnson HM, Thorpe CT, Bartels CM, Schumacher JR, Palta M, Pandhi N, et al. Antihypertensive medication initiation among young adults with regular primary care use. J Gen Intern Med. 2014;29(5):723–31.
44. Jones NR, McCormack T, Constanti M, McManus RJ. Diagnosis and management of hypertension in adults: NICE guideline update 2019. Br J Gen Pract. 2020;70(691):90–1.
45. Macmahon S, Bernstein L, Macdonald G, Andrews G, Blacket R. Comparison of weight reduction with metoprolol in treatment of hypertension in young overweight patients. Lancet. 1985;325(8440):1233–6.
46. Dickerson JC, Hingorani AD, Ashby MJ, Palmer CR, Brown MJ. Optimisation of antihypertensive treatment by crossover rotation of four major classes. Lancet. 1999;353(9169):2008–13.
47. Deary AJ, Schumann AL, Murfet H, Haydock SF, Foo RS-Y, Brown MJ. Double-blind, placebo-controlled crossover comparison of five classes of antihypertensive drugs. J Hypertens. 2002;20(4):771–7.
48. Lund-Johansen P, Omvik P. Acute and chronic hemodynamic effects of drugs with different actions on adrenergic receptors: a comparison between alpha blockers and different types of beta blockers with and without vasodilating effect. Cardiovasc Drugs Ther. 1991;5(3):605–15.
49. Jouven X, Zureik M, Desnos M, Guérot C, Ducimetière P. Resting heart rate as a predictive risk factor for sudden death in middle-aged men. Cardiovasc Res. 2001;50(2):373–8.
50. Gosse P, Roudaut R, Herrero G, Dallocchio M. Beta-blockers vs. angiotensin-converting enzyme inhibitors in hypertension: effects on left ventricular hypertrophy. J Cardiovasc Pharmacol. 1990;16:S145–50.
51. Black MH, Zhou H, Sacks DA, Dublin S, Lawrence JM, Harrison TN, et al. Hypertensive disorders first identified in pregnancy increase risk for incident prehypertension and hypertension in the year after delivery. J Hypertens. 2016;34(4):728–35.
52. Ray JG, Vermeulen MJ, Schull MJ, Redelmeier DA. Cardiovascular Health After Maternal Placental syndromes (CHAMPS): population-based retrospective cohort study. Lancet. 2005;366(9499):1797–803.
53. Regitz-Zagrosek V, Roos-Hesselink JW, Bauersachs J, Blomström-Lundqvist C, Cifkova R, De Bonis M, et al. 2018 ESC guidelines for the management of cardiovascular diseases during pregnancy: the task force for the management of cardiovascular diseases during pregnancy of the European Society of Cardiology (ESC). Eur Heart J. 2018;39(34):3165–241.
54. Magee LA, Von Dadelszen P, Rey E, Ross S, Asztalos E, Murphy KE, et al. Less-tight versus tight control of hypertension in pregnancy. N Engl J Med. 2015;372(5):407–17.
55. Tita AT, Szychowski JM, Boggess K, Dugoff L, Sibai B, Lawrence K, et al. Treatment for mild chronic hypertension during pregnancy. N Engl J Med. 2022;386:1781.
56. Caselli S, Sequì AV, Lemme E, Quattrini F, Milan A, D'Ascenzi F, et al. Prevalence and management of systemic hypertension in athletes. Am J Cardiol. 2017;119(10):1616–22.
57. Pelliccia A, Sharma S, Gati S, Bäck M, Börjesson M, Caselli S, et al. 2020 ESC guidelines on sports cardiology and exercise in patients with cardiovascular disease: the task force on sports cardiology and exercise in patients with cardiovascular disease of the European Society of Cardiology (ESC). Eur Heart J. 2021;42(1):17–96.
58. Niebauer J, Boerjesson M, Carre F, Caselli S, Palatini P, Quattrini F, et al. Recommendations for participation in competitive sports of athletes with arterial hypertension: a position statement from the sports cardiology section of the European Association of Preventive Cardiology (EAPC). Eur Heart J. 2018;39(40):3664–71.
59. https://www.wada-ama.org/sites/default/files/prohibited_list_2018_en.pdf. Accessed 23 Apr 2022.

60. Bailey KR, Grossardt BR, Graves JW. Novel use of Kaplan–Meier methods to explain age and gender differences in hypertension control rates. Hypertension. 2008;51(4):841–7.
61. Persell SD, Karmali KN, Stein N, Li J, Peprah YA, Lipiszko D, et al. Design of a randomized controlled trial comparing a mobile phone-based hypertension health coaching application to home blood pressure monitoring alone: the smart hypertension control study. Contemp Clin Trials. 2018;73:92–7.
62. Persell SD, Peprah YA, Lipiszko D, Lee JY, Li JJ, Ciolino JD, et al. Effect of home blood pressure monitoring via a smartphone hypertension coaching application or tracking application on adults with uncontrolled hypertension: a randomized clinical trial. JAMA Netw Open. 2020;3(3):e200255.
63. McManus RJ, Little P, Stuart B, Morton K, Raftery J, Kelly J, et al. Home and online management and evaluation of blood pressure (HOME BP) using a digital intervention in poorly controlled hypertension: randomised controlled trial. BMJ. 2021;372:m4858.
64. Kario K, Nomura A, Harada N, Okura A, Nakagawa K, Tanigawa T, et al. Efficacy of a digital therapeutics system in the management of essential hypertension: the HERB-DH1 pivotal trial. Eur Heart J. 2021;42(40):4111–22.

Heart Failure with Preserved Ejection Fraction: Current Management and Future Strategies Pre-Heart Failure

30

Enrico Agabiti-Rosei, Stefania Martino, Francesco Torlone, and Claudia Agabiti-Rosei

Abbreviations

ABPM	Ambulatory blood pressure monitoring
ACEI	Angiotensin-converting enzyme inhibitor
AF	Atrial fibrillation
ARB	Angiotensin receptor blocker
ARNI	Angiotensin receptor neprilysin inhibitor
BP	Blood pressure
CAD	Coronary artery disease
CCB	Calcium channel blockers
CHF	Congestive heart failure
CKD	Chronic kidney disease
CV	Cardiovascular
DM	Diabetes mellitus
GFR	Glomerular filtration rate
HBPM	Home blood pressure monitoring
HFpEF	Heart failure with preserved ejection fraction
HFrEF	Heart failure with reduced ejection fraction
HT	Hypertension
LA	Left atrium
LVEF	Left ventricular ejection fraction

E. Agabiti-Rosei (✉) · C. Agabiti-Rosei
Department of Clinical and Experimental Sciences, University of Brescia, Brescia, Italy
e-mail: enrico.agabitirosei@unibs.it; claudia.agabitirosei@unibs.it

S. Martino · F. Torlone
IRCCS Multimedica, Milan, Italy
e-mail: stefania.martino@multimedica.it; francesco.torlone@multimedica.it

M. Dorobantu et al. (eds.), *Hypertension and Heart Failure*, Updates in Hypertension and Cardiovascular Protection,
https://doi.org/10.1007/978-3-031-39315-0_30

LVH	Left ventricular hypertrophy
MRA	Mineralocorticoid receptor antagonist
RAS	Renin-angiotensin system
RCT	Randomized clinical trial
SGLT2	Sodium/glucose cotransporter 2

30.1 Introduction

Chronic heart failure is one of the most frequent causes of death and hospitalization in western countries, and its prevalence is expected to increase due to the population aging [1, 2].

About half of the patients with CHF have a LVEF >50%, i.e., a normal or preserved EF (HFpEF), according to most guidelines, in particular the recent European and American guidelines.

Patients with reduced or preserved EF have similar morbidity and mortality, although CV death seems less frequent in HFpEF. In fact, among patients with HF, as LVEF increases there are more hospitalizations and deaths due to noncardiac causes.

In addition, pharmacological treatments, which have been shown to be effective for improving clinical prognosis in HFrEF, have been less so in patients with HFpEF. These observations may indicate cardiac and noncardiac complex pathophysiological mechanisms in HFpEF, possibly in relation to the burden of comorbidities detectable in these patients as LVEF increases.

Patients with HFpEF are often older and females. Hypertension is highly prevalent in these patients, ranging from about 60% to >90% in specific RCTs. Moreover, AF, CKD, and other noncardiac comorbidities are more common in HfpEF than in HFrEF [3, 4].

30.2 Management of HFpEF

30.2.1 Medical Therapy

Until recently no medical treatment has been demonstrated to consistently reduce mortality in RCT of HFpEF; however hospitalization for HF was reduced in some of them and recent large studies with SGLT2 inhibitors have demonstrated a significantly reduced incidence of the primary end point, i.e., CV death and/or hospitalization for worsening of HF.

Beta Blockers (B-B) are often prescribed in HFpEF, particularly in patients with CAD or AF. However, in a large meta-analysis of RCT in patients with HF, B-B resulted effective in reducing the primary end point only in those with LVEF <50% [5]. Possibly, while some patients with HFpEF will benefit from B-B, in other patients these drugs may even worsen an already impaired chronotropic reserve, with consequent limitation of functional capacity [6, 7]. B-B with vasodilating properties may have differential effects. Nebivolol significantly reduced the primary end point of all-cause mortality or CV hospital admission across a large range of LVEF in the SENIORS trial, but in this study only 15% of patients had an EF >50% [8].

Trials of **ACEIs and ARBs** in HFpEF showed a decreased risk of HF hospitalization, but no significant reduction of mortality.

The **MRA** spironolactone in the TOPCAT trial failed to show a reduction in the primary composite end point, but a small benefit was observed in an exploratory analysis including only patients from the Americas and excluding those from Russia-Georgia who unfortunately in a large proportion did not meet the inclusion criteria. However, in TOPCAT spironolactone was associated with a reduced risk of HF hospitalization [9, 10].

The **ARNI** sacubitril-valsartan, compared to valsartan alone, did not reduce the primary composite end point in patients with HF and EF >45% in the PARAGON-HF trial, but decreased HF hospitalizations. ARNI resulted most effective in two prespecified subgroups, i.e., in females and in patients with LVEF below the median. Moreover, patients with recent HF hospitalization or with elevated troponin or previously using MRAs were more likely to benefit from ARNI [11–13]. This could be an example of how definition of phenotypes could help in improving the efficacy of treatment.

More recently, the use of **SGLT2 inhibitors** reduced the primary outcome of composite CV death or total HF hospitalization in HF with VEF >40%. The results have been clearly and similarly obtained in large RCTs, such as the EMPEROR-Preserved trial, with the use of empagliflozin, and the DELIVER trial, with Dapagliflozin [14, 15]. On the basis of the results, the use of these drugs should be strongly recommended in patients with HFpEF. The DELIVER trial was effective also in patients with previously reduced EF, who subsequently improved with treatment, and also in those admitted to the hospital for symptoms related to worsening HF.

Main characteristics and results of major RCTs in patients with HFpEF are reported in Table 30.1.

30.2.2 Lifestyle-Based Therapy

Diet and exercise training have favourable and additive effects on symptoms and hemodynamic changes in patients with HFpEF. About 80% of patients with HFpEF are overweight or obese, and weight loss has beneficial effects on metabolic profile, cardiac relaxation, and absolute peak oxygen consumption (VO2 peak) [16]. Exercise training also improved VO2 peak, quality of life scores, and 6-min walk distance in a meta-analysis of 8 RCTs including 463 patients with HFpEF [17]. Favourable hemodynamic changes have been also observed with a low sodium diet.

Table 30.1 Major randomized controlled trials in patients with HFpEF

Study, publication year, N participants	Study drug	Patient charateristics	NYHA	Endpoints
CHARM preserved (2203) *N*=3023	Candesartan 25 mg	EF 54% Age 64 Women 40% HT 64% DM 28%	Range = II-IV III-IV = 38%	CVd + fHF Hosp HR = 0.89, p = 0.12 fHF Hosp HR = 0.85, p = 0.07
PEP-CHF (2006) *N*=850	Peindopril 4 mg	EF 64% Age 76 Women 56% HT 69% DM 21%	Range = II-IV III-IV = 24%	All CVd + fHF Hosp HR = 0.92, p = 0.37 fHF Hosp HR = 0.86, p = 0.37
I-PRESERVE (2008) *N*= 4128	Irbesartan 300 mg	EF 59% Age 72 Women 60% HT 88% DM 27%	Range = II-IV III-IV = 80 %	All CVd + fHF Hosp HR = 0.95, p = 0.35 fCV Hosp HR = 0.95, p = 0.44
TOPCAT (2014) *N*=3445	Spironolactone 25 mg	EF 56% Age 69 Women 52% HT 91% DM 33%	Range = II-IV III-IV = 33%	CVd + abort Cardiac arrest + fHF Hosp HR = 0.89, p = 0.19 fHF Hosp HR = 0.83, p = 0.04
PARAGON-HF (2019) *N*=4796	Sacubitril/ valsartan 200 mg	EF 58% Age 73 Women 52% HT 95% DM 43%	Range = II-IV III-IV = 20%	CVd + tot HF Hosp RR = 0.87, p = 0.06 Total HF Hosp RR = 0.85, p = 0.056
EMPEROR-preserved (2021) *N*=5988	Empagliflozin 10 mg	EF 54% Age 72 Women 45% HT 91% DM 49%	Range = II-IV III-IV = 18%	CVd + HF Hosp HR = 0.79, p < 0.001 Total HF Hosp HR = 0.73, p < 0.001

(continued)

Table 30.1 (continued)

Study, publication year, N participants	Study drug	Patient charateristics	NYHA	Endpoints
DELIVER (2022) *N*=6263	Dapagliflozin 10 mg	EF 54% age 72 years Women 44% HT 89% DM 45%	Range = II-IV III-IV = 25%	CVd + HF Hosp HR = 082, $p < 0.001$ Worse HF HR = 0.79, $p < 0.001$

30.3 Heterogeneity in HFpEF

There is a large heterogeneity of the population included in HFpEF trials, related to different demographic characteristics, risk factors, comorbidities, and mortality risk. Main risk factors and associated diseases leading to different pathophysiological mechanisms and morphofunctional alterations, with final development of HFpEF, include hypertension, aging, diabetes, obesity, CKD, anemia, CAD, AF, pulmonary disease, and sleep apnea. Patients with HFpEF are more commonly women, more commonly have HT and AF, and less commonly have CAD. CKD is present in 30–50% of patients, and obesity is both a common comorbidity and a risk factor for HFpEF. These comorbidities and risk factors may coexist in many patients, thus inducing complex combinations which may complicate the whole clinical picture and its specific management. A model accounting for each comorbidity could represent the appropriate approach for a more personalized and effective management [18].

Main risk factors, comorbidities, pathological characteristics, and left ventricular structural changes, which contribute in different combinations to HFpEF phenotype, are reported in Table 30.2.

Table 30.2 Heart failure with preserved ejection fraction (HFpEF) phenotype. *LV* left ventricular

Risk factors	Comorbidities	LV structure	Pathology
• Hypertension • Diabetes • Obesity • Physical inactivity • Ageing	• Atrial fibrillation • Sleep apnea • Coronary artery disease • Valvular heart disease • Pulmonary disease • Renal disease	• Normal • LV hypertrophy • Concentric remodeling • Eccentric remodeling	• Myocardial ischemia • Coronary microvascular inflammation • Interstitial fibrosis • Cardiomyocyte hypertrophy

30.4 Management of Patients with Hypertension and HFpEF

Hypertension represents a major risk factor for HFpEF. HT was detected in about 95% of patients enrolled in the PARAGON-HF trial and in about 91% of those in the EMPEROR-Preserved trial. Pathophysiological characteristics of HT include increased LV afterload, increased arterial stiffness, endothelial dysfunction and systemic low-grade inflammation, autonomic dysfunction, and RAS activation. As a consequence of high BP, CV structural and functional alterations develop, such as LVH, LV fibrosis, increased central aortic pressure and LV filling pressure, LA enlargement, volume expansion, pulmonary HT, and microvascular dysfunction. All these mechanisms contribute to the development of HFpEF; however the variable contribution of different pathophysiological processes may produce different phenotypes, which should be considered for a proper management.

30.4.1 Blood Pressure Targets

The optimal BP target in hypertensive patients with HFpEF has not been directly assessed in a RCT, and therefore any recommendation is extrapolated from observational or retrospective studies in populations with and without HF. In the major RCTs performed so far BP was often controlled at baseline, and this may explain why additional small BP change in the active group did not translate into a further decrease of CV end points. In the PARAGON-HF the lowest risk for CV and renal outcomes was observed in those with baseline and achieved BP in the range 120–129 mmHG [19]. The recent observational study OPTIMIZE-HF suggested that among older patients with HFpEF a worse outcome was observed among patients with SBP < 120 mmHg at discharge from hospitalization for HF [20]. Many large RCTs in hypertension included patients with HMOD and also with early HFpEF and have demonstrated that BP reduction effectively prevents HF events [21]. Thus, also in line with current guidelines, a target of SBP in the range 120–129 mmHg is recommended for patients younger than 65 years and in the range 130–139 mmHg for those with an age > 65 years. However, any decision on treatment should always take into account frailty, tolerability, and comorbidities.

In general, it appears safe to avoid actively decreasing BP to less than 120/70 mmHg. Data indicating the association between CV outcomes and BP values obtained with out-of-office measurements (ABPM-HBPM) are limited and should be obtained in future studies.

30.4.2 Regression of LVH

LVH is the consequence of increased LV afterload and represents a condition of greater risk of future CV events, independently from BP values. Antihypertensive treatment with all major classes of drugs may reduce LVH, although B-B seem to be less effective in this respect [22]. Population studies and several analyses of the LIFE trial have shown that regression of LVH is associated with improvement of LV function and, most importantly, with lower rates of CV events, including HF, independently from BP reduction [23]. Although similar evidence regarding regression of LVH in patients with HFpEF is not yet available, the results obtained in hypertensives with LVH strongly suggest that LV mass reduction should be an important goal of treatment in individuals with HFpEF.

30.4.3 Drug Therapy in Patients with Hypertension and HFpEF

The drug treatment strategy for hypertensive patients with HFpEF should follow the core algorithm of the 2018 ESC/ESH guidelines, with some appropriate adaptations.

30.4.3.1 Diuretics

An intensified use of diuretics to reduce congestion and relieve symptoms is advised. Thiazide/thiazide-like diuretics are indicated for BP control, but loop diuretics are recommended if GFR is below 30 mL/min and often required for volume unloading in order to manage symptoms [24, 25]. Excessive use of diuretics should be avoided, also considering that many patients are particularly sensitive to changes of preload.

30.4.3.2 RAS Blockers

A RAS blocker should be given to hypertensives with HFpEF if not contraindicated. Available evidence indicates that the antihypertensive effect of RAS blockers is associated with improvement of diastolic function, regression of LVH, and also reduction of HF hospitalization in HFpEF patients [26–28]. In addition, these drugs reduce albuminuria and delay renal function worsening in diabetes and CKD, which represent common comorbidities in HFpEF patients [29].

30.4.3.3 ARNIs

Although the demonstration of unequivocal superiority of ARNI over a stand-alone ARB in HFpEF is lacking, they were significantly more effective in women and in those with a LVEF at the lower end of the HFpEF spectrum [30]. In these patient groups ARNIs should be considered as a replacement for RAS blockers in order to reduce HF hospitalizations. ARNIs are also very effective as BP-lowering drugs and may be more effective than RAS blockers in reducing arterial stiffness and LVH [31, 32].

30.4.3.4 MRA

The use of low-dose spironolactone, which is the MRA best studied in hypertensive patients, is recommended in patients with resistant hypertension, but should be considered at an earlier stage in HFpEF to reduce HF hospitalization, particularly in patients who meet the criteria of the TOPCAT trial (i.e., elevated natriuretic peptides or recent hospital admission for HF).The recent RCT HOMAGE evaluated the effect of spironolactone in patients with characteristics of clinically occult HFpEF and found that it decreased SBP and NT-pro-BNP, improved diastolic function, and led to a favorable effect on type-1 collagen turnover [33].

30.4.3.5 Beta Blockers

In the 2018 ESC/ESH Guidelines B-B are recommended at any step of treatment when there is a specific indication, such as angina, post-AMI, and arrhythmias, including AF [29]. An elevated heart rate (>80 b/min) is associated with a grater CV risk in HT and even more in HFpEF. B-B have been considered potentially useful in HFpEF because a reduced heart rate may improve diastolic filling and may reduce oxygen demand. However, caution is advised during their use, since they may aggravate chronotropic incompetence and reduce exercise tolerance. Moreover, prolonged diastolic filling may increase ventricular volume and pressure, enhancing LV wall stress. Central SBP is also reduced less by beta blockers in comparison with other antihypertensives [34]. In a large meta-analysis, including 17,312 patients from 11 RCTs in HF, there was no CV benefit in a small group of 244 patients with EF >50%. Despite these observations B-B remain effective antihypertensive drugs and can be combined with any major drug classes for BP reduction in patients with HFpEF. In hypertensive patients their use is associated with significant reductions of CV end points, as shown in a recent meta-analysis [35].

30.4.3.6 Calcium Channel Blockers

Calcium channel blockers (CCB) are effective BP-lowering drugs which may be used in combination with other compounds in patients with HFpEF, particularly with RAS blockers, in younger patients with no or minimal congestion. Non-dihydropyridine CCB may improve diastolic function and reduce heart rate, but their negative chronotropic and inotropic effects represent a limitation of their use in many cases [29].

30.4.3.7 SGLT2 Inhibitors

The results of recent RCTs have demonstrated their significant effectiveness in reducing the risk of composite CV death or total HF hospitalization in patients with LVEF >40% (up to at least 60%), with or without diabetes. They possess also

BP-lowering and natriuretic effects, which should be taken into consideration in these patients [36, 37].

30.4.4 Recommendations for Treatment of Hypertensive Patient with HFpEF

Lifestyle and pharmacological treatment should be initiated when BP is >140/90 and target of treatment should be obtained according to the recommendations of the 2018 ESC/ESH guidelines. The core algorithm for drug treatment should consider an early use of MRA and/or replacement of RAS blockers with ARNI to reduce HF hospitalizations. According to the recently published results of the EMPEROR-Preserved and DELIVER trials, an SGLT2 inhibitor should be given to all patients with HFpEF.

Drug treatment strategy for patients with hypertension and HFpEF is reported in Table 30.3.

30.4.5 Other Treatment Options

Other medical targets have been explored, but results have usually been disappointing or require additional investigations. These include the use of inorganic nitrates/nitrites, cytokine inhibitors, levosimendan, new antifibrotic, anti-inflammatory, and anti-oxidant agents. Phosphodiesterase-5-inhibitors may be useful for some patients with combined pre- and post-capillary pulmonary hypertension [38, 39].

Device-based therapy is also under investigation. Renal denervation has been able to reduce office and ambulatory BP, to regress LVH and to improve diastolic

Table 30.3 Drug treatment strategy for patients with hypertension and HFpEF. *HFpEF* heart failure with preserved ejection fraction, *ACEi* angiotensin-converting enzyme inhibitor, *ARB* angiotensin receptor blocker, *CCB* calcium channel blockers, *HF* heart failure, *MRA* mineralocorticoid receptor antagonist, *RAS* renin-angiotensin system, *ARNI* angiotensin receptor neprilysin inhibitor, *SGLT2* sodium/glucose cotransporter 2

2018 ESC/ESH HT guidelines			In presence of HFpEF
1 pill	Initial therapy Dual combination	ACEI or ARB + diuretic or CCB	Start with an SGLT2 inibitor
1 pill	Step 2 Triple combination	ACEI or ARB + diuretic + CCB	Consider earlier use of MRA and/or replacement of RAS blocker with ARNI to reduce HF Hospitalization
2 pills	Step 3 Triple combination + Spironolactone or other drug	Resistant hypertension add spironolactone or other diuretic or betablocker	Intensify use of diuretics, including loop diuretics to relieve congestion when required
Beta blockers Consider beta-blockers at any treatment step, when there is a specific indication for their use, e.g. angina, post-myocardial infarction or heart rate control			

function [40]. A small trial did not show a benefit of renal denervation in HFpEF, but more studies are needed to assess its role in these patients [41].

30.4.6 Prevention of HFpEF

Patients with hypertension, diabetes, obesity, and other associated CV risk factors are considered predisposed to HF and therefore should be included in a program of active interventions. Blood pressure, blood glucose, cholesterol, body weight, and smoking should be controlled through nonpharmacological management with interventions to promote a healthy diet and physical activity.

ACEIs should be considered in patients with CV disease to prevent HF. Moreover, an SGLT2 inhibitor is recommended in patients with type 2 DM who are at high CV risk, because of associated atherosclerotic disease, proteinuric CKD, or multiple CV risk factors, to decrease the risk of developing HF. A nonsteroidal MRA, finerenone, may be considered in patients with type 2 DM with proteinuric CKD, on treatment with a RAS blocker, to reduce the risk of developing HF, according to the result of FIDELIO-DKD and FIGARO-DKD trials [42, 43].

30.4.7 Conclusion

HFpEF is the consequence of several risk factors, among which hypertension is probably the most frequent and important. A history of hypertension may be detected in the large majority of patients with HFpEF, and normalization of BP can significantly decrease the incidence of HF. Hypertension is responsible for the development of LVH, impaired relaxation, increased LV filling pressure, LA enlargement, central and peripheral vascular stiffness, and abnormal microcirculation, all of which may lead to HFpEF. Similar alterations may be induced, although with different time and mechanisms, by several risk factors and comorbid conditions, which are often associated with hypertension. The assessment of all these underlying conditions is of great importance in order to understand HFpEF onset, development, and clinical evolution.

Considering the large volume of studies that have demonstrated the efficacy of RAS blockers, MRA, and beta blockers in hypertension, HFrEF, and other CV diseases, it is conceivable that the same drug classes may have a not yet sufficiently clarified possible additional benefit in hypertensive patients with HFpEF. On the other hand, recent trials have shown significant benefit with the use of SGLT2 inhibitors and, in fact, at the present time, these drugs should be considered in every patient with HF, with or without diabetes, whatever the value of EF. Additional benefit could be also given by ARNI and new nonsteroidal MRA.

Furthermore, high-quality studies, with careful patient selection according to different phenotypes, particularly in the presence of hypertension, are needed and will probably provide more specific and definite evidence.

References

1. McDonagh TA, Metra M, Adamo M, Gardner RS, Baumbach A, Böhm M, Burri H, Butler J, et al. 2021 ESC guidelines for the diagnosis and treatment of acute and chronic heart failure. Eur Heart J. 2021;42(36):3599–726.
2. Heidenreich PA, Bozkurt B, Aguilar D, Allen LA, Writing Committee Members, et al. AHA/ACC/HFSA guideline for the management of heart failure. Circulation. 2022;2022(145):e895–e1032.
3. Kasiakogias A, Agabiti-Rosei E, Camafort M, Ehret G, Faconti L, Ferreira JP, Brguljan J, Januszewicz A, Kahan T, Manolis A, Tsioufis K, Weber T, von Lueder TG, Smiseth OA, Wachtell K, Kjeldsen SE, Zannad F, Mancia G, Kreutz R. Hypertension and heart failure with preserved ejection fraction: position paper by the European society of hypertension. J Hypertens. 2021;39(8):1522–45.
4. Gevaert AB, Kataria R, Zannad F, Sauer AJ, Damman K, Sharma K, Shah SJ, Van Spall HGC. Heart failure with preserved ejection fraction: recent concepts in diagnosis, mechanisms and management. Heart. 2022;108(17):1342–50.
5. Cleland JGF, Bunting KV, Flather MD, Altman DG, Holmes J, Coats AJS, et al. Beta-blockers for heart failure with reduced, mid-range, and preserved ejection fraction: an individual patient-level analysis of double-blind randomized trials. Eur Heart J. 2018;39:26–35.
6. Borlaug Barry A, Melenovsky V, Russell Stuart D, Kessler K, Pacak K, Becker Lewis C, et al. Impaired chronotropic and vasodilator reserves limit exercise capacity in patients with heart failure and a preserved ejection fraction. Circulation. 2006;114:2138–47.
7. Meyer M, LeWinter MM. Heart rate and heart failure with preserved ejection fraction: time to slow b-blocker use? Circ Heart Fail. 2019;12:e006213.
8. Flather MD, Shibata MC, Coats AJS, Van Veldhuisen DJ, Parkhomenko A, Borbola J, et al. Randomized trial to determine the effect of nebivolol on mortality and cardiovascular hospital admission in elderly patients with heart failure (SENIORS). Eur Heart J. 2005;26:215–25.
9. Pitt B, Pfeffer MA, Assmann SF, Boineau R, Anand IS, Claggett B, et al. Spironolactone for heart failure with preserved ejection fraction. N Engl J Med. 2014;370:1383–92.
10. Pfeffer Marc A, Claggett B, Assmann Susan F, Boineau R, Anand Inder S, Clausell N, et al. Regional variation in patients and outcomes in the treatment of preserved cardiac function heart failure with an aldosterone antagonist (TOPCAT) trial. Circulation. 2015;131:34–42.
11. Solomon SD, McMurray JJV, Anand IS, Ge J, Lam CSP, Maggioni AP, et al. Angiotensin–neprilysin inhibition in heart failure with preserved ejection fraction. N Engl J Med. 2019;381:1609–20.
12. Gori M, Senni M, Claggett B, Liu J, Maggioni AP, Zile M, et al. Integrating high-sensitivity troponin T and sacubitril/valsartan treatment in HFpEF: the PARAGON-HF trial. JACC Heart Fail. 2021;9(9):627–35.
13. Vaduganathan M, Claggett BL, Desai AS, Anker SD, Perrone SV, Janssens S, Milicic D, Arango JL, Packer M, Shi VC, Lefkowitz MP, McMurray JJV, Solomon SD. Prior heart failure hospitalization, clinical outcomes, and response to sacubitril/valsartan compared with valsartan in HFpEF. J Am Coll Cardiol. 2020;75(3):245–54.
14. Anker SD, Butler J, Filippatos G, Ferreira JP, Bocchi E, Böhm M, EMPEROR-Preserved Trial Investigators, et al. Empagliflozin in heart failure with a preserved ejection fraction. N Engl J Med. 2021;385(16):1451–61.
15. Solomon SD, McMurray JJV, Claggett B, de Boer RA, DeMets D, Hernandez AF, et al. Dapagliflozin in heart failure with mildly reduced or preserved ejection fraction. N Engl J Med. 2022;387(12):1089–98.
16. Kitzman DW, Brubaker P, Morgan T, Haykowsky M, Hundley G, Kraus WE, Eggebeen J, Nicklas BJ. Effect of caloric restriction or aerobic exercise training on peak oxygen consumption and quality of life in obese older patients with heart failure with preserved ejection fraction: a randomized clinical trial. JAMA. 2016;315(1):36–46.
17. Fukuta H, Goto T, Wakami K, Kamiya T, Ohte N. Effects of exercise training on cardiac function, exercise capacity, and quality of life in heart failure with preserved ejection fraction: a meta-analysis of randomized controlled trials. Heart Fail Rev. 2019;24(4):535–47.

18. Gevaert AB, Tibebu S, Mamas MA, Ravindra NG, Lee SF, Ahmad T, et al. Clinical phenogroups are more effective than left ventricular ejection fraction categories in stratifying heart failure outcomes. ESC Heart Fail. 2021;8(4):2741–54.
19. Selvaraj S, Claggett BL, Bohm M, Anker SD, Vaduganathan M, Zannad F, et al. Systolic blood pressure in heart failure with preserved ejection fraction treated with sacubitril/valsartan. J Am Coll Cardiol. 2020;75:1644–56.
20. Tsimploulis A, Lam PH, Arundel C, Singh SN, Morgan CJ, Faselis C, et al. Systolic blood pressure and outcomes in patients with heart failure with preserved ejection fraction. JAMA Cardiol. 2018;3:288–97.
21. Thomopoulos C, Parati G, Zanchetti A. Effects of blood pressure lowering treatment. 6. Prevention of heart failure and new-onset heart failure—meta-analyses of randomized trials. J Hypertens. 2016;34:373–84.
22. Salvetti M, Paini A, Bertacchini F, Stassaldi D, Aggiusti C, Agabiti Rosei C, et al. Changes in left ventricular geometry during antihypertensive treatment. Pharmacol Res. 2018;134:193–9.
23. Okin PM, Devereux RB, Jern S, Kjeldsen SE, Julius S, Nieminen MS, et al. Regression of electrocardiographic left ventricular hypertrophy during antihypertensive treatment and the prediction of major cardiovascular events. JAMA. 2004;292:2343–9.
24. Mullens W, Damman K, Harjola VP, Mebazaa A, Brunner-La Rocca HP, Martens P, et al. The use of diuretics in heart failure with congestion: a position statement from the heart failure association of the European Society of Cardiology. Eur J Heart Fail. 2019;21:137–55.
25. Costanzo MR, Stevenson LW, Adamson PB, Desai AS, Heywood JT, Bourge RC, et al. Interventions linked to decreased heart failure hospitalizations during ambulatory pulmonary artery pressure monitoring. JACC Heart Fail. 2016;4:333–44.
26. Devereux RB, Dahlof B, Gerdts E, Boman K, Nieminen Markku S, Papademetriou V, et al. Regression of hypertensive left ventricular hypertrophy by losartan compared with atenolol: the losartan intervention for endpoint reduction in hypertension (LIFE) trial. Circulation. 2004;110:1456–62.
27. Lund LH, Benson L, Dahlstrom U, Edner M. Association between use of renin-angiotensin system antagonists and mortality in patients with heart failure and preserved ejection fraction. JAMA. 2012;308:2108–17.
28. Khan MS, Fonarow GC, Khan H, Greene SJ, Anker SD, Gheorghiade M, et al. Renin–angiotensin blockade in heart failure with preserved ejection fraction: a systematic review and meta-analysis. ESC Heart Fail. 2017;4:402–8.
29. Williams B, Mancia G, Spiering W, Agabiti Rosei E, Azizi M, Burnier M, et al. 2018 ESC/ESH guidelines for the management of arterial hypertension: the task force for the management of arterial hypertension of the European Society of Cardiology and the European Society of Hypertension. J Hypertens. 2018;36:1953–2041.
30. McMurray JJV, Jackson AM, Lam CSP, Redfield MM, Anand Inder S, Ge J, et al. Effects of sacubitril-valsartan versus valsartan in women compared with men with heart failure and preserved ejection fraction: insights from PARAGON-HF. Circulation. 2020;141:338–51.
31. Suzuki K, Claggett B, Minamisawa M, Nochioka K, et al. Pulse pressure, prognosis, and influence of sacubitril/valsartan in heart failure with preserved ejection fraction. Hypertension. 2021;77(2):546–56.
32. Schmieder RE, Wagner F, Mayr M, Delles C, Ott C, Keicher C, Hrabak-Paar M, Heye T, Aichner S, Khder Y, Yates D, Albrecht D, Langenickel T, Freyhardt P, Janka R, Bremerich J. The effect of sacubitril/valsartan compared to olmesartan on cardiovascular remodelling in subjects with essential hypertension: the results of a randomized, double-blind, active-controlled study. Eur Heart J. 2017;38(44):3308–17.
33. Cleland JGF, Ferreira JP, Mariottoni B, Pellicori P, Cuthbert J, Verdonschot JAJ, et al. The effect of spironolactone on cardiovascular function and markers of fibrosis in people at increased risk of developing heart failure: the heart 'OMics' in AGEing (HOMAGE) randomized clinical trial. Eur Heart J. 2020;42:684–96.

34. Pucci G, Ranalli Maria G, Battista F, Schillaci G. Effects of b-blockers with and without vasodilating properties on central blood pressure: systematic review and meta-analysis of randomized trials in hypertension. Hypertension. 2016;67:316–24.
35. Thomopoulos C, Bazoukis G, Tsioufis C, Mancia G. Beta-blockers in hypertension: overview and meta-analysis of randomized outcome trials. J Hypertens. 2020;38:1669–81.
36. Wiviott SD, Raz I, Bonaca MP, et al. Dapagliflozin and cardiovascular outcomes in type 2 diabetes. N Engl J Med. 2019;380:347–57.
37. McGuire DK, Shih WJ, Cosentino F, et al. Association of SGLT2 inhibitors with cardiovascular and kidney outcomes in patients with type 2 diabetes: a meta-analysis. JAMA Cardiol. 2021;6:148–58.
38. Tschope C, Birner C, Bohm M, Bruder O, Frantz S, Luchner A, et al. Heart failure with preserved ejection fraction: current management and future strategies: expert opinion on the behalf of the nucleus of the "heart failure working group" of the German Society of Cardiology (DKG). Clin Res Cardiol. 2018;107:1–19.
39. Wintrich J, Kindermann I, Ukena C, Selejan S, Werner C, Maack C, et al. Therapeutic approaches in heart failure with preserved ejection fraction: past, present, and future. Clin Res Cardiol. 2020;109:1079–98.
40. Tsioufis C, Papademetriou V, Dimitriadis K, Tsiachris D, Thomopoulos C, Kasiakogias A, et al. Effects of multielectrode renal denervation on cardiac and neurohumoral adaptations in resistant hypertension with cardiac hypertrophy: an EnligHTN I substudy. J Hypertens. 2015;33:346–53.
41. Patel HC, Rosen SD, Hayward C, Vassiliou V, Smith GC, Wage RR, et al. Renal denervation in heart failure with preserved ejection fraction (RDT-PEF): a randomized controlled trial. Eur J Heart Fail. 2016;18:703–12.
42. Bakris GL, Agarwal R, Anker SD, et al. Effect of finerenone on chronic kidney disease outcomes in type 2 diabetes. N Engl J Med. 2020;383:2219–29.
43. Pitt B, Filippatos G, Agarwal R, et al. Cardiovascular events with finerenone in kidney disease and type 2 diabetes. N Engl J Med. 2021;385:2252–63.

Hypertensive Disorders of Pregnancy and Heart Failure

31

Larysa Strilchuk and Renata Cífková

31.1 Introduction

Hypertensive disorders of pregnancy (HDP) are considered to be one of the leading causes of maternal, fetal, and neonatal morbidity and mortality globally. Historically, these disorders were thought to be self-limiting [1], but now it is known that the deleterious impact of HDP is not restricted to the gestation period [2–4].

HDP do not comprise a single entity, but are classified by the 2018 European Society of Cardiology guidelines as follows: pre-existing hypertension, gestational hypertension (GH), preeclampsia (PE), pre-existing hypertension with superimposed GH with proteinuria (Table 31.1). In some cases, when blood pressure (BP) is first recorded after 20 weeks of gestation, reassessment and a proper classification can be made only after 42 days postpartum [5]. The International Society for the Study of Hypertension in Pregnancy (ISSHP) introduced a broader definition of PE, which does not insist on the presence of proteinuria (de novo hypertension after 20 weeks' gestation accompanied by proteinuria and/or evidence of maternal acute kidney injury, liver dysfunction, neurological features, hemolysis or thrombocytopenia, or fetal growth restriction) [6].

L. Strilchuk
Center for Cardiovascular Prevention, Charles University in Prague, First Faculty of Medicine and Thomayer University Hospital, Prague, Czech Republic

Department of Therapy №1, Medical Diagnostics, Hematology and Transfusiology, Lviv Danylo Halytsky National Medical University, Lviv, Ukraine

R. Cífková (✉)
Center for Cardiovascular Prevention, Charles University in Prague, First Faculty of Medicine and Thomayer University Hospital, Prague, Czech Republic

Department of Medicine II, Charles University in Prague, First Faculty of Medicine, Prague, Czech Republic
e-mail: renata.cifkova@ftn.cz

M. Dorobantu et al. (eds.), *Hypertension and Heart Failure*, Updates in Hypertension and Cardiovascular Protection,
https://doi.org/10.1007/978-3-031-39315-0_31

Table 31.1 Classification of hypertensive disorders of pregnancy by the 2018 ESC guidelines [5]

Pre-existing hypertension
• Either preceding pregnancy or developing before 20 weeks' gestation; usually persisting for more than 42 days postpartum,
Gestational hypertension
• Developing after 20 weeks' gestation; usually resolving within 42 days postpartum,
1. Without significant proteinuria
2. With significant proteinuria = preeclampsia (protein excretion >0.3 g/day or ACR ≥ 30 mg/mmol)
Pre-existing hypertension + superimposed gestational hypertension with proteinuria
• Pre-existing hypertension associated with further BP increase and protein excretion >3 g/day after 20 weeks' gestation
Antenatally unclassifiable hypertension
• This term is used when BP is first recorded after 20 weeks of gestation and hypertension is diagnosed; re-assessment is necessary after 42 days postpartum

ACR albumin/creatinine ratio

The total burden of HDP is difficult to estimate. Wang et al. (2021) analyzed the data of the Global Burden of Disease 2019 Study and compared them to similar data from 1990. The authors concluded that the global incidence of HDP had increased from 16.30 million to 18.08 million (by 10.92%), probably due to population growth and multiple gestations, as the age-standardized incidence rates have decreased [7].

31.2 Acute/Subacute Heart Failure in Pregnancy

Acute pulmonary edema is a rare (0.08% to 1.5%) but life-threatening complication of pregnancy, which can affect women ante-, intra- or postpartum. It usually is a complication of PE or eclampsia [8]. Acute pulmonary edema may develop in 3–12% of women with eclampsia [9], whereas for severe PE this proportion may reach 10% [10]. Factors that predispose preeclamptic women to develop pulmonary edema include circulating angiogenic factors, decreased colloid oncotic pressure, increased capillary permeability, diastolic dysfunction, and endothelial cell dysfunction [11, 12]. Iatrogenic fluid overload and the postpartum autotransfusion phenomenon also play significant roles [8]. Nulliparity, multiple gestation pregnancy, and anemia are associated with a higher risk of acute pulmonary edema [12].

The clinical presentation of acute pulmonary edema is characterized by progressive dyspnea, chest discomfort, cough, orthopnea, possible wheezing, and nocturnal paroxysms of dyspnea. As the diagnosis of acute pulmonary edema in pregnancy is mostly clinical, clinicians should be attentive not to misinterpret dyspnea in pulmonary edema with late pregnancy-related dyspnea due to a growing uterus and distension of abdominal muscles [13].

Treatment of stabilized or subacute pulmonary edema in pregnancy is similar to those of nonpregnant patients and includes oxygen therapy, fluid restriction, intravenous loop diuretics, central hemodynamic monitoring, and vasodilators, preferably nitroglycerin, while avoiding potentially fetotoxic agents (angiotensin-converting enzyme inhibitors, angiotensin receptor blockers, angiotensin receptor-neprilysin inhibitors, mineralocorticoid receptor antagonists, and atenolol) [5, 14]. In the absence of pulmonary congestion, diuretics should not be administered, as they may reduce the placental blood flow [15].

According to the Heart Failure Association of the European Society of Cardiology Study Group on Peripartum Cardiomyopathy, pregnant women with acute HF must be immediately referred to an intensive care unit for further assessment of HF severity and fetal status (Fig. 31.1) [14]. A multidisciplinary team (the pregnancy heart team) is required to ensure rapid diagnosis, decision-making, and appropriate therapy. Treatment strategy depends on fetus viability and the possibility to stabilize the mother's condition: hemodynamically unstable women with advanced HF must undergo urgent delivery by cesarean section irrespective of gestation time [14]. Due to this, neonates are exposed to prematurity, leading to an increase in neonatal morbidity and mortality [16]. If stabilization is possible, the delivery may be delayed [14].

As acute pulmonary edema in pregnant women is a secondary condition, measures should be taken to prevent its main causes. To prevent PE and, consequently,

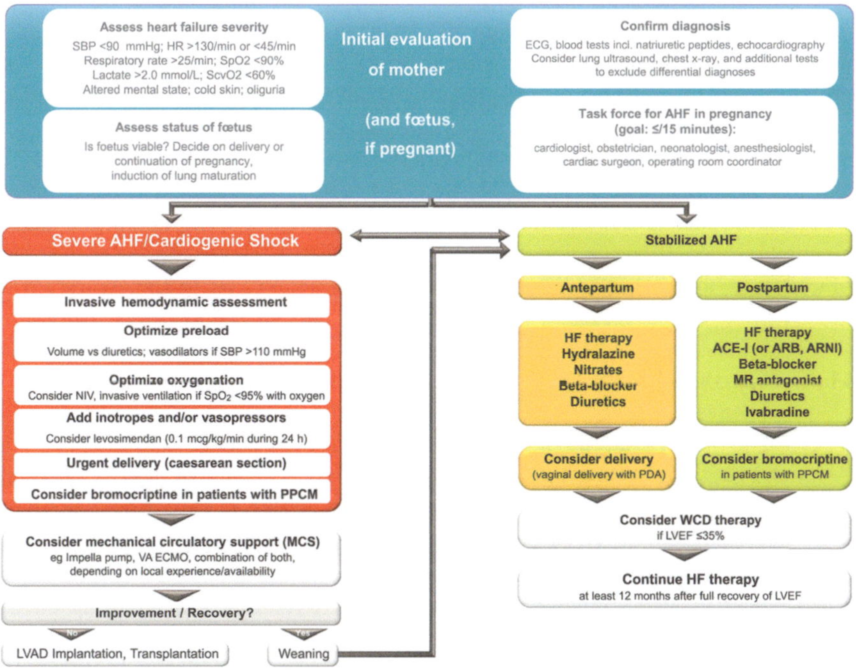

Fig. 31.1 Management of acute heart failure during pregnancy, reproduced with permission from Sliwa et al. [14]. *ACE-I* angiotensin-converting enzyme inhibitor, *AHF* acute heart failure, *ARB* angiotensin receptor blocker, *ARNI* angiotensin receptor-neprilysin inhibitor, *ECG* electrocardiogram, *HR* heart rate, *LVAD* left ventricular assist device, *LVEF* left ventricular ejection fraction, *MR* mineralocorticoid receptor, *NIV* noninvasive ventilation, *PDA* peridural anesthesia, *PPCM* peripartum cardiomyopathy, *SBP* systolic blood pressure, $ScvO_2$ central venous oxygen saturation, SpO_2 blood oxygen saturation, *VA ECMO* veno-arterial extracorporeal membrane oxygenation, *WCD* wearable cardioverter-defibrillator

acute pulmonary edema as its complication, the 2018 ESC guidelines recommend 100–150 mg of aspirin for women at high or moderate risk of PE, administered from week 12 to weeks 36–37 [5]. Other preventive measures include clinical monitoring and restricted fluid administration [17].

31.3 Peripartum Cardiomyopathy

Women with acute pulmonary edema in pregnancy are sometimes diagnosed with peripartum cardiomyopathy (PPCM) [17]. The latter is a rare life-threatening cardiac disease, presenting with symptoms and signs of heart failure (HF) due to the left ventricle (LV) systolic dysfunction and typically occurring in the last month of pregnancy or within 5 months after delivery [5]. A meta-analysis of 22 studies showed that the pooled prevalence of PE in women with PPCM was more than quadruple the average rate (22% vs. 5%, $p < 0.001$) [18]. This supports the hypothesis that PPCM and PE are overlapping pathologic conditions of pregnancy [19].

About half of patients with PPCM recover their LV function, whereas in the other half the impairment persists or severe HF develops [20]. In a retrospective community-based cohort study, PPCM accounted for 68.2% of HF cases [20]. Apart from PPCM, causes of acute HF during pregnancy include pre-existing cardiomyopathy, congenital heart disease, stress/toxic cardiomyopathy, severe valvular heart disease, or myocardial ischemia due to coronary artery dissection or atherosclerosis [19].

If PPCM is treated accordingly, LV ejection fraction improvement is usually seen 6–12 months postpartum, but some women may recover later. Six months of maintaining the LV ejection fraction above 50% usually serves as a criterion for successful treatment; then, a gradual decrease in pharmacotherapy may be considered [21].

31.4 Hypertensive Disorders of Pregnancy and Subsequent Heart Failure

Cardiovascular (CV) consequences of HDP have been widely studied in separate trials and meta-analyses (Table 31.2). Most of these studies have concluded that HDP and heart failure (HF) are closely associated; moreover, this association remains significant after adjustment for many maternal factors.

History of PE was a predisposition for worse left ventricle (LV) diastolic function even 9–16 years after pregnancy, approximately in the fifth decade of life. In a study by Bokstag et al., middle-aged women with a history of PE had significantly higher BP, plasma levels of interleukin-6 and soluble intercellular adhesion molecule-1 (sICAM-1), thicker septal and posterior LV walls, and worse diastolic LV function (reduced mean mitral annular lengthening velocity (E′mean) and a higher ratio of early diastolic mitral flow velocity (E) over E′ mean [E/E′ mean]) [29].

Data on the association of CV pregnancy complications with different HF subtypes (HF with preserved ejection fraction [HFpEF] vs. HF with reduced ejection fraction [HFrEF]) are scarce. According to the survey by Hansen et al. (2021), HDP

Table 31.2 Key trials and meta-analyses evaluating the link between HDP and HF

Type of study	Total no	Major findings	Reference
Rochester epidemiology project, medical record-linkage system	9862	The risk of congestive HF was increased among women with HDP compared with the control group matched by age and parity (HR, 2.11, 95% CI: 1.19–3.76)	[2]
Meta-analysis of 73 studies	>13 mil	Overall RR for HF in women with any HDP compared with women with normotensive pregnancies was 2.87 (95% CI 2.14–3.85; $p < 0.001$)	[22]
Comprehensive review of 21 studies	3,601,192	Women with 1 or more pregnancies accompanied by GH had a greater risk of HF (RR, 1.77; 95% CI, 1.47–2.13)	[23]
Retrospective cohort study	425,649	Compared with women with neither condition, adjusted incident HF risks within 1 year and 5 years after delivery were increased for women with prepregnancy hypertension, HDP, and both conditions	[24]
Retrospective registry-cohort study	>1.4 mil	Women with HDP had double the risk of CV readmission within 3 years after delivery compared to normotensive ones (6.4 vs. 2.5/1000 deliveries; $p < 0.001$). HF was the reason for 84.4% of these readmissions	[25]
Retrospective longitudinal registry, National Health Insurance Research Database (Taiwan)	6475	The risk of congestive HF was increased in PE and eclampsia (HRs 7.39 and 9.07 respectively). The occurrence of HF was independent of post-pregnancy hypertension	[4]
Population-based retrospective cohort study, National Health Insurance research database (Taiwan)	29,186 women with HDP and 116,744 matched controls	The incidence rates of HF in the HDP cohort and the matched comparison cohort were 0.65% and 0.11% (relative risk 5.91). The overall incidence of HF was significantly higher in women with HDP (9.83 versus 1.67 in the control group per 10,000 person-years). Women with HDP had higher HF risk than their counterparts with normotensive pregnancies even at >10 years postpartum	[26]

(continued)

Table 31.2 (continued)

Type of study	Total no	Major findings	Reference
USA Nationwide inpatient sample	>50 mil pregnancy-related hospitalizations	The overall rate of HF was 112 cases per 100,000. The majority (about 60%) of HF cases occurred postpartum, 27.3%—By delivery, and 13.2%—Antepartum. HF was found to be responsible for ≈9% of all maternal deaths before discharge and other adverse outcomes	[27]
Retrospective community-based cohort study (Kaiser Permanente health system)	385,935	HF was diagnosed in 126 cases per 100,000 pregnancies. Women with HF were characterized by more frequent preterm births and cesarean sections, a higher neonatal death rate (1.0% vs. 0.4%, $p = 0.03$), and a 7.7-fold increase in the risk of maternal death	[19]
Prospective observational study of UK biobank data	220,024	HDPs were associated with a greater incidence of HF (age- and race-adjusted HR 1.7, 95%CI 1.04–2.6, $p = 0.03$) and other cardiac disorders	[28]

HDP hypertensive disorders of pregnancy, *HF* heart failure

was significantly associated with HFpEF (OR, 2.06; 95% CI, 1.29–3.27), but not with HFrEF (OR, 1.17; 95% CI, 0.59–2.30) [30]. HFrEF was more common at 1 year postpartum, whereas later cases of HF (at 4 years postpartum) were mostly characterized by normal ejection fraction [31].

31.5 Pathogenesis of Heart Failure in Women with a History of Hypertensive Disorders of Pregnancy

During pregnancy, the woman's body undergoes multiple adaptations (including CV ones) aimed at successful gestation and fetal development. Violations of this process can eventually induce cardiac dysfunction and HF [32]. Cardiac changes in women with PE include impaired myocardial contractility and relaxation; systole-diastolic dysfunction, unfavorable remodeling, and hypertrophy of both ventricles; hemodynamic disorders; localized myocardial ischemia, and fibrosis [33]. These changes may become a substrate for HF (Fig. 31.2).

Nevertheless, the pathophysiology underlying the relationship between HDP and HF is still largely unknown. The existing literature hypothesizes the pregnancy-induced CV risk, pre-conceptional susceptibility toward an increased CV risk, or pregnancy-induced physiological stress superimposed on the pre-pregnancy susceptible background. Shortly after delivery, women with HDP have higher CV risk due to conventional well-known risk factors (hypertension, proatherogenic lipid profile, renal disorders, and diabetes mellitus), and 10–20 years after delivery, they may experience premature CV events [34].

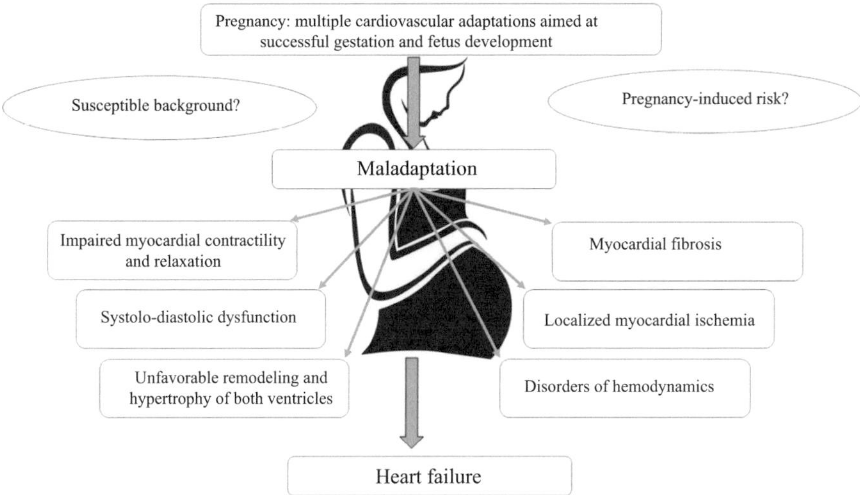

Fig. 31.2 Possible pathophysiological mechanisms of heart failure in women with hypertensive disorders of pregnancy

An extensive literature review by Benagiano et al. concludes that the deleterious impact of HDP on the CV system may be explained by the basic pathophysiologic processes of any CV disease (inflammation, metabolic syndrome, oxidative stress, and hypertension) and the influence of placental biomolecules [3]. Other disease-specific pathophysiologic mechanisms that may explain the long-term CV risks after PE include the increase of sodium and angiotensin II sensitivity, and sympathetic activation [35]. Preeclamptic pregnancy can also result in increased endothelial dysfunction, carotid intima-media thickness [36], coronary artery calcification, and increased pulse wave velocity [37].

31.6 Patterns of Echocardiographic Changes in Women with HDP-Associated Heart Failure

Compared with normotensive control pregnancies, women with HDP demonstrate greater antepartum left atrium (LA) volume, greater LV mass due to increased LV wall thickness, larger inferior vena cava diameter, lower early diastolic septal and lateral mitral annular velocity (e′), and decreased index of long axis contractility during pregnancy/postpartum (Fig. 31.3). Simultaneously, the laboratory tests of women with HDP show higher levels of peripartum brain natriuretic peptide/N-terminal pro-brain natriuretic peptide (BNP/NT-proBNP) and high sensitivity troponin I (hs-TnI). NT-proBNP, BNP, and hs-TnI showed significant positive correlations with LA volume and LV mass indices (LAVi, LVMi). The high peripartum hs-TnI level can be a predictor of poor LV relaxation postpartum, so it may be useful in the detection of women requiring further follow-up of cardiac status [38].

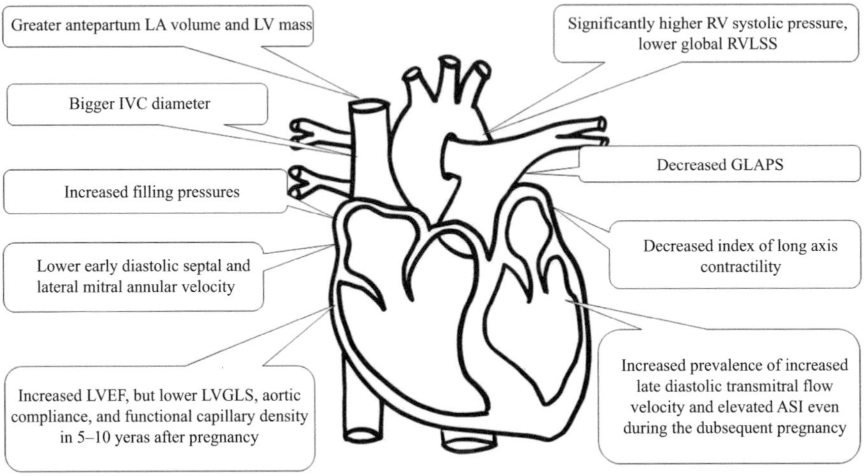

Fig. 31.3 Echocardiographic changes in women with hypertensive disorders of pregnancy. *ASI* arterial stiffness index, *IVC* inferior vena cava, *GLAPS* global left atrial peak strain, *LA* left atrium, *LV* left ventricle, *LVEF* left ventricle ejection fraction, *LVGLS* left ventricle global longitudinal strain, *RV* right ventricle, *RVLSS* right ventricle longitudinal systolic strain

Compared to women with a history of normotensive pregnancies, women with prior HDP still had increased LVMi, LVEF, and LAVi but lower LV global longitudinal strain (LVGLS), E/A, aortic compliance, and functional capillary density 5–10 years after pregnancy [39]. Speckle-tracking echocardiographic data show that PE is accompanied by significantly higher RV systolic pressure, lower global RV longitudinal systolic strain (RVLSS), lower mitral septal e′ velocity, higher septal E/e′ ratio, larger LA area size, and greater posterior and septal LV wall thickness compared to the controls. These differences translate into abnormal diastolic function and abnormal remodeling of left ventricle in women with PE [10].

Early- and late-onset PE may have different effects on heart remodeling. On average, 12 years after delivery (median 12 [9;15]) women with early-onset PE had significantly lower LVGLS magnitude, more restrictive diastolic filling pattern, lower septal e′ velocities, and a subsequently higher E/e′ ratio than controls and women with late-onset PE, so the early-onset PE is more likely to cause the subclinical impairment of LV function [40].

Global left atrial peak strain (GLAPS) is another possible predictor of postpartum persistent hypertension. In hypertensive pregnancies, GLAPS has been found to be significantly decreased compared with normotensive ones. A GLAPS value of ≤23.5% suggests an impaired LA reservoir function and can serve as a marker of severe atrial cardiomyopathy [41].

Estimation of cardiac changes in pregnancy with the help of EchoCG may be challenging. Firstly, indexation may give distorted results as the change in the body surface area is pregnancy-specific; moreover, it is still under debate whether pre-pregnancy or actual weight should be used for calculation. Apart from that, the hearts of pregnant women are more spherical compared to the ellipsoid shape of the female heart beyond the pregnancy, which can influence calculations of the LV volume [42].

31.7 Heart Failure Biomarkers in Women with Hypertensive Disorders of Pregnancy

The European Society of Cardiology guidelines recommend using the NT-proBNP, a conventional marker of HF, for the assessment and monitoring of patients with hypertensive emergencies, including PE [1]. Other authors also support the role of NT-pro-BNP in both PE and diastolic HF in females [43], but this biomarker needs to be validated specifically in women with HDP-associated HF. According to the meta-analysis of 13 studies, BNP levels of 100–500 ng/L demonstrate a positive and negative predictive value of 92 and 97%, respectively, for diagnosing cardiac complications in pregnant women shortly after delivery. Authors claim that BNP had better diagnostic accuracy than NT-pro-BNP; however NT-pro-BNP was evaluated in a smaller number of studies [44].

Glycoprotein growth factor activin A represents one more possible link between HDP and cardiac function. Women with PE or chronic hypertension/GH have significantly higher antepartum levels of activin A than women with normotensive pregnancies. Increased antepartum levels of activin A were closely associated with abnormal GLS before delivery and 1 year postpartum even after adjustment for multiple variables [45]. The activin A hypothesis can at least partly explain the favorable effect of aspirin in hypertensive pregnancies: it was observed that antepartum aspirin administration in the second trimester of pregnancy reduced serum activin A and improved GLS in women with PE [46].

A recent study shows the increased levels of angiotensin II type 1 receptor autoantibodies (AT1R-Ab) in women with PE both during pregnancy and 4 years postpartum compared to women with normotensive pregnancies. The AT1R-Ab increase was linked to prepartum diastolic dysfunction, adverse LV remodeling, and hyperdynamic systolic function in both ventricles. The AT1R-Ab level may be used as a key to potential targeted treatment with angiotensin II receptor blockers [47].

Soluble human leukocyte antigen G (sHLA-G) is significantly lower in early- and late-onset PE groups compared to controls, but after delivery, this difference persisted only in women with early-onset PE. In this group, elevated levels of sHLA-G were observed up to 3 years after PE [48].

Women with PE both during pregnancy and postpartum period show elevated levels of interleukin-6 and tumor necrosis factor-α. Pregnancies complicated with PE are also characterized by increased triglycerides and free fatty acids and, in the postpartum period, by elevated low-density lipoproteins cholesterol and leptin together with a decreased adiponectin level [49].

It has also been shown that PE is associated with elevated plasma levels of vasopressin and copeptin. Current evidence also shows that the excessive stimulation of vasopressin receptors (especially, V1aR) may have an unfavorable impact on both maternal and fetal CV systems [50].

Established and potential biomarkers of HF in pregnancy are presented in Fig. 31.4.

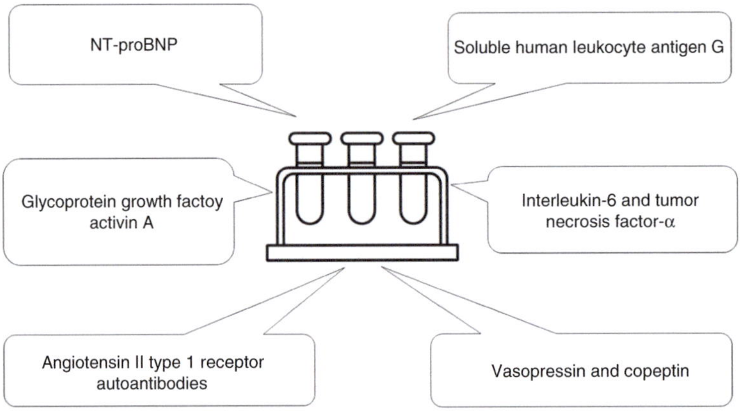

Fig. 31.4 Biomarkers of heart failure in pregnancy. *NT-proBNP* N-terminal pro-brain natriuretic peptide

31.8 Prevention and Treatment of Heart Failure in Women with Current or Prior Hypertensive Disorders of Pregnancy

Strategies for prevention and treatment of women with HDP and HDP-associated HF include two time periods: the pregnancy itself and the postpartum period (early and late phases). Each of these timeframes must have appropriate interventions. For instance, Benschop proposes three timeframes for postpartum CV follow-up with certain measures (Fig. 31.5) [51].

To allow fetal maturation, HDPs are often subject to expectant management with simultaneous monitoring of the maternal condition. However, according to a multicenter open-label randomized controlled trial HYPITAT, induction of labor in women with GH or mild PE after 36 weeks' gestation was associated with improved maternal outcomes (lower maternal mortality and morbidity (eclampsia, HELLP syndrome, pulmonary edema, thromboembolic disease, and placental abruption), less frequent progression of GH or PE to severe hypertension or proteinuria, and lower prevalence of major postpartum hemorrhage) [52]. Apart from that, each week of expectant management of HDP increases the 5-year risk of death or readmission of the woman for acute CV or cerebrovascular disease (hypertensive heart disease, acute myocardial infarction, cardiomyopathy, cardiac arrest, heart failure, stroke, or intracerebral hemorrhage) by 7% [53]. Therefore, PE has to be considered a first CV event in the life of high-risk women, and this event requires secondary prevention and appropriate follow-up [54].

Possible prevention and management measures for women with HDP may include first-trimester screening for PE using the algorithm of The Fetal Medicine Foundation, offering aspirin 100–150 mg/day before 16 weeks of gestation to women at high or moderate risk of PE, early referral to specialist care, lifestyle

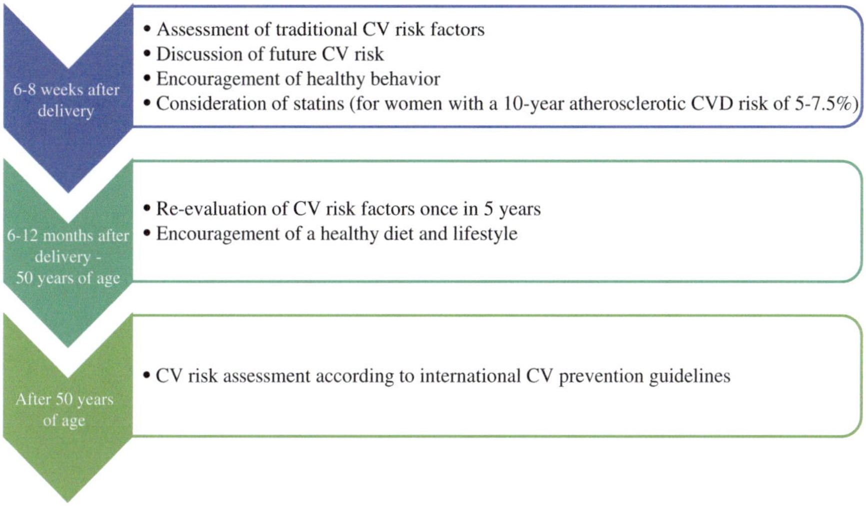

Fig. 31.5 Timeframes of postpartum CV follow-up for women with hypertensive disorders of pregnancy

changes, and regular exercise. It is also recommended to monitor BP, renal functions, and lipid profile annually for the first 5 years after HDP [55].

Animal studies imply that pravastatin may ameliorate LV remodeling, decrease heart fibrosis, and improve cardiac output postpartum independently of the cholesterol-lowering effect [56]. A small pilot randomized clinical trial in 20 women at high risk of PE observed that PE incidence in the pravastatin group was lower than in the placebo group (20% vs. 50%) [57]. The other intervention study with pravastatin in high-risk pregnancies ($n = 1120$) did not show any favorable effect on PE, so further research is needed [58].

The recent single-center Postnatal Enalapril to Improve Cardiovascular Function Following Preterm Preeclampsia (PICk-UP) Study showed that 6 months of postnatal enalapril treatment (titrated to 20 mg) for women with preterm PE improved diastolic function and decreased relative LV wall thickness. Large multicenter randomized clinical trials are required to confirm these findings [59].

In the early postnatal period, women may be non-adherent to any therapy because of neonatal stressors [59], so drug treatment may be challenging. During this period women are also reluctant to visit their doctor for CV check-ups due to the intensive care of their infants [60].

Several professional societies pay attention to HDP and their role in the further follow-up of women, but their recommendations may be confusing, as they differ regarding the time of visits and some other aspects (Table 31.3).

Table 31.3 Key recommendations of different guidelines for the CV follow-up of women with hypertensive disorders of pregnancy

Organization	Advice	References
American College of Obstetricians and Gynecologists	All women affected by HDP need to contact an obstetric care provider within 3 weeks postpartum, a comprehensive postpartum visit by 12 weeks postpartum, and annual CV follow-ups thereafter	[61]
European society of hypertension	Women with prior HDP need annual visits to a primary care physician to check BP and other metabolic factors. Long-term specialist follow-up is recommended. History of HDP should be an integral part of a woman's medical record. Lifestyle modifications need to be employed to reduce future CV risk	[5]
Consensus document from European cardiologists, gynecologists, and endocrinologists	There is a need to establish systematic follow-up recommendations. Self-monitoring of BP and telehealth feedback to the primary care is advised	[62]
National Institute for health and care excellence	Women with prior HDP need to undergo follow-up at 6–8 weeks and 3 months postpartum	[63]

31.9 Conclusions

Acute pulmonary edema is a rare but life-threatening complication of preeclampsia or eclampsia, which can affect women ante-, intra- or postpartum. Pregnant women with acute pulmonary edema often need an urgent pregnancy termination by cesarean section, which exposes neonates to prematurity. In the setting of acute pulmonary edema, women are quite frequently diagnosed with peripartum cardiomyopathy, which can induce persistent severe heart failure.

Raising awareness of the long-term deleterious consequences of HDP is of utmost importance. Women with a history of hypertensive pregnancy need to be encouraged to control their body weight, quit smoking, adopt a healthy diet, and perform some sort of daily physical activity, as these prevention strategies are simple, safe, and cost-effective. As a key point of a woman's history, HDP need to be kept in mind permanently, as numerous studies show unfavorable consequences in women with prior hypertensive pregnancies even up to 25 years after delivery.

During the antepartum and early postpartum periods, women with HDP may benefit from being counseled by multidisciplinary cardio-obstetric teams with a future transition to specially trained general practitioners who can manage the organization of long-term follow-up including CVD risk stratification, monitoring of BP and relevant biomarkers, regular visits to a clinic, continuous lifestyle modification support and, if needed, drug treatment (Fig. 31.6). To aid monitoring in the early postpartum period, telehealth may be employed, as women in this period are often being lost to follow-up due to neonate-associated preoccupation and stressors.

To support the most tailored management strategy, there is a need for separate studies on HDP subtypes, early- and late-onset PE, and different types of HF

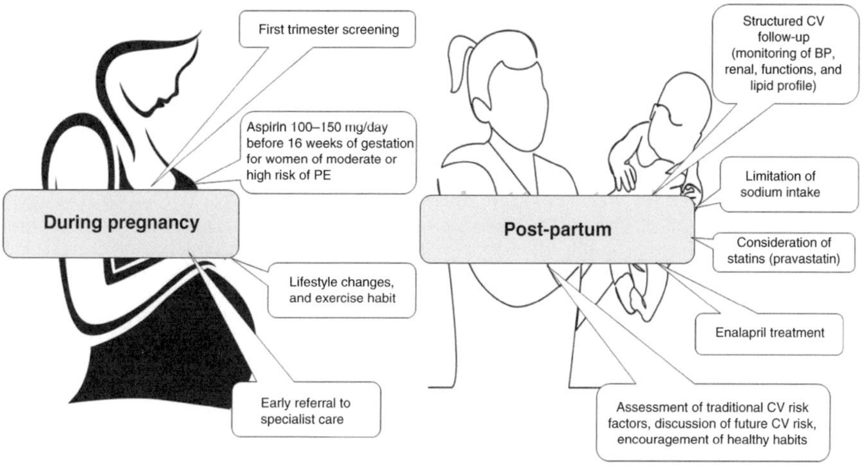

Fig. 31.6 Prevention of post-HDP heart failure. *CV* cardiovascular, *PE* preeclampsia

(HFpEF and HFrEF). Based on the above studies, an algorithmized approach to the CV follow-up of women with prior HDP needs to be developed. This approach must include a time plot of follow-up visits and a clear roadmap for HF prevention in women with a history of HDP.

References

1. Gamble DT, Brikinns B, Myint PK, Bhattacharya S. Hypertensive disorders of pregnancy and subsequent cardiovascular disease: current national and international guidelines and the need for future research. Front Cardiovasc Med. 2019;6(55):55. https://doi.org/10.3389/fcvm.2019.00055.
2. Garovic VD, White WM, Vaughan L, et al. Incidence and long-term outcomes of hypertensive disorders of pregnancy. J Am Coll Cardiol. 2020;75(18):2323–34. https://doi.org/10.1016/j.jacc.2020.03.028.
3. Benagiano M, Mancuso S, Brosens JJ, Benagiano G. Long-term consequences of placental vascular pathology on the maternal and offspring cardiovascular systems. Biomolecules. 2021;11(11):1625. https://doi.org/10.3390/biom11111625.
4. Kuo YL, Chan TF, Wu CY, Ker CR, Tu HP. Preeclampsia-eclampsia and future cardiovascular risk among women in Taiwan. Taiwan J Obstet Gynecol. 2018;57(3):364–9. https://doi.org/10.1016/j.tjog.2018.04.035.
5. Regitz-Zagrosek V, Roos-Hesselink JW, Bauersachs J, Blomström-Lundqvist C, Cífková R, De Bonis M, et al. 2018 ESC guidelines for the management of cardiovascular diseases during pregnancy. Eur Heart J. 2018;39(34):3165–241. https://doi.org/10.1093/eurheartj/ehy340.
6. Brown MA, Magee LA, Kenny LC, Karumanchi SA, McCarthy FP, Saito S, Hall DR, Warren CE, Adoyi G, Ishaku S, International Society for the Study of Hypertension in Pregnancy (ISSHP). The hypertensive disorders of pregnancy: ISSHP classification, diagnosis & management recommendations for international practice. Pregnancy Hypertens. 2018;13:291–310. https://doi.org/10.1016/j.preghy.2018.05.004.
7. Wang W, Xie X, Yuan T, Wang Y, Zhao F, Zhou Z, Zhang H. Epidemiological trends of maternal hypertensive disorders of pregnancy at the global, regional, and national levels: a

population-based study. BMC Pregnancy Childbirth. 2021;21(1):364. https://doi.org/10.1186/s12884-021-03809-2.
8. Pordeus ACB, Katz L, Soares MC, Maia SB, Amorim MMR. Acute pulmonary edema in an obstetric intensive care unit: a case series study. Medicine (Baltimore). 2018;97(28):e11508. https://doi.org/10.1097/MD.0000000000011508.
9. Ghulmiyyah L, Sibai B. Maternal mortality from preeclampsia/eclampsia. Semin Perinatol. 2012;36(1):56–9. https://doi.org/10.1053/j.semperi.2011.09.011.
10. Vaught AJ, Kovell LC, Szymanski LM, Mayer SA, Seifert SM, Vaidya D, Murphy JD, Argani C, O'Kelly A, York S, Ouyang P, Mukherjee M, Zakaria S. Acute cardiac effects of severe pre-eclampsia. J Am Coll Cardiol. 2018;72(1):1–11. https://doi.org/10.1016/j.jacc.2018.04.048.
11. Ramos JGL, Sass N, Costa SHM. Preeclampsia. Rev Bras Ginecol Obstet. 2017;39(9):496–512. https://doi.org/10.1055/s-0037-1604471.
12. Keepanasseril A, Monárrez-Espino J, Vadivelu P, Kumar MD. Risk factors of pulmonary edema in women with preeclampsia from South India: a case-control study. J Hum Hypertens. 2020;34(10):727–34. https://doi.org/10.1038/s41371-020-0300-1.
13. Cypher RL. Pulmonary edema in obstetrics: essential facts for critical care nurses. AACN Adv Crit Care. 2018;29(3):327–35. https://doi.org/10.4037/aacnacc2018958.
14. Sliwa K, van der Meer P, Petrie MC, Frogoudaki A, Johnson MR, Hilfiker-Kleiner D, Hamdan R, et al. Risk stratification and management of women with cardiomyopathy/heart failure planning pregnancy or presenting during/after pregnancy: a position statement from the heart failure association of the European Society of cardiology study group on peripartum cardiomyopathy. Eur J Heart Fail. 2021;23:527–40. https://doi.org/10.1002/ejhf.2133.
15. Hilfiker-Kleiner D, Westhoff-Bleck M, Gunter HH, von Kaisenberg CS, Bohnhorst B, Hoeltje M, Kuehn C. A management algorithm for acute heart failure in pregnancy. The Hannover experience. Eur Heart J. 2015;36:769–70.
16. Dennis AT, Solnordal CB. Acute pulmonary oedema in pregnant women. Anaesthesia. 2012;67(6):646–59. https://doi.org/10.1111/j.1365-2044.2012.07055.x.
17. Parikh P, Blauwet L. Peripartum cardiomyopathy and preeclampsia: overlapping diseases of pregnancy. Curr Hypertens Rep. 2018;20(8):69. https://doi.org/10.1007/s11906-018-0868-9.
18. Bello N, Rendon ISH, Arany Z. The relationship between pre-eclampsia and peripartum cardiomyopathy: a systematic review and meta-analysis. J Am Coll Cardiol. 2013;62(18):1715–23. https://doi.org/10.1016/j.jacc.2013.08.717.
19. Sliwa K, Hilfiker-Kleiner D, Petrie MC, Heart Failure Association of the European Society of Cardiology Working Group on Peripartum Cardiomyopathy, et al. The current state of knowledge on aetiology, diagnosis, management, and therapy of peripartum cardiomyopathy: a position statement from the heart failure association of the European society of cardiology working group on peripartum cardiomyopathy. Eur J Heart Fail. 2010;12(8):767–78. https://doi.org/10.1093/eurjhf/hfq120.
20. Ng AT, Duan L, Win T, Spencer HT, Lee MS. Maternal and fetal outcomes in pregnant women with heart failure. Heart. 2018;104(23):1949–54. https://doi.org/10.1136/heartjnl-2018-313156.
21. Kryczka KE, Demkow M, Dzielinska Z. Peripartum cardiomyopathy—a cardiovascular disease in pregnancy and puerperium. The actual state of knowledge, challenges, and perspectives. Ginekol Pol. 2021;92(2):147–52. https://doi.org/10.5603/GP.a2020.0171.
22. Wu R, Wang T, Gu R, Xing D, Ye C, Chen Y, Liu X, Chen L. Hypertensive disorders of pregnancy and risk of cardiovascular disease-related morbidity and mortality: a systematic review and meta-analysis. Cardiology. 2020;145:633–47. https://doi.org/10.1159/000508036.
23. Lo CCW, Lo ASQ, Leow SH, Fisher G, Corker B, Batho O, Morris B, et al. Future cardiovascular disease risk for women with gestational hypertension: a systematic review and meta-analysis. J Am Heart Assoc. 2020;9(13):e013991. https://doi.org/10.1161/JAHA.119.013991.
24. Malek AM, Wilson DA, Turan TN, Mateus J, Lackland DT, Hunt KJ. Incident heart failure within the first and fifth year after delivery among women with hypertensive disorders of pregnancy and prepregnancy hypertension in a diverse population. J Am Heart Assoc. 2021;10(17):e021616. https://doi.org/10.1161/JAHA.121.021616.

25. Jarvie JL, Metz TD, Davis MB, Ehrig JC, Kao DP. Short-term risk of cardiovascular readmission following a hypertensive disorder of pregnancy. Heart. 2018;104:1187–94. https://doi.org/10.1136/heartjnl-2017-312299.
26. Chen SN, Cheng CC, Tsui KH, Tang PL, Chern CU, Huang WC, Lin LT. Hypertensive disorders of pregnancy and future heart failure risk: a nationwide population-based retrospective cohort study. Pregnancy Hypertens. 2018;13:110–5. https://doi.org/10.1016/j.preghy.2018.05.010.
27. Mogos MF, Piano MR, McFarlin BL, Salemi JL, Liese KL, Briller JE. Heart failure in pregnant women: a concern across the pregnancy continuum. Circ Heart Fail. 2018;11(1):e004005. https://doi.org/10.1161/CIRCHEARTFAILURE.117.004005.
28. Honigberg MC, Zekavat SM, Aragam K, Klarin D, Bhatt DL, Scott NS, Peloso GM, et al. Long-term cardiovascular risk in women with hypertension during pregnancy. J Am Coll Cardiol. 2019;74(22):2743–54. https://doi.org/10.1016/j.jacc.2019.09.052.
29. Bokslag A, Franssen C, Alma LJ, Kovacevic I, Kesteren FV, Teunissen PW, Kamp O, et al. Early-onset preeclampsia predisposes to preclinical diastolic left ventricular dysfunction in the fifth decade of life: an observational study. PLoS One. 2018;13(6):e0198908. https://doi.org/10.1371/journal.pone.0198908.
30. Hansen AL, Søndergaard MM, Hlatky MA, Vittinghof E, Nah G, Stefanick ML, Manson JE, et al. Adverse pregnancy outcomes and incident heart failure in the women's health initiative. JAMA Netw Open. 2021;4(12):e2138071. https://doi.org/10.1001/jamanetworkopen.2021.38071.
31. Breetveld NM, Ghossein-Doha C, Van Kuijk S, Van Dijk AP, Van Der Vlugt MJ, Heidemas WM, et al. Prevalence of asymptomatic heart failure in formerly preeclamptic women: a cohort study. Ultrasound Obstet Gynecol. 2017;49:134–42. https://doi.org/10.1002/uog.16014.
32. Aryan L, Medzikovic L, Umar S, Eghbali. Pregnancy-associated cardiac dysfunction and the regulatory role of microRNAs. Biol Sex Differ. 2020;11:14. https://doi.org/10.1186/s13293-020-00292-w.
33. Melchiorre K, Thilaganathan B, Giorgione V, Ridder A, Memmo A, Khalil A. Hypertensive disorders of pregnancy and future cardiovascular health. Front Cardiovasc Med. 2020;7:59. https://doi.org/10.3389/fcvm.2020.00059.
34. Melchiorre K, Sharma R, Khalil A, Thilaganathan B. Maternal cardiovascular function in normal pregnancy: evidence of maladaptation to chronic volume overload. Hypertension. 2016;67:754–62. https://doi.org/10.1161/HYPERTENSIONAHA.115.06667.
35. Turbeville HR, Sasser JM. Preeclampsia beyond pregnancy: long-term consequences for mother and child. Am J Physiol Renal Physiol. 2020;318:F1315–26. https://doi.org/10.1152/ajprenal.00071.2020.
36. Garovic VD, Milic NM, Weissgerber TL, Mielke MM, Bailey KR, Lahr B, Jayachandran M, et al. Carotid artery intima-media thickness and subclinical atherosclerosis in women with remote histories of preeclampsia: results from a Rochester epidemiology project-based study and meta-analysis. Mayo Clin Proc. 2017;92:1328–40. https://doi.org/10.1016/j.mayocp.2017.05.030.
37. Kirollos S, Skilton M, Patel S, Clare AC. A systematic review of vascular structure and function in pre-eclampsia: non-invasive assessment and mechanistic links. Front Cardiovasc Med. 2019;15:166. https://doi.org/10.3389/fcvm.2019.00166.
38. Umazume T, Yamada S, Yamada T, Ishikawa S, Furuta I, Iwano H, Murai D, et al. Association of peripartum troponin I levels with left ventricular relaxation in women with hypertensive disorders of pregnancy. Open Heart. 2018;5(2):e000829. https://doi.org/10.1136/openhrt-2018-000829.
39. Boardman H, Lamata P, Lazdam M, et al. Variations in cardiovascular structure, function, and geometry in midlife associated with a history of hypertensive pregnancy. Hypertension. 2020;75(6):1542–50. https://doi.org/10.1161/HYPERTENSIONAHA.119.14530.
40. Clemmensen TS, Christensen M, Kronborg CJS, Knudsen UB, Løgstrup BB. Long-term follow-up of women with early-onset pre-eclampsia shows subclinical impairment of the left ventricular function by two-dimensional speckle-tracking echocardiography. Pregnancy Hypertens. 2018;14:9–14. https://doi.org/10.1016/j.preghy.2018.07.001.

41. Sonaglioni A, Lonati C, Lombardo M, Rigamonti E, Binda G, Vincenti A, Nicolosi GL, Bianchi S, Harari S, Anzà C. Incremental prognostic value of global left atrial peak strain in women with new-onset gestational hypertension. J Hypertens. 2019;37(8):1668–75. https://doi.org/10.1097/HJH.0000000000002086.
42. Castleman JS, Shantsila A, Brown RA, et al. Altered cardiac and vascular stiffness in pregnancy after a hypertensive pregnancy. J Hum Hypertens. 2022;37:189. https://doi.org/10.1038/s41371-022-00662-4.
43. Alma LJ, Bokslag A, Maas AHEM, Franx A, Paulus WJ, de Groot CJM. Shared biomarkers between female diastolic heart failure and pre-eclampsia: a systematic review and meta-analysis. ESC Heart Fail. 2017;4(2):88–98. https://doi.org/10.1002/ehf2.12129.
44. Sheikh M, Ostadrahimi P, Salarzaei M, et al. Cardiac complications in pregnancy: a systematic review and meta-analysis of diagnostic accuracy of BNP and N-terminal pro-BNP. Cardiol Ther. 2021;10:501–14. https://doi.org/10.1007/s40119-021-00230-w.
45. Shahul S, Ramadan H, Nizamuddin J, Mueller A, Patel V, Dreixler J, Tung A, Lang RM, Weinert L, Nasim R, Chinthala S, Rana S. Activin a and late postpartum cardiac dysfunction among women with hypertensive disorders of pregnancy. Hypertension. 2018;72(1):188–93. https://doi.org/10.1161/HYPERTENSIONAHA.118.10888.
46. Naseem H, Dreixler J, Mueller A, Tung A, Dhir R, Chibber R, Fazal A, et al. Antepartum aspirin administration reduces Activin A and cardiac global longitudinal strain in preeclamptic women. J Am Heart Assoc. 2020;9:e015997. https://doi.org/10.1161/JAHA.119.015997.
47. Minhas AS, Zakaria S, Fedarko N, Boyer T, Mukherjee M, Sharma G, Everett A, et al. Association of angiotensin receptor autoantibodies with cardiovascular abnormalities in preeclampsia. J Am Heart Assoc. 2021;10:20831. https://doi.org/10.1161/JAHA.121.020831.
48. Jacobsen DP, Lekva T, Moe K, Fjeldstad HES, Johnsen GM, Sugulle M, Staff AC. Pregnancy and postpartum levels of circulating maternal sHLA-G in preeclampsia. J Reprod Immunol. 2021;143:103249. https://doi.org/10.1016/j.jri.2020.103249.
49. Mohseni Z, Derksen E, Oben J, Al-Nasiry S, Spaanderman MEA, Ghossein-Doha C. Cardiac dysfunction after preeclampsia; an overview of pro- and anti-fibrotic circulating effector molecules. Pregnancy Hypertens. 2021;23:140–54. https://doi.org/10.1016/j.preghy.2020.12.001.
50. Szczepanska-Sadowska E, Cudnoch-Jedrzejewska A, Wsol A. The role of oxytocin and vasopressin in the pathophysiology of heart failure in pregnancy and in fetal and neonatal life. Am J Physiol Heart Circ Physiol. 2020;318(3):H639–51. https://doi.org/10.1152/ajpheart.00484.2019.
51. Benschop L, Duvekot JJ, Roeters van Lennep JE. Future risk of cardiovascular disease risk factors and events in women after a hypertensive disorder of pregnancy. Heart. 2019;105(16):1273–8. https://doi.org/10.1136/heartjnl-2018-313453.
52. Koopmans CM, Bijlenga D, Groen H, Vijgen SM, Aarnoudse JG, Bekedam DJ, van den Berg PP, HYPITAT Study Group, et al. Induction of labour versus expectant monitoring for gestational hypertension or mild pre-eclampsia after 36 weeks' gestation (HYPITAT): a multicentre, open-label randomised controlled trial. Lancet. 2009;374(9694):979–88. https://doi.org/10.1016/S0140-6736(09)60736-4.
53. Rosenbloom JI, Lewkowitz AK, Lindley KJ, et al. Expectant management of hypertensive disorders of pregnancy and future cardiovascular morbidity. Obstet Gynecol. 2020;135:27–35. https://doi.org/10.1097/AOG.0000000000003567.
54. Muijsers HEC, Roeleveld N, van der Heijden OWH, Maas AHEM. Consider preeclampsia as a first cardiovascular event. Curr Cardiovasc Risk Rep. 2019;13:21. https://doi.org/10.1007/s12170-019-0614-0.
55. Panaitescu AM, Popescu MR, Ciobanu AM, Gica N, Cimpoca-Raptis BA. Pregnancy complications can foreshadow future disease—long-term outcomes of a complicated pregnancy. Medicina (Kaunas). 2021;57(12):1320. https://doi.org/10.3390/medicina57121320.
56. Kräker K, O'Driscoll JM, Schütte T. Statins reverse postpartum cardiovascular dysfunction in a rat model of preeclampsia. Hypertension. 2020;75:202–10. https://doi.org/10.1161/HYPERTENSIONAHA.119.13219.

57. Constantine MM, West H, Wisner KL, Caritis S, Clark S, Venkataramanan R, Stika CS, et al. A randomized pilot clinical trial of pravastatin versus placebo in pregnant patients at high risk of preeclampsia. Am J Obstet Gynecol. 2021;225:225. https://doi.org/10.1016/j.ajog.2021.05.018.
58. Döbert M, Varouxaki AN, Mu AC, Syngelaki A, Ciobanu A, Akolekar R, De Paco MC, et al. Pravastatin versus placebo in pregnancies at high risk of term preeclampsia. Circulation. 2021;144:670–9. https://doi.org/10.1161/CIRCULATIONAHA.121.053963.
59. Ormesher L, Higson S, Luckie M, et al. Postnatal enalapril to improve cardiovascular function following preterm preeclampsia (PICk-UP): a randomized double-blind placebo-controlled feasibility trial. Hypertension. 2020;76(6):1828–37. https://doi.org/10.1161/HYPERTENSIONAHA.120.15875.
60. Cífková R. Cardiovascular sequels of hypertension in pregnancy. J Am Heart Assoc. 2018;7(10):e009300. https://doi.org/10.1161/JAHA.118.009300.
61. American College of Obstetricians and Gynecologists. ACOG committee opinion no. 736: optimizing postpartum care. Obstet Gynecol. 2018;131(5):e140–50. https://doi.org/10.1097/AOG.0000000000002633.
62. Maas AHEM, Rosano G, Cifkova R, et al. Cardiovascular health after menopause transition, pregnancy disorders, and other gynecologic conditions: a consensus document from European cardiologists, gynecologists, and endocrinologists. Eur Heart J. 2021;42(10):967–84. https://doi.org/10.1093/eurheartj/ehaa1044.
63. National Institute for Health and Care Excellence. Hypertension in pregnancy: diagnosis and management. [NG133]. 2019; https://www.nice.org.uk/guidance/ng133. Accessed 15 Jul 2022.

Hypertension in Cancer Patients

32

Iulia Magdalena Gramaticu, Vlad Mihai Croitoru,
Adina Emilia Croitoru, and Irina Mihaela Cazacu

32.1 Epidemiology and Risk Factors of Hypertension and Cancer

Hypertension is a major global health problem. According to epidemiological data, the prevalence of hypertension in developed countries has more than doubled in the last four decades as a result of population aging [1]. Hypertension is an important risk factor for cardiovascular diseases and is the leading cause of death, accounting for 7.6 million deaths worldwide each year [2].

Cancer morbidity and mortality have also been rising as the population has aged. Cancer is the second leading cause of death worldwide, and cardiovascular disease is the leading cause of late morbidity and death in cancer survivors [3]. There is limited data on the prevalence of hypertension among cancer patients prior to cancer treatment, but cancer patients receiving specific treatments and cancer survivors have a higher prevalence of hypertension than the general population [4]. A large observational cohort study prior to the widespread use of many targeted therapies found that hypertension is the most common comorbidity in cancer registries, with a reported prevalence of 38% [5]. With the development of the novel anticancer therapies which markedly increased survival rates for cancer patients, long-term

I. M. Gramaticu (✉) · V. M. Croitoru
Department of Oncology, Fundeni Clinical Institute, Bucharest, Romania

A. E. Croitoru
Department of Oncology, Fundeni Clinical Institute, Bucharest, Romania

"Titu Maiorescu" University of Medicine, Bucharest, Romania

I. M. Cazacu
Department of Oncology, Fundeni Clinical Institute, Bucharest, Romania

Carol Davila University of Medicine and Pharmacy, Bucharest, Romania

M. Dorobantu et al. (eds.), *Hypertension and Heart Failure*, Updates in Hypertension and Cardiovascular Protection,
https://doi.org/10.1007/978-3-031-39315-0_32

cancer survivors have now a higher risk of cardiovascular disease than of recurrent cancer, and hypertension contributes to this risk [6]. The prevalence of hypertension in childhood cancer survivors exceeds 70% by the age of 50 [7]. It is worth noting that hypertension is the most important modifiable risk factor for poor cardiovascular outcomes among cancer patients [4].

There is a well-known multidimensional relationship between hypertension, cancer, and cardiovascular risk. Smoking, obesity, and diabetes are all risk factors for hypertension, cardiovascular disease, chronic kidney disease, and cancer [8, 9]. Several population studies have found that obesity is at least partially responsible for hypertension in approximately 78% of cases [10, 11], and up to 80% of patients with type 2 diabetes develop hypertension [12].

Many cancer treatments either cause hypertension directly through effects on endothelial function, sympathetic activity, and renin-angiotensin system activity or indirectly through nephrotoxic side effects. Some anticancer therapies cause an acute rise in blood pressure, which may result in deterioration of preexisting cardiovascular conditions and lead to acute hypertension-related complications in severe cases. For example, posterior reversible encephalopathy syndrome is a potentially severe but manageable toxicity of anti-VEGF agents, in patients with uncontrolled arterial hypertension. Thus, cancer patients survival might be impaired as a result of these hypertension-related complications that require a reduction in treatment dosage or even the termination of potentially life-saving anticancer treatment [13, 14]. Apart from the development of rapid-onset hypertension, numerous antineoplastic agents are linked to hypertension many years after the initial exposure. Long-term survivors are more likely to develop hypertension than the general population which increases the risk for cardiovascular disease, long-term end organ damage, and mortality. These detrimental vascular effects are becoming increasingly important as many novel targeted therapies result in long-term anticancer responses, significantly improving cancer survival [4, 15]. Consequently, the prevention, identification, and prompt treatment of hypertension caused by antineoplastic agents are important to avoid both short- and long-term adverse cardiovascular sequelae.

32.2 Pathophysiology of Hypertension and Cancer

The fact that cancer and hypertension frequently co-occur and share multiple risk factors suggests a shared biology for which there is emerging evidence. Some overlapping pathophysiology mechanisms involved in both conditions have been identified, including inflammation and an increase in reactive oxygen species (ROS) and oxidative stress. Inflammatory cells and cytokines are well-known essential elements of the tumor microenvironment, and targeting inflammatory mediators, such as TNF-α, reduces the incidence and development of cancer [16]. Similarly, inflammatory cell infiltration is seen in the renal interstitium and the arterial vascular wall of hypertensive rats [17, 18], and inhibition of this inflammatory process reduces hypertension [19].

According to a recent analysis of GLOBOCAN cancer incidence data, infections caused 2.2 million new cancer cases in 2018, accounting for 13% of all cancer cases (excluding non-melanoma skin cancers) [20]. A meta-analysis of prospective and retrospective cohort studies (fourteen prospective cohort studies, two retrospective cohort studies, and five nested case-control studies, including 142,640 participants) systematically assessed if high levels of circulating inflammation markers (C reactive protein (CRP), high-sensitive CRP (hs-CRP), IL-6 and IL-1β) are associated with an increased risk of hypertension in the general population [21]. A similar relationship has been found between baseline inflammatory status (higher levels of circulating CRP, hs-CRP, and IL-6) and the development of hypertension.

Extensive experimental data from a variety of hypertensive models show that ROS and oxidative stress play a role in the development of hypertension. For example, Endothelin-1(ET-1) has been shown to cause endothelial progenitor cells (EPC) dysfunction in low-renin, salt-sensitive hypertension by increasing oxidative stress through the ET_A/NADPH oxidase pathway [22]. However, the advantages of reducing oxidative stress in cancer patients are unclear. A large Cochrane meta-analysis that included the majority of primary and secondary prevention randomized clinical trials on antioxidant supplements (beta-carotene, vitamin A, vitamin C, vitamin E, and selenium) versus placebo or no intervention did not support the beneficial effects of antioxidant supplements for the prevention of mortality in adults (from various diseases, including CVD and cancer) [23].

32.3 Hypertension and Anticancer Therapy

Many anticancer therapies have been linked to the development and/or aggravation of hypertension.

32.4 VEGF Inhibitors

VEGF inhibitors have improved outcomes in a variety of cancers including renal cell carcinoma, hepatocellular carcinoma, epithelial ovarian cancer, and gastrointestinal stromal cell tumors. The reported incidence and prevalence of hypertension induced by VEGF inhibitors may be inaccurate due to a lack of standardization of hypertension classification and concrete definitions in clinical trials. According to a recent large meta-analysis of 77 studies, 7.4% of patients were diagnosed with severe hypertension, 1.8% with arterial thromboembolism, 1.7% with cardiac ischemia, and 2.3% with cardiac dysfunction. VEGF inhibitors increase the risk of hypertension/severe hypertension, myocardial ischemia, arterial thromboembolism, and cardiac dysfunction by 5.3/5.6, 2.8, 1.5, and 1.4 times, respectively [24].

There are four major classes of agents to inhibit VEGF signaling: (1) monoclonal antibodies directed against circulating VEGF—*bevacizumab*, the first VEGF inhibitor approved by the FDA in 2004; (2) soluble decoy receptors (VEGF-traps)—*aflibercept,* binding to both sides of the VEGF dimer and to the placental growth factor; (3) monoclonal antibodies against the vascular endothelial growth factor receptor (VEGFR)—*ramucirumab*; and (4) TKIs with anti-VEGFR activity such as *sunitinib, sorafenib, pazopanib,* and *axitinib* which act on the intracellular tyrosine kinase domains of VEGFR to limit its activation. Many TKIs interfere also with other important proangiogenic growth factors and receptors, such as the platelet-derived growth factor, fibroblast growth factor, c-Kit, and Flt-3 [25]. VEGF works by attaching to one of its three tyrosine kinase receptors (VEGFR 1, 2, and 3), with VEGFR2 being the major signaling receptor. Activation of the VEGFR leads to multiple downstream effects: increase in capillary permeability, production of the vasodilator nitric oxide (NO), and promotion of vascular endothelial cell survival [26].

Many studies have been conducted to explain the mechanisms involved in the pathogenesis of hypertension related to anti-VEGF treatment. While the exact mechanisms underlying the hypertensive effects of VEGF inhibitors remain unclear, several molecular mechanisms have been proposed.

VEGF is an important modulator of vascular tone and blood pressure. Normally, binding of VEGF to VEGFR-2 on endothelial cells (ECs) results in NO production. NO diffuses to vascular smooth muscle cells (VSMCs), where it induces vasorelaxation by stimulating guanylate cyclase. VEGFR-1/2 stimulation also results in prostacyclin (PGI_2) synthesis, which induces vasorelaxation in VSMCs via adenylate cyclase activation [27, 28]. Thus, blocking VEGFR decreases NO and PGI_2 synthesis and causes vasoconstriction [28]. On the other hand, VEGF inhibitors increase the bioavailability of vasoconstrictors, particularly ET-1, inducing an imbalance between vasodilators and vasoconstrictors in favor of the latter [29]. This enhances vasomotor tone and contributes to the development of hypertension.

Oxidative stress is another important prohypertensive mechanism described. Several studies demonstrated that inhibition of VEGF signaling induces vascular dysfunction through redox-sensitive processes [30].

Microvascular rarefaction, a reduction in microvessel density, has also been observed in patients receiving VEGFI. It has been hypothesized that microvascular rarefaction increases vascular resistance, thus contributing to hypertension induced by VEGF inhibitors [31]. Some authors outline that this rarefaction is most likely functional, given the rapid increase in blood pressure after initiation of therapy and the quick blood pressure normalization after VEGFI withdrawal [32].

Generally, blood pressure increases occur within 1 day of therapy with VEGF inhibitors. A plateau is usually reached within 6 to 10 days, as steady-state concentrations of the drug equilibrate, but significant interindividual variation exists [33]. Moreover, a preclinical study that investigated the salt dependency of sunitinib-induced hypertension showed that VEGF inhibitors-induced hypertension is salt-sensitive [34].

VEGF inhibitors are also associated with nephrotoxic effects through different mechanisms such as thrombotic microangiopathy, minimal change nephropathy, or focal segmental glomerulosclerosis. Thus, the use of VEGF inhibitors may lead to proteinuria and sodium and water retention, causing hypertension [35].

In addition, different clinical studies on cardiotoxicity induced by sunitinib identified hypertension as a risk factor for heart failure [36, 37]. VEGF inhibitors may exert a direct myocardial toxic effect and consequently reduce the heart's capacity to tolerate a rise in afterload as a result of concurrent systemic hypertension. In a large animal model, resting systemic blood pressure has been shown to increase significantly within 1 week of starting sunitinib, along with an increase in systemic and coronary vascular resistance [38]. With an increase in chronic afterload, the heart attempts to adapt through hypertrophic remodeling. In order to support the increased metabolic demand, this hypertrophic response increases oxygen demand, stimulates the HIF-1a-VEGF axis, and eventually leads to vascular neogenesis in the myocardium [39]. In addition, a different study demonstrated not only cardiac hypertrophy at the cardiomyocyte and organ levels, but also an increase in cardiac expression of VEGF-A following thoracic aortic constriction (TAC) in mice, an established model of afterload increase [40]. These data indicate that antiangiogenic agents could be detrimental for heart function in patients who also experience hypertrophic cardiomyopathy.

The role of hypertension as a biomarker in cancer treatment with VEGF inhibitors has been investigated [41]. A retrospective analysis using pooled data from multiple trials showed an association between longer overall survival and the presence of hypertension in patients undergoing treatment with sunitinib for metastatic renal cell carcinoma (mRCC) [42]. In this study, median overall survival was 30.9 months in patients who developed sunitinib-induced hypertension versus 7.2 months in patients without hypertension. Adverse events associated with hypertension were more frequent in patients with a mean systolic blood pressure (SBP) at or above 140 mmHg (overall, 11%; renal events, 5%) compared to patients with a mean SBP below 140 mmHg (overall, 9%; renal events, 3%). In contrast, another study showed no significant association between the development of new or worsening hypertension and deterioration of renal function 30 days after initiating treatment with VEGFR inhibitors for mRCC [43].

32.5 Chemotherapy

32.5.1 Alkylating Agents and Other Alkyl-like Agents

Cyclophosphamide, ifosfamide, busulfan, and cisplatin are frequently used in the treatment of hematologic malignancies (lymphoma, leukexmia) and alkylating agents solid organ malignancies (head and neck cancers, lung, colorectal and genitourinary cancers). In current clinical practice, alkylating drugs are commonly combined with other agents, making it difficult to attribute adverse outcomes to a specific agent. The predominant mechanisms for hypertension induced by alkylating agents are oxidative damage to endothelial cells, increased intimal thickness, and abnormal vascular remodeling [44]. Other postulated mechanisms include nephrotoxicity with associated sodium retention [45]. In contrast to VEGF inhibitor-induced hypertension, platinum therapy-associated hypertension is a long-term effect that can occur years after treatment. Cisplatin is detectable in serum up to 13 years after initial exposure and it may cause chronic endothelial activation. Several studies showed that up to 50% of testicular cancer survivors who received cisplatin-based chemotherapy experienced an increased incidence of hypertension as late effect [46].

32.5.2 Antimetabolites

Gemcitabine has been associated with the development of hypertension in the setting of thrombotic microangiopathy [47], with evidence of endothelial damage in preclinical models of rapidly dividing endothelial cells [48].

32.5.3 Antimicrotubular Agents

Taxanes and vinca alkaloids are among the most commonly used chemotherapeutic agents (e.g., for the treatment of Hodgkin's and non-Hodgkin's lymphoma, breast cancer, non-small cell lung cancer, ovarian cancer, and prostate cancer). These agents inhibit crucial microtubule dynamics and induce cellular responses such as deregulated chromosome formation, abnormal mitotic spindle, metaphase arrest, and eventual cell death.

The true extent of the role of these agents in cardiovascular disease is challenging to determine because antimicrotubular agents are frequently used in conjunction with radiation therapy or other chemotherapy drugs. A comprehensive study evaluated real-world data from the FDA's Adverse Event Reporting System to study cardiovascular adverse events associated with antimicrotubular agents [48]. Over a 30-year period, 13.4% of adverse events related to antimicrotubular agents were cardiovascular (CVAE), with 74.1% caused by taxanes and 25.9% caused by vinca alkaloids. The most common CVAE related to taxanes was arterial hypertension requiring hospitalization (30.2%). For patients treated with vinca alkaloids, hypertension requiring hospitalization was the third most common CVAE. Hypertension requiring hospitalization was significantly lower in monotherapy versus combination therapy (3.4% for taxane monotherapy compared with 5.3% for any taxane combination therapy).

32.5.4 BRAF/MEK Inhibitors

Inhibitors of BRAF (v-raf murine sarcoma viral oncogene homolog B1) and inhibitors of MEK (mitogen-activated protein kinase kinase) are approved for use in patients with different malignancies, including melanoma, colorectal cancer, hepatocellular carcinoma, RCC, and gastrointestinal stroma cell tumors. These agents target BRAF (one of the three RAF kinase family members) and MEK, which are important kinases in the RAS-RAF-MEK-ERK (extracellular signal-regulated kinase) signaling pathway, which plays important roles in promoting cell proliferation, differentiation, and resistance to apoptosis. Cardiovascular adverse events induced by BRAFi/MEKi are rare, but potentially life threatening and need to be monitored. In phase III clinical trials, reduction in left ventricular ejection fraction of any grade was reported in 5.7–11.7%, hypertension in 10.9–29.4%, and QT interval prolongation in 0–4.5% of the patients treated with BRAFi/MEKi, depending on the combinations used [49].

The incidence of hypertension in patients treated with BRAF inhibitors, such as vemurafenib and dabrafenib, varied from 6% to 14% in clinical trials [50]. In another study which evaluated MEK inhibitors versus chemotherapy in metastatic BRAF-mutated melanoma, the incidence of grade 2 to 3 hypertension was 15% in the trametinib group versus 7% in the chemotherapy group [51].

A comprehensive meta-analysis of five pivotal randomized clinical trials, including over 2300 patients treated with combined BRAFi/MEKi therapy, reported an incidence of any-grade hypertension of 20% compared with 14% in controls treated with BRAF inhibitor monotherapy (RR 1.5) [52].

The mechanisms leading to BRAF/MEK inhibitor-induced hypertension are incompletely defined. A complicated molecular loop involving overexpression of CD47, a receptor for thrombospondin-1, has been suggested as the underlying mechanism. Compensatory ERK hyperactivation upon BRAF/MEK inhibition induces CD47 transcription, which in turn inhibits nitric oxide stimulation, a pivotal regulator of the vascular tone [53].

32.5.5 RET Kinase Inhibitors

Several drugs targeting RET have been approved for the treatment of different malignancies, for example cabozantinib and vandetanib for medullary thyroid carcinomas, lenvatinib and sorafenib for differentiated thyroid cancers, or alectinib for lung cancer. Notably, each of these drugs is a multikinase inhibitor that has activity against RET and none of these agents have been approved solely based on their anti-RET kinase action. More recently, the selective RET kinase inhibitors, selpercatinib and pralsetinib, have been approved for use in patients with RET mutations in thyroid and non-small cell lung cancers. In a phase I/II study of 162 patients with RET-mutant medullary thyroid or RET fusion-positive thyroid cancer treated with selpercatinib, 43% developed hypertension of any grade [54]. Similarly, in other study including 105 patients with non-small cell lung cancer, 31% of patients treated with selpercatinib developed any-grade hypertension [55]. Mechanisms underlying RET inhibitor-associated hypertension are not fully understood. However, given the role RET kinase plays within the RAS-RAF-MEK-ERK pathway, RET inhibition may lead to rebound ERK activation similar to that seen with BRAF/MEK inhibition. Thus, CD47 upregulation may also be important in the development of hypertension with RET inhibitors [56].

32.6 PARP (Poly ADP Ribose Polymerase) Inhibitors

From this pharmacological class, only niraparib has been associated with hypertension [57]. In the NOVA trial (Niraparib Maintenance Therapy in Platinum-Sensitive, Recurrent Ovarian Cancer), any-grade and grade 3 or 4 hypertension occurred in 19% and 8% of patients treated with niraparib, respectively, versus 5% and 2% in placebo-treated patients, respectively [58]. Additionally, in the phase II AVANOVA2 study (Niraparib Plus Bevacizumab Versus Niraparib Alone for Platinum-Sensitive, Recurrent Ovarian Cancer), 56% of patients receiving a combination of niraparib and bevacizumab developed hypertension, compared with 22% of patients receiving niraparib monotherapy [59]. Mechanisms underlying niraparib-induced hypertension are poorly understood.

32.7 Proteasome Inhibitors

Bortezomib and carfilzomib, currently used for the treatment of multiple myeloma and mantle cell lymphoma, can cause cardiac toxicity, which is more frequent in patients treated with carfilzomib. In a clinical trial including 929 multiple myeloma patients, carfilzomib led to a longer median progression-free survival than bortezomib, but demonstrated a higher incidence of all-grade (25% versus 14%) and high-grade (9% versus 3%) hypertension. This suggests a correlation between the prohypertensive effects and the efficacy of these agents [60, 61].

32.8 Inhibitors of mTOR (Mammalian Target of Rapamycin)

The phosphatidylinositol 3-kinase (PI3K)/AKT/mammalian target of rapamycin (mTOR) pathway is activated in the majority of human cancers. Activation of the PI3K/AKT pathway in tumor cells can also increase VEGF secretion, both by hypoxia-inducible factor 1 (HIF-1) dependent and independent mechanisms. The PI3K/AKT pathway also modulates the expression of other angiogenic factors such as nitric oxide and angiopoietins. The inhibitors targeting the PI3K/AKT/mTOR pathway have been shown to decrease VEGF secretion and they have prohypertensive effects similar to VEGF inhibitors [62].

In a phase III study (METEOR) which evaluated everolimus versus cabozantinib in patients with advanced renal clear cell carcinoma, a small amount (8%) of patients who received everolimus developed hypertension [63]. Combining everolimus with the VEGF-TKI lenvatinib in patients with metastatic RCC was associated with all-grade and high-grade hypertension in 41% and 14% of patients, respectively, in a randomized phase II study [64].

In addition, 4.1% of patients with pancreatic neuroendocrine tumors (p-NET) and 5.1% of patients with non-p-NET were found to have hypertension after treatment with everolimus in an open-label, multicenter, phase III study [65].

32.9 Endocrine Therapy

Abiraterone acetate, an oral hormonal agent used in the treatment of castration-resistant metastatic prostate cancer, inhibits CYP17, a key enzyme that catalyzes the biosynthesis of androgens, resulting in significant suppression of androgen and cortisol synthesis. The suppression of cortisol leads to increase of ACTH which in turn causes mineral corticoid excess resulting in fluid retention and subsequent hypertension [66].

In a meta-analysis including 5445 patients (from five randomized clinical trials) with metastatic prostate cancer treated with abiraterone and prednisone, the overall incidences of all-grade and high-grade hypertension were 21.9% and 10.2%, respectively. Abiraterone was associated with a significantly increased risk of hypertension of all grade, with a relative risk of 1.80 and high-grade hypertension with a relative risk of 2.11 in comparison with controls [67]. Of note, abiraterone is shown to increase metoprolol effect due to strong CYP2D6 inhibition; thus, increasing metoprolol dose or starting metoprolol is not recommended for treating abiraterone-related hypertension.

32.10 Adjuvant Therapies

Adjuvant therapies, such as erythropoietin, corticosteroids, nonsteroidal anti-inflammatory drugs, and calcineurin inhibitors, may also contribute to the development of hypertension or worsening of previously controlled hypertension.

Erythropoietin use has been associated with systemic hypertension in up to 35% of patients with cancer and anemia secondary to chemotherapy [68]. The possible mechanisms include increased blood viscosity, direct vasoconstricting properties, increased sensitivity to endogenous vasopressors within smooth muscle cells, and vascular remodeling [69].

Glucocorticoid-induced hypertension has been reported in up to 13% of patients [70]. The principal mechanism of corticosteroid-induced hypertension is the overstimulation of the mineralocorticoid receptor, resulting in sodium retention in the kidney. This results in volume expansion and a subsequent increase in blood pressure. Corticosteroids with strong mineralocorticoid effects, such as fludrocortisone and hydrocortisone, produce the greatest amount of fluid retention. Corticosteroid-induced hypertension may respond to diuretic therapy. Mineralocorticoid antagonists and ACEIs/ARBs may also be beneficial [71].

Abdominal radiation has resulted in hypertension due to renal artery stenosis in rare cases [72]. Baroreflex failure manifested as labile blood pressure or hypertensive crisis may develop as a sequela of neck irradiation [73]. Radical nephrectomy for kidney cancer is also associated with the development of hypertension.

32.11 Hypertension Due to Cancer

Hypertension can be a paraneoplastic feature of several malignancies, like hepatocellular carcinoma, renal cell carcinoma, or carcinoid tumors. In hepatocellular carcinoma, paraneoplastic hypertension is due to an excessive production of either renin or angiotensin I [74]. Paraneoplastic hypertension secondary to excessive urinary secretion of catecholamine has been described in some case reports of carcinoid tumors [75]. Paraneoplastic hypertension occurs in approximately 2% of patients diagnosed with RCC and is associated with adverse prognosis [76]. It is noteworthy to mention that hypertension in RCC, reported in 75% of patients, has multiple contributing etiologies: loss of nephron mass post-nephrectomy, treatment with VEGF inhibitors, or secretion of vasoactive peptides, notably endothelin-1.

Malignant pheochromocytoma and paraganglioma (PPGL) are rare tumors and knowledge of the natural history is quite limited. Hypertension described in these cancers is caused by catecholamine hypersecretion. In a retrospective review of patients with malignant PPGL, dopamine hypersecretion has been associated with a more aggressive course and worse prognosis [77]. Adrenocortical carcinomas are rare tumors, frequently presenting with Cushing's syndrome and with features resulting from hypersecretion of glucocorticoid and/or androgens. Presentation with hyperaldosteronism is uncommon [78]. In either case, patients are likely to have hypertension as part of their presenting symptoms.

32.12 Management of Hypertension Related to Anticancer Drugs

32.12.1 Before Cancer Treatment

In addition to hypertension management, clinicians should incorporate a broader ABCDE cardio-oncology framework for cancer patients and management of the cardiovascular disease [78, 79].

The first important step is to identify cancer patients with established atherosclerotic vascular disease, diabetes mellitus, chronic kidney, and traditional cardiovascular risk factors. Cancer itself may be in the next years recognized as a "risk-enhancing factor," given the inflammatory state produced by cancer. Cancer patients who are actively smoking should be offered counseling for tobacco cessation, nicotine replacement therapy, bupropion or varenicline, based on safety of side effect profile. Whenever appropriate, cancer patients should be encouraged to adopt the Dietary Approaches to Stop Hypertension (DASH) eating plan and to maintain a healthy weight (BMI <25 kg/m^2), because even a weight loss of 5% has been shown to reduce blood pressure. Cancer patients should be screened for diabetes mellitus, particularly if they are on long-term or high-dose steroids. Regular aerobic exercise results in reductions in blood pressure of 5–7 mmHg among individuals with hypertension and these reductions translate to a **reduced risk of cardiovascular disease of 20–30%**. Emerging research suggests that **dynamic resistance exercise** may also serve as an efficacious strategy to lower blood pressure to levels similar to aerobic exercise. Any limitations in cardiopulmonary fitness (lymphedema, bony metastasis) should be promptly identified. Clinicians should also review the cardiotoxic effects of cancer therapies [80].

32.13 Blood Pressure Measurement and Monitoring

All cancer patients should undergo BP monitoring. Hypertension guidelines emphasize out-of-office BP measurement for the diagnosis and treatment of hypertension in cancer patients, which is particularly important because they may have more labile BPs and an increased risk for masked hypertension (normotensive in clinic and hypertensive out-of-office) or white coat hypertension (hypertensive in office and normotensive out-of-office) [81]. Patients on anti-VEGF therapy, alkylating agents, or high-dose corticosteroids and patients with preexisting hypertension or those known to be at higher cardiovascular risk may warrant closer BP monitoring. For example, resting BPs should be monitored daily during the first cycle of treatment with VEGF inhibitors.

32.14 Blood Pressure Thresholds to Initiate Treatment and Treatment Targets

From a pragmatic standpoint, patients with active cancer have been excluded from large-scale hypertension trials in the past. Thus, there is little evidence to support antihypertensive therapy and blood pressure treatment thresholds. Based on the best available evidence, the 2017 ACC/AHA Hypertension guidelines outline a series of recommendations that are generally applied to the care of cancer patients. Patients who have an estimated 10-year ASCVD risk of ≥10%, or who have additional cardiovascular comorbidities such as type 2 diabetes mellitus or chronic kidney disease, should be treated to a goal BP < 130/80 mmHg. Patients who have an estimated 10-year ASCVD risk of ≤10% with no additional cardiovascular comorbidities should be treated to a goal BP < 140/90 mmHg, though it may be reasonable to treat these patients to a goal BP < 130/80 mmHg as well [82]. BP targets might be more permissive for cancer patients in the palliative setting, for whom the short-term benefits of anticancer therapy upon their quality of life might outweigh the risk of developing cardiovascular disease in the long term [56].

At present, specific guidance on the choice of antihypertensive agents in the cancer population is lacking. When selecting an antihypertensive agent, it is critical to consider drug pharmacokinetics and pharmacodynamics, as well as the presence of comorbidities and potential side effects. For example, non-dihydropyridine calcium channel blockers (CCB), such as verapamil and diltiazem, should be avoided because they use cytochrome P4503A4, a feature shared by many anticancer agents (e.g., VEGFI sunitinib and sorafenib), potentially increasing chemotherapy toxicity by inhibiting chemotherapy drug metabolism [83]. Dihydropyridine CCBs, which are direct vasodilators, may be the first-line choice in treating patients with cancer, given the role vascular dysfunction plays in the development of hypertension associated with many anticancer agents [84]. A study of 36 patients with ovarian, lung, and colorectal cancer who developed hypertension while on bevacizumab therapy demonstrated the efficacy of amlodipine in achieving BP control [85].

Combination of an angiotensin-converting enzyme inhibitor (ACEI) or an angiotensin II receptor blocker (ARB) with a CCB is generally accepted as the first choice, if it is necessary to combine two or more antihypertensive agents to achieve satisfactory BP levels [82, 86]. ACEI/ARB may be used preferentially in cancer patients with diabetic nephropathy, left ventricular systolic dysfunction (LVSD), and cancer therapy-associated proteinuria.

Cancer patients who have proteinuria (e.g., secondary to VEGF inhibitors) may benefit from renin–angiotensin–aldosterone system (RAAS) inhibitors therapy, because NO production is upregulated by RAAS inhibitors [87].

NO replacement with long-acting nitrate or phosphodiesterase inhibitors might improve VEGF signaling pathway (VSP) inhibitor-associated hypertension [83].

Other antihypertensive agents such as diuretics and β-blockers may be used when ACEI/ARB and CCB are ineffective in normalizing BP [88]. Mineralocorticoid receptor antagonists, such as eplerenone and spironolactone, are increasingly being administered for the management of cancer therapy-associated hypertension that is unresponsive to conventional therapy [89].

Any decision about altering the dose or withholding anticancer therapy should be made via a multidisciplinary approach involving input from all members of the cardio-oncology team. Generally, in patients who are symptomatic or have new evidence of end organ damage, dose reduction or temporary withdrawal of anticancer therapy should be considered. In cases of severe hypertension (≥160/100 mmHg), temporary withdrawal of anticancer therapy until adequate blood pressure control is achieved is recommended.

References

1. Zhou B, Bentham J, Di Cesare M, Bixby H, Danaei G, Cowan MJ, et al. Worldwide trends in blood pressure from 1975 to 2015: a pooled analysis of 1479 population-based measurement studies with 19·1 million participants. Lancet. 2017;389:37–55. https://doi.org/10.1016/S0140-6736(16)31919-5.
2. Arima H, Barzi F, Chalmers J. Mortality patterns in hypertension. J Hypertens. 2011;29:S3–7. https://doi.org/10.1097/01.hjh.0000410246.59221.b1.
3. Abdel-Qadir H, Austin PC, Lee DS, Amir E, Tu JV, Thavendiranathan P, et al. A population-based study of cardiovascular mortality following early-stage breast cancer. JAMA Cardiol. 2017;2:88. https://doi.org/10.1001/jamacardio.2016.3841.
4. Armstrong GT, Oeffinger KC, Chen Y, Kawashima T, Yasui Y, Leisenring W, et al. Modifiable risk factors and major cardiac events among adult survivors of childhood cancer. J Clin Oncol. 2013;31:3673–80. https://doi.org/10.1200/JCO.2013.49.3205.
5. Piccirillo JF. Prognostic importance of comorbidity in a hospital-based cancer registry. JAMA. 2004;291:2441. https://doi.org/10.1001/jama.291.20.2441.
6. Jain M, Townsend RR. Chemotherapy agents and hypertension: a focus on angiogenesis blockade. Curr Hypertens Rep. 2007;9:320–8. https://doi.org/10.1007/s11906-007-0058-7.
7. Gibson TM, Li Z, Green DM, Armstrong GT, Mulrooney DA, Srivastava D, et al. Blood pressure status in adult survivors of childhood cancer: a report from the St. Jude lifetime cohort study. Cancer Epidemiol Biomark Prev. 2017;26:1705–13. https://doi.org/10.1158/1055-9965.EPI-17-0510.
8. Kantor AF, Li FP, Janov AJ, Tarbell NJ, Sallan SE. Hypertension in long-term survivors of childhood renal cancers. J Clin Oncol. 1989;7:912–5. https://doi.org/10.1200/JCO.1989.7.7.912.

9. Cohen JB. Hypertension in obesity and the impact of weight loss. Curr Cardiol Rep. 2017;19:98. https://doi.org/10.1007/s11886-017-0912-4.
10. Garrison RJ, Kannel WB, Stokes J, Castelli WP. Incidence and precursors of hypertension in young adults: the Framingham offspring study. Prev Med. 1987;16:235–51. https://doi.org/10.1016/0091-7435(87)90087-9.
11. Calle EE, Rodriguez C, Walker-Thurmond K, Thun MJ. Overweight, obesity, and mortality from cancer in a prospectively studied cohort of U.S. adults. N Engl J Med. 2003;348:1625–38. https://doi.org/10.1056/NEJMoa021423.
12. Landsberg L, Molitch M. Diabetes and hypertension: pathogenesis, prevention and treatment. Clin Exp Hypertens. 2004;26:621–8. https://doi.org/10.1081/CEH-200031945.
13. Tlemsani C, Mir O, Boudou-Rouquette P, Huillard O, Maley K, Ropert S, et al. Posterior reversible encephalopathy syndrome induced by anti-VEGF agents. Target Oncol. 2011;6:253–8. https://doi.org/10.1007/s11523-011-0201-x.
14. Kim D. Posterior reversible encephalopathy syndrome induced by nivolumab immunotherapy for non–small-cell lung cancer. Clin Case Rep. 2019;7:935–8. https://doi.org/10.1002/ccr3.2122.
15. Armenian SH, Xu L, Ky B, Sun C, Farol LT, Pal SK, et al. Cardiovascular disease among survivors of adult-onset cancer: a community-based retrospective cohort study. J Clin Oncol. 2016;34:1122–30. https://doi.org/10.1200/JCO.2015.64.0409.
16. Mantovani A, Allavena P, Sica A, Balkwill F. Cancer-related inflammation. Nature. 2008;454:436–44. https://doi.org/10.1038/nature07205.
17. Heijnen BF, Van Essen H, Schalkwijk CG, Janssen BJ, Struijker-Boudier HA. Renal inflammatory markers during the onset of hypertension in spontaneously hypertensive rats. Hypertens Res. 2014;37:100–9. https://doi.org/10.1038/hr.2013.99.
18. Harrison DG, Guzik TJ, Lob HE, Madhur MS, Marvar PJ, Thabet SR, et al. Inflammation, immunity, and hypertension. Hypertension. 2011;57:132–40. https://doi.org/10.1161/HYPERTENSIONAHA.110.163576.
19. Rodríguez-Iturbe B, Quiroz Y, Nava M, Bonet L, Chávez M, Herrera-Acosta J, et al. Reduction of renal immune cell infiltration results in blood pressure control in genetically hypertensive rats. Am J Physiol-Ren Physiol. 2002;282:F191–201. https://doi.org/10.1152/ajprenal.0197.2001.
20. de Martel C, Georges D, Bray F, Ferlay J, Clifford GM. Global burden of cancer attributable to infections in 2018: a worldwide incidence analysis. Lancet Glob Health. 2020;8:e180–90. https://doi.org/10.1016/S2214-109X(19)30488-7.
21. Jayedi A, Rahimi K, Bautista LE, Nazarzadeh M, Zargar MS, Shab-Bidar S. Inflammation markers and risk of developing hypertension: a meta-analysis of cohort studies. Heart. 2019;105:686–92. https://doi.org/10.1136/heartjnl-2018-314216.
22. Chen D-D, Dong Y-G, Yuan H, Chen AF. Endothelin 1 activation of endothelin A receptor/NADPH oxidase pathway and diminished antioxidants critically contribute to endothelial progenitor cell reduction and dysfunction in salt-sensitive hypertension. Hypertension. 2012;59:1037–43. https://doi.org/10.1161/HYPERTENSIONAHA.111.183368.
23. Bjelakovic G, Nikolova D, Gluud LL, Simonetti RG, Gluud C. Antioxidant supplements for prevention of mortality in healthy participants and patients with various diseases. Cochrane Database Syst Rev. 2012;2012:CD007176. https://doi.org/10.1002/14651858.CD007176.pub2.
24. Abdel-Qadir H, Ethier J-L, Lee DS, Thavendiranathan P, Amir E. Cardiovascular toxicity of angiogenesis inhibitors in treatment of malignancy: a systematic review and meta-analysis. Cancer Treat Rev. 2017;53:120–7. https://doi.org/10.1016/j.ctrv.2016.12.002.
25. Versmissen J, Mirabito Colafella KM, Koolen SLW, Danser AHJ. Vascular cardio-oncology: vascular endothelial growth factor inhibitors and hypertension. Cardiovasc Res. 2019;115:904–14. https://doi.org/10.1093/cvr/cvz022.
26. Ferrara N. Vascular endothelial growth factor: basic science and clinical progress. Endocr Rev. 2004;25:581–611. https://doi.org/10.1210/er.2003-0027.
27. Robinson ES, Khankin EV, Karumanchi SA, Humphreys BD. Hypertension induced by vascular endothelial growth factor signaling pathway inhibition: mechanisms and

potential use as a biomarker. Semin Nephrol. 2010;30:591–601. https://doi.org/10.1016/j.semnephrol.2010.09.007.
28. Mitchell JA, Ali F, Bailey L, Moreno L, Harrington LS. Role of nitric oxide and prostacyclin as vasoactive hormones released by the endothelium. Exp Physiol. 2008;93:141–7. https://doi.org/10.1113/expphysiol.2007.038588.
29. Kappers MHW, van Esch JHM, Sluiter W, Sleijfer S, Danser AHJ, van den Meiracker AH. Hypertension induced by the tyrosine kinase inhibitor sunitinib is associated with increased circulating endothelin-1 levels. Hypertension. 2010;56:675–81. https://doi.org/10.1161/HYPERTENSIONAHA.109.149690.
30. Neves KB, Rios FJ, van der Mey L, Alves-Lopes R, Cameron AC, Volpe M, et al. VEGFR (vascular endothelial growth factor receptor) inhibition induces cardiovascular damage via redox-sensitive processes. Hypertension. 2018;71:638–47. https://doi.org/10.1161/HYPERTENSIONAHA.117.10490.
31. Mourad J-J, Guetz G d, Debbabi H, Levy BI. Blood pressure rise following angiogenesis inhibition by bevacizumab. A crucial role for microcirculation. Ann Oncol. 2008;19:927–34. https://doi.org/10.1093/annonc/mdm550.
32. Kidoguchi S, Sugano N, Tokudome G, Yokoo T, Yano Y, Hatake K, et al. New concept of onco-hypertension and future perspectives. Hypertension. 2021;77:16–27. https://doi.org/10.1161/HYPERTENSIONAHA.120.16044.
33. Maitland ML, Kasza KE, Karrison T, Moshier K, Sit L, Black HR, et al. Ambulatory monitoring detects Sorafenib-induced blood pressure elevations on the first day of treatment. Clin Cancer Res. 2009;15:6250–7. https://doi.org/10.1158/1078-0432.CCR-09-0058.
34. Lankhorst S, Severs D, Markó L, Rakova N, Titze J, Müller DN, et al. Salt sensitivity of angiogenesis inhibition–induced blood pressure rise: role of interstitial sodium accumulation? Hypertension. 2017;69:919–26. https://doi.org/10.1161/HYPERTENSIONAHA.116.08565.
35. Li W, Croce K, Steensma DP, McDermott DF, Ben-Yehuda O, Moslehi J. Vascular and metabolic implications of novel targeted cancer therapies. J Am Coll Cardiol. 2015;66:1160–78. https://doi.org/10.1016/j.jacc.2015.07.025.
36. Di Lorenzo G, Autorino R, Bruni G, Cartenì G, Ricevuto E, Tudini M, et al. Cardiovascular toxicity following sunitinib therapy in metastatic renal cell carcinoma: a multicenter analysis. Ann Oncol. 2009;20:1535–42. https://doi.org/10.1093/annonc/mdp025.
37. Chu TF, Rupnick MA, Kerkela R, Dallabrida SM, Zurakowski D, Nguyen L, et al. Cardiotoxicity associated with tyrosine kinase inhibitor sunitinib. Lancet. 2007;370:2011–9. https://doi.org/10.1016/S0140-6736(07)61865-0.
38. Kappers MHW, de Beer VJ, Zhou Z, Danser AHJ, Sleijfer S, Duncker DJ, et al. Sunitinib-induced systemic vasoconstriction in swine is endothelin mediated and does not involve nitric oxide or oxidative stress. Hypertension. 2012;59:151–7. https://doi.org/10.1161/HYPERTENSIONAHA.111.182220.
39. Oka T, Akazawa H, Naito AT, Komuro I. Angiogenesis and cardiac hypertrophy: maintenance of cardiac function and causative roles in heart failure. Circ Res. 2014;114:565–71. https://doi.org/10.1161/CIRCRESAHA.114.300507.
40. Izumiya Y, Shiojima I, Sato K, Sawyer DB, Colucci WS, Walsh K. Vascular endothelial growth factor blockade promotes the transition from compensatory cardiac hypertrophy to failure in response to pressure overload. Hypertension. 2006;47:887–93. https://doi.org/10.1161/01.HYP.0000215207.54689.31.
41. Budolfsen C, Faber J, Grimm D, Krüger M, Bauer J, Wehland M, et al. Tyrosine kinase inhibitor-induced hypertension: role of hypertension as a biomarker in cancer treatment. Curr Vasc Pharmacol. 2019;17:618–34.
42. Rini BI, Cohen DP, Lu DR, Chen I, Hariharan S, Gore ME, et al. Hypertension as a biomarker of efficacy in patients with metastatic renal cell carcinoma treated with sunitinib. J Natl Cancer Inst. 2011;103:763–73. https://doi.org/10.1093/jnci/djr128.
43. Boursiquot BC, Zabor EC, Glezerman IG, Jaimes EA. Hypertension and VEGF (vascular endothelial growth factor) receptor tyrosine kinase inhibition: effects on renal

function. Hypertension. 2017;1:HYPERTENSIONAHA.117.09275. https://doi.org/10.1161/HYPERTENSIONAHA.117.09275.
44. Soultati A, Mountzios G, Avgerinou C, Papaxoinis G, Pectasides D, Dimopoulos M-A, et al. Endothelial vascular toxicity from chemotherapeutic agents: preclinical evidence and clinical implications. Cancer Treat Rev. 2012;38:473–83. https://doi.org/10.1016/j.ctrv.2011.09.002.
45. Nuver J, Smit AJ, Sleijfer DT, van Gessel AI, van Roon AM, van der Meer J, et al. Microalbuminuria, decreased fibrinolysis, and inflammation as early signs of atherosclerosis in long-term survivors of disseminated testicular cancer. Eur J Cancer Oxf Engl. 1990;2004(40):701–6. https://doi.org/10.1016/j.ejca.2003.12.012.
46. Valentová M, Mladosievicová B. Coronary heart disease and hypertension as late effects of testicular cancer treatment—a minireview. Klin Onkol. 2011;24:18–22.
47. Izzedine H, Isnard-Bagnis C, Launay-Vacher V, Mercadal L, Tostivint I, Rixe O, et al. Gemcitabine-induced thrombotic microangiopathy: a systematic review. Nephrol Dial Transplant. 2006;21:3038–45. https://doi.org/10.1093/ndt/gfl507.
48. van Hell AJ, Haimovitz-Friedman A, Fuks Z, Tap WD, Kolesnick R. Gemcitabine kills proliferating endothelial cells exclusively via acid sphingomyelinase activation. Cell Signal. 2017;34:86–91. https://doi.org/10.1016/j.cellsig.2017.02.021.
49. Batra A, Patel B, Addison D, Baldassarre LA, Desai N, Weintraub N, et al. Cardiovascular safety profile of taxanes and vinca alkaloids: 30 years FDA registry experience. Open Heart. 2021;8:e001849. https://doi.org/10.1136/openhrt-2021-001849.
50. Long GV, Stroyakovskiy D, Gogas H, Levchenko E, de Braud F, Larkin J, et al. Combined BRAF and MEK inhibition versus BRAF inhibition alone in melanoma. N Engl J Med. 2014;371:1877–88. https://doi.org/10.1056/NEJMoa1406037.
51. Flaherty KT, Robert C, Hersey P, Nathan P, Garbe C, Milhem M, et al. Improved survival with MEK inhibition in BRAF-mutated melanoma. N Engl J Med. 2012;367:107–14. https://doi.org/10.1056/NEJMoa1203421.
52. Mincu RI, Mahabadi AA, Michel L, Mrotzek SM, Schadendorf D, Rassaf T, et al. Cardiovascular adverse events associated with BRAF and MEK inhibitors: a systematic review and meta-analysis. JAMA Netw Open. 2019;2:e198890. https://doi.org/10.1001/jamanetworkopen.2019.8890.
53. Isenberg JS, Roberts DD, Frazier WA. CD47: a new target in cardiovascular therapy. Arterioscler Thromb Vasc Biol. 2008;28:615–21. https://doi.org/10.1161/ATVBAHA.107.158154.
54. Wirth LJ, Sherman E, Robinson B, Solomon B, Kang H, Lorch J, et al. Efficacy of selpercatinib in *RET*-altered thyroid cancers. N Engl J Med. 2020;383:825–35. https://doi.org/10.1056/NEJMoa2005651.
55. Drilon A, Oxnard GR, Tan DSW, Loong HHF, Johnson M, Gainor J, et al. Efficacy of selpercatinib in *RET* fusion–positive non–small-cell lung cancer. N Engl J Med. 2020;383:813–24. https://doi.org/10.1056/NEJMoa2005653.
56. van Dorst DCH, Dobbin SJH, Neves KB, Herrmann J, Herrmann SM, Versmissen J, et al. Hypertension and prohypertensive antineoplastic therapies in cancer patients. Circ Res. 2021;128:1040–61. https://doi.org/10.1161/CIRCRESAHA.121.318051.
57. LaFargue CJ, Dal Molin GZ, Sood AK, Coleman RL. Exploring and comparing adverse events between PARP inhibitors. Lancet Oncol. 2019;20:e15–28. https://doi.org/10.1016/S1470-2045(18)30786-1.
58. Mirza MR, Monk BJ, Herrstedt J, Oza AM, Mahner S, Redondo A, et al. Niraparib maintenance therapy in platinum-sensitive, recurrent ovarian cancer. N Engl J Med. 2016;375:2154–64. https://doi.org/10.1056/NEJMoa1611310.
59. Mirza MR, Åvall Lundqvist E, Birrer MJ, dePont CR, Nyvang G-B, Malander S, et al. Niraparib plus bevacizumab versus niraparib alone for platinum-sensitive recurrent ovarian cancer (NSGO-AVANOVA2/ENGOT-ov24): a randomised, phase 2, superiority trial. Lancet Oncol. 2019;20:1409–19. https://doi.org/10.1016/S1470-2045(19)30515-7.
60. Dimopoulos MA, Moreau P, Palumbo A, Joshua D, Pour L, Hájek R, et al. Carfilzomib and dexamethasone versus bortezomib and dexamethasone for patients with relapsed or refractory

multiple myeloma (ENDEAVOR): a randomised, phase 3, open-label, multicentre study. Lancet Oncol. 2016;17:27–38. https://doi.org/10.1016/S1470-2045(15)00464-7.
61. Laubach JP, Moslehi JJ, Francis SA, San Miguel JF, Sonneveld P, Orlowski RZ, et al. A retrospective analysis of 3954 patients in phase 2/3 trials of bortezomib for the treatment of multiple myeloma: towards providing a benchmark for the cardiac safety profile of proteasome inhibition in multiple myeloma. Br J Haematol. 2017;178:547–60. https://doi.org/10.1111/bjh.14708.
62. Karar J, Maity A. PI3K/AKT/mTOR pathway in angiogenesis. Front Mol Neurosci. 2011;4:4. https://doi.org/10.3389/fnmol.2011.00051.
63. Choueiri TK, Escudier B, Powles T, Tannir NM, Mainwaring PN, Rini BI, et al. Cabozantinib versus everolimus in advanced renal cell carcinoma (METEOR): final results from a randomised, open-label, phase 3 trial. Lancet Oncol. 2016;17:917–27. https://doi.org/10.1016/S1470-2045(16)30107-3.
64. Motzer RJ, Hutson TE, Glen H, Michaelson MD, Molina A, Eisen T, et al. Lenvatinib, everolimus, and the combination in patients with metastatic renal cell carcinoma: a randomised, phase 2, open-label, multicentre trial. Lancet Oncol. 2015;16:1473–82. https://doi.org/10.1016/S1470-2045(15)00290-9.
65. Pavel M, Unger N, Borbath I, Ricci S, Hwang T-L, Brechenmacher T, et al. Safety and QOL in patients with advanced NET in a phase 3b expanded access study of everolimus. Target Oncol. 2016;11:667–75. https://doi.org/10.1007/s11523-016-0440-y.
66. Veccia A, Maines F, Kinspergher S, Galligioni E, Caffo O. Cardiovascular toxicities of systemic treatments of prostate cancer. Nat Rev Urol. 2017;14:230–43. https://doi.org/10.1038/nrurol.2016.273.
67. Zhu X, Wu S. Risk of hypertension in cancer patients treated with abiraterone: a meta-analysis. Clin Hypertens. 2019;25:12. https://doi.org/10.1186/s40885-019-0116-x.
68. Schoen MW, Hoque S, Witherspoon BJ, Schooley B, Sartor O, Yang YT, et al. End of an era for erythropoiesis-stimulating agents in oncology. Int J Cancer. 2020;146:2829–35. https://doi.org/10.1002/ijc.32917.
69. Vaziri ND. Mechanism of erythropoietin-induced hypertension. Am J Kidney Dis. 1999;33:821–8. https://doi.org/10.1016/S0272-6386(99)70413-0.
70. Goodwin JE, Geller DS. Glucocorticoid-induced hypertension. Pediatr Nephrol. 2012;27:1059–66. https://doi.org/10.1007/s00467-011-1928-4.
71. Baid S, Nieman LK. Glucocorticoid excess and hypertension. Curr Hypertens Rep. 2004;6:493–9. https://doi.org/10.1007/s11906-004-0046-0.
72. Fakhouri F, La Batide Alanore A, Rérolle JP, Guéry B, Raynaud A, Plouin PF. Presentation and revascularization outcomes in patients with radiation-induced renal artery stenosis. Am J Kidney Dis. 2001;38:302–9. https://doi.org/10.1053/ajkd.2001.26095.
73. Sharabi Y, Dendi R, Holmes C, Goldstein DS. Baroreflex failure as a late sequela of neck irradiation. Hypertension. 2003;42:110–6. https://doi.org/10.1161/01.HYP.0000077441.45309.08.
74. Arai H, Saitoh S, Matsumoto T, Makita F, Mitsugi S, Yuasa K, et al. Hypertension as a paraneoplastic syndrome in hepatocellular carcinoma. J Gastroenterol. 1999;34:530–4. https://doi.org/10.1007/s005350050309.
75. van der Horst-Schrivers ANA, Wymenga ANM, Links TP, Willemse PHB, Kema IP, de Vries EGE. Complications of midgut carcinoid tumors and carcinoid syndrome. Neuroendocrinology. 2004;80:28–32. https://doi.org/10.1159/000080737.
76. Moreira DM, Gershman B, Lohse CM, Boorjian SA, Cheville JC, Leibovich BC, et al. Paraneoplastic syndromes are associated with adverse prognosis among patients with renal cell carcinoma undergoing nephrectomy. World J Urol. 2016;34:1465–72. https://doi.org/10.1007/s00345-016-1793-7.
77. Hamidi O, Young WF, Iñiguez-Ariza NM, Kittah NE, Gruber L, Bancos C, et al. Malignant pheochromocytoma and paraganglioma: 272 patients over 55 years. J Clin Endocrinol Metab. 2017;102:3296–305. https://doi.org/10.1210/jc.2017-00992.

78. Veron Esquivel D, Batiz F, Farias Vega A, Carrillo Gonzalez PA. Adrenocortical carcinoma, an unusual cause of secondary hypertension. BMJ Case Rep. 2016;2016:bcr2016217918. https://doi.org/10.1136/bcr-2016-217918.
79. Arps K, Pallazola VA, Cardoso R, Meyer J, Jones R, Latina J, et al. Clinician's guide to the updated ABCs of cardiovascular disease prevention: a review part 1. Am J Med. 2019;132:e569–80. https://doi.org/10.1016/j.amjmed.2019.01.016.
80. Marvel FA, Whelton SP, Blumenthal RS. A cardio-oncology cardiovascular prevention framework. JACC CardioOncol. 2019;1:252–5. https://doi.org/10.1016/j.jaccao.2019.11.012.
81. Bamias A, Manios E, Karadimou A, Michas F, Lainakis G, Constantinidis C, et al. The use of 24-h ambulatory blood pressure monitoring (ABPM) during the first cycle of sunitinib improves the diagnostic accuracy and management of hypertension in patients with advanced renal cancer. Eur J Cancer Oxf Engl. 1990;2011(47):1660–8. https://doi.org/10.1016/j.ejca.2011.03.033.
82. Whelton PK, Carey RM, Aronow WS, Casey DE, Collins KJ, Dennison Himmelfarb C, et al. 2017 ACC/AHA/AAPA/ABC/ACPM/AGS/APhA/ASH/ASPC/NMA/PCNA guideline for the prevention, detection, evaluation, and management of high blood pressure in adults: executive summary: a report of the American College of Cardiology/American Heart Association task force on clinical practice guidelines. Hypertens. 2018;71:1269–324. https://doi.org/10.1161/HYP.0000000000000066.
83. Nazer B, Humphreys BD, Moslehi J. Effects of novel angiogenesis inhibitors for the treatment of cancer on the cardiovascular system: focus on hypertension. Circulation. 2011;124:1687–91. https://doi.org/10.1161/CIRCULATIONAHA.110.992230.
84. Abernethy DR, Schwartz JB. Calcium-antagonist drugs. N Engl J Med. 1999;341:1447–57. https://doi.org/10.1056/NEJM199911043411907.
85. Mir O, Coriat R, Ropert S, Cabanes L, Blanchet B, Camps S, et al. Treatment of bevacizumab-induced hypertension by amlodipine. Investig New Drugs. 2012;30:702–7. https://doi.org/10.1007/s10637-010-9549-5.
86. Kalaitzidis RG, Elisaf MS. Uncontrolled hypertension and oncology: clinical tips. Curr Vasc Pharmacol. 2017;16:23. https://doi.org/10.2174/1570161115666170414121436.
87. Umemura S, Arima H, Arima S, Asayama K, Dohi Y, Hirooka Y, et al. The Japanese society of hypertension guidelines for the management of hypertension (JSH 2019). Hypertens Res. 2019;42:1235–481. https://doi.org/10.1038/s41440-019-0284-9.
88. Pasquier E, Street J, Pouchy C, Carre M, Gifford AJ, Murray J, et al. β-Blockers increase response to chemotherapy via direct antitumour and anti-angiogenic mechanisms in neuroblastoma. Br J Cancer. 2013;108:2485–94. https://doi.org/10.1038/bjc.2013.205.
89. Calhoun DA, Jones D, Textor S, Goff DC, Murphy TP, Toto RD, et al. Resistant hypertension: diagnosis, evaluation, and treatment. Circulation. 2008;117:e510–26. https://doi.org/10.1161/CIRCULATIONAHA.108.189141.

Index

M. Dorobantu et al. (eds.), *Hypertension and Heart Failure*, Updates in Hypertension and Cardiovascular Protection, https://doi.org/10.1007/978-3-031-39315-0

T

U

V

FSC
www.fsc.org
MIX
Papier aus verantwortungsvollen Quellen
Paper from responsible sources
FSC® C105338

If you have any concerns about our products,
you can contact us on
ProductSafety@springernature.com

In case Publisher is established outside the EU,
the EU authorized representative is:
Springer Nature Customer Service Center GmbH
Europaplatz 3, 69115 Heidelberg, Germany

Printed by Libri Plureos GmbH
in Hamburg, Germany